AF670370

# Formation and Evolution of Solids in Space

# NATO ASI Series

## Advanced Science Institute Series

*A Series presenting the results of activities sponsored by the NATO Science Committee, which aims at the dissemination of advanced scientific and technological knowledge, with a view to strengthening links between scientific communities.*

The Series is published by an international board of publishers in conjunction with the NATO Scientific Affairs Division

| | |
|---|---|
| **A Life Sciences**<br>**B Physics** | Plenum Publishing Corporation<br>London and New York |
| **C Mathematical and Physical Sciences**<br>**D Behavioural and Social Sciences**<br>**E Applied Sciences** | Kluwer Academic Publishers<br>Dordrecht, Boston and London |
| **F Computer and Systems Sciences**<br>**G Ecological Sciences**<br>**H Cell Biology**<br>**I Global Environment Change** | Springer-Verlag<br>Berlin, Heidelberg, New York, London, Paris and Tokyo |

**PARTNERSHIP SUB-SERIES**

| | |
|---|---|
| **1. Disarmament Technologies** | Kluwer Academic Publishers |
| **2. Environment** | Springer-Verlag / Kluwer Academic Publishers |
| **3. High Technology** | Kluwer Academic Publishers |
| **4. Science and Technology Policy** | Kluwer Academic Publishers |
| **5. Computer Networking** | Kluwer Academic Publishers |

*The Partnership Sub-Series incorporates activities undertaken in collaboration with NATO's Cooperation Partners, the countries of the CIS and Central and Eastern Europe, in Priority Areas of concern to those countries.*

**NATO-PCO-DATA BASE**

The electronic index to the NATO ASI Series provides full bibliographical references (with keywords and/or abstracts) to about 50,000 contributions from international scientists published in all sections of the NATO ASI Series. Access to the NATO-PCO-DATA BASE is possible via a CD-ROM "NATO Science and Technology Disk" with user-friendly retrieval software in English, French, and German (©WTV GmbH and DATAWARE Technologies, Inc. 1989). The CD-ROM contains the AGARD Aerospace Database.

The CD-ROM can be ordered through any member of the Board of Publishers or through NATO-PCO, Overijse, Belgium.

**Series C: Mathematical and Physical Sciences – Vol. 523**

# Formation and Evolution of Solids in Space

edited by

J. Mayo Greenberg
Laboratory Astrophysics,
University of Leiden,
Leiden, The Netherlands

and

Aigen Li
Laboratory Astrophysics,
University of Leiden,
Leiden, The Netherlands

**Springer Science+Business Media, B.V.**

Proceedings of the NATO Advanced Study Institute on
Formation and Evolution of Solids in Space
Erice, Sicily, Italy
10–21 March 1997

A C.I.P. Catalogue record for this book is available from the Library of Congress.

ISBN 978-94-010-6018-9 ISBN 978-94-011-4806-1 (eBook)
DOI 10.1007/978-94-011-4806-1

*Printed on acid-free paper*

# CONTENTS

# PREFACE

The school on Formation and Evolution of Solids in Space, held at the "Ettore Majorana Centre for Scientific Culture" 10-21 March, 1997 was the fourth course of the International School of Space Chemistry held in Erice, Sicily. It was attended by 78 participants from 14 countries. Its intention was to provide a forum for representatives of a variety of diverse fields in astrophysics and astrochemistry to discuss the existence of connections between interstellar and solar system solids. It has long been recognized, although not with the new clarity and focus brought to this subject by the lecturers in this school that interstellar and solar system solids are different stages of the same materials. Sources and evolution of circumstellar dust, interstellar dust, comets, meteorites and interplanetary dust were carefully discussed in the context of their interrelations.

It is perhaps interesting and useful to put the subject in its historical context by bringing out a couple of significant points in its development. First of all, although a connection between interstellar and solar system particles was conjectured it has not always been as we now recognize it. For example, J.L. Greenstein said in an article in 1938: "We can conclude that the extinction of light by an interstellar medium consisting of a reasonable distribution of small dust particles, continuous with that of the hyperbolic meteors, is such as to satisfy the available observational data."

Yes, we do believe that meteors may be related to interstellar dust but not so directly. This topic has provided considerable speculation over the past 25 years. In 1974, A.G.W. Cameron, in the proceedings of the IAU Symposium 52 on Interstellar Dust and Related Topics, wrote a short note entitled "Interstellar Grains in Museums?" To the extent that there are materials in meteorites which isotopically and chemically can be related to stellar and supernova sources this idea has been well supported. However, a major question raised at the school related to the morphology and, more specifically, to the size of the component meteorite materials. How do characteristically tenth micron interstellar dust particles evolve to rocky materials 1000 or more times larger? Until tenth micron silicates are observed in meteorites it will be difficult to understand the transit of materials from stellar and other sources via the interstellar medium to solar system solids. The closest which we appear to be getting in relating interstellar dust to particles in the solar system is the evidence for submicron organic coated particles in IDP's (interplanetary dust particles collected in the upper atmosphere). The so-called GEMS may ultimately offer the clues we have been looking for.

The above examples have the advantages of being directly subjected to laboratory investigation on earth. On the other hand, what we know about comets is either a result of remote observation or of space measurements in situ. Here the connection between comets as solar system objects and interstellar dust seems to be the closest. Comet volatiles compare very well with interstellar ices - as derived from laboratory analogue studies combined with astronomical infrared observations. The laboratory is important here but is still not totally direct. A major input of the results of the in situ investigation of comet Halley by GIOTTO/VEGA space vehicles was the demonstration of the complex organic component of the comet dust. Several contributions in this school are devoted to just this component of interstellar dust and it has become one of the major efforts of future space missions (e.g. the Rosetta mission) to comets to intensify the analyses of the organic fraction of comet nuclei and dust.

The properties and sources of interplanetary dust have occupied the attention of observers for a long time in the interpretation of the zodiacal light and we had several major contribution on this subject. Here, one of the difficult problems has to do with the optical properties of complex structured particles and, if these are related to origins from the interstellar medium we at least may conjecture that the basic units are tenth micron in size. Complexes of such particles provide a major challenge to scattering theory, so that laboratory analogue methods - scaling up the particles for studies at microwave and millimeter wavelength has been the method of choice for many years.

Going back to roots and origins, the formation of dust grains in stellar atmospheres and by supernovae seems to be the ultimate beginnings. But how such particles evolve in interstellar space before becoming part of the ingredients of objects in solar systems is not yet "entirely" solved. One of the key problems is that the solid particles are limited in their composition to the available condensible elements in space. This limited availability must be at least sufficient to provide enough solid particles "of the right sort" to produce the amount and kind of extinction which is generally observed. The optical properties of interstellar dust must conform with the cosmic abundance constraint. There are still several disparate theories on how this is accomplished in reality and the basic evidence for cosmic abundance has itself come into question.

Formation and evolution of interstellar dust must include the question of survival - or destruction. It is believed that the star formation rate in our galaxy is such as to consume all components of the interstellar medium in about 5 thousand million years, so this is the ultimate fate of solid particles as well. But what about other destructive mechanisms? Although only one lecture was devoted to this, it counterbalanced the lectures on the sources of interstellar refractory particles. The bottom line appears to be that when all "stellar" sources are added up there is no way to be able to account for more than 10% of the observed amount of dust because the rates of destruction are about 10 times greater than the source rates.

Nature takes care of this by providing for growth in space just as conjectured by B. Lindblad 50 years ago and followed up by H.C. van de Hulst 40 years ago. The growth here is by accretion of the major condensible elements O, C and N. These atoms go to make up a major fraction of the dust in various molecular forms - from simple ones like water, carbon monoxide, methanol, ammonia, etc. to complex organic molecules which are relatively refractory. These components occupied the attention of a number of lecturers from the point of view of astronomical observations analyzed with the results of laboratory analog studies on the evolution of "ices" at low temperature.

Since the solid particles are bathed in and interact with the gaseous components of the interstellar clouds it was natural to show just how such atomic and molecular species are distributed in the various low and high density regions of space. Of particular interest is the composition of both gas and solid species in regions of pre and post star formation because they can potentially provide the connection between interstellar space and solar system environments. Intensive investigations are underway not only on the O, C, N-based components of the solid particles but also on the refractory Mg, Si, Fe-based components. Interpretation of the astronomical presence and evolution of these solid components has been developed largely on the basis of laboratory experiments on representative analogues of these materials often canonically labeled as silicate but consisting of many other silicon, magnesium and iron compounds as well. How such materials may appear as amorphous or crystalline phases is, for example, one of the key problems engaging astrophysicists. The evolution from crystalline components in stellar atmospheres, to a largely amorphous phase in interstellar space, to the reappearance of a significant crystallinity in comet dust remains an unsolved problem. But the laboratory approach may well offer some possible solutions.

It is clear from the variety of subjects discussed at this school, that solutions to the astrophysical problems bring to here a very wide range of theoretical, observational and laboratory expertise in such fields as astronomy, physics and chemistry and, with the evidence for complex prebiotic organics in meteorites, one must include biology as well because of the undoubted relevance to the origins of life. This aspect is a recurring one in the series of the schools on Chemistry in Space and, although not emphasized here, it is always at the back of our minds. To quote from a previous volume in this series "This is a continuing story. What happens to the solid particles in space after they emerge from stellar sources has important scientific consequence because it ultimately bears on our origins - the origin of solar systems, and, in particular, our own earth, and of life in the universe."

We are grateful for the support of the Scientific Affairs Division of the North Atlantic Treaty Organization (NATO) which made this school possible, as well as for substantial grants from the United States National Science Foundation (NSF), the National Aeronautics and Space Administration (NASA), the Leids Kerkhoven Bosscha Fonds (LKBF), the Netherlands Organization for Scientific research (NWO) and the Space Research Organization Netherlands (SRON). We owe a special debt of gratitude to Beps de Kanter who was involved in all stages of this course. As is always the case, the administration and staff at the "Ettore Majorana Centre for Scientific Culture" provided the logistical and personal support which made it possible for the participants to enjoy a stimulating scientific exchange of ideas with the greatest pleasure.

*J. Mayo Greenberg, Aigen Li*
*Leiden, The Netherlands*
*25 August, 1998.*

# OBSERVATIONAL CONSTRAINTS ON INTERSTELLAR DUST COMPOSITION

THEODORE P. SNOW
*Center for Astrophysics and Space Astronomy, University of Colorado, Campus Box 389, Boulder, CO 80309, USA*

**Abstract.** This paper provides a general review of what is known about the composition of the interstellar dust and how we know it. Several general methods are cited, but the emphasis here is on the measurement of interstellar gas-phase depletions and the inferences drawn from them about the dust. Depletion estimates are derived from measurements of gas-phase abundances, and the first major section of the review describes the methods for determining the composition of the gas in diffuse and translucent interstellar clouds. Following this is a review of depletion results, including a discussion of the so-called 'standard' depletion pattern. Evidence favoring a revision of the assumed reference standard (the 'cosmic' abundances) is then reviewed, along with the implications for models of the dust. Next the dependences of depletions on various environmental factors (cloud velocity, density, and extinction properties) are summarized. Finally, future work in this field is outlined.

## 1. Overview Of Methods For Determining Dust Composition

The composition of the interstellar dust can be inferred be several methods, most of them quite indirect. In this chapter, we provide an overview of the methods, with emphasis on the measurement of interstellar gas abundances and their role in deriving the composition of the dust based on what is missing from the gas. We also provide an update on the most recent results of abundance and depletion studies.

The most direct methods of measuring dust composition either do not distinguish clearly between gas and dust composition, as in the case of X-ray absorption edges; or are limited to regions where the density is high enough that grain mantles have grown to the point where solid-state infrared absorption bands can be detected. Indirect methods include inferences that



*J. M. Greenberg and A. Li (eds.), Formulation and Evolution of Solids in Space,* 1–28.

can be drawn about interstellar grain composition from inclusions found in interplanetary dust particles (IDPs), and deductions about grain composition that are derived from abundance and depletion measures of the interstellar gas.

Measurements of the composition of dust grain mantles that are derived from infrared observations, and deductions based on IDPs, are described elsewhere in this volume. Here we focus on the measurement of gas-phase abundances and depletions, with a brief overview of other techniques such as X-ray absorption-edge observations.

### 1.1. INFERENCES FROM GAS-PHASE DEPLETIONS

The general assumption on which this method is based is that the overall composition of the interstellar medium (gas and dust combined) is the same as the composition of normal stars. Thus the method consists of determining the composition of the gas, through optical and UV absorption lines, and comparing the result with the expected overall composition, which is usually assumed the same as the composition of the Sun. The inferred depletions; i.e. the fractional abundances of elements that are 'missing' from the gas, are assumed to represent the quantities of the elements that are embedded in the dust grains.

This method provides the most complete coverage of the periodic table, largely because the most common elements, in their most populated ionization states, have their ground-state transitions in the optical or, more likely, the ultraviolet, where absorption-line measurements can be made. This method has the further advantage that it refers to global grain compositions, rather than being confined to grain mantles (as in the case of solid-state IR absorption features), and that it provides information on truly *interstellar* grains, something that is problematic for analyzes of inclusions in IDPs. Thus a great deal of what we think we know about the composition of interstellar dust is based on inferences made from gas-phase depletions. This topic is the main theme of this chapter, and will be treated in some detail below.

### 1.2. IR ABSORPTION BANDS

This topic is covered thoroughly elsewhere in this volume. Here we take time only to point out that in principle, IR observations of dust grain mantle features have the potential to provide very useful information on the interaction of gas and dust in diffuse and dense clouds where both the gas phase and the mantle compositions can be measured. If we can measure, for example, water ice, hydrocarbons, solid CO and even silicate dust absorption features in the same lines of sight where we can observe the gas-phase abundances of the elements that make up these compounds, the reward in an improved understanding of the grain growth and mantle accretion processes leading to star and planet formation will be immense. To reap this reward will require pushing the sensitivity limits of both the IR and the UV/optical observations, so that the region of overlap, where the same lines of sight exhibit both grain mantles and measurable gas-phase abundances, will be expanded. Currently it is possible to measure UV absorption lines at high spectral resolution (needed for the analysis of interstellar gas lines) to total extinction values of about $A_V = 2.5$ to 3.0, while infrared data show that the threshold for detecting water ice and hydrocarbon mantles is in the range from 3 to 5 magnitudes of extinction (for detection of solid CO or silicate absorption even more extinction is required).

## 1.3. INCLUSIONS IN INTERPLANETARY DUST PARTICLES

There is ample evidence that interplanetary dust particles include contaminants that are interstellar in origin. The most compelling argument for this comes from inclusions having very non-solar isotope ratios; e.g. the ratio of $^{12}C$ to $^{13}C$ ranges from a few up to several thousand, which clearly implies non-solar origins since the local ratio is about 89. Other elements show similar disparities in some IDP inclusions.

The difficulty in interpreting IDP inclusions as indicators of IS dust properties are: (1) A significant degree of thermal processing is likely to have occurred as the solar nebula formed and went through the transition to planets and interplanetary debris; and (2) individual IS inclusions in grains may not represent typical interstellar grains, but appear in some cases at least to have originated in very specific, chemically-enriched debris from supernova explosions that do not represent the diffuse ISM as a whole.

## 1.4. X-RAY ABSORPTION EDGES

The inner (K-shell and L-shell) electrons of most elements can absorb X-ray photons, providing in principle a unique method for measuring column densities of the elements in IS dust. X-ray absorption edges are observable independently of whether atoms reside in solid or gaseous form, so the full absorptivity of all atoms in the line of sight, whether they are in the gas, grain mantles, or gain cores, is accessible. Therefore in principle, given good absorption cross-sections and high-quality X-ray spectra of distant X-ray sources, observations of X-ray absorption edges can provide a unique handle on interstellar abundances of the common elements.

In practice, to date this method has only been used in reverse, as a means of quantifying the X-ray absorption cross sections (e.g. Ryter et al.

1975; Morrison and McCammon 1983) based on the assumption of standard or 'cosmic' abundances relative to hydrogen. But given reliable cross sections, which can be derived from laboratory measurements, observations of K-shell and L-shell absorption edges against the X-ray spectra of distant sources holds considerable promise as a means to learn about the composition of the dust. The most significant complicating factor is that it is difficult or impossible to distinguish between absorption due to gas-phase atoms and absorption due to atoms in dust grains. But high spectral resolution, such as will be achieved by future instruments, may help discriminate between solid-state and gas-phase absorption, and in any event we can resolve this problem if we also have UV/optical absorption data on gas-phase atoms and ions in the same lines of sight. This latter requirement is made more difficult by the fact that we can obtain X-ray absorption spectra only in lines of sight toward strong X-ray sources, and there may not be coincident UV/optical sources against which we can obtain parallel data on gas-phase abundances.

## 2. Measuring Depletions And Dust Composition

### 2.1. THE UNDERLYING ASSUMPTION: 'COSMIC' ABUNDANCES

Commonly the composition of interstellar dust is inferred indirectly from measurements of depletions, meaning the fractional abundances of the elements that are 'missing' from the interstellar gas. In order to know what is missing, we must know the total composition of the ISM, for the dust and gas combined. Then when we observe the gas-phase abundances from UV/optical absorption lines, we can subtract those in order to deduce the proportions of the elements that are not present in the gas and which therefore must reside in the grains. Of course there are other possible states in which the material may exist, for example in molecules, but the molecular abundances known (from direct observations) or inferred (e.g. from data on the diffuse interstellar bands or from IR emission measures of PAHs) indicate that in the diffuse ISM molecules represent only trace quantities of the common elements. So ignoring the discovery of a reservoir of material in some as-yet unforeseen form, it appears that the vast majority of material not in the atomic gas must be in the dust.

Then the question of determining dust composition devolves into one of measuring gas-phase abundances and knowing what overall, or 'cosmic' abundance standard to apply. Historically we have resorted to the Sun and the solar system to provide the reference standard, on the assumption that the Sun represents the overall composition of our galactic neighborhood. But this has always been known to be somewhat risky, because we know that there are both spatial and temporal variations in stellar composition in the galaxy (e.g Edvardsson et al. 1993). However, in the absence of a better standard, the solar system has been widely cited as the starting point for determining interstellar depletions.

#### 2.1.1. *The solar standard*

Solar abundances of the elements are derived from a combination of spectroscopic observations of the Sun itself (including its corona) and laboratory analyzes of meteorites, which are usually assumed to represent the pristine composition of the solar nebula, the gas cloud from which the entire system formed. One very widely-cited compilation of solar abundances is to be found in Anders and Grevesse (1989), but there are many others (e.g. Lambert 1978; Holweger and Mueller 1974; Grevesse et al. 1992; and Grevesse and Noels 1993), which show, through their variances, how much uncertainty there can be in measuring something as seemingly straightforward as the composition of our local star.

#### 2.1.2. *How it is determined*

Anders and Grevesse (1989) provide data on abundances in meteorites and in the solar photosphere, citing values for carbonaceous chondrites in the meteoritic case, because these meteorites are believed to be very primitive, essentially unaltered by processing during the formation of the solar system. Astronomers, worried that the meteoritic abundances have been altered anyway, tend to rely on the solar photo spheric values. In any case, the agreement is quite good between meteoritic abundances and those derived from spectroscopic measurements of the solar photo sphere.

### 2.2. THE MEASUREMENT OF INTERSTELLAR GAS-PHASE ABUNDANCES

The only observational technique available to us, apart from X-ray absorption-edge measurements (see Section 1.4, above) is the derivation of interstellar column densities from the analysis of absorption lines formed against the continuous spectra of background stars. In this section we review the method and the results.

#### 2.2.1. *Interstellar absorption lines in review*

Depending on the quality of the observational data, there are at least two distinct methods which can be used in the analysis of ultraviolet and visible-wavelength interstellar absorption lines. Before elaborating on those methods, however, it may be instructive to consider the reason why most of the useful lines for analysis arise at ultraviolet wavelengths.

The density in the interstellar gas is very low by any standards. In diffuse clouds, which are most relevant to this review, densities range from about 1 particle/cm$^3$ to perhaps a few hundred particles/cm$^3$. Even in the densest interstellar clouds, the ones that show up as dark regions in photographic images, densities rarely exceed a few times $10^5$ particles/cm$^3$, which is comparable to the best laboratory vacuums here on Earth.

At such low densities, collisions between particles are very rare, and the ionization and excitation of atoms, ions, and molecules in the diffuse ISM are dominated by radiative processes. This means that for all species the electrons will generally be in the ground electronic state, and therefore any optical or UV absorption lines that might be observable will be those that arise from the ground state. For virtually all common elements the transitions from the ground state to any excited electronic level, for atoms, ions, or molecules, require sufficient energy that they arise in ultraviolet wavelengths. Thus in order to observe a low-density medium in absorption, it is generally necessary to make observations in the UV. For this reason, significant progress in measuring interstellar abundances for the common elements did not occur until the 1970s, when ultraviolet spectrographs such as *Copernicus*, with appropriate instrumental capabilities, were placed in orbit.

The detection and analysis of interstellar absorption lines places some constraints on both the types of stars used as background sources and on the qualities of the instruments used to measure the lines. The role of the background stars is to serve as continuum sources, and the stars best suited for this are hot ones, usually of type O or early B, because these stars tend to have relatively few photo spheric absorption lines. Another useful quality is high projected rotational velocity ($vsini$), which broadens the stellar lines and thereby reduces confusion with interstellar lines, which are always quite narrow.

The intrinsic widths of interstellar absorption lines are dominated by Doppler broadening due to gas motions along the line of sight, and are generally broader than the thermal widths appropriate for the gas kinetic temperature. Detailed high-resolution analysis in a few cases has illustrated that nearly every line of sight penetrates a number of discrete clouds, so that the broadening in the integrated line profiles is dominated by the velocity dispersion of the clouds, with a major component as well from internal dispersions within the individual clouds. A line broadening parameter $b$ has been defined, and is equivalent to 0.6 times the FWHM of an observed line (see Spitzer 1978), regardless of whether it is intercloud or intracloud broadening that dominates (so long as the individual contributions are gaussian). Typically for moderately-reddened lines of sight such as those observed with *Copernicus*, the $b$-values are of order 5-10 km/s, whereas for individual clouds, where data have been derived from very high-resolution spectra (often observations of millimeter-wave molecular lines), the $b$-values are of order 1 km/s.

The small intrinsic widths of interstellar absorption lines militate in favor of spectrographs having the highest possible spectral resolution. *Copernicus* had a resolving power of $\lambda/\Delta\lambda = 20,000$ (about 15 km/s) in its highest-resolution mode, while the IUE had a lower resolution, of about 10000 (30 km/s). In its highest-resolution echelle modes, the GHRS on the *HST* had a resolving power of about 100,000 (roughly 3 km/s), and the STIS, recently placed aboard, has a similar resolution. It is important to note that none of these instruments has sufficient resolving power to isolate and measure the intrinsic widths of absorption lines due to individual cloud components. This will be addressed in the next sections, where we discuss analysis techniques.

The high ultraviolet extinction due to dust limits our penetration of interstellar clouds to fairly low levels of extinction, and consequently low total gas column densities. In terms of the total visual extinction $A_V$, *Copernicus* was able to reach extinctions of only about $A_V = 1$, while the GHRS (and the STIS) on the Hubble Space Telescope can do somewhat better, reaching perhaps $A_V = 3$ to 3.5 in the highest-resolution modes. Thus in the use of ultraviolet spectra for interstellar abundance determinations we are limited to diffuse clouds (meaning typical $A_V$ values of 1-2 magnitudes or less) and the high-transmissivity end of the range for translucent clouds ($A_V$ values in the range from 2 to 5 magnitudes).

2.2.2. *The curve-of-growth method*

One measure of the strength of an interstellar absorption line is its equivalent width, which measures the amount of radiant energy that is subtracted from the continuum by absorbing particles along the line of sight from observer to background star. Mathematically the equivalent width can be defined by

$$W_\lambda = \int (I_c - I_\lambda)/I_c, \tag{1}$$

where $I_c$ represents the continuum intensity and $I_\lambda$ represents the intensity at wavelength $\lambda$. The integration is done over the full width of the line. The equivalent width is convenient because to measure it with reasonable accuracy does not require extremely high S/N in the data, and furthermore the quantity is independent of spectral resolving power.

The curve of growth refers to the rate at which the equivalent width of an absorption line grows as a function of the number of absorbing particles along the line of sight. We can identify three distinct phases in this growth: (a) The optically thin case, where $W_\lambda$ is proportional to N, the column density of absorbers (in units of particles/cm$^2$); (b) the saturated regime, where the line strength grows more slowly due to overlapping of absorbers in wavelength space, where $W_\lambda$ is proportional to $\sqrt{\ln N}$; and (c) the damping regime, where natural broadening (due to the uncertainty principle) allows the growth of equivalent width to become more rapid, in proportion to $\sqrt{N}$.

The precise shape of the curve of growth for a given transition depends on the velocity dispersion parameter *b*. The larger the *b*-value, the higher in column density is the point where saturation sets in, because a larger velocity dispersion along the line of sight allows more of the absorbing particles to 'see' the continuum flux and absorb light. Thus in interpreting an equivalent width, one has to consider a family of curves of growth, representing a range in *b*-values. This, along with the very slow dependence of $W_\lambda$ on N in the saturated regime, makes it very difficult to unambiguously interpret the equivalent width of a single line. One can compensate for this if a number of lines of the same species are measured, because then the equivalent widths differ only by a factor related to the line wavelengths and the oscillator strengths (the *f*-values) of the transitions, and a spread of *f*-values translates into a spread of points along the curve of growth. If lines with a sufficiently wide range of *f*-values can be observed, then it is possible to distinguish among curves of growth for different *b*-values and determine an unambiguous column density.

Deriving a column density of absorbers, then, devolves into a task of observing equivalent widths and matching them to theoretical curves of growth. Ideally the lines will be on the linear (optically thin) portion of the curve, or on the damping portion, where the conversion from $W_\lambda$ to N is unambiguous. But as nature would have it, many of the absorption lines that we observe in diffuse and translucent clouds are saturated, lying on the 'flat' portion of the curve. This means that column densities will be uncertain, often by large factors. Not only is it difficult in that case to determine the correct velocity dispersion (i.e. to pick the one among the possible family of curves that best fits the points), but it is also possible for cloud components with small internal velocity dispersions to be hidden entirely within the broader lines formed over the integrated line of sight. There is no easy way around this problem, although statistical means can be used to help narrow the range of allowed fits (Jenkins 1986).

If sufficient information is available about the velocity component structure along a line of sight, for example as might be derived from very high-resolution spectra of specific species (usually either ground-based optical or millimeter-wave radio data), then it is possible to construct a multi-component curve of growth. In principle this would have to be done for each species because the distribution along the line of sight (i.e. as a function of velocity) may vary, but this is impractical, and the normal practice is to assume that all or certain groups of species are probably distributed similarly, and those elements are fitted to the same curve. This technique requires ancillary data on the line-of-sight velocity structure, and has been applied in only a few cases (e.g. Snow, Lamers, and Joseph 1987; Snow et al. 1996).

### 2.2.3. *Line profile fitting*

When data quality is sufficient (i.e. the spectral resolution and the S/N are high), it is feasible to determine abundances through line profile fitting procedures. This is usually accomplished through an iterative process in which trial values for the column density, the velocity dispersion ($b$-value), and the velocities of the observed cloud components are adopted and the synthetic line profile is calculated; then adjustments are made in the input parameters until the best fit to the observed profile is achieved. In most case the velocity structure is frozen first, often based on high-resolution spectra of specific lines, and then only the column density and $b$-value are allowed to float.

The profile-fitting technique can be more accurate than the curve-of-growth method, and has the significant advantage that it allows for the derivation of abundances and depletions separately for distinct clouds, but it shares the uncertainties cited above, due to saturation of lines and overlapping of weak and strong components. As in the case of the curve-of-growth method, results are most accurate when several lines of the same species are available, preferably with a wide range of oscillator strengths.

A variation on the profile-fitting technique is to reconstruct the continuum by determining the optical depth as a function of wavelength offset from line center, and then multiplying the observed profile by $e^{+\tau}$, where $\tau$ is the optical depth. The column density is then adjusted until the reconstructed continuum is level. This technique is especially useful for very strong lines which are damped.

### 2.2.4. *Atomic and molecular hydrogen*

Depletions are always expressed in terms of elemental abundances relative to hydrogen, so in order to measure depletions it is necessary to determine the hydrogen column density. For diffuse clouds H I is usually the dominant form of hydrogen, but in relatively reddened or dense diffuse clouds molecular hydrogen can also be important.

Atomic and molecular hydrogen have numerous transitions in the vacuum ultraviolet, which can be exploited in order to derive hydrogen column densities. In typical reddened lines of sight, the atomic hydrogen lines (the Lyman series) are very strongly saturated and thus are candidates for the profile reconstruction method of column density determination. Atomic hydrogen is nearly always sufficiently abundant for the principal lines (especially Lyman-$\alpha$ at 1216Å) to be damped and therefore well-suited to continuum reconstruction (e.g. Bohlin 1975). In lines of sight having significant total gas column densities (of order $10^{20}/cm^2$ or greater) the molecular hydrogen bands, particularly the lines arising from the J = 0 and 1 rotational states, are usually strong enough to be damped and therefore analyzed by either the profile-fitting or the continuum-reconstruction methods.

## 3. Results: The Depletions Of The Elements

### 3.1. THE 'STANDARD' PATTERN

Until the advent of high-resolution UV spectroscopy with the launch of *Copernicus* in 1972, little was known about the degree to which various elements were depleted from the gas. Apart from some indirect evidence that gaseous calcium might be largely locked up in dust grains in normal diffuse clouds (Routly and Spitzer 1952), no one got an overview of the depletions of the common elements until the first *Copernicus* results were published (Morton et al. 1973). Almost immediately, a comprehensive analysis of a particular line of sight, that toward $\zeta$ Ophiuchi, was completed (Morton 1974, 1975), providing the first detailed view of relative depletions of common elements. This line of sight has continued to serve as a standard for depletion studies, but analyzes of other diffuse-cloud lines of sight have shown that the pattern of depletions (meaning the relative depletions from element to element) is nearly invariant from cloud to cloud. It is instructive to examine the significance of this standard depletion pattern.

#### 3.1.1. *The 'Typical' Line Of Sight To $\zeta$ Ophiuchi*

Since the original studies by Morton (1974, 1975), a number of authors have revised, updated, and extended studies of interstellar abundances and depletions toward $\zeta$ Oph (e.g. de Boer 1979; Snow and Meyers 1979; Savage et al. 1992; Cardelli et al. 1993; Federman et al. 1993; Cardelli et al. 1994; Barlow et al. 1995; Roth and Blades 1995). Snow and Witt (1996) and Meyer (1997) have provided updated summaries of the $\zeta$ Oph depletions. The pattern shown in Fig. 1, representing the most complete and current summary of $\zeta$ Oph depletions now available, may be taken as the 'standard' depletion pattern for the sake of discussion. Later we discuss departures from this pattern.

In general, the standard (i.e. $\zeta$ Oph) depletion pattern shows a tendency for the refractory elements to be more highly depleted than the volatiles. For example, metallic species such as iron are typically depleted by a factor of 100, while others (e.g. calcium) may be depleted by a factor of 1000 or more. On the other hand, volatiles such as carbon, nitrogen, and oxygen are little depleted. This general trend was first interpreted by Field (1974) as suggestive of condensation from the gas phase onto grains under pressure

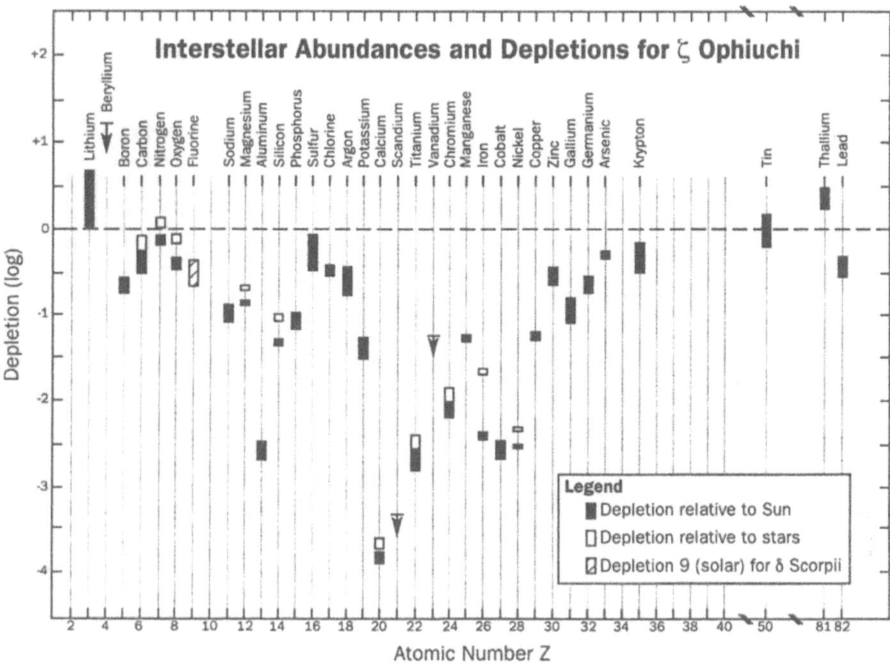


*Figure 1.* The 'standard' depletion pattern. This figure shows a compilation of the most accurate depletions for the $\zeta$ Ophiuchi line of sight, derived from references cited in the text. The filled rectangles represent depletions based on the solar standard, while the open rectangles indicate revised depletions based on stellar composition, where available. For one element, fluorine, the data shown represent another star ($\delta$ Scorpii) because this species has not been measured toward $\zeta$ Oph.

equilibrium, implying that the depletions occurred primarily as a result of processing during star formation or in stellar atmospheres.

The depletion pattern is still interpreted today by many as evidence that the diffuse-cloud depletions were established during the grain-formation process (e.g. Meyer 1997). But this is at odds with the modern view that grains undergo shock processing episodes on a much shorter timescale than their formation time (e.g. Tielens, this volume), which would suggest instead that grain composition should be governed by what survives in solid form during shock events (e.g. Savage and Bohlin 1979). Furthermore, other chemical processes may equally well explain why the refractory elements are more likely to remain in solid form (e.g. Snow 1975), so the general question of why the depletion pattern has the appearance that it does is still open to interpretation.

### 3.1.2. *The Standard Pattern: Other Lines Of Sight*

While few sight lines have been studied in the same detail as the $\zeta$ Oph cloud, a number of other diffuse clouds have been observed for elemental abundance and depletion analyzes. These include several additional stars observed with *Copernicus*; e.g. the lightly-reddened stars $\lambda$ Sco (York 1975) and $\alpha$ Virginis (York and Kinahan 1979), and several other stars with moderate extinction similar to that of $\zeta$ Oph (e.g. o and $\zeta$ Persei, Snow 1976, 1977; several stars in the $\rho$ Ophiuchi cloud, Snow and Jenkins 1980).

More recently the *IUE* has been employed for abundance and depletion studies, yielding depletion patterns similar to the standard set by the early *Copernicus* results (e.g. several stars studied with *IUE* data by Joseph et al. 1986; elemental surveys by Van Steenberg and Shull 1988a, b), but it is important to emphasize that the data quality of the *IUE* was insufficient to reveal even rather substantial depletion variations if they existed. With only a few exceptions (see below), the *IUE* results supported the idea that the diffuse-cloud depletion pattern is invariant. This, along with a compilation of *Copernicus* results by Jenkins, Savage, and Spitzer (1986), led Joseph (1988) to repropose the original idea of Field (1974), that the basic depletions are established early in the lifetimes of dust grains, but with a twist: grain mantles may grow within diffuse clouds, or be destroyed by shock processing in the diffuse ISM, but only by processes that are not selective; i.e. processes that affect all elements equally in an absolute (not proportional) sense.

The Hubble Space Telescope (*HST*), specifically through the use of the Goddard High Resolution Spectrograph (GHRS), has provided abundance and depletion measures with unprecedented accuracies. These studies have led to improved measurements of the depletion pattern in diffuse clouds (e.g. $\xi$ Persei, analyzed by Cardelli et al. 1991, 1993; the improved data on $\zeta$ Oph; Cardelli et al. 1991, 1993), but have not so far been widely applied to more heavily-reddened lines of sight (see comments below). For the diffuse clouds, the *HST*/GHRS data have not significantly altered our perception of the standard depletion pattern, but they have allowed for detailed interpretations of the depletions of specific species having weak lines only accessible to an instrument with the capabilities of the GHRS. But on the whole, even the *HST* results support the notion that relative depletions of the elements are essentially invariant in diffuse clouds.

### 3.1.3. *Significance Of Specific Elements*

So what do we learn from the generally-invariant pattern of diffuse-cloud depletions? We can focus on a few specific species and gain some insight and constraints on the composition of the dust.

Carbon is a key element in most dust models, so the depletion of carbon is a central constraint on those models. Until the availability of *HST*/GHRS data, accurate measurements of carbon column densities (i.e. of the most common ionization stage, C II) were difficult to obtain, but the spin-forbidden C II] line at 2325Å, too weak to be detected previously, has now provided accurate carbon abundances for a few lines of sight (Cardelli et al. 1996; Sofia et al. 1997). Generally the carbon depletions based on these abundances relative to the solar standard are about -0.4 in the log, leaving substantial quantities of carbon available for the dust (but note the discussion below of a possible revision of the cosmic carbon standard).

The depletions of oxygen and silicon are consistent with the formation of silicates (minerals bearing these elements in the form $SiO_4$) in the dust; i.e. the depletions are roughly consistent with an O/Si ratio in the depleted material of 4:1. The silicates may be of many possible forms, but two commonly-considered ones are magnesium orthosilicates ($Mg_2SiO_4$) and olivine ($Fe_2SiO_4$). The Mg/Si ratio in depleted material appears to be approximately 2:1, which suggests that magnesium orthosilicates dominate over olivine (Spitzer and Fitzpatrick 1993, 1995; Snow and Witt 1996). The accuracies of the abundance measurements, and particularly of the inferred depletions, are not sufficient to warrant firm conclusions as to the specific identity of the silicates, however, and indeed this remains an important question for studies of interstellar dust.

#### 3.1.4. *The Initial Conundrum: Where Did All Those Atoms Go?*

When the first comprehensive depletion studies were completed in the mid-1970s, it became apparent that the quantities of the expected major dust constituents available to form dust were more than sufficient. This led Greenberg (1974) to publish a short paper partially entitled "... Where Have All Those Atoms Gone?", in which he pointed out that large quantities of species such as carbon, nitrogen, and oxygen were unaccounted for. Even in a dust model that made generous use of carbon and oxygen, it was impossible to house all of the elements missing from the gas in the same dust that created the observed extinction.

This puzzle appears now to have been resolved, largely through the recognition that the solar standard may not be appropriate to represent the diffuse interstellar medium. This is discussed in the following section.

### 3.2. REEXAMINING THE REFERENCE STANDARD

About three years ago, the first inklings that perhaps the Sun's elemental composition might not represent the interstellar medium began to arise. Sofia, Cardelli, and Savage (1994) and Mathis (1995) first pointed out that the C/H ratio in B stars is smaller than in the Sun, and that if these stars

TABLE 1. Solar and Stellar Abundances and Revised Interstellar Depletion.

| Element | Sun | Stars | $\zeta$ Oph depletion | Dust Abundance |
|---|---|---|---|---|
| Carbon | $400/10^6$H | $214/10^6$H | -0.20 dex | $79/10^6$H |
| Nitrogen | 112 | 66 | +0.07 | 0 |
| Oxygen | 850 | 457 | -0.14 | 126 |
| Magnesium | 38 | 25 | -0.63 | 19 |
| Silicon | 35 | 19 | -1.03 | 17 |
| Iron | 47 | 27 | -2.15 | 27 |

are used instead of the Sun to represent the 'cosmic' carbon abundance, then the carbon depletion is much smaller than previously thought. This stimulus led to further studies of stellar abundances.

### 3.2.1. *Recent Compilations Of Stellar Composition*

Snow and Witt (1995, 1996) carried out exhaustive compilations of published stellar abundances, not only in B stars, but also in field F and G dwarfs and other unevolved stars in and outside of clusters. From these studies a remarkably consistent trend emerged, in which it was found that the Sun has higher abundances (relative to hydrogen) than field and cluster stars in our part of the galaxy, in virtually every heavy element. For carbon, oxygen, nitrogen, magnesium, silicon, and iron the average abundance of stars relative to the Sun is around 60 percent; i.e. the Sun has about a fifty-percent overabundance of these elements compared with other stars (Table 1).

The immediate implication of this result is that depletions based on the solar standard are overestimated by a substantial factor, which implies in turn that less of these materials are available to form dust. The conundrum posed by Greenberg (1974) appears to have been solved; in fact the new constraints on depletions are turning the question around, as we now ask whether enough of the heavy elements are available to explain the extinction due to dust.

It is noteworthy that the revised depletions of magnesium and iron are still roughly consistent with the expected ratio for orthosilicate grains. The carbon to oxygen ratio in the dust now appears to be about $C/O = 0.63$, but with substantial uncertainty.

### 3.2.2. *Is The Solar System Unique?*

One may ask how we went astray by using the solar system as our reference standard for the composition of the interstellar medium; i.e. how did the

Sun come to be overendowed in most or all of the heavy elements?

There are different interpretations possible. One is simply that the Sun represents the upper envelope of a range of compositions for stars of its age and location in the galaxy (e.g. Edvardsson et al. 1993). Another is that the formation of the solar system was accompanied by a nearby supernova explosion, and that the enrichment of heavy elements was a consequence of inclusion in the solar material of debris from that event. This idea has been espoused by a number of authors (Clayton 1977; Cameron and Truran 1977; Olive and Schramm 1982; Cameron, Höflich, and Meyers 1995; Amari, Zinner, and Lewis 1995), but it is not clear whether a supernova can account for all of the species that are enhanced. In any event, for the purpose of estimating interstellar depletions, the cause of the solar overabundances is irrelevant.

But on the other hand, the wide range of stellar compositions found in the reviews by Snow and Witt suggest that it is dangerous to allow any particular star, or perhaps even any particular group of stars, to serve as the standard. Worse, it becomes clear that the reference standard that should be applied may vary from one line of sight to another. For now the best we can do is try to settle on a uniform standard that appears to represent the majority of stars, hence the majority of interstellar clouds.

## 3.3. IMPLICATIONS FOR DUST MODELS

One of the most significant constraints on models of the interstellar dust is that a viable model must be able to account for the observed extinction without requiring more material, or greater quantities of specific elements, than are available in the context of the cosmic abundance reference. The recent revisions in the cosmic abundances have therefore provided new constraints on dust models.

### 3.3.1. *Carbon And Constraints On Dust Models*

Snow and Witt (1995, 1996) review extant dust/extinction models and summarize the new constraints (see Table 1). Regarding carbon, they find that models invoking disordered carbon to explain the general extinction are now in difficulty, as amorphous carbon is too optically inefficient to account for the extinction within the new cosmic carbon budget. On the other hand, extinction models based on carbon in ordered form (e.g. graphite, PAHs) are much more successful, and appear to have little difficulty accounting for the extinction even under the reduced abundance.

Ordered carbon is already required to explain some (perhaps all) of a number of discrete spectral features (e.g. the 2175 Å extinction bump; the infrared emission bands; and possibly the unidentified diffuse interstellar bands), so to adopt ordered forms of carbon to account for the general extinction as well is not a new idea.

### 3.3.2. *Silicate constraints and dust models*

The new constraints on silicate-forming elements again place severe limitations on some models. All models seem now to have a problem with oxygen, particularly the organic refractory mantle models (e.g. Hong and Greenberg 1980). In addition, the revised abundances of magnesium, silicon, and iron are now very close to the minimum needed in most models involving silicate dust.

### 3.3.3. *The dust theorist's response*

To date only a couple of dust models (Mathis 1996; Li and Greenberg 1997) have been produced which reflect the revised cosmic abundance scale and the newer, smaller, depletions. Both rely now on ordered carbon to produce the far-ultraviolet extinction; Mathis also invokes grains with a high porosity (e.g. very 'fluffy' grains) in order to maximize the optical leverage of the material present, while Li and Greenberg have added ordered carbon to help with the UV extinction, but still admittedly face a problem with carbon, which now appears not to be sufficiently abundant to account for the organic refractory grain mantles. The latter authors note that *all* models have a carbon problem, and raise the question of whether the newly-revised cosmic abundance may be incorrect. We comment further on this in our concluding section, below.

## 4. Departures From The Standard Depletions

### 4.1. DEPENDENCE ON CLOUD VELOCITY

One of the very first papers to discuss depletions (Routly and Spitzer 1952) did so in the context of variable depletions with cloud velocity. The ratio of Ca II/ Na I in diffuse clouds was found to vary widely, with a maximum in high-velocity clouds. Later Siluk and Silk (1974) further explored this phenomenon, concluding that shock disruption of grains could explain the enhanced Ca II abundance in rapidly-moving clouds.

At the same time, *Copernicus* data began to reveal substantially reduced depletions in high-velocity clouds for a number of refractory species (e.g. Shull, York, and Hobbs 1977; Shull 1977a; Shull and York 1977), again with the interpretation that shock disruption was responsible (Shull 1977b, c; for a more modern view see also the review by Tielens, this volume, and references cited therein). Further observational work showed that even modest velocities were sometimes associated with reduced depletions, as in the case of $\zeta$ Oph, whose -28 km/sec component was found from *Copernicus* data to have much less refractory depletion than the main low-velocity component (Snow and Meyers 1979).

It is now clear, based on a large body of observational and theoretical work, that dust grains can be eroded or even destroyed by shocks, and that this alters depletions in shocked regions. What is not so clear is whether this process has much to do with the observed general diffuse-cloud depletions, because it is difficult to determine what fraction of the observed material in a typical line of sight has been shocked. As discussed, for example, by Savage and Bohlin (1979) and Joseph (1988), the observed depletion pattern may be governed by shock processing; i.e. the measured gas-phase abundances and depletions may be controlled by what has been *removed* from grains as opposed to what has adhered to them. This question thus becomes embroiled in the larger issue, summarized by Tielens (this volume), of the evolution of interstellar dust grains and whether most grains have been repeatedly shock-processed during their lifetimes.

### 4.2. DEPENDENCE ON CLOUD DENSITY

It has long been supposed, on intuitive grounds, that interstellar depletions ought to be density-dependent; i.e. that the depletions should be enhanced in regions of high density because in such regions the frequency of gas-grain collisions should be enhanced. But this supposition is based on the implicit assumption that the depletions are due to *in situ* gas-grain interactions in which gas-phase material is deposited on the grains. In contrast, if the depletions are governed by the dust formation process, then we would not expect a density dependence, since the conditions under which the depletions were created would be disconnected from the current conditions in the observed interstellar clouds. If the depletions are controlled by shock processing, as discussed above, then we might still expect a density dependence, because shocks cannot penetrate high-density regions as well as low-density or intercloud regimes.

We can consider the density-dependence of depletions in at least two distinct contexts: comparison of lines of sight having contrasting total densities (i.e. total gas column density divided by path length); or comparison of different localized regions (with contrasting densities) along a given line of sight.

#### 4.2.1. *Integrated line-of-sight depletions*

Comparison of contrasting lines of sight (having different total densities) is the more readily accomplished approach, but is also probably the less informative due to ambiguities caused by clumpiness along the lines of sight (high-density regions will dominate the observed column densities, rendering meaningless the averaged density over the full line of sight). Nevertheless, it is useful to consider integrated line-of-sight depletions, especially in circumstances where it appears likely that a single cloud dominates the observed column densities.

One of the first studies aimed at this problem was the survey by Snow and Jenkins (1980) of selected elements toward stars in the $\rho$ Oph cloud. They found that there were no definite variations in *relative* depletions from element to element, while there was a steady progression of increasing *total* depletions with increasing density (i.e. with increasing depth within the cloud). This growth of total depletion could be attributed to non-selective sticking of elements to the grains as the density increased, to non-selective sputtering of elements off of the grains in the less dense portions of the cloud, or to a growing contribution of a pure hydrogen component of the cloud (that is, an interior region with depletions so great that the column densities of the heavy elements remain virtually unchanged with depth in the cloud).

Another approach to the question of possible density-dependent depletions was pursued by Joseph et al. (1986), who used *IUE* data to survey abundances in a group of stars selected for high cloud density according to the presence of strong CN absorption in optical wavelengths. This group of stars was compared with a separate sample having no detected CN absorption, which was taken as a control group of lower-density lines of sight (chemical models indicate that CN is very density-dependent; van Dishoeck and Black 1986). The strong-CN lines of sight were found by Joseph et al. to have systematically higher depletions than the control group, suggesting that depletions are indeed enhanced with increasing density. There was little evidence for selective effects for individual elements, except that it appeared that manganese might be selectively more depleted, relative to other species, in the strong-CN clouds. It is noteworthy that the strong-CN clouds also tend to be distinguished by steep far-UV extinction rises; this is discussed further in a later section.

#### 4.2.2. *Diffuse cloud cores*

In contrast to considering integrated lines of sight, under some circumstances it is possible to compare abundances and depletions in dense regions within a given line of sight with the total column densities and depletions integrated over the entire line of sight. One technique for doing this involves the comparison of lines due to neutral atoms with those of ions of the same elements which are predominantly ionized in diffuse clouds. In these cases, the neutral atomic lines are very density-dependent and arise exclusively in the cores of the clouds (i.e. in the regions of highest density; it is not clear that a simple picture of a cloud with a dense core surrounded by a more diffuse region is a realistic model). But because the neutral atomic lines automatically select the highest-density portion of a line of sight, these lines are useful for comparing depletions in the densest regions with those in the integrated diffuse interstellar material.

The basic equation of ionization equilibrium in the diffuse interstellar medium, where the ionization is predominantly radiative, is

$$n_{i+1} n_e / n_i = \Gamma / \alpha, \tag{2}$$

where $n_{i+1}$ and $n_i$ represent the ionized and neutral forms of an element, $n_e$ is the electron density, $\Gamma$ is the photoionization rate, and $\alpha$ is the radiative recombination rate coefficient. Clearly some of these terms depend on local conditions (e.g. $\Gamma$ is a function of the local radiation field intensity, $\alpha$ depends on temperature, and of course the electron density $n_e$ is a variable). But when this equation is written for two different elements and the results are divided, some of these locally-dependent terms cancel out. We find for elements A and B that

$$(N_{i+1})_A / (N_{i+1})_B = [(N_i)_A/(N_i)_B] [(\Gamma/\alpha)_A / (\Gamma/\alpha)_B], \tag{3}$$

where we have assumed that the ratio of column densities (N) within the neutral atomic zone is approximately equal to the ratio of space densities (n), and that the electron density $n_e$ cancels out. Furthermore, it seems reasonable also to assume that the dependences on the radiation field intensity (in the case of $\Gamma$) and temperature (for $\alpha$) cancel out, at least approximately.

This expression allows one to infer the relative column densities, in the neutral cloud core, of the dominant ions ($N_{i+1}$) based on the observed ratios of the neutral atoms ($N_i$), and thus to assess the depletions in the neutral regions as compared with the integrated line of sight (represented by the observed ion column densities). It is impossible to use hydrogen as the reference element (element B), so another reference standard must be chosen. Typically sulfur (i.e. S I) or sodium (Na I) are used, because both elements are generally little-depleted.

In the only attempt at a survey of cloud core depletions done to date, Snow (1984) found that in most cases the depletions in the cloud cores were remarkably similar to those derived from the integrated lines of sight, based on the ion column densities. In other words, in the sample of clouds surveyed, there was little evidence for dense regions or clumps having enhanced depletions. The one exception was $\zeta$ Oph, the only star in the sample known to have detectable CN absorption and thus the only line of sight known to contain a dense core.

Subsequent studies of neutral atomic lines (Joseph, Snow, and Seab 1989; Cardelli, Federman, and Smith 1991) suggested that in neutral cloud cores of sufficient density (again molecular abundances were used as the criterion), depletions are vastly enhanced over those in more diffuse regions. Both of these studies concluded that the lines of neutrals (Fe I in the Joseph et al. study; Ca I in the case of Cardelli et al.) virtually disappear in cloud cores, presumably due to the very high depletions that occur there.

#### 4.2.3. *Conclusions and questions*

It seems evident that there is some density-dependence of depletions, but questions remain. Perhaps the most basic issue is whether the data represent enhanced depletions inside dense clouds (due to accretion of gas particles onto grains), or diminished depletions in the outer clouds (due to sputtering of material off of grains). This may be a difficult question to settle, since the observed consequences from either mechanism are essentially identical. The specific elements that stick most readily to grains in the first place would likely also be the ones most difficult to dislodge in the second place, so the pattern of relative depletions should be similar in either case.

It may be that both mechanisms are at work, but that there is a threshold in density above which accretion becomes dominant. If so, it will be important to determine at what point (in terms of total extinction, for example) this occurs, and how it relates to other thresholds, such as the point within a cloud where grain mantle materials (such as water ice and hydrocarbons) begin to form detectable infrared absorption features.

### 4.3. DEPENDENCE ON EXTINCTION PROPERTIES

If the observed extinction due to dust is created by the same grains that are responsible for the depletions of gas-phase material into solid form, then clearly it is important to explore the relationship between depletion and extinction. This may provide clues as to the nature of the depletion process, and certainly should provide information on the effect of depletions on the optical properties of the grains where the depleted material resides. This is an area where attention is beginning to be focused, but a lot of work remains to be done.

#### 4.3.1. *Depletions and the far-UV extinction rise*

Surveys of ultraviolet extinction have shown that the far-ultraviolet portion of the extinction curve displays a wide range of behavior (e.g. Bless and Savage 1972; Fitzpatrick and Massa 1986, 1988, 1990). The rise toward short wavelengths that begins around 1500 Å continues, but with highly varying degrees of curvature and slope. There are also variations in the area (equivalent width) and width of the prominent extinction bump at 2175 Å, and it appears that these are linked to the variations in the nature of the far-UV rise (e.g. the series of papers by Fitzpatrick and Massa, cited above).

In a study relating UV, optical, and IR extinction, Cardelli, Clayton, and Mathis (1989) demonstrated that for most lines of sight there is a consistent relationship between the nature of the UV extinction curve (the bump and the far-UV rise) and the ratio of total to selective extinction ($R_V$). This relationship suggests that the grain properties that govern both the UV and visible extinction are linked, and in addition provides a means of predicting the UV extinction behavior for a given line of sight, based on $R_V$. Some deviations from the standard relationship were discussed later by two of the same authors (Mathis and Cardelli 1992).

To date relatively little work has been done to explore the dependence of depletions on the nature of the extinction curve. The paper of Joseph et al. (1986) illustrated that clouds with strong molecular abundances (i.e. CN) appear to have enhanced depletions, and in the same paper it was pointed out that these same clouds typically have unusually steep far-UV extinction rises. Later Cardelli (1988) showed that molecular abundances (specifically $H_2$ and CH) are related to $R_V$ in the sense that small values of $R_V$ (which correlate with a steep far-UV rise) go hand-in-hand with high molecular abundances. This suggests indirectly that depletion and steep far-UV extinction might be related, in view of the studies (e.g. Joseph et al. 1986; Joseph, Snow, and Seab 1989; Cardelli, Federman, and Smith 1991) illustrating a link between high depletions and molecular abundances.

Recently Meyer (1997) showed that the depletion of magnesium correlates strongly with the molecular fraction of hydrogen, particularly when the fraction is greater than about 0.1 (i.e. more than one-tenth of the hydrogen nuclei are in the form of $H_2$). Meyer did not extend his discussion to the possible relationship of magnesium depletion to the nature of the extinction curve, but given the earlier Cardelli (1988) result, we can infer that there must be an inverse correlation between magnesium depletion and $R_V$, i.e. clouds with unusually steep far-UV extinction are likely to have enhanced magnesium depletions. This is consistent with the 1986 results of Joseph et al., cited above.

The various studies suggesting that depletions are enhanced in clouds having steep far-UV extinction and high molecular abundances were summarized and extended by Snow (1992), who developed a possible scenario for the evolutions of clouds with contrasting extinction, molecular, and depletion properties. In that picture, clouds having high far-UV extinction (i.e. low values of $R_V$), high molecular abundances, and high depletions were found to be notably devoid of evidence for recent star formation; whereas the clouds with low far-UV extinction (i.e. high values of $R_V$), low molecular abundances, and normal depletions were found to harbor embedded hot stars and therefore to have undergone recent star formation. The low UV extinction in the latter clouds was attributed to grain coagulation which gives rise to an absence of the small grains normally thought responsible for steep far-UV extinction, and the low molecular abundances were attributed either to diminished grain surface area (for the formation of $H_2$; see also Snow 1983) or to enhanced UV radiation (due to the embedded OB stars) or both. Thus the two types of environments may be distinguished largely by whether or not recent formation of massive stars has taken place.

The question of to what degree depletions are enhanced in clouds having steep far-UV extinction (and high molecular abundances) warrants further exploration, but of course the problem is self-limited, due to the fact that the clouds with unusually steep far-UV extinction are, by definition, the most difficult to observe in the UV where most of the important spectral lines needed for abundances analyzes are to be found. Nevertheless, it is possible to probe more deeply than has been done so far, particularly with the UV spectrographs aboard the *Hubble Space Telescope.*

4.3.2. *The relationships of specific elemental depletions to extinction*

The *HST* affords the opportunity to measure abundances of some specific elements with extremely high accuracy. Thus, apart from studies of the general depletion pattern as a function of extinction curve properties, it has become possible to explore in some detail the behavior of a few specific elements. Among the most important of these are carbon and oxygen, the principal constituents of carbonaceous and silicate grains, respectively.

In a series of papers Cardelli and collaborators (Cardelli et al. 1991, 1993, 1996; Sofia et al. 1997) have reported the abundance and depletion of carbon for several lightly- to moderately-reddened lines of sight, making use of the spin-forbidden line of C II] at 2325 ÅThis line is so weak that it is never saturated; instead the key questions are whether it can be detected and whether the oscillator strength is reliable (of course the latter point has no effect on relative carbon depletions from one line of sight to another). The approach of Cardelli and collaborators has been to observe stars bright enough in the UV to provide extremely high S/N in relatively short exposure times. As a result, most of what we know about carbon depletions is limited to lines of sight having relatively low extinction. It is also important to note that only seven C II] abundances have been unambiguously measured so far, so the sample upon which these conclusions are based is very small.

The most striking result of the published studies is that the carbon depletion appears to be virtually unrelated to the extinction, almost as if carbon has nothing to do with the grains that produce the UV extinction curve. Very similar depletions are found in lines of sight having steep far-

UV extinction and in sight lines having relatively flat far-UV extinction. It is very difficult to interpret this result. Cardelli et al. (1996) conclude that variations in extinction, including features such as the 2175 Å bump that are normally attributed to carbon, are not due to variations in the abundance of carbon in the dust.

A more recent result (Snow et al. 1998) shows that in at least one line of sight (HD 24534), which is substantially more reddened than the stars included in the Cardelli surveys, there is significantly greater carbon depletion than in the less-reddened sight lines. Thus, it is very important to extend measurements of the C II] line to lines of sight having greater extinction, to increase the dynamic range of column densities and extinctions. The Snow et al. result shows that the increased carbon depletion in this one case does not lead to any significant modification of the extinction curve; i.e. the line of sight having the greater carbon depletion has a normal-appearing extinction curve. This leads to the converse of the Cardelli et al. conclusion, because it suggests that not only is the carbon depletion not dependent on the extinction, but that in turn the extinction is unaffected by the carbon depletion.

One very drastic conclusion that could be drawn from these results is that carbon has nothing whatsoever to do with interstellar dust. Perhaps more reasonable interpretations are that the carbon depletion does not affect the grain size distribution enough to modify the extinction curve; that the depletions of all the other elements vary in concert with the carbon depletion, so that the grain composition and optical properties are unaffected by different degrees of depletion; or that the added carbon is in unordered form and therefore does not contribute to the extinction (as noted in Section 3.3.1 above, it appears that carbon must be in some ordered form in order to create the extinction under the revised cosmic carbon abundance). At this point all three of these possible interpretations seem viable.

Another very interesting case is presented by oxygen. Meyer (1997) has compiled O I abundances (based again on a weak spin-forbidden line, this one at 1355 Å) for some 13 stars that have been observed with the GHRS on the *HST*, and finds a result somewhat similar to the carbon situation. There is no significant variation in the gas-phase O/H ratio for these 13 lines of sight, even though there is a substantial range in the molecular fraction of hydrogen and, by inference, in the value of $R_V$ and the nature of the UV extinction. Again, one might conclude that the degree of depletion of an element thought to play a central role in interstellar dust appears to have no relationship to the extinction curve. In this case perhaps one might conclude that silicates are not important constituents of the small grains that are thought to be responsible for the far-ultraviolet extinction and its variation.

It is important to note that both carbon and oxygen behave differently from magnesium, and presumably other refractories, in that the depletions of C and O are nearly invariant (within the limited range of observed reddenings) while magnesium depletion shows a substantial range of depletions. It will be very interesting to see whether this contrasting behavior extends into clouds of higher density and extinction than have so far been studied with the *HST*; i.e. is there a point beyond which even the volatiles begin to undergo enhanced depletion? The Snow et al. (1997) results would appear to suggest that there is such a point.

## 5. Future Work On Depletions

There are many areas where additional observational work is needed in the effort to determine the role of depletions in the evolution of dust, as well as isolating the mechanisms that contribute to depletions. Much more can be done with existing instruments, but some of the new work will require future technology developments.

### 5.1. FURTHER EXAMINATION OF 'COSMIC' ABUNDANCES

The problem of knowing the reference abundances of the elements relative to hydrogen has emerged as a serious one. It is difficult to develop and test grain models without reliable information on how much material is available for the dust. As noted by Meyer (1997), the improved accuracy in measuring gas-phase abundances that is now possible with the *HST* has shifted the limiting factor in depletion measurements from observational errors in the column densities to uncertainties in knowing the relevant reference abundances. In other words, imperfect knowledge of the 'cosmic' abundances of the elements is now the leading source of uncertainty in depletion studies.

One possibility, which is difficult to address, is that stars actually do not represent the composition of the interstellar medium. If during star formation there is any kind of selection process by which some elements are expelled before being accreted onto the star, then those elements would be underrepresented in the star. More specifically, if refractory elements are somehow ejected from pre-stellar cloud cores (perhaps by radiation pressure or magnetic forces acting on grains), then the resulting stellar composition will reflect lower abundances of refractories than are present in the interstellar medium. In that case the use of stellar composition as representative of 'cosmic' abundances is invalid, and depletions derived on that basis are underestimated. This could alleviate the pressure now being brought to bear on dust models, by allowing for larger quantities of carbon, oxygen, silicon iron, etc., than are normally thought to be present. This suggestion would not, however, alter the apparent fact that the Sun has greater abundances of these elements than other stars.

It is difficult to see how to improve the situation, especially in view of evidence that in fact the ISM may have inhomogeneities in composition, over relatively small distance scales. But there is promise in one method for measuring interstellar abundances: the exploitation of inner-shell X-ray absorption edges, as described briefly in Section 1.4 above. The power of this method will be particularly apparent in cases where both X-ray and UV absorption measures can be made toward the same source, because then one can measure the total abundances (in gas and dust) from X-ray spectra, and at the same time, for precisely the same column, also measure the gas-phase abundances alone (from UV absorption lines). This will not be possible for very many lines of sight, but will be extremely useful where it can be done. The *AXAF* mission, to be launched in 1999, may provide the opportunity.

## 5.2. PROBING CLOUDS OF GREATER EXTINCTION

A glaring deficiency in existing depletion measures is the complete lack of data on stars of total extinction much greater than $A_V$ of about 2.5 magnitudes. Because of the steep rise in far-UV extinction, obtaining data for heavily-reddened stars is a difficult challenge. But even with present-day instrumentation a lot more could be done.

To our knowledge, the most heavily-reddened star yet observed with the *HST* for depletion measures is HD 154368 (Snow et al. 1996), which has $E_{B-V} = 0.82$ and $A_V = 2.54$ ($R_V = 3.1$). This, along with the more recent (and less comprehensive) study of HD 24534 (Snow et al. 1998) represent the first (and so far the only) forays into the regime of the translucent clouds. Yet the *HST*, with its current instrumentation (now the STIS, which replaced the GHRS) is capable of doing more, if the appropriate assignment of time could be made. In its highest-resolution mode, the STIS could reach stars having $A_V$ as high as 3.0 mag, and in its moderate-resolution modes $A_V$ values as high as 4 mag could be reached. Yet this has not been done, because the time allocation committees have consistently refused to grant the substantial amounts of time needed.

In principle (and perhaps in practice, pending the outcome of the instrument selection process now under way for the 2002 refurbishment) the *HST* can provide abundance and depletion measures for even larger extinctions. It is not unreasonable to envision a UV spectrograph for the *HST* that is capable of measuring UV absorption lines at moderate resolution (e.g. $\lambda/\Delta\lambda = 30,000$) in stars having $A_V$ values as high as 8.5 (for high values of $R_V$) or 5.5 (for low $R_V$). This would allow depletions to be measured throughout the range of translucent clouds, and even begin to touch on the regime of dark clouds.

One important benefit of penetrating denser clouds, in addition to further exploring the correlations between depletion and other cloud properties such as density, molecular abundances, and extinction curve properties, is to probe the relationship between depletion and the appearance of grain mantles. Infrared observations of solid-state mantle features such as the 3.4 $\mu$m hydrocarbon band and the 3.08 $\mu$m water ice band have been pushed into the translucent cloud regime, and have revealed that thresholds for the appearance of these features arise in the range of $A_V = 3$ to 5 (e.g. Whittet et al. 1988; Whittet 1993; Pendleton et al. 1994). It is obviously important to see how the appearance of these mantles may be related to the depletions, and also to the nature of the UV extinction curve. To fully exploit this will require not only pushing deeper in the UV, for example with the *HST*, but also pushing the limits (going toward low extinction) where the infrared features can be detected. This kind of multispectral approach to the problem of relating depletions to grain properties will prove very fruitful.

## 6. Acknowledgements

It is appropriate to acknowledge with gratitude two late colleagues whose stimulation and inspiration underlie this work: Jason Cardelli, who passed away entirely too soon, and Lyman Spitzer, who pioneered this entire field of research. I am also grateful for the many talks and discussions of dust and depletions that took place at this School, which helped to shape this review. This work has been supported in part by NASA grant NAGW-3600 and NASA/STScI grant GO-2415 to the University of Colorado.

## References

Amari, S., Zinner, E., and Lewis, R. S. 1995, Astrophys. J., 443, 563

Anders, E. and Grevesse, N. 1989, Geochim. Cosmochim. Acta., 53, 327

Barlow, J. M., Crawford, A. I., Diego, F., Dryburgh, M., Fish, A. C.,
Howarth, I. D., Spyromilio, J., and Walker, D. D. 1995, Mon. Not. Roy. Astron. Soc.., 272, 333

Bless, R. C. and Savage, B. D. 1973, Astrophys. J.., 171, 293

Bohlin, R. C. 1975, Astrophys. J.., 200, 402

Cameron, A. G. W., Hflich, P., and Meyers, P. C. 1995, Astrophys. J. Lett., 447, L53

Cameron, A. G. W. and Truran, J. W. 1977, Icarus, 30, 447

Cardelli, J. A. 1988, Astrophys. J., 335, 177,

Cardelli, J. A., Clayton, G. C., and Mathis, J. S. 1989, Astrophys. J., 345, 245

Cardelli, J. A., Federman, S. R., and Smith, V. V. 1991, Astrophys. J. Lett., 381, L17

Cardelli, J. A., Savage, B. D., Bruhweiler, F. C., Smith, A. M., Ebbets, D.C., Sembach, K. R., and Sofia, U. J. 1991, Astrophys. J. Lett., 377, L57

Cardelli, J. A., Mathis, J. S., Ebbets, D. C., and Savage, B. D. 1993, Astrophys. J. Lett., 402, L17

Cardelli, J. A., Sofia, U. J., Savage, B. D., Keenan, F. K., and Dufton, P. L. 1994, Astrophys. J. Lett., 420, L29
Cardelli, J. A., Meyer, D. M., Jura, M. and Savage, BZ. D. 1996, Astrophys. J., 467, 334
Clayton, D. C. 1977, Icarus, 32, 255
de Boer, K. S. 1979, Astrophys. J., 229, 132
Edvardsson, B., Anderson, J., Gustafsson, B., Lambert, D. L., Nissen, P., and Tomkin, J. 1993, Astron. Astrophys., 275, 101
Federman, S. R., Sheffler, Y., Lambert, D. L., and Gilliland, R. L. 1993, Astrophys. J. Lett., 413, l51
Field, G. B. 1974, Astrophys. J., 187, 453
Fitzpatrick, E. L. and Massa, D. 1986, Astrophys. J., 307, 286
Fitzpatrick, E. L. and Massa, D. 1988, Astrophys. J., 335, 703
Fitzpatrick, E. L. and Massa, D. 1990, Astrophys. J. Suppl., 72, 163
Greenberg, J. M. 1974, Astrophys. J. Lett., 189, L81
Grevesse, N. and Noels, A. 1993, in Origin and Evolution of the Elements, eds. N. Prantzos et al., [Cambridge: Cambridge University Press], p. 15
Grevesse, N., Sauval, A. J., and Blomme, R. 1992, in Infrared Solar Physics, eds. D. Rabin and J. T. Jefffries [Dordrecht: Kluwer], p. 539
Holweger, H. and Mueller, E. A. 1974, Solar Phys., 39, 19
Hong, S. S. and Greenberg, J. M. 1980, Astron. Astrophys., 88, 194
Jenkins, E. B. 1986, Astrophys. J., 304, 739
Jenkins, E. B., Savage, B. D., and Spitzer, L. 1986, Astrophys. J., 301, 355
Joseph, C. L. 1988, Astrophys. J., 335, 157
Joseph, C. L., Snow, T. P., and Seab, C. G. 1989, Astrophys. J., 340, 314
Joseph, C. L., Snow, T. P., Seab, C. G., and Crutcher, R. M. 1986, Astrophys. J., 309, 771
Lambert, D. L. 1978, Mon. Not. Roy. Astron. Soc., 182, 249
Li, A. and Greenberg, J. M. 1997, Astron. Astrophys., 323, 566
Mathis, J. S. 1995, in The Diffuse Interstellar Bands, eds. A. G. G. M. Tielens and T.P. Snow [Dordrecht: Kluwer], p. 241
Mathis, J. S. 1996, Astrophys. J., 472, 643
Mathis, J. S. and Cardelli, J. A. 1992, Astrophys. J., 398, 610
Meyer, D. M. 1997, in Astrophysical Implications of the Laboratory Study of Presolar Materials, ed. T. Bernatowicz and E. Zinner [New York: American Institute of Physics], p. 507
Morrison, R. and McCammon, D. 1983, Astrophys. J., 270, 119
Morton, D. C. 1974, Astrophys. J. Lett., 193, L35
Morton, D. C. 1975, Astrophys. J., 197, 85
Morton, D. C., Drake, J. F., Jenkins, E. B., Rogerson, J. B., Spitzer, L., and York, D.G. 1973, Astrophys. J. Lett., 181, L103
Olive, K.A. and Schramm, D. N. 1982, Astrophys. J., 257, 276
Pendleton, Y. J., Sandford, S. A., Allamandola, L. J., Tielens, A. G. G. M., and Sellgren, K. 1994, Astrophys. J., 437, 683
Roth, K. C. and Blades, J. C. 1995, Astrophys. J., 445, 195
Routly, P. M. and Spitzer, L. 1952, Astrophys. J., 115, 227
Ryter, C., Cesarsky, C. J., and Audouze, J. 1975, Astrophys. J., 198, 103
Savage, B. D. and Bohlin, R. C. 1979, Astrophys. J., 229, 136
Savage, B. D., Cardelli, J. A., and Sofia, U. J. 1992, Astrophys. J., 401, 706
Shull, J. M. 1977a, Astrophys. J., 212, 102
Shull, J. M. 1977b, Astrophys. J., 215, 805
Shull, J. M. 1977c, Astrophys. J., 216, 414
Shull, J. M. and York, D. G. 1977, Astrophys. J., 211, 803
Shull, J. M., York, D. G., and Hobbs, L. M. 1977, Astrophys. J. Lett., 211, L139
Siluk, R. S. and Silk, J. 1974, Astrophys. J., 192, 51
Snow, T. P. 1975, Astrophys. J. Lett., 202, L87

Snow, T. P. 1976, Astrophys. J., 204, 759
Snow, T. P. 1977, Astrophys. J., 216, 724
Snow, T. P. 1983, Astrophys. J. Lett., 269, L57
Snow, T. P. 1984, Astrophys. J., 287, 238
Snow, T. P. 1992, Australian J. Phys., 45, 543
Snow, T. P., Black, J. H., van Dishoeck, E. F., Burks, G., Crutcher, R. M., Lutz, B.L., Hanson, M. M., and Shuping, R. Y. 1996, Astrophys. J., 465, 245
Snow, T. P., Black, J. H., van Dishoeck, E. F., Crutcher, R. M., and Lutz, B. L. 1998, Astrophys. J. Lett., 496, L113
Snow, T. P. and Jenkins, E. B. 1980, Astrophys. J., 241, 161
Snow, T. P., Lamers, H. J. G. L. M., and Joseph, C. L. 1987, Astrophys. J., 321, 952
Snow, T. P. and Meyers, K. A. 1979, Astrophys. J., 229, 545
Snow, T. P. and Witt, A. N. 1995, Science, 270, 1455
Snow, T. P. and Witt, A. N. 1996, Astrophys. J. Lett., 468, L65
Sofia, U. J., Cardelli, J. A., and Savage, B. D. 1994, Astrophys. J., 430, 650
Sofia, U. J., Cardelli, J. A., Guerin, K. P., and Meyer, D. M. 1997, Astrophys. J. Lett., 482, L105
Spitzer, L. 1978, Physical Processes in the Interstellar Medium [New York: Wiley]
Spitzer, L. and Fitzpatrick, E. L. 1993, Astrophys. J., 409, 299
Spitzer, L. and Fitzpatrick, E. L. 1995, Astrophys. J., 445, 196
van Dishoeck, E. F. and Black, J. H. 1986, Astrophys. J. Suppl., 62, 109
Van Steenberg, J. M. and Shull, J. M. 1988a, Astrophys. J. Suppl., 67, 335
Van Steenberg, J. M. and Shull, J. M. 1988b, Astrophys. J., 330, 942
Whittet, D. C. B. 1993, in Dust and Chemistry in Astronomy, eds. T. J. Milar and D. A. Williams [London:Institue of Physics], p. 9
Whittet, D. C. B., Bode, M. F., Longmore, A. J., Adamson, A. J., MacFadzean, A. D., Aitken, D. K., and Roche, P. F. 1988, Mon. Not. Roy. Astron. Soc., 233, 321
York, D. G. 1975, Astrophys. J. Lett., 196, L103
York, D. G. and Kinahan, B. F. 1978, Astrophys. J., 228, 127

# THE NATURE AND EVOLUTION OF INTERSTELLAR ORGANICS

JEAN E. CHIAR
*Department of Physics, Applied Physics & Astronomy, Rensselaer Polytechnic Institute, Troy, NY 12180, USA*[†]

AND

YVONNE J. PENDLETON
*NASA Ames Research Center*

**Abstract.** Infrared spectroscopy of solid state absorption features in dense and diffuse interstellar clouds has greatly advanced our understanding of the composition of dust in these regions. However, serious questions regarding the evolution of dust grains have now developed as a result of recent compositional discoveries. The creation of complex organic refractory material on or in icy mantles of grains residing in the protected environs of dense molecular clouds is thought to be the pathway by which organic carriers are formed. Two absorption features in particular, one near 3.4 $\mu$m and another near 4.62 $\mu$m have been attributed to an organic refractory residue which can be produced by ultraviolet photolysis of interstellar ices. However, even though the production site for both features is expected to occur within dense molecular clouds, these features are observed in very different physical regimes of the interstellar medium (ISM). The 3.4 $\mu$m feature is observed in the low density, low temperature portion of the diffuse ISM, and the 4.62 $\mu$m feature is observed towards embedded protostars in high density molecular clouds. Dust grains are thought to cycle between the dense and diffuse clouds fairly efficiently and on rather short timescales, thus the noteworthy absence of the diffuse ISM 3.4 $\mu$m hydrocarbon absorption band and the unique presence of the 4.62 $\mu$m band in the spectra of dense molecular cloud objects presents difficulties for the production site and/or the cycling efficiency. In this paper, we focus on comparisons between the dense and diffuse clouds which specifically address the nature and evolution of the organic component of interstellar dust. A greater understanding of the origin and distribution of interstellar organics is essential

[†]current address: NASA Ames Research Center Moffett Field, CA 94035, USA



*J. M. Greenberg and A. Li (eds.), Formulation and Evolution of Solids in Space,* 29–52.

in any evaluation of the organic inventory available for incorporation into planetary systems and the precursor planetesimal stage.

## 1. Introduction

The basic building blocks of life on Earth originated in stars which formed out of interstellar gas and dust. As discussed by Tielens (this volume), the lifecycle of dust grains includes the injection of the dust into the diffuse interstellar medium (ISM) from their formation site in the circumstellar shells of evolved stars. Upon break-up of dense molecular clouds, primarily due to star formation processes, refractory grain components processed on icy grain mantles during their residence within the cloud (Greenberg 1982) are likely added to the diffuse ISM dust. New generations of dense molecular clouds form out of the swept-up combination of circumstellar dust and the recycled products from previous dense clouds. Given the expected short cycling times between the dense and diffuse ISM (McKee 1989), spectroscopic signatures indicative of refractory organic species should be observable in both the dense and diffuse ISM. In actuality, the organic solid-state features observed at near-infrared wavelengths along sightlines through dense clouds and the diffuse interstellar medium are quite different, revealing little commonality between the locales where key signature organic features are observed.

Attempts to characterize the composition of interstellar dust in both the dense and diffuse ISM have led to observations of bright background sources observed through great quantities of interstellar dust. These features are not intrinsically related to the sources which illuminate the column of dust, but rather are carried by the intervening dust along our line-of-sight. A multitude of absorption and emission features have been observed in this manner (see Whittet & Tielens 1997). Among the many spectroscopic signatures observed in the near infrared are two absorption bands that may be the strongest representatives of the processed organic component of interstellar dust. The 3.4 $\mu$m and 4.62 $\mu$m absorption bands have both been attributed to energetic processing of interstellar ices in dense molecular clouds, yet they appear in mutually exclusive locations in the ISM. The 3.4 $\mu$m absorption feature(s), attributed to the C-H stretching fundamental bands in aliphatic hydrocarbon grains (Sandford et al. 1991), are present along more than a dozen sightlines through the diffuse ISM in our galaxy (Pendleton et al. 1993) and are also seen in dust-embedded Seyfert galaxies (Bridger, Wright, & Geballe 1993). Until recently, the presence of these features in the diffuse ISM has been taken as evidence of the survival of the organic component of interstellar dust from its formation site in the dense cloud to the diffuse interstellar medium environment. However, the noted absence of the 3.4 $\mu$m aliphatic hydrocarbon sub-features in any dense cloud spectra obtained to date raises important questions concerning the origin and/or distribution of the carrier of this band (Allamandola et al. 1992).

Likewise, the near-infrared absorption signature seen at 4.62 $\mu$m in dense cloud spectra towards bright, embedded protostellar objects also presents a problem for the standard picture of dust grain evolution. Possible identifications for this feature include some type of nitrile (CN) or isonitrile (NC) found in the residues produced through energetic processing of typical interstellar ice analog mixtures (Lacy et al. 1984; Tegler et al. 1993, 1995). See reviews by Schutte and Greenberg (this volume) for reviews of laboratory processing of interstellar ice mixtures. In laboratory experiments, the strength of the 4.62 $\mu$m band increases as the temperature is raised after the volatile components of the ices have evaporated (Bernstein, Sandford, & Allamandola 1996). For this reason, it is expected that if the observed 4.62 $\mu$m band is due to a processed ice residue, then it might survive incorporation into the diffuse ISM. However, the 4.62 $\mu$m feature, detected towards embedded protostellar sources, has thus far not been detected along diffuse interstellar medium sightlines.

Constrained by the interfering absorption of the Earth's atmosphere, ground-based infrared observations have been limited to the narrow wavelength regions through which the Earth's atmosphere is relatively transparent. The Infrared Space Observatory, launched in November 1995, is revolutionizing our understanding of the composition of interstellar dust by providing high signal-to-noise, high resolution observations along many sightlines covering the critical portion of the electromagnetic spectrum between 2–30 $\mu$m (Whittet et al. 1996). It is precisely this region which holds the key to our understanding of the composition of interstellar dust. In the diffuse ISM, the incomplete observational database has inhibited unique identifications of the organic materials present, although much has been learned about the functional groups present, their contribution to the overall interstellar carbon budget, and the distribution of solid-state organic features in space. Greater insight has been gained regarding the ices present in dense molecular clouds, primarily because of the larger number of bright sources seen through high column densities of interstellar dust. However, the ultimate goal in our studies of interstellar dust is to understand the complete life-cycle and composition of the grains, and in that regard we have a long way to go.

In this paper, we will address problematic discrepancies in the observations of dense and diffuse clouds, and how they relate to the nature and evolution of interstellar organics. Further discussion of important dust features is covered elsewhere in this volume (see reviews by d'Hendecourt, Schutte, and Greenberg); here we focus on those features specific to the organic component of the ISM, which may have evolved from ice mantles accreted in dense clouds.

## 2. Infrared Spectroscopy

Infrared wavelengths are ideally suited to the study of the composition of interstellar dust, especially in the range between 2 and 30 $\mu$m (5000 to 330 $cm^{-1}$), because this spans the energy range associated with the fundamental inter-atomic vibrations of many molecular bonds associated with the cosmogenically most abundant species (Bellamy 1960; Silverstein & Bassler 1967; Allamandola et al. 1992; d'Hendecourt, Allamandola, & Greenberg 1985). Infrared spectroscopy provides the observational means for detecting absorption features of molecules in the dust. Molecules undergo both vibrational and rotational transitions; rotation is suppressed for molecules in the solid phase. For a diatomic molecule (e.g. CO), there is one allowed fundamental vibrational mode. The fundamental vibrational frequency corresponds to a bond stretch and is given by

$$\nu_{\mathrm{F}} = \frac{1}{2\pi}\left\{\frac{k}{\mu}\right\}^{\frac{1}{2}} \tag{1}$$

where $k$ is the force constant of the chemical bond between the atoms and $\mu$ is the reduced mass of the system (e.g., Banwell & McCash 1994).

For a polyatomic molecule (such as $H_2O$), there are vibrational modes corresponding to either stretching or bending depending on the change of shape of the molecule. These modes are labelled as $\nu_N$ for each of the $N$ degrees of freedom for the molecule, and are observable if the dipole moment of the molecule oscillates during vibration. Overtones occur for non-simple harmonic motion at frequencies near $2\nu_{\mathrm{F}}$, $3\nu_{\mathrm{F}}$, etc. where $\nu_{\mathrm{F}}$ is a fundamental mode. Absorption features are not directly identified with a specific molecule based solely on the wavelength coincidence of a fundamental absorption. Generally, a feature is associated with a certain bond, e.g. the O-H stretch. Laboratory data and observations at wavelengths where corresponding features should appear are necessary to assign a molecule (or a mixture of molecules), such as $H_2O$ or $CH_3OH$ to a particular absorption band.

## 3. Interstellar Dust

Interstellar dust plays a key role in physical and chemical processes in the interstellar medium. Diffuse ($T \sim 30 - 80\,\mathrm{K}$, $n_0 = n_{\mathrm{H}} + 2n_{\mathrm{H_2}} \sim$

$100 - 800\,\mathrm{cm}^{-3}$) and dense ($T \sim 10\,\mathrm{K}$, $n_0 > 10^2\,\mathrm{cm}^{-3}$) clouds are the main contributors to the mass in the ISM. Grain cores composed of refractory materials such as silicate and amorphous carbon provide nucleation sites for the accumulation of icy mantles in dense molecular clouds (for comprehensive reviews of grain models, see e.g., Dorschner & Henning 1995; Whittet 1992). These icy mantles represent the transient interface between gaseous and solid phases. In the diffuse ISM, molecules in the mantles get desorbed back into the gas phase providing an exchange of material between the dust and the gas. Grains catalyze the production of gas-phase molecules by providing a reservoir to absorb the energy released in exothermic reactions. Mantles are also a controlling influence on the abundances of certain gas-phase species. Both theoretical models and observational evidence suggest the existence of at least two stages of mantle growth in molecular clouds, leading to the occurrence of distinct $H_2O$-rich (polar) and CO-rich (nonpolar) ice phases (d'Hendecourt, Allamandola, Greenberg 1985; Tielens et al. 1991).

Computational models show that the conversion of atomic to molecular hydrogen occurs within $10^6$ years of cloud birth and strongly governs the evolution of grain mantles in dense clouds. At the onset of cloud collapse, when the density is low ($n_0 \sim 10^3\ \mathrm{cm}^{-3}$), the mantle formed will be $H_2O$-rich due to the high abundance of atomic hydrogen and oxygen in the gas phase and the ease of forming water on the grain surface. A small amount of methane ($CH_4$) and ammonia ($NH_3$) is also predicted to form on the mantles. The high probability that atomic oxygen and nitrogen will react with the abundant atomic hydrogen on the grains prevents $O_2$ and $N_2$ from forming on the mantles. As cloud collapse progresses and the density increases ($n_0 \sim 10^5\,\mathrm{cm}^{-3}$), a qualitatively different mantle layer forms. The majority of the atomic hydrogen is converted into $H_2$; production of polar species is inhibited, $O_2$ and $N_2$ are produced and CO is a dominant mantle constituent.

Many laboratory groups have studied the effect of energetic processing of interstellar materials (e.g. Greenberg 1963, 1979; Khare et al. 1993; Hagen, Greenberg, & Tielens 1983; d'Hendecourt et al. 1986; Schutte 1988; Allamandola et al. 1988; Ehrenfreund et al. 1991; Bernstein et al. 1995; Gerakines, Schutte, & Ehrenfreund 1996). Several materials have been suggested as candidate carriers of interstellar carbon, as attaching hydrogen atoms create similar $3.4\,\mu$m structure in their spectra. These include Hydrogenated Amorphous Carbon (HAC) (c.f. Ogmen & Duley 1988; Adamson, Whittet, & Duley 1990), Quenched Carbonaceous Composite (QCC) (Sakata et al. 1987), residues produced by the irradiation of ices (Schutte 1988; Allamandola, Sandford, & Valero 1988; Sandford et al. 1991), coal (Papoular et al. 1989, 1991), and plasma processing of polycyclic aromatic hydrocarbons (PAHs) (Lee & Wdowiak 1993; Wdowiak et al. 1995). In many respects these materials are all variations on a theme, and the interstellar medium likely contains many of the proposed materials (see Pendleton 1997, for a review).

Spectra of astrophysical ice analogs can be compared to observational data in order to constrain the types and abundances of mantle constituents. Accurate calculation of the column densities of the carrier of the absorption features depends on an accurate calculation of the band strength ($A$). This is derived from laboratory spectra by taking the integral of the molecular cross section over the band of interest. The strength of the feature can be deduced from the column density ($N$) of the absorber in the laboratory

$$A_i = \frac{\int \tau_i(\nu) d\nu}{N} \tag{2}$$

and the depth of band $i$, $\tau_i$ ($I = I_0 \exp^{-\tau}$, where $I_0$ is the continuum intensity). Tables 1 and 2 contain information on some of the absorption features detected in dense and diffuse clouds, respectively, which are relevant to the discussion presented here.

## 4. Dense and Diffuse Cloud Chemistry

We will briefly discuss the ice mantle, focusing on the primary $H_2O$-ice and CO component. Energetic and/or thermal processing of these simple ices may result in the production of organics (such as hydrocarbons and nitriles). Table 1 compares the column density of $H_2O$ and CO, and the optical depths of absorption features at 3.47 $\mu$m(due to sp$^3$-bonded hydrocarbons), 4.62 $\mu$m (possibly due to nitriles or isonitriles), and 9.7 $\mu$m (silicate) observed towards sources in and behind dense clouds. Table 2 compares optical depths and column densities of aliphatic hydrocarbons observed along several sightlines through the diffuse ISM. The 9.7 $\mu$m absorption feature has been identified with silicate material of crystalline structure in the Beta Pictoris disk (Telesco & Knacke 1991; Knacke et al. 1993) and comets (Hanner et al. 1994) and amorphous structure in dense clouds (Butchart & Whittet 1983; Aitken et al. 1988), the diffuse ISM (Mathis 1990), and dust shells of asymptotic giant branch (AGB) stars (e.g., NML Cyg; Justtanont et al. 1996). As silicates are thought to be a likely grain core component in dense molecular clouds (Greenberg 1982), evolution of the 9.7 $\mu$m absorption feature from dusty circumstellar shells to the diffuse ISM to dense clouds and comets is an important aspect to be addressed. Presently, we can compare the available observational results of some of these regions, but a major step forward awaits more conclusive results from both observations and laboratory experiments.

TABLE 1. Catalog of Dust Column Densities[a] and Visual Extinctions in Nearby Dark Clouds

| Source | $A_V$ (mag.) | $N(CO)$ ($\times 10^{17}$ cm$^{-2}$) | $N(H_2O)$ | $\tau_{3.47}$ | $\tau_{4.62}$ | $\tau_{9.7}$ |
|---|---|---|---|---|---|---|
| Elias 3* | 8.4[a] | 1.73[f] | 9.10[i] | ... | ... | 0.19[i] |
| Elias 13* | 11.2[a] | 0.84[f] | 10.5[i] | < 0.02[l] | ... | 0.54[i] |
| Elias 16* | 22.3[a] | 6.40[f] | 25.4[i] | 0.03[l] | ... | 0.66[i] |
| Tamura 8* | 18.5[b] | 5.55[f] | 24.5[i] | < 0.03[l] | ... | ... |
| HL Tau | 6–8[c] | < 0.4[g] | 14.9[i] | 0.04[l] | ... | 0.71[i] |
| Elias 1 | 8–10[d] | 0.86[f] | 8.93[i] | ... | ... | ... |
| Elias 18 | 17–20[d] | 2.49[f] | 14.0[i] | < 0.03[l] | ... | 0.43[i] |
| W33A | 145[e] | 9.10[h] | > 90[j] | 0.15[j] | 1.34[n] | 7.84[o] |
| NGC 7538 IRS9 | 84[e] | 12.59[h] | 64[j] | 0.10[m] | 0.31[n] | 4.46[o] |
| GL 2136 | | 1.8 | 50 | ... | ... | |
| GL 961 E | 39[e] | 3.01[h] | 40[k] | 0.07[m] | 0.06[n] | 2.11[o] |
| Mon R2 IRS 2 | 21[e] | 2.73[h] | 18[k] | 0.05[m] | 0.29[n] | 2.11[e] |

Note. — Asterisk (*) denotes field star located behind the Taurus Dark Cloud.

References. — (a) averaged from Elias 1978 and Smith et al. 1993. (b) Tamura et al. 1987. (c) Cohen & Kuhi 1979; Cohen 1983. (d) Elias 1978. (e) Tielens et al. 1991. (f) Chiar et al. 1995. (g) Whittet et al. 1989. (h) Chiar 1996. (i) Whittet et al. 1988. (j) Allamandola et al. 1992. (k) Smith et al. 1989. (l) Chiar et al. 1996a. (m) Brooke et al. 1996. (n) Pendleton et al. 1997. (o) Willner et al. 1982.

### 4.1. THE 3 μm SPECTRAL REGION

Icy grains in dense molecular clouds have been detected by the presence of the 3.0 μm absorption band along the line-of-sight towards bright infrared objects in dense molecular clouds (Merrill, Russell, & Soifer 1976; Leger et al. 1976, and many others). This feature is now easily observed toward many sources in and behind dark clouds. Although there are profile differences in terms of the width and shape of the observed 3.0 μm band from source to source, its identification with water ice (as the primary constituent) is well established. In general, the water molecule exhibits an O-H stretching mode at 3.0 μm and an H-O-H bending mode at 6.0 μm (Allamandola & Sandford 1988). Surface reactions involving oxygen atoms are the most likely processes to lead to formation of $H_2O$-ice on grain mantles (e.g. Jones & Williams 1984). The 3.0 μm feature is observable from ground-based telescopes whereas the 6.0 μm feature requires airborne or space-based observations. The shape of the 3.0 μm profile is a sensitive function of the temperature and crystallinity as can be seen by comparing the spectra of the field star Elias 16 (seen through dense cloud dust) and the bright protostar BN (embedded in the Orion Molecular Cloud 1; Smith et al. 1989) in Figure 1. Elias 16 is a highly obscured red giant star located by chance alignment behind the Taurus dark cloud. Grain mantles in this sightline are relatively undisturbed by star formation in the cloud. A broad Gaussian profile (like that of Elias 16; Figure 1, top panel) is produced when water-ice is deposited at a temperature much lower than its melting point, which creates an amorphous structure. A fit to the astronomical spectrum is accomplished with a calculated model for small silicate grains coated with $H_2O$-ice at 23 K (Smith et al. 1989). A sharper profile, such as that observed for the embedded protostar BN (Figure 1, bottom panel), is produced when the ice is warmed: the molecules arrange themselves in a crystalline structure, the most energetically favorable solid. The BN data are fitted with a model based on a mixture of 23, 77, and 150 K (in a ratio of 3.1:1.2:1) ices, showing that there is a range of dust temperatures (as a function of distance from the embedded source) around protostars (Smith et al. 1989). Furthermore, this fit demonstrates that the spectrum is too narrow to be fit with only cold ices and that the warmest grains have temperatures greater than 100 K.

Absorption in excess of what is expected for a pure $H_2O$-ice profile occurs in the 3 $\mu$m spectra of many objects (e.g. Figure 1). The broad, asymmetric 3.0 $\mu$m feature is actually a blend of features due to resonances of $H_2O$-ice and other molecules. Of particular interest is the "long-wavelength wing" between 3.2 $\mu$m and 3.6 $\mu$m. Among the most likely candidates responsible for the wing absorption are: hydrocarbons, alcohols, O-H stretching resonances in $H_2O$ molecules in ammonium hydrate groups (Smith et al. 1989; 1993), and $H_2O_2$ (Meyer et al. 1997). We will compare the $H_2O$-ice and "long-wavelength wing" of dense cloud sources to diffuse ISM observations in §4.3.

### 4.2. THE 4.67 $\mu$m (CO) ABSORPTION

Carbon monoxide is the only molecule abundant enough to collect on the mantle to a significant degree by direct condensation of pre-existing molecules in the gas phase (Duley 1974). Duley was the first to discuss the possibility that CO is a major mantle constituent of interstellar grains with a characteristic frequency that is detectable in the infrared. Solid CO exhibits the C$\equiv$O stretch at 4.67 $\mu$m; this absorption feature is observed in both the intracloud medium and in the dust around embedded objects. Lacy et al. (1984) were the first to detect the 4.67 $\mu$m CO stretch feature toward several embedded objects. Whittet, McFadzean, & Longmore (1985)

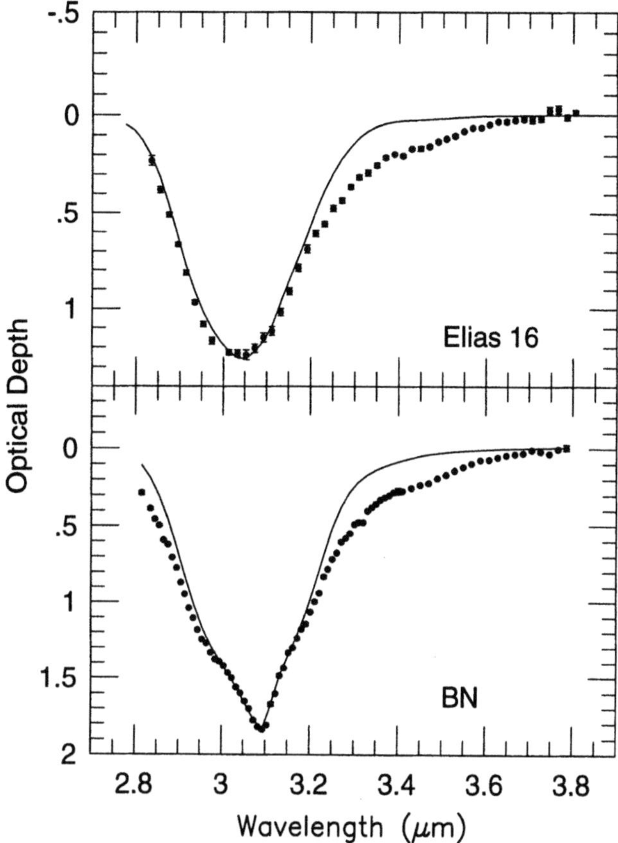


*Figure 1.* [*Top*] $H_2O$-ice spectrum of the field star Elias 16, which samples the quiescent cloud medium in Taurus. The points are the astronomical data and the solid line is the calculated model spectrum for pure $H_2O$-ice mantles at 23 K. [*Bottom*] Spectrum of protostar BN, which samples warm conditions around an embedded object. The solid line is the calculated spectrum for a mixture of pure $H_2O$-ice mantles at temperatures between 23 – 150 K. (Figure from Smith et al. 1989).

were the first to detect solid CO absorption in a dark cloud environment towards background stars. Since then, there have been many studies of solid CO in molecular clouds.

Laboratory studies have shown that the solid CO absorption profile is affected by a number of factors such as the composition of the host matrix, and its thermal history (e.g. Sandford et al. 1988; Palumbo & Strazzulla 1993). Recent studies of solid CO absorption (Tielens et al. 1991; Kerr et al. 1991, 1993; Chiar et al. 1995) show that the feature consists of two overlapping components, (1) a broad, shallow component (at $\sim 4.682\,\mu$m) attributable to CO in a matrix dominated by $H_2O$, and (2) a narrow component (at $\sim 4.673\,\mu$m) produced by either pure CO, CO in a nonpolar matrix (such as CO mixed with $CO_2$, $N_2$ or $O_2$), or CO in a weakly polar matrix (i.e., CO mixed with a small amount of $H_2O$). These components are referred to as the polar and nonpolar components, respectively. In most cases, the nonpolar component dominates the feature. While the two components are blended with each other, they are indicative of the polar and nonpolar phases of mantle growth (see Chiar 1997 for a recent review). The CO profiles observed in different clouds have been compared with laboratory data for both the broad and narrow components in an effort to account for their existence in terms of the chemical and physical state of the cloud (e.g., Tielens et al. 1991; Kerr et al. 1991, 1993; Chiar et al. 1995; Chiar 1996).

## 4.3. THE 3.4 $\mu$m HYDROCARBON REGION

### 4.3.1. *Diffuse Interstellar Medium*

Along the sight-line toward the heavily obscured Galactic Center, the total visual extinction, $A_V$, has been estimated to be $\approx$ 31 mag (Becklin et al. 1978; Henry et al. 1984; Sellgren et al. 1987 and others). While there are indications of some dense molecular cloud absorption along this sightline (Whittet & Tielens 1997), arguments have been made that the majority of interstellar dust, through which the observations are made, arises from the diffuse ISM (Roche and Aitken 1985; Sandford et al. 1991). McFadzean et al. (1989) have shown that while the shape of the 3.0 $\mu$m ice band varies from source to source in the Galactic Center region, the shape of the 3.4 $\mu$m feature does not. This could result from the presence of a molecular cloud close to the Galactic Center, through which we observe the Galactic Center sources. Different sources are seen through varying amounts of the dense cloud. The 3.4 $\mu$m feature, on the other hand, remains constant because for each of the Galactic Center sources, we are looking through approximately the same amount of diffuse ISM material.

As illustrated in Figure 2, 2.8 – 3.8 $\mu$m observations toward the Galactic Center show that these spectra contain a broad ice feature at 3.0 $\mu$m (3300 $cm^{-1}$) and a complex hydrocarbon feature near 3.4 $\mu$m (2950 $cm^{-1}$). Absorption features in the 2.8 –3.70 $\mu$m (3570–2700 $cm^{-1}$) region can be produced by a variety of molecular vibrations. Stretching vibrations from O-H and C-H-bearing molecules are the most likely candidates responsible for the features seen near 3.0 $\mu$m and 3.4 $\mu$m respectively, within cosmic abundance constraints. The diffuse interstellar medium observations indicate the presence of hydrocarbon features which extend from about 3.33 $\mu$m (3000 $cm^{-1}$) to 3.57 $\mu$m (2800 $cm^{-1}$) with sub-features near 3.38, 3.42, and 3.48 $\mu$m (2955, 2925, and 2870 $cm^{-1}$). The positions of the first of these sub-features are characteristic of symmetric and asymmetric C-H stretching

TABLE 2. Hydrogen, Carbon, $-CH_3$, and $CH_2$ Column Densities (in units of $cm^{-2}$), and the Percent of Cosmic Carbon in the Aliphatic Hydrocarbon Component of the Diffuse Medium Dust

| Object | $A_V$ | $\tau^a_{9.7}$ | $N(H)^b$ ($\times 10^{22}$) | $N(C)^b$ ($\times 10^{18}$) | $N(CH_3)$ ($\times 10^{16}$) | $N(CH_2)$ ($\times 10^{16}$) | Percentage Carbon[c] |
|---|---|---|---|---|---|---|---|
| GC IRS 7 | 31 | 3.6±0.4 | 5.9 | 22 | 29 | 57 | 3.9 |
| GC IRS 6E | 31 | 3.6±0.4 | 5.9 | 22 | 31 | 62 | 4.2 |
| GC IRS3 | 31 | 36.6±0.4 | 5.9 | 22 | 35 | 71 | 4.9 |
| HD 229059 | 5.3 | ... | 1.0 | 3.7 | 3.5 | 8.5 | 3.2 |
| HD 194279 | 3.9 | ... | 0.74 | 2.7 | 2.0 | 5.0 | 2.6 |
| BD+40 4220 | 6.2 | ... | 1.2 | 4.4 | 4.0 | 7.5 | 2.6 |
| CygOB2#12 | 10 | 0.58±0.1 | 1.9 | 7.0 | 6.0 | 13 | 2.7 |
| AFGL 2104 | 12.0 | 0.64 ± 0.05 | 2.3 | 8.4 | 5.6 | 13 | 2.2 |
| AS 320 | 5.2 | 0.26 | 0.99 | 3.7 | 2.7 | 6 | 2.4 |
| AFGL 2179 | 12.8 | 0.65 ± 0.05 | 2.4 | 9.0 | 7.0 | 16 | 2.6 |
| Ve 2-45 | 6.5 | 0.36 ± 0.05 | 1.2 | 4.6 | 4.6 | 6.0 | 2.2 |

[a]Silicate optical depths were taken from Roche & Aitken 1984, 1985.

[b]$N(H)$ and $N(C)$ were calculated assuming $N(H) = 1.9 \times 10^{21} A_V$ (Bohlin et al. 1978) and $N(C)/N(H) = 3.7 \times 10^{-4}$ (Allen 1973).

[c]This is a lower limit for the percent of cosmic carbon which is tied up in the $-CH_3$ and $-CH_2$ groups combined, $[N(CH_3) + N(CH_2)]/N(C)$.

frequencies of $-CH_3$ (methyl) and $-CH_2-$ (methylene) groups in saturated aliphatic hydrocarbons (molecules with the formula $C_nH_{2n+2}$). The relative strengths of these sub-features indicate that the average $-CH_2-/-CH_3$ ratio of interstellar hydrocarbons in the diffuse ISM is about 2.5 (Sandford et al. 1991).

Observations of the 3.4 $\mu$m feature along sightlines which have diffuse interstellar extinction values which are substantially lower than that of the Galactic Center are shown in Figure 3. The features are correspondingly weaker, but the peak wavelength and general shape of the bands match very well with the Galactic Center spectra. The data in Figure 3 are the resultant optical depth plots, which are derived from the flux data similar to that shown in Figure 2 (see Sandford et al. 1991 and Pendleton et al. 1994 for a discussion of the choices for a continuum baseline to the flux data).

Recent detection of the 3.4 $\mu$m absorption band complex in other galaxies (Bridger et al. 1993; Wright et al. 1996) has revealed the same substructure and relative strength of the features as found in the spectra of the diffuse ISM in our own galaxy. Figure 4 compares the 3.4 $\mu$m hydrocarbon absorption feature in our own galaxy to that observed in a distant galaxy (the embedded Seyfert galaxy, IRAS 08572+3915). The surprising strength of the overall band complex in the dusty torus of the extragalactic source indicates that dust in other galaxies may provide valuable spectral

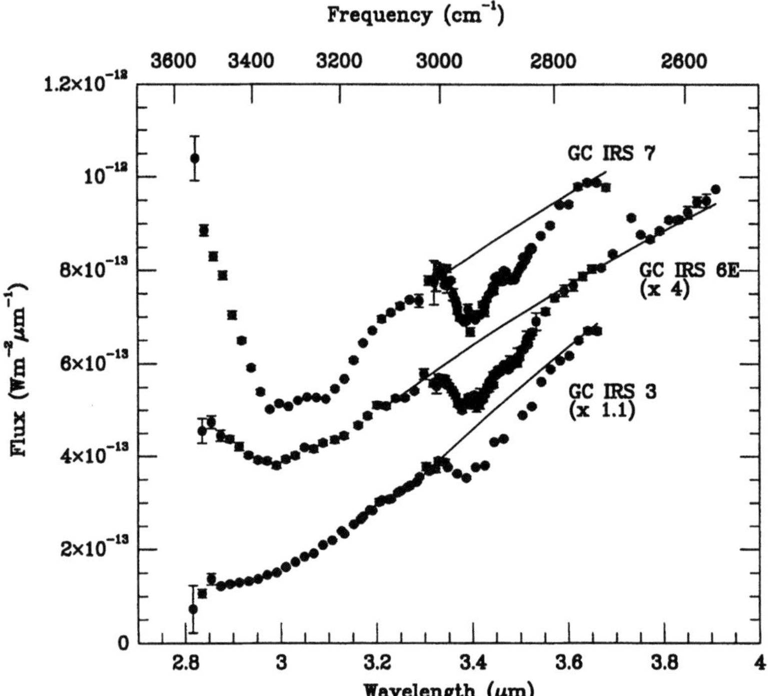


*Figure 2.* The 2.8 - 4.0 $\mu$m (3570 - 2500 cm$^{-1}$) fluxed spectra of sight lines towards Galactic Center sources IRS 7, IRS 6E, and IRS 3. The low resolution data have a resolution of 0.018 $\mu$m per detector $[\lambda/\Delta\lambda]$ = 155 - 220 over the 2.80 - 4.0 $\mu$m (3570 - 2500 cm$^{-1}$) range. High resolution data points between 3.33 and 3.55 $\mu$m (3000 and 2820 cm$^{-1}$) are superposed for GC IRS 6E and GC IRS 7 (taken from Pendleton et al. 1994). Where error bars are not showing, error bars are smaller than the points.

profile information despite the great distances to the illuminating source. In addition, the spectral features may appear at wavelengths that are red-shifted into regions that are more accessible from the ground. Figure 4 also compares the diffuse ISM and extragalactic features to organics found in the Murchison meteorite. As first pointed out by Ehrenfreund et al. (1991), there is a strong similarity between the 3.4 $\mu$m feature seen in the Orgueil and Murchison carbonaceous chondrites and the diffuse ISM spectra. In the meteorite, the aliphatics that produce this feature are primarily cyclic aliphatics (Cronin and Pizzarello 1990). Observationally, we cannot distinguish between cyclic or chain-like aliphatic ($CH_2$ and $CH_3$) groups at this time, but it will be interesting to investigate the nature of the interstellar aliphatics in light of the meteoritic results. At present, we can simply note that the comparison between the three spectra (the diffuse ISM in our galaxy, the dusty region in the distant galaxy, and the Murchison mete-

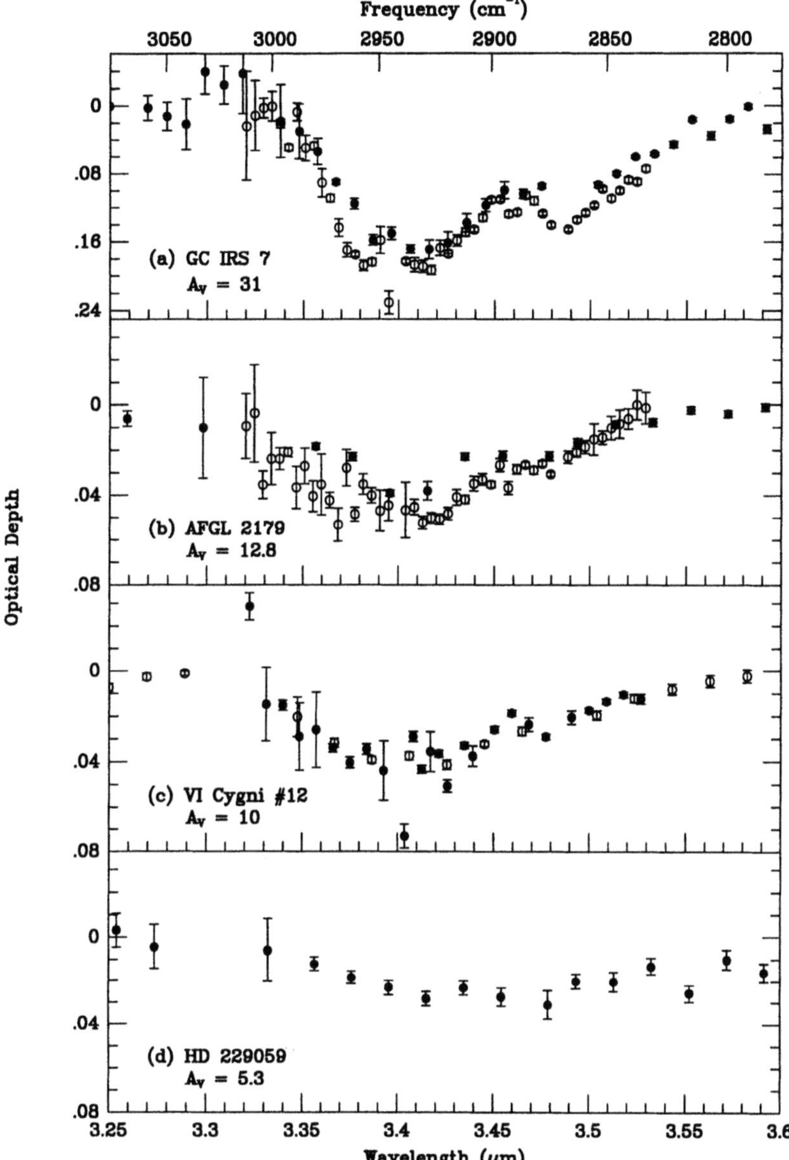


*Figure 3.* Optical depth plots from 3.25 - 3.6 $\mu$m (3077-2778 cm$^{-1}$) for (a) GC IRS 7, (b) the WC star AFGL 2179, (c) the supergiants Cyg OB2 #12, and (d) HD 229059. High resolution data are displayed as open points; low resolution data are displayed as solid points (taken from Pendleton et al. 1994).

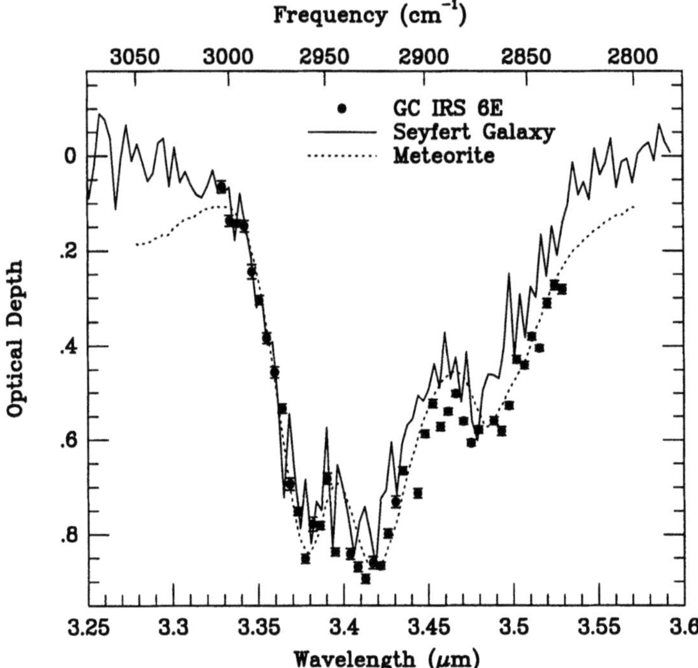


*Figure 4.* A comparison of the optical depth spectrum of the Galactic Center source IRS 6E (solid points), the sublimate from the acid-insoluble component of the Murchison meteorite (dotted line; from DeVries et al. 1993), and the redshift corrected 3.4 $\mu$m feature in the embedded Seyfert Galaxy IRAS 08572+3915 (solid line; taken from Wright et al. 1996).

orite) is remarkable. The spectral features shown in Figure 4 are different than the 3.4 $\mu$m feature observed in dense clouds; we discuss the differences in the following section.

*The 3.28 $\mu$m Aromatic Hydrocarbon Region:* It is important to point out the relationship between the aliphatic hydrocarbon features and the polycyclic aromatic hydrocarbons (PAHs; Leger & Puget 1984; Allamandola et al. 1985, 1989; Joblin, Leger, & Martin 1992). While the aromatic hydrocarbons likely coexist in the diffuse ISM where the $CH_2$ and $CH_3$ group (aliphatic) hydrocarbons are found, and while the C-H stretching absorption feature of aromatic hydrocarbons falls near 3.28 $\mu$m (3050 cm$^{-1}$), only a weak aromatic feature is expected compared to the aliphatic bands, since the aromatics contain fewer hydrogen atoms per carbon atom (H/C$<$ 1) than do aliphatics (H/C$>$ 2), making the intrinsic strength of the aromatic C-H stretching band 2-3 times lower than its aliphatic counterpart. A suggestive detection of the 3.28 $\mu$m feature in absorption towards the Galactic center has been reported (Pendleton et al. 1994). They concluded that the aromatic abundance fraction could be as large as ~10% of the C along

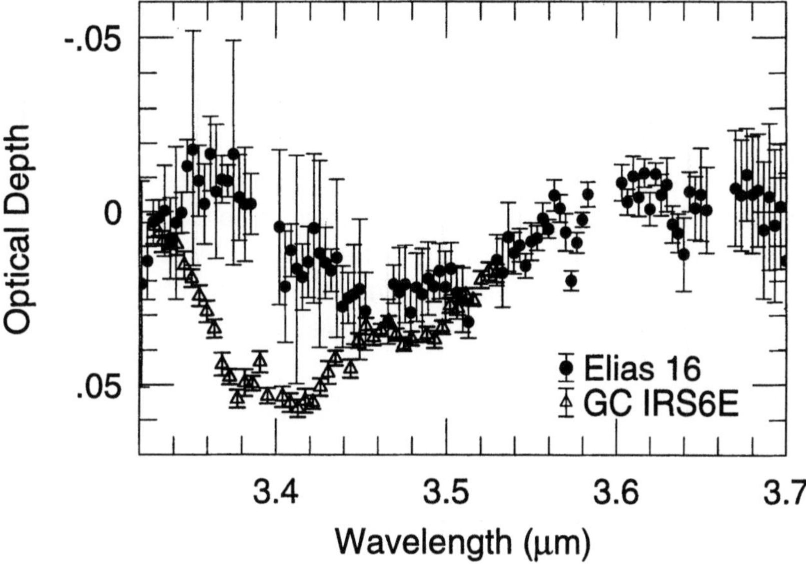


*Figure 5.* The 3.4 μm absorption feature towards the Galactic Center source GC IRS6E (from Pendleton et al. 1994; triangles) and Elias 16 (Chiar et al. 1996; circles) scaled to the depth of the 3.47 μm feature in Elias 16.

the line-of-sight to the Galactic Center. This tentative absorption feature could be due to the absorption counterpart of the infrared emission and/or aromatic hydrocarbon dust grains which are too large to emit in the near IR (cf. d'Hendecourt, this volume; Geballe 1997). In dense clouds, evidence for an absorption feature near 3.25 μm (3080 $cm^{-1}$) which may be due to aromatic material has been reported (Sellgren et al. 1994). The exact position of the C-H stretching absorption varies depending upon the molecule in question, so that a mixture of aromatics would be expected to produce a band centered in the 3.25 μm – 3.28 μm (3080 – 3050 $cm^{-1}$) region. Given the large potential abundance of aromatics implied by these observations, it is clearly important to resolve the presence of the 3.25 μm – 3.28 μm absorption feature in diffuse and dense cloud sources.

### 4.3.2. *Dense Clouds*

In dense molecular clouds, absorption in the 2.8 μm – 3.8 μm region is dominated by the $H_2O$-ice band near 3.0 μm as discussed above. Additional absorption between 3.2 and 3.6 μm, termed the "long wavelength wing," can be seen in Figure 1. Figure 5 shows the spectra of the Galactic Center source IRS6E (Pendleton et al. 1994), which samples diffuse ISM dust, and

Elias 16 (Chiar et al. 1996), which samples dense molecular cloud dust. The structure near 3.4 μm seen in the diffuse ISM profiles is apparently missing in dense cloud spectra. Thus far, none of the dense cloud detections have shown subfeatures like those in the diffuse ISM spectra. Smith et al. (1993) discussed the most likely scenarios for the absence of the aliphatic substructure in the dense cloud spectra. Among the possibilities that were ruled out was the idea that the aliphatic hydrocarbons are part of the grain core which becomes mantled with ice in the dense cloud. Baratta & Strazzulla (1990) have investigated this point, and find that even substantially thick ice layers atop a hydrocarbon residue still reveal pronounced sub-structure near 3.4 μm in the laboratory spectra.

Further evidence that the 3.4 μm wing in dense cloud spectra arises from material in the grain mantle, rather than the core, comes from correlation studies comparing the optical depth of the "long-wavelength wing" with other dense cloud absorption features. Allamandola et al. (1992) detected a broad band at 3.47 μm in luminous protostellar objects, which they attributed to $sp^3$-bonded hydrocarbon material. This structural difference means that in these materials, hydrogen is bonded to a carbon atom which is bonded to three other carbons. This differs significantly from the structure of the diffuse ISM 3.4 μm feature where a carbon atom is attached to either two or three hydrogen atoms (i.e., $CH_2$ or $CH_3$, respectively). The dense cloud 3.47 μm feature has since been detected in other protostellar sources (Sellgren et al. 1994; Brooke et al. 1996), in low-mass embedded objects (HL Tau) and in the quiescent cloud medium in the Taurus dark cloud (Elias 16; Chiar et al. 1996). There is little correlation between the 3.47 μm feature and the 9.7 μm silicate feature; the latter is known to be associated with the grain core. In contrast, the optical depths of the 3.47 μm feature and 3.0 μm $H_2O$-ice feature are strongly correlated (Brooke et al. 1996; Chiar et al. 1996). This strong correlation supports the conclusion that the carrier of the 3.47 μm feature resides in the mantle material, because the 3.0 μm water ice feature is clearly associated with grain mantles. Such a correlation also suggests that the carrier of the 3.47 μm feature does not exist in the core (or an inner mantle layer) beneath the $H_2O$-ice layer. The conclusion that seems most apparent from observations and laboratory experiments is that the material responsible for the 3.4 μm wing in the spectra of dense molecular cloud sources is different from the carrier of the aliphatic hydrocarbon 3.4 μm bands in the diffuse ISM.

### 4.4. THE 4.62 μm (X−C≡N?) BAND

Nitriles (compounds containing the −C≡N group) play a vital role in the formation of prebiotic molecules (Matthews 1992). The discovery of an interstellar absorption feature near 4.62 μm has prompted laboratory comparisons of nitriles and isonitriles (such as $CH_3NC$) to the interstellar observations. This unidentified feature, commonly referred to as the X–C≡N) band, is seen in the spectra of several young stellar objects (e.g. Figure 6; see also Lacy et al. 1984, Tegler et al. 1993; 1995, Weintraub et al. 1994, Pendleton, Tielens, & Tokunaga 1997). Infrared laboratory spectra of ultraviolet photolyzed interstellar analog ice mixtures show a prominent absorption feature at 4.62 μm (2160 $cm^{-1}$). See Pendleton et al. (1997) for a review of possible identifications of the 4.62 μm feature. Although the identification of this feature is by no means secure, it is interesting to contemplate the implications of an identification of the 4.62 μm absorption band with the product of some type of interstellar ice processing inside dense molecular clouds. Laboratory studies have shown that nitriles produced in the residues of UV processed ice mixtures survive temperature increases up to about 200K. While this component is not quite as hardy as the aliphatic hydrocarbon component produced under similar conditions, the survivability of the laboratory band produced near 4.62 μm is consistent with the environment such material would be exposed to in the diffuse ISM. Therefore, it is difficult to understand why the 4.62 μm band is not seen in any of the diffuse ISM observations. Presently, none of the possible identifications for this band can adequately explain this discrepancy, and we are missing another piece of the puzzle of the interstellar dust picture.

The 4.62 μm band detections within dense clouds have thus far been limited to bright, embedded protostellar sources. The lack of detection in the quiescent cloud medium, as evidenced by observations of a bright background field star, Elias 16, seen through large amounts of dense cloud dust, is taken as evidence that ice in the vicinity of embedded sources contains the carrier of the 4.62 μm band whereas the general intracloud material does not (Tegler et al. 1995). If identification of the 4.62 μm band with an a –C≡N molecule is confirmed, then the possible connection to primitive Solar System bodies may become clearer. Detections of the –C≡N stretch first overtone, at 2.2 μm, in some Solar System objects, have been made by several authors (e.g. Jewitt et al. 1982; Bell et al. 1985; Cruikshank et al. 1991). It is important to determine whether these same organic molecules are also present in dust grains in the ISM, as these grains may accrete to form cometary bodies, and comets may have delivered vast quantities of organic materials to the early Earth.

## 5. Summary

The incorporation of organic solids into planetesimals is of prime importance in the evaluation of the materials that were likely delivered to the

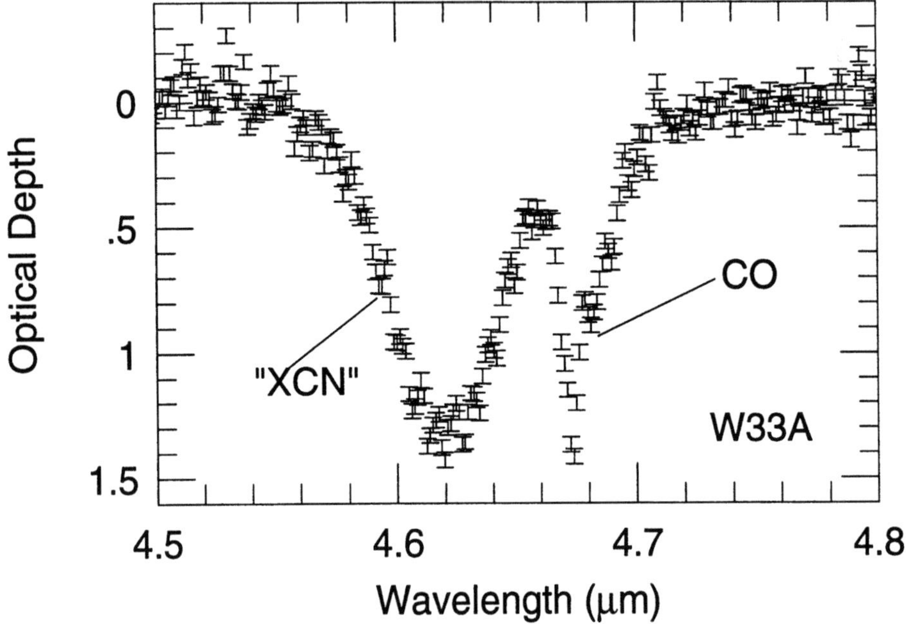


*Figure 6.* The 4.5-4.8 $\mu$m spectrum of the young stellar object, W33A. The feature at 4.62 $\mu$m and the CO feature at 4.67 $\mu$m are shown.

early Earth by comets and asteroids, as these building blocks may have played an important role in the emergence of life on our planet. Planetesimals formed from the material available in the solar nebula, which originated in the interstellar medium. Understanding the production and distribution of organic solids in the ISM, therefore, aids in our understanding of the initial composition of those bodies that would later bombard planets with ices and organic material. The recent discovery of organic solid-state material in distant galaxies, as evidenced by detection of 3.4 $\mu$m hydrocarbon absorption (§4.3) points out the ubiquitous nature of the carrier, which has now been detected along more than a dozen lines-of-sight through the diffuse interstellar medium in our Galaxy. Comparisons of the diffuse interstellar medium 3.4 $\mu$m absorption feature with extracts from carbonaceous meteorites have shown a remarkable similarity in profile and relative strength among the sub-features which arise from $CH_2$ and $CH_3$ groups. The evolution of dust grains and, in particular, the evolution of the organic component of the dust, is central to the issue of the interstellar material incorporated in primitive Solar System bodies. Processing of icy grain mantles in dense molecular clouds and the cycling of the refractory dust components between the dense and diffuse ISM are key aspects of the inventory of material during planetesimal formation. Temperatures in the dense molecular cloud regions are sufficiently cold, and the grains are well enough shielded from the harsh environment outside the cloud, that icy mantles form easily on grain cores. The accretion of ice mantles on dust grains in dense molecular clouds is confirmed by the detection of the 3.0 $\mu$m absorption band, attributed to $H_2O$ and the 4.67 $\mu$m absorption band, attributed to solid CO, in many dense molecular clouds.

Laboratory studies have shown that energetic processing of simple ice mixtures, including $H_2O$ and CO, reproduce many of the spectral features observed in dense molecular clouds. In the laboratory, warm-up of interstellar ice analogs reveals the presence of an organic refractory residue which remains after the volatile ice component has been removed. If the initial laboratory mixture includes some type of nitrogen-bearing component, refractory residues form which exhibit an absorption band in the 4.62 $\mu$m region, a feature which is seen in observations of embedded protostellar objects (Lacy et al. 1984; Tegler et al. 1993, 1995) . This interstellar feature has been named the "X-CN" band, because of the connection to a nitrogen bearing component. The 3.4 $\mu$m aliphatic hydrocarbon feature is also produced in the laboratory through UV photolysis of ices. If these interstellar bands (at 4.62 and 3.4 $\mu$m) are produced through ultraviolet photolysis of ices, the carrier is a hardy, refractory material that would be expected to survive long after the volatile materials are gone. Upon the break-up of a star-forming cloud, surviving refractory material could become incorporated into the surrounding diffuse ISM. Although this theory is supported by observations of the 3.4 $\mu$m absorption feature in the diffuse medium, it is refuted by the absence of this feature in dense clouds and the absence of the 4.62 $\mu$m feature in the diffuse ISM. If these two features are produced via energetic processing of icy grain mantles, as the laboratory studies suggest, then we might expect to see both features in the spectra of dense and diffuse cloud objects. The fact that we do not presents a problem in our current understanding of the evolution of interstellar dust.

Studies presented in this review suggests that the interstellar hydrocarbon feature may be primarily produced via an alternative mechanism, such as in the outflow of carbon stars in the diffuse ISM itself. Another interesting point is that searches for the 4.62 $\mu$m absorption band have so far revealed the feature only in spectra of heavily embedded sources which are associated with the molecular cloud (§4.4). Spectra of background field stars seen through equivalent amounts of dense cloud dust do not reveal the 4.62 $\mu$m absorption band. This suggests that the carrier is not interspersed throughout the dense cloud, but is localized near the embedded object. A more thorough comparison of embedded and background objects must be made in order to confirm this very interesting result, but taken at face value it suggests that close proximity to the forming star is a requirement for production of the 4.62 $\mu$m band. If the identification of the band as a nitrile or isonitrile is confirmed, then the 4.62 $\mu$m absorption band is another aspect of the organic component of interstellar dust that must be better understood in order to evaluate the material that went into the early solar nebula and planetesimals.

Finally, the remarkable comparison of the diffuse ISM feature to that found in the Murchison meteorite underscores the need to fully understand the nature and evolution of interstellar organic solids. Is the similarity an indicator of unaltered interstellar organics in the meteorite? We must know where these materials are made, how they are created, how they are distributed throughout the interstellar medium, and what type of processing they have endured. Only then will we have a chance to learn about the true availability of these organic constituents for incorporation into planetary systems during the star and planet formation processes.

Work on the dense molecular clouds reported here was supported by NASA grant NAGW-3144 and the Department of Education, and is gratefully acknowledged by that author (JEC). JEC currently holds a National Research Council Fellowship at NASA-Ames Research Center. Partial travel support for JEC was kindly provided by J. Mayo Greenberg. Work on the diffuse interstellar medium described here was supported by NASA's Exobiology Program and NASA's IR/Submillimeter Astrophysics Program, and is gratefully acknowledged by that author (YJP).

## References

Adamson, A. J., Whittet, D. C. B., & Duley, W. W. (1990), The 3.4 $\mu$m interstellar absorption feature in Cy g OB2 #12, *Mon. Not. Royal Ast. Soc.*, **243**, 400–404.

Aitken, D. K., Smith, C. H., James, S. D., Roche, P. F., Hyland , A. R., & McGregor, P. J. (1988), 10 $\mu$m spectral observations of SN1987A - The first year, *Mon. Not. Royal Ast. Soc.*, **235**, 19P–31P.

Allamandola, L. J., & Sandford, S. A. (1988), Laboratory simulation of dust spectra, In *Dust in the universe; Proceedings of the Conference, Manchester, England, Dec. 14-1 8, 1987 (A89-39876 17-90)*, Cambridge University Press, Cambridge and New York, pages 229–263.

Allamandola, L. J., Sandford, S. A., Tielens, A. G. G. M., & Her bst, T. M. (1992), Infrared spectroscopy of dense clouds in the C-H stre tch region - Methanol and 'diamonds', *Astrophys. J.*, **399**, 134–146.

Allamandola, L. J., Sandford, S. A., & Valero, G. J. (1988), Photochemical and thermal evolution of interstellar/precometary ice analogs, *Icarus*, **76**, 225–252.

Allamandola, L. J., Sandford, S. A., & Valero, G. J. (1988), Photochemical and thermal evolution of interstellar/precometary ice analogs, *Icarus*, **76**, 225–252.

Allamandola, L. J., Tielens, A. G. G. M., & Barker, J. R. (1985), Polycyclic aromatic hydrocarbons and the unidentified infrared emission bands - Auto exhaust along the Milky Way, *Astrophys.J.*, **290**, L25–L28.

Allamandola, L. J., Tielens, G. G. M., & Barker, J. R. (1989), Interstellar polycyclic aromatic hydrocarbons - The infrared emission bands, the excitation/emission mechanism, and the astrophysical implications, *Astrophys. J. Supp.*, **71**, 733–775.

Banwell, C. N., & McCash, E. M., (1994), *Fundamentals of Molec ular Spectroscopy*, McGraw-Hill, London, fourth edition.

Becklin, E. E., Neugebauer, G., Willner, S. P., & Matthews, K. (1978), Infrared observations of the Galactic Center: IV - The interstellar extinction, *Astrophys. J.*, **220**, 831–835.

Bell, J. F., Cruikshank, D. P., & Gaffey, M. J. (1985), The composition and origin of the Iapetus dark materi al, *Icarus*, **61**, 192–207.

Bellamy, L. J., (1960), *The Infrared Spectra of Complex Molecules* , John Wiley and Sons, New York.

Bernstein, M. P., Sandford, S. A., & Allamandola, L. J. (1996), Hydrogenated polycyclic aromatic hydrocarbons and the 2 940 and 2850 cm $^{-1}$ (3.40 and 3.51 $\mu$m) infrared emission features, *Astrophys. J.*, **472**, L127–L130.

Bernstein, M. P., Sandford, S. A., Allamandola, L. J., Chang, S., & Scharberg, M. A. (1995), Organic compounds produced by photolysis of realistic interstellar and cometary ice analogs containing methanol, *Astrophys. J.*, **454**, 327.

Bridger, A., Wright, G., & Geballe, T. (1993), In McLean, I., editor, *Conference Abstract Volum e of Infrared Astronomy with Arrays: The Next Generation*, UCLA Dept. of Astronomy, Los Angeles, pages 537.

Brooke, T. Y., Sellgren, K., & Smith, R. G. (1996), A study of absorption features in the 3$\mu$m spectr a of molecular cloud sources with H $_2$O ice bands, *Astrophys. J.*, **459**, 209–215.

Butchart, I., & Whittet, D. C. B. (1983), Interstellar grain composition and the infrared spectru m of OH26.50.6, *Mon. Not. Royal Ast. Soc.*, **202**, 971–980.

Chiar, J. E. (1996), *The Nature and Evolution of Interstellar Ices: Stu dies of Methanol and Carbon Monoxide*, PhD dissertation, Rensselaer Polyte chnic Institute, Dept. of Physics, Applied Physics and Astronomy.

Chiar, J. E., Adamson, A. J., Kerr, T. H., & Whittet, D. C. B. (1995), High-resolution studies of solid CO in the Taurus d ark cloud: Characterizing the ices in quiescent clouds, *Astrophys. J.*, **45 5**, 234–243.

Chiar, J. E., Adamson, A. J., & Whittet, D. C. B. (1996), Three micron hydrocarbon and methanol absorption in Ta urus, *Astrophys. J.*, **472**, 665–672.

Cohen, M. (1983), HL Tauri and its circumstellar disk, *Astrophys. J.*, 270, L69–L71.

Cohen, M., & Kuhi, L. V. (1979), Observational studies of pre-main-sequence evolution, *Astrophys. J. Supp.*, **41**, 743–843.

Cronin, J. R., & Pizzarello, S. (1990), Aliphatic hydrocarbons of the Murchison meteorite, ēm Geochim. Cosmochim. Acta, **54**, 2859–2868.

Cruikshank, D. P., Allamandola, L. J., Hartmann, W. K., Tholen, D 2E J., Brown, R. H., Matthews, C. N., & Bell, J. F. (1991), Solid C triple bond N bearing material on outer sol ar system bodies, *Icarus*, **94**, 345–353.

d'Hendecourt, L. B., Allamandola, L. J., & Greenberg, J. M. (1985), Time dependent chemistry in dense molecular clouds. I - Grain surface reactions, gas/grain interactions and infrared spectros copy, *Astron. Astrophys.*, **152**, 130–150.

d'Hendecourt, L. B., Allamandola, L. J., Grim, R. J., & Greenber g, J. (1986), Time-dependent chemistry in dense molecular clouds. II - Ultraviolet photoprocessing and infrared spectroscopy of grain mantle s, *Astron. Astrophys.*, **158**, 119–134.

Dorschner, J., & Henning, T. (1995), Dust metamorphosis in the galaxy, *Astron. Astrophys. Rev.*, **6**, 271.

Duley, W. (1974), Composition of grain mantles in interstellar clouds, ēm *Astrophys. Space Science*, **26**, 199–205.

Ehrenfreund, P., Robert, F., d'Hendencourt, L., & Behar, F. (1991), Comparison of interstellar and meteoritic organic matte r at 3.4 microns, *Astron. Astrophys.*, **252**, 712–717.

Elias, J. H. (1978), A study of the Taurus dark cloud complex, *Astrophys. J.*, **224**, 857–872.

Gerakines, P. A., Schutte, W. A., & Ehrenfreund, P. (1996), Ultraviolet processing of interstellar ice analogs. I - Pure ices., *Astron. Astrophys.*, **312**, 289–305.

Greenberg, J. M. (1963), Interstellar grains, *Annual Rev. Astron. Astrophys.*, **1**, 267–290.

Greenberg, J. M. (1979), Grain mantle photolysis - a connection between the grai n size distribution function and the abundance of complex interstellar molecul es, In *Stars and star systems; Proceedings of the Fourth European Regiona l Meeting in Astronomy, Uppsala, Sweden, August 7-12, 1978*, D. Reidel Publishing Co., Dordrecht, pages 173–193.

Greenberg, J. (1982), Dust in dense clouds - one stage in a cycle, In *Submillimetre wave astronomy*, Cambridge University Press, Cambridge an d New York, pages 261–306.

Hagen, W., Greenberg, J. M., & Tielens, A. G. G. M. (1983), The 3 $\mu$m 'ice' band in grain mantles, *Astron. Astrophys.*, **117**, 132–140.

Hanner, M. S., Hackwell, J. A., Russell, R. W., & Lynch, D. K. (1994), Silicate emission feature in the spectrum of comet Mue ller 1993a, *Icarus*, **112**, 490–495.

Henry, J. P., Depoy, D. L., & Becklin, E. E. (1984), The location of infrared sources in the Galactic Cente r from a deep 1 $\mu$m CCD image, *Astrophys. J.*, **285**, L27–L30.

Jewitt, D. C., Soifer, B. T., Neugebauer, G., Matthews, K., & Danielson, G. E. (1982), Visual and infrared observations of the distant comets P/Stephan-Oterma 1980g, Panther 1980u, and Bowell 1980b, *Astronom. J.* , **87**, 1854–1866.

Joblin, C., Leger, A., & Martin, P. (1992), Contribution of polycyclic aromatic hydrocarbon molecul es to the interstellar extinction curve, *Astrophys. J.*, **393**, L79–L82.

Jones, A. P., & Williams, D. A. (1984), The 3 $\mu$m ice band in Taurus - implications for interstellar chemistry, *Mon. Not. Royal Ast. Soc.*, **209**, 955–960.

Justtanont, K., et al. (1996), The ISO-SWS spectrum of NML Cyg., *Astron. Astrophys.*, _ f 315, L217–L220.

Kerr, T. H., Adamson, A. J., & Whittet, D. C. B. (1991), Sharp infrared CO features in the Rho Ophiuchi mole cular cloud, *Mon. Not. Royal Ast. Soc.*, **251**, 60P–63P.

Kerr, T. H., Adamson, A. J., & Whittet, D. C. B. (1993), Infrared spectroscopy of solid CO - the Rho Ophiuchi molecular cloud, *Mon. Not. Royal Ast. Soc.*, **262**, 1047–1056.

Khare, B. N., Thompson, W. R., Cheng, L., Chyba, C., Sagan, C., Arakawa, E. T., Meisse, C., & Tuminello, P. S. (1993), Production and optical constraints of ice tholin from c harged particle irradiation of (1:6) $C_2H_6/H_2O$ at 77 K, *Icaru s*, **103**, 290–300.

Knacke, R. F., Fajardo-Acosta, S. B., Telesco, C. M., Hackwell, J 2E A., Lynch, D. K., & Russell, R. W. (1993), The silicates in the disk of Beta Pictoris, *Astrophys. J.* , **418**, 440.

Lacy, J. H., Baas, F., Allamandola, L. J., Van De Bult, C. E. P., Persson, S. E., McGregor, P. J., Lonsdale, C. J., & Geballe, T. R. (1984), 4.6 $\mu$m absorption features due to solid phase CO and cyano group molecules toward compact infrared sources, *Astrophys. J.*, **276**, 533–543.

Lee, W., & Wdowiak, T. J. (1993), Origin of the hydrocarbon component of carbonaceous chondrites: The star-meteorite connection, *Astrophys. J.*, **417**, L49- -L51.

Leger, A., & Puget, J. L. (1984), Identification of the 'unidentified' IR emission feat ures of interstellar dust?, *Astron. Astrophys.*, **137**, L5–L8.

Mathis, J. S. (1990), Interstellar dust and extinction, *Annual Rev. Astron. Astrophys.*, **28** , 37–70.

Matthews, C. (1992), In Margulis, L., & Olendzenski, L., editors, *Environmental Evolution*, MIT Press, Cambridge, pages 29.

McKee, C. F. (1989), Photoionization-regulated star formation and the struct ure of molecular clouds, *Astrophys. J.*, **345**, 782–801.

Merrill, K. M., Russell, R. W., & Soifer, B. T. (1976), Infrared observations of ices and silicates in molecula r clouds, *Astrophys. J.*, **207**, 763–769.

Meyer, A., Smith, R., Charnley, S., & Pendleton, Y. (1997), Water-ice in the envelopes of OH-IR stars, *Astronom. J.*. submitted.

Ogmen, M., & Duley, W. W. (1988), Laboratory synthesis of hydrogenated amorphous carbon f ilms, *Astrophys. J.*, **334**, L117–L120.

Palumbo, M. E., & Strazzulla, G. (1993), The 2140 cm$^1$ band of frozen CO - laboratory experiments and astrophysical applications, *Astron. Astrophys.*, **269**, 568- -580.

Papoular, R., Conrad, J., Giuliano, M., Kister, J., & Mille, G. (1989), A coal model for the carriers of the unidentified IR bands, *Astron. Astrophys.*, **217**, 204–208.

Papoular, R., Reynaud, C., & Nenner, I. (1991), The coal model for the unidentified infrared bands. II - the thermal emission mechanism, *Astron. Astrophys.*, **247**, 215–225.

Pendleton, Y. J., Sandford, S. A., Allamandola, L. J., Tielens, A 2E G. G. M., & Sellgren, K. (1994), Near-infrared absorption spectroscopy of interstellar hydrocarbon grains, *Astrophys. J.*, **437**, 683–696.

Pendleton, Y. (1997), Detection of organic matter in interstellar grains, m *Orig. Life Evol. Biosphere*, **27**, 53–78.

Pendleton, Y., Tielens, A., & Tokunaga, A. (1997), Nitriles in protostellar regions, *Astrophys. J.*. submitted.

Roche, P. F., & Aitken, D. K. (1984), An investigation of the interstellar extinction. i - to wards dusty wc wolf-rayet stars, *Mon. Not. Royal Ast. Soc.*, **208**, 481–492.

Roche, P. F., & Aitken, D. K. (1985), An investigation of the interstellar extinction. ii - t owards the mid-infrared sources in the galactic centre, *Mon. Not. Royal Ast. Soc.*, **215** , 425–435.

Sakata, A., Wada, S., Onaka, T., & Tokunaga, A. T. (1987), Infrared spectrum of quenched carbonaceous composite ( QCC). II - A new identification of the 7.7 and 8.6 $\mu$m unidentified in frared emission bands, *Astrophys. J.*, **320**, L63–L67.

Sandford, S. A., Allamandola, L. J., Tielens, A. G. G. M., & Val ero, G. J. (1988), Laboratory studies of the infrared spectral propertries of CO in astrophysical ices, *Astrophys. J.*, **329**, 498–510.

Sandford, S. A., Allamandola, L. J., Tielens, A., Sellgren, K., Tapia, M., & Pendleton, Y. (1991), The interstellar C-H stretching band near 3.4 $\mu$m - constraints on the composition of organic material in the diffuse interstellar medium, *Astrophys. J.*, **371**, 607–620.

Schutte, W. (1988), *The evolution of interstellar organic grain mantle s*, PhD dissertation, Leiden: Rijksuniversiteit.

Sellgren, K., Hall, D. N. B., Kleinmann, S. G., & Scoville, N. Z. (1987), Radial velocities of late-type stars in the Galactic Center, *Astrophys. J.*, **317**, 881–891.

Sellgren, K., Smith, R. G., & Brooke, T. Y. (1994), The 3.2–3.6 $\mu$m spectra of Monoceros R2/IRS-3 an d Elias 16, *Astrophys. J.*, **433**, 179–186.

Silverstein, R. M., & Bassler, G., (1967), *Spectroscopic Identification of Organic Compounds*, John Wiley and Sons, New York. Chapter 3 - Infrared Spectroscopy.

Smith, R. G., Sellgren, K., & Brooke, T. Y. (1993), Grain mantles in the Taurus Dark Cloud, *Mon. Not. Royal Ast. Soc.*, **263**, 749.

Smith, R. G., Sellgren, K., & Tokunaga, A. T. (1989), Absorption features in the 3 $\mu$m spectra of protos tars, *Astrophys. J.*, **344**, 413–426.

Tamura, M., Nagata, T., Sato, S., & Tanaka, M. (1987), Infrared polarimetry of dark clouds. i - magnetic field
structure in heiles cloud 2, *Mon. Not. Royal Ast. Soc.*, **224**, 413–423.

Tegler, S. C., Weintraub, D. A., Allamandola, L. J., Sandford, S. A., Rettig, T. W., & Campins, H. (1993), Detection of the 2165 cm$^{-1}$ (4.619 $\mu$m) XCN b and in the spectrum of L1551 IRS 5, *Astrophys. J.*, **411**, 260–265.

Tegler, S. C., Weintraub, D. A., Rettig, T. W., Pendleton, Y. J.,
Whittet, D. C. B., & Kulesa, C. A. (1995), Evidence for chemical processing of precometary icy gra ins in circumstellar environments of pre-main-sequence stars, *Astrophys. J.*, **439**, 279–287.

Telesco, C. M., & Knacke, R. F. (1991), Detection of silicates in the Beta Pictoris disk, m *Astrophys. J.*, **372**, L29–L31.

Tielens, A. G. G. M., Tokunaga, A. T., Geballe, T. R., & Baas, F. (1991), Interstellar solid CO - polar and nonpolar interstell ar ices, *Astrophys. J.*, **381**, 181–199.

Wdowiak, T. J., Wei, L., Cronin, J., Beegle, L. W., & Robinson , M. S. (1995), Plasma processing of interstellar PAHs into Solar Sy stem kerogen, *Planet. Space Sci.*, **43**, 1175–1182.

Weintraub, D. A., Tegler, S. C., Kastner, J. H., & Rettig, T. (1994), Infrared spectroscopy and imaging polarimetry of the di sk around the T Tauri star RNO 91, *Astrophys. J.*, **423**, 674–680.

Whittet, D. C. B., Adamson, A. J., Duley, W. W., Geballe, T. R., & McFadzean, A. D. (1989), Infrared spectroscopy of dust in the Taurus dark clou ds - Solid carbon monoxide, *Mon. Not. Royal Ast. Soc.*, **241**, 707–720.

Whittet, D. C. B., Bode, M. F., Longmore, A. J., Admason, A. J., McFadzean, A. D., Aitken, D. K., & Roche, P. F. (1988), Infrared spectroscopy of dust in the Taurus dark clou ds - Ice and silicates, *Mon. Not. Royal Ast. Soc.*, **233**, 321–336.

Whittet, D. C. B., McFadzean, A. D., & Longmore, A. J. (1985), Solid CO in the Taurus dark clouds, *Mon. Not. Royal Ast. Soc.*, **216**, 45P–50P.

Whittet, D. C. B., et al. (1996), An ISO SWS view of interstellar ices: first results., *Astron. Astrophys.*, **315**, L357–L360.

Whittet, D. (1992), Dust in the galactic environment, In *The Graduate Series in Astronomy*, Institute of Physics (IOP) Publishing, Bristol.

Whittet, D., & Tielens, A. (1997), Infrared observations of interstellar dust absorption features, In *From Stardust to Planetesimals*, Astronomical Society of the Pacific, San Francisco, pages 161–178.

Willner, S. P., et al. (1982), Infrared spectra of protostars - Composition of the d ust shells, *Astrophys. J.*, **253**, 174–187.

# TRACKING THE ORGANIC REFRACTORY COMPONENT OF INTERSTELLAR DUST

J. MAYO GREENBERG
*Laboratory Astrophysics, University of Leiden, Postbus 9504, 2300 RA Leiden, The Netherlands*
email: greenber@strw.LeidenUniv.nl

**Abstract.** The abundance and composition of complex organic (carbonaceous) material in the interstellar dust is followed as the dust evolves in its cyclic evolution between diffuse and dense clouds. Interstellar extinction, laboratory and space analog experiments, dust infrared absorption spectra, the cosmic abundance of the condensible atoms, and space and ground-based observations of comet dust are used to impose constraints on the organic dust component as mantles on silicate cores.

## 1. Introduction

It has long been accepted that a significant portion of interstellar dust has to be in some form or forms of carbon bearing particles.

When the 217 nm hump was discovered in the interstellar extinction, the proposal was for graphite to provide the feature (Stecher & Donn 1965). F. Johnson (1967) had suggested complex organics bearing a certain resemblance to polycyclic aromatic hydrocarbons in excited states, to account for the unidentified diffuse interstellar bands. Both of these proposals involved what we now call the very small particle/large molecule components of the interstellar dust population.

However, the organics associated with the large tenth micron sized particles which dominate the mass of the interstellar dust and which are responsible for the polarization and the major part of the visual/infrared extinction did not come into their own until three things were realized: (1) that $H_2O$ ice was not always a major component of the dust, (2) that silicates alone could not account for the visual extinction because of cosmic abundance constraints, (3) that the effect of ultraviolet photoprocessing (or other energetic processing) in molecular clouds could lead from simple ices



*J. M. Greenberg and A. Li (eds.), Formulation and Evolution of Solids in Space,* 53–75.

to complex organics. Ultimately, comets which bear many, if not all, the material signatures of interstellar dust, have been shown to be abundant in complex organics (Kissel et al. 1986a, b).

The earliest (indirect) evidence for the existence of an organic component of interstellar dust was the *lack* of evidence for $H_2O$ on grains which had been predicted by the dirty ice model of van de Hulst (1948). Earlier dust theories had been proposed involving metallic grains. There have been many theories which did not involve the $H_2O$ but rather provided the observed extinction in terms of other materials such as graphite and silicate (Mathis, Rumpl & Nordsieck 1977; Draine & Lee 1984). Why, in view of the undeniably reasonable original proposal by van de Hulst that surface reactions between oxygen and hydrogen must lead to ice mantles consisting of water, methane and ammonia, was the strong OH stretch in solid $H_2O$ not observed? In trying to answer this question and incorporating the observations of $H_2CO$ as a surprisingly abundant gas phase molecule (Snyder et al. 1969) I reconsidered the dirty ice grains as being modified in chemical structures by the ever present ultraviolet radiation in space (Greenberg et al. 1971). My reasoning was that the ultraviolet photons would act on the dirty ice in much the same way as the electric spark acts on a gas mixture as in the Miller-Urey experiment (Miller 1955) which was designed to simulate the formation of complex organic (pre-biotic) molecules in a hypothetical primitive earth atmosphere. Actually we now know that the production of formaldehyde in the solid phase with subsequent injection into the gas is not the only source of formaldehyde although it may make a significant contribution (Shalabiea & Greenberg 1994). However there was still a good basis for studying the effects of ultraviolet on molecular dust mantles. This was done more or less simultaneously by two groups: Sagan & Khare (1971) and Greenberg et al. (1971, 1972).

In any case, early laboratory results on the effect of ultraviolet radiation on simple ices showed that complex organic molecules were inevitably produced and that, in the (overall) process, it was certainly reasonable to conjecture that the $H_2O$ ice would be reduced and possibly replaced by other molecules. Initially no specific organic molecules were clearly defined by the experimental results on ultraviolet processing of ices but the fact that they led to the production of molecular weights up to $\sim$ 500 (Greenberg et al. 1971, 1972) was suggested by Greenberg (1973) as the reason why the OH stretch in solid $H_2O$ was not detected in VI Cyg #12 as well as in other sources.

The first *positive* evidence for the presence of an organic dust component came with the observation of the 3.4 $\mu$m absorption feature in the galactic center characterized by CH stretches corresponding to $CH_2$ and $CH_3$ groups in aliphatic hydrocarbons (Wickramasinghe & Allen 1980). When, much later, the 3.4 $\mu$m feature was *also* detected in VI Cyg #12 (Adamson et al. 1990) the problem of why no $H_2O$ was observed would appear to have been resolved. The photon which could break the $H_2O$ molecules could also break the methane and ammonia molecules and, as is always the case where carbon is involved in such energetic processing, the tendency is to create larger carbon rich molecules. Now, the 3.4 $\mu$m feature has been observed in many objects both in our galaxy and in external galaxies and the *idea* of organic mantles on dust has been clearly demonstrated by Pendleton et al. (1994).

However, instead of the subject ending here we are beginning to find that both observationally and theoretically as well as experimentally, much remains to be done before we can gain a thorough understanding of organics in space. Not only are there significant problems related to the exact nature (not only the chemical composition but even the atomic composition) of the organic grain mantles (the 3.4 $\mu$m feature being evidence only for aliphatic hydrocarbons), there also remain unanswered questions of the kinds of material responsible for the hump and the far ultraviolet extinction in the interstellar extinction curve. Although the hump particles are generally believed to be carbonaceous, we now know that they can not be graphite (Greenberg & Chlewicki 1983). The kinds of particles/molecules responsible for the FIR emission features, as well as the far ultraviolet extinction are also largely carbon and are probably closely related to PAH's, polycyclic aromatic hydrocarbons (Joblin, Léger & Martin 1992; also see Li & Greenberg 1997 and references therein).

It is the aim of this paper to provide an evolutionary picture of the formation of the organics in space, considering mostly the mantles on interstellar silicate cores and indicating possible sources of the other carbon components.

Although the 3.4 $\mu$m absorption feature is a clear indication of the presence of aliphatic hydrocarbons as grain mantles the infrared spectra demonstrate that it does not explain the complete mantles of interstellar dust which are needed to account for the observed amount and wavelength dependence of interstellar extinction. It is also not adequate to account for the dust to gas ratio in comets in which at least about half of the dust is organics - the other half being silicates or other refractory material.

It will turn out that the evolutionary picture of interstellar dust provides good reasons for the organic mantle material being different in different regions of space. By combining laboratory chemical analysis and infrared spectra of simulated interstellar organic mantles with infrared spectra of interstellar dust and the atomic composition of cometary organics a coherent picture may be emerging of the how and the why for these differences.

In the core-mantle model of interstellar dust some sort of organic mantles are needed to provide a match to all the major observational constraints of extinction, polarization, albedo and cosmic abundance (Li & Greenberg 1997) so that it is incumbent upon us to resolve their composition.

We shall present observational as well as theoretical and laboratory results bearing on this question. Each of these approaches brings to bear different constraints which help to define the problem.

The major constraints on the amount and composition of organic mantles of interstellar dust involve either directly or indirectly:

a) Sources and sinks of organic evolution;
b) Cosmic abundances;
c) Infrared absorption spectra;
d) Interstellar grain optical modelling;
e) Comets;
f) Laboratory analog methods.

I shall attempt to summarize the present status of each of the above to show how they are interrelated. Because they are so interrelated it is not possible to discuss them totally sequentially.

In Sect. 2 I will discuss some of the constraints on the grain constituents based on possible sources, sinks and atomic abundances. In Sect. 3 some spectral signatures related to organics seen toward the galactic center as well as other diffuse cloud lines of sight will be discussed. Sect. 4 is devoted to a very brief summary of the evolutionary model of interstellar dust. The use of laboratory methods to obtain infrared spectra and composition of organics produced as analogs of interstellar processes will be taken up in Sect. 5. In Sect. 6 the results of interstellar dust optical modelling will be presented. In Sect. 7 the properties of interstellar dust inferred from comet mass spectra and from ground based observations of comet dust and gas will be itemized. Finally in Sect. 8 I will attempt to provide a summary picture of how and why the organics in various spatial contexts differ. The conclusion will be that the organics vary from region to region and that there is need to provide more laboratory data as well as infrared observations to pin down the details.

## 2. Why organic mantles - formation and destruction

Regardless of the details of the composition of organic mantles there are two astrophysical constraints which require the presence of a large fraction of the interstellar dust in the form of carbonaceous mantles on silicate: (1) the cosmic abundance constraint, (2) the grain destruction constraint. The cosmic abundance constraint is simply stated by noting that the relative abundances of the elements in the interstellar medium are such that the observed amount of extinction relative to hydrogen can never be achieved by silicates alone consisting as they do of the rocky elements – Si, Mg, Fe, – but require a large fraction of the ten times more abundant organic elements – O, C, N (Greenberg 1974, 1978). The grain destruction constraint demands that the combination of the organics be refractory molecules dominated by carbon and that the silicates be well shielded from destruction.

### 2.1. SOURCES AND SINKS OF PARTICLES

It has been well documented theoretically that the fraction of silicate grains injected into the interstellar medium which can survive is only 10-20% of what is observed (McKee 1989; Tielens, this volume). The following is a highly simplified treatment. In general, the major source of silicate particles is from M stars and the rate of production is according to Gehrz (1989)

$$(\mathrm{dm/dt})_{\mathrm{sil}}^{\mathrm{prod}} \approx 0.004 - 0.028\,\mathrm{M}_\odot\,\mathrm{yr}^{-1}. \qquad (1)$$

In fact, Habing (this volume) estimates that all AGB stars with silicate envelopes produce a yearly amount of silicate of only $0.0027\,\mathrm{M}_\odot$ and that even adding the possibility of supernova sources leads to at most $0.009\,\mathrm{M}_\odot$ of silicates per year. On the other hand, the typical time to destroy a refractory grain as a result of processes induced in supernova remnants is (see Tielens, this volume)

$$\tau_{\mathrm{dest}} \approx 1 - 5 \times 10^8\,\mathrm{yr}. \qquad (2)$$

Given the fact that the Si, Mg, and Fe atoms are essentially fully depleted in the ISM and the $10\,\mu$m absorption strength (due to the Si-O stretch) is characteristic of their associated abundance being in the form of "silicates", we may roughly estimate the total mass of silicates in the ISM. We start by assuming $10^{-1}$ of the total mass of the galaxy is hydrogen. We then add all the cosmically available (relative to hydrogen) rocky elements in combination with a representative four oxygens to create a generic silicate molecule from which we deduce a mean silicate mass in the galaxy relative to hydrogen

$$(\mathrm{M_{sil}/M_H}) = \sum \mathrm{M_i}\,(\mathrm{n_i/n_H}) + 4\,\mathrm{M_O} \approx 59.6 \times 10^{-3} \qquad (3)$$

where the relative cosmic (solar system) abundances of Grevesse et al. (1996) have been used and $\mathrm{M_i}$, $\mathrm{n_i/n_H}$ are the atomic weights and relative abundances of Mg, Si, Fe. The abundance of oxygen $(\mathrm{n_O/n_H})$ is assumed much larger than all the $\mathrm{n_i/n_H}$ and $\mathrm{M_O}$ is the atomic mass of oxygen (see Greenberg 1998 for details). The total silicate mass in the galaxy is then

$$\mathrm{M_{sil}} \approx 59.6 \times 10^{-4} \times 10^{11}\,\mathrm{M}_\odot = 59.6 \times 10^{7}\,\mathrm{M}_\odot \qquad (4)$$

so that the mean destruction rate is

$$(\mathrm{dm/dt})_{\mathrm{sil}}^{\mathrm{dest}} = \mathrm{M_{sil}}/\tau_{\mathrm{dest}} \approx 0.12 - 0.60\,\mathrm{M_\odot\,yr^{-1}} \quad (5)$$

which is significantly larger than the production rate. Using the maximum production rate prescribed by Habing (this volume) of $0.009\,\mathrm{M_\odot\,yr^{-1}}$ we find that at most $0.009/0.12 = 0.075$ of the observed amount of silicates can possibly be maintained. This general approach is consistent with such detailed discussions as by McKee (1989). The presence of silicates in their known abundance can thus not be provided by any known "stellar" sources. Can they be replenished in the ISM? This appears highly unlikely when one notes the competition with the ten times more abundant organics. One might for example try to imagine silicates being produced by surface reactions but when Si is in competition with the at least ten times more abundant O, C and N, it seems highly unlikely. And to make silicates in the interstellar gas is even more improbable. The O, C and N will much more likely combine with hydrogen and with each other than with Si, Mg or Fe just on the basis of abundances.

What about carbonaceous particles? Can they be replenished by non ISM sources? The injection rate of all carbon rich particles from non ISM sources such as carbon stars, and Wolf-Rayet stars is (Gehrz 1989)

$$(\mathrm{dm/dt})_{\mathrm{carb}}^{\mathrm{prod}} \approx 0.003 - 0.01\mathrm{M_\odot\,yr^{-1}}. \quad (6)$$

The destruction rate of carbon particles and of complex organics is comparable with that of silicate particles (Draine & Salpeter 1979; Greenberg 1982a, b). Given that the depletion of carbon in the ISM is at least as high as 0.2 in the mantles of dust in diffuse clouds (Li & Greenberg 1997) the mass of carbon bearing mantles is at least as large as the silicates so that, again, there is no non-ISM way to maintain the observed dust mass.

A mechanism which could lead to a solution of the puzzle of why there *is* dust in the ISM is that provided by the photo-chemical formation of organic mantles in dust in molecular clouds. These mantles could provide an ablation shield for the silicates when the dust is exposed to the effects of supernova remnants in the diffuse cloud medium if the rate of production of organic mantles in molecular clouds is high enough to compensate for their degradation in diffuse clouds (Greenberg 1982a).

## 2.2. STABILITY OF GRAIN POPULATIONS

The mass creation rate for the organic refractory (OR) in molecular clouds (Greenberg 1986) is

$$(\mathrm{d}\rho_{\mathrm{OR}}/\mathrm{dt}) \approx 8 \times 10^{-42}\,\mathrm{g\,cm^{-3}\,s^{-1}}. \quad (7)$$

Since the mean Galactic mass density of the OR and the silicates is about the same, we see that it is possible to maintain a steady state of OR mantles allowing a destruction rate of $\sim$ 10 times that for silicates. It had earlier been estimated that the erosion rate for organic residues is comparable with that of silicates (Draine & Salpeter 1979; Greenberg 1982a). Thus, although a significant fraction of organics may be destroyed in the diffuse cloud medium, the loss can be compensated by reformation in the molecular clouds (see Sect. 4 on grain cycling). Thus in view of the remaining uncertainties in both the creation and destruction rates the balance required to achieve a steady state is not inconsistent with the data.

## 3. The galactic center and other diffuse cloud lines of sight

The best known – at least the earliest known - feature of the interstellar organics is the 3.4 $\mu$m feature. One of the early infrared spectra towards the galactic center showed this as a quite small feature compared with the very strong silicate absorption at 10 $\mu$m (Willner et al. 1979). It was already clear that, while weak, the 3.4 $\mu$m absorption represented a substantial fraction of the interstellar dust because the silicate abundances were inadequate to account for the extinction.

The 3.4 $\mu$m feature consists of three subfeatures at 2955 $cm^{-1}$ (3.385 $\mu$m), 2925 $cm^{-1}$ (3.420 $\mu$m), and 2870 $cm^{-1}$ (3.485 $\mu$m) corresponding to the C-H stretches in $-CH_2-$ and $-CH_3$ groups in aliphatic hydrocarbons which must be interacting with other chemical groups.

The amount of carbonaceous material responsible for the 3.4 $\mu$m feature is strongly dependent on the nature of the chemical groups attached to the aliphatic carbons. For example each carbonyl (C=O) group reduces its corresponding C-H stretch strength by a factor of $\sim$ 10 (Wexler 1967). Furthermore not every carbon is attached to a hydrogen as in saturated compounds. Aromatic hydrocarbons do not even contribute to the 3.4 $\mu$m feature although they do absorb nearby at $\approx$ 3.28 $\mu$m. Tielens et al. (1996) have made a rather thorough study of the galactic center spectrum concluding that abundance of oxygen in carbonyl groups is constrained to be rather small so that the observed strength (optical depth) of the 3.4 $\mu$m feature provides a reasonable basis for estimating the amount of carbon in aliphatics.

The column number densities of the CH groups can be determined from

$$\mathrm{N} = \int \mathrm{A}(\nu)\mathrm{d}\nu/\mathrm{A} \approx \tau_{\mathrm{max}}\Delta\nu_{1/2}/\mathrm{A} \tag{8}$$

where $\tau(\nu)$ is the frequency dependent optical depth (continuum subtracted), $\Delta\nu_{1/2}$ is the half width of the feature and A is the integrated absorbance in $cm^{-1}$ of the molecular group. Pendleton et al. (1994) have derived such column densities for a number of objects of which we note just two, GC IRS 6E and VI Cyg #12. They find the total number of C's in both $CH_2$ or $CH_3$ to be $\sim 9 \times 10^{17}\,cm^{-2}$ (GC) and $\sim 2 \times 10^{17}\,cm^{-2}$ (VI Cyg #12). The respective masses of carbonaceous (aliphatic) material along these lines of sight are

$$m^{GC}_{carb} = 2.1 \times 10^{-5}\,g\,cm^{-2}, m^{cyg}_{carb} = 0.46 \times 10^{-5}\,g\,cm^{-2}. \tag{9}$$

The relative mass of silicate material along these two sight lines as deduced from the silicate absorption is

$$\tau^{GC}_{sil}/\tau^{cyg}_{sil} = 3.6/0.58 = 6.2 \tag{10}$$

which is a bit higher than that for the carbonaceous material,

$$m^{GC}_{carb}/m^{cyg}_{carb} = 4.5. \tag{11}$$

Does this mean that the dust mantles are smaller towards VI Cyg #12 or that the organic material has a weaker $3.4\,\mu m$ feature per unit mass? At this time we have no obvious observational basis for deciding which of these is closer to reality.

The volume ratio of hydrocarbon (deduced for the $3.4\,\mu m$ feature) and silicates towards the galactic center has been determined by Tielens et al. (1996) to be $V_{HC}/V_{sil} = 0.7/4.3 = 0.16$ (=0.23, including hydrogenated amorphous carbon). But, as will be seen in Sect. 6, the volume of carbonaceous mantle material must be five to six times larger in order for the dust extinction to be as high as it is. It appears that the hydrocarbon component of the organics in diffuse cloud dust mantles must be a relatively minor component even though its absorption feature is the main infrared spectral evidence for the organic mantles.

## 4. A cyclic evolutionary model of interstellar dust

### 4.1. MANTLE EVOLUTION

Interstellar dust is certainly not static as is implied by some modelling of the interstellar extinction. Each dust grain, during its lifetime, passes through a wide variety of environments and interactions both constructive and destructive. I have used Fig. 1 to describe the main steps or events in the life cycle of an average of typical interstellar grain.

The initial phase occurs in the atmospheres of cool evolved oxygen rich stars. Here small silicate particles are produced and ejected into the surrounding interstellar space. This is truly "stardust" - a term which is

DUST IN SPACE

A Brief History of a Dust Grain

START

a) Small silicate particles condense in the atmospheres of cool stars and are ejected into space.

MOLECULAR CLOUDS

b) Silicate particles cool to T=5-15K and act as condensation nuclei for accretion of gas atoms and molecules as a mantle of frost.

c) Complex chemical reactions between gas and solids leads to an $H_2O$ and CO dominated grain mantle.

d) Ultraviolet radiation of mantle breaks simple molecules leading to new combinations and to complex organic molecules (photoprocessing).

e) Star formation occurs leading to dissipation of portions of the collapsed cloud into surrounding low density (diffuse cloud) regions.

DIFFUSE CLOUDS

f) In the low density regions volatile icy mantles are either photoprocessed, evaporated, or destroyed leaving only the complex organic refractory mantles.

These mantles are further subjected to photoprocessing and to destructive processes which can break off small pieces.

BACK TO b, c, d, e, f and repeat many times ($\sim 50$) until the grain is consumed by star formation or becomes part of a comet.

Total mean life time of a dust grain $= 5 \times 10^9$ years $=$ Turnover time for the interstellar medium into an out of stars.

*Figure 1.* Schematic illustration of the main steps or events in the life cycle of an average or typical interstellar grain.

sometimes applied also to the much more complex particles which have actually evolved in the space *between* the stars. Once these silicate particles escape they immediately cool down to temperatures of 10 - 15 K and become coupled to the gas as part of the general interstellar medium. Ultimately, the gas cloud containing the silicate particles becomes denser and the rate at which atoms and molecules collide with the grain is sufficient to grow a mantle. Simple molecules like $H_2O$, $CH_4$, $NH_3$ form initially by surface reactions of the colliding O, C and N with H. Additional molecules - predominantly CO - accrete as such, although surface reactions with hydrogen may also lead to HCO and $H_2CO$ as well as $CH_3OH$. At a density of $n_H = 10^4\,cm^{-3}$ the (condensible) mean atom-grain collision time is $\approx 3 \times 10^4\,s$ ($< 1$ day) so that during the life time of the cloud ($\sim 3 \times 10^7$ yrs) a mantle of frozen molecules is inevitably built up; i.e., a monolayer in $<1$ yr. This build-up is accompanied by bombardment with energetic ultraviolet photons which have a mean collision time of $2 \times 10^5$ s (assuming a flux of $\Phi_{h\nu} \approx 10^4\,cm^{-2}\,s^{-1}$ for $h\nu > 6\,eV$). The collision time for 1 MeV cosmic ray protons ($\Phi_P \leq 3\,cm^{-2}\,s^{-1}$) is at least 1000 times longer; i.e., $\Phi_{h\nu}/\Phi_P \approx 3 \times 10^3$. Higher energy protons ($\geq 1\,MeV$) are not only less abundant, they individually deposit less energy in the grains because their interaction cross sections are smaller. Most of the energetic processing is provided by the photons even though the cosmic rays are individually much more energetic because only about 1 keV of the 1-2 MeV proton energy is deposited in a tenth micron grain so that they contribute (in molecular clouds) at most $\sim 10\%$ of the total energy (Greenberg 1985).

Over the $3 \times 10^7$ years in a molecular cloud the photo-processing produces a complex organic component in the mantle as the molecules are broken and recombine. While the cloud is dispersed - usually as a result of star formation - and subsequently, when the dust grain is in the diffuse cloud phase, it loses its volatile mantle components leaving only the complex organic refractory component. This "first generation" organic material is there subjected not only to the 10,000 as intense (as in the molecular cloud) radiation field it is also subjected to the destructive processes accompanying supernova shocks (Draine & Salpeter 1979; McKee 1989; Tielens, this volume). The resulting and remaining organic material is what is observed in diffuse cloud dust. The destruction by shocks and grain-grain collisions probably result in breaking off pieces of the organics which are then to be observed as very small carbonaceous particles and large molecules (PAH's) contributing to the ultraviolet and far ultraviolet extinction (see Sect. 6). This completes one cycle. Unless the grain is consumed by star formation or has become a part of a planetary/cometary object in a "solar system" it may pass through a series of diffuse cloud/molecular cloud cycles, only the first one starting with "pure stardust". The ultimate fate of a dust grain will be via star formation. The maximum mean lifetime of a grain is then the mean turnover time for the interstellar medium which is about $5 \times 10^9$ years so that each grain, on the average, should have undergone about 50 complete cycles.

### 4.2. SILICATE CORE EVOLUTION

It has now been observed by ISO that the heated silicate emerging from stars is partially crystalline. However, there is no evidence for any degree of crystallinity shown in the $10\,\mu$m interstellar absorption which should show up as an $11.2\,\mu$m structure. Is this an observational problem caused by the mantles on the interstellar silicates or is it real? If it is real there is then the question - to be discussed in the comet dust section - of why or how the silicates seen in emission in the comet dust usually exhibit a degree of crystallinity inconsistent with the concept that the interstellar dust is amorphous. Comets or comet dust could not ever have undergone the degree of heating required to produce crystallinity. If, before comet formation, the protosolar dust had been heated enough to crystallize the silicate ($\sim 800 - 1000\,$K) all the organics as well as the ices would have evaporated. The reconstituted molecular composition of comets would no longer resemble that of the interstellar ices (see Crovisier, this volume and references therein). Furthermore, the ices in comets would be entirely crystalline even if formed at relatively low temperatures (Kouchi et al. 1994). Finally, the ratio of the ortho to para states of water coming off of comets implies that the ice could only have formed at interstellar temperatures of no higher than 30 K (Crovisier, this volume). Even though comet dust grains are heated well above all previous temperatures within the comet nucleus, their temperatures based on observed infrared emission are seldom higher than 600 K (Greenberg & Hage 1990; Li & Greenberg 1998). If one accepts it as a fact that circumstellar silicates and comet dust silicates are partially crystalline and that interstellar silicates are indeed amorphous there is a puzzle to be solved. An evolutionary sequence is suggested in which partially crystalline silicates are first amorphized in the interstellar medium as a result of cosmic ray processing just as crystalline ice is amorphized by energetic processing (Kouchi & Kuroda 1990). We assume that comets are formed and have certainly remained at temperatures less than the $\sim 800 - 1000\,$K required to recrystallize the silicate. If the amorphous state is metastable, some of the crystallinity may be recovered by relatively weak processes which occur in comet dust, involving temperatures of only 500 - 600 K which are created by solar heating as the dust comes off the comet.

## 5. Laboratory analog organics

The use of laboratory methods to produce and study possible organic or carbonaceous constraints of interstellar dust has become very important. A number of potentially useful ideas and methods have been applied to the manufacture of materials to provide the interstellar absorption feature at 220 nm (see for example Mennella et al. 1996; Sakata & Wada 1989). These will not be discussed here. We limit ourselves to: (1) the residues produced from ultraviolet photoprocessing of ice mantles as an analog to the material made in the molecular cloud phase and (2) the (solar) ultraviolet irradiated residues as an analog to the diffuse cloud carbonaceous mantles.

The chemical and physical analogues of first generation organics (molecular cloud phase) suggest that such material is relatively rich in oxygen. It appears to consist of a wide range of aliphatics as well as some aromatics. A starting mixture of $H_2O : CO : NH_3 = 5 : 5 : 1$ led to an estimate of the relative atomic composition of the material as divided into four groups in decreasing degree of volatility.

1) Products evaporated and collected at $100\,K < T < 300\,K$ (Greenberg et al. 1993; Mendoza-Gómez 1992) — $C : O : N : H = 1 : 1.2 : 0.03 : 1.4$;

2) Organic residue material remaining at room temperature and analyzed by GCMS (Briggs et al. 1992) — $C : O : N : H = 1 : 1.06 : 0.1 : 2.2$;

3) Organic residue remaining at room temperature analyzed by mass spectroscopy [relatively more refractory than portion (2)] (Greenberg et al. 1993; Mendoza-Gómez 1992) — $C : O : N : H = 1 : 0.06 : 0.001 : 1.1$;

4) Unanalyzed organic residue (estimates based on the *overall* structure of the background mass spectra) — $C : O : N : H = 1 : 0.06 : >0.001 : 1.1$.

We see that there is quite a range of C : O ratios and that some of what is called organic refractory with respect to the general interstellar medium will sublimate at temperatures $T < 300\,K$ and even at $T \approx 200\,K$.

In any case, the first generation organics do not provide a perfect match to the interstellar $3.4\,\mu m$ feature. In fact, if one examines the infrared spectrum of a residue which is not brought up to room temperature (or one that is) the match is not very good (Allamandola et al. 1988, Pendleton et al. 1994).

In order to approach the true diffuse cloud organic mantle material some of the laboratory residues were subjected to 6 months exposure by solar ultraviolet irradiation on a space platform. These materials provided excellent $3.4\,\mu m$ spectra (Greenberg et al. 1995). While the mass spectral analysis of the material is still preliminary the indications are that it is rich in polycyclic aromatic hydrocarbons. On the other hand the infrared absorption spectra as seen in Greenberg & Li (1996) indicate a rather significant abundance of oxygen as seen by the $3\,\mu m$ absorption feature. Such an absorption is not seen - or is at least rather weak - in diffuse cloud dust if one subtracts the absorption produced by the $H_2O$ in molecular cloud dust along the lines of sight (Schutte & Greenberg 1988). The implication of this is that the mean oxygen abundance in the true organic mantles is even less than in the solar irradiated residues. The strength of the $3.4\,\mu m$ absorption feature per unit mass of laboratory residues (of the first generation type) was determined to be $\sim 440\,cm^2\,g^{-1}$ (see Greenberg 1986). Given the $3.4\,\mu m$ optical depth to the galactic center as $\tau_{3.4} \approx 0.22$ one obtains a column density of organics $M_{OR} = 4.6 \times 10^{-4}\,g\,cm^{-2}$ and assuming a mass density of $\rho_{OR} = 1.8\,g\,cm^{-3}$ (somewhat less than graphite but larger than the ices) one obtains a volume column density of $V_{OR}^{GC} = 2.56 \times 10^{-4}\,cm$. The volume column density of the silicates can be similarly evaluated using a volume absorption coefficient of $\kappa_{sil} = 10^4\,cm^{-1}$ and $\tau_{sil}^{GC} \approx 3.6$ as $V_{sil}^{GC} = 3.3 \times 10^{-4}\,cm$ which suggests approximately equal volumes of cores and mantles rather than the much smaller fraction suggested by Tielens et al. (1996).

## 6. Interstellar dust

### 6.1. DIFFUSE CLOUD DUST

Because silicates alone are of insufficient abundance to produce the observed amount and wavelength dependence of extinction, carbonaceous mantles *must* be involved. A detailed analysis of the material requirements needed to provide the average diffuse cloud dust observations has led to the conclusion that the average mantle to core volume rate is $V_{OR}/V_{sil} = 0.95$ (Li & Greenberg 1997). The relative atomic composition of the mantle in this model was assumed to be $C:O:N:H = 1.0:0.2:0.004:1.0$. Note that the $O:C$ ratio has been taken lower than that given by the first generation organic total taking into account that the diffuse cloud dust mantles have been strongly processed by ultraviolet which reduces all elements with respect to carbon. The cosmic abundance constraint puts rather strict limitations on the carbon depletion in dust. This is shown in Table 1 where it is seen that there are problems in producing the observed amount of extinction with the available carbon atoms. If the $O:C$ ratio in the mantle is reduced the carbon depletion is higher and it becomes increasingly difficult to satisfy the abundance constraint. In addition, with lower oxygen in the mantle there is an even greater excess of oxygen to be accounted for.

### 6.2. MOLECULAR CLOUD DUST

While no $3.4\,\mu m$ feature has been observed in molecular cloud dust, there is indirect evidence that organic mantles exist. This evidence is in the infrared absorption in the neighborhood of the $10\,\mu m$ silicate feature. In fact, the nature of this absorption may suggest why the $3.4\,\mu m$ feature is not observed even though refractory organics are present. It appears that the $10\,\mu m$ and $20\,\mu m$ polarization shapes in the Becklin-Neugebauer (BN) object in Orion can not be satisfied by any silicate materials whether laboratory prepared or natural (see Greenberg & Li 1996, and references therein). It turns out that the measured absorption of the space irradiated residues between 5 and $8\,\mu m$ can modify the shape of the silicate features. Actually this ab-

TABLE 1. The cosmic abundance constraint (in unit of atoms per $10^6$ hydrogen nuclei).

| element | gas phase | 2/3 solar[†] | | $\left[\frac{X}{H}\right]_d^{mod}$ | |
|---|---|---|---|---|---|
| | | | | Diffuse cloud | Silicate/graphite |
| X | $\left[\frac{X}{H}\right]_g^{ISM}$ | $\left[\frac{X}{H}\right]_c^{\frac{2}{3}\odot}$ | $\left[\frac{X}{H}\right]_d^{ava\frac{2}{3}\odot}$ | unified model* | model[‡] |
| Si | ⊕ | 24 | 24 | 20.4 | 32 |
| Fe | ⊕ | 21 | 21 | 20.4 | 29 |
| Mg | ⊕ | 25 | 25 | 20.4 | 35 |
| C | $140 \pm 20$ | 325[a] | 185 | 194 | 300 |
| O | $310 \pm 20$ | 493 | 183 | 103 | 128 |
| N | 61 | $62 \pm 9$ | $1.0 \pm 9$ | 4.3 | not given |

† The "reference abundance" (except for carbon) is assumed to be 2/3 of the solar system abundance as obtained by Grevesse et al. (1996).
* Li & Greenberg (1997).
‡ Draine & Lee (1984).
⊕ Assuming very low gas phase abundance for the rock elements.
[a] This would be ~ 237 if $\left[\frac{C}{H}\right]_c^{\odot} = 355 \times 10^{-6}$ (Grevesse et al. 1996). We assume that (C/O) is 0.6.

sorption is too large while that of the organics in the Murchison meteorite (presumably similar to those of interstellar origin) is too small (see Fig. 3 in Greenberg & Li 1996). If we assume that the actual organic mantle material has an intermediate amount of oxygen it is possible to precisely reproduce the astronomical observation with a mantled pure amorphous silicate. The mantle thickness required to match the BN 10 and 20 $\mu$m polarization is such that $V_{OR}/V_{SIL} \approx 2$ is much *larger* than is characteristic of diffuse cloud dust. This is not too surprising in view of the fact that in the line of sight to the BN object one is seeing molecular cloud dust on which the inner mantle of diffuse cloud organics is likely to be surrounded by an *additional* mantle of incompletely photoprocessed molecular cloud organics. The fact that the 3.4 $\mu$m absorption is not observed may possibly be attributed to hydrogenation *and* oxidation of the inner organics mantle by accretion and photoprocessing in the dense molecular cloud medium. This effect should be investigated in the laboratory by studying the evolution of the 3.4 $\mu$m feature during simultaneous deposition of ices and ultraviolet irradiation. In any case, there is both evidence for an organic mantle and for its being chemically different from the organic mantles of diffuse cloud dust.

I have attempted to show schematically in Fig. 2 how the various organic components of the interstellar dust evolve, starting with the diffuse cloud interstellar dust. In addition to the tenth micron silicate core-organic refractory mantle particles there are thousands of small carbonaceous particles responsible for the 220 nm extinction hump and many more even smaller particles presumed to resemble PAH's which are responsible for the far ultraviolet extinction (Li & Greenberg 1997). As the density of the cloud increases in the transition to the molecular cloud phase, an icy mantle grows and traps some of the hump and PAH particles. The accompanying ultraviolet effects on the ices leads to interactions with the inner organic mantle which modifies it, as indicated by the missing 3.4 $\mu$m feature. It also provides a source of interactions between the radicals produced in the ices and the hump and PAH particles changing *their* character as well. At the same time, the ultraviolet is producing the beginnings of first generation organics in the ices. The transition back to the diffuse cloud phase occurs with a very substantial increase in the ultraviolet photoprocessing. In fact, it could well be that the major processing of the ices leading to first generation organics occurs in the transition phase; i.e., before the ices evaporate or are sputtered away they are subjected to ultraviolet fluxes 1000 - 10,000 times greater than in the molecular cloud. Finally, while the ices are being destroyed in the diffuse cloud, the precursors of the ultimate hump particles and the ultimate PAH's emerge. These are now intensively processed by the ultraviolet and resume their diffuse cloud character as "normal" hump and FUV (PAH) particles. The same intensive ultraviolet modifies the outer first generation organic mantle as well as the inner organic mantle, resulting in the highly processed organic refractory diffuse cloud dust mantle.

The cometary material is schematically indicated as resulting from aggregates of the molecular cloud dust and, as such contains both processed and first generation organics as well as the ices, hump and PAH particles, the latter being trapped in the ices.

## 7. Comet organics

### 7.1. COMET HALLEY MASS SPECTRA

The first direct observational confirmation of the prediction (Greenberg 1979, 1982b) of complex organics in comets came from the mass spectral data obtained for comet Halley dust (Kissel et al. 1986a, b). Further analysis of the data showed that the organic molecules had, on the average, a higher initial energy than the silicate ions which led Krueger & Kissel (1987) to infer a core-mantle structure of the dust particles. Thus according to Jessburger & Kissel (1991) "The existence of the previously postulated (Greenberg 1982b) core-mantle grains seems to be substantiated by data".

Figure 2. Schematic of evolution of the organic component of interstellar dust. DC = diffuse cloud phase; MC = molecular cloud phase; OR = average diffuse cloud organic refractory mantle material; $OR_1$ = "first generation" organics; components with a superscript (') are modified forms. See text and Fig. 1 for additional descriptions.

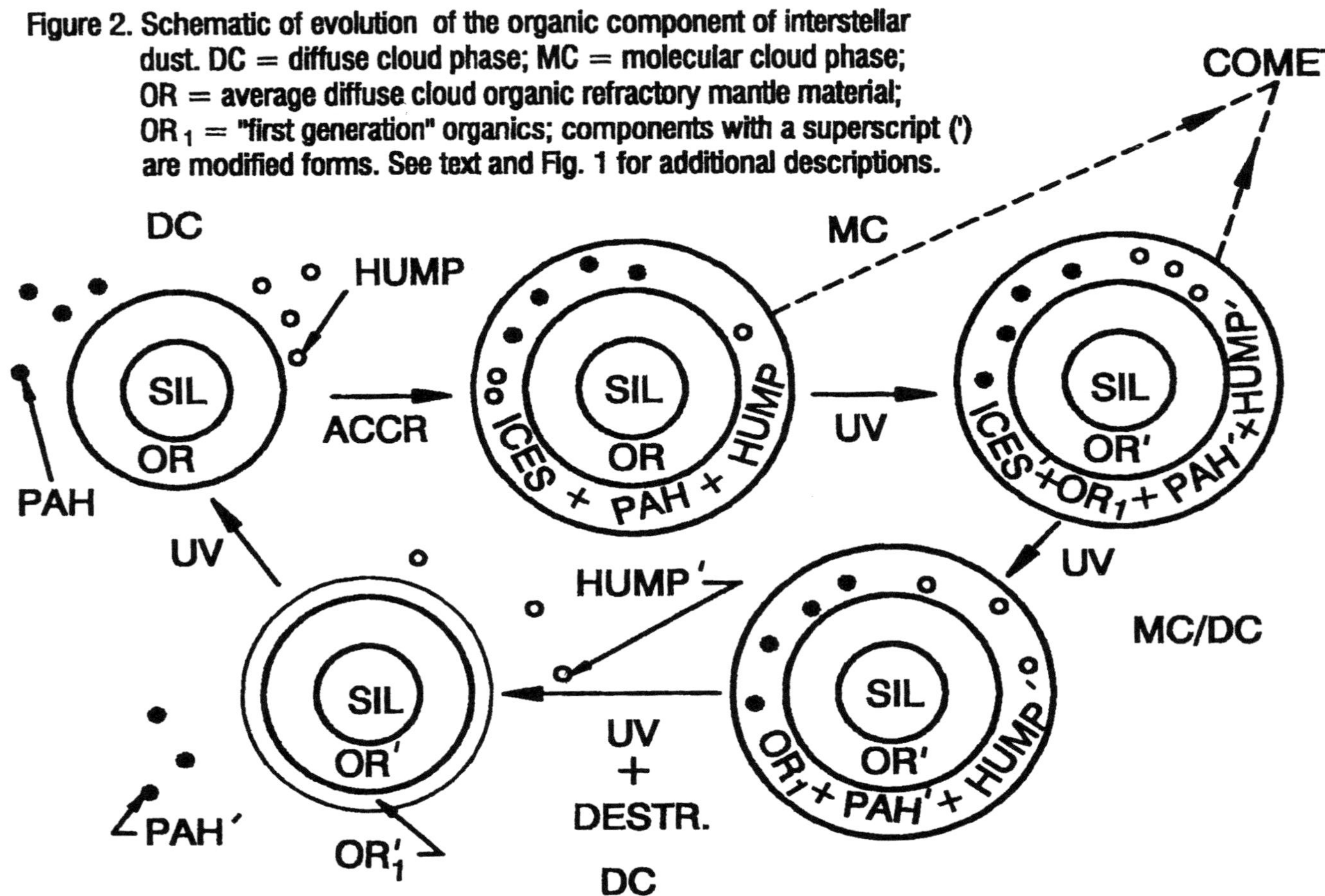

An O : C = 0.6 : 1 ratio deduced from the overall comet Halley organics is much higher than deduced from the Galactic center absorption spectra but is consistent with a mixture of first generation laboratory organics with space irradiated organics as implied by the lab residues returned after being exposed to the sun for 6 months (Greenberg et al. 1995).

If we assume that the comet material contains the solar system abundance of the elements and if we assume that the Si, Mg and Fe are totally bound up in the silicate cores we may derive a mantle to core mass ratio. This turns out to be (Greenberg 1998) $m_{OR}/m_{sil} \approx 1$ which is in agreement with the value obtained by the Giotto/Vega space mission to comet Halley (Krueger & Kissel 1987) as well as with the BN polarization requirement if $\rho_{OR} \simeq 1/2\,\rho_{sil}$.

## 7.2. INFRARED EMISSION BY COMET DUST

I should mention here that the 3.4 $\mu$m emission feature in the comet Halley dust spectrum appears to have come largely from volatile species ($CH_3OH$) rather than from the organic mantles (Reuter & Mumma 1991, Hoban et al. 1991). However, the organic mantles on the silicate cores were required to provide the source of heating of the comet dust - as aggregates of interstellar dust - which made the silicate features visible (Greenberg & Hage 1990). The details of the modelling of comet dust is not emphasized here but only the *physical* requirement that organics must be present. The details may be seen, for comet Halley emission, in Greenberg & Hage (1990) and for comet Hale-Bopp emission, in Li & Greenberg (1998). In the former case the match with observations could best be made with $m_{OR}/m_{sil} = 1$ which implies $V_{OR}/V_{sil} = 2$, similar to the dust towards the YSO (young stellar objects) in the Orion nebula. For Hale-Bopp it appears that $m_{OR}/m_{sil} = 1/2$ rather than 1. In any case, we are led to expect that the comet organics probably resemble the molecular cloud variety - combined oxygen depleted (as inner mantles) and oxygen undepleted organics. The latter would be expected to have been produced in the final stages of cloud contraction leading to the solar system. While the dust seen towards the BN object is not expected to be exactly of this sort, the fact that it is in a region of new star formation suggests that it is post star formation dust which one may presume to have preserved some of the characteristics of the protostellar dust.

## 7.3. DISTRIBUTED CO IN COMET COMAE

An important clue to the abundance of oxygen in comet organics is the presence of CO as a distributed molecule in the comet Halley coma. Eberhardt et al. (1987) found that rather than falling off as $1/R^2$, the CO intensity showed an approximately exponential growth up to about 20,000 km. This was explained in terms of an extended source attributable to the evaporation of the organic fraction of comet dust (Greenberg & Hage 1990; Greenberg & Li 1998). However, to account for the abundance of the source of CO required a very substantial fraction of the organics containing CO groups. For example, using the gas chromatograph analysis of the soluble component of nonvolatile residue of a "first generation" photolysis of the 10 K ice mixture $H_2O : CO : NH_3 = 5 : 5 : 1$ (identified in Sect. 5 as group 2) by Briggs et al. (1992) leads to an O : C ratio of at most 1.2 and probably $O : C \approx 0.6$ for the nonvolatile fraction of the organics as deduced by Krueger & Kissel (1987) for comet Halley organics and by Greenberg & Li (1998) for laboratory analog organics. In fact, adding up all the possible CO groups in the GCMS analyzed molecules gives a fraction by mass of $\sim 0.16$ of that component of the organic refractories. In addition to CO, other simple molecules in comet comae like $C_2$, $C_3$, CN, $H_2CO$ appear to be distributed in such a way that they are neither directly emitted from the nucleus surface nor created as daughter molecules from more complex gas phase species. The only remaining possible source is the organic component in comet dust. The requirements imposed on the comet dust grains by the distributed CO emission are that they be heated sufficiently to evaporate a large fraction of the more volatile fraction of the complex organic refractory molecules and that a large fraction of these contain CO groups. Inferring the *dust/gas* ratio within the mass limits from the comet dust size (mass) distribution obtained by the Giotto spacecraft for comet Halley, and assuming that the refractory organics remaining on the silicate cores are the heating agent by solar radiation in fluffy aggregates of interstellar core-mantle particles, the upper limits of the total amount of CO provided by dust can be approximately determined as a function of porosity (Greenberg & Li 1998). The resulting maximum CO production rate predicted by the comet dust model is significantly less than the observed distributed CO abundance, even using a large O : C ratio. A possible solution lies in the fact that the dust to gas ratio has been underestimated in the dust size distribution employed, or that by neglecting efficient dust fragmentation and sublimation in the innermost coma we have underestimated the fraction of heated dust. On the other hand, it may not be impossible that the extended CO abundance was overestimated due to the crossing of dust jets, the time variation of the comet nucleus activity, and the anisotropic outgassing nature of the nucleus. At this point we can certainly say that, in order to obtain the observed distributed CO production rate from comet dust requires not only the organic refractory mantles and very high porosity, but also it seems that more heatable dust must be available than has been deduced from the space observation.

## 8. Concluding remarks

The varieties of complex organics in grain mantles clearly depend on both the history and the local environment of the dust. However, not only what kind of organics are expected but how much is a key to dust chemical evolution. The following table (Table 2) indicates the wide range of values suggested by various current arguments.

TABLE 2. The wide range of $V_{OR}/V_{sil}$ values.

| Region | $V_{OR}/V_{sil}$ | Comments | Reference |
|---|---|---|---|
| Diffuse cloud | 0.95 | Based on the unified dust model | Li & Greenberg (1997) |
| Diffuse cloud | 0.23 | Based on the infrared absorption spectra of the galactic center | Tielens et al. (1996) |
| Molecular cloud | 2 | Based on matching the 10 and 20 $\mu$m polarization to the BN object | Greenberg & Li (1996) |
| Comet | 2 | Based on Giotto/Vega comet dust mass spectra spectra | Kissel & Krueger (1987) |

The very low value of the organic fraction deduced from IR spectra is difficult to reconcile with that required for interstellar dust grain modeling and it will have to be explained. I suggest that the answer will probably be seen as resulting from a reinterpretation of the spectra; i.e., that the 3.4 $\mu$m strength does *not* provide evidence for *all* of the carbonaceous material but rather only a part. The lack of CO groups in diffuse cloud dust mantles is expected because of the extreme ultraviolet processing while the abundance of CO groups in comet organics is expected because this represents in large part the final stage of cloud chemical evolution before star formation where the ultraviolet is greatly reduced and where the abundance of first generation organics was created and accreted on the dust in the precollapse phase.

There remain many unanswered questions about the nature and abundance of organics as dust mantles but new astronomical observations along with intensified laboratory investigations will help to provide the answers. Perhaps the ultimate and most direct answers must await either in situ studies of comet material or even retrieval of comet material as in the ESA ROSETTA mission or the NASA New Millennium program.

## 9. Acknowledgements

I would like to thank Aigen Li for his great help in the preparation of this paper and Willem Schutte for some valuble discussion and insights.

## References

Adamson, A.J., Whittet, D.C.B., Duley, W.W. (1990), The 3.4 $\mu$m interstellar absorption feature in Cyg OB2 no. 12, Mon. Not. R. Astron. Soc. 243, 400-404.

Agarwal, V.K., Schutte, W.A., Greenberg, J.M., Ferris, J.P., Briggs, R., Connor, S., van de Bult, C.P.E.M., Baas, F. (1985), Photochemical reactions in interstellar grains, photolysis of CO, $NH_3$ and $H_2O$, Origins of Life 16, 21-40.

Allamandola, L.J., Sandford, S.A., and Valerio, G. (1988) Photochemical and chemical evolution of interstellar/pre-cometary ice analogs, Icarus 76, 225-252.

Allen, D.A., Wickramasinghe, D.T. (1981) Diffuse interstellar absorption band between 2.9 and 4.0 $\mu$m, Nature 294, 239-240.

Baas, F., Geballe, T.R., Walther, D.M. (1986) Spectroscopy of the 3.4 $\mu$m emission feature in Comet Halley, Astrophys. J. Lett. 311, L97-L101.

Briggs, R., Ertem, G., Ferris, J.P., Greenberg, J.M., McCain, P.J., Mendoza-Gómez, C.X., Schutte, W.A. (1992) Comet Halley as an aggregate of interstellar dust and further evidence for the photochemical formation of organics in the interstellar medium, Origins of Life and Evolution of the Biosphere 22, 287-307.

Butchart, I., McFadzean, A.D., Whittet, D.C.B, Geballe, T.R., Greenberg, J.M. (1986) Three micron spectroscopy of the galactic centre source IRS 7, Astron. Astrophys. 154, L5-L7.

Chiar, J., Adamson, A.J., Whittet, D.C.B. (1996) Three micron hydrocarbon and methanol absorption in Taurus, Astrophys. J. 472, 665-672.

Crovisier, J. (1998), this volume.

Danks, A.C., Encrenaz, T., Bouchet, P., et al., (1986) Observation of an emission feature at 3.4 $\mu$m in the spectrum of Comet Halley, in 20th ESLAB Symp. on the Exploration of Halley's Comet (B. Battrick, et al., eds.), ESA, Noordwijk, p. 103-106.

Draine, B.T., Salpeter, E.E. (1979) Destruction mechanism for interstellar dust, Astrophys. J. 231, 438-455.

Draine, B.T., Lee, H.M. (1984) Optical properties of interstellar graphite and silicate grains, Astrophys. J. 285, 89-108.

Duley, W.W., Williams, D.A. (1979) Are there organic grains in the interstellar medium? Nature 277, 40-41.

Eberhardt, P., Krankowsky, D., Schulte, W., et al., (1987) The CO and $N_2$ abundance in comet P/Halley, Astron. Astrophys. 187, 481-484.

Gehrz, R.D. (1989) Sources of stardust in the galaxy, in Interstellar Dust (L.J. Allamandola, A.G.G.M. Tielens, eds.), Kluwer, p. 445-453.

Greenberg, J.M., Yencha, A.J., Frisch, H.L. (1971) Growth, distribution and chemical composition of interstellar dust, Bull. Amer. Astron. Soc. 3, 250

Greenberg, J.M., Yencha, A.J., Corbett, J.W., Frisch, H.L. (1972) Ultraviolet effects on the chemical composition and optical properties of interstellar grains, Mém. Soc. Roy. Sci. Liège, 6e série III, p. 425-436

Greenberg, J.M. (1973) Chemical and physical properties of interstellar dust, in Molecules in the Galactic Environment (M.A. Gordon, L.E. Snyder, eds.), Proc. Symp. Interstellar Molecules, Charlottesville, VA, Oct. 1971, Wiley, p. 94-124.

Greenberg, J.M. (1974) The interstellar depletion mystery, or where have all those atoms gone? Astrophys. J. Lett. 189, L81-L85.

Greenberg, J.M. (1978) Interstellar dust, in Cosmic dust (J.A.M. McDonnell, ed.), Wiley, p. 187-294.

Greenberg, J.M. (1979) From interstellar dust to comets to the zodiacal light, in: Cometary Missions (W.I. Axford, H. Fechtig, J. Rahe, eds.), Veröff. Remeis Sternwarte, Bamberg bd. XII nr. 132, p. 119-126.

Greenberg, J.M. (1982a) Dust in dense clouds. One stage in a cycle, in: Submillimetre wave astronomy (J.E. Beckman, J.P. Phillips, eds.), Cambridge University Press, p. 261-306.

Greenberg, J.M. (1982b) What are comets made of - a model based on interstellar dust, in: Comets (L. Wilkening, ed.), University of Arizona Press, p. 131-163.

Greenberg, J.M., Chlewicki, G. (1983) A far ultraviolet extinction law: what does it mean? Astrophys. J. 272, 563-578

Greenberg, J.M. (1985) Evolution of interstellar grains: observations, theory, laboratory experiments, in: Birth and infancy of stars (R. Lucas, et al., eds.), Les Houches, session XLI, 1983, Elsevier, p. 141-203

Greenberg, J.M. (1986) Dust in diffuse clouds: one stage in a cycle, in: Light on dark matter (Proc. IRAS Symp., Noordwijk, 10-14 June 1985) (ed. F. Israel), Reidel, p. 177-188.

Greenberg, J.M., Hage, J.I. (1990) From interstellar dust to comets: a unification of observational constraints, Astrophys. J. 361, 260-274

Greenberg, J.M. (1991) Physical, chemical and optical interactions with interstellar dust, in Chemistry in Space (Greenberg, J.M., Pironello, V., eds), Kluwer, p. 227-261.

Greenberg, J.M., Mendoza-Gómez, C.X., de Groot, M.S., Breukers, R. (1992) Laboratory dust studies and gas-grain chemistry, in: Dust and chemistry in astronomy (T.J. Millar, D.A. Williams, eds.) (Proc. conf. Manchester, Jan. 1992), IOP publ. Ltd., p. 265-288.

Greenberg, J.M., Li, A., Mendoza-Gómez, C.X., Schutte, W.A., Gerakines, P.A., de Groot, M. (1995) Approaching the interstellar grain organic refractory component. Astrophys. J. Lett. 455, L177-L180.

Greenberg, J.M., Li, A. (1996) What are the true astronomical silicates? Astron. Astrophys. 309, 258-266.

Greenberg, J.M. (1998) Making a comet nucleus, Astron. Astrophys. 330, 375-380

Greenberg, J.M., Li, A. (1998) From interstellar dust to comets: the extended CO source in comet Halley, Astron. Astrophys. 332, 374-384.

Grevesse, N., Noels, A., Saural, J. (1996) in: Cosmic Abundances (S. Holt, G. Sonneborn, eds.), ASP conf. series 99, p. 117-176.

Habing, H.J. (1998), this volume.

Hoban, S., Mumma, M., Reuter, D.C., et al., (1991) A tentative identification of methanol ias the progenitor of the 3.52 micron emission feature in several comets, Icarus 93, 122-134.

Jenniskens, P., Baratta, G.A., Kouchi, A., de Groot, M.S., Greenberg, J.M., Strazzulla, G. (1993) Carbon dust formation on interstellar grains, Astron. Astrophys. 273, 583-600.

Joblin, C., Léger, A., Martin, P. (1992) Contribution of polycyclic aromatic hydrocarbons to the interstellar extinction curve, Astrophys. J. Lett. 393, L79-L82.

Johnson, F.M. (1967) Diffuse interstellar lines and chemical characterization of interstellar dust, in: Interstellar Grains (Greenberg, J.M., Roark, T.P., eds), NASA, p. 229-240.

Jones, A.P., Tielens, A.G.G.M., Hollenbach, D.J., et al., (1994) Grain destruction in shocks in the interstellar medium, Astrophys. J. 433, 797-810.

Kissel, J., Brownlee, D.E., Büchler, K., et al. (1986a) Composition of comet Halley dust particles from Giotto observations, Nature 321, 336-337

Kissel, J., Sagdeev, R.Z., Bertaux, J.L., et al. (1986b) Composition of comet Halley dust particles from Vega observations, Nature 321, 280-282

Kissel, J., F.R. Krueger (1987) The organic component in dust from comet Halley as measured by the PUMA mass spectrometer on board Vega 1, Nature 326, 755-760.

Knacke, R., Brooks, T.I., Joyce, R.R. (1986) Observations of 3.2-3.6 micron emission features in Comet Halley, Astrophys. J. Lett. 310, L49-L53.

Kouchi, A., Kuroda, T. (1990) Amorphization of cubic ice by ultraviolet irradiation, Nature 344, 134-135.

Krueger, F.R., Kissel, J. (1987) The chemical composition of the dust of cone P/Halley as measured by "PUMA" on board VEGA-1, Naturwissenschaften 74, 312.

Li, A., Greenberg, J.M. (1997) A unified model of interstellar dust, Astron. Astrophys. 323, 566-584.

Li, A., Greenberg, J.M. (1998) From interstellar dust to comets: infrared emission from comet Hale-Bopp (C/1995 O1), Astrophys. J. Lett. 498, L83-L87.

Mathis, J.S., Rumpl, W., Nordsieck, K.H. (1977) The size distribution of interstellar grains, Astrophys. J. 217, 425-433.

McKee, C.F. (1989) Dust destruction in the interstellar medium, in L.J. Allamandola, A.G.G.M. Tielens (eds.), Interstellar Dust, Kluwer, Dordrecht, p. 431-443.

Mendoza-Gómez, C.X. (1992) Complex irradiation products in the interstellar medium, Thesis, Leiden.

Mennella, V., Colangeli, L., Palumbo, P., Schutte, W.A., Bussoletti, E. (1996) Activation of a UV resonance in hydrogenated amorphous carbon grains by exposure to UV radiation, Astroph. J. Lett. 464, L191-L194.

Moore, M.H., Donn, B. (1982) The infrared spectrum of a laboratory-synthesized residue: implications for the 3.4 micron interstellar absorption feature, Astrophys. J. Lett. 257, L47-L50.

Mumma, M.J., Stern, S.A., Weissman, P.R. (1993) Comets and the origin of the solar system: Reading the Rosetta stone, in Planets and Protostars III (E.H. Levy, J.I. Lunine, M.S. Matthews, eds.), University of Arizona Press, Tucson, p. 1177-1252.

Pendleton, Y.J., Sandford, S.A., Allamandola, L.J., Tielens, A.G.G.M., Sellgren, K. (1994) Near-infrared absorption spectroscopy of interstellar hydrocarbon grains, Astrophys. J. 437, 683-696.

Sagan, C., Khare, B.N. (1971) Science 173, 447

Sakata, A., Wada, S. (1989) Chemical, optical and infrared properties of quenched carbonaceous composites (QCC's), in Interstellar dust (eds. L.J. Allamandola, A.G.G.M. Tielens), Kluwer, Dordrecht, p. 191-196.

Sandford, S.A., Allamandola, L.J., Tielens, A.G.G.M., et al. (1991) The interstellar C-H stretching band near 3.4 microns: constraints on the composition of organic material in the diffuse interstellar medium, Astrophys. J. 371, 607-620.

Schutte, W.A., Greenberg, J.M. (1986) Formation of organic molecules on interstellar dust particles, in: Light on dark matter (Proc. IRAS Symp., Noordwijk, 10-14 June 1985) (F. Israel, ed.), Reidel, p. 229-232.

Schutte, W.A. (1988) The evolution of interstellar organic grain mantles, Thesis, Leiden.

Schutte, W.A., Greenberg, J.M. (1988) The evolution of organic refractory mantles on interstellar grains, in: Dust in the Universe (M.E. Bailey, D.A. Williams, eds.), Cambridge University Press, p. 403-406.

Shalabiea, O.M., Greenberg, J.M. (1994) Two key processes in dust/gas chemical modelling: photoprocessing of grain mantles and explosive desorption, Astron. Astrophys. 290, 266-278

Snyder, L.E., et al. (1969) Microwave detection of interstellar formaldehyde, Phys. Rev. Lett. 22, 679-681.

Stecher, T.P., Donn, B. (1965) On graphite and interstellar extinction, Astrophys. J. 142, 1681-1682

Tielens, A.G.G.M., Allamandola, L.J., Bregman, J., et al. (1984) Absorption features in the 5-8 micron spectra of protostars, Astrophys. J. 287, 697-706.

Tielens, A.G.G.M., Wooden, D.H., Allamandola, L.J., Bregman, J., Witteborn, F.C. (1996) The infrared spectrum of the galactic center and the composition of interstellar dust, Astrophys. J. 461, 210-222.

Tielens, A.G.G.M. (1998), this volume.

Van de Hulst, H.C. (1949) The solid particles in interstellar space, Rech. Astron.Obs. Utrecht 11, pt. 2.

Waelkens, C., Waters, L.B.F.M., de Graauw, M.S., et al. (1996) SWS observations of young main-sequence stars with dusty circumstellar disks, Astron. Astrophys. 315, L245-L248.

Waters, L.B.F.M., Molster, F.J., de Jong, T., et al. (1996) Mineralogy of oxygen-rich dust shells, Astron. Astrophys. 315, L361-L364.

Wexler, A.S. (1967) Integrated intensities of absorption bands in infrared spectroscopy, Appl. Spectrosc. Rev. 1, 29-98.

Whittet, D.C.B., Boogert, A.C.A., Gerakines, P.A., Schutte, W.A., et al., (1997) Infrared spectroscopy of dust in the diffuse interstellar medium toward Cygnus OB2 No. 12, Astrophys. J. 490, 729-734

Wickramasinghe, D.T., Allen, D.A. (1980) The 3.4 $\mu$m interstellar absorption feature, Nature 287, 518-519.

Willner, S.P., Russell, R.W., Puetter, R.C., et al. (1979) The 4 to 8 micron spectrum of the galactic center, Astrophys. J. Lett. 229, L65-L68.

# MID–INFRARED SPECTROPOLARIMETRY AND THE COMPOSITION OF COSMIC DUST

C.M. WRIGHT
*Max–Planck–Institut für extraterrestrische Physik*
*Postfach 1603, D–85740 Garching, Germany*

D.K. AITKEN
*5 Dewsbury Cottages, York YO1 1HB, UK*

C.H. SMITH
*University College, ADFA, UNSW*
*Canberra ACT 2600, Australia*

AND

P.F. ROCHE
*Department of Astrophysics, Oxford University*
*Keble Road, Oxford OX1 3RH, UK*

**Abstract.** Mid–infrared spectroscopic and spectropolarimetric observations of three star formation regions are presented, from which the dust mineralogy is inferred from modelling of the spectra. Crystalline olivine is shown to be a good candidate for the carrier of the 11.2 $\mu$m feature observed toward AFGL 2591, whilst metal oxides are proposed as a possible component of the dust toward the BN Object in Orion. SiC appears to be present in the circumstellar disk of SVS13 (HH7-11 IR).

## 1. Introduction

It is well known that amorphous silicates make up a large fraction of cosmic dust (e.g. Whittet, 1992). This has been established by the observation of broad spectral features at 8–13 and 16–22 $\mu$m toward many astronomical objects, including circumstellar disks around young stars, which show many similarities to those seen in the spectra of terrestrial silicates. Amorphous silicates have also been identified in the spectra of comets and interplanetary dust particles (IDPs) collected in the Earth's stratosphere. However, there are also many trace materials found in IDPs and meteorites, such as crystalline silicates, silicon carbide (SiC) and metal oxides, which have yet



*J. M. Greenberg and A. Li (eds.), Formulation and Evolution of Solids in Space,* 77–83.

to be observed in young circumstellar disks, or are only inferred in a few specific cases. It is of great interest to determine what fraction these trace materials might comprise in circumstellar dust.

Spectropolarimetric measurements in the mid–infrared (here defined as 5–30 $\mu$m), where many minerals possess signature resonances, is an especially powerful tool because the polarization is extremely sensitive to the mineral dielectric function. The polarization is induced by emission or absorption from a medium of spinning, aspheric, paramagnetic dust grains which have had their angular momenta (and hence short axes) mutually aligned through an interaction with the ambient magnetic field (e.g. Aitken 1992). In the mid–infrared both emissive and absorptive polarization components may be observed, but because of their quite unique spectral forms they can usually be separated (e.g. Aitken 1996). In most of the sources we will discuss here we are confident that the dominant, if not sole, polarization mechansim is dichroic absorption.

Examples of the utility of mid–infrared spectropolarimetry can be found in Aitken *et al.* (1988), where the high mass Young Stellar Object (YSO) AFGL 2591 was studied, and Aitken, Smith and Roche (1989), where the BN Object in Orion was studied. However, in the years since that work new dielectric functions have become available in the literature. It is interesting therefore to re–examine the observations to determine how these new functions impact on the polarization spectra. Additionally, further polarimetric observations have been made, and so it is the purpose of this paper to describe some of the preliminary results from these observations and the associated modelling. In Sections 2.1 and 2.2 we describe the up–to–date modelling of AFGL 2591 and the BN Object, whilst in Section 2.3 we present new observations and modelling of SVS13. The papers by Aitken *et al.* (1988) and Aitken, Smith and Roche (1989) contain a description of the observational procedure used to obtain the polarization data, whilst the papers by Martin (1975) and Aitken (1989) contain a detailed explanation of the theory behind deriving dust grain mineralogy from polarimetric measurements, and the modelling technique. Whilst this paper presents some of our primary results, the paper by Wright *et al.* (1997) contains further observations and discussion.

## 2. Circumstellar dust mineralogy

In order to identify mineral components of cosmic dust it is necessary to model the polarization spectra with candidate dust models. The modelling technique has been described in detail elsewhere (e.g. Draine and Lee 1984, Aitken 1989). However, perhaps the most important input parameter in the models is the dielectric function of the candidate material. We have obtained the dielectric function of candidate materials from the literature, and used Effective Medium Theory (EMT) to compute the "average" dielectric function of dust comprised of two of these substances (Bohren and Huffman, 1983). We have used both the Maxwell Garnett and Bruggeman formulae. Our findings are independent of which formula we use, since the results from each are extremely similar.

In all of the models we present here the "astronomical silicate" of Draine and Lee (1984), for which the dielectric function is given in Draine (1985), is used to represent the amorphous silicates responsible for the 10 and 20 $\mu$m features. All models are calculated in the Rayleigh approximation, and we have assumed oblate grains with an axial ratio of 2:1. The output of the models are the absorption and polarization cross sections, which are then directly comparable to the observed flux and polarization spectra respectively. We point out that whilst Greenberg and Li (1996) have questioned the appropriateness of the astronomical silicate, we are not trying to obtain an exact match to the entire 8–13 and 16–22 $\mu$m bands, but rather to "trace" features present in the spectra.

### 2.1. AFGL 2591 AND CRYSTALLINE SILICATES

Figure 1 shows the 8–22 $\mu$m polarization spectrum (solid points) of AFGL 2591. The broad features peaking at around 10.2 and 18 $\mu$m are due to the usual amorphous silicates, whilst the narrow feature at 11.2 $\mu$m has been identified by Aitken *et al.* (1988) with crystalline olivine, based on similarities to terrestrial silicates and a high band strength indicated by the high polarization–to–absorption ratio (i.e. higher than for the 10 $\mu$m feature; see also Martin 1975). Also shown in Figure 1 are two model fits. The dashed line is the spectrum expected from only astronomical silicate grains, whilst the solid line is the spectrum calculated using astronomical silicate with inclusions of crystalline silicate (olivine). The inclusions occupy 17.5% of the total grain volume. The dielectric function for the crystalline olivine was obtained from Mukai and Koike (1990). It can be seen that the EMT curve fits quite well the 8–13 $\mu$m polarization spectrum of AFGL 2591, especially in the 10–13.5 $\mu$m interval, and most notably the 11.2 $\mu$m feature.

The match between the 20 $\mu$m observations and the EMT model is not so good, and certainly not quite as good as that obtained using only astronomical silicate. It is interesting then to pose the question can a real crystalline silicate exist which has a strong 11.2 $\mu$m feature, but only weak features in the 16–22 $\mu$m interval? If so, can this then tell us something about the crystallographic structure, or lack thereof, keeping in mind that the silicate resonances in the 10 and 20 $\mu$m regions are due to Si–O and Si–O–Si stretches and bends respectively?

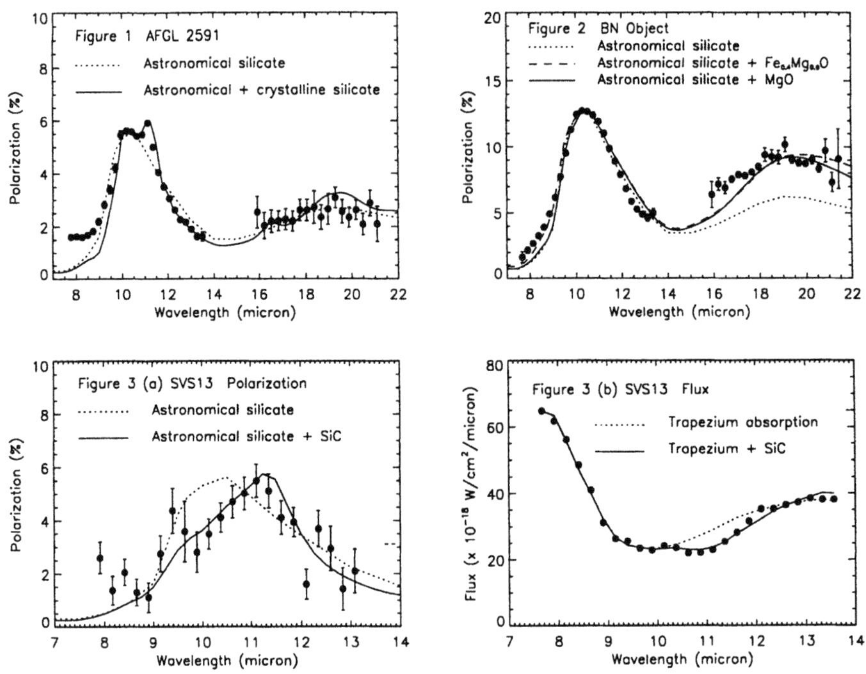


*Figure 1.* Polarization data for AFGL 2591, with representative dust grain model fits. *Figure 2.* Same for the BN Object. *Figure 3.* (a) Same for SVS13. (b) Flux data for SVS13 with model fits. See text for detailed descriptions.

## 2.2. THE BN OBJECT AND METAL OXIDES

Figure 2 shows the 8–22 $\mu$m polarization spectra of the BN Object in the Orion star forming region. Aitken, Smith and Roche (1989) found that the high 20 to 10 $\mu$m polarization ratio could not be matched by any silicate dielectric function then available. Further, Henning and Stognienko (1993) found that porous grains could not increase the ratio without significantly worsening the overall match to the observed spectrum, whilst O'Donnell (1994) found a similar effect using mantles thought to exist on interstellar dust grains. Recently however, Greenberg and Li (1996) were able to adequately fit the BN spectrum with a new silicate dielectric function. We instead propose an alternative model, one in which metal oxides, such as $Fe_xMg_{1-x}O$, are present as inclusions in an amorphous silicate matrix. Such metal oxides have a very strong resonance in the 20 $\mu$m region.

Calculation of such a model is shown as the long dashed line in Figure 2, where $x$=0.4, and the volume fraction of the oxide inclusion is 0.11. The dielectric function was obtained from Henning *et al.* (1995), who considered iron–magnesium oxides with $x$ between 1 and 0.4. Although the position of the 20 $\mu$m polarization peak is not reproduced in the model, the peak does move to shorter wavelengths with decreasing $x$. A value of $x$=0.4 was the minimum considered by Henning *et al.*. A smaller value of $x$, perhaps even pure MgO, might very well provide an even better match. Indeed, this is shown to be the case by the solid curve in Figure 2, where the MgO inclusions volume fraction is 0.075 (dielectric function from Roessler and Huffman 1991).

Such iron/magnesium oxides may be expected to form in the atmospheres of oxygen–rich stars, and so be present in cosmic dust (Henning *et al.* 1995). Also, as noted by Dorschner *et al.* (1995), perfect $SiO_4$ tetrahedra will not always be formed in cosmic silicates, and there will be some Mg and Fe atoms bonded to an O atom, which itself doesn't belong to a tetrahedron. Fe–O and Mg–O vibrations might then be expected. Indeed, one advantage of these oxides is that only a small volume fraction of the grains, $\sim$ 10%, is needed to produce the required 20 to 10 $\mu$m polarization ratio in BN, and so only small variations in this fraction can produce the range of 20 to 10 $\mu$m band ratios commonly observed. However, based on spectroscopy alone it would be extremely difficult to distinguish between the Greenberg and Li (1996) model and that presented here. It would therefore be interesting to search for signatures of iron/magnesium oxides in IDPs, meteorites and comet impact measurements.

### 2.3. SVS13 AND SIC

Figures 3 (a) and (b) show the 8–13 $\mu$m polarization and total intensity (flux) spectra of SVS13 respectively. SVS13 is embedded in the NGC1333 molecular cloud and is the central engine behind the chain of Herbig–Haro objects 7–11. Of the objects studied here it is the most similar, in terms of its mass and luminosity, to our own Sun. The double–trough nature of the intensity spectrum bears a remarkable resemblance to that seen in comets (when inverted), such as Halley (Campins and Ryan 1989), where it is attributed to a mixture of amorphous and crystalline silicates. Without the polarization spectrum as a guide, the temptation to ascribe the spectrum of SVS13 to a similar mixture would be extremely strong.

Several materials known or suspected to be present in cosmic dust, such as crystalline silicates, crystalline water ice, PAHs, SiC or carbonates, may be considered as candidates for the 11.2 $\mu$m feature in SVS13. Without going into the details of the modelling of all these materials, the end result is that both the intensity and polarization spectra can only be satisfactorily modelled using inclusions of SiC in an amorphous silicate matrix. We tried two different dielectric functions of SiC, that determined from bulk measurements and given in Choyke and Palik (1985) (model shown in Figure 3a), and that from particle transmission measurements given in Pegourie (1988). In the former case the volume fraction of the SiC inclusions is 10%, whilst in the latter it is 22.5%. Both provide good matches to the observed spectrum, although the former is slightly better. The ability to essentially "subtract" or "remove" most of the polarization from the amorphous silicate at 10 $\mu$m (disregarding the points around the 9.6 $\mu$m telluric ozone feature), otherwise known as a polarization reversal (Hong and Greenberg 1978), places a very strong constraint on the dielectric function, and hence the responsible material, and is unique to SiC.

The absorption cross section from the SiC models also looks qualitatively like the flux spectrum of SVS13 (when inverted), and a simple two–component model (e.g. Gillett *et al.* 1975) provides a good fit. The model incorporates optically thick (i.e. blackbody) dust emission overlaid by absorption by both Trapezium–like dust and SiC. The Trapezium–like dust simply refers to dust with an emissivity spectrum derived from the Trapezium stars in Orion, whilst the SiC emissivity function was derived from observations of carbon stars. The fit, shown as the solid line in Figure 3b, provides a $\chi^2/N$ of 1.43, an emitting dust temperature of 328 K, and optical depths at 9.7 and 11.2 $\mu$m of 1.33 and 0.18 respectively. The dashed line shows a fit using only Trapezium–like dust in absorption.

The circumstantial evidence in favour of SiC therefore appears strong. This would certainly be a surprising result, given that SiC is typically only observed in emission around evolved carbon stars, where it is created in situ (e.g. Kozasa *et al.* 1996), and has not been observed in the interstellar medium (e.g. Whittet *et al.* 1990). However, for the time being we remain cautious. Although we have modelled three different crystalline silicates, as well as crystalline water ice, and PAHs and carbonates may be ruled out on plausibility arguments, there is still the possibility that the spectrum represents polarized emission rather than absorption, or a combination of the two. However, even if this is the case the spectrum cannot be fit by only astronomical silicate, and the dust composition must be unusual.

## 3. Conclusions

We have shown that mid–infrared spectropolarimetry is an extremely powerful tool in the study of the mineralogy of cosmic dust. However, the proposed detections of crystalline silicates, metal oxides and SiC are confined to only a few objects. It is important to perform similar observations, especially high spectral resolution 10 and 20 $\mu$m spectropolarimetry, to determine how widespread these phenomena are. Observations of more low mass sources, where planetary systems similar to our own might be forming, would be particularly valuable, especially if we are to learn something about the mineral content and chemistry of the circumstellar disk from which our solar system condensed.

## References


Aitken, D.K., 1989, *Spectropolarimetry as a diagnostic of the interstellar medium*, Proc. 22nd ESLAB Symposium on Infrared Spectroscopy in Astronomy, ed. B.H. Kaldeich, ESA Publications division, pp 99–107

Aitken, D.K., 1992, *Observations of magnetic fields*, Aust. J. Phys., **45**(4), 569

Aitken, D.K., 1996, *Spectropolarimetry of solid state features in the infrared*, in Polarimetry of the Interstellar Medium, eds. W.G. Roberge and D.C.B. Whittet, pp 225–242

Aitken, D.K., Roche, P.F., Smith, C.H., James, S.D. and Hough, J.H., 1988, *Infrared spectropolarimetry of AFGL 2591: evidence for an annealed grain component*, Mon. Not. R. Astr. Soc., **230**, 629

Aitken, D.K., Smith, C.H. and Roche, P.F., 1989, *10 and 20 μm spectropolarimetry of the BN Object*, Mon. Not. R. Astr. Soc., **236**, 919

Bohren and Huffman, 1983, *Absorption and Scattering of Light by Small Particles*, Wiley, pp 213–219

Campins, H. and Ryan, E.V., 1989, *The identification of crystalline olivine in cometary silicates*, Astrophys. J., **341**, 1059

Choyke, W.J. and Palik, E.D., 1985, *Silicon Carbide (SiC)*, in Handbook of Optical Constants of Solids, ed. Edward D. Palik, Academic Press, pp 587–595

Dorschner, J., Begemann, B., Henning, Th., Jäger, C. and Mutschke. H., 1995, *Steps toward interstellar silicate mineralogy II. Study of Mg–Fe–silicate glasses of variable composition*, Astron. & Astrophys., **300**, 503

Draine, B.T., 1985, *Tabulated optical properties of graphite and silicate grains*, Astrophys. J. Supp. Ser., **57**, 587

Draine B.T. and Lee, H.M., 1984, *Optical properties of interstellar graphite and silicate grains*, Astrophys. J., **285**, 89

Gillett, F.C., Forrest, W.J., Merrill, K.M., Capps, R.W. and Soifer, B.T., 1975, *The 8–13 μm spectra of compact HII regions*, Astrophys. J., **200**, 609

Greenberg, J.M. and Li, A., 1996, *What are the true astronomical silicates?*, Astron. & Astrophys., **309**, 258

Henning, Th and Stognienko, R., 1993, *Porous grains and the polarization of light: the silicate features*, Astron. & Astrophys., **280**, 609

Henning, Th., Begemann, B., Mutschke, H. and Dorschner, J., 1995, *Optical properties of oxide dust grains*, Astron. & Astrophys. Supp. Ser., **112**, 143

Hong, S.S. and Greenberg, J.M., 1978, *Polarization reversal by small particles over strong resonances*, Astron. & Astrophys., **69**, 341

Kozasa, T., Dorschner, J., Henning, Th. and Stognienko, R., 1996, *Formation of SiC grains and the 11.3 μm feature in circumstellar envelopes of carbon stars*, Astron. & Astrophys., **307**, 551

Martin, P.G., 1975, *Some implications of 10–micron interstellar polarization*, Astrophys. J., **202**, 393

Mukai, T. and Koike, C., 1990, *Optical constants of olivine particles between wavelengths of 7 and 200 μm*, Icarus, **87**, 180

O'Donnell, J.E., 1994, *The effect of grain mantles and grain shape upon the 9.7 and 18 micron silicate features*, Astrophys. J., **437**, 262

Pegourie, B., 1988, *Optical properties of $\alpha$ silicon carbide*, Astron. & Astrophys., **194**, 335

Roessler and Huffman, 1991, *Magnesium Oxide (MgO)*, in Handbook of Optical Constants of Solids II, ed. Edward D. Palik, Academic Press, pp 919–955

Whittet, D.C.B., Duley, W.W. and Martin, P.G., 1990, *On the abundance of silicon carbide dust in the interstellar medium*, Mon. Not. R. Astr. Soc., **244**, 427

Whittet, D.C.B., 1992, *Dust in the Galactic Environment*, IOP Publishing

Wright, C.M., Aitken, D.K., Smith, C.H. and Roche, P.F., 1997, in preparation.

# METHOD OF REGULARIZATION AND MODELS OF INTERSTELLAR DUST

V.G. ZUBKO
*Department of Physics, Technion – Israel Institute of Technology, Haifa 32000, Israel*

**Abstract.** A new, powerful technique based on the method of regularization has successfully been applied for modelling of interstellar dust. Described are the basic of the approach and the first results of its application to studying of interstellar extinction under reduced cosmic abundances.

## 1. Introduction

In spite of the considerable efforts of many investigators the problem of interstellar dust composition has not yet been finally resolved. It is among the most fascinating unsolved problems of modern astronomy. In addition, it has been shown in the recent studies that the real cosmic abundances are around 2/3 of solar ones (Snow, this volume; Greenberg, this volume). This leads to the elemental crisis for many traditional models of interstellar dust (Snow and Witt 1996). Note some last attempts to provide the models to be consistent with the tight abundance constraints. Mathis (1996) has modelled the mean extinction curve using the same reducing factor for all heavy elements and a mixture of small bare graphite and silicate grains together with the porous composite grains containing silicates and various carbonaceous materials: graphite, organic residue, amorphous carbon. However, his limit models ever consistent with observations require $C/H_{dust}$=140–150 p.p.m. (we express abundances as atoms per $10^6$ H atoms) under reducing factor 0.77. In the mean time, the realistic value of $C/H_{dust}$ is around 100 p.p.m. as followed from the study of Cardelli et al. (1996). Li and Greenberg (1997) have proposed a trimodal model of interstellar dust on the base of the cylindrical silicate core–organic refractory mantle grains, very small spherical carbonaceous grains and PAH's. Their model is in good consistency with the average interstellar extinction and



*J. M. Greenberg and A. Li (eds.), Formulation and Evolution of Solids in Space*, 85–90.

polarization laws and some other observational constraints. However, it still requires more carbon in dust: 194 p.p.m., that is, about twice more than expected from the new reference abundance of Cardelli et al. (1996). Note also that both Mathis (1996) and Li and Greenberg (1997) still searched for the grain size distributions in a prescribed form: a power law with an exponential cutoff and a Gaussian function with a cutoff, respectively. Kim et al. (1994) have shown for the first time that the size distribution of interstellar grains in a general form may be deduced from the extinction data by using the Maximum Enthropy Method (MEM). However, for the MEM to work some default (guess solution) should be specified. As a rule, the default is not known reliably. Recently, Zubko et al. (1996) have demonstrated that the method of regularization (MR) may successfully be used to extract the size distributions of interstellar grains from the extinction data. Contrary to MEM, no default is required for MR to work. Below we describe the main ideas of the MR and consider some examples of its application to the analysis of interstellar extinction.

## 2. Statement of a typical problem: modelling of extinction

What we know about the interstellar dust comes from the analysis of a variaty of observational data: the extinction and polarization curves, scattering properties, own radiation and so on. The bulk of data accumulated hitherto may be used to treat the interstellar extinction. We therefore consider a strict formulation of the problem. The optical depth $\tau(\lambda)$ of the dust mattery along a line of sight (coordinate $s$) may be expressed as:

$$\tau(\lambda) = \int \mathrm{d}s \int_0^\infty \pi a^2 Q_{\mathrm{ext}}(\lambda, a) \tilde{f}(a, s)\, \mathrm{d}a, \tag{1}$$

where $Q_{\mathrm{ext}}(\lambda, a)$ is the extinction efficiency for a grain of radius $a$ at wavelength $\lambda$ and $\tilde{f}(a, s)$ is the size distribution function, the number of grains being placed at the distances $s$ to $s$+ds from an observer and having sizes within the interval $a$ to $a$+da. Since it is now difficult to deduce from observations the details of a spatial distribution of dust we introduced a spatially averaged size distribution function $f(a)$ related to a column of unit area:

$$f(a) = \int \mathrm{d}s\ \tilde{f}(a, s). \tag{2}$$

Equation (1) may now be rewritten as

$$\tau(\lambda) = \int_0^\infty \pi a^2 Q_{\mathrm{ext}}(\lambda, a) f(a)\, \mathrm{d}a. \tag{3}$$

Finally, we have found relevant for computational purposes to rewrite equation (3) in the form

$$\tau(\lambda) = \int_0^\infty K(\lambda, a) F(a)\, \mathrm{d}a, \tag{4}$$

where $F(a) = a^3 f(a)$ and $K(\lambda, a) = \pi Q_{\mathrm{ext}}(\lambda, a)/a$. The function $F(a)$ physically reflects the mass distribution of dust grains. It is especially important for consideration of non-spherical grains. Note the limit features of the function $K(\lambda, a)$, following the properties of the Mie extinction efficiencies (Bohren and Huffman 1983): it is practically independent on $a$ for very small grains $a \to 0$ and behaves as $\sim a^{-1}$ for large grains.

For the sake of simplicity the equations (1), (3) and (4) have been written for a single grain constituent. They may be trivially generalized for accounting for the multicomponent composition.

## 3. What is the method of regularization?

Extinction modelling using equation (4) is a typical inverse problem – resolution of a Fredholm integral equation of first order when we need to find the size distribution function $F(a)$ from the observed extinction law $\tau(\lambda)$. In practice, in the integral equation (4) the function $\tau(\lambda)$, extinction, may be deduced from the data of observations, its kernal $K(\lambda, a)$ is calculated using the adopted grain model. We then have to find the size distribution function $F(a)$. Since we deal with an ill-posed problem where a small perturbation of the input data $\tau$ (caused, for example, by various errors) may result in large changes of the solution $F$, a special tool is required for its correct resolution. We have developed the algorithm and the numerical procedure to resolve the equations like (4) on the base of the MR (e.g. Tikhonov et al. 1990). The details of our numerical implementation of MR may be found in Zubko (1997). Here we present the main features of our method. It has been strictly proved that a resolution of the integral equation (4) is equivalent to minimization of the following smoothing functional (Tikhonov et al. 1990):

$$\Phi[\alpha, F] = \Phi_{\mathrm{d}}[F] + \alpha \Phi_{\mathrm{s}}[F] \tag{5}$$

consisting of two parts: the discrepancy functional $\Phi_{\mathrm{d}}[F]$ and Tikhonov's stabilizing functional or stabilizer $\Phi_{\mathrm{s}}[F]$:

$$\Phi_{\mathrm{d}}[F] = \int_{\lambda_{\mathrm{min}}}^{\lambda_{\mathrm{max}}} \left[ \int_0^\infty K(\lambda, a) F(a)\, \mathrm{d}a - \tau(\lambda) \right]^2 \mathrm{d}\lambda, \tag{6}$$

$$\Phi_{\mathrm{s}}[F] = \int_0^\infty [F^2(a) + F_a'^2(a)]\, \mathrm{d}a, \tag{7}$$

where $\lambda_{\min}$ and $\lambda_{\max}$ define the wavelength range of the available extinction $\tau(\lambda)$ data, $F'_a$ is a derivative of $F$ by $a$ and $\alpha$ is the so-called parameter of regularization. The introduction of Tikhonov's stabilizer is decisive in finding a stable solution. It results in some smoothing of the solution, and a certain extent of the smoothing, which is tuned by the parameter of regularization $\alpha$, should be defined by using the uncertainties or errors in both the input data, $\tau$, and numerical algorithms to be used in a computer programme. There are a few rules for finding $\alpha$ in dependence of the nature of the problem under study (Tikhonov et al. 1990). It has been proved that the minimization of the smoothing functional (5) together with a relevant choice of $\alpha$ lead to a mathematically stable solution $F$.

## 4. Features of the computer code

On the base of MR, we are developing a code for a comprehensive modelling of interstellar dust which would take into account as more as possible physical constraints. The code is being written in C language. This results in the executable quite efficient in respect to both CPU time and memory requirements. Currently, the code is able to model interstellar dust 1) using the extinction data, 2) applying the chemical constraints and 3) supposing an artificial mixture of spherical grains:

- homogeneous grains (the optical efficiencies are calculated using the Mie theory, Bohren and Huffman 1983);
- inhomogeneous grains including the composites (the EMT/Mie approach, similar to Mathis 1996);
- multilayer grains (exact theory, the algorithm by Zubko, in preparation).

Note that the mass ratios of the mixture components are not defined as input data, but are found together with the appropriate size distributions. In this sense we get the most optimal solution.

## 5. Example: models of extinction towards the star HD 147165

Zubko et al. (1996) have shown that the star HD 147165 is seen through an apparently single cloud. They have also derived the extinction curve for this star in the wide wavelength range from the ultraviolet (0.13 $\mu$m) to the infrared (3.5 $\mu$m). There is an essential deviation in the UV from the corresponding theoretical extinction curve of Cardelli et al. (1989). To fit the extinction we have built three dust models which are the mixtures of both the bare and inhomogeneous grains. The model O consists of the bare graphite and silicate grains together with the multilayer grains with the silicate core coated by the water ice and refractory organics mantles

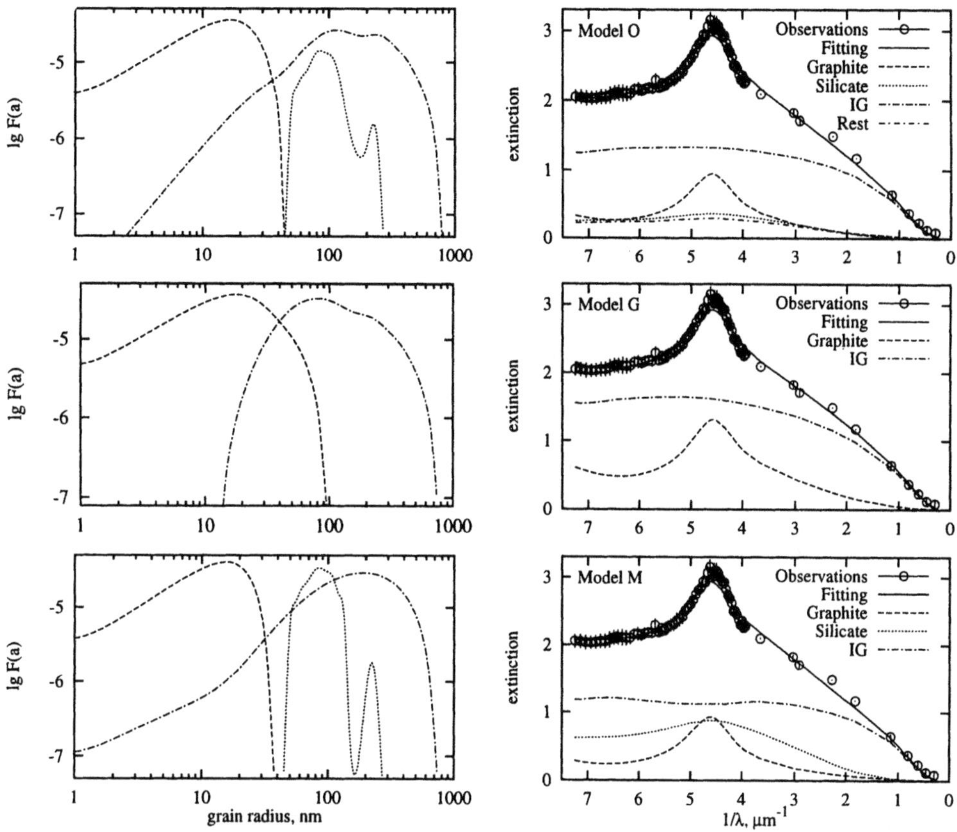


*Figure 1.* Size distributions of dust grains $F(a)$ (left panels) and the respective absolute extinction curves (right panels) with the contributions of each constituents calculated for the star HD 147165. From top to bottom the results are presented for three various models: O, G and M (see the text and Table 1 for a model specification).

(we have found the optimal volume fractions to be: 0.3 for silicate and organic refractory and 0.4 for water ice). It is a development of our previous model containing water ice grains (Zubko et al. 1996). Next model G contains the bare graphite grains and the silicate core–refractory organics mantle grains (the equal volume fractions of the silicates and organic refractory) and, therefore, closely resembles the models of Li and Greenberg (1997). The model M including the bare graphite and silicate grains and the porous composite grains (graphite+amorphous carbon+refractory organics and 45% of vacuum) is similar to Mathis' (1996) one. The optimal size distributions derived by applying the tight chemical constraints (cosmic abundances are around 2/3 Solar) are displayed in Figure 1. For each of the models Table 1 contains the optimal abundances of chemical elements (C, O, Si, Mg and Fe) in dust in p.p.m., the total column mass density in g $cm^{-2}$ of dust and the partial contributions of the bare constituents ($f^M_{\rm Gra}$, $f^M_{\rm Sil}$)

TABLE 1. The model parameters of the line of sight towards the star HD 147165

| Model | C | O | Si | Mg | Fe | $M$ | $f^M_{\rm Gra}$ | $f^M_{\rm Sil}$ | $f^M_{\rm IG}$ |
|---|---|---|---|---|---|---|---|---|---|
| O | 100 | 210 | 24 | 25 | 22 | 3.0–5 | 8.4 | 13.4 | 78.2 |
| G | 150 | 170 | 25 | 25 | 25 | 3.0–5 | 13.4 | | 86.6 |
| M | 170 | 130 | 24 | 24 | 24 | 2.8–5 | 8.5 | 32.20 | 59.3 |

and inhomogeneous grains ($f^M_{\rm IG}$) in per cent. Note that we adopted for the gas-phase abundances C/H$_{\rm gas}$=140 p.p.m., O/H$_{\rm gas}$=310 p.p.m. (Cardelli et al. 1996) and a full depletion of other heavy elements.

All three models provide quite good fits of the observed extinction. They differ, however, by the abundances of C and O required to be in dust. The model O contains the most optimal quantity of carbon (in line with Cardelli et al. 1996), but still requires more oxygen. Another two models needs more carbon with oxygen around the optimal values. Note that all the values of abundances from Table 1 are within their known uncertainties (Greenberg, this volume). On the other hand, the model O contains water ice and it is one of the serious difficulties of this model because the feature at 3.07 $\mu$m is not seen within the diffuse clouds. The model M requires quite more silicates in the bare form. This seems unrealistic as was shown by Greenberg and Li (1996). So, from the analysis of the extinction data we may conclude that the model G seems to be more preferable. However, for more careful analysis we need in more accurate cosmic abundances. In addition, other available observational constraints, e.g. the polarization data, scattering properties, spectra of own radiation and so on should be included in the future models.

## References


C.F. Bohren and D.R. Huffman. *Absorption and Scattering of Light by Small Particles*, Wiley-Interscience, New-York (1983).

J.A. Cardelli, D.M. Meyer, M. Jura and B.D. Savage. *Astrophys. J.* 467 (1996) 334.

J.M. Greenberg and A. Li. *Astron. Astrophys.* 309 (1996) 258.

S.-H. Kim, P.G. Martin and P.D. Hendry. *Astrophys. J.* 422 (1994) 164.

A. Li and J.M. Greenberg. *Astron. Astrophys.* 323 (1997) 566.

J.S. Mathis. *Astrophys. J.* 472 (1996) 643.

T.P. Snow and A.N. Witt. *Astrophys. J.* 468 (1996) L65.

A.N. Tikhonov, A.V. Goncharsky, V.V. Stepanov and A.G. Yagola. *Numerical Methods for the Solution of Ill-Posed Problems*, Nauka, Moscow (1990).

V.G. Zubko. *Mon. Not. Roy. Astron. Soc.* 289 (1997) 305.

V.G. Zubko, J. Krełowski and W. Wegner, *Mon. Not. Roy. Astron. Soc.* 283 (1996) 577.

# MODELS AND OBSERVATIONS OF GAS-GRAIN INTERACTIONS IN STAR-FORMING REGIONS

EWINE F. VAN DISHOECK
*Leiden Observatory P.O. Box 9513*
*2300 RA Leiden, The Netherlands*
*e-mail: ewine@strw.leidenuniv.nl*

**Abstract.** The chemical processes which accompany the birth of new stars from dense molecular clouds are discussed. Basic molecular processes are summarized and recent developments in gas-phase and gas-grain models are reviewed. In dark pre-stellar cores, long carbon-chain molecules produced by low temperature ion-molecule and neutral-neutral reactions are found to be abundant. During the cold collapse phase, most species are depleted onto the grains, where subsequent grain-surface reactions can produce hydrogenated molecules. The amount of depletion is difficult to determine observationally, but new constraints on the abundances and composition of the ices are obtained from the ISO data. Once the young star heats up its surroundings, the ices start to evaporate back into the gas phase, with the most volatile species returned first. These molecules drive a rapid hot core chemistry, resulting in complex organic species such as $CH_3OCH_3$. Shocks created by the interaction of the outflow with the envelope return both refractory and mantle material to the gas. The observational tools for probing each of the phases are discussed.

## 1. Historical background

Observations and models of gas-phase and solid-state interstellar molecules date back at least 60 years. The first gas-phase molecules (CH, $CH^+$ and CN) were detected in 1937–1941 by optical absorption line studies of diffuse clouds. The field came to full bloom in the early 1970's, when millimeter telescopes were constructed and sensitive detectors were developed to search for the rotational lines of molecules. Nearly 120 different gas-phase species have now been identified, not including isotopic varieties, with abundances



*J. M. Greenberg and A. Li (eds.), Formulation and Evolution of Solids in Space,* 91–121.

down to $10^{-11}$ with respect to hydrogen (Ohishi 1997, van Dishoeck 1997). Most of them have been found in chemically active regions, such as the Orion–KL and SgrB2 high-mass star-forming regions, as well as the dark core TMC–1. With the advent of millimeter interferometers, the chemistry in the nearest regions can now be probed on scales of a few 100 AU (Blake 1997).

$H_2O$-ice was identified in 1973, soon after the first infrared detectors became available (Gillett & Forrest 1973). These observations are limited to lines of sight toward bright embedded or background infrared sources, where the vibrational bands of the solids can be seen in absorption. The enormous advances in infrared instrumentation in the last decade, both from the ground and from space with the *Infrared Space Observatory* (ISO), provide a wealth of new information and allow a complete inventory of solid-state species to be made (Tielens & Whittet 1997, Ehrenfreund et al. 1997).

The first models of gas-phase chemistry were constructed by Kramers & ter Haar (1946) and Bates & Spitzer (1951), whereas the formation of $H_2O$, $CH_4$ and $NH_3$ on the surfaces of grains was considered around the same time by van de Hulst and Oort (see van de Hulst 1997 for a historical account). McCrea & McNally (1960), Hollenbach & Salpeter (1971), Watson & Salpeter (1972), Pickles & Williams (1977) and Allen & Robinson (1977) further developed the grain surface chemistry, focussing on the formation of $H_2$. Because such models cannot reproduce the observed abundances of ions and radicals in diffuse and dark clouds, the theoretical emphasis subsequently shifted toward pure gas-phase ion-molecule chemistry (e.g., Herbst & Klemperer 1973, Dalgarno & Black 1976, Watson 1978, Prasad & Huntress 1980, Graedel et al. 1982, Millar et al. 1991, 1997, Lee et al. 1996a).

The observations of large abundances of ices in cold, dense clouds and the detection of complex, saturated organic molecules in hot cores in star-forming regions have revived the interest in models which integrate the gas-phase and grain surface chemistry. The early models considered only the accretion of molecules onto grains, with (e.g., Tielens & Hagen 1982) or without (e.g., Millar & Nejad 1985) any reactions on the surfaces to form molecules. Recent models contain a fully-coupled (time-dependent) gas-grain chemistry, including mechanisms for returning the molecules to the gas phase (e.g., d'Hendecourt et al. 1985, Hasegawa & Herbst 1993, Caselli et al. 1993, Shalabiea & Greenberg 1994, Willacy et al. 1994a,b).

In this chapter, the status of the chemical models will be briefly reviewed and uncertainties in the reaction rates entering the networks will be pointed out. Recent observations and models of molecules in star-forming regions will be discussed, with special emphasis on their role as diagnostics of gas-grain interactions. Finally, the importance of gas-phase and solid-state molecules as indicators of the evolutionary stage, luminosity or outflow activity of young stellar objects is discussed. This chapter is based on the recent reviews by van Dishoeck (1998a) on the chemistry in diffuse and dense clouds, and by van Dishoeck & Blake (1998) on the chemistry in star-forming regions. For an excellent overview of the physics of molecular clouds, see Genzel (1992).

## 2. Basic molecular processes

### 2.1. GAS-PHASE PROCESSES

The networks developed to describe the gas-phase chemistry in star-forming regions contain thousands of reactions between several hundred species (Millar et al. 1997, Lee et al. 1996a). The basic molecular processes have been described in detail in reviews by Dalgarno (1987) and van Dishoeck (1988). At the low densities in the interstellar gas, three-body processes are unimportant, so that only two–body reactions need to be considered. The rate of a reaction between species $X$ and $Y$ is given by $\alpha n(\mathrm{X})n(\mathrm{Y})$ in $\mathrm{cm}^{-3}\ \mathrm{s}^{-1}$, where $\alpha$ is the reaction rate coefficient in $\mathrm{cm}^3\ \mathrm{s}^{-1}$ and $n$ is the concentration in $\mathrm{cm}^{-3}$.

There are two basic mechanisms by which molecular bonds can be formed. The first process is radiative association of gas-phase atoms or molecules, in which the new molecules are stabilized by the emission of a photon. The second mechanism is the formation of molecules on the surfaces of grains, in which the grain carries off the released energy corresponding to the molecular bond. The rate coefficients for the formation processes have considerable uncertainties, often up to an order of magnitude.

Molecules are readily destroyed by the absorption of ultraviolet photons. This process is very effective at the edges of dark clouds and in clouds located close to young stars (so-called photon-dominated regions (PDRs), Hollenbach & Tielens 1997). For molecules such as $H_2$ and CO, the photodissociation can take place only at very short wavelengths between 912 and 1100 Å, whereas for most other species photodissociation can occur by radiation out to 3000 Å. The ultraviolet photons can also ionize atoms with ionization potentials less than 13.6 eV (e.g., C, S, Si, Fe, Mg, ...), thereby increasing the electron abundance inside the cloud. In the unshielded normal interstellar radiation field (e.g., Draine 1978), typical lifetimes against photodissociation or photoionization are only $10^2 - 10^3$ yr.

Inside a cloud, the ultraviolet radiation is reduced because of absorption and scattering by grains (e.g., Roberge et al. 1991). Deep inside dark clouds, little of the ambient radiation penetrates, but a weak ultraviolet field can be maintained by cosmic-ray induced photons, resulting from the excitation of $H_2$ by secondary electrons following cosmic ray ionization of $H_2$ (Gredel et al. 1989). For a typical ionization rate of a few $\times 10^{-17}$ $s^{-1}$, the resulting photodestruction rates are several orders of magnitude lower than those in the unshielded interstellar field, indicating that the two become comparable only at depths corresponding to $A_V \approx 5$ mag or more into the cloud.

Molecular ions are efficiently destroyed by the process of dissociative recombination. Rate coefficients in 10–30 K gas are typically $10^{-7}$ to $10^{-6}$ $cm^3$ $s^{-1}$, with the lower values representative for $H_3^+$ —a key species in the chemistry (Dalgarno 1994, Sundström et al. 1994, Smith & Španel 1993). A major uncertainty in the models is the branching ratio to the various products. Complete experimental data are only just becoming available, and recent results indicate that three–body product channels (e.g., $H_3O^+$ + e $\rightarrow$ OH + H + H or O + $H_2$ + H) have a much larger probability than thought previously (Williams et al. 1996, Vejby–Christensen et al. 1997).

In dense clouds, destruction of neutral molecules can also occur through chemical reactions, in particular with $He^+$ (e.g., $He^+$ + CO $\rightarrow$ He + $C^+$ + O). Collisional dissociation is only important in regions of very high temperature (>3000 K) and density, such as found in shocks.

Once molecular bonds have been formed, they can be rearranged by chemical reactions leading to more complex species. Not only ion–molecule reactions, but also neutral–neutral reactions can have rate coefficients as large as $\sim 10^{-10} - 10^{-9}$ $cm^{-3}$ $s^{-1}$ at low temperatures. (e.g., Smith 1997, Kaiser et al. 1997). Reactions of neutrals and large ions with H or $H_2$ usually have substantial energy barriers, so that they proceed only at higher temperatures (e.g., O + $H_2$ $\rightarrow$ OH + H).

## 2.2. GRAIN SURFACE PROCESSES

The chemistry on the surfaces of interstellar grains has received ample discussion in the literature (e.g., Allen & Robinson 1977, Tielens & Hagen 1982, d'Hendecourt et al. 1985, Tielens & Allamandola 1987, Herbst 1993). Four different steps can be distinguished: (i) accretion; (ii) diffusion; (iii) reaction; and (iv) ejection or evaporation. The time scale for a molecule to collide with a grain and accrete is given by

$$t_{\rm acc} = 2 \times 10^9 / n_{\rm H} y_S \ {\rm yr} \tag{1}$$

where $y_S$ is the sticking probability which is thought to lie between 0.1 and 1.0 for most species (see Williams 1993 for a review). Thus, a grain accretes a few atoms or molecules per day at $n_{\rm H} = 10^4$ $cm^{-3}$. The residence time on the surface depends on the desorption mechanisms. For thermal evaporation, the time scale is

$$t_{\rm evap}^{-1} = \nu_0 \exp(-E_D/kT_D) \tag{2}$$

TABLE 1. Time scales for various grain-surface processes[a]

| Species | $E_D$ (K) | $t_{\rm evap}$ (s) | $t_{\rm hop}$ (s) | $t_{\rm acc}$ (s) |
|---|---|---|---|---|
| H | 350 | 500 | $10^{-12}$ | $7 \times 10^{12}$ |
| $H_2$ | 450 | $10^7$ | $4 \times 10^{-12}$ | $5 \times 10^{12}$ |
| C | 800 | $10^{22}$ | $10^{-2}$ | $2 \times 10^{12}$ |
| $CH_4$ | 2600 | ... | ... | $1 \times 10^{12}$ |

where $\nu_0$ is the oscillation frequency perpendicular to the surface and $E_D$ the binding energy. Typical values for $E_D$ for physisorption on water ice range from 350 K for H to 800 K for C and 2600 K for $CH_4$, resulting in short evaporation times of 500 s for H but very long ($> 10^{15}$ yr) for C and $CH_4$ at $T_D$=10 K. Thus, all neutral species heavier than He can be considered permanent residents at $T_D$=10 K. For chemisorption, typical binding energies are >10,000 K, so that residence times would be infinite even for H.

Once stuck on the grain, a species can hop to the next site on a time scale

$$t_{\rm hop}^{-1} = \nu_0 \exp(-E_b/kT_D) \tag{3}$$

where $E_b$ is the barrier energy, which is typically $0.25 \times E_D$. Thus, light species such as H, $H_2$, C, N and O can hop from site to site, whereas heavier species are immobile. H and $H_2$ are so light that they are expected to tunnel quantum mechanically to the next site on a time scale

$$t_{\rm tun}^{-1} = \nu_0 \exp[-(2a/\hbar)\sqrt{(2mE_b)}], \tag{4}$$

where $t_{\rm tun} \approx 2\times10^{-4}$ s for a 2000 K barrier. A species can scan the complete surface for a reaction partner on a time scale $t_{\rm scan} \approx 10^5 t_{\rm hop}$ determined by a random walk process. Table 1, based on Tielens & Allamandola (1987), summarizes the various timescales for typical species. The values in this table refer to physisorption on $H_2O$ ice at 10 K with $n$=$10^4$ cm$^{-3}$ and a sticking coefficient of 0.5. It is clear that under most circumstances, the rate-limiting step is the accretion of new species, rather than the diffusion over the surface.

## 3. Chemical networks

The gas-phase chemical networks describing the most important formation routes of various species have been discussed extensively in the literature, e.g., by Dalgarno & Black (1976), Watson (1978), Prasad et al. (1987), Herbst & Winnewisser (1987), Dalgarno (1987), Turner (1989), van Dishoeck (1988, 1990a), Millar (1990), Herbst (1995) and van Dishoeck (1998a). Only some of the basic processes will be reviewed here. The abundances of the elements play a key role in the chemistry, especially the absolute and relative amounts of gas-phase carbon and oxygen in atomic or molecular form. It has recently become clear that most of the interstellar elemental abundances are ~30% lower than the solar abundances (Snow & Witt 1996, Cardelli et al. 1996, Meyer et al. 1997, Meyer 1997, see chapter by Snow). The fraction of elements in the gas phase is large (at least 30%) in diffuse clouds, except for the most refractory elements (Si, Fe, ..), which are 99% in solid form. For dense clouds, the depletions of species like C, N, O and S are expected to be more significant, but are still ill constrained quantitatively (Snow et al. 1998).

In the following, the low-temperature networks for molecules containing hydrogen, carbon, oxygen, nitrogen, and sulfur will be discussed. High-temperature chemistry is mentioned in §3.7 and grain-surface chemistry in §3.8.

## 3.1. HYDROGEN AND DEUTERIUM CHEMISTRY

Because $H_2$ has no permanent dipole moment, its gas-phase formation is negligibly slow. The only plausible formation route of $H_2$ is on the surfaces of grains (Hollenbach & Salpeter 1971, Jura 1974). Recent experiments show an unexpectedly low $H_2$ formation rate on silicates (Pironello et al. 1997), although theoretical calculations indicate high sticking values on amorphous water ice (e.g., Takahashi et al. 1998). Further work is needed to better understand these differences. $H_2$ is destroyed mostly by photons through the two step process of spontaneous radiative dissociation. Because the destruction is initiated by discrete absorptions, the lines rapidly become optically thick, thereby shielding molecules deeper in the cloud from the dissociating radiation. Inside dark clouds, chemical reactions also become effective in destroying $H_2$.

$H_2$ is ionized by cosmic rays at a rate $\zeta_o$ of the order of $10^{-17} - 10^{-16}$ $s^{-1}$; 97% of the ionizations produce $H_2^+$. The well–known reaction

$$H_2^+ + H_2 \rightarrow H_3^+ + H \qquad (5)$$

leads to the stable $H_3^+$ ion, which plays a pivotal role in the subsequent ion–molecule chemistry through proton-transfer. Its recent detection in interstellar space provides strong observational support for the gas–phase chemical networks (Geballe & Oka 1996, McCall et al. 1998).

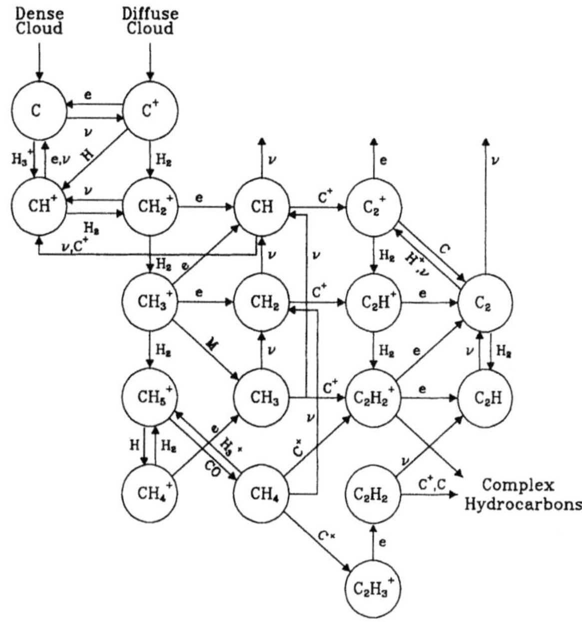


*Figure 1.* Initial steps in the carbon chemistry in diffuse and dark clouds (from: van Dishoeck 1998a).

In contrast with $H_2$, HD can be readily produced by gas–phase reactions, once $H_2$ has been formed

$$D^+ + H_2 \rightleftarrows H^+ + HD. \quad (6)$$

In dense clouds, most of the deuterium is thought to be driven into HD, although a small amount of atomic deuterium of order 10% can be maintained through the dissociative recombination of $DCO^+ + e \rightarrow D + CO$ (Dalgarno & Lepp 1984).

### 3.2. CARBON CHEMISTRY

At the edges of clouds or near a radiation source, atomic carbon is mostly in the form of $C^+$, because its ionization potential is less than 13.6 eV. $C^+$ is very important in the chemistry, because it actively participates in the ion-molecule chemistry leading to more complex hydrocarbon molecules in cold clouds. Formation of $CH^+$ through the reaction

$$C^+ + H_2 \rightarrow CH^+ + H \; - 4640 \text{ K} \quad (7)$$

does not proceed at low temperatures, however, because the reaction is endothermic by about 0.4 eV. Instead, the carbon chemistry is initiated by the slow radiative association reaction

$$C^+ + H_2 \rightarrow CH_2^+ + h\nu \tag{8}$$

with a rate coefficient of $\alpha_8 \approx 1 \times 10^{-15}$ cm$^3$ s$^{-1}$ (Black & Dalgarno 1973, Smith 1989, Gerlich & Horning 1992). $CH_2^+$ reacts rapidly with $H_2$ to form $CH_3^+$, which is followed by dissociative recombination with electrons to produce neutral hydrides like CH and $CH_2$.

At the edges of clouds up to $A_V \approx 2$ mag, the build–up of polyatomic hydrocarbons is limited by rapid photodissociation. At larger depths, reactions of CH, $CH_2$ and $CH_3$ with atomic O leading to species such as CO and $H_2CO$ become the dominant destruction routes. Once the carbon is locked up in the very stable CO molecule, formation of the more complex hydrocarbons ceases.

Inside dense clouds far away from sources of ultraviolet radiation, the amount of $C^+$ is very low. The reaction of C with $H_3^+$ now initiates the carbon chemistry through the reactions $C + H_3^+ \rightarrow CH^+ \rightarrow CH_2^+ \rightarrow CH_3^+$. Because the reaction of $CH_3^+$ with $H_2$ to form $CH_4^+$ is endothermic, only slow radiative association leading to $CH_5^+$ can occur. The rate coefficient for this reaction is uncertain (Gerlich & Horning 1992), but is orders of magnitude slower than that of typical exothermic ion–molecule reactions. Thus, reactions of $CH_3^+$ with other abundant species such as O, CO and HCN become competitive, leading to ions which are the precursors of CO, $CH_2CO$ (ketene) and $CH_3CN$ (methylcyanide) respectively.

The production of complex hydrocarbons in dense clouds occurs via three types of pathways (Herbst 1995): (i) carbon insertion reactions (e.g. $C^+ + CH_4 \rightarrow C_2H_2^+$ or $C + C_2H_2 \rightarrow C_3H + H$); (ii) condensation reactions (e.g. $CH_3^+ + CH_4 \rightarrow C_2H_5^+ + H_2$ or $C_2H + C_2H_2 \rightarrow C_4H_2 + H$); and (iii) radiative association reactions (e.g. $C^+ + C_n \rightarrow C_{n+1}^+ + h\nu$). In general, carbon insertion with $C^+$ is thought to be the dominant route. Since this leads to loss of one hydrogen and since the larger ions $C_nH_m^+$ do not react rapidly with $H_2$, low-temperature gas–phase chemistry produces strongly unsaturated hydrocarbons, in agreement with observations of dark clouds (see §6.3). The necessary $C^+$ ions are produced in small amounts through destruction of CO by $He^+$. Reactions with C may be competitive if they are as rapid as suggested by recent laboratory experiments (Smith 1997).

## 3.3. OXYGEN CHEMISTRY

Because oxygen is mostly neutral even in diffuse clouds, the gas–phase oxygen chemistry is initiated by cosmic ray ionization of H and $H_2$ to form $H^+$ and $H_2^+$, with the latter leading rapidly to $H_3^+$ by reaction (5). In dense clouds, $OH^+$ is then formed through the reaction of O with $H_3^+$. Subsequent hydrogen abstraction reactions and dissociative recombination lead to neutral oxygen–bearing molecules like OH and $H_2O$. Reaction of OH with $C^+$ produces $CO^+$, which reacts with $H_2$ to yield $HCO^+$; $HCO^+$ then dissociatively recombines with electrons, completing one of the major production routes of CO. In cold dense clouds, the reaction $OH + O \rightarrow O_2 + H$ rapidly transforms a significant part of the oxygen into $O_2$.

CO can be destroyed only by photons with wavelengths shortward of 1118 Å. Like $H_2$, the photodissociation of CO occurs through discrete lines, which become optically thick deeper into the cloud. In addition, shielding by dust and coincident lines of H and $H_2$ plays a role (van Dishoeck & Black 1988). The isotopic species $^{13}CO$ and $C^{18}O$ can be partly shielded by $^{12}CO$.

### 3.4. NITROGEN CHEMISTRY

In contrast with the carbon– and oxygen–chemistries, the reactions that initiate the nitrogen chemistry are still not well understood. The reaction of N with $H_3^+$ to form $NH^+$ is endothermic and the exothermic reaction

$$N + H_3^+ \rightarrow NH_2^+ + H \quad (9)$$

is unusual because the transfer of two protons is involved. Recent experiments by Scott et al. (1997), however, suggest a relatively rapid rate coefficient $\alpha_9 \approx 4.5 \times 10^{-10}$ cm$^3$ s$^{-1}$ at room temperatures. Alternatively, the reaction of $N^+$ with $H_2$ may lead to $NH^+$, where $N^+$ is formed by cosmic ray ionization or the destruction of $N_2$ by $He^+$. Once $NH^+$ or $NH_2^+$ are formed, hydrogen abstraction reactions lead to $NH_4^+$, and dissociative recombination to NH, $NH_2$ and $NH_3$.

Neutral–neutral reactions play an important role in the formation of other nitrogen–containing species. For example, reactions of atomic N with CH, OH and NO lead to CN, NO and $N_2$, respectively. Reactions of CN with $C_2H_2$ are probably a major source of the cyanopolyynes ($HC_{2n}CN$) (Fukuzawa & Osamura 1997).

### 3.5. SULFUR CHEMISTRY

Most of the gas-phase formation routes of sulfur hydrides involve at least one endothermic reaction. The reaction of $S^+$ or S with $H_2$ to form $SH^+$ or SH is endothermic, whereas the radiative association reaction is predicted to be slow. S can react with $H_3^+$ to form $SH^+$, but the subsequent reaction of $SH^+$ with $H_2$ is again endothermic. Thus, sulfur hydrides are not expected to be abundant in cold clouds. Their detection in translucent clouds (e.g., Tieftrunk et al. 1994, Turner 1996) and PDRs (e.g., Jansen et al. 1995) signifies the importance of other processes such as grain-surface chemistry.

In contrast, other sulfur–containing species can be readily produced through ion–molecule and neutral–neutral reactions. For example, CS results from reactions of $S^+$ and S with CH and $C_2$, whereas SO stems from reactions with OH. The subsequent reaction of SO with OH is expected to produce copious amounts of $SO_2$. The lack of abundant $SO_2$ in dark clouds indicates an incomplete understanding of either the sulfur chemistry, or the depletion of sulfur (Palumbo et al. 1997), or the physical structure of the clouds.

### 3.6. ISOTOPIC FRACTIONATION

At the low temperatures in interstellar clouds, significant enhancement of various isotopes can occur as a result of a process called 'fractionation'. An extreme example is the deuterium fractionation initiated by the isotope exchange reactions (Wootten 1987, Millar et al. 1989)

$$H_3^+ + HD \rightleftarrows H_2D^+ + H_2 + 227\ \mathrm{K} \tag{10}$$

$$CH_3^+ + HD \rightleftarrows CH_2D^+ + H_2 + 370\ \mathrm{K}, \tag{11}$$

which occur preferentially in the forward direction at low temperatures. These reactions are followed by deuterium transfer reactions such as

$$H_2D^+ + CO \rightarrow DCO^+ + H_2 \tag{12}$$

$$CH_2D^+ + N \rightarrow DCNH^+ + H_2 \tag{13}$$

which lead to $DCO^+/HCO^+$ and DCN/HCN abundance ratios which are up to a factor of 1000 larger than the overall [D]/[H] ratio.

$^{13}C$ can be incorporated preferentially through exchange reactions like

$$^{13}C^+ + {}^{12}CO \rightleftarrows {}^{12}C^+ + {}^{13}CO + 36\ \mathrm{K}, \tag{14}$$

which are most effective in diffuse and translucent clouds where not all carbon is incorporated in CO. Isotope selective photodissociation of $^{13}CO$ counteracts the enhancement at low temperature through reaction (14).

### 3.7. HIGH-TEMPERATURE NETWORKS

At high temperatures of $\sim 200 - 2000$ K such as encountered in the "hot core" regions surrounding massive young stars (see §8.2) and in shocks, gas-phase reactions with H and $H_2$ become significant. Specifically, the reactions

$$O + H_2 \rightleftarrows OH + H - 4480\ \mathrm{K} \tag{15}$$

$$OH + H_2 \rightleftarrows H_2O + H - 2100\,K \quad (16)$$

start to proceed at temperatures of a few hundred K (Ceccarelli et al. 1996, Charnley 1997), and drive most of the available gas-phase oxygen not locked up in CO into $H_2O$. Similar reactions of S with $H_2$ convert the gas-phase sulfur into $H_2S$. However, the back reactions of OH and $H_2O$ with H have comparable energy barriers, so that the balance between O, OH and $H_2O$ depends also on the $H/H_2$ ratio in the warm gas.

Once most of the oxygen has been driven into $H_2O$ at high temperatures, little O, OH and $O_2$ are available in the gas, preventing the formation of, for example, $SO_2$. Since O and $O_2$ also destroy reactive species such as CN, the CN abundance is enhanced at higher temperatures leading to enhanced formation of $HC_3N$ through reaction with $C_2H_2$.

### 3.8. GRAIN SURFACE CHEMISTRY

Because hydrogen is so abundant and mobile (see §2.2), grain surface chemistry is expected to lead primarily to hydrogenated species, such as $H_2O$, $NH_3$ and $CH_4$. Hydrogenation of CO can lead to $H_2CO$ and $CH_3OH$ (Tielens & Charnley 1997). At higher densities, the amount of atomic hydrogen in the gas phase decreases, and reactions with atomic oxygen become important, e.g. $CO + O \rightarrow CO_2$ (Tielens 1983, Tielens & Whittet 1997). Tunneling reactions with $H_2$ also occur competitively once $H_2 >> H$.

Within the icy mantles, photochemical reactions can occur, even if only the cosmic–ray produced photons are available (e.g., Allamandola et al. 1988, Bernstein et al. 1995, Gerakines et al. 1996). For example, photodissociation can lead to radicals (e.g., $H_2O \rightarrow OH + H$ or $O + H_2$), which can subsequently react to form other molecules (e.g., $CO + O$ or $OH \rightarrow CO_2$). In addition, processing by highly energetic cosmic-ray particles can lead to new products (e.g., Moore et al. 1983, Teixeira et al. 1998). Explosive exothermic reactions between stored radicals at higher temperatures produce complex organics (e.g., d'Hendecourt et al. 1982, Greenberg et al. 1995). Further work is needed to determine the relative efficiencies of these processes under various conditions.

## 4. Models

### 4.1. GAS–PHASE MODELS

The calculation of the abundances in star-forming regions requires a physical model in which the temperature, density, radiation field etc. are specified as functions of position and/or time. In general, two different classes of models are considered: (i) Steady–state, depth–dependent models, in which the physical parameters and molecular abundances do not change with time, but are functions of depth into the cloud. Models of the translucent outer envelopes of clouds (e.g., van Dishoeck 1998a) and of dense ultraviolet photon- or X-ray dominated regions (PDRs or XDRs, see e.g., Hollenbach & Tielens 1997, Sternberg et al. 1997, Maloney et al. 1996) fall in this category. (ii) Time–dependent, depth–independent models, in which the concentrations are computed as functions of time at a single position deep into the cloud. Models of dark pre-stellar cores (e.g., Lee et al. 1996a, Millar et al. 1997), collapsing envelopes (e.g., Bergin & Langer 1997) and hot cores near massive young stars (e.g., Charnley et al. 1992) fall into this category. The time scale for reaching chemical equilibrium depends on the density, temperature and ionization fraction and is typically $\sim 10^5 - 10^7$ yr. Some combined time– and depth–dependent models have been made for dark clouds (e.g. Lee et al. 1996b) which are useful to explore the effects of the assumptions.

In both cases, the parameters that enter the models are (1) the elemental abundances of C, O, N, S, metals...; and (2) the cosmic ray ionization rate $\zeta_o$ in $s^{-1}$. In the steady–state, depth–dependent models, additional parameters are (3) the geometry (e.g. plane–parallel, spherical, ...); (4) the density $n_H = n(H) + 2n(H_2)$ as a function of position; (5) the incident radiation field, specified by a factor $I_{UV}$ times the standard interstellar radiation field as given by, e.g., Draine (1978); and (6) the grain parameters, i.e., the extinction curve, albedo and scattering function. The temperatures of the gas and dust as functions of position in the cloud can be obtained self–consistently from the balance of heating and cooling. Alternatively, they can be constrained from observations and provided as additional input parameters.

In the time–dependent models, the additional parameters besides (1) and (2) are: (3) the density, usually taken to be constant with time (so-called pseudo time–dependent models); (4) the visual extinction $A_V$ at the position in the cloud, usually taken to be so large that external photons can be neglected; and (5) the initial abundances of the various species at $t$=0, often taken to be atomic except for $H_2$. The temperature can be obtained from the thermal balance, but is almost always set at 10 K for both the gas and dust, typical of a dark cloud shielded from ultraviolet radiation and heated by cosmic rays only. In these models, the ratios of the local concentrations are taken to be equal to the ratios of the column densities integrated over depth. This assumption is accurate for molecules whose abundances peak in the center of the cloud, but not for species such as radicals whose abundances peak in the outer part of the envelope (see Figure 5 of van Dishoeck 1998a).

### 4.2. THE $C^+ \rightarrow C \rightarrow CO$ TRANSITION

The principal chemical characteristics of the depth– and time–dependent models are governed by the transition of carbon from atomic to molecular form. In the depth–dependent case, $C^+$ recombines to C around $A_V \approx 1$ mag, followed by the conversion to CO around $A_V \approx 2$ mag. The CO photodissociation rate as a function of depth is crucial in this transition, and its calculation requires a careful treatment of the complicated radiative transfer (e.g., van Dishoeck & Black 1988, Warin et al. 1996, Lee et al. 1996b) and the availability of accurate molecular data. The chemistry of other species follows the $C^+ \rightarrow CO$ transition. At the edge, only the simplest diatomic molecules are found. Around $A_V = 1-2$ mag, the presence of both C and $C^+$ and simple hydrides results in an increase in the abundance of hydrocarbon molecules such as CN and $C_2H$ (see Fig. 1). Deeper inside the cloud, atomic carbon is no longer available, and destruction by O becomes more important than photodissociation. Stable species such as $CH_4$, $C_2H_2$ and HCN become dominant.

Many of the same features are observed in the pseudo time–dependent models, if depth is replaced by time. These models usually are based on the assumption that dark clouds originate from diffuse gas, so that all species except $H_2$ are initially in atomic form, with carbon present as $C^+$. On a time scale of $\sim 10^3$ yr, $C^+$ recombines to C, which subsequently transforms to CO after $\sim 10^5$ yr (see Fig. 3). Since the presence of atomic carbon is essential to build up more complex organic species such as $HC_3N$, they are abundant only at early times, but not at steady–state. Recent models by Ruffle et al. (1997) suggest a second peak in the abundances of the long carbon-chains at later times if time-dependent depletion is taken into account.

### 4.3. GAS–GRAIN MODELS

Based on the processes described in §2.2, two different chemical regimes can be distinguished. In the "accretion limited" regime, $t_{\rm hop} << t_{\rm acc}$ so that a species can diffuse on the surface until it finds a co–reactant. The chemistry is limited by the accretion rate of new species. In the "reaction limited" regime, the opposite holds $t_{\rm hop} >> t_{\rm acc}$, so that a species trapped in a site can only react with migrating species that visit that site. Most of the chemical models including gas–grain interactions have been formulated in the "reaction limited" regime using rate equations which mirror those used for gas-phase chemistry (e.g., Hasegawa & Herbst 1993). This approach is only accurate when a large number of reactive species exist on a single grain surface, since only average abundances are calculated. As seen from Table 1, this condition is usually not met, since the accretion times are long and the

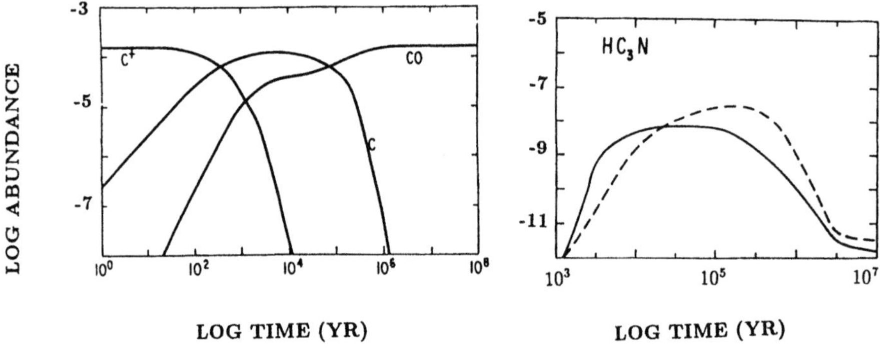


*Figure 2.* Time dependence of the abundances of $C^+$, C and CO (left) and $HC_3N$ (right) for a cloud with $n_H = 2 \times 10^4$ cm$^{-3}$. The latter behavior is typical of all complex carbon–bearing molecules (from: Herbst & Leung 1986).

grains are small, so that only few reactive species are present on a grain at any time. The surface chemistry is therefore likely to be in the accretion–limited regime and can only be properly treated by a Monte–Carlo method which determines the likelihood of two such species arriving from the gas in succession onto a particular grain in a steady-state model (e.g., Tielens & Hagen 1982). Recently, attempts have been made to modify the rate equations to take into account the shortcomings of the reaction–limited approach (Caselli et al. 1998). The consequences for the models have been discussed by Shalabiea et al. (1998). Significant discrepancies between the two formulations occur if desorption is important. So far, no Monte-Carlo method has been developed for use in time-dependent codes.

An important parameter in the gas–grain models is the mechanism for returning the molecules to the gas phase. If no desorption is included, the gas–phase molecules accrete on the grains on a time scale of less than $10^6$ yr at a typical dark cloud density of $10^4$ cm$^{-3}$, inconsistent with the observed widespread molecular emission. The various mechanisms have been discussed by Schutte & Greenberg (1991) and Schutte (1996) (see chapter by Schutte). Thermal evaporation is efficient only at higher temperatures, $T_D > 20$ K. The most effective desorption mechanisms in cold clouds are likely to be cosmic–ray spot heating, which has a time scale of $t_{\rm CR} \approx 4 \times 10^6 - 10^7$ yr, and the explosive desorption mechanism. The latter can be either cosmic–ray induced, giving a time scale $t_{\rm exp-CR} \approx 10^5$ yr, or triggered by grain–grain collisions at velocities greater than $\sim 0.1$ km s$^{-1}$, $t_{\rm exp-gg} \approx 2 \times 10^9/n_H$ yr. These mechanisms are particularly effective for "non–polar" ices containing CO, $O_2$ and $N_2$, but less for $H_2O$ ices, which contain strong hydrogen bonds.

## 4.4. DYNAMICAL MODELS

Dynamical models which go beyond the constant density, pseudo time–dependent models have been developed by various groups. Tarafdar et al. (1985) and Prasad et al. (1991) consider the case in which the density increases from diffuse cloud to dense cloud values, whereas more recent models in this category are by Nelson & Langer (1997) and Shematovich et al. (1997).

Models which describe the gas-grain chemistry during the collapse phase of dark clouds to form stars have been made by Rawlings et al. (1992), Shalabiea & Greenberg (1995) and Ceccarelli et al. (1996), using parametrized fits to the density profiles in simple collapse models such as that of Shu (1977) and Basu & Mouschovias (1994). Bergin & Langer (1997) consider the evolution of a single parcel of gas deep inside the core.

Another class is formed by the dynamical mixing models, in which parcels of gas cycle between the diffuse and dense medium due to turbulence (Boland & de Jong 1982, Chièze et al. 1991, Xie et al. 1995). This process increases the amount of atomic carbon and related molecules inside the clouds. However, Rawlings & Hartquist (1997) have suggested that the microscopic mixing leading to reactions between material from initially different fluid elements is likely to be restricted to the turbulent boundary layers at the edges of clouds in the presence of a magnetic field.

## 4.5. HOT CORE MODELS

A different class of pseudo-time dependent models has been developed to describe the chemistry in "hot cores" near massive young stars. In these regions, the initial composition of the gas is not taken to be atomic but molecular, consistent with that of evaporating ice mantles in the vicinity of young stars. The evaporated molecules subsequently drive a rapid gas-phase chemistry (e.g., Charnley et al. 1992, Caselli et al. 1993, Charnley 1997).

Since methanol is one of the most abundant ice mantle consitutents, it plays a crucial role in the hot core chemistry. Specifically, complex oxygen-containing organics can be formed by protonation of $CH_3OH$ by $H_3^+$, $H_3O^+$ or $HCO^+$, followed by transfer of a -$CH_3$ group to a neutral molecule, e.g., $CH_3OH_2^+ + CH_3OH \rightarrow (CH_3)_2OH^+$. Dissociative recombination of the ion then produces di-methyl ether, $CH_3OCH_3$. The detailed chemistry of interstellar alcohols leading to large esters and ethers has been discussed by Charnley et al. (1995). Nitrogen-containing organics can be formed through reactions of carbon ions with evaporated $NH_3$, or through reactions of CN with evaporated $C_2H_2$.

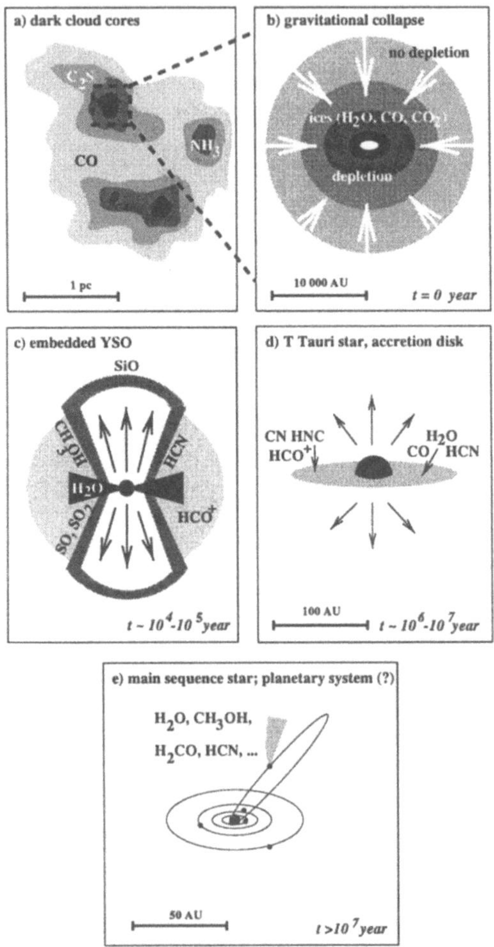


*Figure 3.* A schematic view of low-mass star formation. (a) Dark cloud cores, roughly 1 pc in size, gradually contract until (b) Magnetic support is overcome and inside-out collapse begins at $t = 0$. (c) For $\sim 10^{4-5}$ years a phase of both high accretion and supersonic outflow occurs in deeply embedded young stellar objects. (d) Gradual clearing by the outflow leaves only the young T Tauri star and a residual protoplanetary accretion disk, that, on time scales of $10^{6-7}$ years, leads to the formation of à mature planetary system (e). Characteristic molecules at each of these stages are indicated (Figure by M.R. Hogerheijde, after Shu et al 1987; from: van Dishoeck & Blake 1998).

## 5. Scenario for chemical evolution

### 5.1. CHEMICAL CHARACTISTICS

The star-formation process is accompanied by orders of magnitude changes in the physical parameters of the collapsing material. Figure 3 shows an overview of the various stages in the formation of a single, isolated low mass star from the collapse of a molecular cloud core (Shu et al. 1993). The chemistry responds to the changes in temperature and density. In the cold molecular cores prior to star formation, the chemistry is dominated by low-temperature gas-phase reactions leading to the formation of small radicals and unsaturated molecules such as long carbon-chains. During the collapse phase, the density becomes so high that most molecules accrete onto the grains and form an icy mantle. After the new star has formed, its radiation heats up the surrounding envelope, so that the molecules evaporate back into the gas phase, probably in a sequence according to their sublimation temperatures. In addition, the outflows from the young star penetrate the envelope, creating high temperature shocks and lower temperature turbulent regions in which both volatile and refractory material containing silicon can be returned. These freshly evaporated molecules then drive a rich chemistry in the "hot cores" for a period of $10^5$ yr. Finally, the envelope is dispersed by winds and/or ultraviolet photons, leading to the appearance of photon-dominated regions.

## 5.2. OBSERVATIONAL TECHNIQUES

The above scenario can be tested by a variety of observational techniques, which give complementary information (e.g., Evans et al. 1991, Carr et al. 1995). Submillimeter single-dish observations probe the physics and chemistry in the envelopes surrounding YSOs on scales of $10 - 20''$ (1500-3000 AU at 150 pc). On this scale, many of the different physical regimes (infall, outflow, disks) are blurred together. Millimeter interferometers with typical resolutions of $0.5 - 3''$ may be able to separate the components, but chemical studies are limited to the brightest high-mass regions (e.g., Blake et al. 1996).

Infrared absorption lines studies have the advantage that only a pencil beam line of sight to the young star is sampled. Also, the population distribution over the rotational energy levels can be directly constrained. The excitation temperature provides useful information on the physical parameters of the gas where the molecule is located. Molecules without a permanent dipole moment such as $H_3^+$, $CO_2$, $CH_4$ and $C_2H_2$ can be observed through their ro-vibrational transitions, whereas they have negligible rotational emission. Finally, both gas-phase and solid-state species can be detected at infrared wavelengths (e.g., van Dishoeck et al. 1996). The latter can be distinguished because their spectra consist of a single broad spectral feature that lacks the characteristic ro-vibrational structure of the gas-phase spectrum (see Figure 4). A disadvantage is the low resolving power of most infrared spectrometers, especially those on board ISO.

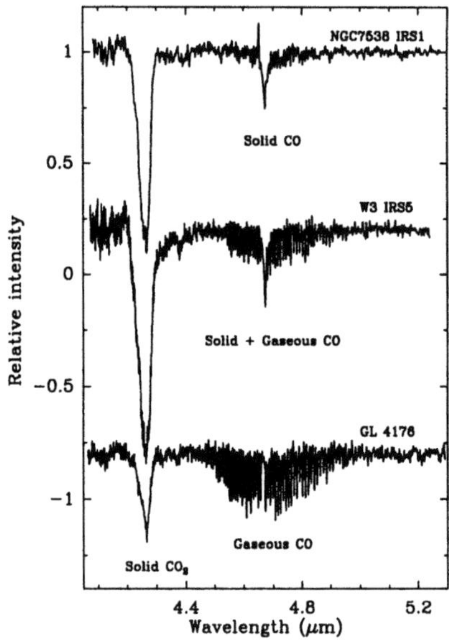


*Figure 4.* Normalized spectra obtained with the *Short Wavelength Spectrometer* on board the *Infrared Space Observatory* toward three massive young stellar objects embedded in dense molecular clouds. The strong, broad absorption at 4.27 $\mu$m is due to solid $CO_2$, whereas the characteristic ro–vibrational P– and R–branch structure at 4.4–4.9 $\mu$m indicates the presence of warm, gaseous CO along the line of sight (van Dishoeck et al. 1998).

This implies that no kinematic information can be obtained, and that only abundances down to $10^{-7} - 10^{-8}$ with respect to hydrogen can be probed, compared with $\sim 10^{-11}$ at millimeter wavelengths.

It should be kept in mind that the derived abundances have considerable uncertainties of up to an order of magnitude, owing to uncertainties in the excitation, optical depth of the lines and unresolved source structure (e.g., Irvine et al. 1987). A good physical model of the source is a prerequisite for deriving accurate abundances. In addition, the $H_2$ column density is often poorly constrained, and determined indirectly from CO isotopic observations. The only direct observation of $CO/H_2=2.7\times10^{-4}$ in the warm, dense cloud toward NGC 2024 IRS2 (Lacy et al. 1994) is a factor of 3 higher than the value of $8\times10^{-5}$ commonly used in dark clouds (Frerking et al. 1982).

## 6. Chemistry in pre-stellar cores

In order to study the effect of YSOs on the chemistry, the abundances in quiescent clouds prior to star formation need to be accurately constrained. Since young stars have powerful outflows which snake through the clouds on parsec scale, care has to be taken to find a parcel of gas that has not yet been affected.

## 6.1. TRANSLUCENT CLOUDS

The best clouds for establishing the reference abundances are probably the translucent clouds with visual extinctions of a few mag and densities of a few thousand $cm^{-3}$. These clouds have been studied by optical absorption lines (e.g., Gredel et al. 1993), by millimeter absorption lines (e.g., Lucas & Liszt 1997) and by millimeter emission observations (e.g., Turner 1996, 1998 and references cited). Detailed models by Turner et al. (1998) show that the observed abundances are well reproduced by the ion-molecule networks. The main exceptions are formed by $NH_3$, $H_2CO$ and $H_2S$, whose formation is probably dominated by grain-surface chemistry.

## 6.2. TMC–1

Most observational effort has been directed toward TMC–1, a small elongated, dense ridge of $\sim 0.6 \times 0.06$ pc in the Taurus dark clouds with a mass less than 40 $M_\odot$. A large chemical gradient is found across the core, with $NH_3$ peaking in the northern part, and long carbon–chain molecules such as $HC_7N$ and $C_4H$ in the southern part (Olano et al. 1988, Pratap et al. 1997). The abundances of the carbon-chains can be fit with pseudo time–dependent models at early times $t \approx 10^5$ yr (Lee et al. 1996a), although Ruffle et al. (1997) point out the existence of a second peak at later times if substantial depletion occurs. The models require that rapid neutral–neutral reactions of the type $C_n$ + O and $C_n$ + N are neglected. Howe et al. (1996) have modeled the chemical gradient across TMC–1 within the dynamical framework of sequential fragmentation by Hanawa et al. (1994), in which the northern part condensed first $\sim 10^6$ yr ago, and the southern part last, $\sim 10^5$ yr ago. They conclude that TMC–1 may be chemically rich compared with other dark clouds because it is undergoing its first cycle of (sudden) collapse starting from very diffuse, atomic carbon–rich gas.

Photons penetrating inhomogeneous dark clouds can also produce atomic carbon inside clouds, as may be indicated by the [C I] 492 GHz observations of Schilke et al. (1995). Alternatively, the observed [C I] emission could reflect the atomic carbon present in the skin or in lower density regions of the (inhomogeneous) cloud. Lee et al. (1996b) have investigated this situation in time– and depth–dependent models with a density gradient from edge to center. The best agreement with observations is again obtained at early times. However, it is difficult to reproduce the data for an ensemble of small, spatially unresolved spherical clouds, because of the large contribution from the envelope material where the build–up of large molecules is prevented by photodissociation of the smaller intermediates.

The model results are sensitive to the $[C]_g/[O]_g$ elemental abundance ratio in the gas as well. If this value is increased from the canonical ratio of 0.4 to higher values, improvement with observations for the complex molecules is obtained (Bergin et al. 1997b). Pratap et al. (1997) show that the observed chemical gradient can also be reproduced by changes in $[C]_g/[O]_g$ along the ridge.

## 6.3. DARK CLOUD CORES

Suzuki et al. (1992) performed a systematic study of a few characteristic molecules ($C_2S$, $HC_3N$, $HC_5N$ and $NH_3$) in a set of dark cores identified by Myers & Benson (1983). Cores with and without stars were selected. The abundances of the carbon chains are found to correlate well with each other, but not with $NH_3$, consistent with the case of TMC–1. The observed $C_2S/NH_3$ abundance can be reproduced quantitatively in models which start from diffuse gas and form dense cores over a period of $10^5$ to $2 \times 10^6$ yr. Thus, this ratio may be a particularly useful tracer of cloud evolution, with TMC-1S among the youngest objects.

High spatial resolution observations of the L1498 pre-stellar core by Kuiper et al. (1996) show a chemically differentiated structure, with $NH_3$ peaking in the inner region and $C_2S$ in the atomic-carbon rich (clumpy) outer part. Further chemical studies of such cores, especially those which have a more centrally concentrated density structure and appear to be on the verge of collapse (e.g., Ward-Thompson et al. 1994, Motte et al. 1998), are warranted.

Caselli et al. (1998) and Plume et al. (1998) have studied the ionization structure in a sample of dark cores, based on $DCO^+/HCO^+$ observations. Plume et al. (1998) find ionization fractions of a few $\times 10^{-7}$, for an adopted cosmic ray ionization rate of $5 \times 10^{-17}$ $s^{-1}$. Surprisingly, no systematic trends between pre-stellar cores and cores with stars have been found, but this may be due to the large scale of the observations (~10,000 AU).

The ice mantles in quiescent dark clouds can be probed through the observation of field stars behind the clouds (e.g., Chiar et al. 1995, Whittet et al. 1998). $H_2O$, CO and $CO_2$ ice have been detected, which indicate that up to 40% of the heavy elements may be frozen out at densities of a few $\times 10^4$ $cm^{-3}$ (Schutte, this volume).

## 6.4. GMC CORES

How does the chemistry in dark clouds such as Taurus compare with that of Giant Molecular Clouds (GMCs) which form high mass stars? Bergin et al.

(1997a) and Ungerechts et al. (1997) have systematically mapped several GMC cores (Orion, M17 and Cepheus A) over a $5' \times 5'$ region in various molecules. The chemical composition was analyzed at several positions away from the regions of star formation. The derived abundances are remarkably uniform (within a factor of a few), both within a single GMC and between different GMCs, in contrast with the situation for some dark cores. Bergin et al. (1997b) show that the measured abundances are best fit by gas–phase models at early times ($\sim 10^5$ yr) if a fairly high elemental abundance ratio $[C]_g/[O]_g \approx 0.8$ is adopted. They argue that the observed CS/SO ratio may be used to constrain the $[C]_g/[O]_g$ ratio, as suggested previously for the L134N dark core (Swade 1989).

## 7. Chemistry in cold collapsing envelopes

The main characteristic of the chemistry in clouds prior to collapse is the formation of ions and unsaturated radicals through ion-molecule and neutral-neutral reactions. How are these abundances modified during the star-formation process?

### 7.1. MODELS

The principal prediction of the collapse models is that most molecules freeze out onto the cold grains at higher densities on time scales of less than $10^6$ yr. The only exceptions are $H_2$, He, $H_3^+$ and perhaps $N_2$. Models appropriate for the cold outer envelopes ($T \approx 10$ K) have been made by Rawlings et al. (1992), Willacy et al. (1994a), Bergin & Langer (1997) and Shalabiea & Greenberg (1995). They differ strongly in the mechanisms that return molecules to the gas, ranging from none in Rawlings et al.'s to explosive desorption in Shalabiea & Greenberg's. The results are also sensitive to the adopted binding energies on the icy mantles, especially whether the outer layer is $H_2O$-rich or CO-rich. The ions $HCO^+$ and $N_2H^+$ are predicted to be good tracers of the envelopes, because their abundances remain high owing to the increase in the $H_3^+$ abundance when its main removal partners (CO, O, ...) are depleted. The use of $HCO^+$ to trace the envelope structure has been demonstrated by Hogerheijde et al. (1997, 1998).

### 7.2. OBSERVATIONS

Observationally, depletion in collapsing cores is very difficult to prove, because every cloud has a "skin" in which the abundances are close to normal (Mundy & McMullin 1997). Even if the skin amounts to only a few % of the total column density, its factor of 10-100 higher abundances can effectively mask any depletions deep inside. The dust obscuration is still so high at this stage that the young stars are too weak at near- and mid-infrared wavelengths for direct observations of ices. Careful modeling of the line and dust emission on various scales appears the only way to probe the abundances in this phase. One of the best-studied cases is that of NGC 1333 IRAS4, where depletions of more than a factor of 10 have been inferred (Blake et al. 1995).

Another well-studied case is the massive star-forming region NGC 2024. A string of deeply embedded YSOs, designated FIR 1–6, has been detected through their submillimeter continuum emission from dust (e.g., Mezger et al. 1992). Surprisingly, the low-lying lines of $C^{18}O$, CS and $NH_3$ do not show any enhancements at the positions of the FIR sources. Interferometer observations by Wilson et al. (1995) and Wiesemeyer et al. (1997) confirm the offset between the continuum and line emission, suggesting depletions of an order of magnitude. However, the analysis is complicated by the fact that the dust emissivity likely changes within the condensations due to coagulation and accretion of a dense ice layer. Also, Chandler & Carlstrom (1996) point out that the dust and gas masses may be comparable if the dust temperatures are substantially higher than assumed by Mezger et al.

At a more evolved, but still deeply embedded stage, infrared absorption line studies of the ice mantles become possible, both for low-mass and high-mass objects (Whittet 1993, Tielens & Whittet 1997, Chiar et al. 1998, Teixeira et al. 1998). ISO allows a nearly complete inventory of these ices to be obtained for deeply embedded sources such as NGC 7538 IRS9, W33 A and RAFGL 7009S (Whittet et al. 1996, d'Hendecourt et al. 1996, Ehrenfreund et al. 1997, see chapter by Schutte). Although only abundances down to $\sim 0.5\%$ of $H_2O$-ice ($\sim 10^{-7}$ w.r.t. $H_2$) can be probed, the inferred mantle composition is remarkably simple, consisting mostly of species resulting from the hydrogenation and oxydation of O, C, N and CO.

## 8. Chemistry in warm envelopes around newly-formed stars

When a young star begins to heat the inner surrounding envelope by radiation and/or shocks above $\sim 100$ K, various chemical phases occur, which can be distinguished into the hot core region, the region of interaction of the outflow with the surrounding envelope, and the more extended warm envelope.

### 8.1. WARM ENVELOPES

Self-consistent models of the thermal balance, chemistry and radiative transfer in spherical envelopes heated by an internal source have been constructed for low-mass YSO's by Ceccarelli et al. (1996) and for high-mass YSOs by Doty & Neufeld (1997). Other calculations of the radial dust temperature distribution include those of Campbell et al. (1995), Kaufman et al. (1998) and van der Tak et al. (1998). For a $\sim 10^5$ $L_\odot$ source, the region where $T_D > 90$ K extends to $\sim 10^{16}$ cm, resulting in evaporation of $H_2O$ ice and an enhancement of the gas-phase $H_2O$ abundance by a factor of 100–1000. The chemistry in this inner hot core is discussed below. The region where $T_D > 20$ K extends to $\sim 10^{17}$ cm, resulting in evaporation of the volatile species, but not of the polar species $H_2O$ and $CH_3OH$. The combined effects of the changing H/CO ratio in the collapsing envelope and the differential outgassing of the volatile and non-volatile species may be responsible for the different "polar" and "non-polar" ice phases which have been observed (e.g., Tielens et al. 1991).

A powerful probe of the physical structure and chemical composition of the envelopes is provided by high spectral resolution infrared observations of the gas. Mitchell et al. (1990) showed the presence of both cold ($T <$ 60 K) and hot ($T = 120 - 1000$ K) gas along the lines of sight toward massive YSO's from $^{12}CO$ and $^{13}CO$ observations . The hot gas contains high abundances of $C_2H_2$, HCN and $CH_4$ (Lacy et al. 1989, 1991, Evans et al. 1991, Carr et al. 1995, Boogert et al. 1998, Lahuis & van Dishoeck 1998), which are enhanced by at least 1–2 orders of magnitude compared with the colder envelope. Gas-phase $CO_2$ has a surprisingly low abundance (van Dishoeck et al. 1996, van Dishoeck 1998b). Clear variations in the gas/solid ratios are seen for various objects, indicating the development of a hot core in the inner envelope in the warmer regions.

Complementary submillimeter observations of a large variety of molecules in these embedded sources have been performed for W 3 IRS5 (Helmich & van Dishoeck 1997, see Figure 5), NGC 2264 IRS1 (Schreyer et al. 1997), AFGL 2591 (van der Tak et al. 1998) and AFGL 2136 (Boonman et al. 1998). Accurate determination of the abundances requires the construction of reliable physical models to disentangle the contributions of changing temperatures and densities from possible radial abundance gradients.

### 8.2. HOT CORE REGIONS

Hot cores are small (<0.1 pc), dense ($n(H_2) > 10^6$ cm$^{-3}$) and warm ($T \approx$ 200 K or higher) regions located close to massive young stars (e.g., Walmsley & Schilke 1993, Ohishi 1997). Observationally, hot cores are characterized by high abundances of fully hydrogenated molecules such as $NH_3$ (e.g., Wyrowski & Walmsley 1996), $H_2S$ (Minh et al. 1991) and $H_2O$ (Cernicharo et al. 1994, Gensheimer et al. 1996) and saturated complex organic molecules such as $CH_3OH$ and $C_2H_5OH$ (e.g., Millar et al. 1995), $CH_3CN$ (e.g., Olmi et al. 1996), $CH_3OCH_3$, $HCOOCH_3$ and $C_2H_5CN$ (e.g., Miao & Snyder 1997, Hatchell et al. 1998). Examples of hot cores include the

Orion hot core and compact ridge (e.g., Blake et al. 1987), SgrB2(N) (e.g., Kuan & Snyder 1996), and objects near ultracompact H II regions such as G34.3+0.15 (Macdonald et al. 1996).

The rich chemistry in hot cores is thought to be driven by the evaporation of icy mantles due to heating near the young star. Charnley et al. (1992) showed that the observed abundances in the Orion hot cores can be reproduced if a mixture of simple ices containing $H_2O$, CO, $CH_3OH$, $NH_3$ and/or HCN is evaporated into the warm gas, followed by gas-phase reactions such as discussed in §4.5. The abundances of the complex species peak after $\sim 10^4$ yr, thus offering a sensitive chemical "clock". Caselli et al. (1993) noted that the different chemical characteristics of the Orion hot core and compact ridge can be explained by a temperature gradient in the original collapsing cloud. More recent hot core models including S-, Si- and P-bearing species have been published by Charnley (1997), MacKay (1995) and Charnley & Millar (1994), respectively. Millar et al. (1997b) have presented a comprehensive time- and depth-dependent treatment of the chemistry in G34.3+0.15 based on a 3-component physical model of the region. Some molecules like $HCO^+$ are present almost entirely in the entended envelope or halo, whereas other species like $CH_3OCH_3$ and $H_2S$ are located primarily in the (ultra-)compact hot core.

The observed high D/H ratios in hot core molecules (e.g., Millar & Hatchell 1998) largely reflect the high D/H ratios in the ices, which can result from two different processes. First, gas-phase deuterium fractionation is effective in the cold collapsing cloud (see §3.6). Second, reactions with D on grain surfaces lead to enhanced abundances of deuterated molecules, because of the high atomic D/H ratio in dense clouds (Tielens 1993).

## 8.3. OUTFLOW REGIONS

The earliest stages of star formation are accompanied by powerful outflows, which originate within the young star/accretion disk boundary region and are essential to shed the high angular momentum content of the original collapsing cloud core (see Reipurth & Bertout 1997 for recent reviews). As the high velocity gas flows outward, it strikes the envelope and creates shocks, which alter the chemistry. Recent reviews of the physical and chemical structure of shocks can be found in Draine & McKee (1993), Hollenbach (1997) and Bachiller (1996, 1997).

If the shocks have high velocities (>50 km $s^{-1}$), the temperature increases to such high values ($> 10^4$ K) that most molecules are collisionally dissociated. In addition, the ultraviolet radiation from such $J$-type shocks can dissociate molecules both within and ahead of the shock. The molecules reform slowly in a lengthy, warm zone in the wake of the shock (Neufeld &

Dalgarno 1989ab).

For lower shock speeds in clouds with low fractional ionization ($x_e < 10^{-6}$), the shock is of $C$-type and the temperatures reach peak values of only 2000-3000 K, insufficient to dissociate the molecules. Reactions with energy barriers such as O + $H_2$ and S + $H_2$ (see §3.7) rapidly drive the available gas-phase oxygen and sulfur into $H_2O$ and $H_2S$.

In addition to these gas-phase processes, grain cores and mantles are affected by shocks. The high-velocity $J$-shocks can cause destruction or thermal sputtering of the cores, resulting locally in much enhanced gas-phase abundances of Si, Fe, ... The lower velocity $C$-shocks can inject a variety of refractory and volatile species into the gas phase through non-thermal sputtering (e.g., Pineau des Forêts & Flower 1997). The released Si can then react with OH and $O_2$ to form SiO, one of the principal diagnostics of shocks (e.g., Schilke et al. 1997, Caselli et al. 1997).

Outflows have been studied observationally in a variety of molecules, including $H_2$ near-infrared emission, CO and SiO. One of the best studied regions chemically is L1157 (Bachiller & Perez-Gutierrez 1997). Both refractory (SiO, SO, ...) and volatile mantle species (HCN, $H_2CO$, $CH_3OH$, ...) are found to be enhanced in the gas by factors ranging from a few to $10^6$. High $H_2O$ abundances in outflow regions have been observed from the ground (e.g., Cernicharo et al. 1994) and with the ISO satellite (e.g., Liseau et al. 1996, Ceccarelli et al. 1998).

## 9. Example: chemical evolution in the W 3 cloud

The W 3 massive star-forming region at $\sim$ 2.3 kpc provides an excellent opportunity to study the chemistry of YSOs at different evolutionary stages originating from the same parent cloud. Helmich et al (1994) and Helmich & van Dishoeck (1997) performed an unbiased submillimeter spectral survey of three YSOs at 15″ (0.15 pc) resolution: IRS5, IRS4 and W 3($H_2O$) (Turner & Welch 1984). Clear physical and chemical differences are found between the three sources, in spite of their similar luminosities (see Figure 5). Toward W 3(IRS5), silicon- and sulfur-bearing molecules such as SiO and $SO_2$ are prominent. This source has a powerful, massive outflow and is presumably the youngest of the three sources, showing also strong ice absorption features in its infrared spectra. Toward W 3($H_2O$), organic molecules molecules like $CH_3OH$, $CH_3OCH_3$ and $CH_3OCHO$ are at least one order of magnitude more abundant, indicating that it is well into the hot core phase. Finally, only simple molecules are found toward IRS4, suggesting that it is more evolved and has already broken free from the parent cloud, setting up an H II region and PDR on the back-side of the cloud.

CHEMICAL EVOLUTION IN THE W 3 MASSIVE STAR-FORMING REGION

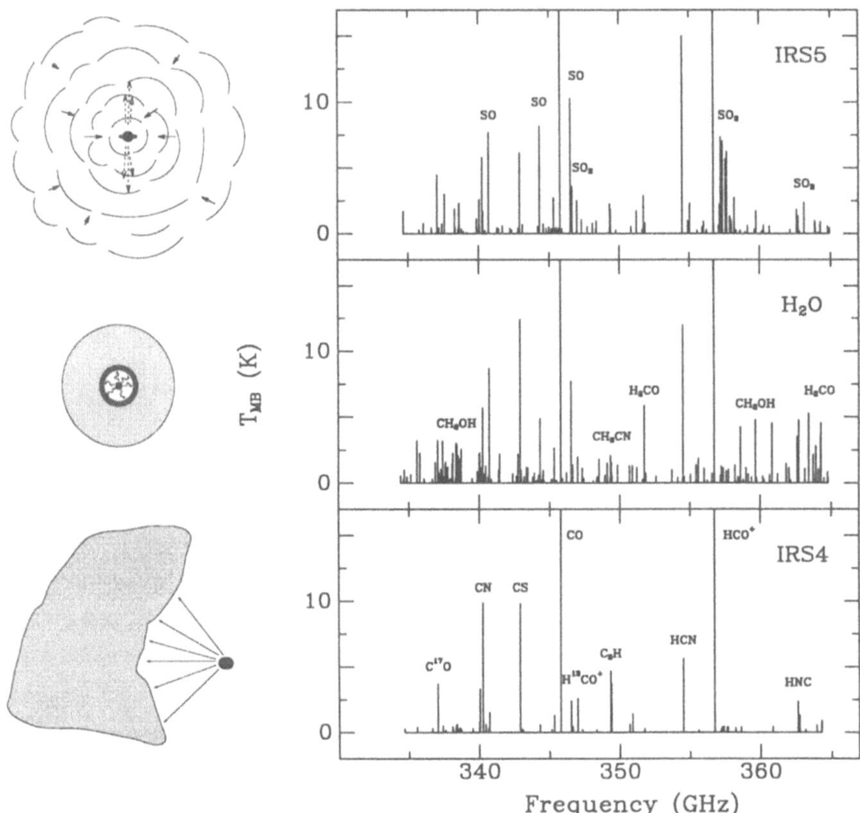


*Figure 5.* Summary of the JCMT 335–365 GHz line survey of three massive star-forming regions in the W 3 molecular cloud. The spectra were constructed from the observed line intensities obtained in double-side band mode. About 90% of this frequency range was covered; the missing 10% was filled in through model excitation calculations, mostly for $SO_2$ and $CH_3OH$. Strong lines in common in the three spectra are labelled in the W 3 IRS4 spectrum only. Large physical and chemical differences are found between the three regions, which are attributed to different evolutionary stages. A possible evolutionary scenario is illustrated by the cartoons: W 3 IRS5 is the youngest object, W 3($H_2O$) is in the "hot core" phase, and W 3 IRS4 is the oldest object with a well-developed H II region which has broken free of the molecular cloud (Based on Helmich & van Dishoeck 1997).

## 10. Concluding remarks

Significant progress has been made in the last decade in the observations of both gas- and solid-state species near young stellar objects, so that detailed chemical inventories can now be made. At the same time, the devlopment of comprehensive gas-grain models of the chemistry in cold and warm clouds has made it possible to put these observations in a coherent framework for the evolution of gas and dust during the process of star formation. Accurate information on the basic molecular processes entering such models remains essential, however. Future high-spatial resolution observations, such as provided by the planned next generation ground-based millimeter arrays and air- or space-borne telescopes such as SOFIA, SIRTF and FIRST, will provide significant tests of these models.

## Acknowledgements

The author is grateful to many colleagues for discussions on interstellar chemistry. This work is supported by the Netherlands Organization for Scientific Research (NWO/NFRA).

## References

Allamandola, L.J., Sandford, S.A., Valero, G.J. 1988, Icarus, 76, 225
Allen, M., Robinson, 1977, ApJ, 212, 396
Bachiller, R. 1996, ARAA, 34, 111
Bachiller, R. 1997, in *Molecules in Astrophysics: Probes and Processes*, IAU Symposium 178, ed. E.F. van Dishoeck, Kluwer, Dordrecht, p. 103
Bachiller, R., Perez-Gutierrez, M. 1997, ApJ, 487, L93
Basu, S., Mouschovias, T.C. 1994, ApJ, 432, 720
Bates, D.R., Spitzer, L. 1951, ApJ, 113, 441
Bergin, E.A., Langer, W.D. 1997, ApJ 486, 316
Bergin, E.A., Ungerechts, H., Goldsmith, P.F., Snell, R.L., Irvine, W.M., Schloerb, F.P. 1997a, ApJ, 482, 267
Bergin, E.A., Goldsmith, P.F., Snell, R.L., Langer, W.D. 1997b, ApJ, 482, 285
Bernstein, M.P., Sandford, S.A., Allamandola, L.J., Chang, S., Scharberg, M.A. 1995, ApJ, 454, 327
Black, J.H., Dalgarno, A. 1973, Astrophys. Lett., 15, 79
Blake, G.A. 1997, in *Molecules in Astrophysics: Probes and Processes*, IAU Symposium 178, ed. E.F. van Dishoeck, Kluwer, Dordrecht, p. 31
Blake, G.A., Mundy, L.G., Carlstrom, J.E. et al. 1996, ApJ, 472, L49
Blake, G.A., Sandell, G., van Dishoeck, E.F., Groesbeck, T.D., Mundy, L.G., Aspin, C. 1995, ApJ, 441, 689
Blake, G.A., Sutton, E.C., Masson, C.R., Phillips, T.G. 1987, ApJ, 315, 621
Boland, W., de Jong, T. 1982, ApJ, 261, 110
Boogert, A.C.A., Helmich, F.P., van Dishoeck, E.F., Schutte, W.A., Tielens, A.G.G.M., Whittet, D.C.B. 1998, A&A, in press
Boonman, A.M., van der Tak, F.P., Helmich, F.P., van Dishoeck, E.F. 1998, A&A, to be submitted
Campbell, M.F., Butner, H.M., Harvey, P.M. et al. 1995, ApJ, 454, 831
Carr, J.S., Evans, N.J., Lacy, J.L., Zhou, S. 1995, ApJ, 450, 667
Caselli, P., Hartquist, T.I., Havnes, O. 1997, A&A, 322, 296
Caselli, P., Hasegawa, T.I., Herbst, E. 1993, ApJ 408, 548
Caselli, P., Hasegawa, T.I. Herbst, E. 1998, ApJ 495, 309
Ceccarelli, C., Hollenbach, D.J., Tielens, A.G.G.M. 1996, ApJ, 471, 400

Cernicharo, J., González-Alfonso, E., Alcolea, J. et al. 1994, ApJ, 432, L59
Chandler, C.J., Carlstrom, J.E. 1996, ApJ, 466, 338
Charnley, S.B. 1997, ApJ, 481, 396
Charnley, S.B, Kress, M.E., Tielens, A.G.G.M., Millar, T.J. 1995, ApJ, 448, 232
Charnley, S.B., Millar, T.J. 1994, MNRAS, 270, 570
Charnley, S.B., Tielens, A.G.G.M., Millar, T.J. 1992, ApJ, 399, L71
Chiar, J.E., Adamson, A.J., Kerr, T.H., Whittet, D.C.B. 1995, ApJ, 455, 234
Chiar, J.E., Gerakines, P.A., Whittet, D.C.B. et al. 1998, ApJ, 498, 716
Chièze, J.P., Pineau des Forêts, G., Herbst, E. 1991, ApJ, 373, 110
Dalgarno, A. 1987, in *Physical Processes in Interstellar Clouds*, eds. G. Morfill and M.S. Scholer, D. Reidel, Dordrecht, 219
Dalgarno, A. 1994, Adv. At. Mol. & Opt. Phys., 32, 57
Dalgarno, A., Black, J.H. 1976, Rep. Prog. Phys., 39, 573
Dalgarno, A., Lepp, S. 1984, ApJ, 287, L47
d'Hendecourt, L.B., Allamandola, L.J., Greenberg, J.M. 1985, A&A, 152, 130
d'Hendecourt, L.B., Allamandola, L.J., Baas, F., Greenberg, J.M. 1982, A&A, 109, L12
d'Hendecourt, L., Jourdain de Muizon, M., Dartois, E. et al. 1996, A&A, 315, L365
Doty, S.D., Neufeld, D.A. 1997, ApJ, 489, 122
Draine, B.T. 1978, ApJS, 36, 595
Draine, B.T., McKee, C.F. 1993, ARAA, 31, 373
Ehrenfreund, P., d'Hendecourt, L., Dartois, E. et al. 1997, Icarus, 130, 1
Evans, N.J., Lacy, J.H., Carr, J.S. 1991, ApJ, 383, 674
Frerking, M.A., Langer, W.D., Wilson, R.W. 1982, ApJ, 262, 590
Fukuzawa, K., Osamura, Y. 1997, ApJ, 489, 113
Geballe, T.R. and Oka, T. 1996, Nature, 384, 334
Gensheimer, P.D., Mauersberger, R., Wilson, T.L. 1996, A&A, 314, 281
Genzel, R. 1992, in *The Galactic Interstellar Medium*, eds. D. Pfenniger and P. Bartholdi, Saas Fee Advanced Course 21, Springer, Berlin, 275
Gerakines, P.A., Schutte, W.A., Ehrenfreund, P. 1996, A&A, 312, 289
Gerlich, D., Horning, S. 1992, Chem. Rev., 92, 1509
Gillett, F.C., Forrest, W.J. 1973, ApJ, 179, 483
Graedel, T.E., Langer, W.D., Frerking, M.A. 1982, ApJS, 48, 321
Gredel, R., van Dishoeck, E.F., Black, J.H. 1993, A&A, 269, 477
Gredel, R., Lepp, S., Dalgarno, A., Herbst, E. (1989, ApJ, 347, 289
Greenberg, J.M., Li, A., Mendoza-Gómez, C.X., Schutte, W.A., Gerakines, P.A., 1995, ApJ, 455, L177
Hanawa, T., Yamamoto, S., Hirahara, Y. 1994, Pub. Astr. Soc. Japan, 420, 318
Hasegawa, T.I., Herbst, E. 1993, MNRAS, 261, 83; 263, 589
Hatchell,J., Thompson, M.A., Millar, T.J., Macdonald, G.H. 1998, A&A, in press
Helmich, F.P., van Dishoeck, E.F. 1997, A&AS, 124, 205
Helmich, F.P., Jansen, D.J., de Graauw, Th., Groesbeck, T.D., van Dishoeck, E.F. 1994, A&A, 283, 626
Herbst, E. 1993, in *Dust and Chemistry in Astronomy*, eds. T.J. Millar and D.A. Williams, IOP, Bristol, 183
Herbst, E. 1995, Ann. Rev. Phys. Chem., 46, 27
Herbst, E., Klemperer, W. 1973, ApJ, 185, 505
Herbst, E., Leung, C.M. 1986, MNRAS, 222, 689
Herbst, E., Winnewisser, G. 1987, Topics in Current Chemistry, 139, 121
Hogerheijde, M.R., van Dishoeck, E.F., Blake, G.A., van Langevelde, H.J. 1997, ApJ, 489, 293
Hogerheijde, M.R., van Dishoeck, E.F., Blake, G.A., van Langevelde, H.J. 1998, ApJ, in press
Hollenbach, D.J. 1997, in *Herbig-Haro Objects and the Birth of Low-mass Stars, IAU Symposium 182*, ed. B. Reipurth & C. Bertout (Kluwer, Dordrecht), p. 181
Hollenbach, D.J., Salpeter, E.E. 1971, ApJ, 163, 155

Hollenbach, D.J., Tielens, A.G.G.M. 1997, ARAA, 35, 179

Howe, D.A., Taylor, S.D., Williams, D.A. 1996, MNRAS, 279, 143

Irvine, W.M., Goldsmith, P.F., Hjalmarson, Å 1987, in *Interstellar Processes*, ed. D.J. Hollenbach and H.A. Thronson Jr. (Reidel, Dordrecht, p. 561

Jansen, D.J., van Dishoeck, E.F., Black, J.H., Spaans, M., Sosin, C. 1995, A&A, 302, 223

Jura, M. 1975, ApJ, 197, 575

Kaiser, R.I., Stranges, D., Lee, Y.T., Suits, A.G., 1997, ApJ, 477, 982

Kaufman, M.J., Hollenbach, D.J., Tielens, A.G.G.M. 1998, ApJ, 497, 276

Kuan, Y.J., Snyder, L.E. 1996, ApJ, 470, 981

Kuiper, T.B.H., Langer, W.D., Velusamy, T. 1996, ApJ, 468, 761

Lacy, J.H., Carr, J.S., Evans, N.J. et al. 1991, ApJ, 376, 556

Lacy, J.H., Evans, N.J., Achtermann, J.M. et al. 1989, ApJ, 342, L43

Lacy, J.H., Knacke, R., Geballe, T.R., Tokunaga, A.T. 1994, ApJ, 428, L69

Lahuis, F., van Dishoeck, E.F. 1998, A&A, to be submitted

Lee, H.-H., Bettens, R.P.A., Herbst, E. 1996a, A&AS, 119, 111

Lee, H.-H., Herbst, E., Pineau des Forêts, G., Roueff, E., Le Bourlot, J. 1996b, A&A, 311, 690

Liseau, R., Ceccarelli, C., Larsson, B. et al. 1996, A&A, 315, L181

Lucas, R., Liszt, H. 1997, in *Molecules in Astrophysics: Probes and Processes*, IAU Symposium 178, ed. E.F. van Dishoeck, Kluwer, Dordrecht, 421

Macdonald, G.H., Gibb, A.G., Habing, R.J., Millar, T.J. 1996, A&AS, 119, 333

MacKay, D.D.S. 1995, MNRAS, 274, 694

Maloney, P.R., Hollenbach, D.J., Tielens, A.G.G.M. 1996, ApJ, 466, 561

McCall, B.J., Hinkle, K.H., Geballe, T.R., Oka, T. 1998, J. Chem Soc. Far. Disc., 109, in press

McCrea, W.H., McNally, D. 1960, MNRAS, 121, 238

Meyer, D.M. 1997, in *Molecules in Astrophysics: Probes and Processes*, IAU Symposium 178, ed. E.F. van Dishoeck, Kluwer, Dordrecht, 407

Meyer, D.M., Cardelli, J.A., Sofia, U.J. 1997, ApJ, 490, L103

Mezger, P.G., Sievers, A.W., Haslam, C.G.T. et al. 1992, A&A, 256, 631

Millar, T.J. 1990, in *Molecular Astrophysics —A volume honoring Alexander Dalgarno*, ed. T.W. Hartquist, Cambridge University Press, 115

Millar, T.J., Nejad, L.A.M. 1985, MNRAS, 217, 507

Millar, T.J., Bennett, A., Herbst, E. 1989, ApJ, 340, 906

Millar, T.J., Rawlings, J.M.C., Bennett, A., Brown, P.D., Charnley, S.B. 1991, A&AS, 87, 585

Millar, T.J., Macdonald, G.H., Habing, R.J. 1995, MNRAS, 273, 25

Millar, T.J., Farquhar, P.R.A., Willacy, K. 1997a, A&AS, 121, 139

Millar, T.J., Macdonald, G.H., Gibb, A.G. 1997b, A&A, 325, 1163

Millar, T.J., Hatchell, J. 1998, J. Chem Soc. Far. Disc., 109, in press

Miao, Y.T., Snyder, L.E. 1997, ApJ, 480, L67

Minh, Y.C., Ziurys, L.M., Irvine, W.M., McGonagle, D. 1991, ApJ, 366, 192

Mitchell, G.F., Maillard, J.-P., Allen, M., Beer, R., Belcourt, K. 1990, ApJ, 363, 554

Moore, M.H., Donn, B., Khanna, R., A'Hearn, M.F. 1983, Icarus, 54, 388

Motte, F., André, P., Neri, R. 1998, A&A, in press

Mundy, L., McMullin, J.P. 1997, in *Molecules in Astrophysics: Probes and Processes*, IAU Symposium 178, ed. E.F. van Dishoeck, Kluwer, Dordrecht, 183

Myers, P.C., Benson, P.J. 1983, ApJ, 266, 309

Nelson, R.P., Langer, W.D. 1997, ApJ, 482, 796

Neufeld, D.A., Dalgarno, A. 1989, ApJ, 340, 869; 344, 251

Ohishi, M. 1997, in *Molecules in Astrophysics: Probes and Processes*, IAU Symposium 178, ed. E.F. van Dishoeck, Kluwer, Dordrecht, 61

Olano, C.A., Walmsley, C.M., Wilson, T.L. 1988, A&A, 196, 194

Olmi, L., Cesaroni, R., Neri, R., Walmsley, C.M. 1996, A&A, 315, 565

Palumbo, M.E., Geballe, T.R., Tielens, A.G.G.M. 1997, ApJ, 479, 839

Pickles, J.B., Williams, D.A. 1977, Astrophys. Spa. Sci., 52, 443

Pineau des Forêts, G., Flower, D.R. 1997, in *Molecules in Astrophysics: Probes and Processes*, IAU Symposium 178, ed. E.F. van Dishoeck, Kluwer, Dordrecht, 113

Pironello, V., Biham, O., Liu, C., Shen, L., Vidali, G. 1997, ApJ, 483, L131

Plume, R., Bergin, E.A., Williams, J.P., Myers, P.C. 1997, J. Chem. Soc. Far. Disc. 109, in press

Prasad, S.S., Huntress, W.J. 1980, ApJS, 43, 1

Prasad, S.S., Tarafdar, S.P., Villere, K.R., Huntress, W.J., Jr. (1987, in *Interstellar Processes*, eds. D. Hollenbach and H.A. Thronson, D. Reidel, Dordrecht, 631

Prasad, S.S., Heere, K.R., Tarafdar, S.P. 1991, ApJ, 373, 123

Pratap, P., Dickens, J.E., Snell, R.L. et al. 1997, ApJ, 486, 862

Rawlings, J.M.C., Hartquist, T.W. 1997, ApJ, 487, 672

Rawlings, J.M.C., Hartquist, T.W., Menten, K.M., Williams, D.A. (1992, MNRAS, 255, 471

Reipurth, B., Bertout, C., ed., 1997, *Herbig-Haro objects and the birth of low-mass stars*, IAU Symposium 182, Kluwer, Dordrecht

Roberge, W.G., Jones, D., Lepp, S., Dalgarno, A. 1991, ApJS, 77, 287

Ruffle, D.P., Hartquist, T.W., Taylor, S.D., Williams, D.A. 1997, MNRAS, 291, 235

Schilke, P., Keene, J., Le Bourlot, J., Pineau des Forêts, G., Roueff, E. 1995, A&A, 294, L17

Schilke, P., Walmsley, C.M., Pineau des Forêts, G., Flower, D.R. 1997, A&A 321, 293

Schreyer, K., Helmich, F.P., van Dishoeck, E.F., Henning, T. 1997, A&A 326, 347

Schutte, W.A. 1996, in *Cosmic Dust Connection*, ed. J.M. Greenberg, Kluwer, Dordrecht

Schutte, W.A., Greenberg, J.M. 1991, A&A, 244, 190

Scott, G.B.I., Fairley, D.A., Freeman, C.G., McEwan, M.J. 1997, Chem. Phys. Lett., 269, 88

Shalabiea, O., Greenberg, J.M. 1994, A&A, 290, 266

Shalabiea, O., Greenberg, J.M. 1995, A&A, 296, 779

Shalabiea, O.M., Caselli, P., Herbst, E. 1998, ApJ, submitted

Shematovich, V.I., Shustov, B.M., Wiebe, D.S., 1997, MNRAS, 292, 601

Shu, F.H. 1977, ApJ, 214, 488

Shu, F.H., Najita, J., Galli, D., Ostriker, E., Lizano, S. 1993, in *Protostars & Planets III*, ed. E.H. Levy and J. Lunine, Tucson, Univ. of Arizona, p. 3

Smith, D., Španel, P. 1993, Int. J. Mass. Spectrom. Ion Proc., 129, 163

Smith, I.W.M. 1989, ApJ, 347, 282

Smith, I.W.M. 1997, in *Molecules in Astrophysics: Probes and Processes*, IAU Symposium 178, ed. E.F. van Dishoeck, Kluwer, Dordrecht, 253

Snow, T.P., Witt, A.N. 1996, ApJ, 468, L65

Snow, T.P., Hanson, M.M., Black, J.H., van Dishoeck, E.F., Crutcher, R.M., Lutz, B.L. 1998, ApJ, 396, L113

Sternberg, A., Yan, M., Dalgarno, A. 1997, in *Molecules in Astrophysics: Probes and Processes*, IAU Symposium 178, ed. E.F. van Dishoeck, Kluwer, Dordrecht, 141

Sundström, G., Mowat, J.R., Danared, H. et al. 1994, Science, 263, 785

Suzuki, H., Yamamoto, S., Ohishi, M. et al., 1992, ApJ, 392, 551

Swade, D.A. 1989, ApJ, 345, 828

Takahashi, J., Masuda, K., Nagaoka, M. 1998, ApJ, submitted

Tarafdar, S.P., Prasad, S.S., Huntress, W.T., Villere, K.R., Black, D.C. 1985, ApJ, 289, 220

Teixeira, T.C., Emerson, J.P., Palumbo, M.E., 1998, A&A, 330, 711

Tieftrunk, A., Pineau des Forêts, G., Schilke, P., Walmsley, C.M. 1994, A&A, 289, 579

Tielens, A.G.G.M. 1983, A&A, 119, 177

Tielens, A.G.G.M., Allamandola, L.J. 1987, in *Interstellar Processes*, eds. D. Hollenbach and H.A. Thronson, D. Reidel, Dordrecht, 379

Tielens, A.G.G.M., Charnley, S.B. 1997, Origin of Life, 27, 23

Tielens, A.G.G.M., Hagen, W. 1982, A&A, 114, 245

Tielens, A.G.G.M., Tokunaga, A.T., Geballe, T.R., Baas, F. 1991, ApJ, 381, 181
Tielens, A.G.G.M., Whittet, D.C.B. 1997, in *Molecules in Astrophysics: Probes and Processes*, IAU Symposium 178, ed. E.F. van Dishoeck, Kluwer, Dordrecht, 45
Turner, B.E. 1989, Space Science Rev., 51, 235.
Turner, B.E. 1995, ApJ, 449, 635.
Turner, B.E. 1996, ApJ, 468, 694.
Turner, B.E. 1998, ApJ, 495, 804
Turner, B.E., Lee, H.H., Herbst, E. 1998, ApJS, 115, 91
Turner, J.L., Welch, W.J. 1984, ApJ, 287, L81
Ungerechts, H., Bergin, E.A., Goldsmith, P.F., Irvine, W.M., Schloerb, F.P. and Snell, R.L. 1997, ApJ, 482, 245.
van de Hulst, H. 1997, in *Molecules in Astrophysics: probes and processes*, IAU Symposium 178, Dordrecht, Kluwer, p. 13
van der Tak, F., van Dishoeck, E.F., Evans, N.J., Blake, G.A. 1998, ApJ, submitted
van Dishoeck, E.F. 1988, in *Millimetre and Submillimetre Astronomy*, ed. R.D. Wolstencroft and W.B. Burton, Kluwer, Dordrecht, 117
van Dishoeck, E.F. 1990, in *Molecular Astrophysics —A volume honoring Alexander Dalgarno*, ed. T.W. Hartquist, Cambridge University Press, 55
van Dishoeck, E.F., ed., 1997, *Molecules in Astrophysics: probes and processes*, IAU Symposium 178, Dordrecht, Kluwer; For links to up-to-date lists of detected molecules and molecular data bases, see `http://www.strw.leidenuniv.nl/~iau34`
van Dishoeck, E.F. 1998a, In *The Molecular Astrophysics of Stars and Galaxies*, ed. T.W. Hartquist and D.A. Williams, Oxford, Oxford University
van Dishoeck, E.F. 1998b, J. Chem. Soc. Far. Disc., 109, in press
van Dishoeck, E.F., Black, J.H. 1988, ApJ, 334, 771
van Dishoeck, E.F., Blake, G.A. 1998, ARAA, 36, in press
van Dishoeck, E.F., Blake, G.A., Draine, B.T., Lunine, J.I. 1993, in *Protostars and Planets III*, eds. E.H. Levy and J.I. Lunine (Univ. of Arizona, p. 163
van Dishoeck, E.F., Helmich, F.P., de Graauw, Th. et al. 1996, A&A, 315, L349
Vejby–Christensen, L., Andersen, L.H., Heber, O., Kella, D., Pedersen, H.B., Schmidt, H., Zajfman, D. 1997, ApJ, 483, 531
Walmsley, C.M., Schilke, P. 1993, in *Dust and Chemistry in Astronomy*, eds. T.J. Millar and D.A. Williams, IOP Publishing, Bristol, 37
Ward-Thompson, D., Scott, P.F., Hills, R.E., André, P. 1994, MNRAS, 268, 276
Warin, S., Benayoun, J.J., Viala, Y.P. 1996, A&A, 308, 535
Watson, W.D. 1978, ARAA, 16, 585.
Watson, W.D., Salpeter, E.E. 1972, ApJ, 174, 321; 175, 659
Whittet, D.C.B. 1993, in *Dust and Chemistry in Astronomy*, eds. T.J. Millar and D.A. Williams, IOP Publishing, Bristol, 9.
Whittet, D.C.B., Gerakines, P.A., Tielens, A.G.G.M. et al. 1998, ApJ, 498, L159
Whittet, D.C.B., Schutte, W.A., Tielens, A.G.G.M. et al. 1996, A&A, 315, L357
Wiesemeyer, H., Güsten, R., Wink, J.E., Yorke, H.W. 1997, A&A, 320, 287
Willacy, K.R., Rawlings, J.M.C., Williams, D.A. 1994a, MNRAS, 269, 921
Willacy, K.R., Williams, D.A., Duley, W.W. 1994b, MNRAS, 267, 949
Williams, D.A. 1993, in *Dust and Chemistry in Astronomy*, eds. T.J. Millar and D.A. Williams, IOP Publishing, Bristol, 143
Williams, T.L., Adams, N.G., Babcock, L.M., Herd, C.R. and Geoghegan, M. 1996, MNRAS, 282, 413
Wilson, T.L., Mehringer, D.M., Dickel, H.R. 1995, A&A, 303, 840
Wootten, A. 1987, in *Astrochemistry*, eds. M.S. Vardya and S.P. Tarafdar, Kluwer, Dordrecht, 311
Wyrowski, F., Walmsley, C.M. 1996, A&A, 314, 265
Xie, T.L., Allen, M. and Langer, W.D. 1995, ApJ, 440, 674

# MODIFIED MODELS OF GAS/GRAIN INTERACTIONS

OSAMA M. SHALABIEA
*Department of Physics, The Ohio State University, Columbus, OH 43210, USA and Department of Astronomy, Faculty of Science, Cairo University, Cairo, Egypt*

ERIC HERBST
*Departments of Physics and Astronomy,
The Ohio State University, Columbus, OH 43210, USA*

AND

PAOLA CASELLI
*Osservatorio Astrofisico di Arcetri, Largo E. Fermi 5, I-50125 Firenze, Italy*

## 1. INTRODUCTION

It is customary to use rate equations for grain chemistry in gas-grain chemical models of interstellar clouds (e.g. d'Hendcourt et al. 1985; Hasegawa et al. 1992; Hasegawa & Herbst 1993; Shalabiea & Greenberg 1994). In this technique, the surface reaction rates can be expressed in terms of rate coefficients analogous to those in the gas phase. Work by Tielens (1995; see also Charnley et al. 1997) showed that, at least for a simple system, the rate equation method (hereafter RE) does not reproduce the Monte Carlo (hereafter MC) approach in selected instances for which the latter approach must be regarded as essentially exact. Recently, Caselli et al. (1998) duplicated the work of Tielens (1995) and suggested several modifications to the rate equations, which allow solutions that approach the MC values in the limit when the latter is correct. Caselli et al (1998) confined their attention to rather simple, unrealistic systems. The main aim of this work is to apply their modified rates to a more complex gas-grain model first studied by Hasegawa & Herbst (1993) and to determine if the results are any different from those previously obtained.



*J. M. Greenberg and A. Li (eds.), Formulation and Evolution of Solids in Space,* 123–130.

## 2. MODELS AND MODIFICATIONS OF THE RATE COEFFICIENTS

### 2.1. SIMPLE MODELS

The simplest system investigated by Caselli et al. (1998) contains a gas phase which consists entirely of atomic hydrogen and oxygen, with fixed total abundances $n(H)$ and $n(O)$. Upon sticking to grain surfaces, these atoms can undergo the following reactions only:

$$\begin{aligned} \mathrm{H{:}gr} + \mathrm{H{:}gr} &\longrightarrow \mathrm{H_2{:}gr} \\ \mathrm{O{:}gr} + \mathrm{O{:}gr} &\longrightarrow \mathrm{O_2{:}gr} \\ \mathrm{H{:}gr} + \mathrm{O{:}gr} &\longrightarrow \mathrm{OH{:}gr}. \end{aligned}$$

The three molecules formed are assumed to remain on the grains and to be unreactive. With the rate equation approach:

$$\begin{aligned} \frac{dN(H)}{dt} &= \kappa_{acc}(H)n(H) \quad -t_{evap}^{-1}(H)N(H) - \kappa_{\mathrm{H,H}}N(H)N(H) \\ &- \kappa_{\mathrm{H,O}}N(H)N(O), \end{aligned}$$

$$\frac{dN(O)}{dt} = \kappa_{acc}(O)n(O) - \kappa_{\mathrm{O,O}}N(O)N(O) - \kappa_{\mathrm{H,O}}N(H)N(O),$$

$$\frac{dN(H_2)}{dt} = 0.5\kappa_{\mathrm{H,H}}N(H)N(H),$$

$$\frac{dN(OH)}{dt} = \kappa_{\mathrm{H,O}}N(H)N(O),$$

$$\frac{dN(O_2)}{dt} = 0.5\kappa_{\mathrm{O,O}}N(O)N(O),$$

where the $N(I)$ are the number of species I per grain, and

- $\kappa_{acc}$ ($\mathrm{cm^3s^{-1}}$), the accretion rate for species I, is given by the cross section of the grain multiplied by the average speed of the species,
- $\mathrm{t}_{evap}^{-1}(\mathrm{H})$ is the inverse of the H evaporation time $\mathrm{t}_{evap}(\mathrm{H})$,
- $\kappa_{\mathrm{I,J}}$, the so called rate coefficients, are sums of inverse diffusion times for sweeping of the entire grain ($t_s$) by species $I$ and $J$ via tunneling or thermal hopping (Hasegawa et al. 1992):

$$\kappa_{\mathrm{I,J}} = t_s^{-1}(I) + t_s^{-1}(J).$$

Specific values of $\kappa_{acc}(\mathrm{H})$, $\kappa_{acc}(\mathrm{O})$, $\mathrm{t}_s^{-1}(\mathrm{H})$, $\mathrm{t}_s^{-1}(\mathrm{O})$, and $\mathrm{t}_{evap}^{-1}(\mathrm{H})$ for a 10 K cloud and a standard grain size are given by Hasegawa et al. (1992). These values have been used here. A comparison between the RE and MC results for this simple model is shown in Fig. 1. In this figure, the steady-state fractional abundance relative to the total number of surface species (mole fraction; X) of surface $O_2$, $H_2$, and OH is plotted versus the gas phase number density $n(H)$ with a fixed value of $n(O) = 1.0$ cm$^{-3}$. Fig. 1(a) presents the RE results, while Fig. 1(b) presents the MC results, in which H and O atoms land randomly on an individual grain according to the grain collision frequency.

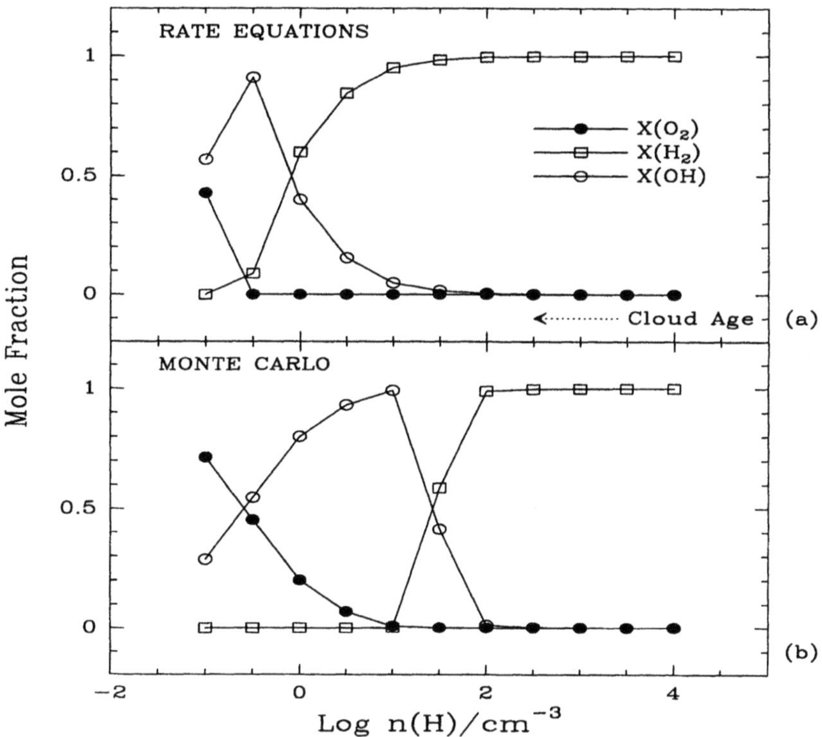


*Figure 1.* Fractional abundances relative to the total number of surface species (mole fraction; X) of surface $O_2$, $H_2$ and OH versus gas phase number density of atomic hydrogen, $n(H)$, from (a) rate equation calculations, and (b) Monte Carlo calculations. Reproduced with permission from The Astrophysical Journal by the American Astronomical Society.

For both sets of results, it can be seen that at very low $n(H)$, the dominant species is $O_2$ (note that lower values of $n(H)$ are required for the RE approach than depicted here); at intermediate $n(H)$, the dominant species is OH; while at high $n(H)$, the dominant species is $H_2$. The "phase transitions" – the values of $n(H)$ at which one molecule takes over from another as dominant species – are very different for the RE and MC approaches, with the values for the RE approach occurring at much smaller values of

$n(H)$. Clearly as shown by Tielens (1995) and Caselli et al. (1998) there is a large difference between these two approaches. The RE approach tends to make hydrogenation overly efficient compared with the MC approach.

Caselli et al. (1998) made several modifications to the above rate equations to bring the results for the simple system into agreement with the MC results. The modifications are mainly to those surface rate coefficients for reactions involving H atoms. The first modification to these rate coefficients is an adjustment so that the probability of reaction between an adsorbed H and a second atom or molecule X in a time period $\Delta t$,

$$P = \kappa_{\mathrm{H,X}} \Delta t$$

cannot be greater than unity, where $\Delta$t is the smaller of the accretion and evaporation times of H. The accretion time $t_{acc}$ for species H is defined by the equation

$$t_{acc}(H) = \frac{1}{\kappa_{acc}(H)n(H)}.$$

Letting $x$ be the ratio between the inverse evaporation and accretion times for atomic hydrogen:

$$x = t_{evap}^{-1}(\mathrm{H})/t_{acc}^{-1}(H)$$

the modified rate coefficients of Caselli et al. (1998) have different values depending upon whether or not $x$ is greater than or less than unity. In the former case, the inverse grain diffusion rate for H – $t_s^{-1}(H)$ – is replaced, in the rate coefficient expressions, by the inverse evaporation time, whereas for the latter case, it is replaced by the inverse accretion time.

Secondly, Caselli et al. (1998) strongly modified the rate coefficient for $H_2$ formation in the limit that $x \geq 1$ to the following:

$$\kappa'_{\mathrm{H,H}} = \kappa_{\mathrm{H,H}} \times x \times exp(-x).$$

The modification to the $H_2$ formation rate adjusts for the fact that one H atom can evaporate before another lands on the same grain.

A third (minor) modification corrects for an apparent overly strong reduction in the rate of hydrogenation induced by the first correction when $x \geq 1$. This is shown below for the specific case of the reaction between H and O:

$$\kappa'_{\mathrm{H,O}} = \kappa_{\mathrm{H,O}} + \kappa_{\mathrm{H,O}} \times \frac{t_{acc}^{-1}(O)}{t_{acc}^{-1}(H)}.$$

In addition to these three modifications, there are additional modifications provided for cases of reactions involving an abundant grain species (e.g. CO, $H_2$) when the activation energy is low but non-negligible.

When the modifications to the RE approach are applied and used for the simple O-H system, the agreement with the MC approach is excellent. A high level of agreement between the modified RE and MC treatments was also found by Caselli et al. (1998) for several other simple systems.

## 2.2. COMPLEX MODELS

In order to investigate how much the previous suggested modifications affect the calculated abundances for a large and much more complex model, we have constructed a gas-grain model by combining the new standard gas-phase model of Lee et al. (1996) with the grain reactions of Hasegawa & Herbst (1993). This model has a total of 656 species with 4742 reactions. Desorption via thermal evaporation and cosmic ray heating is included. Hasegawa & Herbst (1993) considered two activation energies for grain-surface reactions involving atoms or radicals and $H_2$; we have used the higher activation energy here, which has the effect of reducing the importance of molecular hydrogen in grain surface chemistry. As initial conditions, we have chosen to have hydrogen predominantly in its molecular form: viz., $n(H_2) \simeq 10^4$ cm$^{-3}$ and $n(H) = 1$ cm$^{-3}$. We have run time-dependent gas-grain models using the unmodified grain surface rate equations (hereafter UM) and with the appropriate modifications of Caselli et al. (1998; hereafter MM). Our main goal has been to determine whether or not there are any differences rather than to treat the modifications in great detail. Given the small amount of gaseous atomic hydrogen expected at all times, the parameter $x$ will be significantly greater than unity, so that the H atom diffusion rate will be replaced by its evaporation rate in the modified rate equations. With such a large value of $x$, $x \times \exp(-x)$ is essentially zero, so that there is no $H_2$ formation from two H atoms and $H_2$ can only be generated by reactions such as

$$H + HX \longrightarrow H_2 + X.$$

Our results for both the UM and MM models at three different times are shown in Table 1, where we list fractional abundances of assorted species with respect to $n(H_2)$ both in the gas and on the surfaces of grains. The three times listed are an "early time (ET)" of 100 yr, a "middle time (MT)" of $10^5$ yr, and a "late time (LT)" of $10^7$ yr. Central to understanding the results are the gas-phase abundances of atomic oxygen and hydrogen; these are plotted vs. time in Fig. 2. Although both O and H start off with appreciable (and reasonably similar) abundances, as time progresses, gas-phase atomic O begins to disappear along with most heavy species, and at late times, there is much more H in the gas than O. Based on the results of Caselli et al. (1998), we expect therefore that there should be less difference between the MM and UM models at late times, since we are in a situation where the H atomic abundance dominates that of heavy atoms such as O, which corresponds to the far right of Fig. 1. Here hydrogenation will dominate the grain-surface chemistry. At early times, however, we expect that the modifications of Caselli et al. (1998) will lead to substantial changes, generally in the direction towards less hydrogenation.

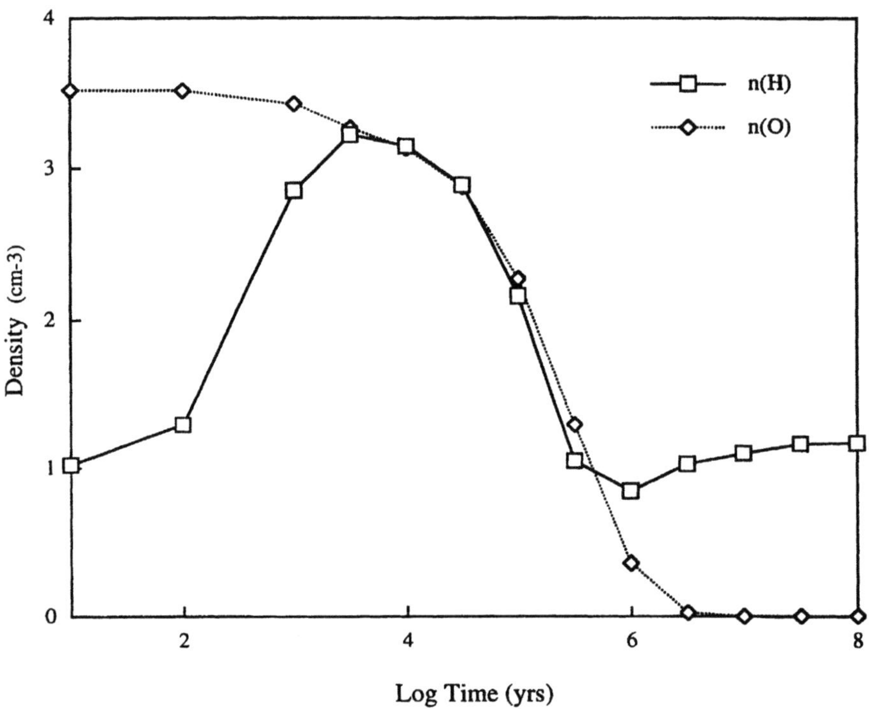


*Figure 2.* Atomic O and H abundances ($cm^{-3}$) plotted against evolution time for the so-called modified model (MM).

If one looks carefully at the results in Table 1, one can indeed see that the largest discrepancies for surface abundances of synthesized molecules between the MM and the UM models are at early time, as expected, and that the discrepancies tend to become much smaller as the time progresses. A full analysis of these and other results will be published later.

TABLE 1. The relative $n(X)/n(H_2)$ abundances of some species for the UM and MM models at early (ET, $10^2$yrs), middle (MT,$10^5$yrs) and late (LT, $10^7$yrs) times.

| Species | ET | | MT | | LT | |
|---|---|---|---|---|---|---|
| | gas | surface | gas | surface | gas | surface |
| $N_2$ | *1.64E-13 | 5.051E-13 | 6.05E-08 | 1.46E-06 | 1.55E-06 | 1.09E-05 |
| | **2.14E-13 | 3.157E-09 | 5.82E-08 | 1.79E-06 | 1.48E-06 | 1.28E-05 |
| $O_2$ | 4.33E-13 | 1.277E-16 | 8.40E-07 | 9.58E-12 | 1.54E-11 | 1.22E-14 |
| | 4.33E-13 | 1.096E-12 | 8.29E-07 | 9.27E-12 | 1.63E-11 | 5.40E-14 |
| $H_2O$ | 3.78E-11 | 5.299E-08 | 9.00E-07 | 4.10E-05 | 5.40E-10 | 2.30E-04 |
| | 3.77E-11 | 4.755E-08 | 8.70E-07 | 3.90E-05 | 4.40E-10 | 2.30E-04 |
| $CH_4$ | 2.79E-10 | 2.454E-09 | 1.00E-06 | 1.20E-05 | 3.40E-09 | 1.40E-06 |
| | 2.79E-10 | 1.319E-09 | 1.00E-06 | 9.40E-06 | 1.20E-08 | 4.50E-06 |
| $NH_3$ | 2.55E-14 | 6.886E-09 | 5.20E-08 | 2.00E-06 | 8.10E-09 | 1.50E-07 |
| | 6.46E-16 | 2.801E-11 | 9.10E-09 | 2.40E-07 | 1.30E-08 | 1.70E-07 |
| $CO_2$ | 3.05E-13 | 2.444E-17 | 1.70E-07 | 4.80E-08 | 8.00E-16 | 9.80E-07 |
| | 3.05E-13 | 3.203E-13 | 1.60E-07 | 2.70E-07 | 1.80E-15 | 1.60E-06 |
| $H_2CO$ | 1.39E-12 | 5.641E-12 | 6.80E-08 | 2.10E-06 | 1.00E-11 | 4.10E-05 |
| | 1.39E-12 | 1.602E-10 | 6.60E-08 | 3.50E-06 | 2.30E-11 | 6.10E-05 |
| $CH_3OH$ | 8.71E-18 | 5.559E-12 | 1.10E-09 | 4.20E-06 | 4.00E-13 | 6.15E-05 |
| | 8.71E-18 | 4.498E-10 | 1.00E-09 | 3.90E-06 | 4.60E-13 | 6.22E-05 |

UM is the Unmodified model
** MM is the Modified model

## 3. CONCLUSIONS

The modifications to the grain rate equations suggested by Caselli et al. (1998) so that the rate equation method obtains the same abundances as the Monte Carlo approach for simple systems has been used in a complex gas-grain model of dense interstellar clouds. Differences between calculated surface abundances of assorted species for a model in which the unmodified rate equations are used and a modified model are significant mainly at early and middle times, when the gas phase contains large abundances of heavy species that are reactive when they land on grain surfaces. At later times, atomic H becomes the only major gas-phase species capable of rapid reaction when it lands on grain surfaces so that hydrogenation is dominant; a condition which appears to be accurately handled by the unmodified treatment. Much more work is needed to verify our conclusions for a wide range of gas-grain chemical models with varying initial conditions and more efficient desorption mechanisms. It is expected that more efficient desorption will lead to larger differences between the UM and MM approaches because there will be more heavy reactive species available to land on grains at late times.

## 4. Acknowledgements

EH acknowledges the support of the National Science Foundation for his research programme in astro-chemistry. Funds from this source of support have been used to pay for the visit of OMS to The Ohio State University.

## References

Caselli, P., Hasegawa, T. I., & Herbst, E. 1998, ApJ, in press
Charnley, S. B., Tielens, A. G. G. M., & Rodgers, S. D. 1997, ApJ, 482, L203
Hasegawa, T.I., Herbst, E., & Leung, C.M. 1992, ApJS, 82, 167
Hasegawa, T.I., & Herbst, E. 1993, MNRAS, 261, 83
Hasegawa, T. I., Herbst, E., & Leung, C. M. 1992, ApJS, 82, 167
d'Hendecourt, L.B., Allamandola, L.J., & Greenberg, J.M. 1985, A&A, 152, 130
Lee, H.-H., Bettens, R. P. A., & Herbst, E. 1996, A&AS, 119, 111
Shalabiea, O.M. & Greenberg, J.M. 1994, A&A, 290, 266
Tielens, A. G. G. M. 1995, Lecture at a conference held at Leiden University

# THE ORGANIC CHEMISTRY OF TMC-1 REVISITED

S.B. CHARNLEY
*Space Science Division, NASA Ames Research Center Moffett Field, CA 94035, and Astronomy Department, University of California, Berkeley, CA 94720.*

**Abstract.**
I reconsider the formation of complex interstellar molecules in the dark cloud TMC-1.

## 1. Introduction

The dark molecular cloud TMC-1 has been the focus of intense observational and theoretical effort (van Dishoeck et al. 1993). The correct explanation for both the origin of its rich organic chemistry (e.g. Ohishi et al. 1992), and the spatial gradients in its chemical abundances (e.g. Olano et al. 1988), is one of the central problems in astrochemistry. Table 1 lists the molecules detected at the so-called 'cyanopolyyne peak' (CP) where organic species have their peak abundances. Other notable positions in the cloud are the 'ammonia peak' and the '$DCO^+$ peak' which are offset to the northeast and where the organic emission is weaker; conversely, the $NH_3$ and $DCO^+$ emission progressively declines as one moves down the ridge towards the CP. As no other dark cloud is currently known to exhibit the organic diversity seen in TMC-1, it is likely that the organic chemistry and the abundance gradients have a common cause. The challenge is to explain the observations with a theory possessing the miniumum number of assumptions. Many chemical schemes and cloud models have been proposed to account for the TMC-1 chemistry, however, no concensus has been reached on either issue.

In this contribution, I qualitatively outline a chemical scenario, based on work aimed at explaining the organic chemistry of star-forming cores, that accounts for the molecular diversity observed in TMC-1 with a minimal number of reaction mechanisms.



*J. M. Greenberg and A. Li (eds.), Formulation and Evolution of Solids in Space,* 131–137.

TABLE 1. Molecules detected at the TMC-1 'cyanopolyyne peak'

| | | | | | | |
|---|---|---|---|---|---|---|
| $CO$ | $C_2O$ | $C_3O$ | $CH_2C_2$ | $CH_2C_3$ | $CH_2C_5$ | |
| $CH$ | $C_2H$ | $C_3H$ | $C_4H$ | $C_5H$ | $C_6H$ | $C_8H$ |
| $HC_3N$ | $HC_5N$ | $HC_7N$ | $HC_9N$ | $HC_{11}N$ | | |
| $CN$ | $HCN$ | $CH_2CN$ | $C_3N$ | $CH_3CN$ | $CH_2CHCN$ | |
| $CH_3CCH$ | $CH_3C_4H$ | $CH_3C_3N$ | | | | |
| $H_2CO$ | $CH_3OH$ | $HNCO$ | $HCOOH^{\dagger}$ | $CH_2CO$ | $CH_3CHO$ | $HC_2CHO$ |

Based on Ohishi et al. (1992) and Ohishi (1996).
† upper limit only.

## 2. Organic reaction pathways

Perhaps the most intriguing facet of Table 1 is that the 'cyanopolyyne peak', ostensibly the richest interstellar source of C-bearing molecules, also contains several O-bearing organic molecules. Apart from methanol, all these molecules contain the carbonyl (C=O) group.

### 2.1. O-BEARING MOLECULES: GRAIN-SURFACE REACTIONS

Charnley (1997a,b) has proposed a theory for the formation of oxygen-bearing organic molecules on grain surfaces involving C-atom additions to *stable* organic radicals. This scheme is shown schematically in Figure 1 where the intermediate radicals are produced by H atom addition to carbon double and triple bonds, beginning with the formation of HCO from CO (Tielens & Hagen 1982; Tielens & Allamandola 1987). Carbon chain growth leads to a ketene backbone: large linear molecules with a carbonyl and an acetylenic functional group at either end. Hydrogen atom additions evolve the mantle composition from ketenes through aldehydes and on to alcohols, in line with a previous suggestion that mantle alcohols are the saturated end point of grain surface chemistry (Charnley et al. 1995).

Comparing Table 1 and Figure 1, it is difficult to escape the conclusion that grain mantles have recently been disrupted at the TMC-1 'cyanopolyyne peak'. The non-detection of $NH_2CHO$ in TMC-1 (Irvine et al. 1989), as well as the lower abundances of other species (e.g. $CH_3OH$ , Ohishi et al 1992), suggests that dark cloud mantles are less reduced relative to those evaporated in hot cores. This route to methanol ice formation has been considered previously (see Tielens & Allamandola 1987), however, experiments (Hiraoka et al. 1994) and observations (Chiar et al. 1996) indicate that cold H atom addition to $H_2CO$ is an inefficient process on 10K grains.

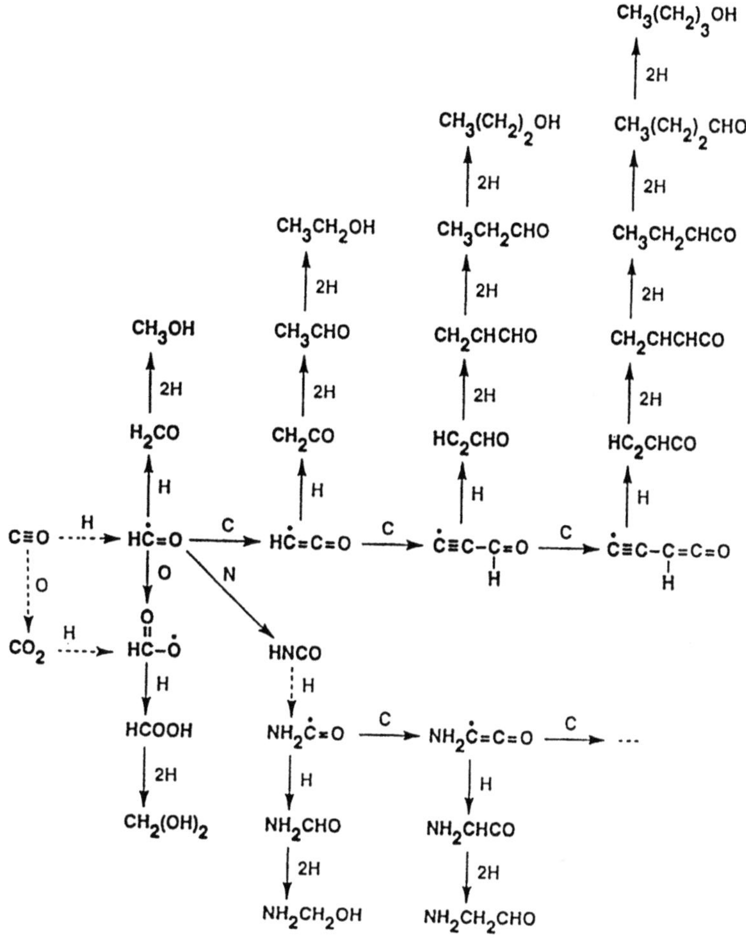


*Figure 1.* The composition of interstellar grain mantles. The pristine mantle comprises of simple molecules accreted from the gas or formed by exothermic H atom additions. Energetic processing allows H atom addition to unsaturated molecules and a rich organic chemistry seeded by CO ensues. Broken arrows indicate reactions with activation energy barriers; where 2H is shown, a barrier penetration reaction followed by an exothermic addition is implicitly indicated (adapted from Charnley 1997a)

Indeed, the experiments of Hiraoka et al. yield a solid $H_2CO:CH_3OH$ ratio of 5:1 which is larger than that found at the TMC-1 CP (10:1, Ohishi et al. 1992). This may therefore be the highest ratio attainable in quiescent dark clouds by cold H atom tunnelling, it indicates that, as the lowest $H_2CO:CH_3OH$ ratios ($>$ 1:100) and highest solid and gas phase methanol abundances are observed only towards luminous protostellar regions, a connection exists between the local energy density of the environment and the conversion efficiency of CO to $CH_3OH$ on dust grains. Figure 1 suggests that HCOOH and $C_2H_5OH$ should be present at the CP and, as propioaldehyde ($HC_2CHO$) is detected, suggests that acrolein ($CH_2CHCHO$) and propionaldehyde ($CH_3CH_2CHO$) may also be detectable.

### 2.2. CYANOPOLYYNES AND POLYACETYLENES: NEUTRAL-RADICAL GAS PHASE REACTIONS

Based on laboratory data, Herbst & Leung (1990) proposed that reaction of CN with $C_2H_2$ and $C_2H_4$ would lead to $HC_3N$ and $CH_2CHCN$ in TMC-1. Cherchneff & Glassgold (1993) considered the formation of cyanopolyynes and polyacetylenes in the envelope of the carbon star IRC+10216 and proposed a generalized network of reactions based upon laboratory experiments:

$$C + C_nH_m \longrightarrow C_{n+1}H_{m-1} + H$$

$$C_{2n+1}N + C_{2m}H_2 \longrightarrow HC_{2(n+m)+1}N + H \quad (n \leq 4; m \leq 5)$$

$$C_2H_k + C_mH_n \longrightarrow C_{m+2}H_{n+k-1} + H \quad (k = 0, 1)$$

Application of a similar neutral-radical reaction network to complex molecule formation in TMC-1, however, lowered the computed abundances of complex organics, particularly because the high neutral carbon abundances suppressed acetylene formation, as well as the inclusion of many hypothesised O-atom destruction reactions (Herbst et al. 1994).

Early in the chemical evolution of TMC-1, one million years or so, high abundances of $C_2H_2$ and $C_2H_4$ should form and be accreted into grain mantles (Brown & Charnley 1990; Herbst et al. 1994). Hydrogenation of solid $C_2H_2$ could also lead to $C_2H_4$ (Tielens 1992). Disruption and injection of these mantles would dramatically increase the gas phase abundances of $C_2H_2$ and $C_2H_4$ and perhaps transiently boost gas phase production of the cyanopolyynes and other carbon chain species, despite the efficacy of reactions involving atomic oxygen and carbon in inhibiting the development of complexity.

### 2.3. METHYLATED CHAINS: ION-NEUTRAL GAS PHASE REACTIONS

The methanol abundance in TMC-1 ($\sim 2 \times 10^{-9}$, Ohishi et al. 1992) is much lower than observed in regions of massive star formation hot cores ($\sim 10^{-7} - 10^{-6}$, e.g. Menten et al. 1988) but only a factor of ten less than observed in dark clouds containing low-mass protostars ($\sim 10^{-8}$), such as B1 (Bachiller et al. 1990). As remarked upon above, this is probably due in large measure to the degree of mantle saturation, however, this does not rule out the possibility that the mantles originally disrupted at the CP contained more $CH_3OH$ and that gas phase reactions have 'eroded' this pristine abundance. A similar picture has been proposed for the ethanol chemistry of hot cores (Charnley et al. 1995); in TMC-1 the methanol excess may have driven reaction sequences in which methyl cation transfer from protonated methanol produced several of the observed organics:

$$C_2H_2 \xrightarrow{CH_3OH_2^+} CH_3CCH_2^+ \xrightarrow{e^-} CH_3CCH$$

$$HC_3N \xrightarrow{CH_3OH_2^+} CH_3C_3NH^+ \xrightarrow{e^-} CH_3C_3N$$

$$C_2H_2 \xrightarrow{H_3^+} C_2H_3^+ \xrightarrow{e^-} C_2H \xrightarrow{C_2H_2} C_4H_2$$

$$C_4H_2 \xrightarrow{CH_3OH_2^+} CH_3C_4H_2^+ \xrightarrow{e^-} CH_3C_4H$$

Methylation of the higher cyanopolyynes should therefore produce species such as $CH_3C_5N$ and $CH_3C_7N$.

### 2.4. CUMULENE HYDROCARBONS AND $C_NO$: SURFACE CHAIN FRAGMENTATION

If carbon chain growth is favoured over hydrogenation in the putative TMC-1 mantles, then, upon injection into the gas phase, chemical fragmentation of the large ketene molecules, i.e. by protonation followed by dissociative electron recombination, could be a source of organics. Although highly speculative, various interesting product branchings could occur in principle, for example

$$[H - C \equiv C - (CH) = C = C = O\,]H^+ \xrightarrow{e^-} CH_2C_2 \;+\; C_2O \;+\; H$$

The products and associated branching ratios are unknown.

## 3. Conclusions

There is strong circumstantial evidence that grain mantles have been removed at the 'cyanopolyyne peak' in TMC-1. The existence of a distribution of activation energy barriers involving peneration of $C \equiv C$, $C = C$, as well as $C = O$ bonds, naturally explains the differences in mantle composition between cold dark clouds and energetic protostellar environments, as being due to varying degrees of mantle saturation. In TMC-1, mantle disruption may be due to the action of MHD waves - the relative drift of charged and neutral grains could be high enough to lead to collision energies where icy mantles exceed the critical temperature for radical explosion to occur (Greenberg 1997). There is some evidence that MHD waves are affecting the chemistry of TMC-1 (Charnley & Butner 1995).

The sudden release of previously accreted $C_2H_2$ and $C_2H_4$ may lead to the production, by previously considered chemical reactions, of the observed cyanopolyynes and polyacetylenes. Methylation of neutral molecules and the gas phase fragmentation of large surface-formed molecules, having the structure $H - C \equiv C - (CH) = C = C = C = ... = C = C = O$, are additional pathways to the observed organics. The gas phase schemes presented in this paper are qualitative, detailed chemical kinetic modelling is presently underway to test the feasibility of these ideas (Millar & Charnley 1997).

## Acknowledgements

Theoretical astrochemistry at NASA Ames is supported by NASA's Origins of Solar Systems and Exobiology Programs (RTOP nos. 344-37-22-04 & 344-38-12-14). Funds for the support of this study have been allocated by NASA Ames under Interchange No. NCC2-1592.

## References

Bachiller, R. et al. 1990, Anatomy of a dark cloud - A multimolecular study of Barnard 1 , *A&A*, **236**, 461

Brown, P.D. and Charnley, S.B. 1990, Chemical Models of Interstellar Gas-Grain Processes I. , *MNRAS*, **244**, 432

Charnley, S.B. 1997a, On the Nature of Interstellar Organic Chemistry, in C.B. Cosmovici, S. Bowyer and D. Wertheimer (Eds.), *Astronomical & Biochemical Origins and the Search for Life in the Universe*, (Editrice Compositori; Bologna), 89

Charnley, S.B. 1997b, in preparation

Charnley, S.B. and Butner, H. 1995, Molecular Signatures of MHD Waves , *Astrophys. Space. Sci.*, **224**, 443

Charnley, S.B., Kress, M.E., Tielens, A.G.G.M. and Millar, T.J. 1995, Interstellar Alcohols , *ApJ*, **448**, 232

Cherchneff, I., Glassgold, A.E. 1993, The Formation of Carbon Chain Molecules in IRC +10216 , *ApJ*, **419**, L41

Chiar, J. et al. 1996 , Three Micron Hydrocarbon and Methanol Absorption in Taurus , *ApJ*, **472**, 665

Greenberg, J.M. 1997, private communication

Herbst, E. and Leung, C.M. 1990 , The gas phase production of $CH_2CN$ and other organo-nitrogen species in dense interstellar clouds , *A&A*, **233**, 177

Herbst, E., et al. 1994, The Effect of Rapid Neutral-Neutral Reactions on Chemical Models of Dense Interstellar Clouds , *MNRAS*, **335**, 344

Hiraoka, K., Ohashi, N., Kihara, Y., Yamamoto, K., Sato, T., and Yamashita, A., 1994, Chem. Phys. Letters 229, 408

Hiraoka, K., Ohashi, N., Kihara, Y., Yamamoto, K., Sato, T., and Yamashita, A., 1994, Formation of Formaldehyde and Methanol from the Reactions of H atoms with Solid

CO at 10-20K , *Chem. Phys. Lett.*, **229**, 408

Irvine, W.M. et al. 1989, Observations of some oxygen-containing and sulfur-containing organic molecules in cold dark clouds , *ApJ*, **342**, 871

Menten, K. et al. 1988, Methanol in the Orion region. I - Millimeter-wave observations , *A&A*, **198**, 253

Millar, T.J. and Charnley, S.B. 1997, in preparation

Ohishi, M., Irvine, W.M. and Kaifu, N. 1992, Molecular Abundance Variations Among and Within Cold, Dark Molecular Clouds, in P.D. Singh (ed.), *Astrochemistry of Cosmic Phenomena*, (Kluwer, Dordrecht), 171

Ohishi, M. 1996, Observations of Hot Cores, in E.F. van Dishoeck (ed.), *Molecules in Astrophysics: Probes and Processes*, (Kluwer, Dordrecht), 61

Olano, C.A. et al. 1988, The relative distribution of $NH_3$, $HC_7N$ and $C_4H$ in the Taurus Molecular Cloud 1 (TMC-1) , *A&A*, **196**, 194

Tielens, A.G.G.M. 1992, Grain Surface Chemistry, in N. Kaifu (ed.), *Chemistry and Spectroscopy of Interstellar Molecules*, (University of Tokyo Press), 237

Tielens, A.G.G.M. and Allamandola, L.J., 1987, Composition, Structure and Chemistry of Interstellar Dust, in D. Hollenbach and H. Thronson (eds.), *Interstellar Processes*, (Reidel, Dordrecht), p.397

Tielens, A.G.G.M. and Hagen, W. 1982, Model Calculations of the Molecular Composition of Interstellar Grain Mantles , *A&A*, **114**, 245

van Dishoeck, E.F., Blake, G.A., Draine, B.T. and Lunine, J.I. 1993, The Chemical Evolution of Protostellar and Protoplanetary Matter, in E.H. Levy and J.I. Lunine (eds.), *Protostars and Planets III*, (Tucson:University of Arizona), 163

# MOLECULAR GAS PHASE COUNTERPARTS TO SOLID STATE GRAIN MANTLE FEATURES

*implication for gas/grain chemistry*

E. DARTOIS, L. D'HENDECOURT, F. BOULANGER AND
J.-L. PUGET *IAS-CNRS, Université Paris XI, Orsay Cedex, France*

M. JOURDAIN DE MUIZON
*DESPA, Observatoire de Paris, Meudon, France*
*LAEFF-INTA, ESA Satellite tracking Station, Madrid, Spain*

M. BREITFELLNER
*ISO Science Operations, ESA/VILSPA, Madrid, Spain*

AND

H. J. HABING
*Leiden Observatory, Leiden, The Netherlands*

**Abstract.** ISO SWS (AOT6) observations of a protostellar object RAFGL 7009S are presented. This source is extremely promising for the chemistry arising from grain surfaces and mantles, since many solid state features of simple molecules are detected such as $H_2O$, CO, $CO_2$, $^{13}CO_2$ and $CH_4$, together with weaker ones not yet unambiguously identified. Gas phase $H_2O$, $CO_2$, $CH_4$ and CO can be seen superimposed on the solid state features. We derive an estimate for the gas phase temperature around 50 K. The gas to solid ratios of these simple molecules is found to be $\sim$0.2 for $H_2O$ and $CH_4$ and an order of magnitude lower for $CO_2$. We discuss the possible implications for interstellar chemistry in such objects.

## 1. Introduction

Star formation takes generally place in molecular clouds where the environment is dense and well radiation shielded, developping a cold temperature gas-grain chemistry. The new star formed provides a background source which allows to probe cold matter in absorption both in the gas and on grains. Surface reactions and UV photoprocessing of icy mantles adsorbed on the grains are important processes for interstellar chemistry. Such processes can be analysed and understood by comparing the observations to laboratory simulations which allow to reproduce precisely the interstellar



*J. M. Greenberg and A. Li (eds.), Formulation and Evolution of Solids in Space,* 139–145.

spectra. Simple interstellar gas phase molecules such as $H_2O$ and $CO_2$ have not been extensively observed in the infrared with ground based telescopes since their rovibrational transitions are hidden by strong telluric absorptions. Furthermore, linear and symetric molecules present weak radio transitions and their abundances can only be inferred from chemical models or through reactions involving other species detectable at radio wavelength. In particular, prior to ISO observations, $CO_2$ gas phase abundance was inferred from observations of the related ion $HCO_2^+$. The first results we obtained on the protostellar object RAFGL 7009S with ISO (d'Hendecourt et al.,1996) were the identification of solid state molecular features : $H_2O$, CO, $CO_2$, $^{13}CO_2$, $CH_4$, "XCN" as well as a deep and broad absorption feature at 6.8 $\mu$m. The "silicate" bands at 9.7 (saturated) and 18 $\mu$m are also prominent in this object. Higher resolution observations show rovibrational lines of gas phase $H_2O$, CO, $CO_2$ and $CH_4$. We use the extensive HITRAN rovibrational lines database (Rothman et al, 1996) as well as our laboratory spectra which help us to disentangle the gas phase lines from the solid ones. We then derive the different abundances of these molecules both in the gas and solid phases and discuss the implications for the chemistry in protostellar environments.

## 2. Results

*Gas phase $CO_2$ band.* The $\nu_2$ bending mode of $CO_2$ is detected in the ISO spectrum of RAFGL 7009S slightly shifted compared to the $CO_2$ ice band as previously published by van Dishoeck et al. (1996). The result is shown in Fig. 1A. In addition to P and R branches, the $CO_2$ $\nu_2$ mode presents a Q branch which presents a single feature at SWS resolution (~1500) with a fairly strong integrated absorption coefficient. Furthermore, due to the important opacity present in the source which results in a high gas column density, P and R branches are also seen although the P branch suffers from the lower S/N in the region of the solid phase absorption. A model was constructed using the HITRAN 96 database constants for the $\nu_2$ bending mode line positions (Rothman et al 1996.) and Reichle and Young's integrated absorption coefficient (1972). We first use a voight profile and assume an excitation temperature $T_{ex}$ with LTE conditions for the gas. We convolve the model spectrum with a gaussian having the ISO-SWS06 resolution to allow comparison. The spectrum is then compared to the observation after extraction of the continuum level and assuming a given temperature (40K). We have little constrain on the temperature reached by the gas but we can exclude a 200 K excitation temperature from the asymmetry it would produce on the Q branch as shown in the Fig. 1B. A turbulent velocity of $v = 3$ km s$^{-1}$ is taken based on various CO isotopes ra-

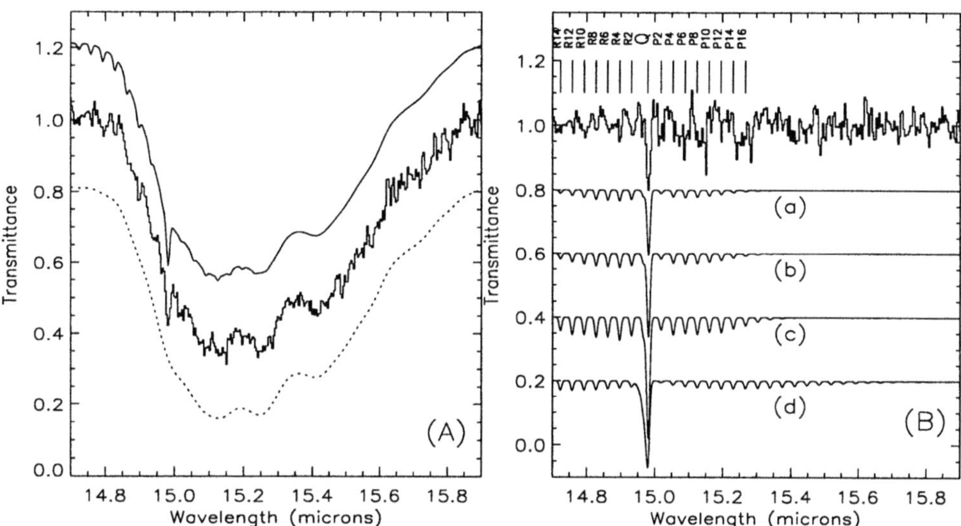


*Figure 1.* Gas phase $CO_2$ $\nu_2$ band Observed in RAFGL 7009S and model spectra. Panel (A): (upper curve) model spectra made using gas (1x$10^{17}$ molecule.cm$^{-2}$) and solid phase estimate, (middle curve) ISO SWS06 data and adopted solid phase transmittance (lower curve). Panel (B): ISO transmittance spectrum after removal of the solid phase "baseline" and model spectra with $v = 3$ km s$^{-1}$, $T$=40K and (a) 5x$10^{16}$ molecule.cm$^{-2}$, (b) 1x$10^{17}$ molecule.cm$^{-2}$ , (c) 5x$10^{17}$ molecule.cm$^{-2}$, (d) $T$=200K and 1x$10^{17}$ molecule.cm$^{-2}$. The vertical lines indicate the different transitions involved.

dio lines ($^{13}C^{16}O$, $^{12}C^{18}O$) (McCutcheon et al. 1991) and is typical for such environments. The determination of the gas column density is performed in the following manner : we compute gas transmittance spectra with increasing number of $CO_2$ molecules and divide the ISO reduced spectrum by the expected transmittance from the model. We then calculate the variance of the division. The variance diminishes until it reaches a minimum governed by the signal-to-noise, column density and model. The best fit is obtained with $N_{CO_2} \approx 1^{+1}_{-0.5}.10^{17}$ molecules.cm$^{-2}$ (see Fig. 1.). The final result is given in Fig. 1A (upper trace) built using the solid phase estimation and the gas phase model spectrum presented just above. To derive the $CO_2$ ice column density we use the band strength given by Gerakines et al. (1995) and derive 2.5x$10^{18}$ cm$^{-2}$ (d'Hendecourt et al. 1996). Thus, the derived $CO_2$ gas to solid ratio is about 5x$10^{-2}$. This number is similar to the one given by van Dischoek et al (1996) in other objects. Clearly, $CO_2$ is much more abundant in the solid than in the gas phase.

*Gas phase $H_2O$ band* . As for $CO_2$, the $H_2O$ $\nu_2$ band is detected in our spectra of RAFGL 7009S, but the rotation-vibration bands seen in absorption, compared to carbon dioxide span a wider range ($\sim$ 5.5 to 6.9$\mu$m) since the molecule is lighter. We proceed in the same manner as for $CO_2$. The temperature estimate from individual lines seems rather difficult to obtain. When choosing the strongest lines to make a rotation diagram one is lim-

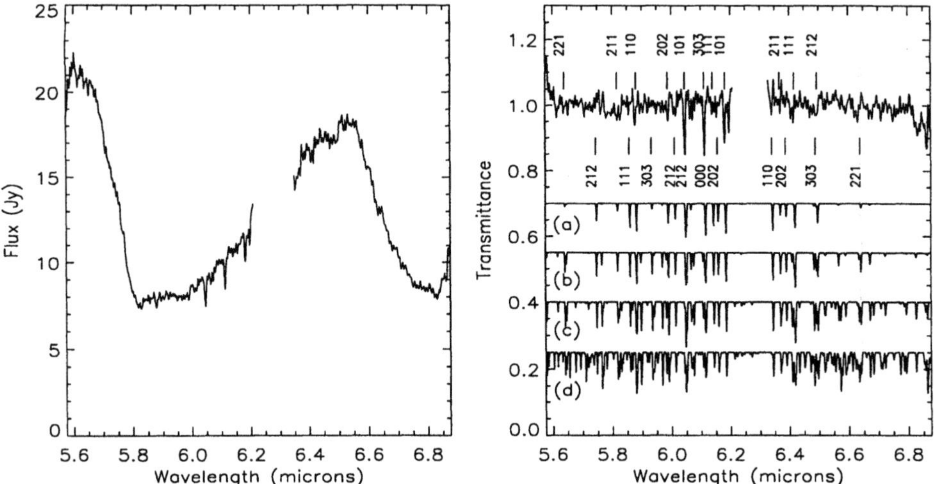


*Figure 2.* Gas phase $H_2O$ $\nu_2$ band Observed in RAFGL 7009S and model spectra with $v$= 3 km s$^{-1}$ and 2x10$^{18}$cm$^{-2}$ at temperature (a) $T$=25K (b) $T$=50K (c) $T$=100K (d) $T$=300K. Assignment of the pure rotationnal levels from which transitions comes from have been made for the strongest bands occurring at $T$=50K, showing that we mainly deal with a cold component of the gas.

ited by saturation effects and, taking the non-saturated ones, one is limited by the signal-to-noise ratio. Nevertheless, the tentative assignment of the observed lines can give information on the range of temperatures reached in the gas. In Fig. 2 it can be seen that the strongest observed transitions coincide with low levels rotational lines (ground, $J$=1 or 2 levels). The comparison of the data and the model spectra allow to exclude dominant absorption from gas hotter than about 50 K. The spectrum between e.g. 5.65 and 5.75 or 5.9 and 6 $\mu$m would present many more lines than observed if the gas was at a higher temperature. We then propose that most of the water vapor temperature lies between 20 and 50K for what we call the cold component. This cannot exclude the possibility of a second hot component that could be responsible for few lines that seems to be present (with a high $J$), however with a much lower column density (at least a factor of 3 below the cold one). The signal-to-noise ratio does not allow to draw a firmer conclusion.

*Gas phase $CH_4$ $\nu_2/\nu_4$ dyad* . Methane is inactive in its rotational ground state mode and thus not observable in the radio range. $CH_4$ can then only be clearly seen through the rovibrational modes (Boogert et al., 1997) such as the nu$_2$/nu$_4$ dyad falling in the 7.7 $\mu$m range. We present the $CH_4$ analysis in Fig. 3 using a slighlty different approach than the ones presented for $CO_2$ and $H_2O$. Indeed, we rather use a real laboratory UV photolysed ice spectrum obtained from a mixture ($H_2O$,$CO$,$CH_4$,$NH_3$,$O_2$:15,8,3,2.75,5) to

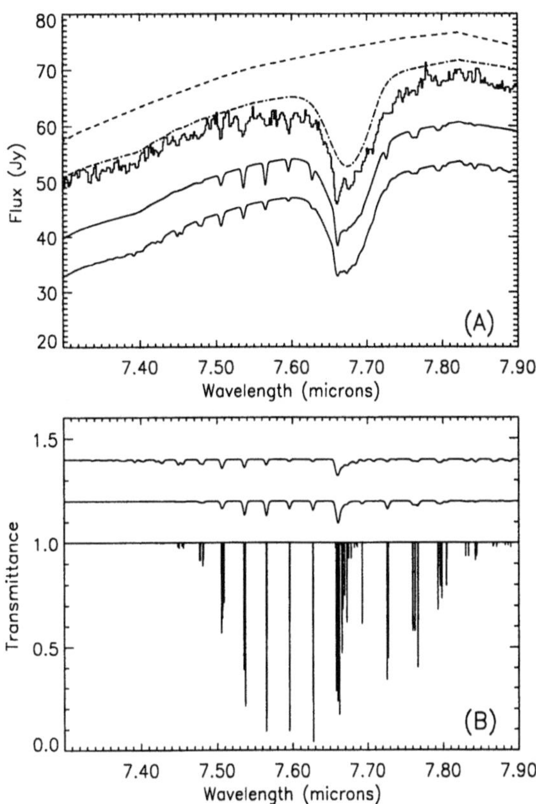


*Figure 3.* Gas phase $CH_4$ of the $\nu_2/\nu_4$ dyad. Panel (A) : From upper to lower curve, continuum adopted, resultant spectrum with the $CH_4$ laboratory ice feature, ISO spectrum of RAFGL 7009S, and spectrum when both gas phase model and laboratory spectrum are taken into account. Even the broad absorption around 7.4 $\mu$m seems to be reproduced by laboratory experiment. Panel (B) : Model spectrum with $T = 50$ K, doppler parameter b = 3 km s$^{-1}$ and a column density of $10^{17}$cm$^{-2}$ at $R$=1800 and full resolution.

produce a synthetic spectrum (lower curve of panel A). The gas phase model spectrum is generated using the exhaustive list of weighted squared transition moments for the $nu_2/nu_4$ dyad given in the HITRAN database. Assuming a doppler parameter and a line profile as explained for $H_2O$, a model can be built and compared to observations. The best spectrum is obtained with a column density of $1.2^{+0.5}_{-0.3}.10^{17}$cm$^{-2}$ with a temperature around 50 K, shown in Fig. 4B. The ratio of gas to solid methane is then 0.23, a number similar to the one for $H_2O$ but significantly different from $CO_2$. Finally, although we choose not to discuss the solid state effects here, we wish to emphasize that, for $CH_4$, a non-polar molecule which is non disturbed by its environment, a quasi-exact match between the laboratory data and the observed features is obtained, showing that grain size effects or shapes are unimportant for this molecule, thus validating our laboratory simulations.

## 3. Discussion

TABLE 1. Gas to solid ratios in RAFGL 7009S.

| | Gas $cm^{-2}$ | Solid $cm^{-2}$ (d'Hendecourt et al. 1996) | Gas/Solid |
|---|---|---|---|
| $H_2O$ | $\geq 2 \times 10^{18}$ | $1.1 \times 10^{19}$ | $\geq 0.18$ |
| $CO_2$ | $1.0^{+1}_{-0.5} \times 10^{17}$ | $2.5 \times 10^{18}$ | $\sim 0.04$ |
| $CH_4$ | $1.2^{+0.5}_{-0.3} \times 10^{17}$ | $4.3 \times 10^{17}$ | $\sim 0.28$ |

In Table 1 the column densities estimated are summarized. These results allow a quantitative insight on the role of grains in interstellar chemistry. The absorption spectra of gas phase species imply relatively low temperatures (~50K). From the $H_2O$ lines we do not observe a significantly abundant warm component. This is in contrast with other observations of young embedded stellar objects for which water vapor is observed at about 300 K (van Dishoeck et al. 1996). An important question is to know whether water forms on grains or in the gas. The formation of $H_2O$ molecule in the gas phase involves reactions with activation energies. The high abundance of water in dark clouds such as Taurus (10% of the cosmic oxygen abundance, Whittet et al. 1988) is not reproduced by chemistry models assuming formation in the gas followed by accretion on dust. Formation on grain surfaces seems required. Most models including water formation onto grain surfaces predict $H_2O$ rich mantles (Tielens and Hagen 1982, d'Hendecourt et al. 1985, Hasegawa et al. 1992, Bergin et al. 1995). It is generally accepted that chemical gas phase models cannot reproduce the observed amounts of $CO_2$ on grains. Laboratory experiments (d'Hendecourt et al. 1986) show that $CO_2$ is efficiently produced by UV photolysis of ices containing CO, $H_2O$ and/or $O_2$ as well as cosmic ray irradiation. Surface reactions have also been proposed as an important formation process (Tielens and Hagen, 1982). The grain/gas abundance ratio is much higher for $CO_2$ than $H_2O$ and $CH_4$. This result might be taken as evidence that $CO_2$ is essentially produced on grains and is rapidly destroyed in the gas phase. Indeed $CO_2$ reacts with $H^+$,$H_2^+$ or $H_3^+$ to form $HCO^+$ or $HCO_2^+$ ions. High abundances of these ions in some sources may be a direct evidence of the release of grain species in the gas phase as noted by Breukers (1991). If the observed molecules are effectively formed on grains, the grain/gas abundance ratio provides information on the desorption mechanism. The fact that this ratio is comparable for methane and water while these molecules have different condensation temperatures suggests that this mechanism might be non thermal. Cosmic ray induced desorption (Leger et al, 1985) together with chemical explosions (d'Hendecourt, 1982) are plausible and must be investigated by further laboratory work.

## References

Bergin E. A., Langer W. D., Goldsmith P. F., A&A 441, 1995, 222.
Boogert A. C. A., Schutte W. A., Helmich F. P. et al., 1997, A&A 317, 929.
Breukers R. J. L. H., thesis, University of Leiden, 1991.
Van Dishoeck E. F. and Helmich F. P., 1996, A&A 315, L177.
Van Dishoeck E. F., Helmich F. P., de Graauw Th. et al., 1996, A&A 315, L349.
Gerakines P. A., Schutte W. A., Greenberg J. M. et al., 1995, A&A 296, 810.
Hasegawa T. I., Herbst E., Leukng C. M. , 1992, ApJS 82, 167.
D'Hendecourt L. B., Allamandola L. J., Baas F., Greenberg J. M., 1982, A&A 109, L12.
D'Hendecourt L. B., Allamandola L. J., Greenberg J. M., 1985, A&A 152, 130.
D'Hendecourt L. B., Allamandola L. J., Grim R., Greenberg J. M., 1986 A&A 158, 119.
D'Hendecourt L., Jourdain de Muizon M., Dartois E. et al., 1996, A&A 315, L365.
Leger A., Jura M., Omont A., 1985, A&A 144, 147.
McCutcheon W. H., Dewdney P.E., Purton C.R., Sato T., 1991, AJ 101, 1435.
Reichle H. G., Young C., 1972, Canadian Journal of Physics 50, 2662.
Rothman L. S., Gamache R. R., Tipping R. H. et al., 1996, J.Q.S.R.T. 48, 469.
Tielens A. G. G. M. and Hagen W., 1982, A&A 114, 245.
Whittet D. C. B., Bode M. F., Longmore A. J. et al., 1988, MNRAS 233, 321.

# DIFFUSE INTERSTELLAR BANDS – THE $75^{TH}$ ANNIVERSARY

J. KRELOWSKI
*Center for Astronomy*
*N. Copernicus University*
*Gagarina 11*
*Pl-87-100 Toruń*
*Poland*

**Abstract.** This paper reviews the 75 year long history of observations of diffuse interstellar bands (DIBs), the puzzling features which appear in the spectral range from the near UV to near IR, in spectra of reddend stars. They apparently form absorption spectra of dark, HI clouds, together with other spectroscopic features such as atomic and molecular lines or continuous extinction and polarization. DIB intensity ratios vary evidently from cloud to cloud in the interstellar matter (ISM), strongly suggesting that several carriers must be identified to account for the whole DIB spectrum. The recent surveys discovered hundreds of very weak interstellar features; the pattern of them is clearly related to the well-known, strong DIBs and thus they may facilitate the identification of the DIB carriers. These unidentified species are most likely big molecules or very small dust grains which can produce only one strong and several very weak features each. In single cloud cases, the DIB intensity ratios clearly split into several groups, most probably characterizing different cloud environments. The paper discusses different types of HI interstellar clouds and gives a suggestion of how to classify them and put into an evolutionary sequence.

## 1. Introduction

Diffuse Interstellar Bands (DIBs) were originally discovered at Lick Observatory; the original report on the discovery of two "stationary" spectral features, centered near 5780 and 5797 Å in spectra of some spectroscopic binaries was published by Heger in 1922 – exactly 75 years ago. The interstellar natures of several strong DIBs were established by Merrill (1934,



*J. M. Greenberg and A. Li (eds.), Formulation and Evolution of Solids in Space,* 147–166.

1936), Beals & Blanchet (1937), York (1971) and Herbig (1975). Despite many efforts the features remain unidentified. The number of known interstellar features keeps growing – we still discover more and more weak interstellar lines or bands of unknown origin with the growing S/N ratio in our spectra. The solid–state detectors improved the quality of astrophysical spectra so much that it became possible to abandon the observations of heavily reddened, distant stars, towards which the interstellar absorptions are averaged over unknown number of individual clouds, characterized by unknown physical parameters. Instead, we can observe now the DIBs in spectra of slightly reddened, bright stars which are most likely to be obscured by only single clouds. In the latter case we can believe that the obscuring medium is homogeneous *i.e.* characterized by rather narrow range of physical parameters. The importance of such an approach became evident after it has been proved that the optical properties of individual clouds differ substantially.

To the present day no reliable identification of the carrier(s) of DIBs has been found. Almost all conceivable forms of matter have been proposed as the possible carriers, ranging from dust grains, to free molecules of very different sizes and structures, to even the hydrogen negative ion. It is commonly believed now that DIBs originate in "something in between" of the identified molecules and dust grains. Such compounds or clusters can share some quantum as well as classical spectral properties, *i.e.* they can be responsible both for DIBs and some segments of the interstellar extinction curve. Their internal structures can be very complicated and the possible presence of many species makes the task of identification very difficult.

The most recently discovered DIBs are typically very weak (Herbig 1988, Herbig & Leka 1991, Jenniskens & Dèsert 1994, Krełowski, Sneden & Hiltgen 1995). The growing S/N ratio makes it possible to discover hundreds of new features; all of them are, however, very shallow and thus – barely seen in spectra of slightly reddened stars obscured by single clouds.

The problem of DIB carrier identification is complicated by the possible existence of multiple overlapping features comprising each major DIB (first mentioned by Herbig 1975). For example, the prominent 5780 DIB (henceforth, major bands will be labeled by their approximate central wavelengths in Å) is situated inside a very broad ($\simeq$ 20 Å) but extremely shallow DIB usually called the 5778 band. Further structure within the general profile of this broad band, already suggested by Herbig (1975) was demonstrated convincingly by Krełowski, Schmidt & Snow (1997).

The commonly observed Doppler splitting of the sharp profiles of ISM atomic lines (reflecting the fact that the interstellar medium is not continuous but built of separate clouds) exists also in DIB profiles. Herbig & Soderblom (1982) showed that the extremely narrow 6196 DIB has a complex, Doppler–split profile toward some lines-of-sight. Since that paper it has been clear that the DIBs observed in spectra of heavily reddened, distant stars may be composed of components originating in individual clouds of different radial velocities. In cases of broader DIBs, Doppler splitting is not seen directly, but the resultant profiles are much broader than in single-cloud cases (*e.g.* Westerlund & Krełowski 1988).

Even twenty years ago the interstellar medium had a clear connotation: "the same everywhere". However, the first survey of ISM extinction curves in the vacuum UV (Bless & Savage 1972) showed large differences between extinction curves derived from spectra of bright, nearby stars – the objects very likely to be obscured by only single, individual clouds.

It is important to analyse the DIB strengths towards the stars in which we observe very sharp ISM atomic lines, in order to restrict the considerations to pretty homogeneous media which can permit to reveal possible relations of DIB spectra to other ISM parameters.

Absorption interstellar features are usually observed in spectra of early type, reddened OB stars because of their high luminosity and the scarcity of stellar lines; the spectra allow to observe a variety of interstellar absorptions such as:

- continuous extinction - which affects the broad-band colors of stars by the selective attenuation of starlight; for review see e.g. Mathis (1990)
- polarization of starlight, caused by the same grains, which must be at least partially aligned by some mechanism, most probably – magnetic field
- spectral resonant lines of rarefied atomic gases (especially in far-$UV$) (Morton 1975); their pattern changes from object to object – *e.g.* Crawford (1992); in the visual wavelength range these are Fraunhofer $D_1$ and $D_2$ of $NaI$ as well as $H$ and $K$ lines of $CaII$ and the weaker lines of $KI$ and $CaI$
- absorption features of simple molecular species ($CH$, $CH^+$, $CN$, $C_2$, $H_2$, $CO$), some of them known from late 1930's; their intensity ratios to $E_{B-V}$ vary strongly from cloud to cloud (Crawford, 1989; Krełowski *et al.*, 1992; Snow, 1992)
- the unidentified diffuse interstellar bands: the rapidly growing set of the heroes of this article. For a recent review see Herbig (1995).

These absorption features are very likely to be dependent on the global physical parameters of individual clouds. As the well–known atomic or molecular features can allow us to determine the above mentioned parameters, it seems very important to establish possible relations between the unidentified spectrum of diffuse bands and the other absorptions.

The present review does not repeat all the details which can be found in the recent reviews by Herbig (1995) or Krełowski (1988). In the coming sections we will deal mostly with the most recent observational results.

## 2. Diffuse Bands – their profiles and stellar contaminations

The heroes of this review are considered as interstellar features. It is because they do not participate in the stellar line oscillations, clearly seen in spectra of spectroscopic binaries, and their profiles remain unchanged in spectra of fast rotators. Such spectral lines, known from the very beginning of the present century, are called "stationary". On the other hand the DIB profiles are usually much broader than those of atomic gas lines like e.g. the sodium doublet despite they are much shallower. This is why they are called: diffuse.

The identification of diffuse interstellar bands is the longest standing unsolved problem in all of spectroscopy. It is clear that they are not atomic lines, as it was believed right after the first discovery of Heger. However, it is still impossible to find a laboratory spectrum of any substance which would match the astrophysical spectra. The difficulties follow some observational problems such as still too low S/N ratio which prevents precise measurements of weak features. The same factor complicates a determination of complete spectra of single agents – it is now well–known that DIBs originate in more than one agent but we see the spectra of many of them mixed together. Also the DIB profiles are often blended *i.e.* DIBs of evidently different origin (like 5795 and 5797) cannot be resolved does not matter the spectral resolution. In many cases we observe some stellar contaminations inside DIB profiles (*e.g.* the *FeII* line inside the 5850 DIB). The latter problem can be very severe in a case of very broad feature like 4430 which is usually blended with tens of stellar features. In cases of rapid stellar rotation it is impossible to resolve stellar and interstellar contributions. A high resolution may be helpful but only inside a certain range of apparent rotational velocities.

Profiles of the majority of DIBs are relatively narrow and situated in the wavelengths much longer than 4430 Å. This is why they are only seldom contaminated with telluric lines. The DIB profiles are analyzed in the recent publication of Krełowski & Schmidt (1997). It demonstrates the lack of symmetry in profiles of all strong diffuse bands. Also there are some substructures, seen in at least some of the profiles not contaminated with the Doppler splitting. The latter have been demonstrated convincingly in ultra high resolution spectra by Hibbins, Miles & Sarre (1995). However, Krełowski & Schmidt (1997) have shown that the same substructures can be revealed inside the 6614 DIB profile in the resolution 60,000 as well as with the ultra high one – 600,000. Apparently interstellar clouds are turbulent with a characteristic turbulence velocity of a few km/s.

## 3. Families of diffuse bands

Since the publication of Chlewicki *et al.* (1986) it is clear that a single carrier cannot account for all DIBs. The publications of Krełowski & Walker (1987) and of Krełowski & Westerlund (1988) revealed the varying strength ratios of the major diffuse bands (5780 and 5797) from object to object. This phenomenon, confirmed in other papers (e.g. Josafatsson & Snow,1987; Herbig, 1993), is now well–known. The ratio of 5780 to 5797 varies greatly in different lines-of-sight; similar phenomena can be presented in cases of many other DIBs.

It is to be emphasized that these phenomena are seen in spectra of nearby stars, characterized by narrow interstellar sodium lines. It is very unlikely to find a varying DIB strength ratio towards heavily reddened stars, most probably obscured by several clouds. However, such the "average galactic diffuse spectrum" cannot be interpreted in terms of any physical theory.

Krełowski & Walker (1987) proposed to divide the observed DIBs into several "families" *i.e.* sets of features observed always together and with the same strength ratios. Every such "family" can be created by a single carrier. The limited amount of high-quality (adequate resolution and S/N) spectra makes the lists of "family members" very unstable.

The very recent survey of the measurements of stronger DIBs done by Moutou & al. (1997) clearly suggests that usually any pair of two DIBs is correlated up to some level. However, some of these pairs are correlated very poorly (*e.g.* 5850 and 5780) whereas some others correlate up to a very high degree (*e.g.* 6614 and 6196). Generally the degree of correlation is higher when the widths of the two features are similar.

## 4. Weak unidentified interstellar features

The phenomenon described in the above Section clearly requires a search for more interstellar features which could be originated in the same species as the strong diffuse bands. Any spectroscopic identification of a molecule or a solid particle requires the match of several spectral features. Also any relation of the DIBs to any other identified spectral features may be very helpful. The recent survey of Krelowski, Sneden & Hiltgen (1995) showed the presence of many weak interstellar features which clearly behave in the same fashion as the well–known, strong ones and thus may form "families" which can facilitate the future identification. They are shown in Fig. 2.

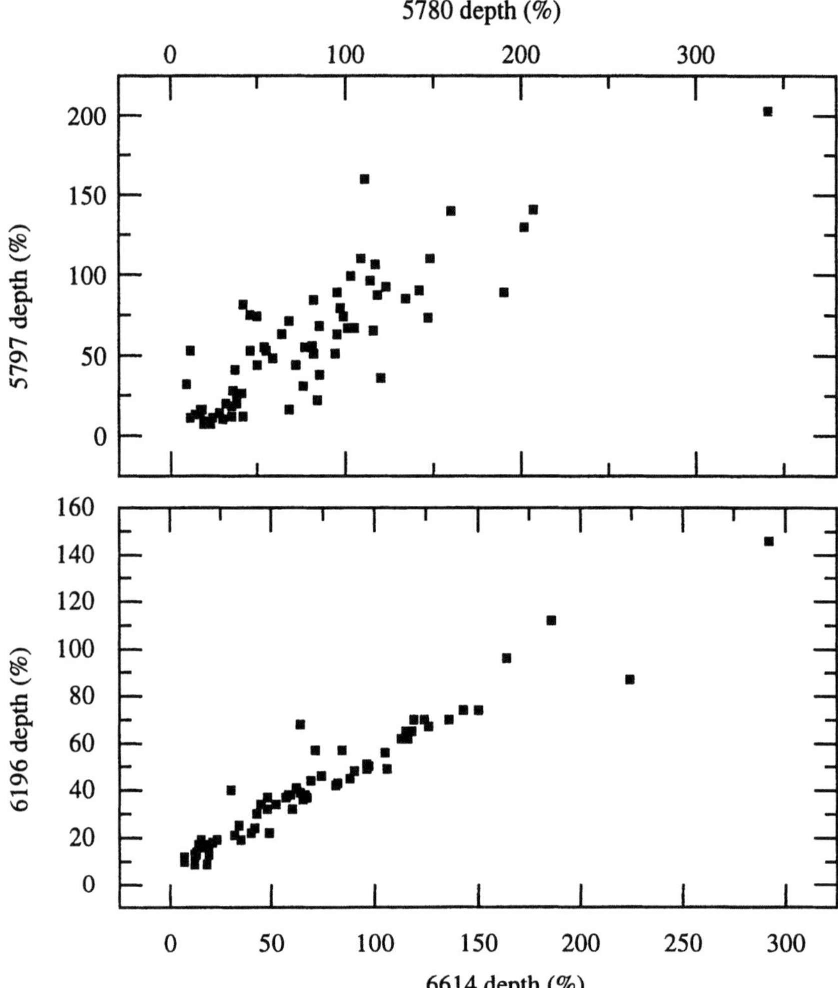


*Figure 1.* The correlations between the DIBs of very different widths (upper frame) and those of similar widths (lower frame). Data from Moutou et al. (1997).

Apparently hundreds of weak interstellar features can be observed in the visual spectral range. Some of them are detectable only in the spectra of clouds characterized by relatively very strong 5780 diffuse band, others – only while the above mentioned DIB is very weak in relation to a color excess or to the neighbor DIB (5797). This phenomenon suggests their possible common origin with some strong DIBs which can facilitate the identification of all the unidentified interstellar features.

The recent survey of Krełowski, Schmidt & Snow (1997) proposes a method to extract the DIB "families" from the whole spectrum of strong and weak features. When one spectrum is scaled up or down to match

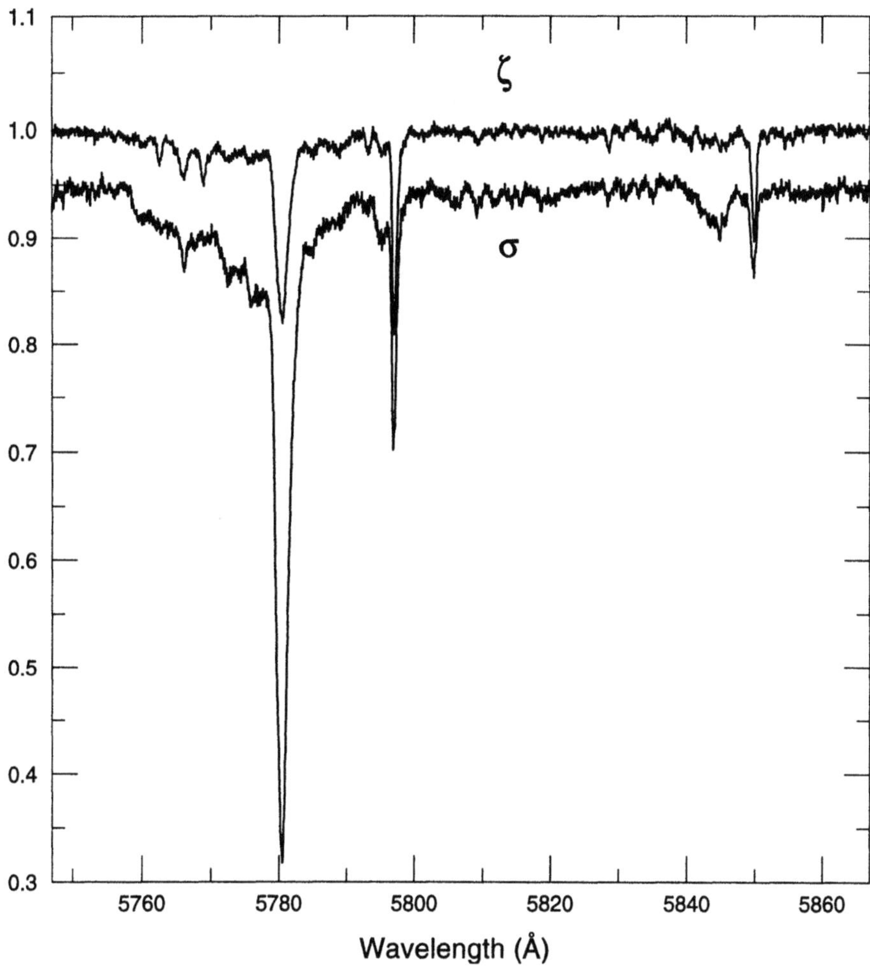


*Figure 2.* The very high S/N average spectra of "zeta" and "sigma" type. All the average stars are fast rotators and thus the narrow weak features are of interstellar origin. Note that their pattern changes together with the major DIBs 5780/5797 strength ratio. Data from Krełowski, Schmidt & Snow (1997).

central depth of some well–known diffuse band in another type of spectrum, one can observe identical depths of some weak features. It is evident that strong and weak broad features can be strictly proportional which may suggest their common origin.

## 5. Diffuse bands vs. other interstellar features

The quantitative parameters of the components of absorption spectra of individual *HI* clouds seemingly change together. According to Joseph, Snow,

& Seab (1989) and Cardelli, Federman, & Smith (1991), gas phase depletions, implied by neutral atomic abundances, are linked to the shape of the UV extinction curve. The intensity ratios of the diffuse interstellar bands, described in the previous subsection, also seem to be related to the shape of the extinction curve and to other spectral features of interstellar origin (Krełowski & al. 1992).

### 5.1. RELATION BETWEEN THE CONTINUOUS EXTINCTION AND DIFFUSE BANDS

It has been suggested for the first time by Krełowski & Westrelund (1988) that the strongly different ratio of 5780 and 5797 diffuse bands towards $\sigma$Sco and $\zeta$Oph was related to the different shapes of vacuum–UV extinction curves found already by Bless & Savage (1972).

Figure 3 demonstrates that the strength ratio of the major diffuse bands clearly depends on the ratio of the far–UV color excess and $E_{B-V}$. It may suggest that the carriers of the narrow DIBs, such as *e.g.* 5797 are related in a way to the carrier of the far–UV extinction rise, most probably – the small, bare silicate grains. It is, however, hard to say whether the grains give only a protection against the far–UV photons to the DIB carriers or they also facilitate their formation.

Krełowski *et al.* (1987) showed that also the ratio of famous 4430 DIB absorption to $E_{B-V}$ is related to a parameter characterizing the far-UV rise of the extinction curve. This band is, however, stronger when the far–UV rise is not observed in the extinction curve.

The extinction curves, as those of Fig 3, are considerably different. The differences, well–known since the Bless & Savage (1972) "classical" paper, reliably show that the physical conditions inside individual clouds may be considerably different. If very distant stars are observed these differences are smoothed and we get finally the galactic average extinction curve (Krełowski & Papaj 1992). We can also expect some "average" DIB ratios in spectra of distant objects. Such interstellar spectra are, however, useless – we cannot interpret properly such blends.

### 5.2. INTERSTELLAR MOLECULES AND DIFFUSE BANDS

The inter–correlations between ISM cloud parameters can help identification of the DIB carriers. Krełowski *et al.* (1992) suggest that the $CN$ features around 3875 Å as well as those of the $CH$ between 3878 and 3890 Å appear only in certain cloud environments: the above mentioned "zeta"

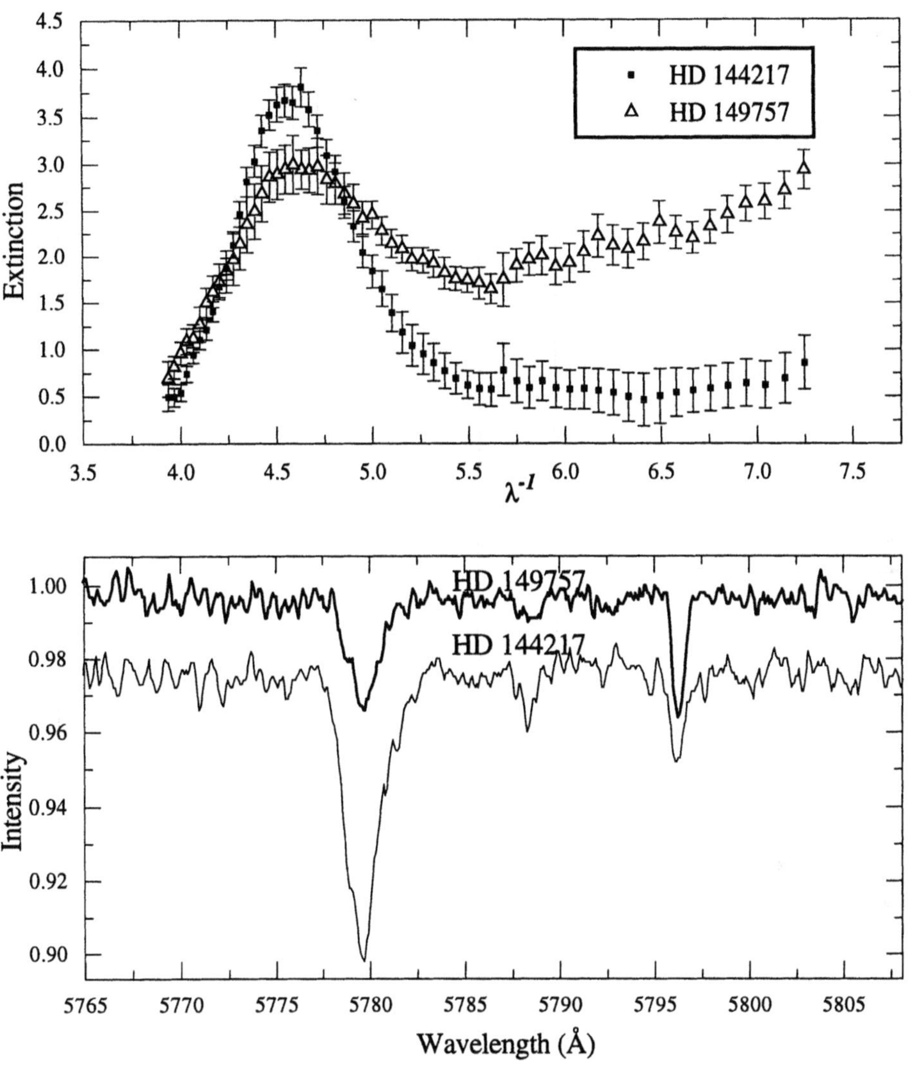


*Figure 3.* The extinction curves observed together with the major DIBs in the spectra of "zeta" type (HD 149757) and "sigma" type (HD 144217), type objects. The vacuum–UV spectra are these of TD–1 satellite; those, showing the DIBs are from the Russian Special Astrophysical Observatory.

clouds. In such objects the narrow DIBs, such as 5797, 6379 or 5850 are relatively strong. The extinction curve shows the steep far–UV rise.

Fig. 4 compares two spectra: one of them belongs to the "sigma" type cloud where the 5780 DIB is much deeper than the neighbor 5797 (HD184915), the second – to the "zeta" type object with the 5797 DIB deeper than 5780 (HD149757). Both stars are very fast rotators: their stellar lines are so flattened that they are hardly visible except the strong $H8$ line. In HD149757 we can observe clearly the nice sequence of molecular lines: the sharp features must not be originated in the stellar atmosphere. In the second object, despite its very similar reddening, the molecular lines remain below the level of detection. Apparently the presence or absence of molecular features in spectra of interstellar clouds is related to the ratio of the two major diffuse bands. Let's emphasize that the broad 5780 DIB is twice as deep in HD184915 as in HD149757. The strength of the latter can apparently grow when the molecular features disappear.

However, the existing samples of spectra containing the molecular features are very scarce. It is not only because of the lack of observational activity of astronomers. The accessible early type stars are usually either very distant or very slightly reddened. Following this the observed molecular features are either formed in several clouds along any sightline or are too weak to be measured with a proper precision. We know that interstellar clouds form several types but we do not know how many such types really exist in the space. When several clouds are situated along one sightline any analysis is very difficult. Trying to understand physical processes in HI clouds as well as to determine their chemical compositions we have to deal with very limited sample of nearby, bright $OB$ stars.

## 5.3. ATOMIC LINES AND DIFFUSE BANDS

It is generally very difficult to investigate relations between diffuse bands and interstellar atomic lines. The former are observed practically only in the wavelength range available from the ground–based observatories, the latter – mostly from the extraatmospheric ones. The most well–known interstellar lines, $H$ and $K$ or the sodium $D$ doublet are almost always saturated in spectra of reddened stars and thus it is difficult to construct any relation between relatively shallow DIBs and those saturated atomic features.

The very recent observations, conducted at the Special Astrophysical Observatory in Russia, allowed to accumulate a pretty big sample of targets in which the $KI$ doublet (7664 and 7698 Å) is observed. The former line is usually blended with the telluric features, but the latter is easily accessible. The $KI$ lines are strong enough to be measurable in all reddened targets but they are not saturated in any of our sample spectra. Fig 5 illustrates the unexpected positive correlation between the intensity of the $KI$ line and the narrow 5797 DIB. This relation was confirmed using the sample of 38 spectra. The same figure shows that the neighbor 5780 broad DIB is not correlated with the $KI$ line.

The only way in which any of the interstellar atomic lines can be atten-

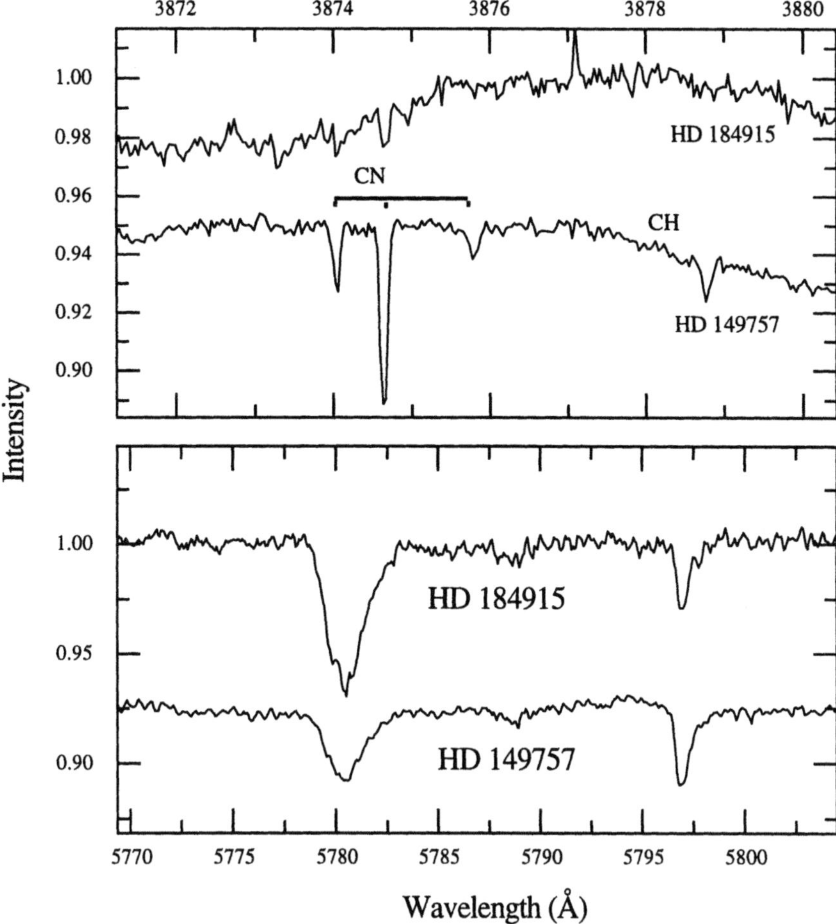


*Figure 4.* The spectral range of the sharp molecular features of $CN$ and $CH$ observed in the CFHT spectra of HD184915 ("sigma" type) and HD149757 ("zeta" type). The lower frame shows the major DIBs observed in the same spectra.

uated is the photoionization. Apparently the strength ratio of the major DIBs is related in a way to the ionization state of potassium. Let's remind that $KI$ has the binding energy very low – only 4.64 eV and thus can be ionized by every photon of $\lambda < 2800$ ÅThis may happen in the environment of neutral hydrogen cloud.

As similar phenomena seem to be observed also for the weak lines of CaI (4226 Å) and LiI (6707 Å) we can generally conclude that the spectrum of diffuse interstellar bands depends strongly on the ionization state of many heavy elements which exist inside $HI$ clouds. The result has been recently described by Krełowski, Galazutdinov and Musaev (1997). It seems

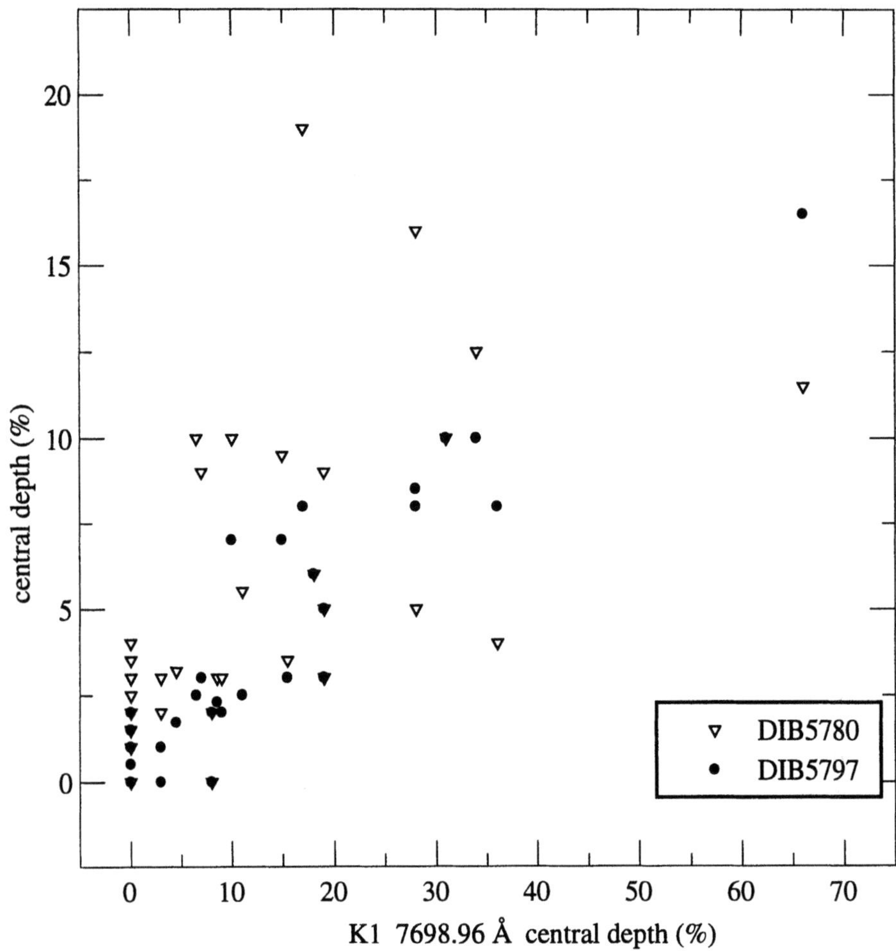


*Figure 5.* The correlation between the neutral potassium interstellar absorption line and the major DIBs. The correlation is good for the narrow feature and very poor in the case of the broad one. Apparently the narrow DIBs are sensitive to the ionization level of some neutral elements. The result taken from Krełowski, Galazatdinov & Musaev (1997).

necessary to conduct a broad survey of interstellar atomic lines observed using the *HST*.

## 6. Molecular origin of diffuse bands

### 6.1. SELECTIVE ATTENUATION AND MATRIX SHIFTS IN DIB PROFILES

It has been demonstrated by Krełowski & Sneden (1994) that the diffuse bands can be selectively attenuated in spectra of some targets. *e.g.* in the spectrum of HD37022 the 5797 narrow DIB is below the level of detection whereas the 5780 DIB is evidently present but very weak; the still broader 6284 DIB is of "normal" strength in relation to $E_{B-V}$. In the same objects Porceddu, Benvenuti & Krełowski (1992) found exceptionally broad 5780 profiles. The effect of broadening in these cases could not have been created by the Doppler effect as the weak, Doppler shifted sodium line components seen in the spectra of observed stars, could not be responsible for such a substantial broadening of the feature which is already broad.

The question we shall here address is whether one can relate such shape changes with any other phenomena and, in particular, with some sort of interaction with interstellar dust. The evidence that the sharper (narrow) bands are more reduced in regions of high local extinction has been attributed to the possibility that the molecules, being the DIB carriers, are relatively small and may be attacked by ultraviolet radiation after they accrete in the icy mantles of dust. However, such an effect is less likely to occur for large molecules because they may readily disperse the absorbed energy into vibrational modes and are therefore less likely to be partially dissociated within the dust mantle. The larger molecules responsible for DIB's are expected to provide broadened band shapes.

Analyzing the high resolution and high S/N spectra acquired at McDonald Observatory Krełowski & Greenberg (1998) found the additional effect in the Orion Trapezium. It is demonstrated in Fig. 6 which compares the spectra of HD37022 and HD24912. The objects are of the same $E_{B-V}$=0.50, but the former is one of the Orion objects. Fig. 6 compares the narrow (5797 and 6270), broad (5780) and very broad (6284) features in the spectra of HD37061 and HD24912. In the presence of the same reddening (color excess) the broad feature is of the same strength. The narrow DIBs: 5797 and 6270, are below the level of detection in the spectrum of our Orion star while the 5780 DIB is weakened but evidently present.

In all profiles of the diffuse bands seen in the Orion we observe both broadening and red–shift. In most of cases the DIBs are attenuated in comparison to $E_{B-V}$ while related to other places in the Galaxy. This is evidently not true in the case of 6284. Let us emphasize that this band is almost twice as broad as 5780. Our spectra clearly suggest that the narrow bands like 5797 disappear as the first, whereas those of intermediate width still remain visible, but much weaker. The broadest diffuse feature shows only changes of the profile and the red–shift, but the depth remains the same.

Apparently the mantles which are formed involve the carriers of diffuse interstellar bands which are almost certainly some molecules, most probably large, carbon bearing ones. The molecules are of different sizes. The smallest ones produce the narrowest and most numerous DIBs. They loose their

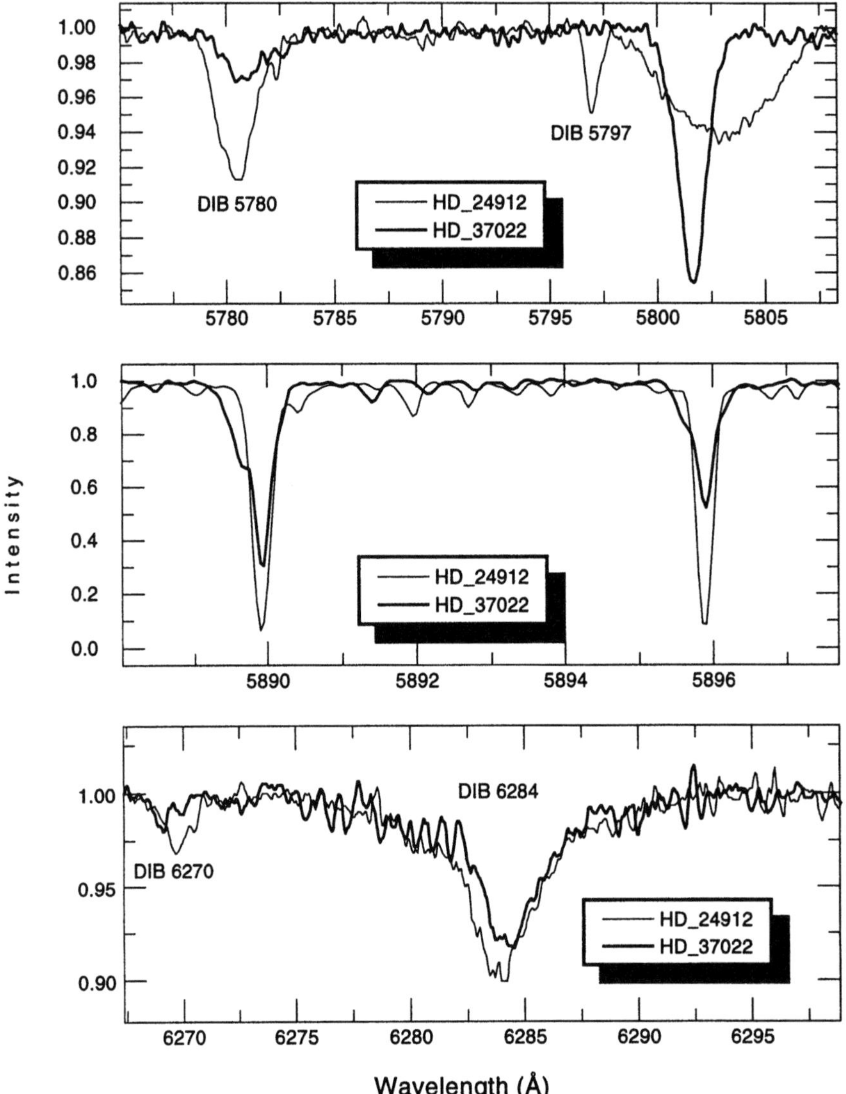


*Figure 6.* The attenuation of the narrow DIBs and the redshift of the broad ones observed in the spectrum of the brightest Orion Trapezium star: HD 37022. The comparison star, HD 24912 is of the same reddening. The result is of Krełowski & Greenberg (1998), the spectra from the Russian Special Astrophysical Observatory.

identity right after they stick the grain surfaces and thus the narrowest DIBs are not observed in the Orion $OB1$ association. The broader DIBs have the strengths reduced as their carriers slowly are being incorporated into the grain mantles, but they are still observed. The molecules, residing on grain surfaces still produce some broad DIBs, but weaker and red-shifted. It is to be emphasized that all DIBs observed in the Orion $OB1$ association are red-shifted. The broadest DIBs, created in the largest molecules are not attenuated. But even they are red-shifted which the fact proves that their carriers are residing on grain surfaces. The fact that the large, carbon bearing molecules are observed residing on grain surfaces strongly supports the idea of organic grain mantles which should exist on all evolved grains (Krełowski & Greenberg (1998).

## 6.2. DIFFUSE BANDS IN SPECTRA OF BE STARS

It has recently been found (Krełowski & Sneden, 1995) that diffuse interstellar bands are very weak or completely absent in the spectra of Be stars. The result is based on high resolution spectra obtained at the McDonald Observatory. These spectra contain only three strong stellar lines which can be seen in all targets: $H_\alpha$, the Fraunhofer $D_3$ 5876 Å and 6678 Å HeI lines. The presence of the first line in emission does not influence the diffuse spectrum ($H_\alpha$ is clearly seen in emission in the famous, heavily reddened star HD183143, frequently observed because of high reddening and strong diffuse bands. The helium lines are, however, rarely seen in emission (if yes – together with $H_\alpha$). Only while HeI lines are observed in emission and their profiles clearly suggest the origin in circumstellar disks, the diffuse interstellar bands are always much weaker in relation to $E_{B-V}$.

We illustrate this fact in Fig.7 using the spectra of HD 23180 and HD 24534, belonging, in fact to the same association: Per OB2. The major diffuse bands 5797 and 5780 are exactly of the same strength, suggesting the obscuration effects of the same cloud. However, the color excess of HD24534 is twice as big as that of HD23180. Apparently HD 24534 is additionally obscured by some cloud (circumstellar disk) producing the reddening but not containing the DIB carriers. This star is characterized by $D_3$ and 6678 Å emissions revealing the profiles typical for circumstellar shells.

Are the observed reddenings of Be stars caused by a circumstellar extinction of grains being quite different from those in the diffuse interstellar clouds or their colors are altered by some local free–free emission? The recent review of Krełowski and Papaj (1993) suggests that the intrinsic flux distributions of both B and Be stars are identical; the observed ones are altered by different extinction laws. Additional considerations are necessary to understand the physics of circumstellar shells.

It is to be emphasized that the shapes of extinction curves derived from the spectra of Be stars with helium emission lines are also peculiar. They typically lack the prominent 2200Å "bump", their far–UV branches are flat, typical for a "gray" extinction in this spectral range, characteristic of

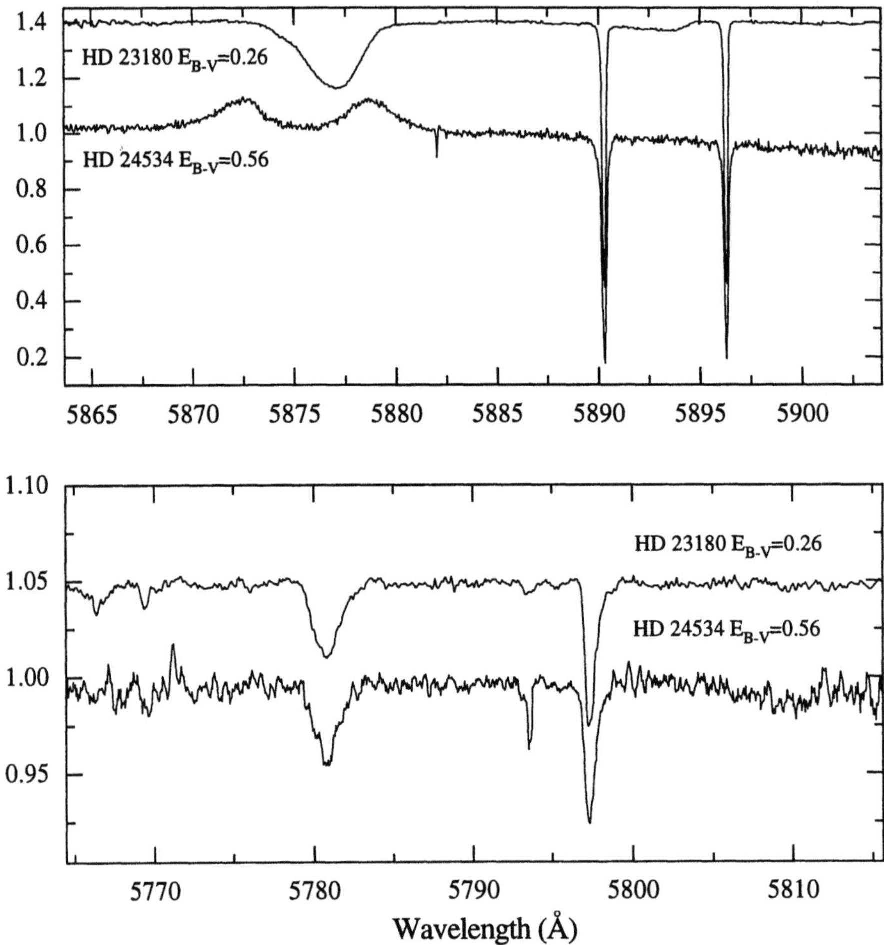


*Figure 7.* The interstellar sodium doublet and the helium stellar lines observed in HD24534 and HD23180. The upper D3 profile clearly indicates that HD24534 is surrounded by a circumstellar disk. Spectra from the McDonald Observatory.

rather big grains and their infrared segments are suggesting very high total–to–selective extinction ratio; the latter is probably due to circumstellar infrared emissions originating in the disks (Papaj *et al.*, (1991). It should be mentioned that the total extinction of such circumstellar shells is limited to $E_{B-V}$ not exceeding 0.2 – 0.3 mag. The spectra suggest the accretion going farther than in. *e.g.* the Orion association.

The DIB carriers, accreted in the grain mantles inside the circumstellar disks do not produce DIBs any more. No observations of simple molecular species exist in "clear" circumstellar spectra such as HD45725 or HD41335; however in spectra of "mixed" cases *i.e.* when both circumstellar and interstellar reddening is observed, as in HD202904, HD200120 and HD10516 the features of $CN$ and $CH$ in the near UV range are below the level of detection (Szczerba *et al.* (1994).

## 7. Evolution of interstellar clouds

The spectra, considered in this paper are apparently the absorption spectra of clouds composed mostly of neutral hydrogen and helium *i.e.* $HI$ clouds. The spectra of individual clouds are apparently different proving that physical conditions inside $HI$ clouds may be quite different. The clouds can be possibly divided into several spectral types as suggested for the first time by Krełowski & Sneden (1995). We do not know how many such types can be defined basing on spectral characteristics. However, the changes of the observed spectra of individual clouds may be related to a kind of the cloud evolution.

It is thus the interesting question: do we have enough of data to attack the problem of cloud evolution? Is it possible to define any parameter(s) which can trace a cloud evolution?

### 7.1. THE MAJOR DIB RATIO AS A PARAMETER OF THE EVOLUTIONARY STATE

It seems attractive to put the spectra of different clouds into a sequence using a the parameter the 5797/5780 strength ratio. This parameter gets the value of zero in the Orion Trapezium spectra. It's maximum value is observed in the Per $OB2$ stars such as HD23180 or HD24398. It is always a matter of some debate how many Doppler components can be found in the interstellar sodium line profiles. The answer usually depends on the spectral resolution.

For the narrow sodium lines we observe many objects with the varying DIB ratio. This can be very likely the effect of the evolution of the clouds situated inside $OB$ associations such as globules or elephant trunks. Let's remind that such objects are typically limited with a bright rim: the limit which the Lyman continuum photons can reach. This bright edge can be, however, easily crossed by the photons of $\lambda > 912$Å. They can ionize heavy elements such as $K$, $Ca$, $Li$ and others which makes their spectral features attenuated in spectra of "sigma" type objects. As the degree of ionization can be different, depending on the local conditions, the intraassociatory clouds can evolve pretty quickly.

This evolution apparently starts from the matter containing a lot of small dust particles and small molecules – the true interstellar cloud. The degree of ionization is low due to the lack of Lyman continuum photons in the general ISM environment. In the course of the cloud irradiation and collapse the grains grow, changing the shape of the extinction curve. The far–UV rise disappears because of the lack of small grains which become core–mantle ones. The DIB carriers stick to grain surfaces as well as gas atoms and simple molecules. However, it is possible, that this process may create a larger population of big molecules being the carriers of broad DIBs.

In very dense intraassociatory clouds all the DIB carriers stick to grain surfaces. Bigger molecules still do not loose their identity but their spectral features are red–shifted and usually weaker, except the broadest ones. In this phase the grains are covered with so thick mantles that even the prominent 2200 Å extinction feature is barely visible (Fitzpatrick & Massa 1991). Possibly all the DIBs disappear together with the 2200 Å extinction bump in circumstellar disks.

We are, however, facing one more problem. The diffuse bands have been observed as emission features in the spectrum of the Red Rectangle (Sarre & al.). The features are apparently observed in a circumstellar shell of a very special shape. Most of them are the features corresponding to the narrow DIBs such as 5797; the feature centered near 5813 Å was first observed in emission and only later found in absorption. The features are also red–shifted.

Are the DIBs seen in emission originated in mantles of hot grains? The red–shift may suggest so. Also the fact that the Red Rectangle is apparently a circumstellar shell. However, if the above is true, we have to conclude that the DIB carriers, after they stick to dust particles are not incorporated in the sense of reacting with the bulk material but just "frozen" into the grain mantles. Such molecules can apparently be expelled into the space as soon as the grain temperature is high enough. Why this phenomenon takes place so rarely – we do not know.

## 8. Conclusions

In 1997, 75 years after the diffuse interstellar bands have been discovered, they still remain unidentified. However, it seems generally accepted that their carriers are some complicated molecules, most probably – carbon bearing.

The varying pattern of the diffuse spectrum clearly proves that interstellar clouds are not all identical – they differ in spectral characteristics which, in turn, reflect the different physical conditions and/or evolutionary stages of intervening clouds. The problem of the evolution of clouds can apparently be addressed using the rich DIB spectrum. A cloud evolution involves the formation of more and more complicated molecules as well as the accretion of atoms and molecules onto grain surfaces. Dust grains in dense clouds should contain a lot of organic matter – complicated molecules, including the DIB carriers. How far they can be found in *e.g.* cometary matter – remains an open question.

# Acknowledgements

The author wants to express his gratitude to the Polish National Committee for Scientific Research for the financial support under the grant PB–0913/P3/94/07 and to Dr. Gazinur Galazutdinov for the very helpful discussions and suggestions.

## References

Beals, C.S. & Blanchet, G.H. (1937), *Publs. Astr. Soc. Pacific*, **49**, 224.
Bless, R.C., & Savage, B.D. (1972), *Astrophys. J.*, 171, 293.
Cardelli, J.A., Federman, S.R., & Smith, V.V. (1991), *Astrophys. J. Lett.*, **381**, L17.
Chlewicki, G., van der Zwet, G.P., van IJzendoorn, L.J. & Greenberg, J.M. (1986). *Astrophys. J.*, **305**, 455.
Crawford, I.A. (1989), *Mon. Not. R. astr. Soc.*, **241**, 575.
Crawford, I.A. (1992), *Mon. Not R. astr. Soc.*, **254**, 264.
Fitzpatrick, E.L., & Massa, D. (1990), *Astrophys. J. Suppl.*, **72**, 163.
Heger, M.L. (1922), *Lick Obs. Bull.*, **10**, 146.
Herbig, G.H. (1975), *Astrophys. J.*, **196**,129.
Herbig, G., H. and Soderblom, D. R. (1982), *Astrophys. J.*, **252**, 610.
Herbig, G.H. (1988), *Astrophys. J.*, **331**, 999.
Herbig, G.H. & Leka, (1991), *Astrophys. J.*, **382**, 193.
Herbig, G.H. (1993) *Astrophys. J.*, **407**, 142.
Herbig, G.H. (1995), *Ann. Rev. Astr. Astrophys.*,**33**, 19.
Jenniskens, P. and Dèsert, X. (1994), *Astr. Ap. Suppl. Ser.*, **160**, 39.
Josafatsson, K. & Snow, T.P., 1987. *Astrophys. J.*, **319**, 436.
Joseph, C.L., Snow, T.P., & Seab, C.G. (1989), *Astrophys. J.*, **340**, 314.
Krełowski, J., & Walker, G.A.H. (1987), *Astrophys. J.*, **312**, 860.
Krełowski, J., Walker, G.A.H., Grieve, G.R., & Hill, G.M. (1987), *Astrophys. J.*, **316**, 449.
Krełowski,J. & Westerlund, B.E. (1988), *Astr. Ap.*, **190**, 339.
Krełowski, J. (1988), *Publs. Astr. Soc. Pacific*, **100**, 896.
Krełowski, J., Snow, T.P., Seab, C.G. & Papaj, J. (1992), *Mon. Not. R. astr. Soc.*, **258**, 693.
Krełowski, J., & Papaj, J. (1992), *Acta Astr.*, **42**, 211.
Krełowski, J., Papaj, P. (1993), *Publs. Astr. Soc. Pacific*, **105**, 1209.
Krełowski, J. and Sneden, C. (1995), in *The Diffuse Interstellar Bands*, A.G.G.M. Tielens and T.P.Snow (eds.), Kluwer Acad. Publ., p.13.
Krełowski J., Sneden, C. & Hiltgen D. (1995), *Planetary Space Sci.*, **43**, 1195.
Krełowski, J., Schmidt, M. (1997), *Astrophys. J.*, **477**, 209.
Krełowski, J., Schmidt, M. & Snow, T.P. (1997), *Publs. Astr. Soc. Pacific*, (submitted).
Krełowski, J., Galazutdinov, G.A., and Musaev, F.A. (1997), *Astrophys. J.*, (submitted).

Krełowski, J. & Greenberg, J.M. (1998), *Astr. Ap.*, (submitted).
Mathis, J.S. (1990), *Ann. Rev. Astron. Astroph.*, 28, 37.
Merrill, P.W. (1934), *Publs. Astr. Soc. Pacific*, **46**, 206.
Merrill, P.W. (1936), *Publs. Astr. Soc. Pacific*, **48**, 179.
Morton, D.C. (1975), *Astrophys. J.*, **197**, 85.
Moutou, C., Krełowski, J., d'Hendecourt, L., Ehrenfreund, P. (1997) *Astr. Ap.*, (submitted).
Papaj, J., Wegner, W., & Krełowski, J. (1991), *Mon. Not. R. Astr. Soc.*, 252, 403.
Porceddu, I., Benvenuti, P., **Krełowski**, J. (1992), *Astr. Ap.*, **260**, 391.
Snow, T. P. (1992), *Australian J. Phys.*, **45**, 543.
Szczerba, R., Krełowski, J., Walker, G.A.H., Kennelly, E.T., Sneden, C., Volk, K., Hill, G. (1994), in *The Diffuse Interstellar Bands: Contributed Papers*, NASA Conference Publication 10144, A.G.G.M. Tielens (ed.), p. 33.
Westerlund, B., E. and Krełowski, J. (1988), *Astr. Ap.*, **203**, 134.
York, D.G. (1971), *Astrophys. J.*, **166**, 65.

# OBSERVATIONAL CONSTRAINTS ON THE CARRIERS OF THE DIFFUSE INTERSTELLAR BANDS (DIBS)

J. CAMI[1], P. EHRENFREUND[2] AND B.H. FOING[3]

[1]*Space Research Organization of the Netherlands (SRON)*
*P.O. Box 800, 9700 AV Groningen, The Netherlands*
[2]*Leiden Observatory*
*P.O. Box 9513, 2300 RA Leiden, The Netherlands*
[3]*ESA Space Science Department, Solar System Division*
*ESTEC/SO, 2200 AG Noordwijk, The Netherlands*

**Abstract.** In this contribution we describe the techniques used to infer properties of the carriers of the Diffuse Interstellar Bands (DIBs) from the analysis of astronomical data with emphasis on the recent results obtained from so-called single cloud observations. All results seem consistent with large carbonaceous gas phase molecules like PAHs and fullerenes as carriers of the DIBs. We discuss some of the puzzling properties of the DIBs like the variability of DIB strengths and DIB correlations, and summarize recent results which provide constraints for the identification of the carriers.

## 1. Introduction

Whenever human beings observe a phenomenon they cannot identify as something for which there may be an easy and logical explanation, they call it a UFO, a miracle or a mystery. Whenever astronomers observe a phenomenon for which they cannot find an easy and logical explanation, they call it a problem. The mystery associated with the identification of the carriers of the Diffuse Interstellar Bands (DIBs) is indeed the longest standing problem in astronomical optical spectroscopy.
Although the strongest DIB was detected as early as 1919 in the spectrum of a heavily reddened B supergiant (Code 1958), it was only years later that the interstellar nature of the DIBs was established from the observation that those bands do not show radial velocity variations in spectroscopic



*J. M. Greenberg and A. Li (eds.), Formulation and Evolution of Solids in Space,* 167–176.

binaries and that there is a correlation of band intensities with reddening. Nowadays, the term "Diffuse Interstellar Bands (DIBs)" is the name commonly used to describe a set of ~ 200 absorption lines in the visible and NIR spectral range showing similar characteristics but they still remain unidentified (see Herbig 1995 for a review). One notable exception may be the positive identification of two DIBs with lines originating from $C_{60}^+$ (Foing & Ehrenfreund 1994, 1997).
A definite identification of DIBs requires a good match of astronomical spectra with spectroscopic data obtained in laboratory experiments. However, a wealth of possible carriers has been proposed, ranging from solid state particles (from the correlation with $E_{B-V}$) over impurities embedded in grains to gas phase molecules. To guide the laboratory research in this field the astronomical data are analyzed to provide reasonable candidates as carriers for the DIBs.
Arguments against a solid state origin or impurities (the constancy of the wavelengths of DIBs and the absence of polarization structure within a DIB profile) favour strongly the idea that the carriers of the DIBs are gas phase molecules. It is clear that the carrier molecules in that case should show a rotational fine structure and be able to survive the harsh interstellar conditions. The most promising classes of molecules seem to be Polycyclic Aromatic Hydrocarbons (PAHs), fullerenes and their ions (Léger & d'Hendecourt 1985, van der Zwet & Allamandola 1985, Kroto 1987, Léger et al. 1988, Kroto & Jura 1992). These are suggested to be the most abundant free organic molecules with a stable carbon skeleton against photodissociation which also show strong transitions in the UV and the visible.
Some properties of the carrier molecules can be inferred from a careful analysis of astronomical data. In the next sections we discuss the three most important methods used and focus on the recent results obtained from observations of so-called *single cloud* lines-of-sight.

## 2. Mutual DIB correlations and Families

### 2.1. METHOD AND EARLIER RESULTS

One of the most striking properties of the DIBs is that the relative strength of different DIBs can change drastically from one line-of-sight to another, even for lines-of-sight having the same reddening $E_{B-V}$. A well-known example is the ratio of the $\lambda\lambda$5780 and 5797 DIBs. Clearly this indicates that the properties of the carrier molecules (and hence the carrier molecules themselves) are not the same for all DIBs. It is thus useful to group the DIBs that do show similar variations in strength together in *families* and search for tight mutual DIB correlations that would predict a same carrier molecule for different DIBs.

In the past, several people attempted to group a limited number of strong DIBs into families (Clewicki et al. 1986, Krelowski & Walker 1987, Josafatsson & Snow 1987, Krelowski & Westerlund 1988) but with contradictory results. The contradictions arise from a variety of reasons. First of all, the DIB sample studied was very limited and not completely the same in all papers. Second, the authors use different approaches and interpretations of correlations or non-correlations. But what is probably the most important factor is the choice of environment to study : the results of a correlation study seem to depend on the lines-of-sight which are studied. Stars with a rather high reddening are obscured likely by several intervening interstellar clouds. Therefore, the resulting band strengths are ill-defined averages over the contributing components and interpretation of the results is very difficult, if not impossible. The most promising results were thus to be expected from a study of so-called *single-clouds* : nearby stars that are only obscured by one single intervening interstellar cloud.

## 2.2. RESULTS FROM SINGLE CLOUD OBSERVATIONS

We recently made a survey of band correlations (Cami 1996, Cami et al. 1997) using the ELODIE spectrograph ($R \sim 42000$; spectral range 3906 to 6811 Å) at the *Observatoire de Haute Provence* (OHP) with several objectives.

First of all, the 13 observed lines-of-sight were chosen in single clouds, representing very different physical environments (Sco OB2, Per OB2 and Orion). This increases the sensitivity of the measured DIB strengths to the different properties of the carriers. Second, special care was taken to have a consistent set of DIB equivalent width measurements with minimal errors. For that reason we preferred not to include broad DIBs where a large error on the measurements is possible, due to mispositioning of the continuum baseline. DIBs that suffer from atmospheric contamination as well as DIBs blended by a stellar component were also rejected. The resulting DIB sample still contains 44 DIBs.

The most important problem remaining is what procedure should be used to find objective and reliable results. How well the strengths of different DIBs are correlated in the different lines-of-sight can be easily quantified by calculating the linear Pearson correlation coefficient. It is the interpretation of these coefficients and an assessment of the reliabilty of the results which are often done in a very subjective way. We applied a statistical approach to this problem. First we used the measured equivalent widths and the estimated errors as input for a Monte Carlo simulation yielding 100 times two simulated sets of "estimates" of DIB strengths in the observed lines-of-sight. We then calculated the correlation coefficient $r_i$ for every two

sets and finally calculated the average $r$ of the 100 correlation coefficients and the standard deviation $\sigma$ on this average. It is clear that without errors $r = 1$ and $\sigma = 0$. We found however that the inaccuracies on the measurements not only influence the standard deviation $\sigma$ (and hence the reliability) on the autocorrelation $r$, but also decrease the value of $r$. The errors on the measurements thus generate a "decorrelation bias" which we measured to be $\sim 1.8\sigma$.
We used the same kind of approach to calculate the mutual correlation coefficients and corrected for the observed decorrelation bias. Assuming now that DIBs originating from the same carrier should have $r = 1$, we can assess for which DIBs we can expect that the real value of $r$ is indeed very close to 1. From our results we have to conclude however that in our sample of DIBs there are *no DIBs for which* $r > 0.9$ *with a confidence* > 60%, supporting the 1 DIB - 1 carrier hypothesis in the observed wavelength range and for the 44 DIBs considered. However, this should not be regarded as a definite conclusion : more observations with a high S/N and a larger sample of DIBs can improve the reliability of the results and even allow tight correlations to be dragged out of the noise.
Although there may be no DIBs that originate from the *same* carrier, a closer look at the reliable ($\sigma/r$ <10 %) correlation coefficients shows that some DIBs probably originate from *similar* carriers : two sets of DIBs (Families I and II) are found that all correlate well with each other :

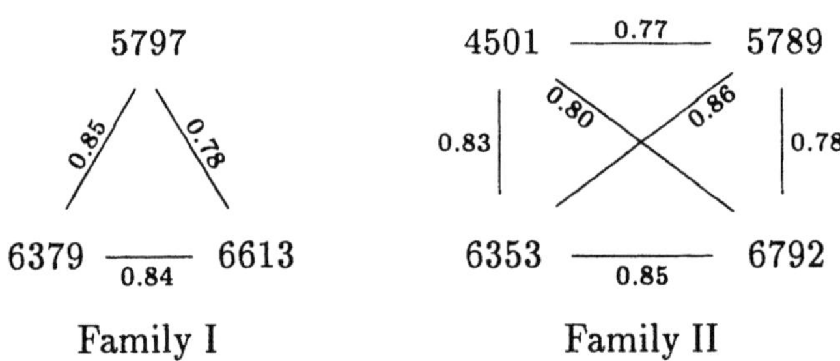


There is more evidence to believe that the DIBs in Family I are indeed originating from a very similar carrier. In the Red Rectangle these are the DIBs that are found in emission (Scarrott et al. 1986) together with the $\lambda$5850 DIB. Our sample does not allow to group the latter with Family I, but future observations may confirm that also this DIB belongs to the same family. Also, the line profiles of the 3 DIBs from Family I show similar (but not the same) properties for the corresponding carriers (see Sect. 4).

## 3. Environmental behaviour

Searching for mutual DIB correlations or classifying the DIBs into families may be useful for future laboratory work, but one can learn much more about the carriers by studying how the strength of the DIBs changes in different environments. The strength of DIBs is not only influenced by the local chemistry (e.g. in circumstellar shells around evolved stars) but also - and in the interstellar medium by large - by the physical conditions. The most promising results can again be expected from a single cloud study, for the same reasons as before.

### 3.1. THE IONIZATION HYPOTHESIS

The same sample of DIBs and stars that was used for the correlation study can now be used for investigating the influence of the local surrounding interstellar UV field on the DIB strength.

The strength of the UV field is therefore estimated using the ratio $\frac{\text{H I}}{\text{H I}+2\text{H}_2}$ of hydrogen column densities towards the studied lines of sight. Almost all the DIBs show the same qualitative behaviour (see Fig. 1) : in the presence of weak UV fields, the DIBs are generally weak but they get stronger with increasing UV flux until they reach a maximum after which the DIB-strengths decrease again. The main differences between different DIBs - especially when not belonging to the same family - seem to be the required UV flux to reach maximum strength and the slopes of both the rise and the decline.

This behaviour strongly suggests that the DIB carriers undergo photo-ionization. The rise and fall of the DIB strength could just express the displacement of ionization stage as a function of the UV field : if the carrier of a DIB is a molecule in ionization stage $n$, then we can expect that for low UV fluxes most of the molecules remain in stage $n-1$ and hence the DIBs are weak; the stronger the UV field grows, the more molecules get into stage $n$, so the DIB strength increases up to a maximum. An increase of the UV flux then causes the molecules to excite into stage $n+1$, which causes a decrease of DIB strength. This effect of UV penetration and photo-ionization has been further quantified to estimate the ionization potential of the carrier molecules of the $\lambda\lambda$5780, 5797, 6379 and 6613 DIBs (Sonnentrucker et al. 1997).

The observed differences in behaviour of DIBs not belonging to the same family thus only reflect that the carrier molecules do not have the same ionization and recombination properties. Members of the same family all have very similar values for these molecular parameters, indicating their similar nature.

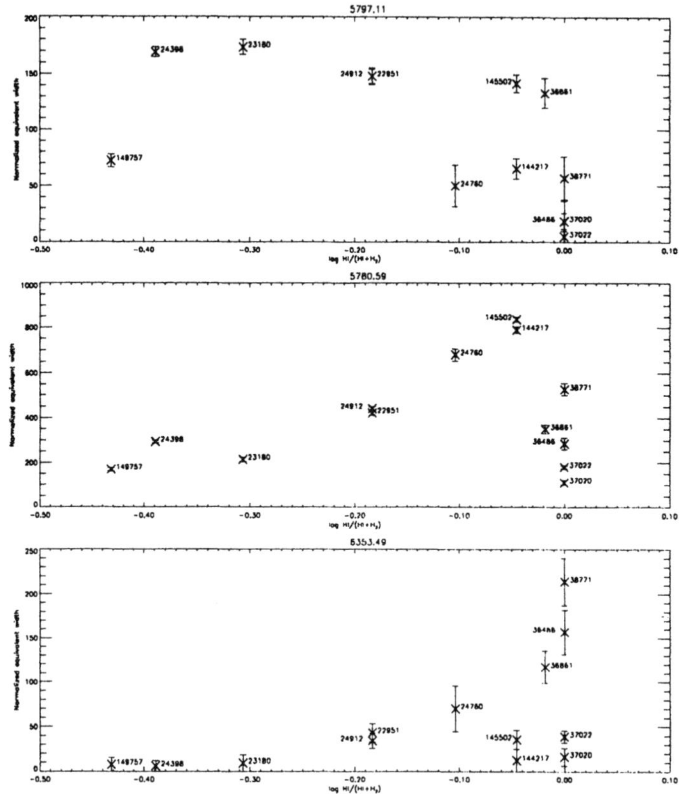


*Figure 1.* The normalized equivalent widths of the λλ5797 (top), 5780 (middle) and 6353 (bottom) DIBs as a function of the $\frac{\mathrm{HI}}{\mathrm{HI}+2\mathrm{H}_2}$ column density ratio. One notes a rise and fall of the DIB strength at different values of $\frac{\mathrm{HI}}{\mathrm{HI}+2\mathrm{H}_2}$ indicating different photo-ionization and recombination properties (taken from Cami et al. 1997)

## 3.2. CLOUD TYPES : A SEQUENCE IN STRENGTH OF THE UV FIELD

The ionization behaviour of the DIBs can also help to explain the rather puzzling division of clouds into families. This division into the so-called $\sigma$ and $\zeta$ type clouds (after their prototypes in the lines-of-sight towards $\sigma$ Sco, HD147165, and $\zeta$ Oph, HD149757) and also two more types called the *Orion* and *CS* types originated from the observation that the ratio of the strengths of two of the strongest DIBs, the λλ5780 and 5797, either has a very high value or either a low value. Lines-of-sight with intermediate ratios seem to be rather scarce. But not only does the ratio of these DIB strengths change, also other interstellar features have a different appearance (see Table 1). This led to the conclusion that there exists only a discrete spectrum of cloud types, and not a continuous one. Lines-of-sight with intermediate parameters were assumed to be composed of two (or more) components. A good example is the molecular cloud complex towards Per OB2 which

| | $\zeta$ | $\sigma$ | *Orion* | *CS* |
|---|---|---|---|---|
| $\lambda$5780 | comparable to $\lambda$5797 | very strong | weak but present | hardly visible |
| $\lambda$5797 | very deep | shallow | absent | hardly visible |
| Far-UV extinction | high | low | flat | gray |
| 2200 Å bump | broad | narrow | very weak | absent |
| $R_V$ | ~3.1 | ~4.5 | ~5 | hard to obtain |
| Simple molecules (CH, CN) | strong | below level of detection or very weak | not detected | not detected |
| Example | $\zeta$ Oph | $\sigma$ Sco | HD37022 | HD45725 |

TABLE 1. The division of clouds into families as summarized in Krelowski & Sneden (1995)

was said to consist of 2 overlapping clouds of different optical properties (Krelowski et al. 1996). With our knowledge about the ionization behaviour of the DIBs, we can now understand the differences between these cloud types.

As we have seen, the main differences between the DIB carriers are their ionization and recombination properties which manifest themselves when investigating the influence of the UV field on the DIB strength. As one of the main diagnostics for the classification of cloud types is the variable strength of the different DIBs, it is clear that the different cloud types are in fact lines-of-sight where the UV properties do strongly differ. It turns out that all the $\zeta$ like lines-of-sight are characterized by relatively low UV fluxes, while for the typical $\sigma$ lines-of-sight the UV field is much stronger. The *Orion* type clouds have even stronger UV fields. The carrier of the $\lambda$5797 DIB is easily ionized; the DIB reaches its maximum strength already at the typical $\zeta$ environment and decays only slowly. The $\lambda$5780 DIB on the other hand is very weak for the typical $\zeta$ objects and rises sharply to a maximum for typical $\sigma$ objects. Due to this *sharp* rise, there are only a few lines-of-sight with intermediate value for the ratio of the $\lambda$5780/5797.

We can also explain the other characteristics of the different cloud types. Cardelli (1988) notes that the total-to-selective extinction $R_V$ can be used as a parameter to estimate the strength of the surrounding UV-field. A low value of $R_V(\sim 3.1)$ corresponds to weak UV fields, say, the $\zeta$ type of clouds. Stronger UV fields imply higher $R_V$; we see that $\sigma$ and *Orion* types indeed have a subsequently higher $R_V$ (respectively $\sim$ 4 and $\sim$ 5.5) and are thus indeed characterized by stronger UV fields. Small molecules like CH and CN are easily destroyed by moderate UV fluxes; hence they are absent in the $\sigma$ and *Orion* types. The far-UV non-linear rise absorption feature correlates with the abundance of molecular hydrogen (Jenniskens et al. 1992). This makes the possibility that the far-UV non-linear rise is caused by neutral molecules very likely. However, neutral PAHs of sizes less than about 60 C-atoms have no strong absorptions in the visible and probably do not account for DIBs. But ionized PAHs do (see Salama et al. 1996 for a review). If DIBs are due to some ionized PAH species and the far-UV non-linear rise is due to neutral PAHs, a low degree of ionization will result in relatively weak DIBs and a large far-UV non-linear rise which is consistent with the observations and is also supported by an earlier study by Désert et al. (1995). The final conclusion is that the apparent division of clouds into $\zeta$, $\sigma$ and *Orion* types is not a physical one, but only represent a gradient in strength of the surrounding interstellar UV radiation field.

## 4. Line profile analysis

Another important tool to get information on the carriers of the DIBs, comes from the study of DIB line profiles observed at very high resolution and comparing them with theoretically calculated line profiles of molecules with different geometries, like C-chains, PAHs and fullerenes. This technique has so far only been applied to 3 DIBs, which happen to be the DIBs in our Family I. But although the $\lambda\lambda$5797, 6379 and 6613 DIBs seem to form a family, their line profiles are not identical. The profiles of the $\lambda\lambda$5797 and 6379 DIBs are both double-peaked, while the $\lambda$6613 DIBs shows a triple peak (see Ehrenfreund & Foing 1996).

If we now assume that the DIB carriers are some large molecules, then it is not unreasonable to assume that the substructures are unresolved ro-vibrational lines, with the observed peaks in the line profiles corresponding to the P, Q, and R branches. In such a case, the peak separation between the P and R branch is mainly determined by the largest of the rotational constants $B$ and the excitation temperature $T$. Measuring the peak separation in the line profile of a given DIB and assuming a reasonable value for the temperature, it is then possible to estimate a value for $B$ which corresponds to the smallest moment of inertia for the carrier of this DIB. Assuming a certain geometry it is then possible to obtain an estimate for the mass of the carrier by comparison with calculated rotational constants. This has been done for the three DIBs in Family I (see Ehrenfreund & Foing 1996).

The measured rotational constants for both double-peaked DIB line profiles correspond to linear C-chains of 12-18 C-atoms, or PAHs of about 50 C-atoms (scaling from pyrene or ovalene gives similar results). Remarkable however is the rotational constant derived for the $\lambda$6613 DIB, which is within errors consistent with the rotational constant for the buckminsterfullerene, $C_{60}$.

## 5. The two DIBs at 9577 and 9632 Å

The $C_{60}$ molecule and its ions has been proposed as a good candidate to account for at least some of the DIBs (Kroto 1987, Léger et al. 1988). A laboratory spectrum of $C_{60}^+$ in a Ne matrix shows two strong transitions in the NIR. Foing & Ehrenfreund (1994, 1997) searched in astronomical spectra for the presence of DIBs corresponding to these transitions and found two DIBs at $\lambda\lambda$9577 and 9632 Å, the right wavelengths if one allows for a matrix induced wavelength shift of 10 $cm^{-1}$ for both transitions.
The two DIBs have a constant band ratio in all observed lines-of-sight (within the errors) and are clearly enhanced in UV dominated regions like HD37022, consistent with an ionized carrier of high stability. Both DIBs have a width of 3 $cm^{-1}$, compatible with theoretical ro-vibrational line contour calculations of $C_{60}$ at rotational temperatures of 50-80 K. Moreover the bands show a red wing excess for high excitation temperatures, as is expected for ro-vibrational bands.
All these arguments strongly favour the idea that those DIBs are indeed absorptions arising from $C_{60}^+$, but the definite identification has to come from comparison of astronomical spectra with a *gas-phase* laboratory spectrum of $C_{60}^+$ which is not available yet at present days.

## 6. Conclusions

This contribution indicates that although the DIB carriers remain unidentified, important knowledge about the carriers can be obtained from careful analysis of astronomical observations.
The constancy of the wavelengths of the DIBs in all lines-of-sight, the substructures for some DIBs, the absence of a polarization structure within the bands and the ionization behaviour of the DIBs as observed in single cloud stars strongly support the DIB carriers to be gas-phase molecules. An analysis of the mutual DIB correlations shows that most DIBs are originating from different carriers with different properties, although some families can be distinguished, whose carriers are probably very similar. The ionization behaviour of the DIBs explains changes in the ratios of DIB strengths as different ionization and recombination properties of the carriers while the variability of the DIB strengths from one line-of-sight to another can be explained by differences in the flux of the surrounding UV field. These results also allow to explain the apparent division of clouds into $\zeta$, $\sigma$ and *Orion* types and shed some light on the nature of the far-UV non-linear rise of extinction curves. As the carriers must be quite abundant, the DIB carriers are very likely carbon bearing molecules; as they must be very stable to survive the harsh conditions encountered in the interstellar medium, the most promising classes of molecules to account for the DIBs seem to be

PAHs, fullerenes and their ions. An analysis of the DIB line profiles and comparison with theoretical ro-vibrational contours of PAHs and fullerenes reveals that the DIB carriers are large molecules, $\sim$ 50 C-atoms.
Laboratory experiments on PAHs, fullerenes and especially their ions in gas phase will be indispensable to come up with a definite identification for at least the strongest of the DIBs. The positive identification of the two DIBs at $\lambda\lambda$9577 and 9632 Å with $C_{60}^{+}$ illustrates the progress made in the recent years. Present day high quality astronomical observations coupled with new insights and ideas will eventually be the necessary guide for laboratory research and thus be an important tool to unravel the mystery of the DIBs and solve a major problem in optical astronomy.

## References


1. Cami, J., 1996, "Band correlations and Line Profiles of Diffuse Interstellar Bands (DIBs) in single-clouds and DIBs in Reflection Nebulae", Masters Thesis, Leiden Observatory
2. Cami J., Sonnentrucker P., Ehrenfreund P., Foing B.H., 1997, A&A, in press
3. Cardelli J., 1988, ApJ 335, 177
4. Chlewicki G., van der Zwet G.P., Van Ijzerdoorn L.J., Greenberg J.M., 1986, ApJ 305, 455
5. Code A.D., 1958, PASP 70, 407
6. Désert F.-X., Jenniskens P., Dennefeld M., 1995, A&A, 303, 223
7. Ehrenfreund P., Foing B.H., 1996, A&A, 307, L25
8. Foing B.H., Ehrenfreund P., 1994, Nature 369, 296
9. Foing B.H., Ehrenfreund P., 1997, A&A 317, L59
10. Herbig G.H., 1995, ARA&A 33, 19
11. Jenniskens P., Ehrenfreund P., Désert F.X., 1992, A&A 265, L1
12. Josafatsson K., Snow T.P., 1987, ApJ 319, 436
13. Krelowski J. & Sneden C., 1995, in "Diffuse Interstellar Bands", eds. A.G.G.M. Tielens & T.P. Snow, Kluwer Academic Publisher, 13
14. Krelowski J., Walker G.A.H., 1987, ApJ 312, 860
15. Krelowski J., Westerlund B.E., 1988, A&A, 190, 339
16. Krelowski J., Megier A., Strobel A., 1996, A&A 308,908
17. Kroto H.W., 1987, in "Polycyclic Aromatic Hydrocarbons and Astrophysics", eds. A. Léger & L. d'Henedecourt, Reidel Publishing Company, 197
18. Kroto H.W., Jura M., 1992, A&A 263, 275
19. Léger A., d'Hendecourt, L., 1985, A&A 146, 81
20. Léger A., d'Hendecourt L., Verstraete L., Schmidt W., 1988, A&A 213, 351
21. Salama F., Bakes E.L.O., Allamandola L.J., Tielens A.G.G.M., 1996, ApJ 458, 621
22. Scarrott S.M., Watkin S., Miles J.R., Sarre P.J., 1992, MNRAS, 255, 11
23. Sonnentrucker P., Cami J., Ehrenfreund P., Foing B.H., 1997, A&A, in press
24. van der Zwet G.P., Allamandola L.J., 1985, A&A 146, 76

# LABORATORY SIMULATION OF PROCESSES IN INTERSTELLAR ICES

W. A. SCHUTTE
*Leiden Observatory Laboratory*
*P. O. Box 9513, 2300 RA Leiden, the Netherlands fax:(31)-715275819, E-mail: schutte@strwchem.strw.LeidenUniv.nl*

**Abstract.** Laboratory applications to the study of interstellar ices are reviewed. An overview is given of the state-of-the-art in infrared spectroscopy of astrophysical ice analogs. Laboratory results on chemical processes relevant to the evolution of interstellar ice mantles are summarized. These are grain surface chemistry involving activationless reactions between atoms and molecules, activationless acid-base reactions, cryogenic reactions involving formaldehyde, and the energetic processing of the ices by ultraviolet radiation or energetic ions. Furthermore, the sublimation properties of pure and mixed ices of astrophysical relevance are summarized. Some areas in which the laboratory results have stimulated essential progress in the study of interstellar and comet ices are pointed out. Finally, some avenues are discussed through which the combined observation/laboratory simulation approach could clarify a number of standing issues on the role of ices in dense cloud chemistry.

## 1. Introduction

The objectives of laboratory experiments in the field of the study of interstellar ices are twofold. The first objective is to obtain infrared spectra (generally between 4000 – 400 $cm^{-1}$; 2.5 – 25 $\mu m$) of relevant species embedded in astrophysical ice analogs. Such data are necessary to identify absorption bands that are present in the spectra of infrared sources which suffer extinction from dense interstellar regions where ices are formed. The second objective is the simulation of physical and chemical processes which are likely to play a role in the evolution of interstellar ices. Examples are photoprocessing, ice surface chemistry, and thermal and non-thermal sub-



*J. M. Greenberg and A. Li (eds.), Formulation and Evolution of Solids in Space,* 177–201.

limation processes. This paper reviews the present state of laboratory research in these areas.

This paper is laid out as follows. Sect. 2 gives a concise description of the experimental techniques to produce interstellar ice analogs. Sect. 3 reviews infrared spectroscopy of astrophysical ice analogs. In Sect. 4 an overview is given of laboratory simulation studies of chemical and physical processes involved in the evolution of icy grain mantles. Sect. 5 gives some highlights of the application of the laboratory results to the study of interstellar ices. Sect. 6, finally, discusses how experimental work may continue to play a vital role in the interpretation of data gathered on interstellar dust and comets.

## 2. Experimental procedures

Figure 1 shows a schematic outline of a basic set-up to create interstellar ice analogs. It consists of a vacuum chamber equipped with a substrate mounted on a cold finger which can be cooled to $\sim$ 10 K. The substrate generally consists of a soft infrared transmitting window, e.g., CsI, which is sandwiched in a metal holder using indium seals to obtain good thermal conductivity. Deposition of the gas mixture is performed through one or two deposition tubes at a very low rate ($< 5$ $\mu$m / hour). Ices that are formed in this way have the amorphous structure typical for the interstellar medium. Having a double deposition facility can be useful in many cases, e.g., for depositing mixtures of species which would react with each other at room temperature, or for depositing comparable abundances of species of widely different volatility. Generally, the vacuum chamber is placed in the beam of an infrared spectrometer. The beam passes through two outside windows (generally KBr) as well as the ice sample and the CsI substrate window, thus allowing in situ transmission spectroscopy.

Experimental details were described by Hudgins et al. (1993) and Gerakines et al. (1995). A comparison between typical experimental conditions and the conditions in interstellar dense clouds is given in table 1.

Measuring the integrated band strength per molecule is essential to determine column densities of the components of interstellar ice. To this end a layer of pure ice of the desired molecule is deposited. The layer thickness is measured by registering with a photomultiplier tube the variation during deposition in the intensity of a HeNe laser beam (0.6328 $\mu$m) which is reflected of the surface of the sample (Fig. 1). While the thickness of the ice sample increases, interference between the light reflected from the ice surface and reflected from the substrate surface is in turn constructive or destructive, causing the reflected intensity to trace a fringing pattern over time. From the number of recorded fringes m, the sample thickness can

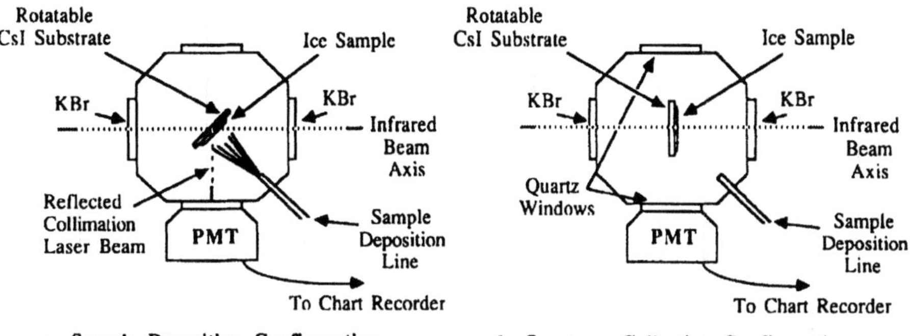


*Figure 1.* Schematic of the vacuum chamber of a basic set-up to produce interstellar ice analogs. The pictures show the system in the configurations used (a) during sample deposition and (b) during infrared spectral measurement (From Hudgins et al. 1993).

TABLE 1. Interstellar vs. Laboratory conditions

| Parameter | Interst. | Laboratory |
|---|---|---|
| Grain Temp. (K) | ~10–70 | $\geq 10$ |
| Condensation rate (molec. $cm^{-2}$ $s^{-1}$) | $10^5$–$10^7$ | $\sim 10^{15}$ |
| Condensing species | molec.: CO,$N_2$,$O_2$,... | no problem |
| | atomic: H,O,C,... | generally not |
| UV field (Photons $cm^{-2}$ $s^{-1}$;> 6eV) | $10^3$–$10^8$ | $10^{15}$ |
| Particle field | MeV $H^+$ & $He^{2+}$ | KeV $He^+$ |

than be obtained from:

$$d \; = \; (m\lambda_0)/(2n_{ice}cos\Theta_{ice}) \tag{1}$$

where d is the ice thickness, $\lambda_0$ is the wavelength of the laser, $n_{ice}$ is the visible index of refraction of the ice, and $\Theta_{ice}$ is the angle of reflection within the ice sample. The integrated band strength A is than found from

$$A \; = \; (\tau_{int} \times m)/(d \times \rho) \tag{2}$$

where $\tau_{int}$ gives the optical depth integrated over the feature, m is the molecular mass, and $\rho$ is the ice density.

To simulate interstellar chemical processes, the basic set-up shown in figure 1 can be extended in several ways. For vacuum-UV processing, a lamp can be mounted on the set-up with a transmitting window, usually $MgF_2$. Generally a hydrogen discharge lamp is used, which produces most of its output between 7 and 10.5 eV, roughly equivalent to a $\sim$ 20,000 K black body. Irradiation by charged particles is usually performed with an "ion-gun" or van de Graaff accellerator (Strazzulla & Baratta 1991; Moore & Hudson 1992), yielding protons and $He^+$ ions in the KeV-MeV energy range. Grain surface chemistry involving H atoms has been simulated by dissociating molecular hydrogen in a discharge tube, leading the atomic gas through a cooled region to obtain temperatures of $\sim$ 20 K, and squirting the H atoms on the ice sample (Hiraoka et al. 1994).

## 3. IR spectroscopy of interstellar ice analogs

The objectives of laboratory spectroscopy of interstellar ice analogs is threefold. First, the solid state spectra strongly depend on the ambient ice conditions, and identification of observed ice features therefore requires comparison to amorphous ice samples of appropriate composition and temperature. Second, to derive the column density from an observed infrared feature, the integrated band strength must be determined (Sect. 2). Third, careful studies of the obtained band shape and position for ices of varying composition and temperature may yield important information on the matrix in which an interstellar ice component is embedded.

Table 2 gives an overview of the molecules with their infrared features that are expected to have significant abundances in interstellar ices. Some of the listed species have already been detected (Sect. 5.1), while others are predicted to be present on the basis of theoretical modelling or laboratory simulation of grain chemistry, or because they are produced by comets.

In the following we will review the existing literature on spectroscopy of molecules in astrophysical ice analogs. A broad overview of the overall spectroscopic properties of a large number of astrophysically relevant species was given by d'Hendecourt & Allamandola 1986.

### 3.1. $H_2O$

$H_2O$ was the first molecule to be extensively studied in the laboratory in the context of astrophysical ices. This research was stimulated by the early detection of the 3 $\mu$m absorption band of amorphous $H_2O$ ice towards the embedded protostellar BN object (Gillett & Forrest 1973; Gillett et al. 1975).

The first infrared spectroscopy of amorphous water ice deposited at 10 K were published by Hagen et al. (1981), which lead to the identification of the interstellar feature. Optical constants for amorphous water ice after deposition at $\sim$ 10 K and subsequent annealing up to 150 K can be found in Greenberg et al. 1983, Kitta & Krätschmer 1983, Léger et al. 1983, Bergren et al. 1978, and Hudgins et al. 1993.

TABLE 2. Features of (possible) interstellar ice components

| Molecule | Mode | $\nu$ $cm^{-1}$ | $\lambda$ $\mu m$ | A cm | ref. |
|---|---|---|---|---|---|
| $H_2O$ | OH stretch | 3280 | 3.05 | 2.0 (-16) | 1,2 |
| | OH bend | 1670 | 6.00 | 8.4 (-18) | 1,2 |
| $CH_3OH$ | $\nu_3$ | 2830 | 3.54 | 6.6 (-18) | 1,3 |
| | combination modes | 2560 | 3.91 | 2.6 (-18) | 18 |
| | $CH_3$ deform./ OH bend | 1460 | 6.85 | 1.2 (-17) | 1,2 |
| | $\nu_7,\nu_{11}$ | 1130 | 8.85 | 2.4 (-18)[b] | 18 |
| | $\nu_8$ | 1030 | 9.75 | 1.8 (-17) | 1,3 |
| CO | CO stretch | 2140 | 4.67 | 1.0 (-17) | 4 |
| $CO_2$ | $CO_2$ stretch | 2340 | 4.27 | 7.6 (-17) | 5 |
| | $CO_2$ bend | 655 | 15.2 | 1.5 (-17) | 6 |
| $^{13}CO_2$ | $CO_2$ stretch | 2280 | 4.39 | 7.6 (-17) | 6 |
| $H_2CO$ | $\nu_5$ | 2884 | 3.47 | 2.7 (-18) | 7 |
| | $\nu_1$ | 2822 | 3.54 | 3.7 (-18) | 7 |
| | $\nu_2$ | 1720 | 5.81 | 9.6 (-18) | 7 |
| | $\nu_3$ | 1495 | 6.69 | 3.9 (-18) | 7 |
| $CH_4$ | $\nu_3$ | 3010 | 3.32 | 7.7 (-18) | 8 |
| | $\nu_4$ | 1300 | 7.68 | 7.3 (-18) | 8 |
| $CH_3CH_2OH$ | $\nu_a(CH_3)$ | 2980 | 3.36 | 6 (-18) | 16 |
| | $\nu(CO)$ | 1045 | 9.57 | 1.1 (-17) | 16 |
| $C_2H_6$ | $\nu_7$ | 2980 | 3.36 | 1.6 (-17) | 16 |
| | $\nu_8+\nu_{11}$ | 1460 | 6.85 | 6.0 (-18) | 16 |
| $NH_3$ | $(\nu_1,\nu_3)$ | 3370 | 2.97 | 2.2 (-17) | 9 |
| | $\nu_2$ | 1070 | 9.35 | 1.7 (-17) | 9 |
| OCS | OCS stretch | 2025 | 4.93 | 1.5 (-16) | 10 |
| $OCN^-$ | $\nu_3$ | 2165 | 4.62 | 1.1 (-16) | 11 |
| HCOOH | $\nu(CO)$ | 1710 | 5.85 | 6.7 (-17) | 12 |
| | $\delta(OH)$ | 1220 | 8.20 | 2.6 (-17) | 17 |
| $O_2$ | OO stretch | 1550 | 6.45 | {0.07–3} (-18)[a] | 13 |
| $O_3$ | $\nu_3$ | 1040 | 9.62 | 1.4 (-17) | 14 |
| $H_2$ | HH stretch | 4140 | 2.42 | $\sim$ 1 (-19) | 15 |

[a] Band strength varies with matrix [b] For $CH_3OH$ in $H_2O$ rich matrix
1. d'Hendecourt & Allamandola 1986; 2. Hagen et al. 1981; 3. Schutte et al. 1991; 4. Jiang et al. 1975; 5. Yamada & Person 1964; 6. Gerakines et al. 1995; 7. Schutte et al. 1993; 8. Boogert et al. 1997; 9. Sandford et al. 1993a; 10. Hudgins et al. 1993; 11. Keane & Schutte 1998; 12. Schutte et al. 1996b; 13. Ehrenfreund et al. 1992; 14. Smith et al. 1985; 15. Sandford et al. 1993; 16. Boudin et al. 1998; 17. Maréchal 1987; 18. Sandford & Allamandola 1993b

Admixtures of strong bases, for example $NH_3$, causes the formation of water-base complexes, giving rise to a red shoulder adjacent to the OH stretching mode. This effect was studied by Hagen et al. (1983).

Ehrenfreund et al. (1996a) looked into the infrared properties of $H_2O$ ice isolated in apolar ice matrices (CO, $O_2$, $N_2$, $CO_2$). This could be relevant for interstellar ices for which a separate phase containing predominantly apolar molecules has been shown to exist (Sect. 5.2). The brake up of the hydrogen bonding network results in a strong blue shift of the OH stretching mode, giving rise to a series of sharp bands between 3800 and 3300 $cm^{-1}$ (2.63 – 3.03 $\mu m$) due to $H_2O$ monomers, dimers, trimers etc. (Hagen & Tielens 1981). The position of these bands depends sensively on the composition of the apolar ice matrix. They could therefore, if detected, act as a probe of the matrix composition, e.g., the infrared inactive molecules $O_2$ and $N_2$.

Far-IR studies of amorphous water ice were performed by Moore & Hudson (1992, 1994). Water ice has one strong feature in the far-IR near 220 $cm^{-1}$ (45 $\mu m$) For amorphous water ice this feature is very broad ($\sim$ 100 $cm^{-1}$), but upon transition to cubic ice around 140 K, it considerably sharpens and a second, weaker, band, developes near 160 $cm^{-1}$ (62 $\mu m$). This double structure has been observed in emission towards the M giant IRAS 09371+1212 (Omont et al. 1991).

### 3.2. CO

After $H_2O$, the next molecule discovered in interstellar ice was CO (Lacy et al. 1984). CO has a single infrared mode near 2140 $cm^{-1}$ (4.67 $\mu m$) due to the CO stretching vibration. Extensive laboratory research has been aimed at determining its spectroscopic properties in astrophysically relevant ice matrices. Sandford et al. (1988) were the first to make a systematic study. They found a clear difference between the CO band for CO diluted in water ice or CO diluted in apolar ice (e.g., CO itself). When diluted in $H_2O$, the CO feature becomes broadened and shifts redwards. The broadening may be understood in terms of the high polarity of the amorphous $H_2O$ matrix, which provides a variety of distinct sites with various degrees of attraction between the CO and the matrix. The redshift can be understood in terms of the generally high attractive interaction between CO and the polar $H_2O$. These results showed that the interstellar CO feature actually consists of two components, one of CO embedded in apolar ice, and one of CO diluted in $H_2O$ ice (Sect. 5.2). This finding sparked extensive further research in this area, generally with the aim to better constrain the nature of the apolar component in which most of the CO is found. Palumbo & Strazzulla (1992,1993) investigated the properties of CO in various polar and apolar mixtures, before and after irradiation by 3 keV He ions, as an analog of the processing that may take place by Cosmic Ray particles in dense clouds. Elsila et al. (1997) investigated CO in simple and more complex apolar ice mixtures of $N_2$, $O_2$ and $CO_2$. Ehrenfreund et al. (1996, 1998) provided an even more extensive database of such ice mixtures, particularly ices containing various amounts of $CO_2$. Optical constants were provided in both the Elsila and Ehrenfreund studies. The Ehrenfreund data are accessible on the www (http://www.strw.leidenuniv.nl/ ehrenfreund/isodb).

The integrated band strength of the CO molecule in mixed ices was measured by Gerakines et al. (1995). This study introduced a new method which involved separate deposition of the CO and the diluting molecule, in order to eliminate systematic errors of earlier studies which relied on determining the composition of a pre-mixed gas. In contrast to earlier studies it was found that the CO band strength varies little with the ice matrix.

### 3.3. $CO_2$

Sandford et al. (1990) studied the properties of $CO_2$ ice diluted in mixtures with $H_2O$, $O_2$, CO, and $H_2O/CH_3OH$. A very thorough study of $CO_2$ in a wide variety of mixtures with the apolar species CO, $O_2$, $N_2$ including optical constants has been recently published (Ehrenfreund et al. 1996, 1998).

Gerakines et al (1995) investigated the dependence of the band strength of the stretching and bending mode of $CO_2$ on the ice matrix. It was found that the bending mode shows a limited ($\sim$ 30 %) variation with ice matrix, while the strong stretching mode has a nearly constant integrated strength.

### 3.4. $CH_4$

Boogert et al. (1997) did spectroscopy of $CH_4$ in binary mixtures with $H_2O$, CO, $CO_2$, $O_2$, $N_2$, $NH_3$ and $CH_3OH$, as well as $CH_4$ diluted in $H_2O/CH_3OH$. Dilutions were varied between highly diluted (dilutant: $CH_4$ $\approx 50/1$) to mixtures with roughly equal abundances. Published results focused on the $\nu_4$ feature near 1302 $cm^{-1}$ (7.68 $\mu m$), since this band has been observed both by airborne and space observatories (Boogert et al. 1996, 1997).

As for CO, the $CH_4$ feature showed broadening and redshift in polar matrices ($H_2O$, $CH_3OH$, $NH_3$). Besides this, experiments with $CO/CH_4$ and $H_2O/CH_4$ mixtures at various mixing ratio's showed that there is a general trend of the width of the feature to reach a maximum close to the 1:1 ratio. This trend can apparently be ascribed to the high complexity of matrices consisting of two species of comparable abundances, providing a maximum of sites into which a molecule can become trapped.

### 3.5. $CH_3OH$

As much as three absorption features observed towards massive protostellar sources can be ascribed to methanol ice (see Sect. 6). These are the $\nu_3$ band near 3.54 $\mu$m (2829 $cm^{-1}$), the weak combination band at 3.91 $\mu$m (2560 $cm^{-1}$) and the $\nu_8$ feature at 9.75 $\mu$m (1026 $cm^{-1}$). Good fits can be obtained with mixtures of methanol and water. Schutte et al. (1991) studied the $\nu_8$ feature of highly and mildly diluted methanol in water ice. It was found that the band shifts strongly to the red with increasing dilution, apparently due to the attractive interaction of $CH_3OH$ with $H_2O$. On the other hand, the position and width of the $\nu_3$ band shows little variation between pure and $H_2O$ diluted $CH_3OH$ (Schutte et al. 1996a). The double 3.91 $\mu$m (2560 $cm^{-1}$) combination band, finally, shows a blue-shift upon dilution with $H_2O$ (Allamandola et al. 1992).

### 3.6. OCS

OCS has a very strong feature near 2040 $cm^{-1}$ (4.90 $\mu$m) which has been detected towards the massive embedded protostar W33A. Due to the extremely large intensity of this band (table 1), as well as the happy circumstance that it lies in a spectral region which is free of absorption features of the main ice components, this molecule can be detected even at very low abundances in the ice mantles. A detailed study of its spectroscopic properties in matrices of $H_2O$ and $CH_3OH$ at various mixing ratio's was performed by Palumbo et al. (1995). The band strength of the OCS feature was measured by Hudgins et al. (1993).

### 3.7. $OCN^-$

UV photolysis of ice mixtures containing CO and $NH_3$ leads to the production of a strong band near 2167 $cm^{-1}$ (4.61 $\mu$m). This feature gives a good match to the XCN absorption band which has been observed towards a number of high- and low-mass embedded objects (Sect. 5.1). The $^{13}C$, $^{15}N$ and $^{18}O$ isotopic shifts of this feature obtained when using labelled CO and $NH_3$ in the original ice mixture as well as the presence of additional weak bands shows that the laboratory band can be ascribed to the $OCN^-$ ion (Grim and Greenberg 1987; Schutte and Greenberg 1997). This species is presumably produced after prior photochemical formation of the HNCO acid, followed by the acid-base reaction with $NH_3$ (Grim et al. 1989). Indeed, in a recent study the $OCN^-$ ion was produced without UV radiation by deposition and warm-up of mixtures of iso-cyanic acid (HNCO), $NH_3$ and $H_2O$ (Keane & Schutte 1998). This clearly demonstrates the viability of the acid-base reaction at cryogenic temperatures.

The laboratory studies show that the position of the $OCN^-$ feature depends on ice composition and temperature. In mixtures of CO and $NH_3$ it falls at about 10 $cm^{-1}$ lower frequency than in a $H_2O$ dominated matrix (Grim & Greenberg 1987; Schutte & Greenberg 1997). In $H_2O$ dominated matrices warm-up to 120 K causes a 2 $cm^{-1}$ redshift of the band (Tegler et al. 1993).

### 3.8. MISCELLANEOUS STUDIES

A number of laboratory studies were performed into the spectroscopic properties of molecules which sofar have not been positively identified in interstellar ices. Much of this research was in anticipation of the observations by the Infrared Space Observatory (ISO). Ehrenfreund et al. (1992) investigated the stretching feature of solid $O_2$ at 1550 $cm^{-1}$ (6.45 $\mu$m) in ices containing $H_2O$, CO and $CO_2$. They found that the strength of this very weak feature, which only becomes infrared active due to perturbation by the ice matrix, strongly depends on the composition of the mixture. The strongest $O_2$ bands were obtained in mixtures containing a lot of $CO_2$ ice. Schutte et al. (1996a) studied the infrared properties of $H_2CO$ in matrices of $H_2O$ and $CH_3OH$. The published results focused on the $\nu_5$ feature near 2880 $cm^{-1}$ (3.47 $\mu$m), whose presence was indicated by observations of GL 2136. Boogert et al. (1997) studied the properties of the asymmetric stretching mode of $SO_2$ near 7.55 $\mu$m (1324 $cm^{-1}$), in a variety of mixtures with $H_2O$, CO and $CH_3OH$. This study was motivated by the presence of a shallow feature near 7.62 $\mu$m (1312 $cm^{-1}$) towards some protostellar objects (Boogert et al 1996, 1997). Bernstein et al. (1997) studied the intense CN stretching mode of a large number of organic nitriles and iso-nitriles (e.g., HCN) which falls in the region 2300 – 2050 $cm^{-1}$ (4.35 – 4.90 $\mu$m).

## 4. Simulation of chemical and physical processes in interstellar ices

A number of chemical and physical processes that play a role in the evolution of interstellar ices have been studied in the laboratory. These are surface chemistry, activationless acid-base reactions, cryogenic formaldehyde chemistry, photochemistry and radiation chemistry as well as the sublimation properties of pure and mixed ices.

In the following a summary is given of these research areas.

### 4.1. SURFACE REACTIONS

Grain surface chemistry occurs because light atomic species H, C, O and N can migrate over the grain surface by thermal hopping or through tunnelling, allowing them to react with accreted radicals and molecules. These reactions are generally thought to play a pivotal role in determining the composition of interstellar ices (Tielens & Hagen 1982, d'Hendecourt et al. 1985, Hasegawa et al. 1992) . It is still unclear whether light radicals such as $CH_3$ and OH are also able to diffuse over the grain surface. While Tielens & Allamandola (1987a) argued that only the above atomic species will show appreciable surface diffusion, some models include surface chemistry by heavier radicals (e.g., Hasegawa & Herbst 1993). One interesting aspect is that the nature of the surface chemistry may sensitively depend on temperature; while at the temperature of dust in dense clouds of about 10 K diffusion by non-atomic species may be inhibited, at temperatures of 20 – 30 K which may occur in protostellar regions, surface diffusion of heavier species will become important, allowing a more complex grain surface chemistry (Tielens & Allamandola 1987a, Hasegawa & Herbst 1993).

A number of potentially important grain surface reactions have been studied in the laboratory. These are summarized in Table 3, together with a number of important reactions which have been proposed, but for which no laboratory studies have yet been performed.

TABLE 3. Laboratory studies of grain surface chemistry

| | react. 1 | react.(s) 2 | product 1 | product 2 | references |
|---|---|---|---|---|---|
| 1 | CO | H | HCO | | 1,2 |
| 2 | HCO | H | $H_2CO$ | | 1,2 |
| 3 | $H_2CO$ | 2H | $CH_3OH$ | | 2 |
| 4 | $H_2CO$ | H | HCO | | 2 |
| 5 | CO | O | $CO_2$ | | 3 |
| 6 | $O_2$ | O | $O_3$ | | 3 |
| 7 | N | 3H | $NH_3$ | | 4 |
| 8 | C | 4H | $CH_4$ | | |
| 9 | O | 2H | $H_2O$ | | |
| 10 | $O_2$ | 2H | $H_2O_2$ | | |
| 11 | $H_2O_2$ | H | $H_2O$ | OH | |
| 12 | $N_2$ | 2H | $N_2H_2$ | | |
| 13 | $CO_2$ | 2H | HCOOH | | |
| 14 | N | O | NO | | |
| 15 | NO | C | OCN | | |
| 16 | OCN | H | HNCO | | |

1. van IJzendoorn et al. 1983; 2. Hiraoka et al. 1994; 3. Grim & d'Hendecourt 1986; 4. Hiraoka et al. 1995.

One set of important surface reactions start with the abundant molecule

CO, forming the more complex species $H_2CO$, $CH_3OH$ and $CO_2$. $CH_3OH$ and $CO_2$ are important constituents of interstellar ice, while the indications for solid $H_2CO$ are less clear (Allamandola et al. 1992, de Graauw et al. 1996, Schutte et al. 1996a). Van IJzendoorn et al. (1983) photolysed a mixture of $CH_4$ and CO embedded in argon and subsequently studied the diffusion and reactions of the produced H atoms upon subsequent warm-up by a few K. While HCO was efficiently produced (reaction 1, Table 3), formation of $H_2CO$ was not observed. Hiraoka et al. (1994) bombarded CO ice with atomic hydrogen cooled to $\sim$ 20 K. It was found that the conversion of CO to $H_2CO$ and $CH_3OH$ is very low in these experiments. They attributed this to efficient destruction of $H_2CO$ by hydrogen abstraction (reaction 4, Table 3). Thus laboratory evidence that $H_2CO$ and $CH_3OH$ may be efficiently produced by grain surface reactions of CO and H is still lacking.

Grim & d'Hendecourt 1986 studied the production of $CO_2$ by reactions of CO with atomic O. A mixture of CO and $O_2$ in an Argon matrix was UV photolysed to produce atomic O. During subsequent warm-up by a few K, promoting the diffusion of O, formation of $O_3$ was observed but little or no $CO_2$ was produced. This experiment clearly shows that reaction 6 should proceed. In this paper it was concluded that the production of $CO_2$ by reaction 5 would not proceed. However, perhaps the $CO_2$ production was quenched by the presence of large quantities of $O_2$, and it can not be excluded that the reaction of CO with O could proceed at more pure CO ice surfaces.

Hiraoka et al. 1995 bombarded an ice mixture of atomic and molecular nitrogen with H atoms. The formation of $NH_3$ showed the importance of reaction 7 in grain surface chemistry. Another important result of this study was the observation that H atoms may penetrate into the ices and react inside the matrix.

### 4.2. ACID–BASE REACTIONS

Laboratory studies have shown that a number of acid–base reactions often have no or very small activation barriers and thus may efficiently proceed on grain surfaces. Such acid–base reactions may thus modify the classical scheme of grain surface chemistry. Table 4 summarizes the acid–base reactions for which laboratory studies showed that they proceed at very low temperature ($T \leq 30$ K).

### 4.3. FORMALDEHYDE REACTIONS

In the presence of traces of $NH_3$ formaldehyde can become involved in a large suite of reactions under cryogenic conditions. This may involve simple

TABLE 4. Activationless acid-base reactions of astrophysical relevance

| acid | base | neg. ion | pos. ion | reference |
|---|---|---|---|---|
| $HNO_3$ | $NH_3$ | $NO_3^-$ | $NH_4^+$ | Ritzhaupt & Devlin 1977 |
| HCOOH | $NH_3$ | $HCOO^-$ | $NH_4^+$ | Schutte et al. 1998, in preparation |
| HCOOH | $N_2H_4$ | $HCOO^-$ | $N_2H_5^+$ | Boudin et al. 1998 |
| HNCO | $NH_3$ | $OCN^-$ | $NH_4^+$ | Keane & Schutte 1998 |

polymerization (i.e., formation of polyoxy methylene) but also reactions involving other species like $NH_3$ and $CH_3OH$ (Schutte et al. 1993). These reactions occur when the ice is warmed above $\sim$ 60 K, indicating that they possess a finite, though small, activation energy. Table 5 summarizes the products that were observed in various astrophysical ice analogs.

TABLE 5. Products of formaldehyde polymerization reactions in astrophysical ice analogs

| Mixture composition | Products |
|---|---|
| Pure $H_2CO$ | No polymers observed |
| $H_2O/H_2CO/NH_3$ | $HCO[-O-CH_2]_n-H$ |
| | $HOCH_2NH_2$ |
| | $HOCH_2-O-CH_2NH_2$ |
| | $HOCH_2OH$ |
| | $HOCH_2-O-CH_2OH$ |
| $H_2CO/NH_3/H_2O/CH_3OH$ | $H_3C-CH(OH)-O-(CH_3)$ |

## 4.4. PHOTOCHEMICAL PROCESSES

A number of studies on photochemical processes in astrophysical ice analogs have been published. A database on UV irradiated ices created at the Leiden Observatory can be accessed on the www (http://www.strw.leidenuniv.nl/ lab). Some of the main photochemical processes are summarized in Table 6 . Important photoproducts of astrophysical ice analogs consisting of $H_2O$, CO, $NH_3$, $O_2$ and $CH_3OH$ are $CO_2$, $CH_4$, $O_3$, $HCONH_2$, $OCN^-$, HCOOH, $HCOO^-$, $HCOOCH_3$, and $CH_3CH_2OH$. Furthermore, photolysis of ice mixtures containing $H_2O$, $NH_3$ and CO or $CH_3OH$ also leads to the production of a non-volatile residue, which consists of a variety of oxygen and nitrogen-rich organic molecules. (Agarwal et al. 1985; Briggs et al. 1992; Bernstein et al. 1995).

TABLE 6. Examples of different processes induced by UV photolysis (from Hagen 1982; d'Hendecourt et al. 1986; Allamandola et al. 1988; Grim et al. 1989; Bernstein et al. 1995; Gerakines et al. 1996; Schutte & Greenberg 1997)

| Reactant 1 | Reactant 2 | Product 1 | Product 2 |
|---|---|---|---|
| photodissociation | | | |
| $H_2O$ | | $OH$ | $H$ |
| $NH_3$ | | $NH_2$ | $H$ |
| $NH_3$ | | $NH$ | $H_2$ |
| $CH_3OH$ | | $H_2CO$ | $H_2$ |
| $CH_3OH$ | | $CH_4$ | $O$ |
| $CO_2$ | | $CO$ | $O$ |
| addition | | | |
| $CO$ | $H$ | $HCO$ | |
| $CO$ | $NH$ | $HNCO$ | |
| $CO$ | $OH$ | $CO_2$ | $H$ |
| $HCO$ | $H$ | $H_2CO$ | |
| $HCO$ | $OH$ | $HCOOH$ | |
| $HCO$ | $NH_2$ | $HCONH_2$ | |
| acid-base interaction | | | |
| $HCOOH$ | $NH_3$ | $HCOO^-$ | $NH_4^+$ |
| $HNCO$ | $NH_3$ | $OCN^-$ | $NH_4^+$ |
| formaldehyde reactions | | | |
| $H_2CO$ | $H_2CO$ | $HCOOCH_3$ | |
| 6 $H_2CO$ | 4 $NH_3$ | $C_6H_{12}N_4$ | 6 $H_2O$ |

Typical for photochemical processes is the cleavage of molecular bonds and formation of free radicals. These radicals can then recombine to form larger organic species. In pure $CH_4$ ice, this process was observed to produce carbon chains containing up to $\sim$ 7 C atoms (Gerakines et al. 1996). Moreover, the energy provided by the UV photons enables acid-base reactions and reactions involving formaldehyde. The latter process is probably responsible for the formation of the bulk of the non-volatile species found in the organic residue (Schutte 1988, Schutte et al. 1993; Bernstein et al. 1995). Particularly, the complex molecule Hexamethylene Tetramine (HMT) has been found in large quantities in the organic residue created by irradiation of $H_2O/CH_3OH/NH_3$ mixures (Bernstein et al. 1995). This molecule is formed by reactions between $H_2CO$ and $NH_3$, where the $H_2CO$ is formed in large quantities by photodissociation of $CH_3OH$. Due to the available excess energy, the formaldehyde reactions promoted by UV photolysis are of a different nature than the ones promoted by simple warm-up (Sect. 4.3; Bernstein et al. 1995).

### 4.5. RADIATION CHEMISTRY

A limited number of laboratory studies have been performed into the effects of irradiating interstellar ice analogs with charged particles analogous to cosmic rays. Irradiating 10 K ice mixtures of $H_2O$ and $CO_2$ with 1.5 MeV $He^+$ produced $H_2CO$ by reactions of liberated C with the $H_2O$ molecules (Pirronello et al. 1982). Palumbo & Strazzulla (1993) irradiated a number of samples with 3 KeV $He^+$. Having similar specific energy loss, the radiation effects should be similar to $\sim$ 1 MeV cosmic ray protons. Irradiation of 10 K methanol ice produced $CO_2$ and CO, but, contrary to UV irradiation, no $H_2CO$ was produced (Strazzulla, private communication). Irradiation of CO produced $CO_2$, $C_2O_2$ and CO. In a recent study, Hudson & Moore (1997) produced ethane ($C_2H_6$) and ethylene ($C_2H_4$) by irradiation mixtures of $H_2O$ with $C_2H_2$ and/or $CH_4$ by 0.8 MeV protons. Kaiser & Roessler (1997), observed the production of complex polycyclic aromatic hydrocarbons when irradiating $CH_4$ ice with 9 Mev $He^{2+}$ ions. Finally, the efficient conversion of carbon containing ices to refractory organic materials has been reported by several authors (Moore & Donn 1982; Foti et al. 1984; Strazzulla et al. 1984; Lanzerotti et al. 1987; Strazzulla & Baratta 1992).

### 4.6. SUBLIMATION

Studying the sublimation behaviour of astrophysical ice analogs is important for understanding the composition of ices and the gas phase in warm protostellar regions. Furthermore, studying non-equilibrium sublimation behaviour caused by impulsive heating events, e.g., a cosmic ray hit, is important to understand the exchange between gas and solid phase in the general dense cloud medium.

For pure ices, the sublimation rate ($s^{-1}$) of a surface molecule is given by:

$$R_s = \nu_0 . exp(-\Delta H_s/kT) \tag{3}$$

where $\nu_0$ is the lattice vibrational frequency of the molecule within its surface matrix site, $\Delta H_s$ is the binding energy of the molecule on the ice surface, k is the Boltzmann constant, and T is the ice temperature. From this equation sublimation rates in the laboratory on timescales of, say, hours, can be used to determine sublimation rates at much larger timescales in the astrophysical situation. Table 7 gives, for some pure ices of astrophysical relevance (Sect. 6.2), the temperature at which an ice layer of 0.01 $\mu$m thicknness sublimes in $10^5$ years (Note that this temperature, due to the steep dependence of sublimation rate on temperature changes very little when the layer thickness or sublimation timescale is varied by a factor ~10). $\nu_0$ and $\Delta H_s$ were adapted from Sandford & Allamandola 1993. The final entry in table 7 shows that pure $H_2$ ice will not form under astrophysical conditions, while other ices are stable in the general dense cloud medium where grain temperatures are ~ 10 K (Greenberg 1971; Mathis et al. 1983). In hot cloud cores temperatures can range up to ~ 100 K (Goldsmith 1987), and sublimation of the ice is expected in this case.

TABLE 7. Temperature at which a 0.01 $\mu$m layer of pure ice sublimes in $10^5$ year

| Molecule | Temp. (K) |
|---|---|
| $H_2O$ | 94 |
| $CH_3OH$ | 79 |
| $CO_2$ | 50 |
| $CO$ | 18 |
| $H_2$ | 2 |

For mixed ices of volatile and less volatile components the sublimation behaviour is rather complicated since in this case the sublimation rate of the volatile component will not be dominated by the diffusion rate, but by the thermal evolution of various lattice sites in the non-volatile matrix in which the volatile species can be trapped. For example, for CO higly diluted in $H_2O$ sublimation takes place only at ~ 125 K (Sandford et al. 1988, Schmitt et al. 1988; it must be noted that this temperature is still well below that sublimation temperature for water ice of ~ 160 K). However, if the CO abundance exceeds ~ 16 % of that of $H_2O$, the excess CO quickly sublimes near 30 K (Schmitt et al. 1988), similar to the behaviour of pure CO ice.

For $CO_2$ diluted in $H_2O$, a $CO_2$ abundance up to ~ 10 % can remain trapped in the ice up to ~ 130 K. The excess $CO_2$ sublimes in the temperature interval 85 – 130 K (Schmitt et al. 1988; Sandford & Allamandola 1990; ~ 85 K is the sublimation temperature of pure $CO_2$ ice in the laboratory).

An interesting case is presented by mixtures of $CH_3OH$ and $H_2O$ ice. For mixtures with more than 7 % $CH_3OH$, separation of the excess $CH_3OH$ from the $H_2O$ ice occurs at 120 K. After the separation, the relatively volatile $CH_3OH$ ice rapidly sublimes (Blake et al. 1991; Sandford et al. 1993). The remaining $\sim$ 7 % $CH_3OH$ ice remains trapped until the $H_2O$ ice itself evaporates at about 160 K.

It must be noted that the various temperature regions in which the volatile species leave the $H_2O$ matrix can be related to structural transitions in the ice and the related loss of sites in which volatile species can be trapped. A thorough discussion of this effect can be found in Jenniskens & Blake (1994).

Finally, a limited number of laboratory studies were done on non-thermal sublimation processes. These kind of processes dominate the sublimation of ices in general dense clouds, where due to the low grain temperatures ($\sim$ 10 K) thermal sublimation is negligible. Benit et al. (1987) irradiated $H_2O$ ice by ions with energy in the KeV – Mev range. It was found that this process leads to evaporation of the ices along the entire track of the particle. d'Hendecourt et al. (1982), Léger et al. (1985) and Schutte & Greenberg (1991) studied the sublimation caused by thermal runaway which can occur in photoprocessed ices which contain potential chemical energy in the form of stored free radicals. In space, such a runaway can for example be triggered by the passage of a cosmic ray particle. It was found that this process may be highly efficient in subliming volatile ices like CO, but does not cause significant sublimation of refractory ices such as $H_2O$ and $CH_3OH$.

## 5. Astrophysical applications of the laboratory studies

We summarize highlights of the application of the laboratory simulation studies which have greatly enhanced our understanding of solid state processes in interstellar dense clouds.

### 5.1. IDENTIFICATION OF INTERSTELLAR ICE ABSORPTION BANDS

Comparison between the spectra of infrared sources obscured by dense cloud material and the laboratory spectra of interstellar ice analogs have been succesfully used to identify many of the observed absorption lines with molecules frozen in icy mantles. Table 8 summarizes the lines and molecules which sofar have been identified, together with the relative abundance as derived from the observed integrated optical depth using the integrated band strength as found in the laboratory (Table 2).

TABLE 8. Absorption features of interstellar ice components

| Feature μm | cm$^{-1}$ | molecule | Abundance % | references |
|---|---|---|---|---|
| 2.42 | 4128 | $H_2$ | $\geq 1$ | 1 |
| 3.05 | 3280 | $H_2O$ | 100 | 2,3 |
| 3.54 | 2826 | $CH_3OH$ | 0 – 8 | 4,5,26 |
| 3.91 | 2560 | $CH_3OH$ | | 5 |
| 4.27 | 2342 | $CO_2$ | 12 – 16 | 6,24 |
| 4.39 | 2278 | $^{13}CO_2$ | ~ 0.2 | 6 |
| 4.62 | 2165 | XCN($OCN^-$ ?) | 0 – 0.8 | 7,8,9,10,14,19 |
| 4.67 | 2140 | CO | 1 – 40 | 11,12,13,14 |
| 4.91 | 2139 | OCS | 0.04 | 15,16 |
| 5.85[a] | 1710 | HCOOH[b] | 3 | 17 |
| 6.0 | 1670 | $H_2O$ | | 18,17 |
| 7.68 | 1302 | $CH_4$ | 0.5 – 2 | 20,21,22 |
| 8.90 | 1120 | $CH_3OH$ | | 23 |
| 9.75 | 1030 | $CH_3OH$ | | 23 |
| 15.2 | 660 | $CO_2$ | | 6,25 |

[a] Blended with the $H_2O$ 6.0 μm band.
[b] Larger organic acids (R-COOH) may also provide a good fit.
1.Sandford et al. 1993b; 2. Willner et al. 1982; 3. Smith et al. 1989; 4. Grim et al. 1991; 5. Allamandola et al. 1992; 6. de Graauw et al. 1996; 7. Lacy et al. 1984; 8. Tegler et al. 1993; 9. Grim & Greenberg 1987; 10. Schutte & Greenberg 1997; 11. Lacy et al. 1984; 12. Tielens et al. 1991; 13. Kerr et al. 1993; 14. Chiar et al. 1998; 15. Geballe et al. 1985; 16. Palumbo et al. 1995; 17. Schutte et al. 1996b; 18. Tielens & Allamandola 1987b; 19. Tegler et al. 1995; 20. Lacy et al. 1991; 21. Boogert et al. 1996; 22. Boogert et al. 1997; 23. Skinner et al. 1992; 24. Gürtler et al. 1996; 25. d'Hendecourt & Jourdain de Muizon 1989; 26. Schutte et al. 1996a.

## 5.2. PHASES OF THE INTERSTELLAR ICE

Information on the composition of the ice matrix in which a molecule is embedded can be obtained by comparing the band profile in various matrices with the observed feature. Such comparisons have shown that interstellar ices are not homogeneously mixed but consist of at least four independent phases. The approximate compositions of these phases are listed in table 9.

The first evidence for the existence of separate interstellar ice phases was obtained from the CO band near 2140 cm$^{-1}$ (4.67 μm; Sandford et al. 1988; Tielens et al. 1991). This band clearly consists of two components; a sharp feature at 4.674 μm (2139 cm$^{-1}$) and an underlying broader band centered at 4.680 μm (2137 cm$^{-1}$). The broader feature is clearly due to CO diluted in $H_2O$ ice. The narrow component points to CO inside an apolar

TABLE 9. Phases of interstellar ices

| Phase | molecule | appr. abundance | derived from | references |
|---|---|---|---|---|
| I | $H_2O$ | 100 | $H_2O$ OH str. | 1 |
| | CO | 1 – 10 | CO str. | 2,3,4 |
| | $CO_2$ | $\sim 8$ | $CO_2$ str. | 5 |
| | $(CH_4)^a$ | (0.5 – 2) | $CH_4$ $\nu_4$ | 6 |
| II | $H_2O$ | $\sim 8$ | $CH_3OH$ $\nu_8$ and $CH_3OH$ 3.90 μm f. | 7,8 |
| | $CH_3OH$ | 5 | $CH_3OH$ $\nu_3$ | 8,9,10 |
| | $(CH_4)^a$ | (0.5 – 2) | $CH_4$ $\nu_4$ | 6 |
| | OCS | 0.04 | OCS str. | 11 |
| III | CO | 0 – 40 | CO str. | 2,3,4 |
| | $(O_2, N_2, CO_2)^b$ ? | ? | CO stretch | 4,12 |
| IV | $CO_2$ | $\sim 4$ | $CO_2$ bend | 5 |

[a] It is unclear whether $CH_4$ resides in the phase I or phase II ice (Boogert et al. 1996).
[b] Indirect indications of the presence of these molecules have been reported
1. Smith et al. 1989; 2. Sandford et al. 1988; 3. Tielens et al. 1991; 4. Chiar et al. 1998; 5. de Graauw et al. 1996; 6. Boogert et al. 1996; 7. Skinner et al. 1992; 8. Allamandola et al. 1992; 9. Grim et al. 1991; 10. Schutte et al. 1996a; 11. Palumbo et al. 1995; 12. Ensila et al. 1997

ice matrix. Thus it was found that CO is not homogeneously mixed with the dominant $H_2O$ ice, but is predominantly concentrated in a separate ice phase (phase III ice).

Fitting the $\nu_8$ band of solid methanol observed towards the massive embedded protostar GL2136 required a $H_2O/CH_3OH \leq 1.5$ (Skinner et al. 1992). Since the ratio of the column densities of $H_2O$ and $CH_3OH$ is $\sim 13$, this clearly showed that methanol, like CO, must be predominantly concentrated in a separate phase away from the dominant $H_2O$ ice component (phase II). On the other hand pure methanol ice gave a poor fit to the 3.90 $\mu$m feature seen towards W33A, while a good fit can be obtained with $H_2O/CH_3OH \approx 2$. Thus the methanol enriched ice component should still contain a considerable quantity of $H_2O$.

The broad $CO_2$ bending mode near 15.2 $\mu$m (658 cm$^{-1}$) was first observed by ISO. While the bulk of the feature is fitted by $CO_2$ diluted in $H_2O$ ice, the band displays relatively sharp substructure at 15.10 and 15.25 $\mu$m (662 and 656 cm$^{-1}$). This double structure is well matched by (almost) pure

$CO_2$ ice (de Graauw et al. 1996), the existence of a fourth, $CO_2$ dominated phase of interstellar ice (phase IV).

### 5.3. ORGANIC MOLECULES IN COMETS

The PUMA and PIA impact mass spectrometers on board the Vega and Giotto spacecraft detected that refractory organic species constitute a large fraction of the dust of Halleys comet. From the distribution of the impact energy transferred to the organic and silicate component it was deduced that the particles consist of an organic coating on a silicate core (Jessberger et al. 1987). This is consistent with an origin as core-mantle interstellar dust particles (Greenberg 1982). In this model, organic species are formed by processing of the icy mantles inside dense clouds and in the presolar nebula. At least part of the cometary organic dust should be rich in oxygen, as indicated by the high oxygen abundance of the refractory grain component (Jessberger & Kissel 1991). This is consistent with the laboratory results in which highly oxygen-rich organic molecules are produced from photoprocessing of interstellar ice analogs (Sect. 4.4, 4.5).

## 6. Future work

In the near future the study of laboratory produced interstellar ice analogs will remain essential for the interpretation of the new spectroscopic data provided by ISO. Spectroscopy of components of interstellar ices in matrices of different composition may show how molecules are distributed over the various ice phases (Sect. 5.2). This could clarify the ambient chemical conditions at the time when the accretion of these phases took place and thus inform about the chemical evolution of dense clouds and protostellar regions.

A long standing question in the study of interstellar ice is the importance of UV and particle processing. Unfortunately, the most abundant products of processing as seen in the laboratory (Sect. 4.4 and 4.5) may, according to theoretical models, also be produced by surface chemistry (Sect. 4.1). It therefore seems unlikely that ISO could detect a “smoking gun” species which could be unambiguously traced to processing. Rather, the effects of energetic processing should be quantitatively studied and compared to the composition of the interstellar ice to see whether processed ices are able to match the *overall* observed abundance patterns.

Infrared spectrocopic observations are generally not very sensitive to complex species which could be present in interstellar ice as trace components ($< 0.5$ % of $H_2O$). Such molecules, consisting of at least 2 C atoms, are for example produced by photoprocessing (Sect. 4.4). However, when ice mantles sublime, e.g., in hot dense cores, the composition of the gas phase can be probed at very high sensitivity by observing the mm and sub-mm emission lines produced by molecular rotational transitions. In this way complex trace molecules may be retrieved. Such species could for example be $CH_3CH_2OH$ and $HCOOCH_3$. These molecules are observed in considerable quantities in hot cores (Millar et al. 1995; Charnley et al. 1995; Millar 1997; Ohishi 1997; Helmich et al. 1998). Both are produced by photolysis of ices containing $CH_3OH$ (Bernstein et al. 1995; Gerakines et al. 1996; Sect. 4.3). However, gas phase destruction and formation processes makes determining the original abundances of these molecules in the icy mantles somewhat complicated (Charnley et al. 1995; Helmich et al. 1998). Nevertheless, complex molecules of this kind should be very good probes of processing, since grain surface chemistry likely produces them in only minor quantities (Hasegawa & Herbst 1993). Thus, sensitive measurements of the sublimation products of interstellar ice analogs which have been processed by UV or particle bombardment could be essential for modelling the gas phase composition of hot cores in dense clouds.

Another important project would be the analysis of the organic residues produced by energetic (UV or particle) processing of interstellar ice analogs. A database of this kind, together with the information already available on thermal production of organic species by cryogenic formaldehyde reactions (Sect. 4.3), could be essential for understanding the origin of the organic component of comets. In the not too distant future the composition of comets will be investigated in situ, for example by ROSETTA. Comparison with the laboratory results could give important information on the conditions in the pre-solar nebula.

Finally, while theoretical models indicate that grain surface chemistry plays a dominant role in producing the molecules found in interstellar ices, little laboratory simulation in this area has so far been performed. Indeed laboratory evidence for the main surface chemical pathways which have been proposed for the production of the most abundant molecules in interstellar ices is still lacking. A deeper understanding of the chemical evolution of interstellar ice and the relationship between the ice composition and the chemical conditions in the gas phase at the time of condensation awaits further studies in this area.

## References

Agarwal V.K., Schutte W., Greenberg J.M., Ferris J.P., Briggs R., Connor S., van de Bult, C.P.E.M., Baas F. 1985, Photochemical reactions in interstellar grains. Photolysis of CO, $NH_3$, and $H_2O$, *Origins of Life* **16**, 21–40.

Allamandola L. J., Sandford S. A. and Valero G. J. 1988, Photochemical and thermal evolution of interstellar/precometary ice analogs, *Icarus* **76**, 225–252.

Allamandola L. J., Sandford S. A., Tielens A. G. G. M. and Herbst T. M. 1992, Infrared spectroscopy of dense clouds in the C–H stretching region: Methanol and "diamonds", *ApJ* **399**, 134–146.

Benit J., Bibring J.-P., Della-Negra S., Le Beyec Y., Mendenhall M., Rocard F., Standing K. 1987, *Nuclear Instruments and Methods in Physics Research* **B19/20**, 838–842.

Bergren M.S., Schuh D., Sceats M.G., Rice S.A. 1978, The OH stretching region infrared spectra of low density amorphous solid water and polycrystalline ice Ih, *J. Chem. Phys.* **69**, 3477–3482.

Bernstein M.P., Sandford S.A, Allamandola L.J., Chang S., Scharberg M.A. 1995, Organic compounds produced by photolysis of realistic interstellar and cometary ice analogs containing methanol, *ApJ* 454, 327–344.

Bernstein M.P., Sandford S.A, Allamandola L.J. 1997, The infrared spectra of nitriles and related compounds frozen in Ar and $H_2O$, *ApJ* **476**, 932–942.

Blake D., Allamandola L., Sandford AA., Hudgins D., Freund F. 1991, Clathrate hydrate formation in amorphous cometary ice analogs in vacuo, *Science* **254**, 548–551.

Boogert, A.C.A., Schutte, W.A., Tielens, A.G.G.M., Whittet, D.C.B., Helmich, F.P., Ehrenfreund, P., Wesselius, P.R., de Graauw, Th., and Prusti, T. 1996, Solid methane towards deeply embedded protostars, *A&A* **315**, L377-L380.

Boogert A.C.A., Schutte W.A., Helmich F.P, Tielens A.G.G.M., Wooden D.H. 1997, Infrared observations and laboratory simulations of interstellar $CH_4$ and $SO_2$, *A&A* **317**, 929–941.

Boudin N., Schutte W.A., Greenberg J.M. 1998, Constraints to the abundance of a number of molecules in interstellar ice: Laboratory studies and astrophysical implications, submitted to *A&A*.

Briggs R., Ertem G., Ferris J. P., Greenberg J. M., McCain P. J., Mendoza-Gomez C. X. and Schutte W. 1992, Comet Halley as an aggregate of interstellar dust and further evidence for the photochemical formation of organics in the interstellar medium, *Origins of Life and Evolution of the Biosphere* **22**, 287–307.

Charnley S.B., Kress M.E., Tielens A.G.G.M., Millar T.J. 1995, Interstellar alcohols, *ApJ* **448**, 232–239.

Chiar J. E., Adamson A. J., Kerr T. H. and Whittet D. C. B. 1994, Solid carbon monoxide in the Serpens dark cloud, *ApJ* **426**, 240–248.

Chiar J.E., Gerakines P.A., Whittet D.C.B., Pendleton Y.J., Tielens A.G.G.M., Adamson A.J., Boogert A.C.A. 1998, Processing of icy mantles in protostellar envelopes, *ApJ*, in press.

de Graauw, Th., Whittet, D.C.B., Gerakines, P.A., Bauer, O.H., Beintema, D.A., Boogert, A.C.A., Boxhoorn, D.R., Chiar, J.E., Ehrenfreund, P., Feuchtgruber, H., Helmich, F.P., Heras, A.M., Huygen, R., Kester, D.J.M., Kunze, D., Lahuis, F., Leech, K.J., Lutz, D., Morris, P.W., Prusti, T., Roelfsema, P.R., Salama, A., Schaeidt, S.G., Schutte, W.A., Spoon, H.W.W., Tielens, A.G.G.M., Valentijn, E.A., Vandenbusshe, B., van Dishoeck, E. F., Wesselius, P.R., Wieprecht, E., Wright, C.M. 1996, SWS observations of solid $CO_2$ in molecular clouds, *A&A* **315**, L345-L348.

d'Hendecourt L. B., Allamandola L. J., Baas F., and Greenberg J. M. 1982, Interstellar grain explosions: Molecule cycling between gas and dust, *A&A* **109**, L12–L14.

d'Hendecourt L. B., Allamandola L. J. and Greenberg J. M. 1985, Time dependent chemistry indense molecular clouds: I. Grain surface reactions, gas/grain interactions and infrared spectroscopy, *A&A* **152**, 130–150.

d'Hendecourt L. B., Allamandola L. J., Grim R. J. A. and Greenberg J. M. 1986, Time dependent chemistry indense molecular clouds: II. Ultraviolet photoprocessing and infrared spectroscopy of grain mantles, A&A 158, 119–134.

d'Hendecourt L. B. and Allamandola L. J. 1986, Time dependent chemistry in dense molecular clouds. III. Infrared band cross sections of molecules in solid state at 10 K, A&AS, 64, 453–467.

d'Hendecourt L. B. and Jourdain de Muizon M. 1989, The discovery of interstellar carbon

dioxide, A&A 223, L5–L8.
Ehrenfreund P., Breukers R., d'Hendecourt L. and Greenberg J. M. 1992, On the possibility of detecting solid $O_2$ in interstellar grain mantles, A&A 260, 431–436.
Ehrenfreund, P., Gerakines, P. A., Schutte, W. A., and van Dishoeck, E. F. 1996a, Infrared properties of isolated water ice, A&A 312, 263–274.
Ehrenfreund, P., Boogert, A.C.A., Gerakines, P.A., Jansen, D.J., Schutte, W.A., Tielens, A.G.G.M., van Dishoeck, E.F. 1996b, A laboratory database of solid CO and $CO_2$ for ISO, A&A 315, L341-L344.
Ehrenfreund, P., Boogert, A.C.A., Gerakines, P.A., Tielens, A.G.G.M., van Dishoeck, E.F. 1998, Infrared spectroscopy of interstellar apolar ice analogs, A&A, in press.
Elsila J., Allamandola L.J., Sandford S.A. 1997, The 2140 $cm^{-1}$ (4.673 $\mu$m) solid CO band: The case for interstellar $O_2$ and $N_2$ and the photochemistry of non-polar interstellar ice analogs, ApJ 479, 818–833.
Foti G., Calcagno L., Sheng K.L., Strazzulla G., Micrometre-sized polymer layers synthesized by MeV ions impinging on frozen methane 1984, Nature 310, 126–128.
Geballe T. R., Baas F., Greenberg J. M. and Schutte W. 1985, New infrared absorption features due to solid phase molecules containing sulfur in W 33 A, A&A 146, L6–L8.
Gerakines P. A., Schutte W. A., Greenberg J. M. and van Dishoeck E. F. 1995, The infrared band strenghts of $H_2O$, CO and $CO_2$ in laboratory simulations of astrophysical ice mixtures, A&A 296, 810–818.
Gerakines, P. A., Schutte, W. A., Ehrenfreund, P., and van Dishoeck, E. F. 1996, Ultraviolet processing of interstellar ice analogs. I. Pure ices, A&A 312, 289–305.
Gillett F. C. and Forrest W. J. 1973, Spectra of the Becklin–Neugebauer point source and the Kleinmann–Low nebula from 2.8–13.5 microns, ApJ 179, 483–491.
Gillett F. C., Jones T. W., Merrill K. M. and Stein W. A. 1975, Anisotropy of constituents of interstellar grains, A&A 45, 77–81.
Goldsmith P.F. 1987, Molecular clouds: An overview, in *Interstellar Processes*, D.J. Hollenbach and H. Thronson eds., Reidel, Dordrecht, pp. 51–70.
Greenberg J.M. 1971, Interstellar grain temperatures: Effects of grain materials and radiation fields, A&A 12, 240–249.
Greenberg J.M. 1982, What are comets made of ? A model based on interstellar dust. In *Comets*, L. Wilkening, Ed., Univ. of Arizona press, Tucson, pp. 131–163.
Greenberg J.M., van de Bult C.E.P.M., Allamandola L.J. 1983, Ices in space, J. Phys. Chem. 87, 4243–4260.
Grim R. J. A. and d'Hendecourt L. B. 1986, Time dependent chemistry in dense molecular clouds: IV. Interstellar grain surface reactions inferred from a matrix isolation study, A&A 167, 161–165.
Grim R. J. A. and Greenberg J. M. 1987, Ions in grain mantles: The 4.62 micron absorption by $OCN^-$ in W33A, ApJ 321, L91–L96.
Grim R. J. A., Greenberg J. M., de Groot M. S., Baas F., Schutte W. A., and Schmitt B. 1989, Infrared spectroscopy of astrophysical ices: New insights in the photochemistry, A&AS 78, 161–186.
Grim R. J. A., Baas F., Geballe T. R., Greenberg J. M. and Schutte W. 1991, Detection of solid methanol toward W 33A, A&A 243, 473–477.
Gürtler J., Henningg Th., Kömpe C., Pfau W., Krätschmer W., Lemke D. 1996, Detection of an absorption band at the position of the 4.27-$\mu$m band of solid $CO_2$, A&A 315, L189–L192.
Hagen W., Tielens A.G.G.M., Greenberg J.M. 1981, The spectrum of amorphous solid water and ice $I_c$ between 10 and 140 K, Chem. Phys. 56, 367–379.
Hagen W. and Tielens A. G. G. M. 1981, Infrared spectrum of $H_2O$ matrix isolated in CO at 10 K: Evidence for bifurcated dimers, J. Chem. Phys. 75, 4198–4207.
Hagen W. 1982, *Chemistry and infrared spectroscopy of interstellar grains*, PhD thesis, University of Leiden, the Netherlands.
Hagen W., Tielens A.G.G.M., Greenberg J.M. 1983, A laboratory study of the infrared spectra of interstellar ices, A&AS 51, 389-416.

Hasegawa T. I., Herbst E. and Leung C. M. 1992, Models of gas–grain chemistry in dense interstellar clouds with complex organic molecules, ApJS 82, 167–195.

Hasegawa T. I. and Herbst E. 1993, New gas–grain chemical models of quiescent dense interstellar clouds: the effects of $H_2$ tunnelling reactions and cosmic ray induced desorption, MNRAS 261, 83–102.

Helmich F.P., Millar T.J., van Dishoeck E.F. 1998, in preparation.

Hiraoka K., Yamashita A., Yachi Y., Aruga K., Sata T., Muto H. 1995, Ammonia formation from the reaction of H atoms with N atoms trapped in a solid $N_2$ matrix at 10–30 K, ApJ 443, 363–370.

Hiraoka K., Ohashi N., Kihara Y., Yamamoto K., Sata T., Yamashita A. 1994, Formation of formaldehyde and methanol from the reaction of H atoms with solid CO at 10–20 K, Chem. Phys. Letters 229, 408–414.

Hudgins D. M., Sandford S. A., Allamandola L. J. and Tielens A. G. G. M. 1993, Mid– and Far–infrared spectroscopy of ices: Optical constants and integrated absorbances, ApJS 86, 713–870.

Hudson R.L., Moore M.H. 1997, Hydrocarbon radiation chemistry in ices of cometary relevance, Icarus 126, 233–235.

Jenniskens P. and Blake D. F. 1994, Structural transitions in amorphous water ice and astrophysical implications, Science 265, 753–756.

Jessberger E.K., Kissel J., Cehmical properties of cometary dust and a note on carbon isotopes, In *Comets in the Post-Halley era*, R.L. Newburn, M. Neugebauer, and J. Rahe, Eds., Kluwer Academic Publishers, Dordrecht, pp. 1075–1091.

Jiang G.J., Willis B.P., Brown K.G. 1975, Absolute infrared intensities and band shapes in pure solid CO and CO in some solid matrices, J. Chem. Phys. 62, 1201–1211.

Kaiser R.I., Roessler K. 1997, Theoretical and experimental studies of the interaction of cosmic ray particles with frozen matter in space. I. Synthesis of polycyclic aromatic hydrocarbons by cosmic ray induced multi center mechanism, ApJ, in press.

Keane J., Schutte W.A. 1998, in preparation.

Kerr T. H., Adamson A. J. and Whittet D. C. B. 1993, Infrared spectroscopy of solid CO: the Ophiuchi molecular cloud, MNRAS 262, 1047–1056.

Kitta K., Krätschmer W. 1983, Status of laboratory experiments on ice mixtures and on the 12 $\mu$m $H_2O$ ice feature, A&A 122, 105–110.

Lacy J. H., Baas F., Allamandola L. J., Persson S. E., McGregor P. J., Lonsdale C. J., Geballe T. R. and van de Bult C. E. P. 1984, 4.6 micron absorption features due to solid phase CO and cyano group molecules toward compact infrared sources, ApJ 276, 533–543.

Lacy J. H., Carr J. S., Evans N., Baas F., Achtermann J. M. and Arens F. 1991, Discovery of interstellar methane: Observations of gaseous and solid $CH_4$ absorption toward youngg stars in molecular clouds, ApJ 376, 556–560.

Lanzerotti L.J., Brown W.L., Marcantonio K.J. 1987, Experimental study of erosion of methane ice by energetic ions and some considerations for astrophysics, ApJ 313, 910–919.

Léger A., Gauthier S., Defourneau D., Rouan D. 1983, Properties of amorphous $H_2O$ ice and origin of the 3.1 $\mu$m absorption, A&A 117, 164–169.

Léger A., Jura M., Omont A. 1985, Desorption from interstellar grains, A&A 144, 147–160.

Maréchal Y. 1987, IR spectra of carboxylic acids in the gas–phase: A quantitative reinvestigation, J. Chem. Phys. 87, 6344–6353.

Mathis J.S., Mezger P.G., Panagia N. 1983, Interstellar radiation field and dust temperatures in the diffuse matter and in giant molecular clouds, A&A 128, 212–229.

Millar T.J., Macdonald G.H., Habing R.J. 1995, The detection of hot ethanol in G34.3+0.15, MNRAS 273, 25–29.

Millar T.J. 1997, Models of hot molecular cores, In *I.A.U. symposium no. 178: Molecules in astrophysics: Probes and processes*, E. F. van Dishoeck, Ed., Kluwer, Dordrecht, pp. 75–88.

Moore M.H., Donn B. 1982, The infrared spectrum of a laboratory synthesized residue: Implications for the 3.4 micron interstellar absorption feature 1982, ApJ 257, L47–L50.

Moore, M. H. Hudson R. L. 1992, Far–infrared spectral studies of phase changes in water ice produced by proton irradiation, ApJ 401, 353–360.

Moore, M. H. Hudson R. L. 1994, Far–infrared spectra of cosmic–type and mixed ices, A&AS 103, 45–56.

Ohishi M. 1997, Observations of "hot cores", In *I.A.U. symposium no. 178: Molecules in astrophysics: Probes and processes*, E. F. van Dishoeck, Ed., Kluwer, Dordrecht, pp. 61–74.

Omont A., Moseley S.H., Forveille T., Glaccum W.J., Harvey P.M., Likkel L., Loewenstein R.F., Lisse C.M. 1990, Observations of the 40–70 micron bands of ice in IRAS 09371+1212 and other stars, ApJ 355, L27–L30.

Palumbo M. E. and Strazulla G. 1992, The 2140 $cm^{-1}$ band of frozen CO in ion–irradiated and unirradiated mixtures with methanol and water, A&A 259, L12–L14.

Palumbo M. E. and Strazulla G. 1993, The 2140 $cm^{-1}$ band of frozen CO: laboratory experiments and astrophysical applications, A&A 269, 568–580.

Palumbo M.E., Tielens A.G.G.M., Tokunaga A.T. 1995, Solid carbonyl sulphide (OCS) in W33A, ApJ 449, 674–680.

Pironello V., Brown W. L., Lanzerotti L. J., Marcantonio K. J., Simmons E. J. 1982, Formaldehyde formation in a $H_2O/CO_2$ ice mixture under irradiation by fast ions, ApJ 262, 636–640.

Ritzhaupt G., Devlin J.P., 1977, Ionic vs. molecular nature of monomeric ammonium and hydronium nitrate. Infrared spectra of $H_3O^+NO_3^-$ and $NH_4^+NO_3^-$ solvated in argon matrices, J. Phys. Chem. 81, 521–525.

Sandford S. A. and Allamandola L. J. 1988, The condensation and vaporization behavior of $H_2O$:CO ices and the implications for interstellar grains and cometary activity, Icarus 76, 201–224.

Sandford S. A., Allamandola L. J., Tielens A. G. G. M., and Valero L. J. 1988, Laboratory studies of the infrared spectral properties of CO in astrophysical ices, ApJ 329, 498–510.

Sandford S. A. and Allamandola L. J. 1990, The physical and infrared spectral properties of $CO_2$ in astrophysical ice analogues, ApJ 355, 357–372.

Sandford S. A. and Allamandola L. J. 1993a, $H_2$ in interstellar and extragalactic ices: Infrared characteristics, ultraviolet production and implications, ApJ 409, L65–L68.

Sandford S. A., Allamandola L. J. and Geballe T. R. 1993, Spectroscopic detection of molecular hydrogen frozen in interstellar ices, Science 262, 400–402.

Sandford S. A. and Allamandola L. J. 1993b, Condensation and vaporization studies of $CH_3OH$ and $NH_3$ ices: Major implications for astrochemistry, ApJ 417, 815–825.

Schmitt B., Grim R. J. A. and Greenberg J. M. 1988, Molecular diffusion in ices – Implications for the composition of interstellar grain mantles and comet nuclei, in *experiments on cosmic dust analogues*, eds. E. Bussoletti, C. Fusco and G. Longo (Dordrecht, Kluwer), p. 259–269.

Schmitt B. 1994, Physical and chemical processes in icy grain mantles, in *AIP conference proceedings 312: Molecules and grains in space*, ed. I. Nenner (New York, AIP press), 735–757.

Schutte, W. A. 1988, *The evolution of interstellar organic grain mantles*, PhD thesis, University of Leiden, the Netherlands.

Schutte W. A. and Greenberg J. M. 1991, Explosive desorption of icy grain mantles in dense clouds, A&A 244, 190–204.

Schutte W. A., Tielens A. G. G. M. and Sandford S. A. 1991, 10 micron spectra of protostars and the solid methanol abundance, ApJ 382, 523–529.

Schutte W. A., Allamandola L. J. and Sandford S. A. 1993, An experimental study of the organic molecules produced in cometary and interstellar ice analogs by thermal formaldehyde reactions, Icarus 104, 118–137.

Schutte, W. A., Gerakines, P. A., Geballe, T. R., van Dishoeck, E. F., and Greenberg, J. M. 1996a, Discovery of solid formaldehyde towards the protostar GL 2136: Observations and laboratory simulation, A&A 309, 633–647.

Schutte, W.A. 1996, Formation and Evolution of Interstellar Icy Grain Mantles, *in: "The Cosmic Dust Connection"*, ed. J. M. Greenberg, pp. 1-42.

Schutte, W.A., Tielens, A.G.G.M., Whittet, D.C.B., Boogert, A., Ehrenfreund, P., de Graauw, Th., Prusti, T., van Dishoeck, E. F., and Wesselius, P. 1996b, The 6 and 6.8 $\mu$m absorption features in the spectrum of NGC7538:IRS9, A&A 315, L333-L336.

Schutte W.A., Greenberg J.M. 1997, Further evidence for the $OCN^-$ assignment to the XCN band in astrophysical ice analogs, A&A 317, L43–L46.

Skinner C. J., Tielens A. G. G. M., Barlow M. J., and Justtanont K. 1992, Methanol ice in the protostar GL 2136, ApJ 399, L79-L82.

Smith M. A. H., Rinsland C. P., Fridovich B., Rao K. N. 1985, in *Molecular Spectroscopy: Modern research*, Ed. K. N. Rao (Academic Press), Vol. III, p. 111.

Smith R. G., Sellgren K. and Tokunaga A. T. 1989, Absorption features in the 3 micron spectra of protostars, ApJ 344, 413–426.

Strazzulla G., Calcagno L., Foti G. 1984, Build up of carbonaceous material by fast protons in Pluto and Triton, A&A 140, 441–444.

Strazzulla G. and Baratta G. A. 1991, Laboratory study of the IR spectrum of ion-irradiated frozen methane, A&A 241, 310–316.

Strazzulla G. and Baratta G. A. 1992, Carbonaceous Material by ion irradiation in space, A&A 266, 434–438.

Tegler S.C., Weintraub D.A., Allamandola L.J., Sandford S.A., Rettig T.W., Campins H. 1993, Detection of the 2165 $cm^{-1}$ inverse centimeter (4.619 micron) XCN band in the spectrum of L1551 IRS5, ApJ 411, 260–265.

Tegler S.C., Weintraub D.A., Rettig T.W., Pendleton Y.J., Whittet D.C.B., Kulesa C.A. 1995, Evidence for chemical processing of precometary ice grains in circumstellar environments of pre–main-sequence stars, ApJ 439, 279–287.

Tielens A.G.G.M. and Hagen W. 1982, Model calculations of the molecular composition of interstellar grain mantles, A&A 114, 245–260.

Tielens A.G.G.M., Allamandola L.J. 1987a, Composition, structure, and chemistry of interstellar dust, in *Interstellar Processes*, D.J. Hollenbach and H. Thronson eds., Reidel, Dordrecht, 397–469.

Tielens A. G. G. M. and Allamandola L. J. 1987b, Evolution of interstellar dust, in *Physical processes in interstellar clouds*, eds. G. E. Morfill and M. Scholer, p. 333–376.

Tielens A. G. G. M., Tokunaga A. T., Geballe T. R. and Baas F. 1991, Interstellar solid CO: Polar and nonpolar interstellar ices, ApJ 381, 181–199.

van IJzendoorn L.J., Allamandola L.J., Baas F., Greenberg J.M. 1983, Visible spectroscopy of matrix isolated HCO, J. Chem. Phys. 78, 7019–7018.

Whittet, D.C.B., Schutte, W.A., Tielens, A.G.G.M., Boogert, A.C.A., de Graauw, Th., Ehrenfreund, P., Gerakines, P.A., Helmich, F.P., Prusti, T., and van Dishoeck, E. F. 1996, An ISO view of interstellar ices - First results, A&A 315, L357-L360.

Willner S. P., Gillett F. C., Herter T. L., Jones B., Merrill K. M., Pipher J. L., Puetter R. C., Rudy R. J., Russell R. W. and Soifer B. T. 1982, Infrared spectra of protostars: Composition of the dust shells, ApJ 253, 174–187.

Yamada H., Person W.B. 1964,, Absolute infrared intensities of the fundamental absorption bands is solid $CO_2$ and $N_2O$, J. Chem. Phys. 41, 2478–2487.

# COSMIC DUST AND LABORATORY SIMULATION: WISHES, RESULTS AND OPEN PROBLEMS

L. COLANGELI, V. MENNELLA
*Astronomical Observatory of Capodimonte, Via Moiariello 16, Napoli*

AND

P. PALUMBO, A. ROTUNDI
*Istituto Universitario Navale, Via De Gasperi 5, Napoli*

**Abstract.** Cosmic dust plays an active role in several processes occurring in the interstellar medium, in star formation regions and around evolved stars. Thus, the identification of its properties is one of the major goals of modern astrophysics. In this framework, laboratory experiments are a fundamental tool, which allows us to interpret observations and to validate theoretical models on a quantitative basis. This is the reason why, in the recent years, a new generation of "laboratory astrophysics" has grown up, aimed at performing systematic experiments under carefully controlled and reproducible conditions. The goal is to make comparative analyses between samples produced and processed under different ambient conditions, which may reproduce - at the best we can do in a laboratory on Earth - the actual space environment. In this paper we intend to give an overview of recent results obtained in the Cosmic Physics Laboratory in Naples on several soot samples. A key of interpretation which correlates dust optical behaviour to structural properties in a self-consistent scenario will be discussed. This is a sort of evolutionary progression, which is linked both to production conditions and to processing mechanisms, such as thermal annealing, UV irradiation and ion bombardment. The derived general principles may be applicable to interpret spectral behaviours observed in different astronomical environments for carbon-based materials: interstellar and circumstellar media, cometary and interplanetary environments. Of course, several uncertainties still remain, especially concerning both the mass budget available in space as solid condensed forms and the aggregation status of cosmic dust. From the experimental point of view, many points are not fully clarified and require further analyses to remove several question marks. In this sense, a new plan of laboratory activities, based on innovative techniques, has to be



*J. M. Greenberg and A. Li (eds.), Formulation and Evolution of Solids in Space,* 203–228.

developed to perform significant steps ahead. This in the view of pursuing new frontiers along the line that should bring to a consistent interpretation of physical and chemical phenomena involving cosmic dust.

## 1. To start

Carbonaceous materials, together with silicates, are one of the two main species forming cosmic dust. Several observational hints do exist about the presence of various forms of carbon-based materials in the interstellar space, the circumstellar medium and the interplanetary and cometary environments. On the other hand, it is nowadays widely accepted, and observations confirm, that dust in space changes its intimate nature through different evolutionary phases, as long as it is present in various cosmic environments. Therefore, it is no more sufficient to state that: carbon is present in cosmic dust. Further steps ahead are necessary in order to distinguish in more detail the actual physical, chemical, structural and morphological properties of materials that, generally speaking, fall in the class of carbonaceous materials. They are actually characterised by significant intrinsic differences in various space environments.

Attempts to identify the properties of the refractory cosmic dust are based both on theoretical modelling and on laboratory simulations. Until some years ago, several researchers have focused their attention at reproducing typical fingerprints evidenced in astronomical observations by using spectra synthetically produced and/or obtained in laboratory. The main drawback of this approach is that several free parameters are often chosen, so as to properly reproduce astronomical data, without a careful check of their real physical meaning. On the other hand, the so-called laboratory simulations have frequently been limited to a restricted number of samples, without a serious control of boundary constraints, such as probability of formation in space and cosmic elemental abundance. These approaches have shown several limitations. Very often, as long as astronomical observations have improved in number and in quality, the new bits of information have demonstrated that suggested species were totally inappropriate in the light of new observational evidences and constraints.

More recently, models have been produced which are able to incorporate, in a self-consistent way, constraints coming from several astronomical observations. As a consequence, the number of free parameters has been reduced, even if new freedom degrees have been gained by considering, for example, the status of aggregation of cosmic grains [1]. From the experimental point of view, a number of research groups have started to apply a more systematic approach to their experimental activity. A wide variety of experimental techniques, generally used for solid state physical analyses, is applied to cosmic dust "analogue" materials. Different kinds of samples, pertaining to the same class, are considered.

In other words, a new interdisciplinary effort has been started, in which astrophysical competencies have to be properly mixed with solid state knowledge: the so-called "solid-state" astrophysics has started to play a renewed significant role. It must be stressed that such laboratory activity is only apparently less directly connected to astrophysical questions. The goal is not any more to simply match one or more features observed in astronomical spectra in a striking way, but to produce a sort of general framework, in which similarities and differences of samples can be interpreted in a self-consistent way. This scenario can be extrapolated, if not directly applied, to actual space conditions. We are convinced that this is the root to correctly interpret astronomical "unidentified" or - sometime - "over-identified" features (e.g. the UV interstellar/circumstellar extinction "bumps" and several IR features) and new signatures that modern instruments (e.g. ISO) have disclosed.

In the line of the elements reported above, systematic laboratory studies are performed in several laboratories on properly selected cosmic dust analogues: this is the case of silicate [2,3] and carbonaceous materials [4,5]. As far as carbon-based compounds are concerned, a large variety of natural and laboratory synthesised samples exists, i.e. coals, kerogens, soot grains. They present many similarities but also several differences, in terms of morphology, structure, physical and optical behaviour. Therefore, an in depth analysis is required to understand such properties. For several years, our group has performed a systematic laboratory study of soot particles synthesised and processed under carefully controlled conditions. The results obtained by means of spectroscopic, morphological and structural analyses have been presented in various papers [6-15] and offer a clear reference frame.

In section 2 we report a review of the experimental techniques applied in our laboratory to produce, process and analyse soot samples. Section 3 is devoted to an overview of the most significant results and of relevant correlation found. In section 4 the evolutionary trends, generally valid and applicable to cosmic dust, are evidenced. Finally, section 5 is aimed at a critical analysis of what has been done so far in our laboratory experiments in comparison with astronomical evidences.

## 2. What to do in the laboratory

In this section we recall the experiments performed at the Cosmic Physics Laboratory in Naples. We give a summary of methods applied, during several years, in order to make clear the framework in which the different analyses have to be considered in the line of a progressively deeper study of soot (i.e. "disordered" carbon) grains. We wish to mention that the way our approach and results are presented here is not always consistent with the chronological evolution of our experimental activity. Often, negative (more than positive) results have induced to explore new lines of approach and to apply new experimental techniques, which, at the end, have shed light on the complete set of obtained results.

### 2.1. PRODUCTION

Different forms of soot are condensed by arc discharge (applied voltage = 100 V, current = 10 A) between amorphous carbon electrodes in a 10 mbar atmosphere of pure Ar (ACAR sample) or pure $H_2$ (ACH2 sample) and by burning benzene in air (BE sample) at normal conditions (see Table 1) [6]. The ACAR grains are formed by pure carbon, but traces of oxygen and hydrogen, acquired from electrodes and/or during exposure to air after production, are present. The ACH2 grains are "hydrogen rich", as they maintain a significant fraction of hydrogen linked to free bonding carbon sites of the "disordered" structure. The BE grains should be, in principle, different from the other two samples, because they are condensed by burning benzene in air. In fact, the intrinsic structure of benzene contains the basic building blocks (benzene rings) which form graphitic materials. On the contrary, ACAR and ACH2 are obtained from vaporisation of amorphous carbon rods.

The ACAR, ACH2 and BE samples are - so to say - a starting point. They roughly represent three points of reference that we have used for several measurements to understand how production conditions (e.g., ambient pressure and atmosphere) can affect the properties of condensed grains. In particular, ACAR and ACH2 samples represent two extremes on a scale in which the hydrogen content can be varied. A finer tuning in this sense is obtained by producing carbon grains by arc discharge in a mixed atmosphere of Ar and $H_2$. Samples have been produced with different relative abundance of $H_2$ in Ar (see Table 1). They offer the opportunity of studying in more detail how the hydrogen content can influence the properties of the condensed products.

TABLE 1. Production techniques.

| Sample | Raw material | Method | Atmosphere | P† (mbar) | D‡ (cm) |
|---|---|---|---|---|---|
| ACAR | ACE• | AD* | 100% Ar | 10 | 5 |
| ACH2 | ACE | AD | 100% $H_2$ | 10 | 5 |
| BE | Benzene | Burning | Air | 1013 | 5 |
| ACH_0.4 | ACE | AD | 0.4% $H_2$ in Ar | 10 | 5 |
| ACH_0.6 | ACE | AD | 0.6% $H_2$ in Ar | 10 | 5 |
| ACH_0.8 | ACE | AD | 0.8% $H_2$ in Ar | 10 | 5 |
| ACH_1.0 | ACE | AD | 1.0% $H_2$ in Ar | 10 | 5 |
| ACH_10 | ACE | AD | 10% $H_2$ in Ar | 10 | 5 |
| ACH_30 | ACE | AD | 30% $H_2$ in Ar | 10 | 5 |

Notes: † P = pressure; ‡ D = collecting distance; • ACE = Amorphous carbon electrodes; * AD = Arc discharge.

### 2.2. PROCESSING

A way to further increase the variety of soot samples is to process those produced with the techniques described in section 2.1. This approach offers the advantage to follow the physical-chemical evolution of condensed samples under different kinds of treatment. The processing methods are chosen to simulate mechanisms active in space. In our experiments we have applied thermal annealing, UV irradiation and ion bombardment, within the ranges reported in Table 2.

The thermal annealing was considered as a first, rough way to simulate grain processing active in space [16,17]. As long as the temperature increases, a progressive effusion of hydrogen from the samples is produced, so that spectral and structure changes can be analysed and correlated to hydrogen release in a systematic way [8]. For ACH2 [7] and ACAR samples [10], the annealing has been performed for 3 hours at a pressure $< 10^{-5}$ mbar in an isothermal stainless steel cavity, at fixed temperatures (see Table 2). The temperature, measured by a thermocouple, was controlled by a microprocessor to better than 1%. For each annealing process, a freshly produced sample was used and an average heating rate of about $10\,^{\circ}\mathrm{C\,min^{-1}}$ was applied. In the following, the samples will be identified with a label ACAR$Tt$ and ACH2$Tt$, with $t =$ temperature values.

To reproduce more closely actual processing active in space, we have exposed fresh ACH2 samples to three different UV radiation doses (see Table 2) [11]. The processing was performed at room temperature in a vacuum

TABLE 2. Processing experiments performed on ACH2 and ACAR samples.

| Starting Sample | Technique | Parameter | Values |
|---|---|---|---|
| ACH2 | Thermal annealing | Temperature | 250, 325, 400, 415, 450, 500, 550, 600, 700, 800 °C |
| ACAR | Thermal annealing | Temperature | 250, 325, 415, 515, 700, 780 °C |
| ACH2 | UV irradiation by $(2 \pm 1) \times 10^{15}$ photons $cm^{-2}\,s^{-1}$ | UV photon dose | $5 \times 10^{20}$, $1 \times 10^{22}$, $4 \times 10^{22}$ eV $cm^{-2}$ |
| ACH2/ACAR | Bombardment by 3 keV $He^+$ ions | Ion dose | 6.6, 66, 660 eV/C-atom |

chamber at a pressure $< 10^{-6}$ mbar. A hydrogen flow discharge lamp, with an $MgF_2$ window, operating at 2 mbar and 100 V, was used as source. Its spectrum is characterised by molecular and atomic Lyman emission. The flux at the sample is $(2 \pm 1) \times 10^{15}$ photons $cm^{-2}\,s^{-1}$, with an average energy per photon = 10 eV. To achieve the UV doses reported in Table 2, the samples were exposed to the UV source for different time intervals: 7, 172 and 550 hours, corresponding to $5 \times 10^{20}$, $1 \times 10^{22}$, $4 \times 10^{22}$ eV $cm^{-2}$, respectively. In the following, the UV irradiated samples will be recalled as ACH2UV$\tau$, where $\tau$ is the time of exposure.

In addition, freshly produced ACAR and ACH2 grains have been subject to ion bombardment [13]. Samples have been treated at room temperature, in a vacuum chamber at $< 10^{-7}$ mbar. The beam of a 3 keV He+ Varian ion gun has been spread across a 1 $cm^2$ surface of the target samples. A low current of 1.5 $\mu A\,cm^{-2}$ ensured a very low sample heating. Different fluences were used (from $10^{15}$ to $10^{17}$ ions $cm^{-2}$) to obtain the deposited energy doses given in Table 2. In the following, the samples will be recalled as ACARI$n$ and ACH2I$n$, with $n = 15, 16, 17$, i.e. the power of the fluence, $10^n$ (ions $cm^{-2}$) applied to the samples, to obtain doses of 6.6, 66 and 660 eV/C-atom, respectively.

### 2.3. CHARACTERISATION

A detailed analysis of condensed (section 2.1) and processed (section 2.2) samples is absolutely necessary, in order to correlate their evolution to production and treatment parameters. With this aim, several techniques

TABLE 3. Methods applied to characterise our samples and derived information.

| Technique | Information |
|---|---|
| Field Emission Scanning Electron Microscopy (FESEM) | 3-D aggregation status, surface morphology, grain shape and size |
| Transmission electron microscopy (TEM) | 2-D grain distribution, grain internal structure |
| Electron diffraction of selected areas (SAED) | crystalline degree |
| Visible spectroscopy | optical gap |
| Raman spectroscopy | disorder degree |
| Energy dispersive X-ray (EDX) analysis | elemental distribution |
| Micro combustion calorimetry (MCC) | hydrogen content |
| Far UV - UV spectroscopy | electronic transitions |
| Near IR spectroscopy | C-H, C-O, C-C resonances |
| Far IR spectroscopy | structure, morphology |

are used to study the morphology, structure and optical behaviour of grains.

In the experiments performed at the Cosmic Physics Laboratory of Naples, we have progressively increased, in number and sensitivity, the analytical methods applied to our samples. In Table 3, we summarise the experimental approaches used to gain as much information as possible.

Sample preparation varies for the different analyses listed in Table 3. As a baseline, we have tried to reduce as much as possible the sample manipulation, in order to limit sample modification induced after condensation. We collected, during condensation, the grains on substrates appropriate for subsequent analyses. Samples deposited on UV-grade fused silica disks have been used for thermal annealing and UV and ion bombardment. These substrates are appropriate for Vis, UV and far-UV spectroscopy. The sample morphology can be studied by FESEM on the same substrates, attached to Al pin stubs with a conductive silver paste. This means that several analyses can be performed exactly on the same sample, a powerful tool to have a self-consistent characterisation of dust. Furthermore, the same dust, once scraped off from the collecting quartz disk, can be properly prepared for other analyses. High resolution FESEM, EDX and TEM measurements are usually performed on small dust fractions dispersed respectively onto smooth silicon wafers (attached to the Al stubs), onto smooth carbon pin stubs, and onto holey-carbon thin films supported by a Cu grid [14]. SAED and TEM analyses have been performed also on ultra-thin sections of dust embedded in epoxy. For IR spectroscopy (up to about 50 $\mu$m) and for Raman analyses, the deposition can be done directly on IR windows, such as KBr or CsI. Dust is also collected on a glass screen and scraped off for other preparations (e.g., embedding in polyethylene matrix for far-IR spectroscopy).

Usually, spectroscopic properties are derived by comparing the transmission spectrum of a "blank" substrate (e.g. clean window or pure matrix pellet) with that of a "sample" (substrate including dust). The mass extinction coefficient, $K(\lambda)$, is:

$$K(\lambda) = S/M \ln [1/T(\lambda)] \tag{1}$$

where S = sample cross-section, M = mass deposited onto the substrate or embedded in the matrix and $T(\lambda)$ = [T of the "sample" /T of the "blank"]. This approach requires a careful determination of the sample surface and of the dust mass deposited onto / embedded in it. The use of a micro-balance (sensitivity at least $\pm 1\,\mu$g) is mandatory.

In the case of far-UV, where no transparent material can be found, the approach shown in Figure 1 is used. The function $T(\lambda)$ is obtained by:

$$T(\lambda) = [(I_s/I_{os}) / (I_b/I_{ob})]^{1/2} \tag{2}$$

where, as shown in Figure 1, $I_{os}$ and $I_{ob}$ are the incoming light beams and $I_s$ and $I_b$ are the outgoing beams. Note that $I_b/I_{ob}$ is the reflectivity R of the substrate, and, with good approximation, $I_s/I_{os} = T^2R$. Different experimental set-up are used according to the spectral range, as synthesised in Table 4.

The combination of FESEM, TEM and SAED observations provides a detailed morphological and structural analysis at different levels. FESEM is appropriate for a 3-dimensional study of grain aggregation at various magnifications. TEM and SAED are devoted to the identification of the inner structure of single particles.

Spectroscopy from the far-UV to the mm range is needed to analyse spectral features which are directly related to the intimate electronic structure and/or to the presence of heterogeneous elements linked to free standing bonds. For carbons, typical absorption features at 80-90 nm of wavelength are due to $\sigma$-$\sigma^*$ electronic transitions, while bands due to $\pi$-$\pi^*$ electronic transitions fall in the range 200 - 260 nm. The Vis range is generally

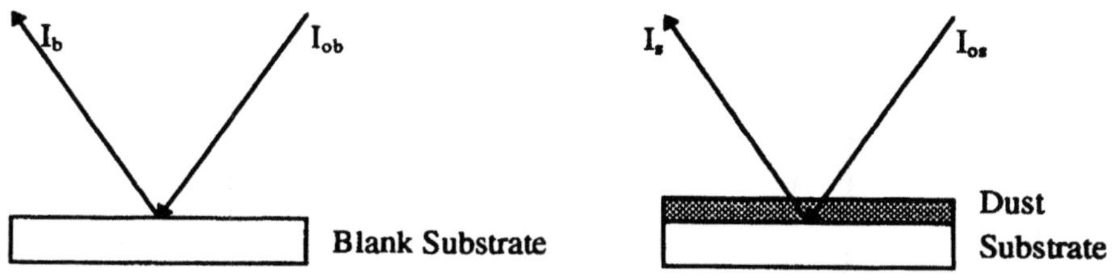


*Figure 1.* Scheme of spectral measurements in the far-UV.

TABLE 4. Experimental set-up used for spectroscopic measurements.

| Spectral range | Set-up | Sample preparation |
|---|---|---|
| 40 nm - 300 nm | BESSY Synchrotron light facility McPherson spectrometer | deposition onto quartz, LiF or goad-coated quartz substrates |
| 190 nm - 3.2 μm | P.E. Lambda 9 and 19 double beam spectrophotometers | deposition onto quartz or LiF substrates |
| 2.5 - 40 μm | P.E. 883 double beam spectrophotometers | deposition onto KBr/CsI windows embedding into KBr/CsI matrices |
| 20 μm - 2 mm | Bruker IFS66V interferometer | embedding in polyethylene matrix |

characterised by a featureless trend and can be fitted by the Tauc relation [8]:

$$(\alpha E)^{1/2} = B\ (E - E_g) \tag{3}$$

with $\alpha$ = absorption coefficient, E = energy and B = constant. From this, it is possible to derive information on the optical gap, $E_g$, i.e. on the distance between valence and conduction bands. In the medium IR (3 - 20 μm) the typical stretching and bending bands of heteroatoms linked to the disordered structure fall. At longer wavelengths, down to the mm range, no main feature exists and the spectral trend can be represented by a power law, $\lambda^{-\beta}$. The spectral index, $\beta$, depends both on the structure and on the grain size and shape. Thus, spectroscopy is a powerful tool not only to study the optical behaviour of carbon-based materials, but also a source of information about their electronic structure and elemental composition.

Finally, Raman spectroscopy usually allows one to identify the two main Raman-active features of carbons, due to the "disorder" (D-line) and "graphitic" (G-line) content. Their relative intensity should be related to the intimate order degree of grains.

## 3. To obtain what?

According to the elements reported in section 1, the aim of our studies is not only to reproduce spectral features observed in space but also to correlate their behaviour to the intimate nature of potential carriers. Therefore, the main efforts are oriented to determine if and how boundary production parameters, as well as processing methods, may influence morphology, structure and optical properties of soot particles.

### 3.1. MORPHOLOGY

A systematic study of morphological properties has been recently performed on ACH2, ACH_0.4, ACH_0.8, ACH_10, ACH_30, ACAR, ACH2T415, ACH2T700 and ACH2T800 samples [14]. Several inherent forms of carbon have been observed thanks to FESEM, TEM and SAED techniques. In synthesis we have identified:

a. Chain-like aggregates, amorphous carbon smoke formed by spherical (sphericity $0.9 \div 1.0$) grains (diameter from 7 to 15 nm);

b. Poorly graphitised carbon, isolated domains of loops and rings delineated by a single graphite layer on smooth amorphous carbon;

c. Bucky-onions (diameter $= 10 \div 40$ nm) and bucky-tubes ($100 \times 10$ nm in size);

d. Very rare graphitic carbon ribbons and single-crystal carbon platelets.

The amorphous carbon smoke is the most abundant form in all examined samples and is the typical product of condensation. Poorly graphitised carbon is a result of auto-annealing during coagulation. "Bucky-carbons", formed by arc-discharge [18], are products of the condensation process "proper". The graphitic carbon ribbons, present only in thermally annealed samples, are probably the result of a graphitisation process, just beginning at the considered temperatures [19]. The origin of the very rare and well structured single-crystal carbon platelets could be the result of peculiar chemical-physical conditions occurred during soot production.

A comparative analysis indicates that all the detected carbon forms are inherent to the considered samples. All, but the chain-like carbon, do not vary with production conditions and thermal treatment, i.e. are not affected by the C/H ratio. On the contrary, some structural variations seem to occur within the chain-like aggregates [15].

## 3.2. OVERALL SPECTRAL PROPERTIES

The use of different spectroscopic instruments and techniques has allowed us to characterise several of our samples over most of the range from 40 nm to 2 mm. In Figure 2 we report the general spectral behaviour of the three reference samples: ACH2, ACAR and BE. The main spectral features are summarised in Table 5, while the IR main bands are shown in Figure 3.

For all the samples mentioned in section 2.1 and 2.2, a summary of the information derived is reported in Table 6. The main spectral features will be discussed in the following sections.

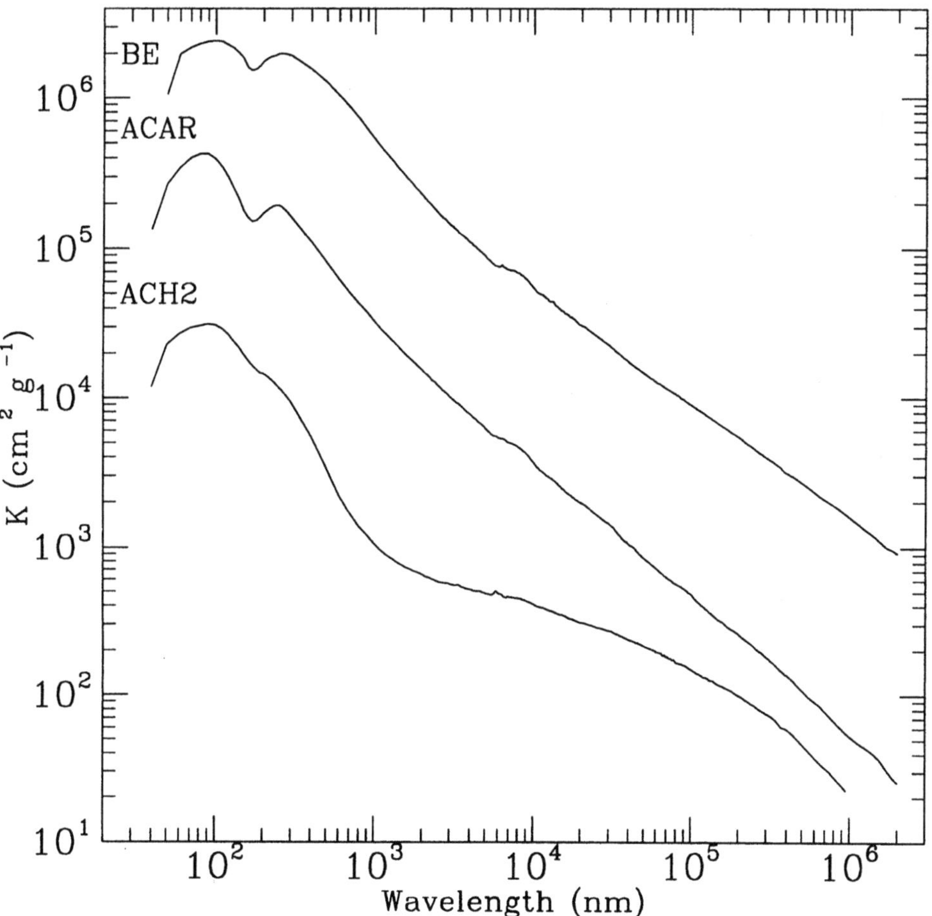


*Figure 2.* Mass extinction coefficient, K(λ) vs. wavelength of ACAR, ACH2 and BE samples (BE and ACH2 spectra are upper and lower shifted by 1 order of magnitude, respectively) [6].

TABLE 5. Main spectral features of ACH2, ACAR and BE samples.

| Sample | FUV bump (nm) | UV bump (nm) | Spectral index $\beta$ | | |
|---|---|---|---|---|---|
| | | | 0.6 - 1 $\mu$m | 1 - 5 $\mu$m | 15 $\mu$m - 2 mm |
| ACH2 | $93 \pm 2$ | lacking | ... | $0.3 \pm 0.1$ | $0.5 \div 1.1$ |
| ACAR | $82 \pm 2$ | $242 \pm 2$ | $1.3 \pm 0.1$ | $1.0 \pm 0.1$ | $0.9 \pm 0.1$ |
| BE | $95 \pm 2$ | $260 \pm 2$ | $1.3 \pm 0.1$ | $1.1 \pm 0.1$ | $0.8 \pm 0.1$ |

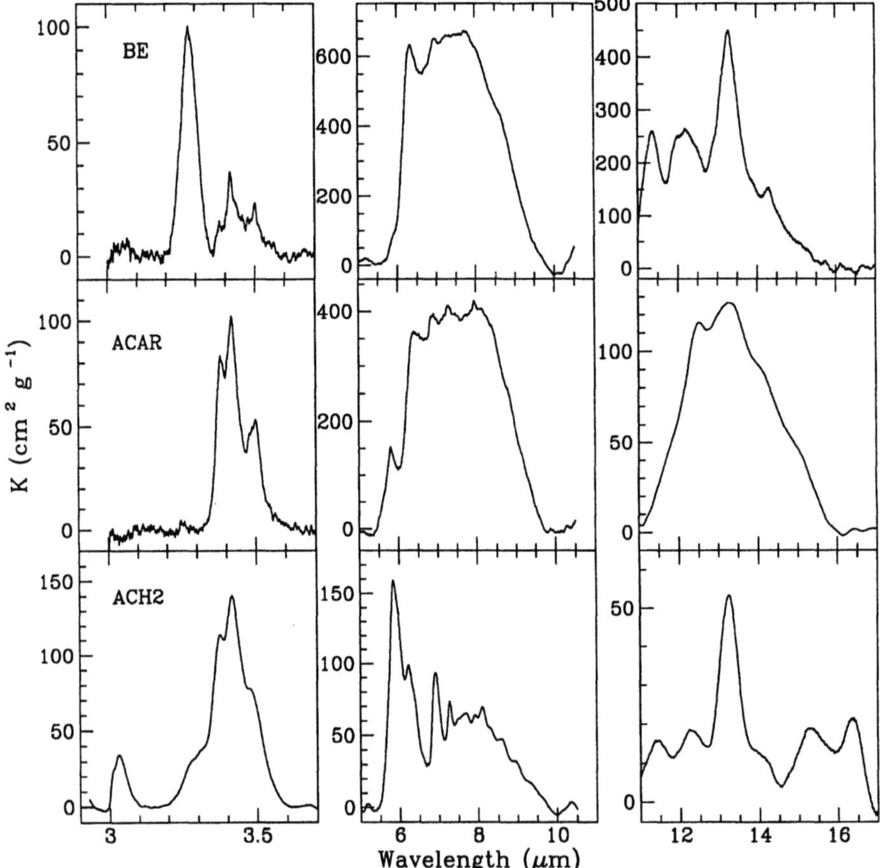


*Figure 3.* Infrared bands detected in ACH2, ACAR and BE samples. The continuum contribution has been subtracted [6].

### 3.3. HYDROGENATION AND IR BANDS

The study of the IR absorption bands (see Figure 3) is important to derive information about:

a) the kind of heteroatoms linked to the carbon structure;
b) the hydrogenation degree;
c) the structural arrangement of carbon.

All our samples present hydrogen as dominating heteroatom and, in some cases, a very little oxygen contamination. Values of hydrogen content for some of the analysed samples are reported in column 3 of Table 6. The most direct way to determine the sample hydrogen content is by Micro-Combustion Calorimetry, MCC [7]. On the other hand, the integrated band intensity, A, of the C-H stretching bands, falling in the 3400 - 2700 $cm^{-1}$ (i.e. 3.0 - 3.7 $\mu$m) range, is proportional to the hydrogenation degree. Thus, it is possible to use IR spectroscopy to derive the hydrogen content of dust, by scaling the values of A for different samples to that measured by MCC for ACH2. This normalisation is effective only under the hypothesis that all the hydrogen measured by MCC is actually linked in C-H bonds.

Furthermore, bands between 3400 and 2700 $cm^{-1}$ are due to different resonant modes (see Table 7), so that a proper determination of their relative integrated intensities can be used to derive the $sp^2/sp^3$ abundance ratio (see column 7 in Table 6). To properly separate the different spectral components, a careful deconvolution of measured bands by Lorentzian oscillators is necessary. The C-H stretching bands evolve as a function of hydrogen content in the arc discharge chamber. For the examined cases (see Table 6) the $sp^3$ structure dominates, but a progressive increase in $sp^2$ content is observed as long as the atmosphere becomes more and more hydrogen rich (see column 7 in Table 6). This could be interpreted as an effect of saturation of free $sp^3$ sites by hydrogen and a progressive involvement of $sp^2$ active sites. On the other hand, going from ACH2 to ACAR and to BE samples (see Figure 3) and by annealing ACH2 samples (see Figures 4 and 5, panels a and b) the $sp^2/sp^3$ ratio increases. The BE sample seems to be an extreme case in which the $sp^2/sp^3$ ratio is $> 1$.

### 3.4. OPTICAL GAP

For all analysed samples a broad optical absorption edge is observed in the Vis range, so that the optical gap by the Tauc has been determined [8,11,13]. As reported in Table 6 (fourth column), on most of the considered grains the evaluation of the optical gap, $E_g$, has been performed. This parameter is sensitive to the production conditions and the processing. In particular, we notice that:

1. $E_g$ decreases from ACH2 (1.34 eV) to ACAR (0.52 eV) and to BE

TABLE 6. Summary of information on analysed samples.

| Sample | Processing | H/C ratio | $E_g$ (eV) | UV peak (nm) | $L_a$ (Å) | $sp^2/sp^3$ |
|---|---|---|---|---|---|---|
| ACH2 | none | 0.6$^{\bullet}$ | 1.34 | – | 6 | 0.1 |
| ACH_0.4 | none | – | 0.70 | 230 | 11 | 0.003 |
| ACH_0.6 | none | – | 0.78 | 223 | 10 | – |
| ACH_0.8 | none | – | 0.83 | 218 | 9 | 0.048 |
| ACH_1.0 | none | – | 0.92 | 203 | 8 | 0.35 |
| ACH2T250 | 250 °C$^a$ | 0.66$^{\star}$ | 1.22 | 194 | 6 | 0.15 |
| ACH2T325 | 325 °C | 0.5$^{\star}$ | 1.20 | 198 | 6 | 0.2 |
| ACH2T400 | 400 °C | – | – | 208 | – | – |
| ACH2T415 | 415 °C | 0.2$^{\bullet}$<br>0.35$^{\star}$ | 0.92 | 214 | 8 | 0.4 |
| ACH2T450 | 450 °C | – | – | 216 | – | – |
| ACH2T500 | 500 °C | – | – | 230 | – | – |
| ACH2T515 | 515 °C | – | – | – | – | 4 |
| ACH2T550 | 550 °C | 0.17$^{\star}$ | 0.35 | 241 | 20 | – |
| ACH2T600 | 600 °C | 0.0$^{\bullet}$ | 0.19 | 246 | 40 | – |
| ACH2T700 | 700 °C | – | -0.06 | 252 | – | – |
| ACH2T800 | 800 °C | – | -0.13 | 259 | – | – |
| ACH2UV7 | $5 \times 10^{20\,b}$ | – | 1.34 | – | 6 | – |
| ACH2UV172 | $1 \times 10^{22}$ | – | 1.29 | 215 | 6 | – |
| ACH2UV550 | $4 \times 10^{22}$ | 0.3$^{\dagger}$ | 1.23 | 215 | 6 | – |
| ACH2I15 | ion 6.6$^c$ | – | 1.23 | 205 | 6 | – |
| ACH2I16 | ion 66 | – | 1.11 | 206 | 7 | – |
| ACH2I17 | ion 660 | – | 0.69 | 215 | 11 | – |
| ACAR | none | – | 0.52 | 242 | 15 | – |
| ACART250 | 250 °C | – | 0.53 | 242 | 15 | – |
| ACART325 | 325 °C | – | 0.52 | 241 | 15 | – |
| ACART415 | 415 °C | – | 0.50 | 243 | 15 | – |
| ACART515 | 515 °C | – | 0.37 | 244 | 21 | – |
| ACART700 | 700 °C | – | 0.07 | 250 | – | – |
| ACART780 | 780 °C | – | 0.0 | 253 | – | – |
| ACARI15 | ion 6.6 | – | 0.42 | 245 | 18 | – |
| ACARI16 | ion 66 | – | 0.52 | 219 | 15 | – |
| ACARI17 | ion 660 | – | 0.62 | 218 | 12 | – |
| BE | none | – | 0.15 | 260 | 51 | 2.6 |

$^{\bullet}$ – from MCC measurements; $^{\star}$ – from IR spectroscopy of C-H stretching bands; $^{\dagger}$ – from comparison with annealing measurements; $^a$ – annealing temperature; $^b$ – UV fluence in eV cm$^{-2}$; $^c$ – ion fluence in eV/C-atom.

TABLE 7. Reference wavenumbers for stretching C-H features.

| $sp^2$ CH aromatic stretching | $sp^3$ $CH_3$ asymmetric stretching | $sp^3$ $CH_2$ asymmetric stretching | $sp^2$ CH symmetric stretching | $sp^3$ $CH_2$ symmetric stretching |
|---|---|---|---|---|
| 3040 - 3060 $cm^{-1}$ | 2960 - 2970 $cm^{-1}$ | 2925 $cm^{-1}$ | 2870 $cm^{-1}$ | 2855 - 2865 $cm^{-1}$ |

(0.15 eV) samples;

2. $E_g$ decreases with decreasing the hydrogen content during production by arc discharge.

Moreover, a reduction of the optical gap is observed with:
1. increasing of the annealing temperature (see Figure 5c);
2. increasing of the UV dose;
3. increasing of the ion dose for ACH2.

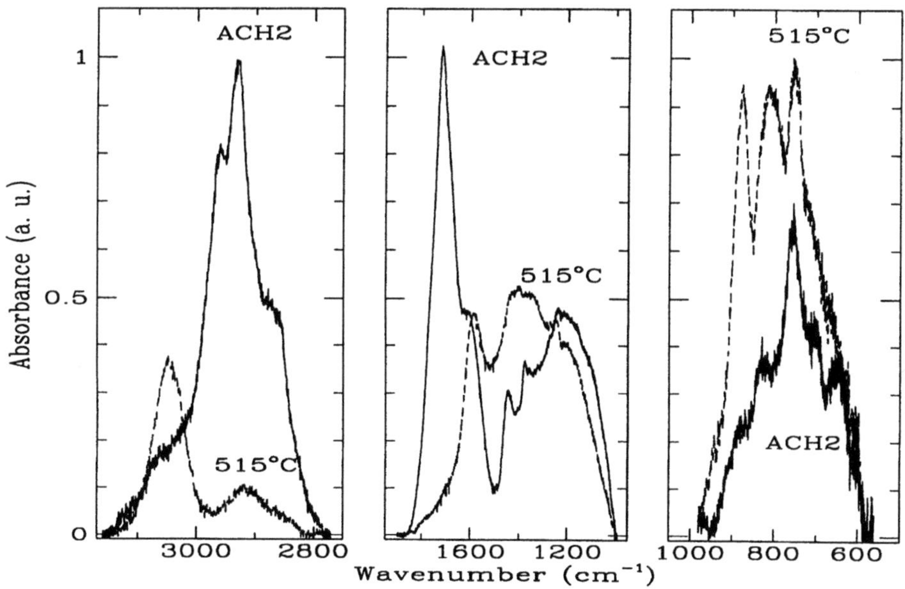


*Figure 4.* Evolution of the C-H stretching bands for thermal annealing of ACH2 up to 515 °C [12].

We recall that in carbon-based materials the gap between valence and conduction band is driven by the amount/size of $sp^2$ hybridised carbons [20]. In graphitic materials the $sp^2$ hybridisation dominates and the energy

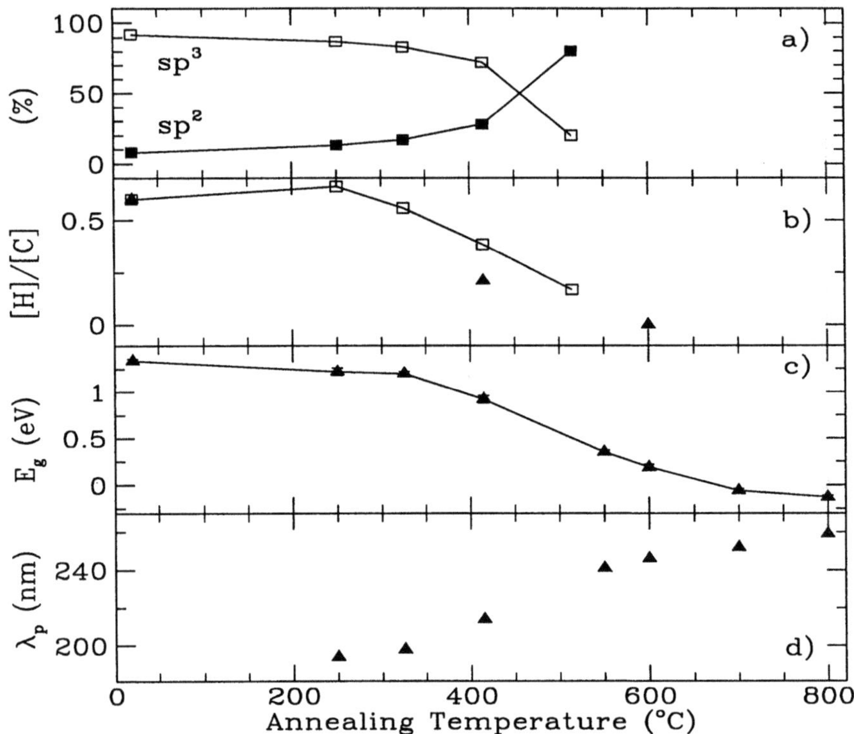


*Figure 5.* Dehydrogenation of ACH2 grains by thermal annealing [12]: a) $sp^2$ and $sp^3$ concentration; b) H/C abundance ratio (triangles: from MCC; squares: from IR spectroscopy); c) optical gap behaviour; d) UV $\pi$-$\pi^*$ bump position.

gap is rather small; in diamond-like materials the carbon atoms are mainly $sp^3$ hybridised and the gap is larger. Our results suggest that, no matter the production or processing method, as long as the hydrogen content in the grains is reduced, the optical gap decreases, i.e. a more graphitic-like structure is obtained.

It is interesting to note that ion bombardment of ACAR grains produces a progressive increase of $E_g$, from 0.42 to 0.62. In this case processing acts to destroy the $sp^2$ aromatic organisation present in the sample.

### 3.5. THE UV "BUMP"

Another feature which appears sensitive to the production conditions and to the processing mechanisms is the so-called UV "bump". This wide feature is sometime lacking (only a weak shoulder around 200 nm is observed in ACH2) and shifts in wavelength position from about 194 nm (ACH2T250) up to 260 nm (in BE and ACH2T800 samples). According to the data

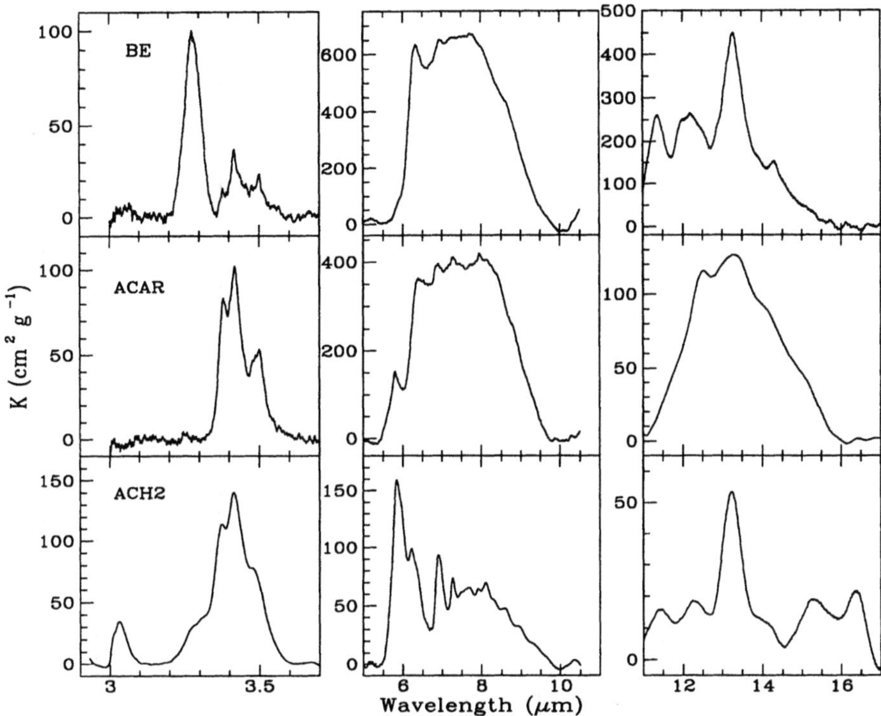


*Figure 6.* UV-Vis spectra of samples produced by arc discharge in mixed $H_2$-Ar atmospheres. Curves from bottom to top refer to 0.4, 0.6, 0.8 and 1.0% of $H_2$ in Ar.

summarised in Table 6 (fifth column), the peak position tends to shift towards longer wavelengths in function of:

a) decreasing of $H_2$ content in the arc discharge chamber (Figure 6);

b)increasing of the annealing temperature of ACH2 (Figure 5d) and ACAR grains;

c) UV irradiation of ACH2 samples;

d) ion bombardment of ACH2 samples (Figure 7).

It is quite evident that, no matter what is the active mechanism, as long as grain hydrogen content is reduced (either due to production conditions or to processing) the bump moves to longer wavelengths. This trend is consistent with the attribution of the resonance to a $\pi$-$\pi^*$ interband transition. The presence of hydrogen promotes $sp^3$ bond formation and localisation of the p bonds. The hydrogen loss produces a rearrangement of the intimate grain structure, in favour of $sp^2$ bonds and of the $\pi$-$\pi^*$ interband transition activation [20,21]. A growth in number and size of $sp^2$ clusters within the grain structure is consistent with this behaviour [8,11,13].

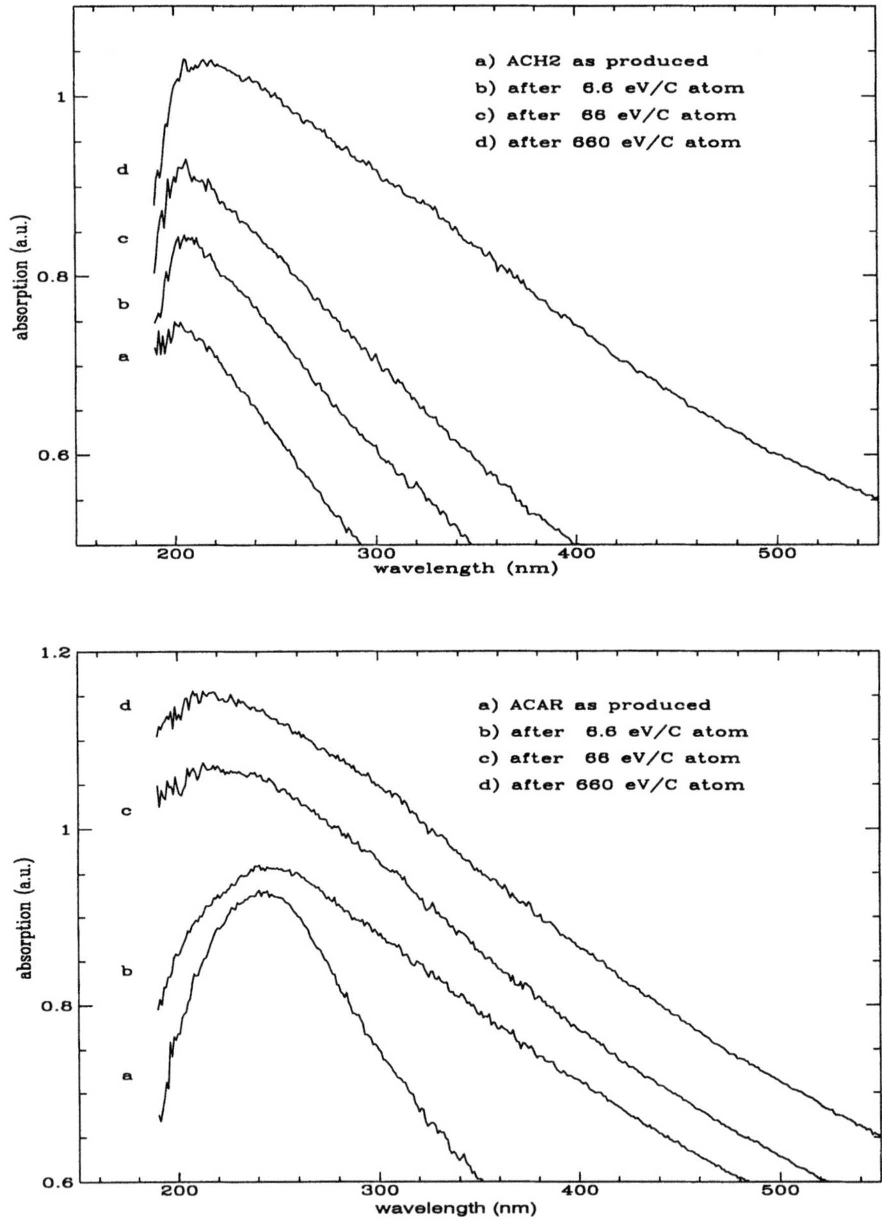


*Figure 7.* UV-Vis spectra of ion bombarded ACH2 (top) and ACAR (bottom) samples [13].

Again, the ion bombardment of ACAR grains appears to produce an opposite effect: i.e. the shift of the bump position towards shorter wavelengths (Figure 7). This is a sign that ion bombardment produces a (partial) disruption of $sp^2$ clusters in the ACAR sample.

## 4. What does it mean?

### 4.1. THE EVOLUTION OF CARBON GRAINS

Spectral, structural and morphological data summarised in the previous section indicate that the behaviour of the carbon grains, produced under different conditions or exposed to various processes, are driven by common factors. A possible analysis is based on the grain structural order degree. We recall that, in principle, a larger $sp^2/sp^3$ ratio, i.e. a larger fraction of $sp^2$ "clusters", implies a more "ordered" (i.e. more graphitic) structure. The ambient conditions, mainly the hydrogen amount in the condensation environment, can trigger the production of more or less ordered samples, as illustrated below.

Production conditions (pure $H_2$ atmosphere) suggest, and measured data (see Table 6) confirm, that ACH2 is the most hydrogenated and the most disordered sample in our set. The UV $\pi$-$\pi^*$ electronic transition lacks (see Figure 2), the aliphatic $sp^3$ bonding dominates in the C-H stretching IR region (Figure 3 and Table 6), the energy gap is large ($E_g = 1.34\,eV$) and the far IR spectral trend is quite flat (spectral index $\beta$ varying between 0.5 and 1.1). These elements confirm that hydrogen, present during condensation, inhibits the formation of large aromatic assemblies of C atoms, the short range order dominates and electronic carbon bonds are localised. The typical "disordered" internal structure is formed by small $sp^2$ "clusters" in a $sp^3$ dominant matrix. The cluster size is given by the so-called aromatic coherence length, $L_a$ (Å) $\sim 7.7/E_g$ (eV) [21]. For ACH2, $L_a \sim 6$ Å, as expected for very disordered carbons [4].

According to data available in the literature for films [20], the hydrogen content of samples affects the order degree of carbon. On the basis of our results, the same concept appears applicable to soot particles. Grains produced in $H_2$-Ar mixed atmosphere, show a clear dependence on the $H_2$ content. A similar behaviour is promoted by thermal annealing, UV irradiation and ion bombardment. The progressive closing of the energy gap, the appearance and shift of the UV $\pi$-$\pi^*$ electronic transition towards a sort of upper edge (260 nm) typical of graphitic materials [7,22], the increase of the $sp^2/sp^3$ ratio are all self-consistent results. The relative intensity of "D" and "G" Raman-active bands for annealed [8] and ion bombarded samples [23] varies consistently. All these results confirm that the structure and the optical properties of soot grains change simultaneously. As long as hydrogen content decreases, the "aromatisation" process brings to a progressive growth, in number and size, of plane $sp^2$ clusters, the structural units of soot grains. The increase in aromatic coherence length, $L_a$, is a direct evidence of such changes. In Figure 8, we put together optical (the UV $\pi$-$\pi^*$ bump position) and structural (the energy gap) parameters. A clear general trend appears typical of carbon grains.

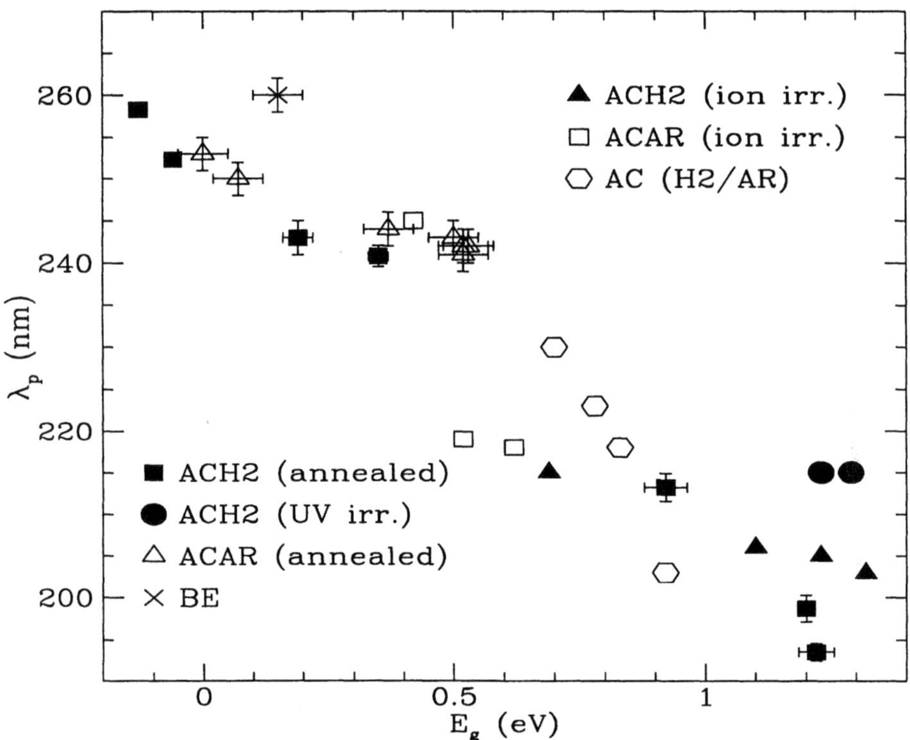


*Figure 8.* The dependence of the $\pi$-$\pi^*$ bump position on the optical gap, for different kinds of carbon grains. The various symbols refer to different samples and processing.

Interesting enough, when ACAR samples are subject to ion bombardment, a sort of reverse processing occurs. This appears quite reasonable. In fact, optical and structural data for ACAR (see Table 6) indicate that this sample has a clustering degree ($L_a$ of the order of 15 Å) higher than ACH2. Thus, the applied ion doses appear sufficient to induce a "de-clustering" process, as indicated by the UV peak shift and optical gap increase. It must be noted that the same ion dose, used on a "disordered" (ACH2) and a "more ordered" (ACAR) sample, brings to an intermediate "partially disordered" status, as testified by parameters reported in Table 6.

Clearly, a confirmation from electron microscopy observations about the overall behaviour of the inner structure of grains would further strengthen our interpretation. As mentioned in section 3.1, the lack of information

[4] about the different morphological typologies present in soot grains has been overcome [14]. Observational evidences suggest that amorphous carbon grains, forming chain-like structures, are the only component structurally sensitive to the hydrogen content [15]. This result seems to confirm that modifications in the intimate nature of amorphous particles are directly connected to observed spectral changes.

### 4.2. COMPARISON WITH OTHER C-BASED MATERIALS

The general trend, found for our samples, appears common to other carbon-based materials. As an example, coals are a class of natural disordered carbons, containing hydrogen and oxygen, which have been studied for decades [4]. They have been classified in "ranks", according to their graphitic content (i.e. their age), and range from very young sub-bituminous to graphitic materials. The progression of properties for these well studied bulk materials is in line with the interpretation of the results obtained for our soot particles. Again, a driving parameter is the "ordering" degree.

Many analogies do exist between the thermal evolution of ACH2 grains and vapour-deposited hydrogenated carbon films studied by several authors [20]. A striking similarity exists between the trends shown in Figure 5 and those reported for hydrogenated carbon film [24]. Finally, we note significant links between our data and those obtained for UV irradiated organic refractory residues [25].

## 5. And in space?

According to the items reported in section 1, we wish to stress that the study of samples in the laboratory does not imply a direct application of the obtained results to actual space conditions. In other words, we do not claim that our "analogues" are actually similar to cosmic dust. It is, however, true that the general results reported in section 4 must have a general validity. They can be used to interpret observations pertaining to carbon-based cosmic grains.

### 5.1. COMPARISON WITH ASTROPHYSICAL CONDITIONS

As mentioned in section 2, thermal annealing is only a rough way to simulate cosmic dust processing. On the contrary, UV irradiation and ion bombardment are active in space. It is, then, useful to check if the doses applied in our experiments are representative of rates expected in space.

According to Hagen et al. [26], the interstellar radiation flux is $F \sim 9.6 \times 10^8 \exp(-2\,A_V)$ [$\mathrm{eV\,cm^{-2}\,s^{-1}}$], with $A_V$ = visual extinction [27]. In the diffuse medium $A_V \sim 0.6$ so that in a typical residence time of $3 \times 10^7$ yrs, the total UV dose should be around $3 \times 10^{23}\,eV\,cm^{-2}$ [28]. This value is a factor of 7 larger than the maximum dose applied in our experiments. By extrapolating our results, we conclude that the typical $E_g$ of interstellar carbon grains should be about 1.2 eV. It is evident that hydrogenated carbon grains would never be able to be converted into pure graphite grains under the above mentioned conditions.

As far as ion bombardment is concerned, low energy (1 MeV) protons are the most effective in grain processing in space conditions. The energy they are able to release by crossing C-based grains should be of the order of $10^{-14}\,eV\,cm^2/atom$ [29]. For a 1 MeV proton flux of $3\,cm^{-2}\,s^{-1}$ [29], the energy expected to be deposited in a carbon grain is of the order of $3 \times 10^{-14}\,eV\,s^{-1}/atom$, which, for the residence time of carbon grains in the interstellar medium, yields 30 eV/atom. Since our 3 keV $He^+$ ions have a stopping power similar to that of 1 MeV protons [25], a direct comparison with doses applied in our tests are possible. According to our data, it looks that ion processing of hydrogenated carbon grains should not be significant, but on time scales of the order of $5 \times 10^8$ yrs, corresponding to several evolutionary cycles between diffuse and dense clouds [30-32].

In summary, UV photon processing should be the dominant mechanism affecting hydrogenated grains. On the other hand, ion irradiation could be effective in "de-clustering" more aromatised grains (in a time scale of 1 or 2 diffuse/dense cloud cycles). Thus, UV and ion processing are mechanisms which may act on carbon grains in a combined or antithetical way, depending on the starting disorder degree of the cosmic carbon grains. The neat effect could be a sort equilibrium stage, in terms of ordering of carbon intimate structure. Of course, this equilibrium could vary depending on the local space environment conditions.

### 5.2. COMPARISON WITH ASTROPHYSICAL FEATURES

It is not our intention to fit observed features, commonly attributed to carbon dust, with our experimental results. On the contrary, we believe that a lot of information can be gained by evidencing the most striking differences between laboratory results and astronomical observations. This in the view of improving laboratory simulation to get closer and closer to the actual space conditions. On the other hand, the critical comparison of laboratory results and astronomical observations could suggest lines of interpretation of observed features.

A prime element of comparison can be the study of the so-called UV extinction "bump", observed both in interstellar and in circumstellar space. Several papers [33-36] have dealt with the problem of finding carbon-based grains whose structure and morphology is adequate to produce an UV spectral behaviour matching the observations. The peak observed in the circumstellar environment of C-rich stars falls at 240-250 nm and its shape is very close to that measured for our ACAR grains. It looks reasonable to conclude that its origin is similar to that determined for our laboratory "analogues", i.e. electronic transitions in $sp^2$ clusters forming carbon-based grains. The observed peak position suggests that grains should have a non negligible internal order degree ($L_a > 20$ Å, according to data in Table 6).

The interstellar bump is observed at 217.5 nm, in several directions through the diffuse medium, suggesting that its carriers must be stable. Moreover, the band presents width variations along different lines of sight and is often sharper than the circumstellar feature. The differences in profile and position of interstellar vs. circumstellar bumps may suggest that the carrier materials are different in the two cases. More probably, they are different evolution stages of similar carbon-based materials. Under this hypothesis also the interstellar bump should be attributed to electronic transitions in graphitic clusters. This suggestion seems compatible with the trend of laboratory carbon grains. As mentioned above, the extrapolation of UV irradiation experiments brings to an optical gap of 1.2 eV for grains in the diffuse medium, which should be only partially graphitised. In the view of the discussion reported in section 5.1, the stable peak position at 217.5 nm should be the result of a balancing of various processes (e.g. UV and ion irradiation). Interstellar grains should present a higher disorder ($L_a \sim 10$ Å, according to data in Table 6) than circumstellar ones.

A serious problem concerns the possibility of simulating the interstellar "bump" bandwidth. This is not matched by any of the samples we have produced. Recent measurements on matrix isolated carbon grains [37] evidence a dependence of bandwidth from carbon particle clumping. Tentatively, we can infer that circumstellar carbon grains are clumped (wide bump), while in the interstellar medium they are mainly isolated (sharp feature).

Observational data in the 3.3-3.4 $\mu$m spectral region show that: a) the aliphatic 3.4 $\mu$m bands dominate in interstellar medium and comets [38,39] and b) the aromatic 3.3 $\mu$m emission feature prevails in regions rich in UV flux [40]. Consistently with our results, grains present a more graphitic character in media where UV radiation may have processed them.

A puzzling question is why the 3.4 $\mu$m band, typical of less processed materials, and the 217.5 nm bump, characteristic of some processing, coexist in the diffuse interstellar medium. Tentatively, different populations (size ranges?) of carbon-based compounds could be invoked to interpret the two spectral features.

Last but not least, the cosmic elemental abundance budget must be carefully considered when models or laboratory data are applied to reproduce observations. Snow & Witt [41] have compared updated carbon abundances with requirements of various models. Several of them face problems to be reconciled with the new abundance constraints. All models, whether core-mantle or based on clustered grains [1,5,42-44], seem to confirm that carbon grains play a significant role in the overall cosmic dust budget. However, according to the approach declared in section 1, our results maintain their general validity as have allowed us to evidence some evolutionary trends true for laboratory "analogues" and applicable to space materials.

## References


1. Dominik, C., and Tielens, A.G.G.M. (1997) The physics of dust coagulation and the structure of dust aggregates in space, Astrophys. J. 480, 647-673.
2. Jäger, C., Mutschke, H., Begemann, B., Dorschner, J., and Henning, Th. (1994) Steps toward interstellar silicate mineralogy I, Astron. Astrophys. 292, 641-655.
3. Dorschner, J., Begemann, B., Henning, Th., Jäger, C., and Mutschke, H. (1995) Steps toward interstellar silicate mineralogy, II, Astron. Astrophys., 300, 503-520.
4. Papoular, R., Conard, J., Guillois, O., Nenner, I., Reynaud, C., and Rouzaud, J.-N. (1996) A comparison of solid-state carbonaceous models of cosmic dust, Astron. Astrophys. 315, 222-236.
5. Greenberg, J.M., Li, A., Mendoza-Gómez, C.X. et al. (1995) Approaching the interstellar grain organic refractory component, Astrophys. J. 455, L177-L180.
6. Colangeli, L., Mennella, V., Palumbo, P., Rotundi, A., and Bussoletti, E. (1995) Mass extinction coefficients of various submicron amorphous carbon grains: tabulated values from 40 nm to 2 mm, Astron. Astrophys. Suppl. Ser. 113, 561-577.
7. Mennella, V., Colangeli, L., Blanco, A., Bussoletti, E., Fonti, S., Palumbo, P., and Mertins, H.C. (1995) A dehydrogenation study of cosmic carbon analogue grains, Astrophys. J. 444, 288-292.
8. Mennella, V., Colangeli, L., Bussoletti, E., Monaco, G., Palumbo, P., and Rotundi, A. (1995) On the electronic structure of small carbon grains of astrophysical interest, Astrophys. J. Suppl. Ser. 100, 149-157.
9. Mennella, V., Colangeli, L., and Bussoletti, E. (1995) The absorption coefficient of cosmic carbon analogue grains in the wavelength range 20 - 2000 $\mu m$, Astron. Astrophys. 295, 165-170.
10. Mennella, V., Colangeli, L., Bussoletti, E., Merluzzi, P., Monaco, G., Palumbo, P., and Rotundi, A. (1995) Laboratory experiments on cosmic dust analogues: the structure of small carbon grains, Planet. Spa. Sci. 43, 1217-1221.
11. Mennella, V., Colangeli, L., Palumbo, P., Rotundi, A., Schutte, W.A., and Bussoletti, E. (1996) Activation of a UV resonance in hydrogenated amorphous carbon grains by exposure to UV radiation, Astrophys. J. 464, L191-L194.
12. Mennella, V., Colangeli, L., Cecchi Pestellini, C., Palomba, E., Palumbo, P., Rotundi, A., and Bussoletti E. (1997) The role of hydrogen in small amorphous carbon grains: the IR spectrum, From Stardust to Planetesimals (contributed papers), eds. Y.J. Pendleton and A.G.G.M. Tielens, NASA CP3343, 109-112.
13. Mennella, V., Baratta, G.A., Colangeli, L., Palumbo, P., Rotundi, A., Bussoletti, E., and Strazzulla, G. (1997) Ultraviolet spectral changes in amorphous carbon grains induced by ion irradiation, Astrophys. J. 481, 545-549.
14. Rotundi, A., Rietmeijer, F.J.M., Colangeli, L., Mennella, V., Palumbo, P., and Bussoletti, E. (1998) Evidence of inherent forms of carbon in soot materials of astrophysical interest, Astron. Astrophys. 329, 1087-1096.
15. Rotundi, A., Rietmeijer, F.J.M., Colangeli, L., Mennella, V., Palumbo, P., and Bussoletti, E. (1997) Chain-like aggregates of amorphous carbon grain: modifications induced by hydrogen content, in preparation.

16. Hecht, J.H. (1986) A physical model for the 2175 Å interstellar extinction feature, Astrophys. J. 305, 817-822.
17. Sorrell, W.H. (1990) The $\lambda$ 2175 Å feature from irradiated graphitic particles, MNRAS 243, 570-587.
18. Harris, P.J.F, Green, M.L.H., Chi Tsang, S. (1993) J. Chem. Soc. Faraday Trans. 89(8), 1189-1192.
19. Oberlin, A. (1984) Carbonization and graphitization, Carbon 22, 521-541.
20. Robertson, J. (1986) Amorphous carbon, Adv. Phys. 35, 317-374.
21. Robertson, J. (1991) Hard amorphous (diamond-like) carbons, Prog. in Solid State Chem. 21, 199-334.
22. Marchand, A., (1987) Various kinds of solid carbon: structure and optical properties, Polycyclic Aromatic Hydrocarbons and Astrophysics, Reidel, Dordrecht, pp. 31-54.
23. Baratta, G.A., Arena, M.M., Strazzulla, G., Colangeli, L., Mennella, V., and Bussoletti, E. (1996) Raman spectroscopy of ion irradiated amorphous carbons, Nuc. Inst. Meth. Phys. Res. 116, 195-199.
24. Dischler, B., Bubenzer, A., and Koidl, P. (1983) Bonding in hydrogenated hard carbon studied by optical spectroscopy, Solid State Comm. 48, 105-108.
25. Jenniskens, P., Baratta, G.A., Kouchi, A., de Groot, M.S., Greenberg, J.M., and Strazzulla, G. (1993) Carbon dust formation on interstellar grains, Astron. Astrophys. 273, 583-600
26. Hagen, W., Allamandola, L.J., and Greenberg, J.M. (1979) Interstellar molecule formation in grain mantles: the laboratory analog experiments, results and implications, Astrophys. Space Sci. 65, 215-240.
27. Mathis, J.S., Mezger, P., and Panagia, N. (1983) Interstellar radiation field and dust temperatures in the diffuse interstellar matter and in giant molecular clouds, Astron. Astrophys. 128, 212-229.
28. Jenniskens, P. (1993) Optical constants of organic refractory residue, Astron. Astrophys. 274, 653-661.
29. Strazzulla, G. and Baratta, G.A. (1992) Carbonaceous material by ion irradiation in space, Astron. Astrophys. 266, 434-438.
30. Seab, C.G. (1987) Grain destruction, formation, and evolution, Interstellar Processes, Reidel, Dordrecht, pp. 491-512.
31. McKee, C.F. (1989) Dust destruction in the interstellar medium, Interstellar Dust, Kluwer, Dordrecht, pp. 431-443.
32. Greenberg, J.M. (1989) Interstellar dust: an overview of physical and chemical evolution, Evolution of interstellar dust and related topics, North Holland, Amsterdam, pp. 7-52.
33. Draine, B.T. and Malhotra, S. (1993) On graphite and the 2175 Åextinction profile, Astrophys. J. 414, 632-645.
34. Kim, S., Martin, P.G., and Hendry, P.D. (1994) The size distribution of interstellar dust particles as determined from extinction, Astrophys. J. 422, 164-175.
35. Mathis, J.S. (1994) The origin of variations in the 2175 Å extinction bump, Astrophys. J. 422, 176-186.
36. Ugarte, D. (1995) Interstellar graphitic particles generated by annealing of nanodiamonds and their relation to the 2175 Å peak carrier, Astrophys. J. 443, L85-L88.
37. Schnaiter, M., Mutschke, H. , Henning, Th., Lindackers, D., Strecker, M., and Roth, P. (1996) Ultraviolet spectroscopy of matrix isolated amorphous carbon particles, Astrophys. J. 464, L187-190.
38. Tokunaga, A.T. and Brook, T.Y. (1990) Did comets form from unaltered interstellar dust and ices ? The evidence from infrared spectroscopy, Icarus 86, 208-219.
39. Pendleton, Y.J., Sandford, S.A., Allamandola, L.J., Tielens, A.G.G.M., and Sellgren, K. (1994) Near-infrared absorption spectroscopy of interstellar hydrocarbon grains, Astrophys. J. 437, 683-696.
40. Bregman, J.D., Dinerstein, H.L., Goebel, J.H., Lester, D.F., Witteborn, F.C., and

Rank, D.M. (1983) Observations of NGC 7027 from 5.2 to 7.5 microns: detection of Ni II and additional dust features, Astrophys. J. 274, 666-670.
41. Snow, T.P. and Witt, A.N. (1995) The interstellar carbon budget and the role of carbon in dust and large molecules, Science 270, 1455-1460.
42. Mathis, J.S. (1996) Dust models with tight abundance constraints, Astrophys. J. 472, 643-655.
43. Li, A., and Greenberg, J.M. (1997) A unified model of interstellar dust, Astron. Astrophys. 323, 566-584.
44. Dwek, E. (1997) Can composite fluffy dust particles solve the interstellar carbon crisis?, Astrophys. J. 484, 779-784.

# STARDUST MINERALOGY

*The Laboratory Approach*

JOHANN DORSCHNER
*Astrophysikalisches Institut und Universitäts-Sternwarte*
*Schillergäßchen 3, D-07745 Jena, Germany*

**Abstract.** Some basic mineralogical problems which are not always adequately considered in the astrophysical investigation of solids in circumstellar envelopes around evolved stars (stardust) are discussed: adequacy of mineralogical terms, identification problems, universality claim, amorphousness vs. crystallinity, heterogeneous grain structure, and spectral masking. The state of the art in the laboratory simulation of the main groups of stardust solids detected in oxygen-rich and carbon-rich circumstellar solids is critically reviewed: silicates, oxides, carbides, nitrides, sulfides and diamonds. Relevant laboratory data on presolar grains in meteorites and interplanetary dust particles are compared with the astrophysical results.

## 1. Introduction

Astrophysical modelling of cosmic objects and processes requires knowledge of material parameters. They can be easily provided for gases because atoms and molecules in cosmic environments are identical with those measured in terrestrial laboratories. However, the solids immersed as dust grains in circumstellar and interstellar gas are, as a rule, quite different from terrestrial solids.

There are only some singular cases, in which such exotic cosmic dust solids can be studied by direct measurements. Some highly refractory and chemically resistant presolar grains contained in the cloud from which the solar system formed have been preserved in primitive meteorites and can be isolated from them [1, 96]. Refractory grains frozen in cometary nuclei, which are released by the sublimation of the ices through the solar radiation and contribute to the interplanetary dust particle (IDP) population, penetrate the earth's atmosphere and can be collected by planes and studied in the laboratory. In such IDPs indications for presolar dust components have



*J. M. Greenberg and A. Li (eds.), Formulation and Evolution of Solids in Space*, 229–264.

been found [15, 16]. Cometary dust grains have also been studied via in-situ impact mass-spectrometry during the perihelion passage of P/Comet Halley [61, 62]. The results formed the base for detailed mineralogical modelling [103, 104]. The most fascinating possibility is the direct investigation of interstellar grains passing the solar system. They have been detected by the dust analysers on board of the spacecraft Ulysses [51]. The problem of determining their chemical composition will be tackled for the first time during the Cassini mission launched in 1997; the collection of cometary and passing interstellar dust will be attempted within the NASA project "Stardust" scheduled to be launched in 1999.

Space experiments are expensive, extremely time-consuming and restricted to very small quantities of matter. For instance, the total mass of dust investigated by the Halley impact mass-spectrometry was not larger than that of a 10 $\mu$m IDP! At present, the most effective way to provide the astrophysicists with urgently needed realistic data of cosmic dust solids is the production of laboratory analogues [19], i.e. the experimental "imitation". Primitive solar system solids [25, 60, 77] and the aforementioned exceptional cases of in-situ and laboratory studies of genuine cosmic dust are important guides to make the laboratory simulation products as similar as possible to the cosmic originals.

The new branch of dust research via experimental simulation found its technical articulation in the terms "laboratory astrophysics" (mostly used) and "solid-state astrophysics" introduced on international conferences in 1968 ("International Conference on Laboratory Astrophysics" in Lunteren, Holland) and 1974 ("Symposium on Solid-state Astrophysics" in Cardiff, Wales). In the past twenty years, experimental groups working in this field have grown up in USA, the Netherlands, Italy, France, Japan, and Germany. Today this research field forms a constitutive element within the interdisciplinary Cosmic Dust Connection initiated by J. Mayo Greenberg (see, e.g., [47, 48]). The courses of the International School of Space Chemistry held at the "Ettore Majorana" International Centre for Scientific Culture are impressive manifestations of the work of this interdisciplinary connection.

The Astrophysical Institute of the University of Jena belongs to the "early birds" in laboratory astrophysics. More than twenty years ago, the experimental work was performed in collaboration with chemical institutions. After the German unification, the Max-Planck-Gesellschaft acknowledged the experience of the Jena dust group by the foundation of the research unit "Dust in Star-forming Regions" at this Institute. This unit headed by Thomas Henning started its work in 1992 and was generously supported by the MPG for five years. In the frame of this research unit the author of this review paper had the great opportunity to establish three laboratories (chemical, spectroscopic and dust lab) and to work with an efficient team of physicists and chemists. Main scientific goals have been preparation, spectroscopy and analytical characterization of dust analogues; determination of optical constants in the wavelength range 0.2-2000 $\mu$m; production of particle beams of analogues and study of the particles by matrix-isolation spectroscopy; investigations of presolar dust grains; sticking and aggregation experiments of dust grains in the laboratory and within the space experiment CODAG (COsmic Dust AGgregation) in order to get realistic data on the initial growth of planetesimals in the solar nebula.

## 2. Basic mineralogical stardust problems

### 2.1. ARE STARDUST SOLIDS MINERALS?

In the outflows and ejecta of evolved stars the largest number of different dust species among the cosmic dust populations has been found [27, 92]. These dust species are commonly summarized by the figurative term "stardust" [83]. The two most efficient dust builders among the chemical elements, oxygen and carbon, provide a natural chemical division of the stardust solids. In oxygen-rich envelopes grains consisting of silicates and oxides of the most abundant metals are formed, whereas in the reducing conditions of carbon-rich envelopes carbon, carbides, sulfides, and nitrides are expected to condense . The most basic question to be discussed first is: Can these stardust solids be considered minerals or not?

In mineralogical textbooks (see, e.g., [76]), minerals are defined as *natural, physically and chemically homogeneous solids.* They are characterized by a crystallographic class and a clearly limited compositional range. Chemical composition and lattice structure are narrowly connected with each other. Almost all minerals known so far are crystalline inorganics.

There are few problems to extend the mineral definition to cosmic rocks, e.g. lunar and martian rocks. As a rule, even very primitive solar system solids (carbonaceous chondrites, IDPs) and the stardust grains isolated from meteorites are crystalline and fulfil the mineral definition automatically. However, prominent circumstellar dust bands have wide and smooth profiles presenting strong evidence for the absence of long-range order and translational symmetry in these solids. Therefore, they are commonly considered amorphous solids.

In contrast to a crystalline phase that represents a sharp physical state exactly defined by microphysical energetic relations, amorphousness comprehends a wide transitional field of metastable structures. Lacking long-range order does not mean that the structure of the solids is completely disordered. In many cases, amorphous solids are characterized by short-range order, i.e., by a fixed coordination number and by average values of bond lengths and angles of the coordinated atoms. Well-ordered molecular building blocks are present, but they form a disordered fabric, e.g., the $SiO_4$ tetrahedra in the continuous random network of a silicate glass. Some rare examples of glassy solids are even present in terrestrial mineralogy, e.g. opal. In some textbooks [76], these amorphous natural solids are discriminated from ordinary minerals by the technical term "mineraloids".

In other cases, high-resolution electron microscopy reveals that the amorphous solids in question consist of randomly distributed micro-crystallites.

On the other hand, in cosmic environments amorphous solids are thinkable that exhibit chaotic structure, in which already the molecular building blocks have serious defects, so that non-stoichiometric composition and even dangling bonds are present. Silicate and oxide smokes are experimentally produced representatives of this type [88, 87, 100]. Such solids are surely beyond the limits of the mineraloid definition. Unfortunately, at present we are not yet in a position to classify sharply which kind of structure is present in a given cosmic dust solid. It can be expected that the refractory interstellar solids are more heavily disordered than those are in circumstellar envelopes.

## 2.2. SOME IMPORTANT QUESTIONS OF STARDUST MINERALOGY

### 2.2.1. *Reliable identification?*

The most basic problem is the reliable identification of an observed band. Solid-state bands in stellar spectra prove the presence of certain chemical bonds in the circumstellar dust material being the carrier of IR-active vibrations. Formulations, such as "to have identified circumstellar silicates" by their fundamental IR vibrations at 10 and 18 $\mu$m, are inexact. What has been found is merely the fact that the spectra exhibiting bands at 10 and 18 $\mu$m are due to solid grains containing Si–O bonds in such a spatial arrangement that stretching and bending vibrations can take place. Identification of silicates in the strict sense would mean that information is available on the average number of bridging oxygen atoms per tetrahedron, which connect neighbouring tetrahedra, and on the cations incorporated in the silicate structure. This is not the case. We do not even know whether all silicon ions are perfectly coordinated by four oxygen ions or not, whether the central tetrahedra-forming ions are in all cases Si ions or have been partly replaced by Al or $Fe^{3+}$. However, the wide and smooth profiles give clear evidence for disordered structure so that we consider these solids amorphous. All other statements are based on additional, more or less plausible assumptions [58].

Silicates represent an exceptionally favourable case because there are two bands being due to the same solid. The normal "identification" stories of circumstellar species rest on single bands only. The good fit of an observed band by data of a laboratory analogue is a necessary but not a sufficient condition for a correct identification. Astrophysical and astrochemical plausibility can support the assignment, but not ultimately prove it. The strongest confirmation is given by finding unambiguous profile characteristics and, if possible, by detecting additional spectral features in the observed spectra.

### 2.2.2. *Universal stardust solids?*

The Low Resolution Spectrometer (LRS) catalogue of the IRAS satellite [91] has made available a few thousand spectra in the wavelength range 8–22 $\mu$m; many of them containing "fingerprints" of stardust silicates, oxides, silicon carbide, and probably also silicon disulfide. The strongest features observed up to now, i.e. the silicate bands and the SiC feature at 11.3 $\mu$m, exhibit great diversity in their profiles (see Figures 1 and 9).

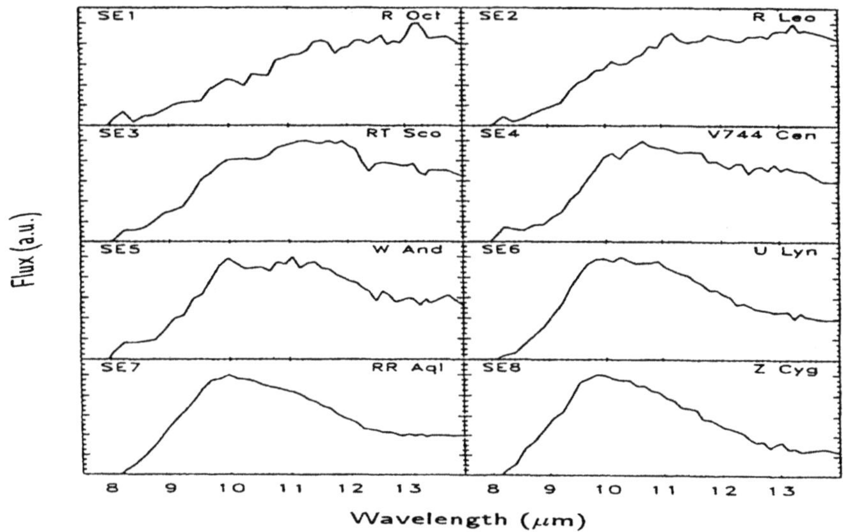


*Figure 1.* Classification system resting on 8 silicate emission (SE) indices applied to the 10 $\mu$m silicate feature [108].

The ubiquity of silicate grains, which have been detected in all dusty oxygen-rich cosmic environments, has promoted the idea of a unique cosmic silicate type, often called "astronomical silicate". However, this term is not used in the same meaning by different authors. Theoretical constructions like the "astronomical silicate" by [33] are based on compilations of data of very different origin and are useful tools for model calculations of dusty objects. Obviously, these data compiled for the practical application in modelling are not helpful in investigations of stardust mineralogy. Considering the spectral diversity visible in evolved stars (Fig. 1), we agree with the statement by Goebel et al. [45] that *"a universal astronomical silicate does not exist in oxygen-rich circumstellar shells"*.

Quite different is the situation in the interstellar silicate component. Here, long-time modification, e.g. by cosmic-ray irradiation and by photochemical modification, could have indeed levelled the optical properties of solids and really provided a unified "astronomical silicate". This interstellar silicate problems are out of the scope of this paper. We point to the interesting concept of the "true astronomical silicate" proposed by [49].

Silicates are not the only stardust species showing spectral diversity (cf., e.g., Fig. 9). Observable spectral variations can be due to morphological variety (size and shape distributions), and due to compositional and structural differences of the grain material. Compositional variety can be caused by chemical effects, e.g., the substitution of Mg by Fe in silicates, oxides and sulfides or of Si by Al in the $SiO_4$ tetrahedra. Variety of spectral features can also be induced by structure modifications leading, e.g., to the occurrence of different proportions between crystalline and amorphous dust grain components. Special consequences of such structural inhomogeneity will be discussed in the next Sections.

### 2.2.3. *Amorphousness from the very beginning?*

Altogether, there is little experimental experience with the nucleation of refractory particulates in conditions comparable to circumstellar environments. Nuth [87] stressed that in most cases the vapour-phase condensate formed in such experiments is not the most thermodynamically-favoured form of a mineral, *"while more often the condensate is found to be a glass or an even more chemically heterogeneous disordered material"*. From a theoretical standpoint, considerations of microphysical time scales suggest that the primary stardust condensates should be rather amorphous than crystalline solids [105]. These conclusions form the base for the widely held conviction that circumstellar solids, in particular silicates, are principally amorphous. There are, however, some facts that admonish to be somewhat cautious with such generalizations. As shown in Fig. 1 by far not all of the circumstellar silicate bands have really structureless profiles. Recent observations of oxygen-rich circumstellar envelopes with the Short Wavelength Spectrometer (SWS) of the satellite observatory ISO show several relatively narrow bands in the wavelength range 20–45 $\mu$m, which were suggested to be due to crystalline silicates [79, 114]. This finding does not necessarily contradict the concept of amorphous primary condensates. The initially amorphous material could have been subjected to thermal metamorphism by annealing followed by crystallization.

Apart from a possible thermal evolution of the grain material, it even cannot be completely ruled out that highly refractory materials, e.g. $Al_2O_3$

and SiC in oxygen-rich and carbon-rich envelopes, respectively, are formed as nanometer-sized monomineralic crystallites from the first. They could serve as nuclei, onto which in cooler layers massive condensation of amorphous material takes place. There are indications that minor crystalline phases must have been present when the main stardust components condensed (see Section 2.2.4).

Generally, it is a great challenge to experimenters to simulate particle condensation in the laboratory and to prepare this way analogues that represent solid states of different degree of order which populate the whole transitional field from amorphous to crystalline solids. Performing this is not only a problem of having the right preparational techniques at hand, but also a problem of the controlling analytics confirming that the irregular structure desired was really reached and that this did not happen more or less by chance, but rather in a reproducible way. An important point to be considered in future nucleation experiments is the fact that in contrast to the terrestrial laboratory in cosmic environments condensation occurs in conditions of drastically reduced gravity. That means, future progress requires performing space experiments.

2.2.4. *Heterogeneity vs. homogeneity?*

The traditional laboratory approach to the interpretation of circumstellar dust spectra is based on the assumption that the grains consist of a homogeneous amorphous solid. However, if grain formation takes place during the outward motion of a cooling gas, onion-layered grains should be expected, the layering of which reflects the temperature development during the condensation process [23].

Apart from onion-layered grains that are piece-by-piece homogeneous, irregularly inhomogeneous grains are thinkable. They could be formed, if a growing host grain incorporates chemically different subgrains already immersed in the gas from condensation processes at higher temperatures. Bernatowicz et al. [10, 11] found nm-sized Ti- and Zr-Mo carbide crystallites embedded in stardust graphite grains. Other imaginable cases would be grain aggregates formed by coagulation and grains containing small crystalline islands, phase separations, and holes (porous grains). Via mixture rules and discrete dipole approximations, theorists have developed tools to compute the optical properties of such heterogeneous grain models (see Fig. 2)

2.2.5. *Spectral masking by superposition?*

Apart from true band profile structuring due to crystallinity, spurious structuring can occur if, for instance, a relative narrow band of a minor dust component superposes a broad band of an amorphous main component. A

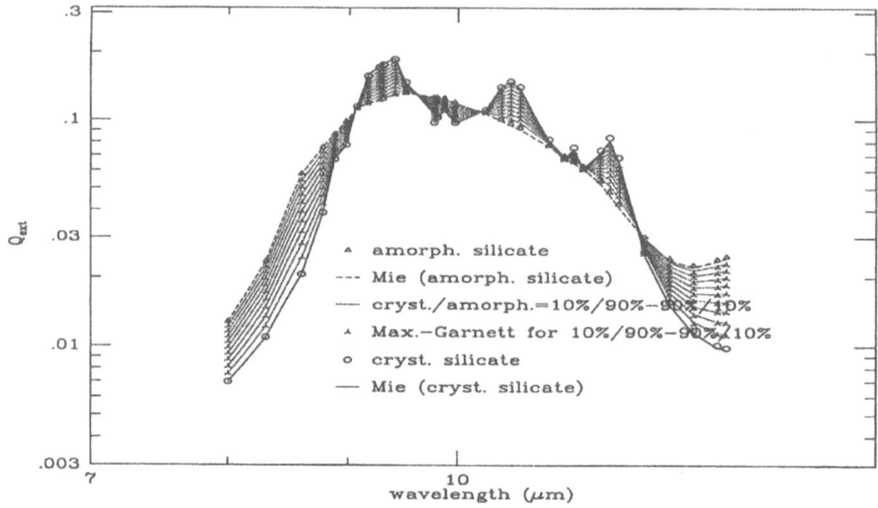


*Figure 2.* Change of the 10 $\mu$m band of amorphous bronzite with gradual admixture of crystalline bronzite, calculated with the Maxwell-Garnett mixture rule [62].

well-known example of this kind is the narrow 13 $\mu$m band riding on the "red wing" back of the broad 10 $\mu$m silicate band in many oxygen-rich stars. Its tentative identification with $Al_2O_3$ [95] has been recently confirmed [8].

Superpositions of bands can also take place in the continuum. If a weak superposing band has a broad profile, it strengthens the continuum and is hardly recognized as a discrete feature. In oxygen-rich stars broad features of amorphous $Al_2O_3$ and other metal oxides could strengthen the circumstellar opacity in the "trough" between the main silicate bands and, thus, be responsible for the relatively large opacity in this wavelength range, which is commonly attributed to the silicate dust alone. The opacity increase due to additional components would give a simple solution to overcome the trough opacity problem of the silicate laboratory analogues. Up to now, all silicate analogues fail to reproduce the observed trough opacity (see Section 3.2.3). Arguments against oxides of metals much less abundant than Mg and Fe, which result from cosmic abundance constraints, are not necessarily valid because many of these evolved stars exhibit peculiar abundances interpreted a consequences of dredge-up processes. In this connection, we point to the well-known fact in stellar spectroscopy that in spite of the low abundances of Ti, Zr, and V strong bands of the oxides of these elements are typical of late-type stars.

## 3. Circumstellar solids in oxygen-rich sources

### 3.1. SILICATES

#### 3.1.1. *The early laboratory work*

The stardust research era began in 1968 by the detection of the 10 $\mu$m emission band of silicate grains in oxygen-rich giants and supergiants [39]. Dramatic progress has been made by the IRAS observations in 1983. About 1800 IRAS LRS spectra covering the wavelength range 8–22 $\mu$m show circumstellar silicate emission features with peaks at about 10 and 18 $\mu$m [91]. In particular, the 10 $\mu$m band profile displayed considerable mineralogical diversity, and many authors introduced classification schemes [72, 73, 74, 108, 109]. In addition to the conspicuous wide silicate bands, further circumstellar features in O-rich stars have been found, in particular, in the spectra of Mira stars (Table 1). At present, the satellite observatory ISO is establishing a new FIR spectral data base considerably extending the wavelength range.

TABLE 1. Bands detected in oxygen-rich envelopes

| Wavelength ($\mu$m) | Band width Sharp | Band width Broad | Recent references |
|---|---|---|---|
| 7.15 | x | | [45] |
| $\approx$ 10 | | x | [108, 109] |
| 10 | x | | [74] |
| 11.3 | x | | [74] |
| 13.1 | x | | [8, 110] |
| $\approx$ 18 | | x | [108] |
| 19.5 | | x | [74] |
| 23.5 | x | | [79, 114] |
| 29.5 | x | | [79, 114] |
| 33.5 | x | | [79, 114] |
| 40.5 | x | | [79, 114] |
| 43.5 | x | | [79, 114] |

The observations of the silicate vibrational bands strongly impacted laboratory astrophysics. More than a decade ago, the experimental efforts were concentrated to the reproduction of the broad structureless 10 $\mu$m silicate band which was observed in emission in evolved stars and in the Orion trapezium nebula and in absorption in Galactic Centre sources and in deeply embedded young stellar objects. Important preparational techniques applied to producing silicate analogues with structureless broad bands have been [67]:

1. Evaporation and subsequent recondensation (arc discharge, laser pulses, thermal evaporation)
2. Reactive sputtering producing silicatic smokes
3. Amorphization of silicate surfaces by ion irradiation
4. Quenching silicate melts in order to get glasses
5. Producing amorphous silicates by sol-gel methods

Many authors immediately compared the measured spectra of the analogues with the observed spectra and were satisfied if they could reproduce the broad structureless 10 $\mu$m profiles. Studies of the structure of the analogues, confirmation of their experimental reproducibility, exact derivation of the optical constants, and studies of specific effects of small particles etc. were seldom performed. One of the noteworthy exceptions was Huffman's thorough study of spectra of small particles [12, 56] and his permanent warnings of experimental and theoretical pitfalls [55, 57]. Basic laboratory work, especially in the field of particle formation and also in analytics, was performed by Donn, Nuth and coworkers [24, 89, 87, 100].

### 3.1.2. *Studies of Silicate Glasses*

Systematic studies of glasses as dust analogues began in Jena more than 10 years ago and have been continuously performed up to now [29, 30, 31, 58, 82]. The glass conception for stardust silicates was based on the following assumptions:

- Silicate dust consists of homogeneous grains
- The glass state is a good approximation to the amorphous structure
- Mg and Fe are the predominant cations
- The stoichiometric ratios correspond to those of olivine and pyroxene

An important principle of the experimental work done in Jena was to obtain a sufficient characterization of the analogue material by the techniques of physical and chemical analytics available in the own lab and attainable by "laboratory tourism" to other chemical and physical institutes. From these results relations between chemical and spectral parameters ("diagnostic relations") have been derived in order to use them to the interpretation of the observations [31, 82]. The silicate glasses studied so far actually showed the band peaks at the right positions, but failed to reproduce the large band widths of the stardust silicates and the high absorption in the trough between the bands (Fig. 3).

The mineralogical aspects discussed in Section 2. forced to modify the laboratory philosophy. As a first step, in addition to Mg and Fe, the metals Al, Na, and Ca have been included in the experiments by preparing several

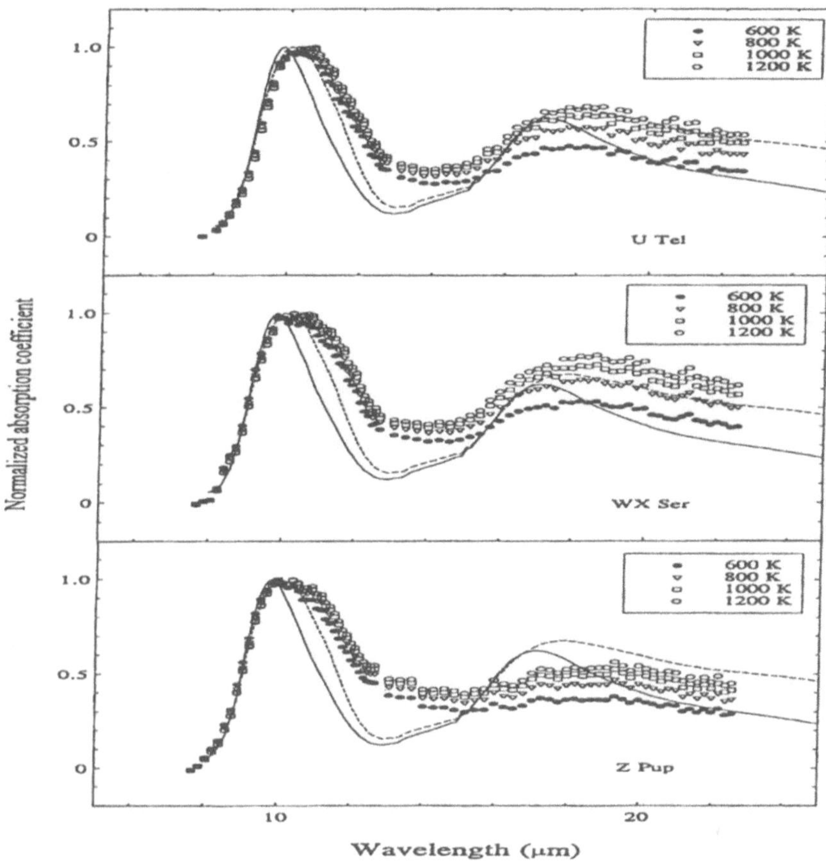


*Figure 3.* Normalized absorption coefficient derived from IRAS LRS spectra of three Mira stars compared with spectra of Rayleigh spheres (solid line) and a CDE (dashed line) calculated for the olivine glass with the cation ratio Mg/Fe=1. The observational curves differ by the temperature assumed at the inner edge of the silicate-containing envelope [31].

series of aluminosilicate glasses (ASG, Table 2). They consist of a continuous network of $SiO_4$ tetrahedra, in which Si is partly substituted by Al. In addition to the usual bending vibrations of the bridging oxygens normal to the Si(Al)–O–Si(Al) plain, which are the cause of the 18 $\mu$m band, ASG also show in-plain bending vibrations of the bridging oxygens, which cause an additional weak band or at least a distinct shoulder in the 13-14 $\mu$m range [38]. A comparison of an observed spectrum with Mg-Fe-ASG grains is shown in Fig. 8. In order to increase the trough opacity of the laboratory spectrum, a small percentage of amorphous $Al_2O_3$ was added.

TABLE 2. Diagnostic parameters of the characteristic bands of aluminosilicate glasses [81] calculated for Rayleigh spheres. Symbols: $\lambda_{peak}$ peak position of a band, $S = Q/a$ absorption coefficient, $W$ full width of half maximum of a band, $B = S(20)/S(10)$ band ratio, * no evaluation possible, ** shoulder at about 14 $\mu$m. For the sake of comparison, two pyroxene glasses are added.

| Formula parameter | 10 $\mu$m band | | | 20 $\mu$m band | | | 13 $\mu$m band | band ratio |
|---|---|---|---|---|---|---|---|---|
| $x$ | $\lambda_{peak}$ ($\mu$m) | $S_{peak}$ ($\mu m^{-1}$) | $W$ ($\mu$m) | $\lambda_{peak}$ ($\mu$m) | $S_{peak}$ ($\mu m^{-1}$) | $W$ ($\mu$m) | $\lambda_{peak}$ ($\mu$m) | $B$ |
| $NaAlSi_xO_{2x+2}$: | | | | | | | | |
| $x$=2 | 9.36 | 2.33 | 1.26 | 20.8 | 0.78 | 2.3 | 13.9 | 0.33 |
| 3 | 9.19 | 2.43 | 1.24 | 20.8 | 0.87 | 2.1 | 13.3 | 0.36 |
| 4 | 9.08 | 2.57 | 1.19 | 20.8 | 0.92 | 2.0 | 12.6 | 0.36 |
| $Mg_xAlSi_xO_{3x+1.5}$: | | | | | | | | |
| $x$=2 | 9.65 | 1.67 | 1.92 | 17.8 | 0.50 | 11.5 | ** | 0.30 |
| 3 | 9.61 | 1.73 | 1.80 | 17.7 | 0.56 | 9.0 | ** | 0.32 |
| 4 | 9.53 | 1.70 | 1.77 | 17.7 | 0.57 | 8.0 | ** | 0.34 |
| 9 | 9.48 | 1.79 | 1.63 | 17.6 | 0.62 | 6.6 | ** | 0.35 |
| $Mg_2Si_2O_6$: | 9.32 | 1.72 | 1.64 | 17.7 | 0.66 | 5.95 | - | 0.38 |
| $(MgFe)_{0.5x}AlSi_xO_{3x+1.5}$: | | | | | | | | |
| $x$=2 | 9.80 | 1.28 | 2.48 | 17.8 | 0.47 | 11.7 | ** | 0.37 |
| 3 | 9.62 | 1.37 | 2.36 | 17.8 | 0.50 | 10.1 | ** | 0.37 |
| 4 | 9.63 | 1.43 | 2.26 | 17.8 | 0.53 | 9.7 | ** | 0.37 |
| 9 | 9.60 | 1.48 | 2.12 | 17.8 | 0.58 | 7.0 | ** | 0.39 |
| $MgFeSi_2O_6$: | 9.25 | 1.32 | 2.27 | 17.9 | 0.57 | 6.64 | - | 0.44 |
| $Ca_2Mg_{0.5}AlSi_{1.5}O_7$: | 10.04 | 1.66 | 1.35 | 19.1 | 0.49 | * | 14.5 | 0.29 |
| $Ca_2Al_2SiO_7$: | 10.10 | 1.28 | 2.32 | 20.0 | 0.39 | * | 14.1 | 0.30 |

## 3.2. OXIDES

### 3.2.1. *Aluminium oxide*

Aluminium oxide ($Al_2O_3$, generally called alumina) is a longstanding candidate for a stardust solid [116]. Grains of $\alpha$–$Al_2O_3$ (corundum) together with the mixed Mg-Al oxide $MgAl_2O_4$ (spinel) have been considered as being the earliest high-temperature condensate in the solar nebula [50, 69], in circumstellar envelopes of M stars [37], and in supernova shells [20, 64, 65]. Al-enriched inclusions (CAI: calcium aluminium inclusions [75]) detected in primitive meteorites represent the high-temperature condensate of the solar nebula, which can be studied in the laboratory. Among the presolar grains extracted from primitive meteorites micrometer-sized corundum and spinel grains have been found [85]. The isotope signature of the corundum grains points to AGB star origin.

Evidence for alumina grains has been found in IRAS LRS spectra of Mira and other evolved stars, which show a weak, but distinct band at 13 $\mu$m. It has been tentatively assigned to the Al–O vibrations of crystalline $Al_2O_3$ [95, 40]. However, only recently the handicap for a reliable identification of this band, namely the lack of a sufficiently accurate band profile derived from the observations, was overcome by making available such data [8, 110]. In both papers the silicate emission underlying the 13 $\mu$m band was reconstructed by a polynomial fit. The resulting profiles shown in Fig. 4 agree very well with each other in the short-wavelength wing and the peak position, but show different band widths (1.05 $\mu$m in [8], 0.6 $\mu$m in [110]).

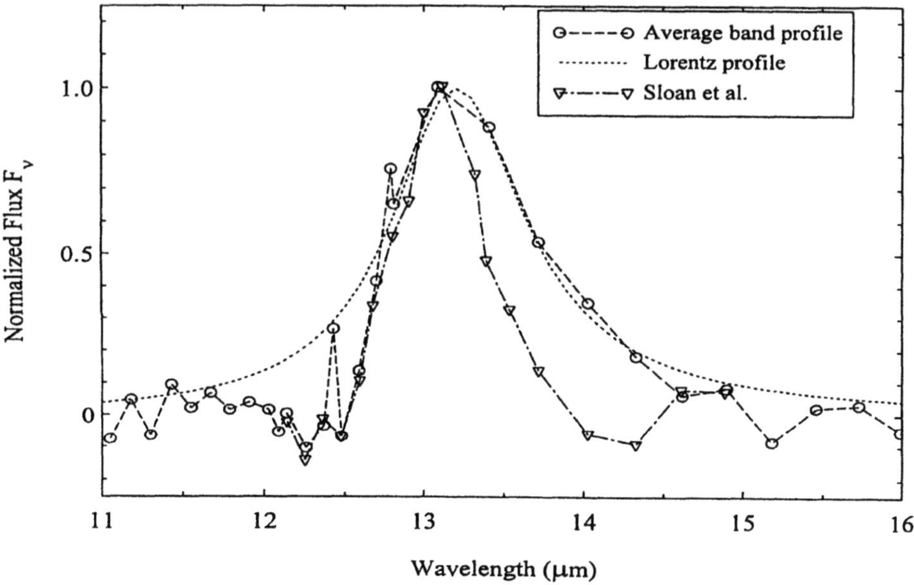


*Figure 4.* Average profile of the 13 $\mu$m band derived from 51 selected IRAS LRS spectra (dashed curve, open circles) [8]. The dotted curve is a Lorentz profile of the same width and peak position. The dash-dotted curve (triangles) is the profile derived by Sloan et al. [109].

Fig. 5 shows the comparison of the observed 13 $\mu$m band after [8] with calculated profiles for Rayleigh particles of crystalline and amorphous alumina taken from the literature and from own laboratory measurements, respectively. Amorphous alumina (a-$Al_2O_3$) has been produced by a sol-gel technique. The comparison shows that a-$Al_2O_3$ can be ruled out as the band carrier. However, such grains would strengthen the long-wavelength wing of the 10 $\mu$m silicate band and considerably contribute to the opacity of the trough region. Rayleigh particles of $\alpha$-$Al_2O_3$ produce a band approximately at the position of the observed band, but with a profile much sharper than the observations. However, the profil of the $\alpha$-$Al_2O_3$ band is sensitively depending on the grains' shapes. For instance, it considerably widens if a continuous distribution of ellipsoids (CDE) is admitted (see Fig. 5). Calculations of CDE were introduced by Bohren and Huffman [12] as a criterion to test the sensitivity of bands against shape effects. Since the profile of the Al–O vibrational bands in $\alpha$-$Al_2O_3$ sensitively depends on the grains' shapes, the reproduction of the observed profile seems to be only a question of finding an appropriate shape distribution. An ultimate proof of the identification of the 13 $\mu$m band with $\alpha$-$Al_2O_3$ could be the detection of the second band at 21 $\mu$m (see Fig. 5).

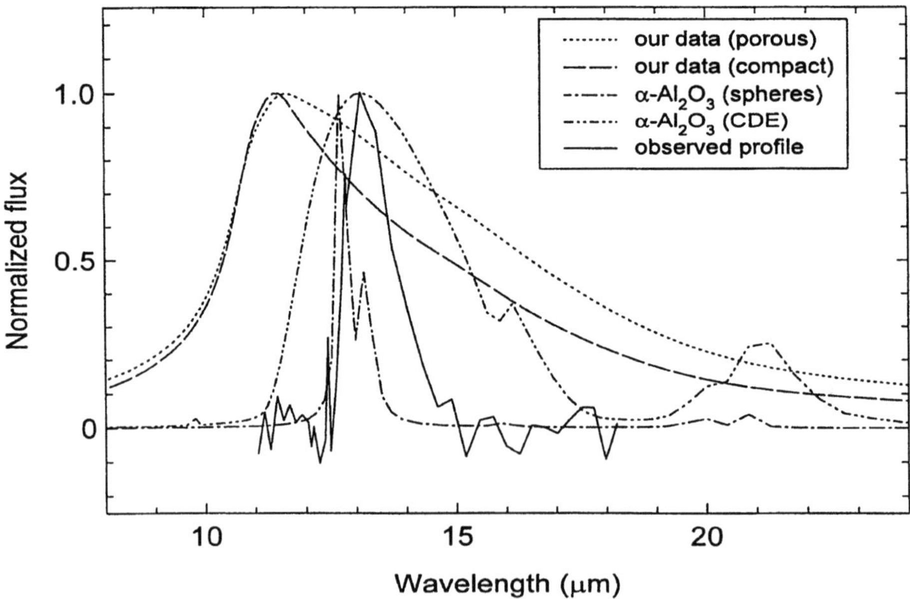


*Figure 5.* Profiles of the Al–O vibrational bands for Rayleigh spheres of different alumina analogues (porous and compact a-$Al_2O_3$ [8] and $\alpha$-$Al_2O_3$ [3]) and the observed profile [8]. The CDE-curve for $\alpha$-$Al_2O_3$ shows a much wider 13 $\mu$m band and a distinct second band at 21 $\mu$m.

In spite of this convincing arguments in favour of the alumina hypothesis, it cannot be completely excluded that the 13 $\mu$m band is caused by another carrier. The tight correlation of the strength of this band with that of the 10-$\mu$m silicate band (Fig. 6) points to the possibility that the 13 $\mu$m carrier is somehow connected with the silicates; in which way is unknown.

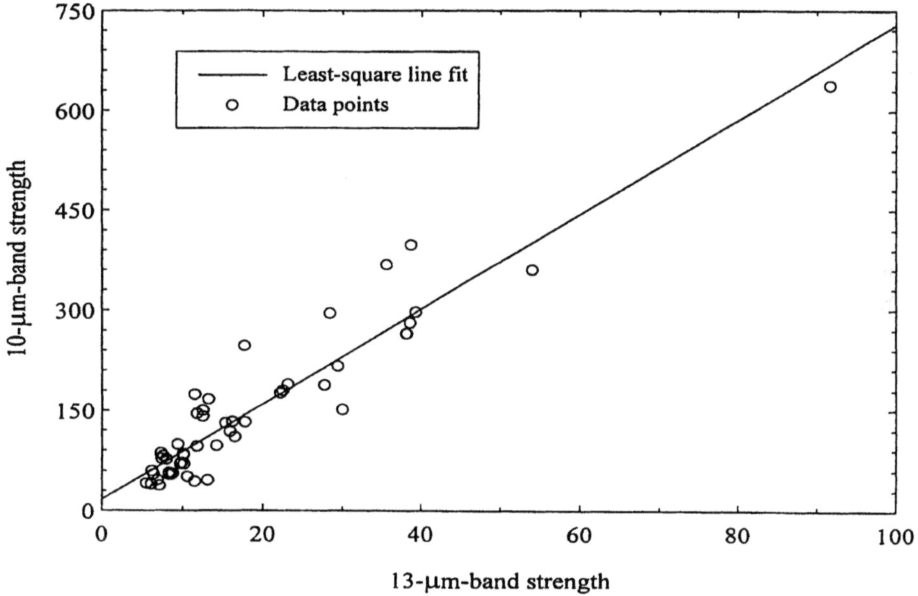


*Figure 6.* Correlation between the band strengths (in Jy) of the 10 and the 13 $\mu$m bands [8]

### 3.2.2. *Iron and magnesium oxides*

Apart from aluminium oxide, there is no conclusive observational evidence for the presence of other metal oxides in oxygen-rich circumstellar envelopes. There are, however, some indirect arguments. From an experimental point of view Nuth [87] has stressed the decisive role of metal oxide clusters in circumstellar condensation. Chemical arguments presented in [99] support the assumption that under highly oxidizing conditions in post-AGB stars magnetite ($Fe_3O_4$) and ferric iron oxide ($Fe_2O_3$) could be formed. This nonequilibrium processes could explain the depletion of Fe observed in some post-AGB stars [13].

If, on the other hand, Nuth and Hecht [88] are right with their argument that the initial circumstellar condensate should not be completely oxidized because Si and Al tend to reduce other metal oxides and, thus, keep the oxidation degree low, then not the ferric but rather the ferrous oxide should be dominant. For this reason, ferrous oxide (wüstite) and its solid solutions with MgO (magnesiowüstite) could also play an important role. Since optical constants of such mixed oxides ($Fe_xMg_{1-x}O$) had not been available in the literature, in the Jena labs a series of such samples was prepared, from which optical constants in the wavelength range 0.2–500 $\mu$m have been derived [6, 53].

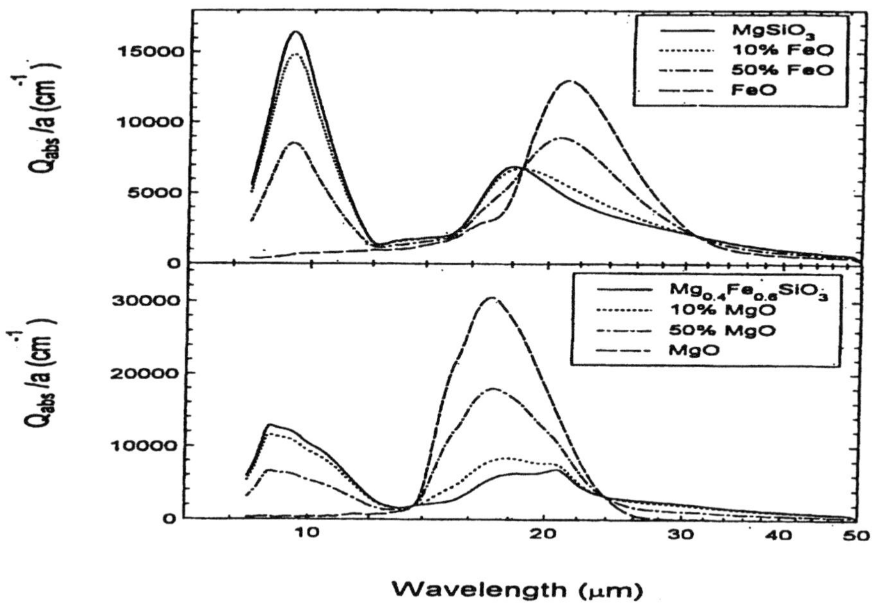


*Figure 7.* Superposition of Mg-Fe oxide spectra and silicate spectra. Growing oxide content drastically strengthens the 20 $\mu$m band and, thus, produces unusual band ratios [52].

There are some indications in IRAS LRS spectra that Fe and Mg oxides could really be present. Candidates are sources with weak or lacking 10 $\mu$m bands, but exhibiting strong band absorption in the range 17-25 $\mu$m ( see Fig. 7). Such objects have been attributed to the class $\lambda$ in [44].

### 3.2.3. *The trough opacity problem*

As a rule, silicate laboratory analogues fail to reproduce the relatively high emission observed in the trough region 12–17 $\mu$m. As it was already indicated in section 2.2.5., the silicate absorption coefficient can have been enhanced by the superposition of contributions due to other absorbers. For instance, amorphous alumina like the laboratory analogue product discussed in section 3.2.1. would effectively strengthen the trough opacity, and a small share of such grains could remove the discrepancy (Fig. 8). This simple proposal which would solve the trough opacity problem throws up some interesting astrophysical questions. The strengthening of the continuum between the silicate bands would require that a considerable part of the alumina grains should be a-$Al_2O_3$, whereas the 13-$\mu$m band is caused by crystalline $\alpha$-$Al_2O_3$. If $Al_2O_3$ condenses in the amorphous form, then in the envelopes exhibiting the 13 $\mu$m band a considerable fraction of it must have passed an annealing stage leading to crystallization. At present, little is known about grain evolution within circumstellar envelopes. Since not all envelopes with silicate emission show the 13 $\mu$m band, crystallization seems not to take place in every circumstellar envelope. If a-$Al_2O_3$ is actually responsible for the increased trough opacity and $\alpha$-$Al_2O_3$ for the 13 $\mu$m band, then sources with strong 13 $\mu$m bands should show reduced trough opacity and vice versa. The detection of such a relation would prove the proposed grain evolution picture.

From the viewpoint of the elemental abundance, aluminium compounds seem to be the most promising candidates accounting for the trough opacity. Considering the peculiar abundances observed in Mira stars and their interpretation by dredge-up processes, a noticeable contribution to the trough opacity could also stem from oxides of elements like Ti, Cr, Mn and some others. The possible role of Ti is indicated by the observation of strong photospheric TiO bands in late-type stars. In Table 3 some candidates are listed.

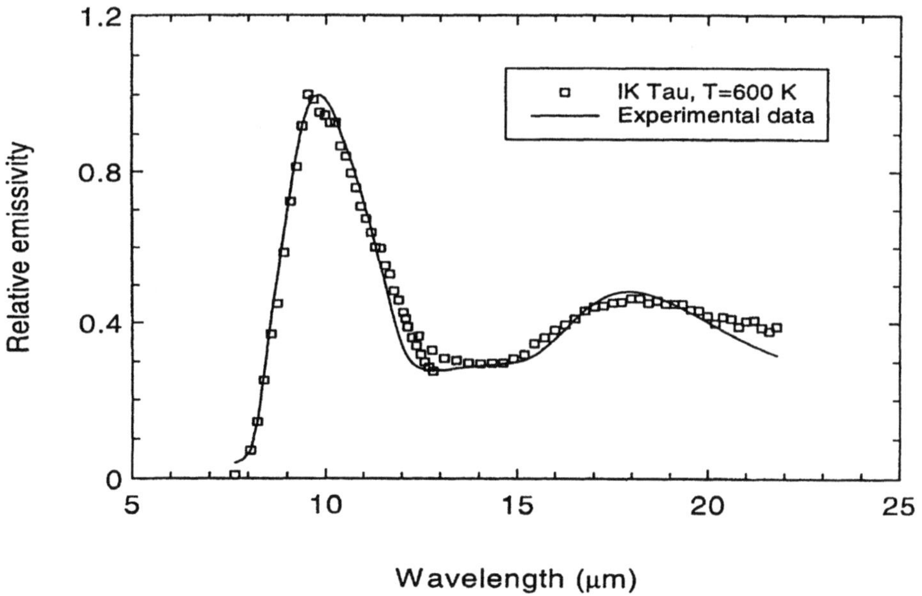


*Figure 8.* Reproduction of the 8–22 $\mu$m spectrum of the Mira star IK Tauri (IRAS LRS spectrum) by Mg-Fe-ASG + 16 vol.-% amorphous alumina [81].

TABLE 3. Potential absorbers in the trough (13-17 $\mu$m)

| Mineral | Formula | Absorption range |
|---|---|---|
| Cyanite, Disthene | $Al_2SiO_5$ | Bands in the whole range |
| Spinel | $MgAl_2O_4$ | Strong band at 14.7 $\mu$m |
| Sapphirine | $Mg_2Al_2[O_6/SiO_4]$ | Absorption in the whole range |
| Chromite | $FeCr_2O_4$ | Strong band at 15.5 $\mu$m |
| Rutile, Anatas, Brookite | $TiO_2$ | Bands in the whole range |
| Ilmenite | $FeTiO_3$ | Bands in the whole range |
| Pyrolusite | $\beta$-$MnO_2$ | Band at 14.6 $\mu$m |

### 3.3. POSSIBLE OTHER STARDUST COMPONENTS

Because of the relatively large abundance of sulfur and the presence of CO molecules, it makes sense to consider sulfates and carbonates of the most abundant metals as candidates of stardust components. At high temperatures, before sulfur is oxidized in the gas-phase, it could form the highly refractory magnesium sulfide MgS, which is oxidized to magnesium sulfate at lower temperatures. The vibrational bands of sulfates are positioned at about 8.7 and 16 $\mu$m, those of carbonates at about 7, 11.5, and 13.7 $\mu$m At the sulfate positions no conspicuous features have been found in IRAS LRS spectra, those of carbonates approximately coincide with three sharp bands listed in Table 1. At present, a detailled study that proves or disproves this association is still lacking.

## 4. Circumstellar solids in carbon-rich sources

### 4.1. CARBIDES

Carbon in elemental form and with heteroatoms, e.g., H and N, in its structure is adopted to be the main dust component in the carbon-rich envelopes of evolved stars. The wide field of experiments with carbon and carbonaceous matter has been excluded from the present review. However, in an appendix section of this review, new experiments on meteoritic nanodiamonds are presented, which probably represent a special stardust or interstellar dust component somehow related to type-II-supernovae. For the work on carbon dust analogues see the review by Colangeli and the contribution by Schnaiter et al. in this volume.

#### 4.1.1. *Silicon carbide*

In 1972, four years after the detection of silicate emission bands in oxygen-rich stars, Hackwell [52] found a strong emission band in the carbon stars V Hya and CIT 6 (IRC +30 219). The first tentative attribution of this band with its peak in the range 11-12 $\mu$m to silicon nitride [117] failed. In 1974, Treffers and Cohen [113] assigned this band to silicon carbide (SiC) grains. However, in this connection they demonstrated that the SiC band profile sensitively depends on grain shapes, so that, in principle, the band can peak at any position between the transverse and logitudinal optical phonon wavelengths 10.75 and 12.7 $\mu$m, respectively. In spite of these warnings, early laboratory studies of embedded $\alpha$-SiC grains [28] were considered as confirmation of this identification. To which extent the SiC feature can be modified by shape effects is demonstrated in Fig. 11.

The detection of SiC stardust was not a completely unexpected event. SiC condensation in carbon stars had been predicted by Friedemann [35] five years before the observational detection. Since the SiC band like the silicate band could also be observed by ground-based astronomy in the 8–13 $\mu$m window, the stardust investigation of O-rich and C-rich envelopes developed simultaneously in the 1970ies [78, 117, 113].

The IRAS LRS catalogue contains more than 500 stars showing the 11 $\mu$m emission feature of SiC [91]. The band profiles show variations of peak positions and widths from source to source. Recent ground-based observations have strikingly confirmed this ([111], Fig. 9). In analogy to the silicate bands, the spectral variety formed the base for classification schemes [46].

Concerning the identification of the emission feature positioned on the average at about 11.3 $\mu$m, there was an interesting wavelength coincidence. One of the strong unidentified IR bands (UIR bands), first detected in planetary and reflection nebulae, however, later on also in cool carbon stars (see, e.g., [106]), has its absorption peak at the same position. However, the 11.3 $\mu$m UIR band is always accompanied by other members of the UIR band series and can, therefore, be discriminated from the SiC feature.

Speck et al. [111] also canceled a mystery connected with cosmic SiC. They proved that the SiC band does appear in absorption (Fig. 10). Up to this time, only one (unconfirmed) observation existed that showed weak indications for SiC absorption. The apparent absence of SiC absorption bands had evoked the search for possible mechanisms explaining this strange behaviour [66, 115]. Now, the existence of SiC absorption in optical thick envelopes has been demonstrated beyond any doubt. However, at present spectral evidence for the presence of SiC in the diffuse interstellar clouds does not yet exist. The occurrence of small quantities of stardust SiC in interstellar matter is solely supported by the fact that such grains have been found among the presolar particles extracted form unequilibrated me-

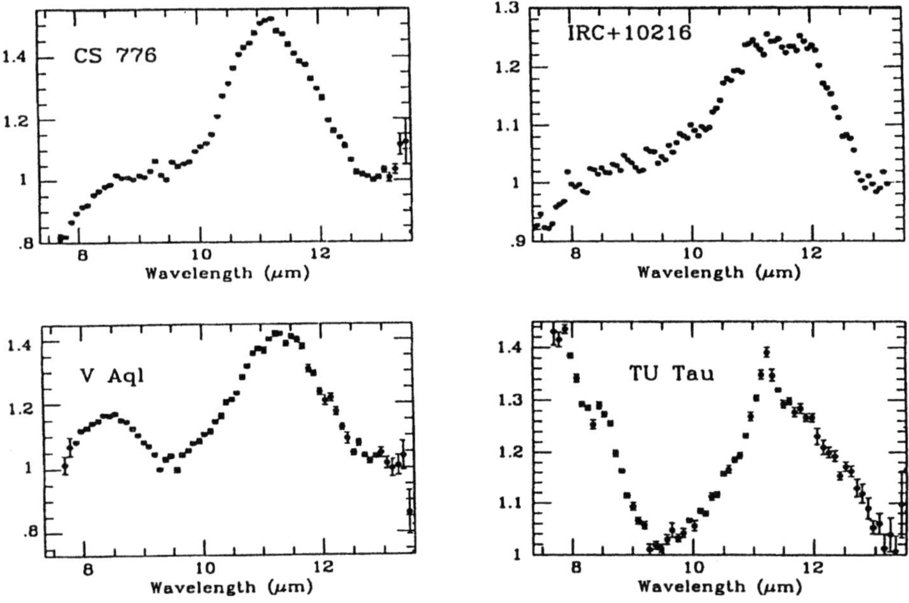


*Figure 9.* Four typical SiC emission profiles selected out of the UKIRT observations by Speck et al. [110].

teorites [1, 96]. Isotope measurements yielded striking evidence that these grains actually are stardust. Most of the grains studied so far are in the size range 1–10 $\mu$m and belong to the cubic modification ($\beta$-SiC).

The experimental base for the interpretation of the observed SiC band at 11.3 $\mu$m are absorption spectra of KBr-embedded size-separated SiC grains obtained from commercial mixtures of polytypes of $\alpha$-SiC grains [28, 36]. The most extensive laboratory work of this kind was done by Borghesi et al. [14] who measured several size-separated powder samples of $\alpha$-SiC and one sample of $\beta$-SiC. The band of $\beta$-SiC turned out to be positioned at somewhat shorter wavelength and to have a remarkably narrower profile than the corresponding findings of the $\alpha$-SiC samples.

As already stressed, the direct comparison of laboratory spectra of SiC powder with observed spectra meets serious objections because it ignores the sensitive dependence of the SiC band profile on the grain shapes, which is due to surface modes occurring at the Fröhlich wavelength $\lambda_F$. At this value, for very small grains the dielectric function $\epsilon(\lambda_F)$ fulfils the condition $\epsilon(\lambda_F) = -2\epsilon_m(\lambda_F)$, where $\epsilon_m$ is the dielectric function of the surrounding medium [12]. As already demonstrated in [113], with a suitable shape distribution almost any profile peaking between the phonon band wavelengths $\lambda_T$ and $\lambda_L$ can be obtained. This shape-dependence is illustrated in Fig.

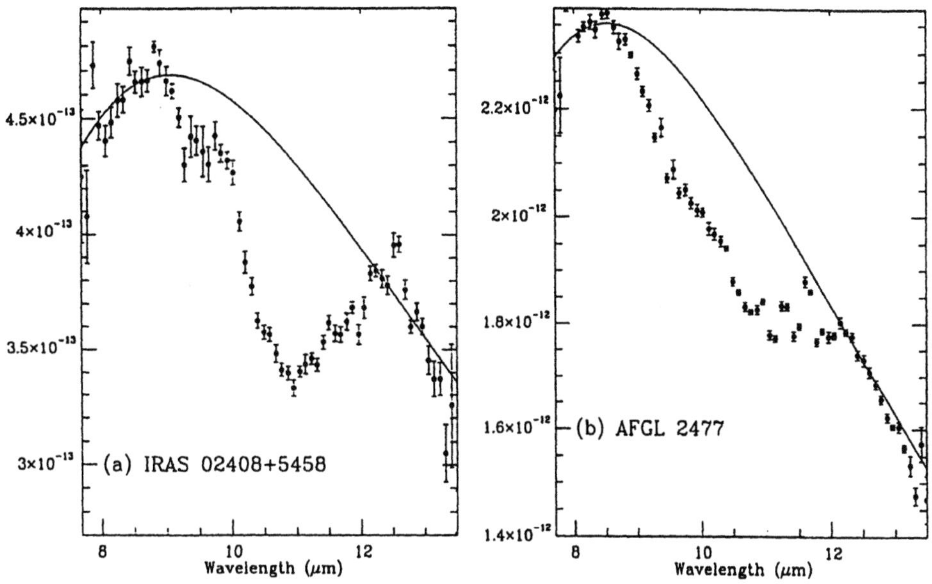


*Figure 10.* Two sources with SiC absorption (Flux in W $m^{-2}$ $\mu m^{-1}$). The solid curves are the blackbodies used for normalization [110].

11.

The laboratory bands of the SiC grains show typical shoulders (at both phonon wavelength for $\alpha$-SiC, only at $\lambda_T$ for $\beta$-SiC; [14]), which are not visible in the astronomical profiles. This was considered as evidence for predominantly amorphous circumstellar SiC particles [4]. With the aforementioned shape effects in mind, this argument is no longer conclusive. Recently, in Jena an experimental study of amorphous SiC was started [21]. Samples of a-SiC were produced by pyrolysis of dimethylpolysilanes in an inert atmosphere. Calculations with the optical constants derived from these amorphous samples suggest that the observed stardust SiC band profiles are narrower than those of the new a-SiC data. The experiments will be continued by the incorporation of some astrophysically interesting heteroatoms in the SiC structure, e.g. N, Al, and S. It is a well-known fact in the material sciences that nitrogen can be bonded into SiC.

### 4.1.2. *Other carbides*

Silicon carbide is evidently not the only carbide present in carbon-rich stars, even if no spectral evidence for other carbides is available at present. However, there is clear evidence for their existence from meteoritic presolar graphite spherules. In such spherules with diameters ranging from 0.3 to 9 $\mu$m nanometre-sized crystallites of titanium carbide and zirconium-

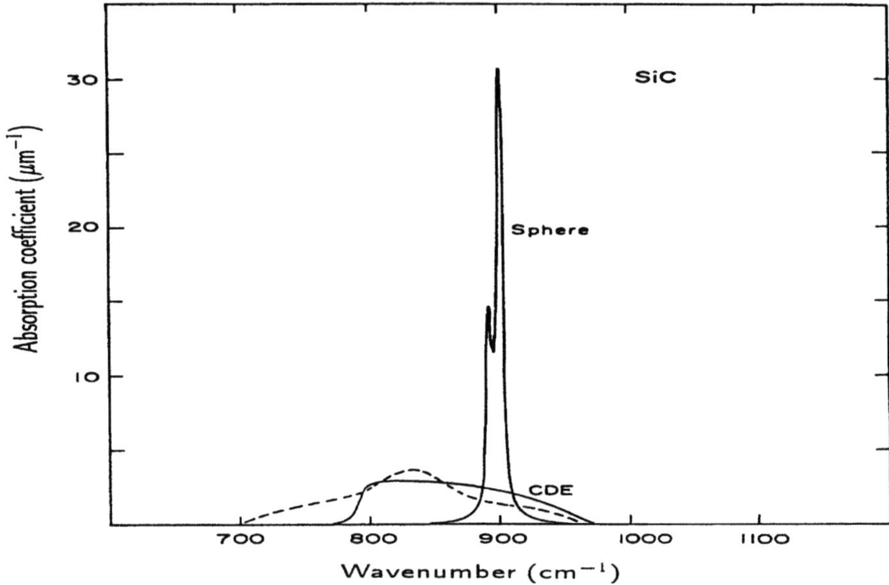


*Figure 11.* Demonstration of shape effects of SiC. The sharp band labeled with "Sphere" is the result of a calculation for small spherical grains, using bulk SiC data [96]. The curve labelled CDE is the calculation for a continuous distribution of ellipsoids, the dashed profile corresponds to the laboratory powder measurements in KBr. After [12].

molybdenum carbide have been found in the centres of the spherules as well as incorporated between the graphene sheets. The metal carbides evidently served as heterogeneous nucleation centres for the carbon condensation, and they were entrained by carbon as the spherule grew in the carbon-rich environment [10, 11]. Apparently, the carbon condensation preceded the SiC-formation, for only one of 67 grains investigated in [11] contained silicon carbide at all.

In the Jena labs, a series of carbide analogues supposed to be of astrophysical significance has been prepared: TiC, $Fe_3C$, $Na_2C$, $MgC_2$, $Mg_2C_3$, $CaC_2$. The goal is to make available optical constants of these solids for their potential use in improved envelope modelling and in order to search for spectral features in the observed spectra of carbon-rich sources.

## 4.2. NITRIDES AND SULFIDES

### 4.2.1. *Silicon nitride*

Apart from carbides, the reducing conditions present in carbon-rich objects allow the formation of other oxygen-free solids, too. From the viewpoint of cosmic elemental abundances [98], nitrides and sulfides are the most proba-

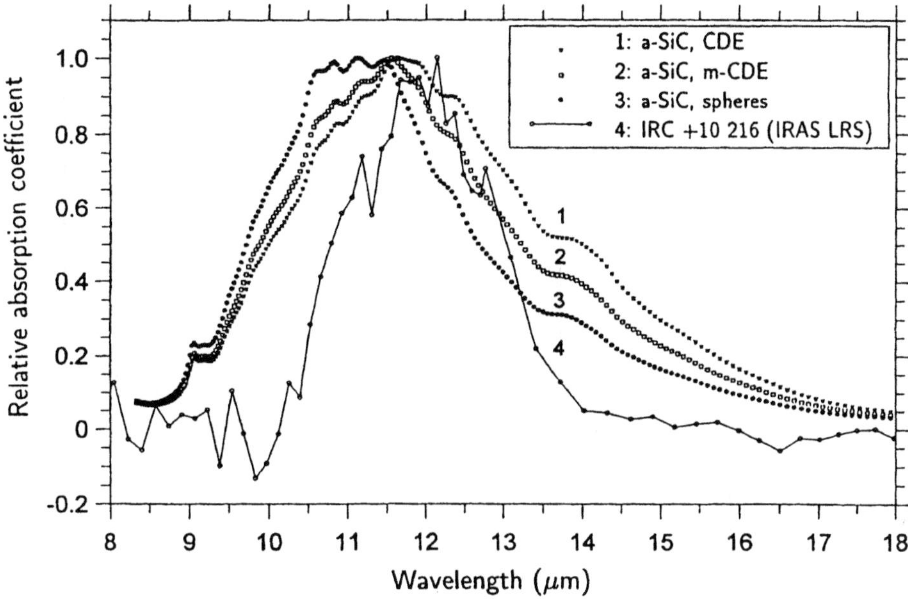


*Figure 12.* Comparison of experimental data of amorphous SiC [21] with the profile of IRC +10 216. This carbon star has a wide, flat-top SiC profile.

ble candidates. Among the nitrides, $Si_3N_4$ could be of interest. It has rarely been discussed in the stardust literature, and the tentative assignments of observed features to $Si_3N_4$ were wrong [41, 117]. Concerning laboratory measurements, Russell et al. [101] prepared crystalline (mixture of $\alpha$ and $\beta$ $Si_3N_4$) and amorphous samples of silicon nitride and measured their emissivities. After these results, grains of $Si_3N_4$ would contribute to the emission in the spectral range 9-12 $\mu$m and, thus, could modify the SiC grain profiles (Fig. 13). It cannot be entirely excluded that the structure observed within the SiC absorption profiles (Fig. 10) is influenced by $Si_3N_4$ absorption. Evidence for the existence of silicon nitride grains as a minor stardust component comes from meteorites, from which single $Si_3N_4$ grains could be isolated, which exhibited an isotopic pattern pointing towards a supernova origin [86].

4.2.2. *Sulfides*

Sulfur belongs to the abundant elements in the sun and in primitive Solar System condensates [98]. Ordinary chondrites contain about 5 % troilite (FeS) Iron-nickel sulfides are commonly found in interplanetary dust particles. The total content of iron sulfides in the Comet Halley dust, determined by impact mass spectrometry, was estimated to 10–15 % [103, 104].

Some types of carbon-rich evolved stars show solid-state bands at posi-

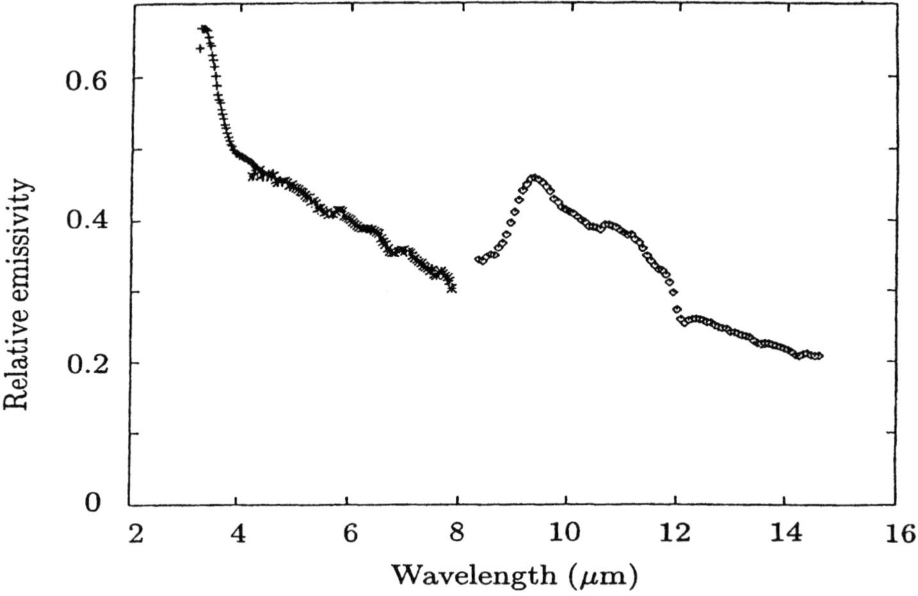


*Figure 13.* Silicon nitride emissivity measured by Russell et al. 1989 [100].

tions typical of sulfide grains. An extremely strong band centred at about 30 $\mu$m found in KAO spectra of carbon-rich AGB stars, proto-planetary nebulae and planetary nebulae [22, 94] has been tentatively attributed to the vibrations of magnesium sulfide [42, 90]. More recent laboratory measurements [80] confirmed this assignment and stressed that in addition to pure MgS also mixed Mg-Fe sulfides deserve attention. The complex refractive indices of the mixed Mg-Fe sulfide in dependence on the Mg/Fe ratio are represented in Fig. 14. At present, the Mg–S vibrations of $Mg_{1-x}Fe_x$ with $x < 0.5$ offer the best quantitative explanation of the observed 30-$\mu$m band profile. The observed strength is in agreement with the solar values of the abundances of S and Mg [5, 94].

In contrast to Mg-rich sulfide grains, the iron-rich series members and pure FeS are chemically stable in interstellar conditions and, thus, should be able to occur both as interstellar dust component and as presolar ingredient in primitive Solar System solids [5, 26]. The sulfide-bearing GEMS in interplanetary dust particles, which are considered remainders of interstellar dust [16], give strong support to this expectation. In consequence, it makes sense to search for the spectral signature of high iron sulfides (Fig. 15) in cosmic sources with high dust content, e.g., in deeply embedded molecular cloud sources (protostars). Fig. 15 shows an interesting temperature effect. Since the strengths of the single components of the band group depend on

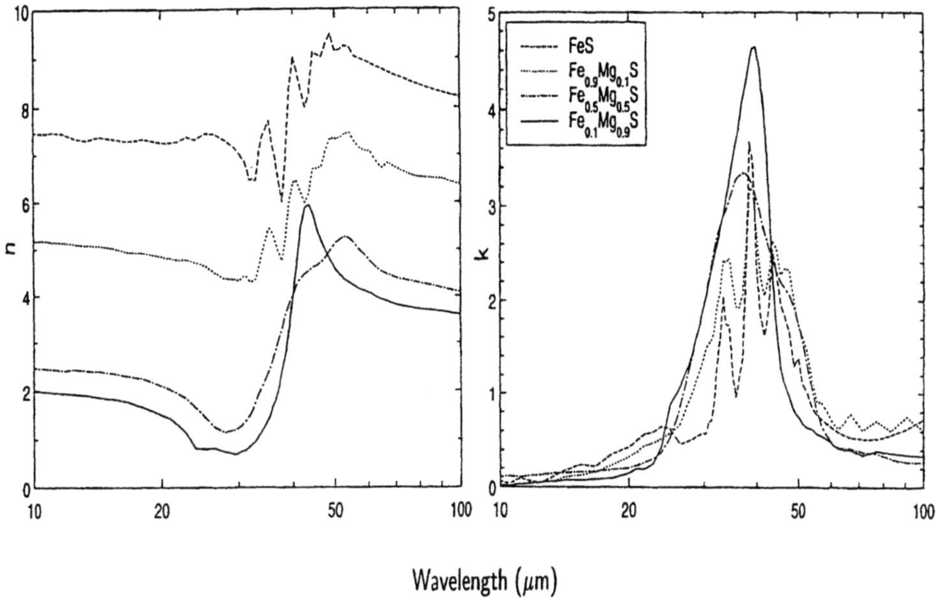


*Figure 14.* The wavelength dependence of $n$ and $k$ of mixed Mg-Fe sulfides. With increasing Fe content the strong single band of the insulator MgS changes to the characteristic band group of FeS [79] (see also Fig. 15).

the temperature in a different way, the FeS band group could serve as a "dust thermometer".

A further interesting candidate of sulfides in carbon-rich objects is silicon disulfide ($SiS_2$) suggested by Goebel [43] as a stardust component. The observational background of this proposal was the detection of a strong emission band at 21 $\mu$m by Kwok et al. (1989) in IRAS LRS spectra of carbon-rich proto-planetary nebulae [68]. Omont et al. [93] concluded that its carrier is a transient species related to the extreme conditions to which dust is exposed during this stellar phase. Apart from $SiS_2$, other carrier materials of this band have been suggested, e.g., carbonaceous dust [18].

In order to lay a reliable experimental base for the sulfide assignment, optical constants of amorphous $SiS_2$ have been derived by Begemann et al. [7]. The data show a strong band that can satisfactorily reproduce the peak position if the shape effects on the profile are considered and if the grains are sufficiently cool (see Fig. 17). A crucial experiment for the $SiS_2$ assignment would be the detection of the weak band present at about 16.5 $\mu$m in the laboratory spectrum. A first inspection of ISO SWS spectra of proto-planetary nebulae showing the 21 $\mu$m band yielded indications for the presence of a weak band-like excess at the wavelength in question [79]. Some chemical speculations on possible formation mechanisms of sulfides

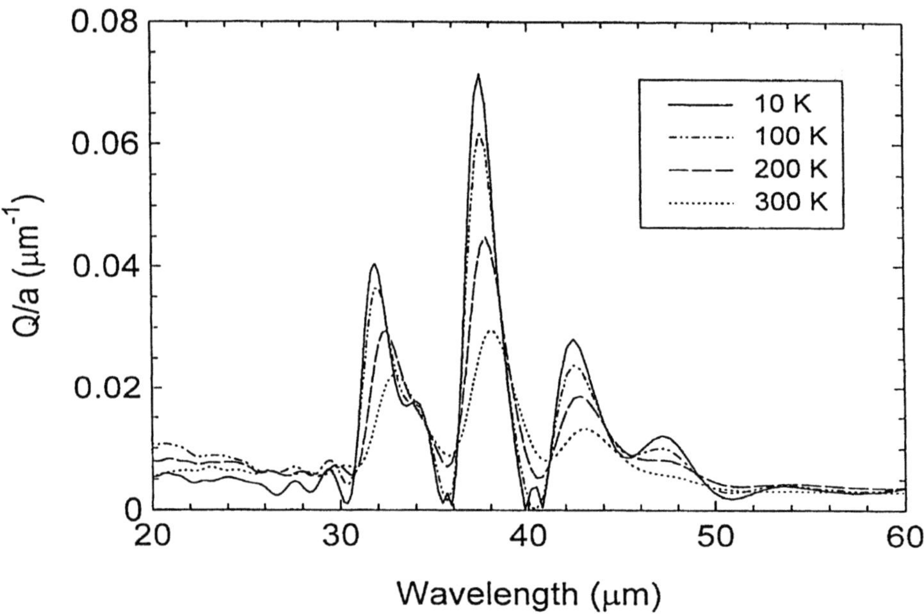


*Figure 15.* The absorption coefficient of Rayleigh grains of FeS grains in dependence on the temperature.

**Chemical Reaction leading to the Formation of Stardust Sulfides**

1. Mg-Fe sulfides (X = Mg, Fe):

$$X(g) + CS(g) \rightleftharpoons XS(s) + C(s) \quad (1)$$

$$X(g) + H_2S(g) \rightleftharpoons XS(s) + H_2(g) \quad (2)$$

2. Silicon disulfide:

$$Si(g) + 2CS(g) \rightleftharpoons SiS_2(s) + 2C(s) \quad (3)$$

$$SiS(g) + CS(g) \rightleftharpoons SiS_2(s) + C(s) \quad (4)$$

$$SiS(g) + H_2S(g) \rightleftharpoons SiS_2(s) + H_2(g) \quad (5)$$

$$SiC(s) + 2H_2S(g) \rightleftharpoons SiS_2(s) + CH_4(g) \quad (6)$$

*Figure 16.* Possible ways to the formation of sulfides in C-rich objects

in C-rich objects are given in Fig. 16. Like MgS, $SiS_2$ would not be able to survive a longer stay in the interstellar environment.

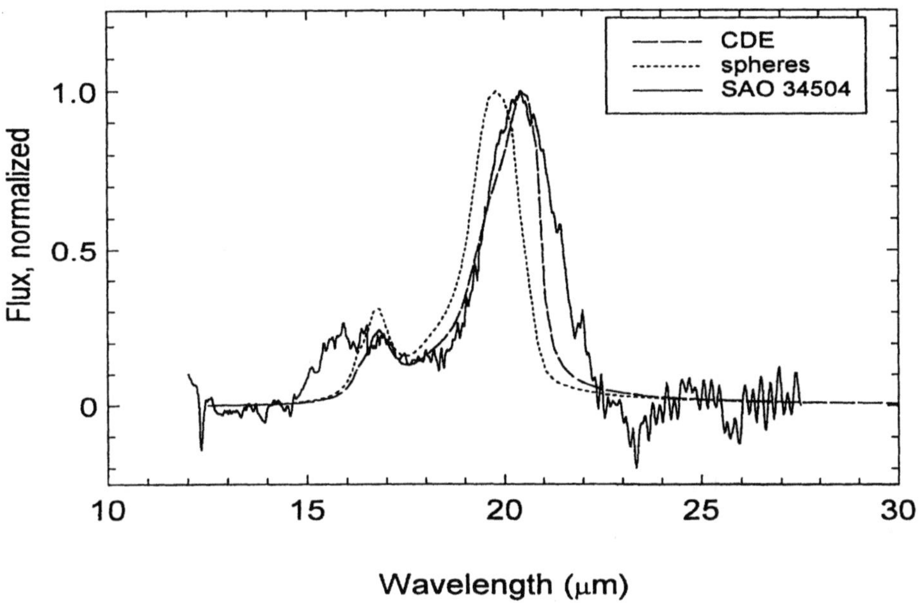


*Figure 17.* ISO SWS spectrum of the source SAO 34504 (≡ IRAS 22272 +34504) compared with calculated spectra with the optical constants of a-$SiS_2$ for spheres and a CDE.

## 5. Meteoritic nano-diamonds – a stardust component?

Diamonds, suggested to be a cosmic dust component already in 1969 [102], were surprisingly found ten years ago by Lewis et al. [70] as a finely dispersed component in unequilibrated meteorites (content on the average 400 ppm). The extraction procedure mainly rests on destructive acid dissolution of all components that can be dissolved ([112], Fig. 18).

The presolar character of this component is proven by isotopic anomalies, diamond being the carrier of the well-known Xe-HL anomaly in meteorites [70], which is generally attributed to the r- and p-processes of nucleosynthesis in supernovae.

For laboratory astrophysicists, it is a great challenge to study almost unaltered dust coming from sources outside the Solar System. The particular fascination of the meteoritic nano-diamonds consists in the expectation that they could give us the key for understanding the cosmic carbon dust, at all. Carbon is a main component of the interstellar dust, but its structure is hardly understood. In contrast to presolar SiC and graphite grains, the tiny sizes of the diamonds (mass-weighted mean diameter of the crystallites: 1.6 nm [34]) prevents extracting single grains and performing microanalytical and isotopic studies of isolated grains. The chemical activity of the sur-

**Steps for the Extraction of Meteoritic Nano-diamonds**

1. Solution of the silicates by alternating treatment with 10m HF/1m HCl and 6m HCl at room temperature and at 60-65°C
2. Extraction of precipitated sulfur with $CS_2$
3. Oxidation with $K_2Cr_2O_7$ (80°C)
4. Treatment with concentrated $H_3PO_4$ (3 hrs, 180°C)
5. Oxidation with $HClO_4$ (3 hrs, 190°C)
6. Second treatment with HF/HCl (4 hrs, 80°C)
7. Washing with $CH_2Cl_2$
8. Addition of water. Ultrasonification at this step produced a colloidal suspension in the water containing the bulk of the diamonds, with the water-immiscible organic liquid at the bottom of the test tube together with residual spinels and SiC.
9. Separation of the diamonds and further purification by colloidal extraction in 0.1m $NH_4OH$.

*Figure 18.* Chemical extraction procedure after Tang and Anders [111]

faces of such tiny grains dramatically masks their optical properties. On the other hand, the content of the nano-diamonds is much higher than that of the coarser presolar dust components (SiC, graphite, corundum, spinel, silicon nitride) and permits spectroscopic investigation.

For the spectrometry in the Jena labs, a diamond sample of 7 mg was available, which had been extracted by Ulrich Ott at the Max-Planck-Institut für Chemie in Mainz from the Murchison meteorite. Electron diffraction measurements of the sample strikingly confirmed the results first presented in [70] that the grains mainly consist of face-centered cubic (fcc) diamond. In spite of this unquestionable fact, the UV and the Raman spectra of the Murchison diamonds show indications for the presence of an aromatically bonded ($sp^2$-hybridized) carbon component. The steep rise of the UV absorption coefficient with decreasing wavelength sets in with the meteoritic diamonds already at larger wavelengths than it is the case with terrestrial fcc diamond, and in contrast to terrestrial diamond, the meteoritic diamonds show a weak, but distinct absorption feature at the position of the interstellar extinction bump (217.5 nm) (Fig. 19). The latter feature can be attributed to the $\pi$–$\pi^*$ transition typical of the $\pi$ electrons of $sp^2$-hybridzed carbon.

In the Raman spectra, the characteristic diamond line at 1332 $cm^{-1}$ could not be detected in our sample. Instead two weak and very wide Raman bands at about 1150 and 1530 $cm^{-1}$ were found. The comparison with Raman spectra of CVD (chemical vapour deposition) diamond films [84] and films of $sp^2$-dominated hydrogenated amorphous carbon films [107] indicates a diamondlike structure containing $\pi$-bonded carbon. In this connection, the fact is important that the Raman scattering cross-section of aromatically bonded carbon is larger by a factor of about 100 than that of carbon in a regular diamond structure.

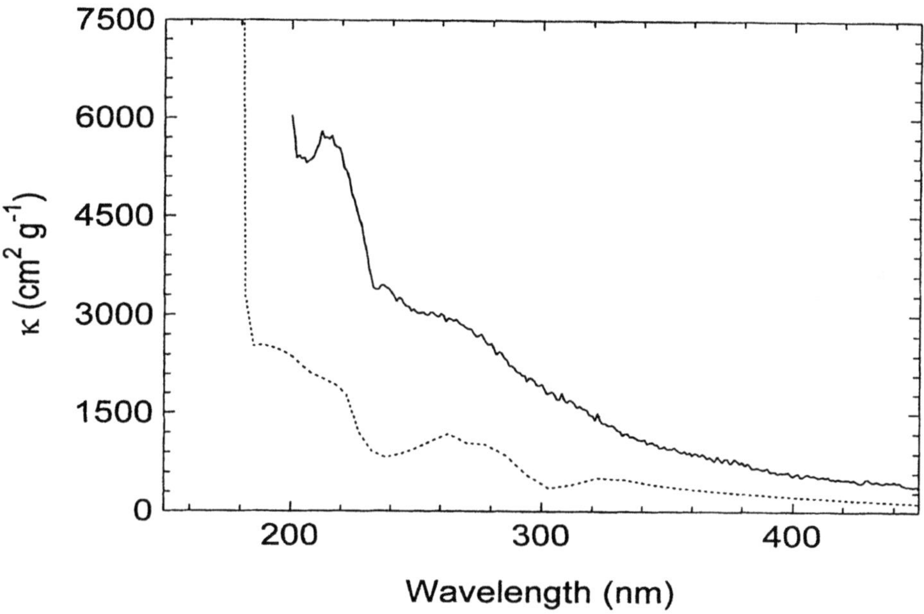


*Figure 19.* UV absorption spectrum of Murchison diamonds. The dotted curve is the absorption coefficient for terrestrial diamonds [70].

The IR spectrum of the Murchison diamonds confirmed the presence of prominent absorption features due to chemical groups attached to surface atoms [71]. Since our measurements [81] were extended up to submillimeter wavelengths, we could demonstrate that this spectral behaviour is continued into the far IR (Fig. 20). At 120 $cm^{-1}$ (83 $\mu$m) a particularly strong FIR band was found. If it is a genuine interstellar diamond feature, it could be useful for astronomical FIR observations of interstellar diamonds. Concerning the strong bands in the middle IR, our measurements corroborated the suspicion that, at least, some of them are artifacts due to chemical alterations during the extraction procedure and/or due to preparational contamination [54, 81].

The electron energy loss (EEL) spectra allowed the quantitative evaluation of the ratio of $sp^2$- to $sp^3$-bonded carbon, determined independently from core losses as well as plasma losses. Values between 0.12 and 0.2 have

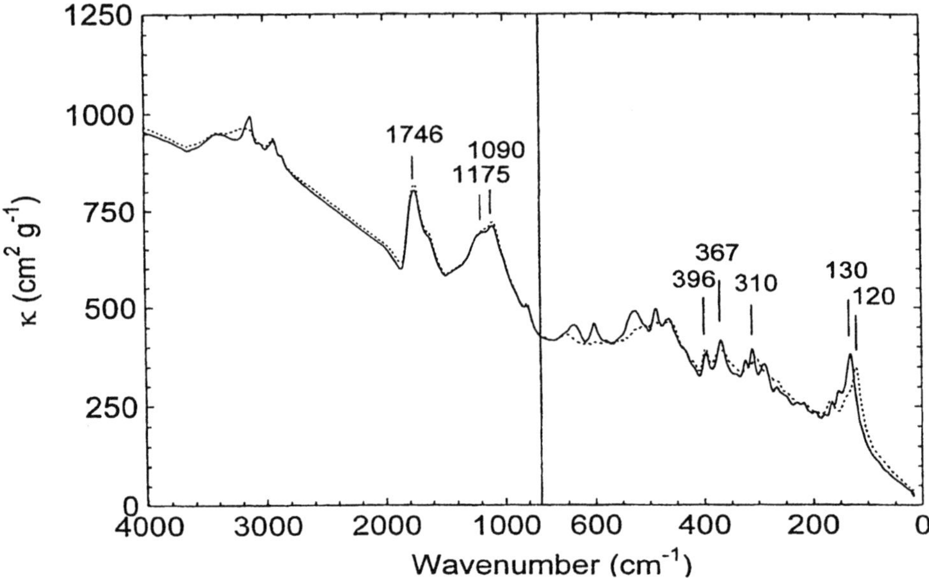


*Figure 20.* IR absorption spectrum of Murchison diamonds at different sample temperatures (solid line: 10 K, dotted line: 300 K). The identification of the bands and the reasons for the suspicion of artificial origin are discussed in [80].

been obtained [32, 59]. They confirmed earlier results obtained on Murray and Allende diamonds [9]. A simple interpretation of these values is that the diamond crystallites are covered by a chemically modified surface layer containing $sp^2$-bonded carbon. It cannot be completely excluded that this aromatically bonded carbon is a laboratory effect as well. Simulation experiments are needed to clarify this point.

At present, it is not positively known, in which cosmic environment the nano-diamonds are formed. The isotopic anomalies indicate a close relation to SN explosions. Generally, chemical vapour deposition in a carbon-rich environment has been suggested as formation mechanism. The problem is that CVD reqires a substrate for the diamond formation and that the relatively high nitrogen content (about 1 % by mass) suppresses the CVD diamond formation. Recently, new possibilities, e.g., homogeneous nucleation [17] and formation within carbon onions [2], have come within scope and deserve further attention.

## 6. Acknowledgments

I am grateful to the director of the International School of Cosmochemistry, J. Mayo Greenberg, for inviting me to present this lecture and for supporting my participation in the 4th course of this school. I further thank several

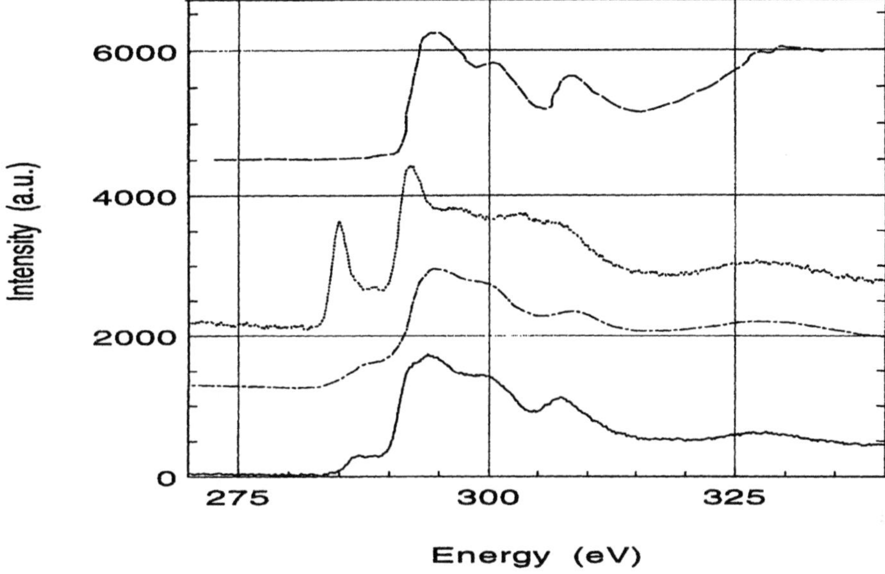


*Figure 21.* Comparison of EEL spectra (core loss) of Murchison diamonds (Jäger et al. [58], solid line), Murray diamonds (Bernatowicz et al. [9], dash-dotted line), terrestrial diamonds (dashed line) and graphite (dotted line). In addition to the peaks typical of fcc diamond, the meteoritic diamonds show a shoulder at the energy value where the graphite spectrum has a characteristic loss peak due to $\pi$-electrons.

colleagues who made available new data to me during the preparation of this lecture: Cornelia Jäger, Takashi Kozasa, Frank Molster, Harald Mutschke, and Angela Speck.

## References


1. Anders, E. and Zinner, E. (1993) Interstellar grains in primitive meteorites: diamond, silicon carbide, and graphite, *Meteoritics* **28**, 490-514.
2. Banhart, F. and Ajayan, P.M. (1996) Carbon onions as nanoscopic pressure cells for diamond formation, *Nature* **382**, 433–435.
3. Barker Jr., A. S. (1963) Infrared lattice vibrations and dielectric dispersion in corundum, *Phys. Rev.* **132**, 1474–1481.
4. Baron, Y., de Muizon, M., Papoular, R., and Pégourie, B. (1987) An analysis of the emission features of the IRAS low-resolution spectra of carbon stars, *Astron. Astrophys.* **186**, 271–279.
5. Begemann, B., Dorschner, J., Henning, T., Mutschke H., and Thamm, E. (1994) A laboratory approach to the interstellar sulfide dust problem, *Astrophys. J.* **423**, L71–L74.
6. Begemann, B., Henning, Th., Mutschke, H., and Dorschner, J. (1995) Magnesium-iron oxides - astrophysical origin and optical constants, *Planet. Space Sci.* **43**, 1257–1261.
7. Begemann, B., Dorschner, J., Henning, Th., and Mutschke, H. (1996) Optical properties of glassy $SiS_2$ and the 21 micron feature, *Astrophys. J.* **464**, L195–L198.
8. Begemann, B., Dorschner, J., Henning, Th., Mutschke, H., Gürtler, Kömpe, C. and

Nass, R. (1997) Aluminum oxide and the opacity of oxygen-rich circumstellar dust in the 12-17 micron range, *Astrophys. J.* **476**, 199–208.

9. Bernatowicz, T.J., Gibbons, P.C., and Lewis, R.S. (1990) Electron energy loss spectrometry of interstellar diamonds, *Astrophys. J.* **359**, 246–255.
10. Bernatowicz, T.J., Amari, S., Zinner, E.K., and Lewis, R.S. (1991) Interstellar grains within interstellar grains, *Astrophys. J.* **373**, L73–L76.
11. Bernatowicz, T.J., Cowsik, R., Gibbons, P.C., Lodders, K., Fegley, Jr., B., Amari, S., and Lewis, R.S. (1996) Constraints on stellar grain formation from presolar graphite in the Murchison meteorite, *Astrophys. J.* **472**, 760–782.
12. Bohren, C. F. and Huffman, D. R. (1983) *Absorption and Scattering of Light by Small Particles*, Wiley, New York.
13. Bond, H.E. (1991) Chemical composition of post-AGB stars, in G. Michaud and A. Tutukov (eds.), *Evolution of Stars: The Photospheric Abundance Connection*, Kluwer Academic Publisher, Dordrecht, pp. 341–349.
14. Borghesi, A., Bussoletti, E., Colangeli, L., and De Blasi, C. (1985) Laboratory Study of SiC submicron particles at IR wavelengths: a comparative analysis, *Astron. Astrophys.* **153**, 1–8.
15. Bradley, J. and Ireland, T. (1996) The search for interstellar components in interplanetary dust particles, in B.A.S Gustafson and M.S. Hanner (eds.), *Physics, Chemistry, and Dynamics of Interplanetary Dust*, A.S.P. Conf. Ser. **104**, San Francisco, pp. 275–282.
16. Bradley, J. (1997), this volume
17. Bürki, P.R. (1996) Low-pressure formation routes for interstellar microdiamonds: chemical vapour deposition vs. homogeneous nucleation, *Meteoritics* **31**, A24-A25.
18. Buss Jr., R. H., Cohen, M., Tielens, A. G. G. M., Werner, M. W., Bregman, J. D., Witteborn, F. C., Rank, D., and Sandford, S.A. (1990) Hydrocarbon emission features in the infrared spectrum of warm supergiants, *Astrophys. J.* **365**, L23–L26.
19. Bussoletti, E., Fusco, C., Longo, G. (eds.) (1988) *Experiments on Cosmic Dust Analogues*, ASS Library **149**, Kluwer Academic Publishers, Dordrecht.
20. Clayton, D. D. (1982) Cosmic chemical memory: a new astronomy, *Q. J. Roy. Astron. Soc.* **23**, 174-212.
21. Clément, D. (1996) Spectroskopische Untersuchungen an amorphem Siliciumcarbid und Vergleich mit astronomischen Infrarotquellen, *Diploma Thesis*, Gesamthochschule Kassel und Max-Planck-Gesellschaft, AG "Staub in Sternentstehungsgebieten" an der Universität Jena.
22. Cox, P. (1993), Far-infrared spectroscopy of solid state features, in S. Kwok (ed.), *Astronomical Infrared Spectroscopy: Future Observational Directions*, A.S.P. Conf. Ser. **41**, San Francisco, pp. 163–170.
23. Dominik, C., Sedlmayr, E., and Gail, H.-P. (1993) Dust formation in stellar winds. VI. Moment equations for the formation of heterogeneous and core-mantle grains, *Astron. Astrophys.* **277**, 578–594.
24. Donn, B. (1986) Experimental investigations relating to the properties and formation of cosmic grains, in J. A. Nuth III and R. E. Stencel (eds.) *Interrelationships Among Circumstellar, Interstellar, and Interplanetary Dust*, NASA Conf. Publ. 2403, pp. 109-134.
25. Dorschner, J. 1967. Theoretische Untersuchungen über den interstellaren Staub. I. Vorschlag eines Staubmodells aus meteoritischen Silikaten, *Astron. Nachr.* **290**, 171–181.
26. Dorschner, J. and Schulze, H. (1993) Refractory cometary sulphides and their possible stellar/interstellar origin. *Astronomische Gesellschaft Abstract Ser.* **9** 119.
27. Dorschner, J. and Henning, Th. (1995) Dust metamorphosis in the galaxy, *Astron. Astrophys. Rev.* **6**, 271–333.
28. Dorschner, J., Friedemann, C., and Gürtler, J. (1977) Silicon carbide and the infrared excess of carbon stars, *Astron. Nachr.* **298**, 279–283
29. Dorschner, J., Friedemann, C., Gürtler, J., and Henning, T. (1988) Optical prop-

erties of glassy bronzite and the interstellar silicate bands, *Astron. Astrophys.* **198**, 223-232.
30. Dorschner, J., Gürtler, J., and Henning, Th. (1989) Pyroxene glasses – candidates for the interstellar silicate dust component, *Astron. Nachr.* **310**,303–309.
31. Dorschner, J., Begemann, B., Henning, Th., Jäger, C., and Mutschke, H. (1995) Steps toward interstellar silicate mineralogy. II. Study of Mg-Fe-silicate glasses of variable composition *Astron. Astrophys.* **300**, 503–520.
32. Dorschner, J., Henning, Th., Jäger, C., Mutschke, H., and Ott, U. (1996) Spectrum and solid-state structure of Murchison diamonds, *Meteoritics* **31**, A37–A38.
33. Draine, B. T. and Lee, H. M. (1984) Optical properties of interstellar graphite and silicate grains, *Astrophys. J.* **285**, 89–108.
34. Fraundorf, P., Fraundorf, G., Bernatowicz, T., Lewis, R., and Tang, M. (1989) Stardust in the TEM, *Ultramicroscopy* **27** 401–412.
35. Friedemann, Chr. (1969) Evolution of silicon carbide particles in the atmospheres of carbon stars, *Physica* **41**, 139–143.
36. Friedemann, C., Gürtler, J., Schmidt, R., and Dorschner, J. (1981) The 11.5 micrometer emission from carbon stars: Comparison with IR spectra of submicrometer-sized silicon carbide grains, *Astrophys. Space Sci.* **79**, 405–417.
37. Gail, H.-P., and Sedlmayr, E. (1986) The primary condensation process for dust around late M-type stars, *Astron. Astrophys.* **166**, 225–236.
38. Gervais, F., Blin, A., Massiot, D., Coutures, J. P., Chopinet, M. H., Naudin, F. (1987) Infrared reflectivity spectroscopy of silicate glasses, *J. Non-Cryst. Solids* **89**, 384–401.
39. Gillett, F.C., Low, F.J., and Stein, W.A. (1968) Stellar spectra from 2.8 to 14 microns, *Astrophys. J.* **154**, 677–pp.
40. Glaccum, W. (1995) Infrared dust features of late-type stars and planetary nebulae, in M.R. Haas, J.A. Davidson, and E.F. Erickson (eds.), *Airborne Astronomy Symposium on the Galactic Ecosystem: From Gas to Stars to Dust*, ASP Conf. Ser. **73**, San Francisco, p. 395.
41. Goebel, J.H. (1986) New dust species in circumstellar shells, *Bull. Amer. Astron. Soc.* **18**, 1003–1004.
42. Goebel, J. H., and Moseley, S.H. (1985) MgS grain component in circumstellar shells, *Astrophys. J.* **290**, L35-L39.
43. Goebel, J. H. (1993) $SiS_2$ in circumstellar shells *Astron. Astrophys.* **278**, 226–230.
44. Goebel, J. H., Volk, K., Walker, H., Gerbault, F., Cheeseman, P., Self, M., Stutz, J., and Taylor, W. (1989) A Bayesian classification of the IRAS LRS Atlas, *Astron. Astrophys.* **222**, L5–L8.
45. Goebel, J. H., Bregman, J. D., and Witteborn, F. C. (1994) A 7 micron dust emission feature in oxygen-rich circumstellar shells *Astrophys. J.* **430**, 317–322.
46. Goebel, J.H., Cheeseman, P., and Gerbault, F. (1995) The 11 micron emissions of carbon stars, *Astrophys. J.* **449**, 246-257.
47. Greenberg, J. M. (1989). The core-mantle model of interstellar grains and the cosmic dust connection, in L. J. Allamandola and A. G. G. M. Tielens (eds.) *Interstellar Dust: Proc. IAU Symp. 135*, Kluwer Academic Publishers, Dordrecht, pp. 345–355
48. Greenberg, J.M. (ed.) (1996) *The Cosmic Dust Connection* NATO ASI Ser. **C 487**, Kluwer Academic Publishers, Dordrecht.
49. Greenberg, J. M., and Li, A. (1996) What are the true astronomical silicates? *Astron. Astrophys.* **309**, 258-266.
50. Grossman, L., and Larimer, J. W. (1974) Early chemical history of the solar system. *Rev. Geophys. Space Phys.* **12**, 71–101.
51. Grün, E., Gustafson, B., Mann, I., Baguhl, M., Morfill, G. E., Staubach, P., Taylor, A., and Zook, H. A. (1994) Interstellar dust in the heliosphere. *Astron. Astrophys.* **286**, 915–924.
52. Hackwell, J. A. (1972) Long wavelength spectrometry of M, S and C-stars *Astron. Astrophys.* **21**, 239–248.

53. Henning, Th., Begemann, B., Mutschke, H., and Dorschner, J. (1995) Optical properties of oxide dust grains, *Astron. Astrophys. Suppl. Ser.* **112**, 143–149.
54. Hill, H.G.M., d'Hendecourt, L.B., and Perron, C. (1996) Can any infrared absorption bands be linked to microdiamonds?, *Meteoritics* **31**, A61-A62.
55. Huffman, D. R. (1988) Methods and difficulties in laboratory studies of cosmic dust analogues, in E. Bussoletti, C. Fusco, and G. Longo (eds.) *Experiments on Cosmic Dust Analogues*, ASS Library Vol. 149, Kluwer Academic Publishers, Dordrecht, pp. 25–42.
56. Huffman, D. R. (1988) The applicability of bulk optical constants to small particles, in P. W. Barber and R. K. Chang (eds.) *Optical Effects Associated with Small Particles*, Advanced Series in Applied Physics, Vol. 1, World Scientific Publ. Co., Singapore, pp. 279–323.
57. Huffman, D. R. (1989) Pitfalls in calculating scattering by small grains, in L. J. Allamandola and A. G. G. M. Tielens (eds.) *Interstellar Dust: Proc. IAU Symp. 135* Kluwer Acad. Publ., Dordrecht, pp. 329–336.
58. Jäger, C., Mutschke, H., Begemann, B., Dorschner J., and Henning, Th. (1994) Steps toward interstellar silicate mineralogy. I. Laboratory results of a silicate glass of mean cosmic composition. *Astron. Astrophys.* **292**, 641–655
59. Jäger, C., Dorschner, J., and Mutschke, H. (1997) Properties of Murchison diamonds, in preparation
60. Jones, A. P. and Williams, D. A. (1987) Interplanetary material as a guide to the composition of interstellar grains, *Mon. Not. R. Astron. Soc.* **224**, 473–479.
61. Kissel, J., et al. (1986a) Composition of comet Halley dust particles from Vega observations, *Nature* **321**, 280–282.
62. Kissel, J., et al. (1986b) Composition of comet Halley dust particles from Giotto observations, *Nature* **321**, 336–338.
63. Kozasa, T. (1995), unpublished results
64. Kozasa, T., and Hasegawa, H. (1987) Grain formation through nucleation process II: Nucleation and grain growth accompanied by chemical reaction, *Prog. Theor. Phys.* **77**, 1402–1410.
65. Kozasa, T., Hasegawa, H., and Nomoto, K. (1991) Formation of dust grains in the ejecta of SN 1987A. II. *Astron. Astrophys.* **249**, 474–482.
66. Kozasa, T., Dorschner, J., Henning, Th., Stognienko, R. (1996) Formation of SiC grains and the 11.3 $\mu$m feature in circumstellar envelopes of carbon stars, *Astron. Astrophys.* **307**, 551-560.
67. Krätschmer, W. (1988) Laboratory production of amorphous silicates, in E. Bussoletti, C. Fusco, and G. Longo (eds.) *Experiments on Cosmic Dust Analogues*, ASS Library Vol. 149, Kluwer Academic Publishers, Dordrecht, pp. 95–102.
68. Kwok, S., Volk, K. M., Hrivnak, B. J. (1989) A 21 micron emission feature in four proto-planetary nebulae, *Astrophys. J.* **345**, L51–L54.
69. Lattimer, J. M., Schramm, D. N., and Grossman, L. (1978) Condensation in supernova ejecta and isotopic anomalies in meteorites. *Astrophys. J.* **219**, 230–249.
70. Lewis, R. S., Tang, M., Wacker, J. F, Anders, E., and Steel, E. (1987) Interstellar diamonds in meteorites, *Nature* **326**, 160–162.
71. Lewis, R. S., Anders, E., and Draine, B. T. (1989) Properties, detectability and origin of interstellar diamonds in meteorites *Nature* **339**, 117–121.
72. Little-Marenin, I.R. and Price, S. D. (1986) The shapes of circumstellar "silicate" features. In: D.J. Hollenbach and H.A. Thronson (eds.),*Summer School on Interstellar Processes*, NASA Tech. Memo. 88342, pp. 137-138
73. Little-Marenin, I. R., and Little, S. J. (1988) Emission features in IRAS low-resolution spectra of MS, S, and SC stars, *Astrophys. J.* **333**, 305–315.
74. Little-Marenin, I. R., and Little, S. J. (1990) Emission features in IRAS LRS spectra of M Mira variables, *Astron. J.* **99**, 1173–1186.
75. MacPherson, G. J., Wark, D. A., and Armstrong, J. T. (1988) Primitive Material surviving im chondrites: Refractory inclusions, in J. F. Kerridge and M. S. Matthews

(eds.), *Meteorites and the Early Solar System*, Univ. of Arizona Press, Tucson, pp. 746–807.

76. Matthes, S. (1990) *Mineralogie*, 3rd edition, Springer-Verlag, Berlin.
77. McDonnell, J.A.M. (1988) Solar system dust as a guide to interstellar matter, in M.E. Bailey and D.A. Williams (eds.), *Dust in the Universe*, Cambridge University Press, Cambridge, pp. 169–181.
78. Merrill, K.M., and Stein, W.A. (1976) 2–14 $\mu$m stellar spectrophotometry. I. Stars of the conventional spectral sequence, *Publ. Astron. Soc. Pacific* **88**, 285–293; II. Stars from the 2 $\mu$m Infrared Sky Survey, *Publ. Astron. Soc. Pacific* **88**, 293–307.
79. Molster, F.W. 1997, personal communication.
80. Mutschke, H., Begemann, B., Dorschner, J., and Henning, Th. (1994) Infrared data of sulphides of interstellar dust importance, *Infrared Phys. Technol.* **35**, 361–374.
81. Mutschke, H., Dorschner, J., Henning, Th., Jäger, C., and Ott, U. (1995) Facts and artifacts in interstellar diamond spectra, *Astrophys. J.* **454**, L157–L160.
82. Mutschke, H., Begemann, B., Dorschner, J., and Henning, Th. (1997) Steps toward interstellar silicate mineralogy. III. The role of aluminium in stardust silicates. Submitted to *Astron. Astrophys.*.
83. Ney E.P. (1977) Star dust *Science* **195**, 541–546.
84. Nistor, L. C., Van Landuyt, J., Ralchenko, V. G., Kononenko, T. V., Obraztsova, E. D., and Strelnitsky, V. E. (1994) Direct observations of laser-induced crystallization of a-C:H films. *Appl. Phys.* A 58, 137–144.
85. Nittler, L., Alexander, C. M. O'D., Gao, X., Walker R. M., and Zinner E. (1995a) Oxygen-rich stardust in meteorites, in M. Busso, R. Gallino, and C. M. Raiteri (eds.), *Nuclei in the Cosmos: 3rd International Symposium on Nuclear Astrophysics*, AIP Conf. Proc. No. 327, New York, pp. 585–589.
86. Nittler, L. R., Hoppe, P., Alexander, C. M. O'D., Amari, S., Eberhardt, P., Gao, X., Lewis, R. S., Strebel, R., Walker, R. M., and Zinner, E. (1995b), Silicon nitride from supernovae, *Astrophys. J.* **453**, L25-L28.
87. Nuth III, J.A. (1996) Grain formation and metamorphism, in J.M. Greenberg (ed.), *The Cosmic Dust Connection*, NATO ASI Ser. C, Vol. 487, Kluwer Academic Publisher, pp. 205-221.
88. Nuth III, J. A., and Hecht, J. H. (1990) Signatures of aging silicate dust, *Astrophys. Space Sci.* **163**, 79–94.
89. Nuth, J.A. and Donn, B. (1983) Laboratory studies of the condensation and properties of amorphous silicate smokes, *J. Geophys. Res.* **88** Suppl., A847–A852.
90. Nuth, J. A., Moseley, S. H., Silverberg, R. F., Goebel, J. H., and Moore, W. H. (1985) Laboratory infrared spectra of predicted condensates in carbon-rich stars, *Astrophys. J.* **290**, L41–L43.
91. Olnon, F. M., Raimond, E. (eds.), Neugebauer, G., van Duinen, R.J., Habing, H.J. et al. (1986) IRAS catalogues and atlases. Atlas of low-resolution spectra, *Astron. Astrophys. Suppl. Ser.* **65**, 607–1065.
92. Omont, A. (1993) Circumstellar matter: Cool IRAS sources. in S. Kwok (ed.), *Astronomical Infrared Spectroscopy: Future Observational Directions*, A.S.P. Conf. Ser. **41**, San Francisco, pp. 87–95.
93. Omont, A., Cox, P., Moseley, S. H., Glaccum, W., Casey, S., Forveille, T., Szczerba, R., and Chan, K.-W. (1995a) Mid- and far-infrared emission bands in C-rich protoplanetary nebulae, in M. R. Haas, J. A. Davidson, and E. F. Erickson (eds.), *Airborne Astronomy Symposium: From Gas to Stars to Dust*, A.S.P. Conf. Ser. **73**, San Francisco, pp. 413–417.
94. Omont, A., Moseley, S. H., Cox, P., Glaccum, W., Casey, S., Forveille, T., Chan, K.-W., Szczerba, R., Loewenstein, R. F., Harvey, P. M., and Kwok, S. (1995b) The 30 micron emission band in carbon-rich pre-planetary nebulae, *Astrophys. J.* **454**, 819–825.
95. Onaka, T., de Jong, T., and Willems, F.J. (1989) A study of Mira variables based on IRAS LRS observations. I. Dust formation in the circumstellar shell, *Astron. As-*

*trophys.* **218**, 169–179.
96. Ott, U. (1993) Interstellar grains in meteorites, *Nature* **364**, 25–33.
97. Palik, E. D. (1985) *Handbook of Optical Constants of Solids*, Academic Press, Orlando, pp. 587–595.
98. Palme, H., and Beer, H. (1993) Abundances of elements in the solar system. Landolt-Börnstein (New Series), Group VI, Vol. 3, Subvolume A, pp. 196–221.
99. Rietmeijer, F. J. M. (1992) Nonequilibrium iron oxide formation in some low-mass post-asymptotic giant branch stars, *Astrophys. J. Lett.* **400**, L39–L41.
100. Rietmeijer, F.J.M., Nuth, J.A., and Mackinnon., I.D.R. (1986) Analytical electron microscopy of Mg-SiO smokes: a comparison with infrared and XRD studies, *Icarus* **66**, 211–222.
101. Russell, R.W., Chatelain, M.A., Hecht, J.H., and Stephens, J.R. (1989) $Si_3N_4$ emissivity and the unidentified infrared bands, in A.G.G.M. Tielens and L.J. Allamandola (eds.) *Interstellar Dust: Contributed Papers*, NASA CP-3036, pp. 157–162.
102. Saslaw, W. C. and Gaustad, J. E. (1969) Interstellar dust and diamonds, *Nature* **221**, 160–162.
103. Schulze, H., 1992: Untersuchungen zur stofflichen Charakterisierung der Staubkomponente des Kometen P/Halley auf der Grundlage von Staubimpakt-Massenspektren, Ph. D.-Thesis, Humboldt-Universität Berlin
104. Schulze, H., Kissel, J., and Jessberger, E.K. 1997 Chemical heterogeneity of the dust particles of comet P/Halley and hints to their mineralogy, in Y.Pendleton and A.G.G.M.Tielens *From Stardust to Planetesimals: Review Papers*, ASP Conf. Ser., San Francisco, in press.
105. Sedlmayr, E. (1989) Dust condensation in stellar outflows, in L. J. Allamandola and A. G. G. M. Tielens (eds.), *Interstellar Dust*, Proc. IAU Symp. **135**, Kluwer Academic Publishers, Dordrecht, pp. 467–477.
106. Sellgren, K. (1990) The infrared interstellar emission features, in E. Bussoletti and A. A. Vittone (eds.), *Dusty Objects in the Universe* Kluwer Academic Publisher, Dordrecht, pp. 35–47.
107. Shroder, R.E., Nemanich, R.J., and Glass, J. T. (1990) Analysis of the composite structures in diamond thin films by Raman spectroscopy. *Phys. Rev.* **B 41**, 3738-3745.
108. Simpson, J.P. (1991) IRAS low-resolution spectral observations of the 10 and 18 micron silicate emission features, *Astrophys. J.* **368**, 570-579.
109. Sloan, G.C., and Price, S.D. (1995) Silicate emission at 10 microns in variables on the asymptotic giant branch, *Astrophys. J.* **451**, 758–767.
110. Sloan, G.C., LeVan, P.D., and Little-Marenin, I.R. (1996) Sources of the 13 micron feature associated with oxygen-rich circumstellar dust, *Astrophys. J.* **463**, 310–319.
111. Speck, A. K., Barlow, M. J., and Skinner, C.J. (1997) The nature of silicon carbide in star outflows, *Monthly Not. R. Astron. Soc.* in press.
112. Tang, M. and Anders, E. (1988) Isotopic anomalies of Ne, Xe, and C in meteorites. II. Interstellar diamond and SiC: Carriers of exotic noble gases, *Geochim. Cosmochim. Acta* **52**, 1235–1244.
113. Treffers, R. and Cohen, M. (1974) High-resolution spectra of cool stars in the 10- and 20-micron region *Astrophys. J.* **188**, 545–552.
114. Waters, L. B. F. M., Molster, F. J., de Jong, T., et al. (1996) Mineralogy of oxygen-rich dust shells, *Astron. Astrophys.* **315**, L361–L364.
115. Whittet, D. C. B., Duley, W. W., and Martin, P. C. (1990) On the abundance of silicon carbide dust in the interstellar medium, *Mon. Not. Roy. Astron. Soc.* **244**, 427-431.
116. Wildt, R. (1933) Kondensation in Sternatmosphären *Zeitschr. f. Astrophys.* **6**, 345–354.
117. Woolf, N. J. (1973) Circumstellar infrared emission, in J. M. Greenberg and H. C. van de Hulst (eds.), *Interstellar Dust and Related Topics*, Reidel, Dordrecht, pp. 485–504.

# INFRARED SPECTROSCOPY OF COSMIC DUST ANALOGUES AT LOW TEMPERATURES

H. MUTSCHKE AND TH. HENNING
*Astrophysical Institute and University Observatory (AIU), Friedrich Schiller University Jena, Schillergäßchen 2-3, D-07745 Jena, Germany*

**Abstract.** Cosmic dust analogue materials including an amorphous and a crystalline silicate, and ferrous oxide have been studied by reflectance spectroscopy of bulk surfaces as well as by transmission spectroscopy of particulates at temperatures down to 10 K. From the reflectance measurements optical data have been derived. The spectra of the crystalline materials (bronzite and FeO) show a sharpening and strengthening of the vibrational bands with decreasing temperature. The metal ion vibrations observed in the silicate glass of approximately cosmic composition in the far infrared have been found to be strongly temperature-dependent. The FIR continuum absorption caused by free charge carriers in the semiconducting FeO and by phonon difference processes in the crystalline silicate, respectively, showed the expected decrease with decreasing temperature.

## 1. Introduction

Infrared extinction data of cosmic dust analogue materials measured in the laboratory are widely used in modelling circumstellar envelopes and proto-planetary accretion disks. They are critical parameters in the determination of total dust masses in accretion disks, molecular cloud cores and galaxies from their thermal dust radiation (Henning *et al.*, 1995a). In most cases, however, the possible temperature dependence of these infrared optical data is not taken into account simply because temperature-dependent laboratory data for relevant materials are lacking. Laudable exceptions are measurements by Day (1976) and Agladze et al. (1994; 1996) on a number of crystalline and amorphous silicates, respectively, but in limited spectral ranges.



*J. M. Greenberg and A. Li (eds.), Formulation and Evolution of Solids in Space*, 265–272.

This paper presents low-temperature infrared spectra of some of our previously studied dust analogue materials: an amorphous silicate of pyroxene stoichiometry with a cosmic composition of the metals (Jäger *et al.*, 1994), a crystalline silicate of pyroxene type (bronzite) (Dorschner *et al.*, 1988) and ferrous oxide (FeO) (Henning *et al.*, 1995b). Points of interest are first the lattice vibrational bands of these materials, and second the far-infrared continuum absorption. For these puposes, specular reflectance spectra of polished bulk surfaces as well as extinction spectra of powders dispersed in polyethylene pellets have been measured. From the former, we derived the complex refractive index $m=n+ik$ of the materials and calculated the absorption behaviour for grains small compared to the wavelength (Rayleigh limit).

Before discussing the results, we give a short introduction to the temperature effects which are expected in the spectra of cosmic dust analogues. A more comprehensive description of the experiments and results of further measurements are contained in (Henning and Mutschke, 1997).

## 2. Temperature effects on infrared spectra of solids

### 2.1. LATTICE VIBRATIONS

At room temperature, thermal energies are comparable to radiation energies in the infrared spectral range. Therefore, lattice vibrations that can be excited by infrared photons usually are not in their ground state so that photon-induced transitions will occur between higher energy levels. Anharmonicity of the potential then leads on one hand to a somewhat lower transition energy and on the other hand to a stronger damping because of more efficient coupling to other modes. As a consequence, it has to be expected that at higher sample temperature vibrational absorption bands will be broadened and shifted to smaller energies (larger wavelengths). This has been studied for some crystalline solids like MgO already in the sixties (Jasperse *et al.*, 1966). For amorphous solids, much weaker effects are expected because the shapes of their broad bands are determined rather by a superposition of different vibrational modes than by the properties of one single mode.

Anharmonicity of the lattice potential is also the reason for the occurrence of multi-phonon processes. Difference processes (e.g. destruction of one phonon, creation of one of higher energy) determine the far-infrared (FIR) continuum absorption of crystalline solids. These processes of course require the presence of phonons. Their probability, thus, strongly decreases if a material is cooled down, which leads to the excellent transparency of e.g. halide crystals at very low temperatures (Stolen & Dransfeld, 1965; Hadni, 1970). This effect cannot be observed with an amorphous solid because disorder-induced single-phonon processes completely dominate over the multiphonon absorption in this case. The FIR continuum absorption, therefore, does not show a significant temperature dependence for amorphous materials. At millimetre wavelengths, low-energy transitions have been observed in glasses (Mon *et al.*, 1975; Strom & Taylor, 1977; Bösch, 1978). These transitions are also of great interest for the emission of cold interstellar dust (Agladze *et al.*, 1994; Agladze *et al.*, 1996) but could not be studied with the experimental setup used for our measurements.

### 2.2. FREE-ELECTRON EXCITATION

A considerable number of the proposed cosmic-dust analogue materials show semiconducting behaviour, for instance: iron oxides, silicon carbide, and amorphous carbon. At room temperature, these solids contain a more or less large number of thermally excited electrons in their conduction bands which give rise to an absorption continuum increasing with the wavelength. If the number density of the free charge carriers is very high, also plasma reflection can be observed for a bulk material at frequencies lower than the plasma frequency. For small particles, surface plasmon resonances occur at wavelengths where the real part of the dielectric function is negative.

The number density of the free electrons follows a Boltzmann law with the gap energy as the parameter. Therefore, cooling of the material results in a drastic decrease of the number of free charge carriers and, therefore, in a strong shift of the absorption/reflection phenomena towards much lower frequencies.

## 3. Experimental

The measurements have been carried out by means of a Bruker 113v FTIR Spectrometer equipped with a continuous-flow liquid helium cryostat with KRS5 and polyethylene infrared optical windows. In the cryostat, the samples are cooled via helium contact gas (1–5 mbar) which ensures a uniform temperature of the samples. Both the temperatures of the liquid helium cooled walls of the sample chamber and of the sample have been measured by means of Si-diode detectors. The sample holder allowed switching between the sample and the reference without warming up. As the references, we used a gold-coated mirror for the reflection and pure polyethylene pellets for the transmission measurements, respectively.

For measuring the specular reflectance of polished sample surfaces, we fitted the cryostat to the standard reflectance accessory of the spectrometer (for quasi-normal incidence of radiation). Reflection at the cryostat windows leads to a “zero signal” which has been measured by turning the sample out of the beam. It was subtracted both from the measured sample and reference spectra. From the reflectance spectra, optical data (complex refractive index $m=n+ik$) have been derived by Kramers-Kronig analysis. Subsequently, absorption spectra of small particles in the Rayleigh limit have been calculated in order to allow a comparison of the results of reflectance and transmittance measurements.

The pyroxene glass of "cosmic composition" as well as the iron oxide sample were prepared by melting in an arc furnace and subsequent quenching on a water-cooled copper plate. For the FeO, the preparational details and the results of chemical and structural analytics are described in a previous paper (Henning *et al.*, 1995b). In the case of the pyroxene, a natural hypersthene mineral was used as the "raw material" for the production of a glass (Dorschner *et al.*, 1989) with the following cationic composition determined by energy-dispersive X-ray analysis (in at.%): Mg 28.1%, Fe 17.3%, Al 1.1%, Ca 2.2% and Si 51.3%.

The crystalline bronzite sample has a much lower iron content than the glassy pyroxene (about 5%) and is polycrystalline with crystallite sizes of about 1mm. Since the spot size of the spectrometer beam is about 5 mm in diameter, crystallites of different orientation contribute to the reflection signal. In a first approximation, we assume that the different orientations are equally averaged as it is the case in powder samples.

For the reflection measurements, the samples have been embedded into epoxy resin for polishing. For the transmission measurements, the material was ground in a ball mill. The powder fraction smaller than one micron in diameter was separated by sedimentation in acetone and embedded in polyethylene pellets with mass concentrations ranging between $10^{-3}$ and $10^{-2}$.

## 4. Results

### 4.1. PYROXENE GLASS

Fig.1a shows the reflectance spectrum of the silicate glass of "cosmic composition" in the wavenumber range 1500–20 $cm^{-1}$. In the silicate bands at 10 and 20 $\mu m$ (1000 and 500 $cm^{-1}$) obviously there is no temperature dependence above the experimental accuracy limits. In the far infrared, however, the relatively flat behaviour of the reflectance at room temperature changes during cooling down into a rise of the reflectance towards small wavenumbers. Such a FIR rise has been observed by Gervais et al. (1987) for many commercial glasses at room temperature and has been attributed to vibrations of network-modifier cations such as calcium or sodium relative to the silicate network. In the case of our pyroxene glass, the content of these species (calcium 2.2%) is very low. Magnesium and iron cause similar vibrations which occur, however, in the immediate vicinity of the silicate bending mode at about 20 $\mu$m and cannot be resolved (Gervais *et al.*, 1987). Therefore, it is not surprising that the rise in the reflectance is not yet visible at room temperature. The occurrence of this feature during cooling-down indicates a large degree of anharmonicity which is understandable since the environment of the metal ions in the glass is irregular.

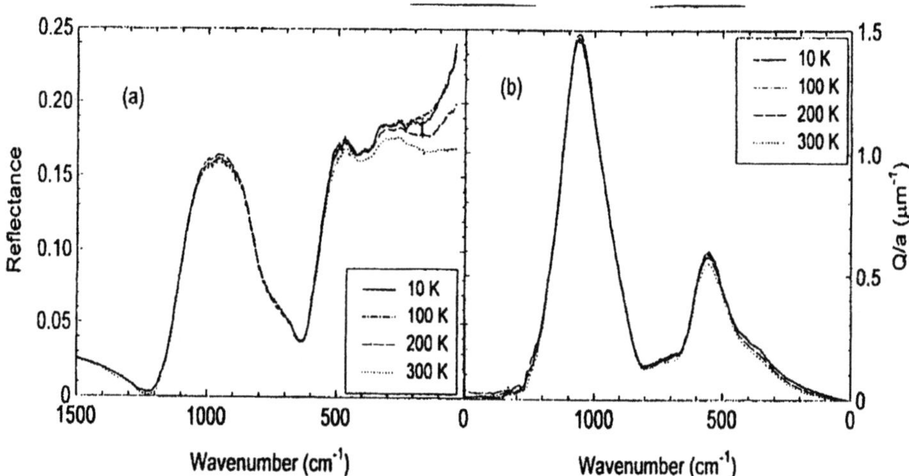


*Figure 1.* Bulk reflectance (a) and calculated small-particle extinction (Rayleigh spheres in vacuum) (b) for a pyroxene glass of approx. cosmic metal composition at different temperatures.

Fig.1b shows the effect of these metal ion vibrations on the (calculated) FIR continuum absorption of grains. Obviously, the "quasi-continuum" produced by these excitations dominates the total absorption at wavenumbers smaller than about 450 $cm^{-1}$ and results in a considerable increase of the absorption with falling temperature (factor of 3–4 at $\lambda$=200–500 $\mu$m for 10 K compared to room temperature).

## 4.2. BRONZITE

In Fig.2a the reflectance spectra of the crystalline bronzite are given. They show the expected strengthening during cooling and in some cases also a small shift towards larger wavenumbers (cf. 2.1). However, with about 10–20% the increase in band strength is smaller than expected. Day (1976) reported a much stronger effect, especially in the region of the 20 $\mu$m band, for another crystalline pyroxene (enstatite). The measured transmittance spectra (Fig.2b) also contradict the result by Day (1976). However, there may be a strong dependence on the sample. Further measurements on crystalline silicates are required to clarify this point.

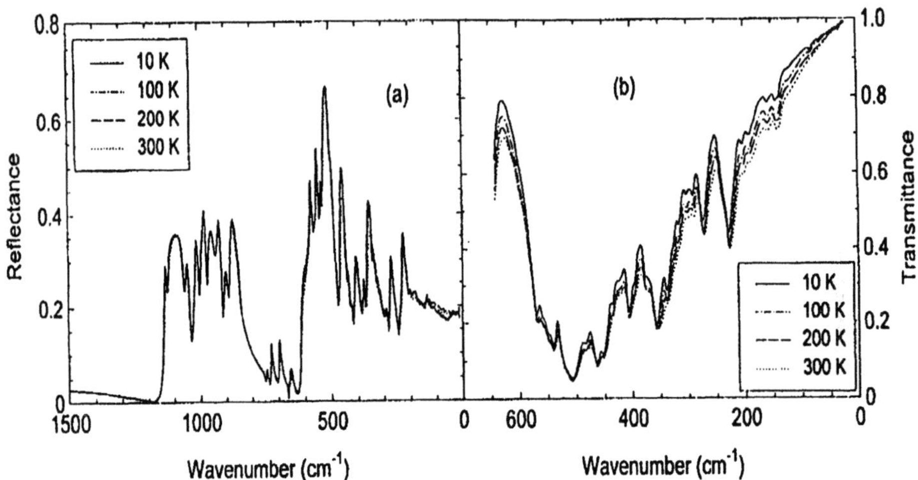


*Figure 2.* Bulk reflectance (a) and measured small-particle transmission (b) spectra for a crystalline silicate of pyroxene stoichiometry (bronzite) at different temperatures.

The transmittance spectra show also the expected decrease of the continuum absorption with falling temperature which is due to the freezing-out of phonon-difference processes. The reflectance spectra are insensitive to this effect because they are determined mainly by the real part of the refractive index.

## 4.3. IRON OXIDE

Fig.3a shows the reflectance of bulk FeO. In contrast to the silicates, the spectra are not only determined by the single-phonon transition which occurs at about 500 cm$^{-1}$ but also by free charge carriers. The latter cause a rise in the reflectance in the FIR which weakens, as expected, if the sample is cooled.

In Fig.3b the results of the small-particle calculations are presented. They show that the strengthening of the vibrational band is very pronounced (factor of 2–3 stronger at 10 K compared to room temperature). The reason for the strength of this effect compared to the silicates is that the absorption is dominated by a geometrical resonance (surface mode, see e.g. (Bohren & Huffman, 1983)). Therefore, it is influenced by the behaviour of both the real and imaginary part of the refractive index which reflect in that case also the temperature dependence of electronic excitations.

The free charge carrier absorption is better shown in a logarithmic plot (inset in Fig.3b). It becomes important for wavenumbers smaller than 200 cm$^{-1}$. Its temperature dependence may decrease the absorbance by a

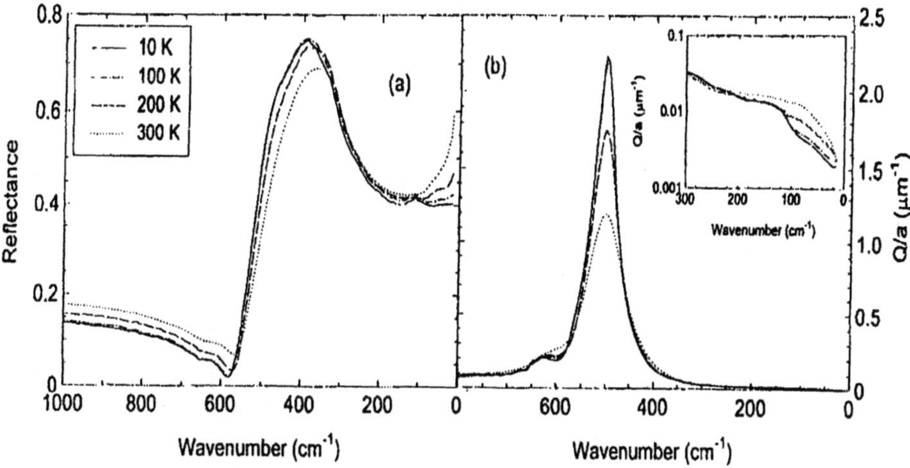


*Figure 3.* Bulk reflectance (a) and calculated small-particle extinction (Rayleigh spheres in vacuum) (b) for iron oxide (FeO) at different temperatures.

factor of 5 at about 100 $cm^{-1}$.

## 5. Conclusion

The relatively weak strengthening of the crystalline silicate's vibrational bands with falling temperature is in contradiction to the results of Day (1976). Measurements on a variety of silicate minerals are required to clarify this point. The temperature dependence of phonon-difference processes, free charge carrier excitation and metal ion vibrations influences heavily the far-infrared continuum absorption in crystalline materials, semiconductors, and glasses, respectively. These effects should be studied further as well. Transmission measurements on slabs of bulk material should improve the measurement sensitivity considerably and allow the extension of the measurements into the millimetre wavelength range.

## 6. Acknowledgment

We thank the Jena laboratory team: C. Jäger, B. Begemann, G. Born, W. Teuschel, and J. Dorschner.

## References

Agladze, N.I., Sievers, A.J., Jones, S.A., Burlitch, J.M. and Beckwith, S.V.W. (1994) Reassessment of Millimeter-wave Absorption Coefficients in Interstellar Silicate Grains, *Nature*, **372**, p. 243

Agladze, N.I., Sievers, A.J., Jones, S.A., Burlitch, J.M. and Beckwith, S.V.W. (1996) Laboratory Results on Millimeter-wave Absorption in Silicate Grain Material at Cryogenic Temperatures, *Astrophys. J.*, **462**, p. 1026

Bohren, C. F. and Huffman, D. R. (1983) *Absorption and Scattering of Light by Small Particles*, John Wiley, New York.

Bösch, M.A. (1978) Anomalous Absorption in Amorphous Solids at Low Temperature, *Phys. Rev. Lett.*, **40**(13), p. 879

Day, K.L. (1976) Temperature Dependence of Mid-Infrared Silicate Absorption, *Astrophys. J.*, **203**, p. L99

Dorschner, J., Friedemann, C., Gürtler, J. and Henning, Th. (1988) Optical Properties of Glassy Bronzite and the Interstellar Silicate Bands, *Astron. Astrophys.*, **198**, p. 223

Dorschner, J., Gürtler, J. and Henning, Th. (1989) Pyroxene Glasses - Candidates for the Interstellar Silicate Dust Component, *Astron. Nachr.*, **310**, p. 303

Gervais, F., Blin, A., Massiot, D., Coutures, J.P., Chopinet, M.H. and Naudin, F. (1987) Infrared Reflectivity Spectroscopy of Silicate Glasses, *Journal of Non-Crystalline Solids*, **89**, p. 384

Hadni, A. (1970) Lattice Absorption in the Very Far Infrared, in S. S. Mitra and S. Nudelman (eds.), *Far-Infrared Properties of Solids*, Plenum Press, New York, pp. 561–587

Henning, Th., Michel, B. and Stognienko, R. (1995) Dust Opacities in Dense Regions, *Planet. Space Sci.*, **43**, p. 1333

Henning, Th., Begemann, B., Dorschner, J. and Mutschke, H. (1995) Optical Properties of Oxide Dust Grains, *Astron. Astrophys. Suppl. Ser.*, **112**, p. 143

Henning, Th. and Mutschke, H. (1997) Low-Temperature Infrared Properties of Cosmic Dust Analogues, *Astron. Astrophys.*, submitted

Jäger, C., Mutschke, H., Begemann, B., Dorschner, J. and Henning, Th. (1994) Steps toward Interstellar Silicate Mineralogy — I. Laboratory Results of a Silicate Glass of Mean Cosmic Composition, *Astron. Astrophys.*, **292**, p. 641

Jasperse, J.R., Kahan, A., Plendl, J.N. and Mitra, S.S. (1966) *Phys. Rev.*, **146**, p. 526

Mon, K.K., Chabal, Y.J. and Sievers, A.J. (1975) Temperature Dependence of the Far-Infrared Absorption Spectrum in Amorphous Dielectrics, *Phys. Rev. Lett.*, **35**(20), p. 1352

Stolen, R. and Dransfeld, K. (1965) Far-Infrared Lattice Absorption in Alkali Halide Crystals, *Phys. Rev.*, **139**, p. A1295

Strom, U. and Taylor, P.C. (1977) Temperature and Frequency Dependences of the Far-Infrared and Microwave Optical Absorption in Amorphous Materials, *Phys. Rev. B*, **16**(12), p. 5512

# MATRIX ISOLATION OF AMORPHOUS CARBON GRAINS IN BORON-OXIDE GLASS

*Far UV Spectroscopy*

M.S. ROBINSON, L. COLANGELI AND V. MENNELLA
*Osservatorio Astronomico di Capodimonte*
*Via Moiariello 16, I-80131, Napoli, Italy*

AND

E. BUSSOLETTI AND A. ROTUNDI
*Istituto di Fisica Sperimentale, Istituto Universitario Navale*
*Via A. DeGaspaeri 5, I-80133, Napoli, Italy*

**Abstract.** Small carbon particles less than 20 nm in diameter are widely regarded as an essential element in models to fit the interstellar extinction curve, particularly in regard to the 217.5 nm "bump". The techniques used in the laboratory to make amorphous carbon cosmic dust analogues produce aggregates of small (<15 nm) particles. Due to the difficulties in isolating the individual grains, spectroscopic investigations to date have been on these agglomerates not on the smaller grains. In this paper we discuss efforts underway in our laboratory to isolate, in a boron-oxide glass matrix, individual grains of amorphous carbon produced by arc discharge in an argon atmosphere. Isolation in the boron-oxide glass matrix permits investigations from 160 nm to the near ir. Results presented here give preliminary indication that the amorphous carbon material being isolated in our glass matrix is less agglomerated.

## 1. Introduction

An important source of information about the nature of the dust component of the interstellar medium comes from the interpretation of the interstellar extinction curve. From the wavelength dependence of the curve and the general properties of light scattering from small, homogeneous spheres it can be concluded that there must be an appreciable population of grains with diameter $a \leq 20$ nm [1]. Models of the dust component of the interstellar medium, generally require small, carbonaceous grains, especially for the



*J. M. Greenberg and A. Li (eds.), Formulation and Evolution of Solids in Space,* 273–279.

217.5 nm "bump". ([2] for example). No specific carbonaceous material has been identified, however, as the carrier of this feature. For these reasons this laboratory has been involved in a program to investigate the characteristics of amorphous carbon analogues to these grains [3], [4], [5]. Laboratory analogues studied in this laboratory have been produced by arc discharge from amorphous carbon rods in an atmosphere of 10 mbar of either argon or hydrogen. This material consists of small amorphous carbon spheres with $a \sim 7$-15 nm which form into aggregates prior to being collected for spectroscopic study. These aggregates show a complex range of morphologies [6], but are invariably much larger than the $a \sim 20$ nm size range appropriate for astrophysical situations. Given the astrophysical importance of particles with $a \leq 20$ nm, we have begun to develop a technique for isolating individual (7-15 nm) grains for spectroscopic study in the visible to far uv.

## 2. Experimental Technique

The technique under development is to entrap the isolated grains in a matrix of boron-oxide glass. The technique of matrix isolation using boron-oxide glass has been used previously to study PAH molecules and their ions [7], [8]. Isolation in a glass is an attractive option for investigating the far uv characteristics of amorphous carbon grains for a number of reasons. Boron-oxide glass is transparent down to 160 nm, permitting the study of the spectra of these materials beyond the region of the 217.5 nm "bump". Samples are prepared and studied at room temperature so that cryogenic equipment is not needed. The samples can be handled and examined outside the spectrometer for close visual inspection. Finally, the samples are stable, and may be stored for several weeks allowing repeated examination of the same samples.

Because boron-oxide glass is extremely fragile, it cannot be handled unless it is sandwiched between rigid plates, which must also be transparent over the wavelength range to be studied. For this investigation disks of spectral grade fused silica (12 mm diameter) have been used. To make a boron-oxide glass suitable for spectroscopy in the uv and far uv range a quantity of boric acid ($H_3BO_3$) is first placed between two disks. This "sandwich" is then heated on a conventional laboratory hot plate at 300°$C$. At 240°$C$ the boric acid forms a melt according to the following reaction: $6H_3BO_3 \longrightarrow 3(B_2O_3 \bullet \frac{2}{3}H_2O) + 7H_2O$ After removing the "sandwich" from the hot plate the melt solidifies into a glass which is transparent from 160 nm into the near infrared.

Amorphous carbon particles produced at 10 mbar aggregate prior to being collected. By reducing the pressure to 1 mbar, aggregation does not occur until the particles are collected. In attempting to isolate individual grains we have been producing material in an atmosphere of 1 mbar of argon.

Two techniques have been used to attempt to isolate carbon grains in the matrix. The first and simplest, referred to hereafter as the mixing technique, is to produce a quantity of the amorphous carbon material, collected on a glass sheet and subsequently scraped off, and mix a portion of this with boric acid prior to forming the melt. Agglomerates of amorphous carbon are expected to form on the glass sheet during deposition. This technique will result in isolated grains in the matrix only if the material disaggregates in the melt. The results described below indicate that disaggregation did not occur.

In the second technique, hereafter referred to as the layering technique, a layer of amorphous carbon is first deposited on each of two quartz substrates for 2-3 minute. After the first 2-3 min deposition approximately 10-20 $\mu$g of boric acid is placed on each disk and melted to form a glass. The expectation is that little agglomeration will take place on the substrate over the short deposition time and that particles collected on the quartz will become dispersed in the glass during the melt. Amorphous carbon is then deposited again on the boron-oxide glass/quartz disks for 2-3 min. Again the disks are removed, 10-20 $\mu$g of boric acid is placed on the disks and melted. The disks are returned to the deposition chamber for a third deposition. Finally the two disks are combined together to form a single "sandwich".

When samples are prepared by either of these methods, a reference is also prepared consisting of boron-oxide glass between quartz substrates with no amorphous carbon in the matrix. Spectra are taken of the sample disk and of the reference. The extinction of the amorphous carbon is determined by subtracting the extinction of the reference from the extinction of the sample.

Samples have been prepared using both the mixing and the layering techniques with amorphous carbon produced in an argon atmosphere at 1 mbar. Samples have also been prepared using the mixing technique where the amorphous carbon has been produced in an argon atmosphere at 10 mbar.

## 3. Instrumentation and Analysis

Spectra were taken of the samples over the range 160-1000 nm using two instruments, a Perkin-Elmer lambda 19 for the uv-vis (190-1000 nm) and a McPherson 0.5 m evacuable scanning monochromator for the far uv (160-310 nm). The extinction determined for all samples by the two instruments agreed well over the range from 220-310 nm. Single spectra for the whole

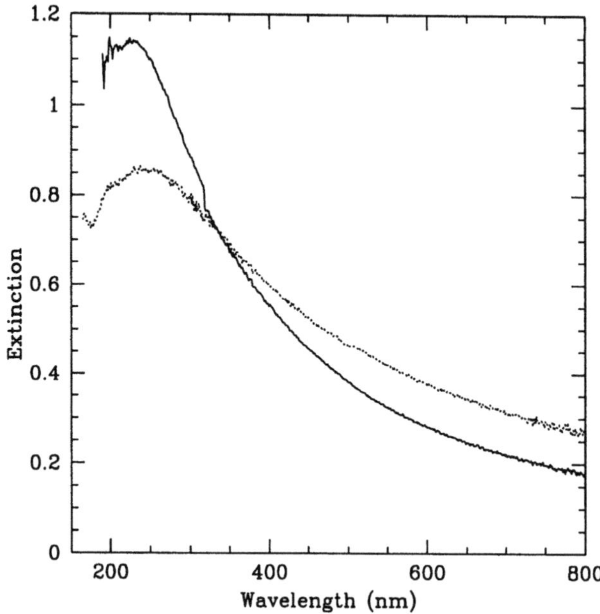


*Figure 1.* Extinction of amorphous carbon produced in 10 mbar of argon and collected directly on a quartz disk (solid line) and amorphous carbon produced in 1 mbar of argon and embedded in a boron-oxided glass matrix by means of the mixing technique (dotted line).

range were obtained by adding a constant factor of less than 0.5 abs units in all cases, to the extinction measured in the far uv to obtain agreement at 300 nm. The value of 300 nm was selected because it was a convenient value, not near the extreme edge of either instruments range of measurement, and within the region of good agreement between the instruments. The spectra shown in the figure are the data from 160-300 nm obtained with the McPherson vacuum uv instrument and the data from 300-1000 nm from the Perkin-Elmer lambda 19.

## 4. Results and Discussion

A comparison of the extinction of amorphous carbon produced at 10 mbar and collected on a quartz disk vs. the extinction of amorphous carbon produced at 1 mbar and embedded in a boron-oxide glass matrix by means of the mixing technique is shown in figure 1. The peak positions in the two cases are nearly the same, which is taken as evidence that no significant degree of disagglomeration was achieved by the boron-oxide glass technique used here.

Figure 2 shows a graph of the extinction of amorphous carbon produced

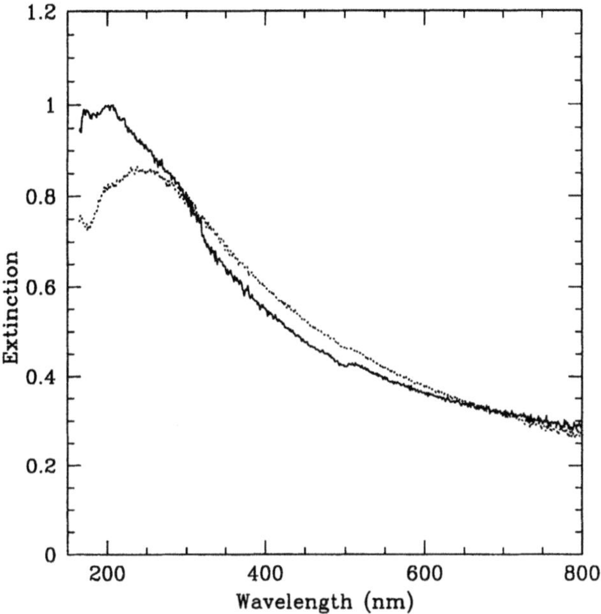


*Figure 2.* Extinction of amorphous carbon produced in 1 mbar of argon and embedded in a boron-oxide glass matrix by the layering technique (solid line) and by the mixing technique (dotted line).

in an atmosphere of 1 mbar of argon embedded in a boron oxide glass matrix by each of the techniques described above. The effect of this technique, developed to reduce the extent of agglomeration of the particles, appears to shift the peak to shorter wavelengths (from 240 nm to 210 nm) and to narrow the width of the peak. The peak shift toward shorter wavelength and a decrease in the peak width are in qualitative agreement with the results of [9] obtained using an entirely different technique.

When considering the results of matrix isolation experiments there may be a shift in the wavelength of the peak, due to interactions between the matrix and the material being studied. A solid neon matrix will generally perturb the spectrum least, and result in a spectrum closest to that of free particles, while argon and glass matrices will perturb the spectrum to progressively greater degrees. A reasonable estimate of the magnitude of the matrix shift due to boron-oxide glass as compared to solid argon or neon matrices can be made by comparing the matrix shift of PAH molecules such as pyrene and benzo[ghi]pyrene in boron-oxide glass and in rare-gas matrices. A comparison of the positions of absorption maxima of pyrene and benzo[ghi]perylene in neon and argon matrices reported by Salama *et al.* [10] with the positions of the same absorption features of pyrene and benzo[ghi]perylene in a boron-oxide glass matrix (see [11] and [8]) shows that the peak is shifted by approximately 4-5 nm redward in argon as compared to neon and 4-5 nm redward in boron-oxide glass as compared to argon. Therefore it is not unreasonable to expect a matrix shift due to the boron-oxide glass to be on the order of 10 nm as compared to free particles.

Our most recent experiments have reproduced the shift in peak wavelength from 240 nm to 210 nm when the particles are isolated in the matrix using the layering technique as compared to using the mixing technique. An important test of these results will be made by obtaining the extinction characteristics of amorphous carbon produced at 10 mbar and embedded in boron-oxide using the layering technique. Because the amorphous carbon produced at 10 mbar agglomerates prior to deposition, no change in the degree of agglomeration, and consequently the peak position, should be observed. If it is confirmed that the amorphous carbon particles are isolated as individual grains in the matrix this may prove to be a very useful technique for studying the spectral characteristics of isolated carbon grains.

*Acknowledgements* This work has been supported under MURST, ASI, and CNR contracts. Michael S. Robinson is supported by the North Atlantic Treaty Organization under a Grant awarded in 1996 (NSF number DGE-9633963).

## References

1. B.T. Draine (1995) Properties of Dust in Diffuse Clouds in A. Ferrara, C.F. McKee, C. Heiles, and P.R. Shapiro (eds.), *The Physics of the Interstellar Medium and Intergalactic Medium* Astronomical Society of the Pacific Conference Series Vol. 80, 133-147.
2. Li, A. and Greenberg, J.M. (1997) A Unified Model of Interstellar Dust, *A&A*, **323**, 566-584.
3. Mennella, V., Colangeli, L., Palumbo, P., Rotundi, A., Schutte, W., and Bussoletti, E. (1996) Activation of an Ultraviolet Resonance in Hydrogenated Amorphous Carbon Grains by Exposure to Ultraviolet Radiation, *ApJ*, **464**, L191-L194.
4. Mennella, V., Colangeli, L., Bussoletti, E., Merluzzi, P., Monaco, G., Palumbo, P., and Rotundi, A. (1995) Laboratory Experiments on Cosmic Dust Analogues: the Structure of Small Carbon Grains, *Planet. and Space Sci.*, **43**, 1217-1221.
5. Colangeli, L., Mennella, V., Bussoletti, E., Merluzzi, P., Palumbo, P., and Rotundi, A. (1995) Extinction Signatures of Amorphous Carbon Grains from the Vacuum UV to the Far-IR, *Planet. and Space Sci.*, **43**, 1263-1269.
6. Rotundi, A., Rietmeijer, F.J.M., Colangeli, L., Mennella, V.,Polumbo, P., and Bussoletti, E. (in press) Evidence of Inherent Forms of Carbon in Soot Material of Astrophysical Interest, submitted to *A&A*.
7. Lee, W. and Wdowiak, T. J. (1993) Laboratory Spectra of Polycyclic Aromatic Hydrocarbon Ions and the Interstellar Extinction Curve, *ApJ*, **410**, L127-L130.
8. Robinson, M. S., Beegle, L. W., and Wdowiak, T. S. (1997) Inference of a 7.75 eV Lower Limit in the Ultraviolet Pumping of Interstellar Polycyclic Aromatic Hydrocarbon Cations with Resulting Unidentified Infrared Emissions, *ApJ*, **474**, 474-478.

9. Schnaiter, M., Mutschke, H., Henning, Th., Lindackers, D., Strecker, M., and Roth, P. (1996) Ultraviolet Spectroscopy of Matrix-Isolated Amorphous Carbon Particles, *ApJ*, **464**, L187-L190.
10. Salama, F., Joblin, C., and Allamandola, L.J. (1995) Neutral and Ionized PAHs: Contribution to the Interstellar Extinction, *Planet. and Space Sci.*, **43**, 1165-1173.
11. Robinson, M.S. (1996) *Laboratory Spectra of Polycyclic Aromatic Hydrocarbon Cations in the Vacuum Ultraviolet: Astrophysical Implications*, Ph.D. Thesis, The University of Alabama at Birmingham, Birmingham.

# MATRIX-ISOLATED NANO-SIZED SOOT GRAINS AND THEIR RELATION TO SOLID CARBON IN SPACE

M. SCHNAITER, H. MUTSCHKE, J. DORSCHNER AND TH. HENNING
*Astrophysikalisches Institut und Universitäts-Sternwarte*
*Friedrich-Schiller-Universität Jena*
*Schillergäßchen 2-3, D-07745 Jena, Germany*

**Abstract.** The effect of particle shape or particle clustering onto the extinction behavior of nano-sized hydrogenated carbon grains was investigated experimentally. The particles were extracted by a molecular beam technique at different condensation and clustering states and were isolated in an argon matrix for UV spectroscopy. The state of clustering in the samples was controlled by TEM analysis. A clear correlation was found between the measured UV feature width and the degree of particle clustering in agreement with theoretical investigations for compact clusters. Therefore, the results show that the optical properties of carbonaceous grain material are strongly influenced by the particle shape and the clustering degree. For particles produced in hydrogen-containing atmospheres, the UV extinction peak was shifted blueward to a position close to the position of the 217.5 nm hump. This shift was found to be nearly independent of the amount of hydrogen in the condensation zone. As in the case of the interstellar 217.5 nm hump the UV features of the matrix isolated soot grains can fitted very well by a "Drude" profile plus a $\sim \lambda^{-1}$ background. Our experimental results show the most narrow UV profiles ever measured on carbonaceous dust analogues in the laboratory and are comparable to the width of the interstellar 217.5 nm hump.

## 1. Introduction

Beside silicates solid carbonaceous material is the most important component of cosmic dust. That is why models of the interstellar dust composition discussed at the moment are based mainly on these materials (for a recent review see [1]). The main goal of these models is to give more or less satis-



*J. M. Greenberg and A. Li (eds.), Formulation and Evolution of Solids in Space,* 281–289.

factory representations of the observed spectral characterisctics (interstellar extinction curve with its superimposed spectral features, FIR emission, polarization, etc.). They also have to take into account the cosmic elemental abundance constraints [2].

The dominating and most controversially discussed feature of the interstellar extinction curve is the 217.5 nm hump which is very stable in position but varies in strength and width with respect to different interstellar environments [3], [4]. In the most dust models small ($a \leq 10$ nm) graphite grains in the sense of spheroidal particles formed of stacked plane graphene sheets were included as hump particles [5]. This is more a theoretical construct than a real grain model because graphite is a highly anisotropic material and therefore can hardly be found in form of such small spherical particles in nature. Real carbonaceous dust grains produced in laboratories or found in meteorites are highly disordered. In these grains, the carbon atoms are bound in different hybridization states ($sp^1$, $sp^2$, and $sp^3$) and with varying short-range and medium-range order [6]. The often used term "amorphous" for this solid state structure is misleading because of the lacking crystalline counterpart (in contrast to a-Si for instance) [7].

The UV absorption features of carbonaceous solids located around 80 nm and between 200 and 260 nm can be attributed to interband transitions of the $\sigma$ and $\pi$ electrons, respectively.

A lot of experimental work was done in order to derive the relationship between the internal structure and the optical absorption characteristics of soot grains and carbon thin films (for review see [7]). Much less experimental work was done in studying the influence of the particle shape onto the optical behavior in the case of small carbonaceous grains. Indeed, this effect could play a fundamental role in the interpretation of the observed narrow band width of the 217.5 nm hump with its environmental variations.

It is well known from the Mie theory that in the case of spheres having sizes small compared to the wavelength of light ($2\pi a/\lambda \ll 1$) surface resonances (*Fröhlich* modes) are expected in spectral regions of strong absorption features of the grain material [8]. As a consequence of this, these material absorption bands become stronger, narrower, and get a blue-shift in the case of small isolated spheres. For arbitrarily shaped particles or for clusters of particles this effect is reduced depending on the shape distribution and the state of clustering [9]. It was a long-standing open question if the dielectric function of soot fulfils the *Fröhlich* conditions for the occurrence of surface resonances in the spectral region of the $\pi - \pi^*$ transition. The difficulty in answering this question is due to the fact that soot condenses always as heavily coagulated chain-like clusters of primary particles [10]. If surface modes occur, the dielectric function cannot be derived reliably from the absorption spectrum measured on such samples because

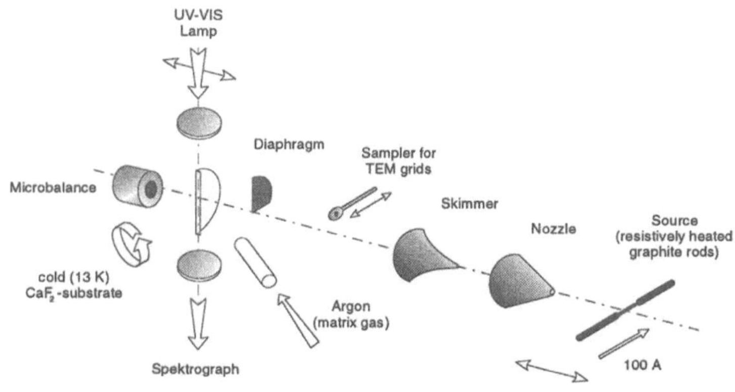


*Figure 1.* Experimental setup

of their complex cluster morphology. Therefore, the only way to get the answer is to measure the absorption characteristics directly on isolated primary soot grains. The problem here is the small primary particle size ($a \leq$ 10 nm) and the strong coalescence in the condensing clusters, so that a deagglomeration (e.g. by mechanical treatment) is hopeless. That is why we developed a new experimental technique for the extraction of primary soot grains out of different soot sources (flames, bulk evaporation) and the subsequent isolation of these particles in noble gas ice matrices for *in-situ* spectroscopy. Indeed, first experiments on small carbon grains ($a \approx 3$ nm) produced in low-pressure acetylene/oxygen flames have shown that the $\pi - \pi^*$ absorption profile varies strongly with particle shape and clustering degree [11]. A more thorough investigation was performed for particles produced by evaporation of carbon rods in quenching gas atmospheres at varying hydrogen content.

## 2. Experimental setup

The device we used combines two experimental techniques well established in modern molecular physics and cluster physics, namely the matrix-isolation spectroscopy and the molecular beam technique. In Fig. 1 the experimental setup is sketched. In the source chamber carbon rods are resistively heated up to the sublimation temperature. The generated carbon vapour is quenched in a cooling gas atmosphere (argon at 10 mbar) to form small solid grains. The quenching gas is expanded through a nozzle (orifice 0.8 mm) into a second chamber pumped separately to pressures of about $10^{-3}$ mbar. Due to the drop in pressure during this expansion the particles are decoupled from the gas and a further particle growth is stopped. A sharp-edged skimmer peels a particle beam out of the free jet into a third chamber pumped separately to pressures of about $10^{-5}$ mbar. In this chamber a sampler is located for collecting particles from the beam for transmission electron microscopy (TEM) analysis. In a fourth chamber at a pressure of about $10^{-7}$ mbar a cold $CaF_2$ substrate (previously cooled down to 10 K) is mounted rotatably. The substrate covers only half of the beam cross section. The other half is used for the mass rate monitoring by a microbalance. The beam particles reaching the cold surface of the substrate are isolated there in a growing argon ice matrix. For each sample prepared, an area of pure ice for reference is generated on the substrate by shutting a part of the beam with a diaphragm. After the isolation of a sufficient amount of particles in the ice, UV-VIS spectroscopy is performed in transmission after rotating the cold substrate into the spectrometer beam.

## 3. Results and discussion

Two sample series were prepared and analyzed, one at constant cooling gas conditions (pure argon at 10 mbar) but at varying distance between the source and the extraction nozzle. A second one was performed at a constant short extraction distance but at varying content of hydrogen in the quenching gas. In Fig. 2, representative examples of the TEM results are shown. The beam consists completely of isolated small primary particles with diameters of about 5-15 nm when extracting out of a condensation zone 15 mm away from the evaporation source (Fig. 2a). In the case of 25 mm extraction distance the beam is dominated by small clusters formed of a few to about 50 primary particles (Fig. 2b). Apparently, a clear growth in cluster size with increasing distance or with decreasing gas temperature is typical for the used carbon source indicating stable thermal conditions in the condensation process and in the particle growth.

The corresponding extinction characteristics for the clusters of Fig. 2a,b isolated in solid argon are plotted in Fig. 3. The spectra show a systematic broadening and a slight red-shift of the $\pi - \pi^*$ absorption feature which is obviously correlated with the increase of the clustering in the samples. This result is in agreement with the theoretical expectations for small compact clusters [9].

Compared to vacuum, the ice matrix in which the particles are embedded induces a red-shift and a slight broadening of each of these features [8]. In the case of solid argon, the maximum values of this matrix effect are 11 nm and 6 nm for the peak position and the feature width (FWHM), respectively [11].

Keeping this in mind, the UV feature of the isolated sample extracted at 15 mm distance get a strong red-shift and broadening after the evaporation of the ice matrix (Fig. 3). It seems that this spectrum represents some sort

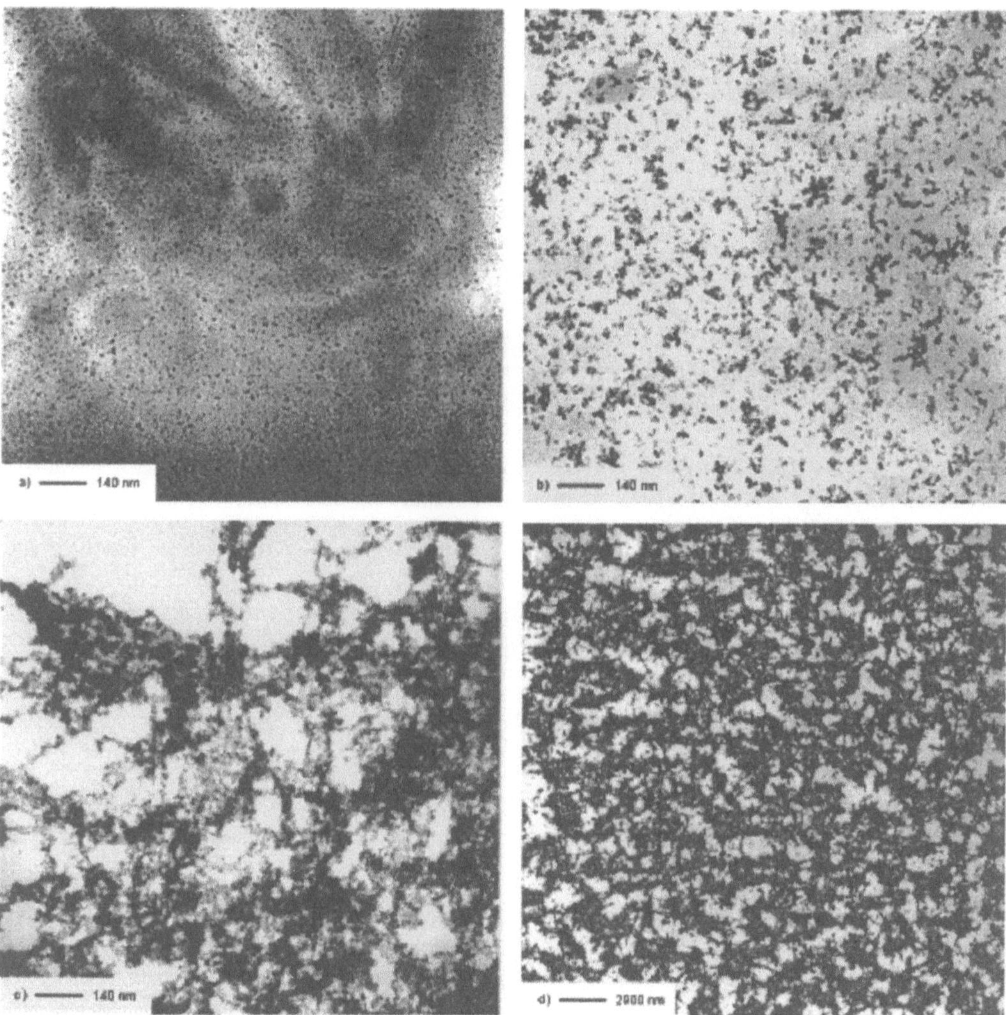


*Figure 2.* TEM images of soot grains collected from the beam at an extraction distance relative to the source of 15 mm (a) and 25 mm (b). TEM images of the residue left on the substrate after the evaporation of the ice matrix, at the same magnification as in the upper part (c), at a lower magnification to show the long-range ordered structure in these residues (d).

of upper limit for the peak profile variations due to the cluster growth. It is comparable to the spectra of the heavily clustered particle samples investigated so far. To confirm this, the result reported by Colangeli et al. [10] for soot produced in an arc discharge in argon atmosphere at 10 mbar (ACAR) is also plotted in Fig. 3. The corresponding TEM results indeed

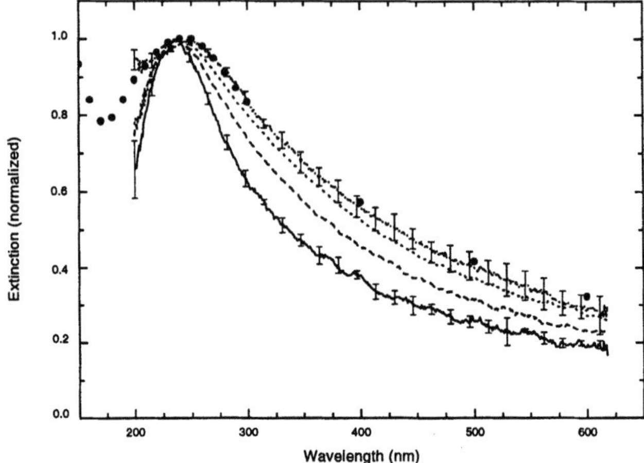


*Figure 3.* UV-VIS absorption spectra of soot grains extracted at different distances and isolated in solid argon at 13 K: 15 mm (solid line), 20 mm (dashed line) and 25 mm (dotted line). The spectrum of the 15 mm sample after the evaporation of the ice matrix is also plotted (dashed-dotted line). It fits accurately the ACAR spectrum reported by Colangeli et al. [9] (filled circles).

show that the residue consists of strongly coagulated primary particles (Fig. 2c) which cover in a chain-like structure the whole substrate (Fig. 2d) similar to the morphological results reported for the ACAR particle sample [10]. These results prove that the $\pi - \pi^*$ absorption feature is influenced by surface modes, and, therefore, morphological effects have to be taken into account when discussing the extinction characteristics of (coagulated) soots.

To investigate the effect of hydrogen incorporation onto the $\pi - \pi^*$ transition, a second sample series was prepared at varying hydrogen content in the quenching gas. The extraction distance in this run was 15 mm from the source to ensure that the freshly condensed nano-particles were extracted and isolated in a non-coagulated state only. This was proved by TEM analysis as in the samples before. The mixing ratio between hydrogen and argon in the quenching gas was successively enhanced from 1:100 to 5:10 for the individual samples while the pressure was held constant at 10 mbar. The result of this investigation is shown in Fig. 4.

For the sample prepared at a ratio of 1:100, there was nearly no change found in the absorption charateristics compared to the pure argon case. A large blue-shift of about 15 nm (from 237 to 222 nm) was observed when the hydrogen content was enhanced to a ratio of 5:100. This blue-shift is in agreement with the results generally found for hydrogenated amorphous carbon particles and thin films. It is usually attributed to an increased $sp^3$ site content in these grains leading to smaller and more separated $sp^2$ clusters. Due to this structural change the $\pi - \pi^*$ transition energy and the optical gap energy are enhanced and a blue-shift of the UV absorption

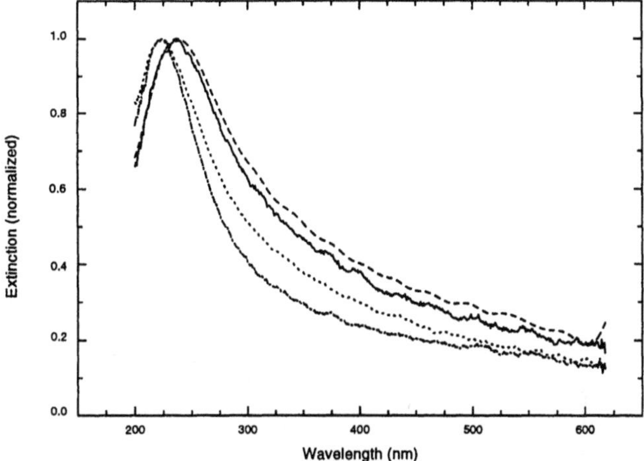


*Figure 4.* UV-VIS absorption spectra of isolated soot grains extracted at a 15 mm distance to the source in a quenching gas atmosphere at 10 mbar with varying hydrogen content: pure Ar (solid line), $H_2$:Ar=1:100 (dashed line), $H_2$:Ar=5:100 (dotted line), and $H_2$:Ar=5:10 (dashed-dotted line).

feature is expected [6].

Another important result taken from Fig. 4 is that once the peak had shifted it remained stable at that wavelength position even for higher hydrogen contents while the peak width was decreasing slightly. A possible explanation for this result could be a limited hydrogen incorporation at the hot condensation zone from which the primary particles are extracted. We suggest that due to the high temperature in that zone a thermal equilibrium state between hydrogen incorporation and hydrogen loss exists in the condensation process of these grains. Furthermore, the decrease in feature width shown in Fig. 4 suggests that there is an enhanced structural order present in the hydrogenated grains (in terms of a narrow $sp^2$ cluster size distribution). However, these suggestions will be the subject of further investigations by high-resolution TEM and electron energy loss spectroscopy (EELS).

The mean feature width calculated from the spectra of the hydrogenated soot samples is 130 (± 19) nm FWHM which does not agree very well with the value of 48 (± 12) nm for the the "Drude profile" fit of the 217.5 nm hump of the average interstellar extinction curve published by Fitzpatrick & Massa [3]. In this comparison, one has to keep in mind that a $\sim \lambda^{-1}$ continuum background has been substracted from the interstellar extinction curve to get this "Drude" representation, and it is not yet known to certainty what part of this background is due to the hump particles themselves. Indeed, the extinction features of isolated particles can be fitted very well by a "Drude" profile plus a $\sim \lambda^{-1}$ background whereas attempts failed in fitting the UV features of clustered samples in the same way. For example, a value of 73 nm FWHM was found for the "Drude" profile of the isolated particle sample prepared at a ratio of 5:10. However, we emphasize that the UV-VIS extinction results presented here for the matrix-isolated nano-sized soot grains show the most narrow $\pi - \pi^*$ features ever measured on carbonaceous dust analogues in the laboratory.

Concerning the peak positions of the matrix-isolated samples a red shift has to be taken into account due to the refractive index of the argon matrix ($\sim$ 1.29). This red shift is of about 10 nm for graphite and should be somewhat less for soot grains due to the weaker absorption strength. So the extinction curves for hydrogenated nano-sized soot grains in vacuum are expected to peak between 212 nm and 222 nm.

## 4. Summary and conclusions

In this work we have shown that particle shape and clustering strongly influence the absorption characteristics of small carbon soot particles. By matrix-isolation the effects of clustering and hydrogenation onto their UV-VIS spectral behaviour could be investigated separately. In any case a narrowing of the $\pi - \pi^*$ feature was found for the isolated grain samples compared to the coagulated soot samples investigated so far in the laboratory.

Small changes in the clustering state effect mainly the feature width and less the peak position in agreement with theoretical investigations for compact particle clusters.

Above a certain lower limit of hydrogen content in the quenching gas, hydrogen is incorporated into the atomic structure of the soot grains. The corresponding change in the internal bonding structure results in a blue-shift of the UV feature towards a peak position of 222 nm which is comparable (after the correction of the matrix shift) to the position of the 217.5 nm hump of the interstellar extinction curve. Further changes in the hydrogen content of the grains also result mainly in a change of the UV feature width whereas the peak position is quite stable.

*Acknowledgements* We thank Prof. J. M. Greenberg for the financial support which has enabled the participation in the school for two of us (M. Schnaiter and H. Mutschke). We also thank Prof. H.-W. Meyer and his staff at the Institut für Ultrastrukturforschung Jena for providing the TEM and for their help in operating it. This work has been supported by the Deutsche Forschungsgemeinschaft (DFG).

## References

1. Dorschner, J., Henning, Th. (1995) Dust metamorphosis in the galaxy, *Astron. Astrophys. (Rev.)*, **6**, 271

2. Henning, Th. (1997) Interstellar Dust Grains - An Overview, in E. van Dishoeck (ed.), *Molecules in Astrophysics*, IAU Symp. 178, Kluwer Acad. Publ., Dordrecht, in press
3. Fitzpatrick, E.L., Massa, D. (1986) An analysis of the shapes of ultraviolet extinction curves. I. The 2175 Å bump, *Astrophys. J.*, **307**, 286
4. Cardelli, J., Clayton, G., Mathis, J. (1989) The relationship between infrared, optical, and ultraviolet extinction, *Astrophys.J.*, **345**, 245
5. Draine, B.T. (1989) On the interpretation of the $\lambda$ 2175 Å feature, in L.J. Allamandola, A.G.G.M. Tielens (eds.), *Interstellar Dust*, IAU Symp. 135, Kluwer Acad. Publ., Dordrecht, p. 313
6. Robertson, J., O'Reilly, E., (1987) Electronic and atomic structure of amorphous carbon, *Phys. Rev. B*, **35**, 2946
7. Papoular, R., Conard, J., Guillois, O., Nenner, I., Reynaud, C., Rouzaud, J.-N., (1996) A comparison of solid-state carbonaceous models of cosmic dust, *Astron. Astrophys.*, **315**, 222
8. Bohren, C.F., Huffman, D.R. (1983) *Absorption and Scattering of Light by Small Particles*, Wiley, New York
9. Rouleau, F., Henning, Th., Stognienko, R. (1996) Constraints on the properties of the 2175 Åinterstellar feature carrier, *Astron. Astrophys.*, **468**, 834
10. Colangeli, L., Mennella, V., Palumbo, P., Rotundi, A., Bussoletti, E. (1995) Mass extinction coefficients of various submicron amorphous carbon grains: tabulated values from 40 nm to 2 mm *Astron. Astrophys. (Suppl. Ser.)*, **113**, 561
11. Schnaiter, M., Mutschke, H., Henning, Th., Lindackers, D., Strecker, M., Roth, P. (1996) Ultraviolet extinction of matrix-isolated amorphous carbon particles, *Astrophys. J.*, **464**, L187

# TEMPERATURE DEPENDENCE OF THE FIR ABSORPTION COEFFICIENT FOR CARBON AND SILICATE GRAINS

J.R. BRUCATO, V. MENNELLA AND L. COLANGELI
*Osservatorio Astronomico di Capodimonte*
*via Moiariello 16 I-80131 Napoli, Italy*

AND

E. BUSSOLETTI
*Istituto di Fisica Sperimentale,*
*Istituto Universitario Navale*
*via A. De Gasperi 5, I-80133 Napoli, Italy*

**Abstract.** In this paper we present some preliminary results concerning the measure of the mass absorption coefficient of carbonaceous and silicates grains at various temperatures in the far-infrared spectral region. We have performed transmittance measurements in the 20-200 $\mu$m spectral range at temperature between 24 and 300 K. Data show significant spectral index variations with the temperature. In particular when the temperature decreases, $\beta$ increases from 0.8 to 1.0 for the amorphous carbon, from 1.4 to 1.6 for amorphous fayalite and from 1.8 to 2.3 for crystalline forsterite. These studies are relevant for astrophysical applications since in the far-infrared, information on the total mass and, in some cases, discrimination among dust models, can be derived from opacity measurements of interstellar grains.

## 1. Introduction

The knowledge of the optical propreties of different materials of astronomical interest is important when we want to determine, for example the mass of young objects with circumstellar disks which are candidates to form planets [1], or the mass of galaxies [2]. Since the propreties of the interstellar dust have mostly been deduced by theoretical and observational analysis [3], it is necessary to support the results by laboratory measurements. The



*J. M. Greenberg and A. Li (eds.), Formulation and Evolution of Solids in Space,* 291–296.

study in the laboratory of the optical property of cosmic dust analougues at room temperature is appropriate to investigate the properties of the dust in the circumstellar sources; but if we consider, for example, that the dust in molecoular clouds could be colder it is necessary to take into account that the optical property could change with the temperature. Moreover, recent observations of some Seyfert galaxies showed mid and far infrared emission being due to two different dust components: one warm and another cold. The temperatures were estimated to be in the range of 177-161 K for the warmer, and 27-36 K for the colder one [4].

Up to now few works have been devoted to the study of the absorption coefficient when the temperature varies [5], [6], [7]. So, in this paper our purpose is to analyse the temperature dependence of the mass absorption coefficient in the far infrared (20-200 $\mu$m) for carbon and silicate grains.

## 2. Experimental

In this work we have studied some representative materials of different classes like carbon-based materials and silicates. The amorphous carbon sample (BE) was produced by burning benzene in air. This method produces spherical grains with an average diameter of 30 nm and which aggregate in a chain-like fractal structure. We have chosen two examples of silicates, fayalite and forsterite, in the two states amorphous and crystalline respectively. The choice of crystalline forsterite is supported by recent observations with the ISO satellite that have shown the presence, in oxygen-rich shells, of a dust mixture of amorphous and crystalline silicates such as olivine and pyroxenes [8]. The forsterite grains ($Mg_2SiO_4$) were obtained by grinding a piece of natural rock in a mortar, until it became a fine powder with average size of 0.2 $\mu$m. The amorphous fayalite ($Fe_2SiO_4$) sample was obtained by laser ablation of a natural rock in $O_2$ atmosphere (10 mbar) so obtaining particles with average size of 13 nm.

Samples for extintion measurements were prepared by embedding the materials in polyethylene matrixes by means of the standard hot pellet technique. The sample masses were determined by means of a microbalance (sensitivity = 1 $\mu$g). The two components were mixed together obtaining an homogeneous mixture. With a hot pressing film maker (Specac Mod. 15620), a circular (13 mm) pellet of constant thickness (2 mm) was obtained. The dust mass embedded in the sample was finally determined by weighing the pellet and taking into account the dust to polyethylene mass ratio of starting mixture.

Transmission measurements were carried out with a FTIR Bruker 66v spectrophotometer operating at a pressure of $10^{-1}$ mbar to reduce the water vapor absorption in the air. This allows us to better control the experimen-

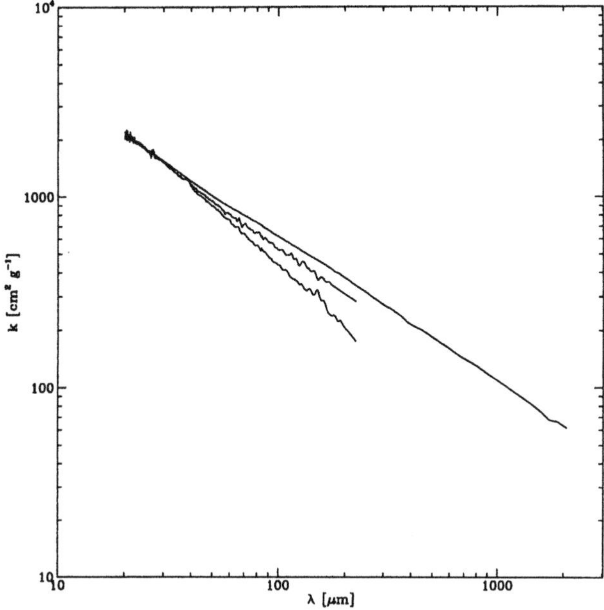


*Figure 1.* Mass absorption coefficient of amorphous carbon BE at different temperatures. The temperatures, at which the spectra shown were taken, are in order from the top to bottom: 295, 160, and 24 K.

tal conditions during the measurements. The spectral resolution was 2 $cm^{-1}$ over the whole spectral range considered, 20-200 $\mu$m. Several experimental configurations of the spectrophotometer were used. Two mylar beamsplitters with different thickness were used to cover the whole spectral range. A high sensitivity room temperature DTGS thermal element detector was used. The sample was mounted between two copper rings, with a central hole of 6 mm, on a alluminium cold finger inside a high-vacuum chamber ($10^{-7}$ mbar). The temperature of the sample was controlled by a water-cooled closed-cycle helium cryostat (10-300 K). The sample temperatures attainable were previously calibrated using a silicon diode temperature sensor. The minimum temperature reached was 24 K. A heater, mounted on the cold finger allows the sample temperature to be maintained or changed by a programmable controller from the low-temperature limit to room temperature.

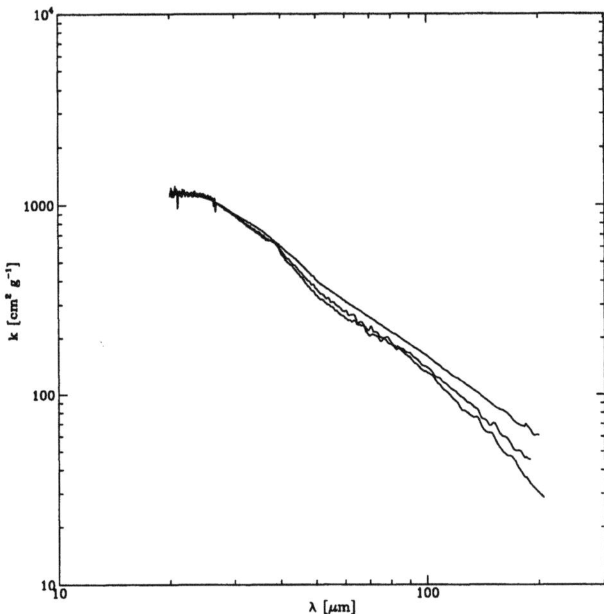


*Figure 2.* Mass absorption coefficient of amorphous fayalite at different temperatures. The temperatures, at which the spectra shown were taken, are in order from the top to bottom: 295, 160, and 24 K.

## 3. Results

The mass extinction coefficient was determined from the relation:

$$K(\lambda, T) = \frac{S}{M} ln \frac{t_p(\lambda, T)}{t(\lambda, T)}$$

$t(\lambda, T)$ represents the transmission of the sample with mass $M$ at different temperatures and $S$ is its cross-sectional surface. We used blank polyethylene pellets as reference; in this case $t_p(\lambda, T)$ represents the transmission of polyethylene at different temperatures. Since the extinction of the analysed materials decreases as the wavelength increases, we used samples with higher masses in longer wavelength spectral region in order to improve the signal to noise ratio.

In the FIR-mm spectral range the absorption coefficient is featureless, and it is assumed to have a power law wavelength dependence $\lambda^{-\beta}$ with spectral index $\beta$. We have estimated the spectral indices, by means of a least square fit of the experimental data. The variations of the mass absorption coefficient with the temperature for the materials studied in this work are shown in Figures 1-3. The behaviour of the mass absorption coefficient of

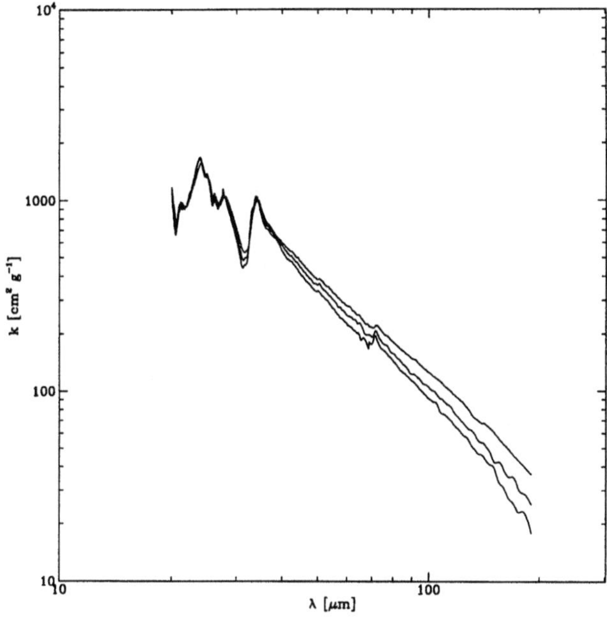


*Figure 3.* Mass absorption coefficient of crystalline forsteryte at different temperatures. The temperatures, at which the spectra shown were taken, are in order from the top to bottom: 295, 160, and 24 K.

the BE sample has been extented up to 2 mm, by using the data reported by Mennella *et al.* [9]. The mass absorption coefficient decreases when the temperature is lower, and at 200 $\mu$m it is reduced of a factor 2. The BE has a spectral index $\beta$=0.8 at room temperature, this is in agreement with the results obtained by other authors [10], [11], [12], in a different spectral region. At the temperature of 160 K $\beta$ is 0.9 and increases to 1.0 at 24 K. The spectral index for amorphous fayalite sample, in the spectral region 50-200 $\mu$m, takes on the following values: 1.4 at 300 K, 1.5 at 160 K and 1.6 at 24 K; at 200 $\mu$m the mass absorption coefficient decreases of a factor 1.8 (Figure 2). For crystalline forsteryte we have the same behaviour and the spectral index assumes, in the spectral region 50-200 $\mu$m, the following values: 1.8 at 300 K, 2.1 at 160 K and 2.3 at 24 K; at 200 $\mu$m the mass absorption coefficient decreases of a factor 2 (Figure 3). It is noted that the far-infrared absorption coefficient, for crystalline materials, has more or less an inversely quadratic dependence on wavelength in agreement with theoretical work [13]. However, for small amorphous grains, with a radius less than 100 Å, and for large wavelength, greater than 100 $\mu$m, a spectral index of 1 is expected; so the origin of this discrepancy has to be investigated. Some hints about this point can also be obtained from the extention of measurements to the millimeter spectral region. This will be the subject of a future work.

## 4. Conclusions

The results of the temperature dependence of the mass absorption coefficient of carbonaceous and silicate grains between 300 and 24 K, and 20 to 200 $\mu$m have been presented and discussed. The mass absorption coefficient provides the only direct means to assess the total mass in many interstellar regions. Our measurements indicate that, at low temperature, the mass absorption coefficient is lower and the spectral slope is steeper than they are at room temperature; therefore estimates of the mass of cold interstellar dust may be increased accordingly.

*Acknowledgements.* This work has been supported by ASI, CNR and MURST research contracts.

## References

1. Beckwith S.V.W. and Sargent A.I. (1991) Particle emissivity in circumstellar disks, *ApJ* **381**,250-258.
2. Blain A.W. and Longhair M.S. (1993) Submillimetre cosmology, *MNRAS* **264**, 509-521.
3. Li A. and Greenberg J.M. (1997) A Unified model of interstellar dust, *A&A* **323**, 566-584.
4. Rodríguez Espinosa J.M., Pérez García A.M., Lemke D. and Meisenheimer K. (1996) Bimodal dust emission in three classical Seyfert galaxies: NGC 3227, NGC 4051 and NGC 4151, *A&A* **315**, L129-L132.
5. Zaikowski A., Knacke R.F. and Porco C.C. (1975) On the presence of phyllosilicate minerals in the interstellar grains, *Ap&SS* **35**, 97-115.
6. Day K.L. (1976) Further measurements of amorphous silicates, *ApJ* **210**, 614-617.
7. Agladze N.I., Sievers A.J., Jones S.A., Burlitch J.M. and Beckwith S.V.W. (1996) Laboratory results on millimeter-wave absorption in silicate grain materials at criogenic temperatures, *ApJ* **462**, 1026-1040.
8. Waters L.B.F.M. *et al.* (1996) Mineralogy of oxygen-rich dust shells, *A&A* **315**, L361-L364.
9. Mennella V., Colangeli L. and Bussoletti E. (1995) The absorption coefficient of cosmic carbon analogue grains in the wavelength range 20-2000 $\mu$m, *A&A* **295**, 165-170.
10. Blanco A., Fonti S. and Martino F. (1991) Far infrared extinction measurements on carbonaceous particles of astrophysical relevance, *Infrared Phys.* **31**, 167-171.
11. Bussoletti E., Colangeli L., Borghesi A. and Orofino V. (1987) Tabulated extinction efficiencies for various types of submicron amorphous carbon grains in the wavelength range 1000 Å- 300 $\mu$m, *A&ASS* **70**, 257-268.
12. Koike C., Hasegawa H. and Manabe A. (1980) Extinction coefficients of amorphous carbon grains from 2100 Åto 340 $\mu$m, *Ap&SS* **67**,495-502.
13. Seki J. and Yamamoto T. (1980) Amorphous interstellar grains: wavelength dependence of far-infrared emission efficiency, *Ap&SS* **72**, 79-86.

# MORPHOLOGY, COMPOSITION AND MIR-FIR SPECTROSCOPY OF SILICATES OF ASTROPHYSICAL INTEREST

G. FERRINI[a,b], A. ROTUNDI[b,c], L. COLANGELI[b] AND V. MENNELLA[b]
[a] *Università degli Studi di Napoli Federico II*
*Dipartimento Scienze Fisiche*
*Mostra d'Oltremare pad. 19, 80125 - Napoli*
[b] *Osservatorio Astronomico di* Capodimonte
*Via Moiariello 16, 80131 - Napoli*
[c] *Istituto Universitario Navale*
*Via A. De Gasperi 5, 80133 - Napoli*

**Abstract**. An important step in the characterisation of cosmic analogues is represented by the study and comparison of morphological, chemical and spectroscopical properties of "crystalline" and "amorphous" materials. In order to achieve this aim we have determined the mass extinction coefficient in the infrared region, between 2.5 and 200 μm, of Fayalite ($Fe_2SiO_4$) in the two forms. Amorphous materials were obtained by laser vaporisation of a crystalline target. We have studied the morphology of the grains, computed their dimensional distribution and average diameter and analysed the elemental composition, using a field emission Scanning Electron Microscope equipped with an Energy Dispersive X-ray detection system.

## 1. Introduction

Two broad emission/absorption bands near 10 and 20 μm are often present in the spectra of several infrared sources, like HII regions [1,2], oxygen-rich late-type stars and protostellar objects [3-8] and toward the galactic centre [9]. Similar bands are observed in the spectra of comets [10-12] which are believed to contain the most primitive material in the solar system. Recent observations performed by means of the Infrared Space Observatory (ISO), confirm this identification [13-16] and reveal the presence of many new features [17, 18]. The wide bands at 10 and 20 μm are widely believed to be due to the stretching mode of Si-O and the bending mode of Si-O-Si bonds respectively, in small "amorphous" silicate grains. Most of the other features, being all narrow compared to the amorphous silicate bands, suggest the presence of "crystalline" forms of silicates.



*J. M. Greenberg and A. Li (eds.), Formulation and Evolution of Solids in Space,* 297–303.

At present the nature of the silicate materials responsible for these features is still uncertain, so that laboratory measurements, performed on natural and synthetic silicates, either in crystalline or amorphous form, are of great importance for the interpretation of astronomical data.
Laboratory studies and simulations of cosmic dust have to take into account a number of constraints linked to the spectroscopic properties of grains, cosmic abundance of condensable elements and material depletion in space environments. According to recent cosmic abundance evaluation [19], silicates in space are expected to contain iron and magnesium, since these elements are present in sufficient amount to efficiently combine with silicon. Silicates as pyroxenes ($[Mg,Fe]SiO_3$) and olivine ($[Mg,Fe]_2SiO_4$) are considered good candidates to reproduce astronomical observations.
In this paper preliminary results about the analysis on crystalline and amorphous Fayalite ($Fe_2SiO_4$), the iron-rich member of the olivine group, are presented. We report: a) the mass extinction coefficient in the mid infrared (MIR between 2.5 and 20 μm) and in the far infrared (FIR between 20 and 200 μm); b) the dimensional distribution of the grains and their average dimension; c) the elemental composition.

## 2. Experimental

The fayalite sample used in this work is a mineral of terrestrial origin coming from the iron mine of Forsythe in Quebec, Canada. Since it is expected that the grains responsible for the 10 and 20 μm bands are submicron in size, to compare observations with laboratory results it is necessary to produce particles with similar average dimensions.
Fine powder of crystalline fayalite was obtained by grinding 200 mg of the bulk material for 1 hour in an agate mill. Amorphous fayalite was produced by vaporisation and condensation in $O_2$ atmosphere at 10 mbar pressure, of a bulk crystalline target using a Nd-YAG Q-switched LASER (mod. Continuum Surelite II) working at a wavelength of 1064 nm.
In order to perform laboratory measurements on the produced materials a suitable preparation of the samples is necessary for each kind of analysis.
The morphological characterisation of the samples (crystalline and amorphous) was made using a field emission Scanning Electron Microscope (SEM mod. Stereoscan FE360) with a maximum resolution of 2 nm. The crystalline sample was deposited on a silicon wafer chip fixed on an aluminium stub, by means of the evaporation of an ethanol and powder solution. The amorphous material was directly collected during the laser bombardment on a similar substrate. An Energy Dispersive X-ray (EDX) detection system linked to the SEM, has been used to find the elemental composition of the materials as well as to test the purity of the mineral. For the EDX analysis amorphous materials have been directly deposited on carbon stubs, while direct analysis of the mineral was performed by fixing bulk pieces of material on carbon stubs.

The MIR transmission spectra of the crystalline and amorphous grain samples, embedded in KBr pellets and deposited on a KBr window respectively, have been obtained using a spectrophotometer (mod. Perkin-Elmer 880) with a resolution of 2.4 $cm^{-1}$.

The FIR transmission spectra of the samples, embedded in polyethylene pellets, have been obtained using a FT-IR spectrophotometer (mod. Bruker IFS 66v) with a resolution of 2 $cm^{-1}$.

## 3. Results and Discussion

In Figure 1 the MIR and FIR mass extinction coefficient of the crystalline and amorphous fayalite are shown. They have been computed by the relation:

$$K(\lambda) = \frac{S}{M} ln\left(\frac{1}{T(\lambda)}\right)$$

where $S$ is the cross-surface of the sample, $M$ is the dust mass, $T(\lambda)$ is the measured transmission normalised to that of a blank substrate or pellet.

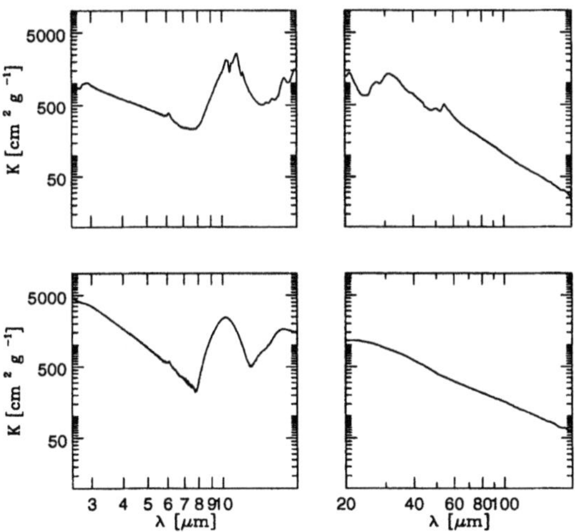


*Figure 1.* MIR (left) and FIR (right) mass extinction coefficient of crystalline (top) and amorphous (bottom) fayalite.

It can be seen that in the spectra of the amorphous sample the richness of features that are present in the crystalline one is lost. This is due to the total loss, in the amorphous grains, of long range order typical of a crystalline lattice. Two broad bands near 10 and 20 μm in the mid infrared region survive, while bands lack in the far infrared region, where only a smooth trend is visible.
It is important to note that the reported spectra must be corrected for the matrix effects, which may affect the peak positions [20]. This will be the subject of further analysis.

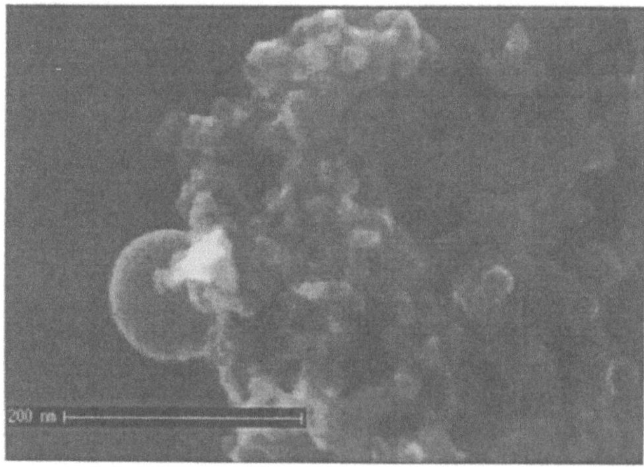

*Figure 2.* SEM images of crystalline (top) and amorphous (bottom) fayalite grains.

In Figure 2 two SEM images of the crystalline and amorphous samples are reported. They show the morphology of the two different forms of the material. We have found that crystalline grains show a power law distribution of dimensions [21]:

$$n(d) \propto d^{-p}$$

where $n$ is the number of grains, $d$ their dimension and $p$ a characteristic power index. We have measured a power index $p \sim 2.2$ and an average diameter $<d> \sim 0.2$ µm.
In the amorphous sample we have observed two types of morphology, resulting from the laser-target interaction and thermodinamical production conditions [22,23]:

1. *spherical isolated particles;*
2. *aggregates of particles.*

The former also show a power law distribution of dimensions with a power index $p \sim 2.6$ and an average diameter $<d> \sim 35$ nm. The latter, that frequently are organised in chain like-structures [22], have dimensions normally distributed around an average diameter $<d> \sim 13$ nm.

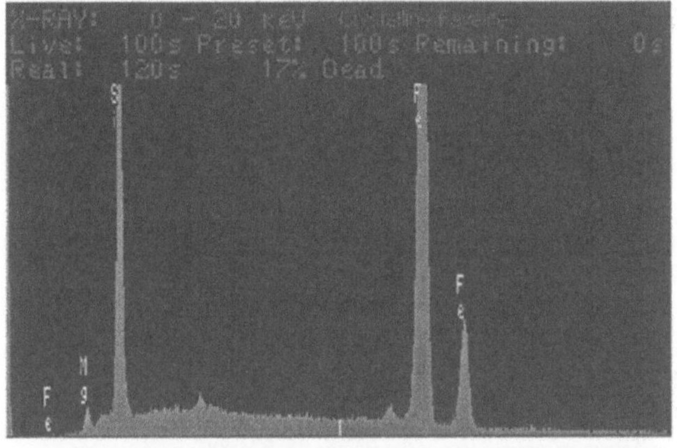


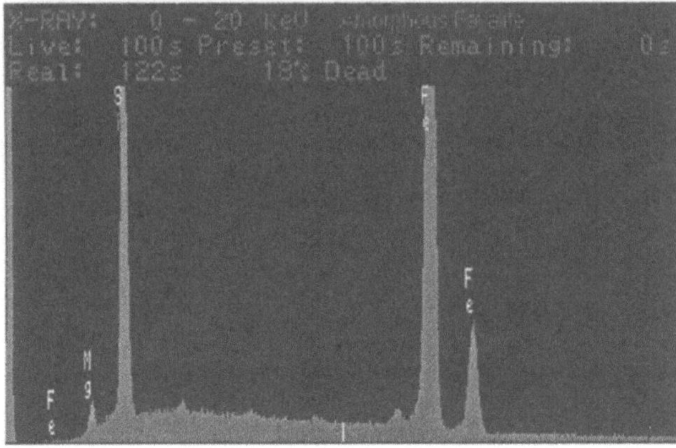


*Figure 3.* EDX qualitatively results about composition of crystalline (top) and amorphous (bottom) fayalite samples.

The EDX analysis (Figure 3) has shown that the elemental composition of the condensed grains is qualitatively the same as the target materials. This is due to the rapid laser-assisted evaporation of crystalline target and condensation of produced vapour [22].
Moreover, we can conclude that either unprocessed or processed fayalite samples are quite pure: a very low concentration of magnesium and other trace elements are observed.

## 4. Conclusions

In this paper we have presented preliminary results of the optical, morphological and chemical properties of amorphous and crystalline fayalite. The comparison of different forms of silicates is fundamental in order to disentangle between amorphous and crystalline silicates, as candidates for reproduce astronomical observations. The comparative analysis reported here is of main importance to interpret IR features recently observed by ISO from a number of astronomical sources.

## References

1. Wynn-Williams, C. G. and Becklin, E. E. (1974) Infrared emission from HII regions, *Publications of the Astronomical Society of the Pacific* **86**, 5-25.
2. Roche, P. F. (1989) Spectroscopy of galaxies in the mid infrared, in ESA, *Infrared Spectroscopy in Astronomy*, pp. 447-453.
3. Merrill, K. M. and Stein, W. A. (1976) 2-14 μm M stellar spectrophotometry I. Stars of the conventional spectral sequence, *Publications of the Astronomical Society of the Pacific* **88**, 285-293.
4. Merrill, K. M. and Stein, W. A. (1976) 2-14 μm M stellar spectrophotometry II. Stars from the 2 μm M infrared sky survey, *Publications of the Astronomical Society of the Pacific* **88**, 294-307.
5. Merrill, K. M. and Stein, W. A. (1976) 2-14 μm stellar spectrophotometry III. AFCRL sky survey objects, *Publications of the Astronomical Society of the Pacific* **88**, 874-887.
6. Willner, S. P., Gillet, F. C., Herter, T. L., Jones, B., Krassner, J., Merrill, K. M., Pipher, J. L., Puetter, R. C., Rudy, R. J., Russell, R. W. and Soifer, B. T. (1982), Infrared spectra of protostars - Composition of the dust shells, *Astrophysical Journal* **253**, 174-187
7. Olnon, F. M., Raimond, E., Neugebauer, G., van Duinen, R. J., Habing, H. J., Aumann, H. H., Beintema, D. A., Boggess, N., Borgman, J., Clegg, P. E., Gillett, F. C., Hauser, M. G., Houck, J. R., Jennings, R. E., de Jong, T., Low, F. J., Marsden, P. L., Pottasch, S. R., Soifer, B. T., Walker, R. G., Emerson, J. P., Rowan-Robinson, M., Wesselius, P. R., Baud, B., Beichman, C. A., Gautier, T. N., Harris, S., Miley, G. K. and Young, E. (1986), IRAS catalogues and atlases - Atlas of low-resolution spectra, *Astronomy and Astrophysics Supplement Series* **65**, 607-1065.
8. Omont, A. (1993), Circumstellar matter: cool IRAS sources, in S. Kwok (eds.), *Astronomical Infrared Spectroscopy: Future Observational Directions*, ASP Conference Series, San Francisco, vol. 41, pp. 87-95.
9. Woolf, N. J. (1975), Circumstellar dust, in G. B. Field and A. G. W. Cameron (eds), *The Dusty Universe*, Neale Watson Academic Publications, Inc., New York, pp. 59-87.
10. Ney, E. P. (1974), Multiband photometry of Comets Kohoutek, Bennett, Bradfield, and Encke, *Icarus* **23**, 551-560.
11. Krishna Swamy, K. S., Sandford, S. A., Allamandola, L. J., Witteborn, F. C. and Bregman, J. D. (1989), Infrared emission from comets, *Astrophysical Journal* **340**, 537-549.
12. Hanner, M. S., Lynch, D. K. and Russell, R. W. (1994), The 8-13 micron spectra of comets and the composition of silicate grains, *Astrophysical Journal* **425**, 274-285.

13. Lutz, D., Feuchtgruber, H., Genzel, R., Kunze, D., Rigopoulou, D., Spoon, H. W. W., Wright, C. M.; Egami, E., Katterloher, R.; Sturm, E. Wieprecht, E.; Sternberg, A., Moorwood, A. F. M. and de Graauw, T. (1996), SWS observations of the Galactic center, *Astronomy and Astrophysics* **315**, L269-L272.
14. Crovisier, J.; Brooke, T. Y., Hanner, M. S.; Keller, H. U., Lamy, P. L.; Altieri, B., Bockelée-Morvan, D., Jorda, L.; Leech, K. and Lellouch, E. (1996), The infrared spectrum of comet C/1995 O1 (Hale-Bopp) at 4.6 AU from the Sun, *Astronomy and Astrophysics* **315**, L385-L388.
15. van der Hucht, K. A., Morris, P. W., Williams, P. M., Setia Gunawan, D. Y. A., Beintema, D. A., Boxhoorn, D. R., de Graauw, T., Heras, A., Kester, D. J. M., Ahuis, F., Leech, K. J., Roelfsema, P. R., Salama, A., Valentijn, E. A. and Vandenbussche, B. (1996), ISO-SWS spectrophotometry of galactic Wolf-Rayet stars: preliminary results, *Astronomy and Astrophysics* **315**, L193-L196.
16. Gürtler, J., Henning, T., Kömpe, C., Pfau, W., Krätschmer, W. and Lemke, D. (1996), Detection of an absorption feature at the position of the 4.27 μm band of solid $CO_2$, *Astronomy and Astrophysics* **315**, L189-L192.
17. Waelkens, C., Waters, L. B. F. M., de Graauw, M. S., Huygen, E., Malfait, K., Plets, H., Vandenbussche, B., Beintema, D. A., Boxhoorn, D. R., Habing, H. J., Heras, A. M., Kester, D. J. M., Lahuis, F., Morris, P. W., Roelfsema, P. R., Salama, A., Siebenmorgen, R., Trams, N. R., van der Bliek, N. R., Valentijn, E. A. and Wesselius, P. R. (1996), SWS observations of young main-sequence stars with dusty circumstellar disks, *Astronomy and Astrophysics* **315**, L245-L248.
18. Waters, L. B. F. M., Molster, F. J., de Jong, T., Beintema, D. A., Waelkens, C., Boogert, A. C. A., Boxhoorn, D. R., de Graauw, T., Drapatz, S., Feuchtgruber, H., Genzel, R., Helmich, F. P., Heras, A. M., Huygen, R., Izumiura, H., Justtanont, K., Kester, D. J. M., Kunze, D., Lahuis, F., Lamers, H. J. G. L. M., Leech, K. J., Loup, C., Lutz, D., Morris, P. W., Price, S. D., Roelfsema, P. R., Salama, A., Schaeidt, S. G., Tielens, A. G. G. M., Trams, N. R., Valentijn, E. A., Vandenbussche, B., van den Ancker, M. E., van Dishoeck, E. F., van Winckel, H., Wesselius, P. R. and Young, E. T. (1996), Mineralogy of oxygen-rich dust shells, *Astronomy and Astrophysics* **315**, L361-L364.
19. Snow, Th. P. and Witt, A. N. (1996), Interstellar Depletions Updated: Where All the Atoms Went, *Astrophysical Journal Letters* **468**, L65-L68.
20. Dorschner, J., Friedemann, C. and Gürtler, J (1978), Laboratory spectra of phyllosilicates and the interstellar 10-micrometer absorption band, *Astronomische Nachrichten* **299**, 269-282.
21. Mathis, J. S., Rumpl, W. and Nordsieck, K. H. (1977), The size distribution of interstellar grains, *Astrophysical Journal* **217**, 425-433.
22. Stephens, J. R. and Kothari, B. K. (1978), Laboratory analogues to cosmic dust, *The Moon and the Planets* **19**, 139-152.
23. Sankur, H. and Cheung, J. T. (1988), Formation of Dielectric and Semiconductor Thin Films by Laser-Assisted Evaporation, *Applied Physics A* **47**, 271-284.

# IR REFLECTANCE SPECTROSCOPY OF MARTIAN ANALOGUES

F.ESPOSITO[1,2], E.PALOMBA[1], L.COLANGELI[1]
*(1) Osservatorio Astronomico di Capodimonte-Napoli*
*(2) Università degli Studi di Napoli Federico II*

**Abstract.** The purpose of this paper is to show some preliminary results about IR reflectance spectroscopic properties of Martian analogues. We focus on IR measurements because most of Mars radiance is contained in this spectral range. Samples we analyze are candidate for Martian regolith. We perform two kind of reflectance measurements: specular reflectance and diffuse reflectance. The former allows us to compute the optical constants while the latter can be used to predict emittance. We present spectra in transmittance, diffuse and specular reflectance for montmorillonite and the deduced optical constants. It is important to note how spectral features in reflectance vary with particle size.

## 1. Introduction

In the last years Mars has become again one of the most important objectives for spatial missions. In fact both USA and Russia are engaging themselves in missions like Mars Global Surveyor, Mars Pathfinder and Mars'96, even if this last unfortunately failed. The MGS orbiter is equipped with a Thermal Emission Spectrometer (TES). This instrument works in the spectral range 2000-250 $cm^{-1}$ with a spectral resolution of 5 $cm^{-1}$ and it will allow us to obtain a Mars detailed spectroscopic map. The correct interpretation of TES data involves a systematic laboratory activity consisting in a complete characterization of Martian analogue spectroscopic properties. For our laboratory work we have selected samples to be analysed among possible components of Mars planet regolith. They are: montmorillonite, calcite, magnesite, hematite and palagonite.
Regolith is formed by unconsolidated layers of soils, sediments and rock fragments that overlie bedrock. These layers have albedo $\sim 0.26$ (bright areas) and cover regions like *Hellas, Arabia, Lunae Planum, Elysium* (ref.[1]).



*J. M. Greenberg and A. Li (eds.), Formulation and Evolution of Solids in Space,* 305–310.

## 2. Why reflectance measurements in the infrared?

Most of Mars radiance is contained in the IR range where most minerals and atmospheric molecules exhibit strong rotational and rotovibrational bands. Radiance is the sum of two terms: a reflectivity term and an emissivity term so that:

$$L_{tot} = L_{ref}(\sigma) + L_{em}(\sigma) = A\frac{r^2}{D^2}B(T_S, (\sigma)) + \epsilon B(T_M, (\sigma))$$

where
$L_{ref}$= reflected sunlight, $L_{em}$= grey body emission, A = bo lometric albedo (fraction of solar radiation reflected in all direction), r = radius of the Sun ($4.66 \times 10^{-3}$A.U.), D = Mars-Sun mean distance (1.52 A.U.), $T_S$= Sun photosphere temperature (5800 K), $T_M$= Martian surface temperature, $\epsilon$= emissivity.

The emissivity prevailes at shortest wavenumbers, while reflectivity dominates between 2000 and 6000 $cm^{-1}$. Our laboratory measurements are focused on: 1) diffuse reflectance, 2) specular reflectance.
Diffuse reflectance data can be used for a direct comparison with observational data. They can also used to predict emittance behaviour through the Kirchhoff's law:

$$E = 1 - R$$

where: E = total emissivity, R = total reflectance.
This law is valid only at thermal equilibrium, an approximation applicable to the Martian surface.
Specular reflectance data allow us to calculate optical constants of selected minerals through, for example, the Lorentz model (ref.[3]).
In this model the dielectric function is:

$$\epsilon(\lambda) = \epsilon_\infty + \sum_{i=1}^{N} \frac{S_i}{1 - i\gamma_i\frac{\nu}{\nu_i^2} - (\frac{\nu}{\nu_i^2})^2} = \epsilon' + i\epsilon''$$

The theoretical specular reflectance (Fresnel's law) to fit the experimental one is given by:

$$R(\lambda) = \frac{1}{2}(|r_\pi|^2 + |r_S|^2)$$

with

$$r_\pi = \frac{\epsilon cos(\phi) - Z}{\epsilon cos(\phi) + Z}; \qquad r_S = \frac{Z - cos(\phi)}{Z + cos(\phi)}; \qquad Z = (\epsilon - sin^2(\phi))^{\frac{1}{2}}$$

The optical constants are then calculated from:

$$n = \sqrt{\frac{\sqrt{\epsilon'^2 + \epsilon''^2} + \epsilon'}{2}}; \qquad k = \sqrt{\frac{\sqrt{\epsilon'^2 + \epsilon''^2} - \epsilon'}{2}}$$

## 3. Experimental

For specular measurements, bulk samples are crushed in an agate mill and pressed to obtain pellets (diameter 13 mm, thickness 1 mm) with highly reflective surfaces.

$\epsilon_\infty = 2.01 \pm 0.03$

| Oscillator | Strength S | $\frac{\gamma}{\nu^2}(cm)$ | $\frac{1}{\nu^2}(cm^2)$ |
|---|---|---|---|
| 1 | $0.01 \pm 0.01$ | $0 \pm 10^{-5}$ | $(4.33 \pm 0.07) \cdot 10^{-7}$ |
| 2 | $0.053 \pm 0.002$ | $(5.9 \pm 0.1) \cdot 10^{-5}$ | $(7.917 \pm 0.008) \cdot 10^{-7}$ |
| 3 | $0.30 \pm 0.05$ | $(7.5 \pm 0.8) \cdot 10^{-5}$ | $(4.59 \pm 0.02) \cdot 10^{-6}$ |
| 4 | $0.012 \pm 0.01$ | $(2.7 \pm 0.1) \cdot 10^{-5}$ | $(9.12 \pm 0.01) \cdot 10^{-7}$ |
| 5 | $0.18 \pm 0.01$ | $(3.8 \pm 0.2) \cdot 10^{-3}$ | $(9.59 \pm 0.02) \cdot 10^{-7}$ |
| 6 | $(9 \pm 6) \cdot 10^{-3}$ | $(2.4 \pm 0.8) \cdot 10^{-5}$ | $(1.189 \pm 0.009) \cdot 10^{-6}$ |
| 7 | $(5 \pm 5) \cdot 10^{-3}$ | $(2.1 \pm 0.8) \cdot 10^{-5}$ | $(1.16 \pm 0.01) \cdot 10^{-6}$ |
| 8 | $(8 \pm 2) \cdot 10^{-3}$ | $(4 \pm 1) \cdot 10^{-5}$ | $(1.29 \pm 0.01) \cdot 10^{-6}$ |
| 9 | $(3.1 \pm 0.8) \cdot 10^{-3}$ | $(2.6 \pm 0.8) \cdot 10^{-5}$ | $(1.578 \pm 0.008) \cdot 10^{-6}$ |
| 10 | $(0 \pm 6) \cdot 10^{-3}$ | $0.0 \pm 0.2$ | $(3 \pm 2) \cdot 10^{-6}$ |
| 11 | $(2.6 \pm 0.4) \cdot 10^{-2}$ | $(1.3 \pm 0.1) \cdot 10^{-4}$ | $(2.67 \pm 0.01) \cdot 10^{-6}$ |
| 12 | $0.21 \pm 0.01$ | $(9.0 \pm 0.4) \cdot 10^{-5}$ | $(3.621 \pm 0.005) \cdot 10^{-6}$ |
| 13 | $0.14 \pm 0.01$ | $(2.1 \pm 0.1) \cdot 10^{-4}$ | $(3.24 \pm 0.02) \cdot 10^{-6}$ |
| 14 | $(7 \pm 3) \cdot 10^{-2}$ | $(6 \pm 1) \cdot 10^{-5}$ | $(4.37 \pm 0.04) \cdot 10^{-6}$ |
| 15 | $0.21 \pm 0.04$ | $(1.4 \pm 0.1) \cdot 10^{-4}$ | $(4.96 \pm 0.06) \cdot 10^{-6}$ |

TABLE 1. Oscillator parameters for montmorillonite

For diffuse reflectance measurements, after crushing, the particle samples are sieved to 5 different particle-size ranges:
$> 200$, 200-100, 100-50, 50-20, $< 20$ $\mu$m.
However scanning electron microscope shows that, after seaving, lots of

small particles (diameter of about 5 $\mu$m) remain linked to the largest. To eliminate this problem sedimentation is needed. This work is in progress.

For our reflectance measurements, we use a Fourier transform Michelson interferometer (Bruker Equinox 55) with a diffuse/specular accessory.

## 4. Results and discussion

Specular and diffuse reflectance measurements of montmorillonite are shown in fig.1 for the 400-2000 and 2000-6000 cm$^{-1}$ ranges respectively. A transmittance measurement is also shown for comparison. In fig.2 the computed real index "n" and imaginary index "k" for montmorillonite are reported,respectively. The oscillator parameters used in fit are shown in table 1.

In the range 400-2000 cm$^{-1}$, see fig.1, the strongest vibration bands, the "reststrahlen" bands (ref.[2]), appear as specular reflectance peaks. These peaks are due to a first surface reflection, i.e. the all returned radiation is reflected from the mineral surface and none passes through it. When we move to the diffuse reflectance from powder, the reststrahlen bands become weaker and weaker with particle size decreasing.

In the range 1300-1600 cm$^{-1}$ the specular reflectance has no evident structures while overtone and combination bands become more important.

A low absorption coefficient at these wavenumbers implies high reflectance, due to scattering, which increases with particles size decreasing.

In summary, fundamental bands are the mineral fingerprints between 400 and 2000 cm$^{-1}$ while overtone and combination bands are the typical characteristics between 1300 and 6000 cm$^{-1}$.

## 5. Conclusions and future developments

We have developed a method to measure diffuse and specular IR reflectance of minerals. This method will be applied to study all selected minerals.

It is interesting to note that we have been able to observe different spectral behavior with varying grain dimensions. So it is very important a systematic study which also includes sedimentation.

Finally, we have developed an algorithm to determine optical constants through the Lorentz model. This algorithm makes a non-linear fit of experimental specular reflectance with a theoretical one given by the Fresnel's low. The dielectric function is calculated via the Lorentz relations. We have obtained the real and imaginary parts of refraction index for montmorillonite in the range 400- 1250 cm$^{-1}$.

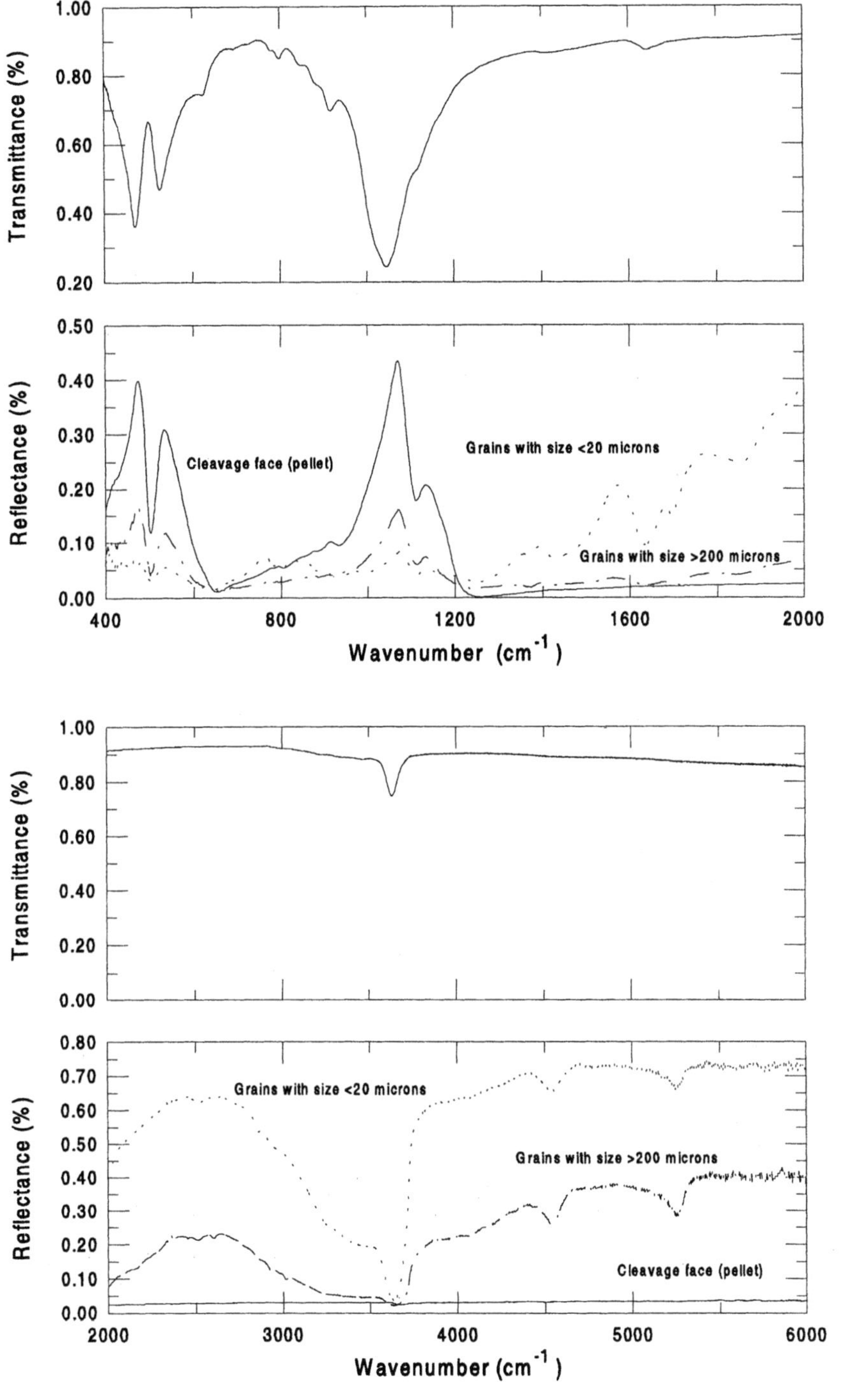


*Figure 1.* Transmittance and reflectance of montmorillonite.

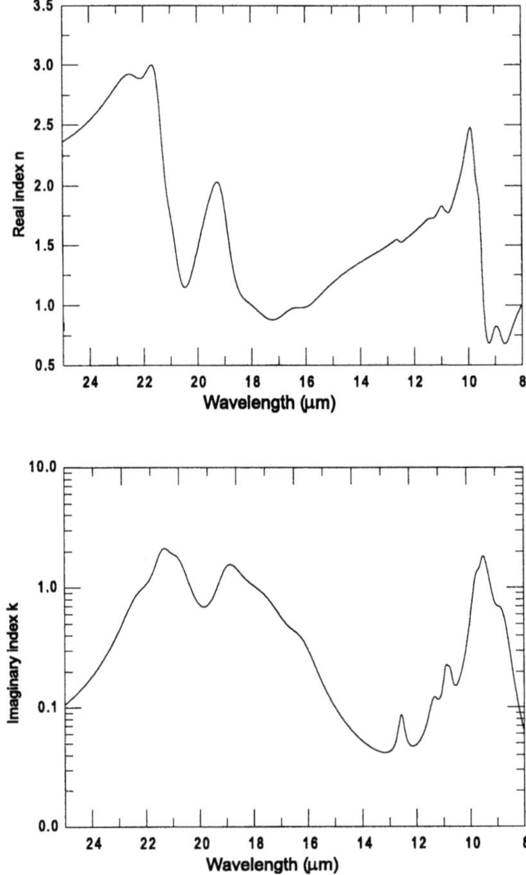


*Figure 2.* Real and imaginary parts of refraction index for montmorillonite.

## 6. References

[1] H.H. Kieffer, B.M. Jarosky, C.W. Snyder, M.S. Matthews "Mars" *University of Arizona Press, 1992.*

[2] W. Salisbury "Mid-Infrared Spectroscopy: Laboratory Data" in "Remote Geochemical Analysis: Elemental And Mineralogical Composition", Carle M. Pieters, Peter A.J. Englert (eds.) *Cambridge University Press, 1993* (p. 79-83).

[3] C.F. Bohren, D.R. Huffman "Absorption and Scattering of Light by Small Particles", *John Wiley & Sons, Inc., 1983* (p. 228-244).

# PROPERTIES OF DUST PRODUCING STARS

H.J. HABING
*Sterrewacht*
*Huygens Laboratorium*
*Postbus 9513*
*2300 RA Leiden*

**Abstract.** Almost all long–period variables with periods longer than 300 days eject matter at a high rate ($10^{-6}$ $M_{\odot}$/yr) but at a low outflow velocity (10 km/s). These stars have initially a modest mass (1 to $6M_{\odot}$) and they are in one of the very last and very short phases of their evolution (the Asymptotic Giant Branch or AGB). At some distance from the star small solid particles condense in the outflow. Long-period variables of spectral type M produce silicate grains and carbon stars produce the carbon dust. Together these two types of stars produce enough dust to explain all that is seen in interstellar space. There are many M-type variables in the inner Galaxy and very few carbon stars. In the outer parts of the galactic disk, say beyond the solar circle, the carbon stars have the majority. Thus one expects a strong increase in the ratio of carbonaceous to silicate dust when one travels from the inner Galaxy to the outside.

## 1. Dusty envelopes

On the topic of this presentation I have published an extensive review (Habing, 1996).

Some stars produce dust, solid particles of submicron size, particles that are lost by the star to the interstellar medium. There are three different facts that, when put together, lead to this conclusion. The first fact: these stars radiate a strong continuum in the infrared, a continuum that is many times stronger than expected from the stellar photosphere; the smoothness of the continuum suggests solid particles as the origin of the radiation; the radiation is strong in the infrared and one concludes that these solid particles have a temperature of a few hundred Kelvin and must be located close



*J. M. Greenberg and A. Li (eds.), Formulation and Evolution of Solids in Space,* 311–323.

the star. The second fact concerns spectral lines emitted by gas surrounding these stars; the line shape proves that the gas either contracts onto or expands away from the star; a detailed analysis of the OH maser lines shows that the motion is outward; the velocities are so high that gravitational attraction cannot stop the flow. And, third fact, the dust particles absorb light photons easily and are propelled outward by light pressure. Because the motion of the dust is tied to the gas by friction the gas is also propelled outwards. Thus: the stars loose mass in the form of gas and dust and at a rapid rate; the stars are continuously producing dust particles and eject them in insterstellar space.

The infrared continuum is marked by a few broad emission and/or absorption bands and these help to distinguish between two brands of mass–losing stars: in one group of stars the particles are made of some sort of silicate mineral, in the other group the paricles are made of SiC and some kind of carbonaceous material. This division into two groups corresponds to the spectroscopic division of red giants into M–type stars and carbon stars. An astrophysical explanation for this division has been given by Gilman (1969).

In practically all cases where one can see the star through the dust envelope it is a very luminous ($L_*$ between 3,000 and 20,000 $L_\odot$), long–period (300 to $500^d$) variable with a large pulsational amplitude ($\Delta m_{\rm bol} \approx 1$ mag): a Mira variable or, in a minority of cases, a Semi–Regular variable. In addition to these visible stars there are infrared objects, equally variable, equally luminous but where the envelope is optically thick and the star is too faint at visual wavelengths to be seen directly. Visible and invisible objects differ only in the optical depth of the dust envelope; otherwise they have the same structure. It has been thought that the invisible objects are more luminous than those visible, but there is no good evidence for this conclusion. I will call visible and invisible objects together by one acronym: "LALPVs": "large–amplitude, long–period variables". Although almost all stars with a "dusty envelopes" are LALPVs the reverse is only partially true: Mira variables with $P < 300^{\rm d}$ do not show signs of mass loss, and only Miras with longer periods loose matter fairly rapidly, that is $|\dot{M}| > 10^{-7}$ $M_\odot$ per year.

In the argumentation above several times words like "most objects ..." and "a few objects..." have been used because there are exceptions to every definition. It is in favour of the argumentation that all exceptions, e.g. RCrB and RV Tauri stars, can be understood within the same framework and that they are rare in comparison with Mira and SR variables. And Miras and SRs themselves are rare objects among other red giants: in a magnitude–limited sample of red giants the Miras constitute no more than 1 % of the total (Blanco, 1965).

In addition to these variable, evolved stars there is a completely separate, small class of nearby main-sequence stars that are surrounded by a disk of dust. The examples are well-known: $\beta$ Pictoris, Vega, Fomalhaut and others. This disk must be a left-over from the time the star formed. This is supported by the fact that young stars, e.g. T Tau and Herbig AeBe stars, have similar disks, but in these young stars the disks are thicker. Carsten Dominik will discuss these main-sequence stars in another lecture during this course. I restrict myself to the very luminous red giants and their relatives, that show dust formation in action.

## 2. Most dust-producing stars are on the AGB

The high luminosity and the cool surface identifies these stars with a very short, but spectacular episode during stellar evolution called the Asymptotic Giant Branch, or AGB. This is the last phase of a star during which the luminosity is caused by nuclear fusion; after the AGB phase a star develops into a planetary nebula and next into a white dwarf that then slowly fades away. Define $M_{\rm ms}$ as the mass of a star when it is on the main-sequence and has not yet suffered mass loss. Only the least massive stars ($M_{\rm ms} < 0.9 M_{\odot}$?) and the most massive ($M_{\rm ms} > 8.0 M_{\odot}$) have a different fate. A bit of theory of the evolution of single stars with a mass between 0.9 and 8 $M_{\odot}$: After leaving the main sequence a single star (double stars are counted as single when the exchange of mass is too low to influence the evolution of either component) means that the star is not influenced in its evolution by a companion) follows a theoretically well prescribed path through the HR-diagram. The actual path depends on $M_{\rm ms}$; two different paths can be distinguished: that for low-mass stars ($M_{\rm ms} < 2M_{\odot}$), and that for intermediate-mass stars ($2M_{\odot} < M_{\rm ms} < 8M_{\odot}$). These boundaries of 2 and 8 $M_{\odot}$ have been taken from a set of models calculated by the Padua group, in particular those published by Bressan et al. (1993); other groups find the same boundary at a numerically somewhat different stellar mass.

Start with the low-mass stars: during the main-sequence episode they burn hydrogen into helium. This will continue untill the mass of helium exceeds some limit; then the helium contracts into a core where the density is so high and the temperature so low that the electrons become a degenerate gas; this core contains a sizeable fraction of the stellar mass. The energy production continues by hydrogen burning in a thin shell surrounding the helium core. A peculiar property of stars with such a core is a rapid increase in luminosity: for example, a star with an initial mass $M_{\rm ms} = 1.0 M_{\odot}$ will remain on the main sequence for a time of 12 Gyr and slowly increase its luminosity to 10 $L_{\odot}$, and then, within 0.27 Gyr it will reach 2500 $L_{\odot}$. During this quick rise in luminosity the star is a red giant and in the HR–diagram it populates the so–called Red Giant Branch (RGB).

When the stellar core reaches a mass of about 0.46 $M_{\odot}$, and its luminosity about 2500 $L_{\odot}$, the core collapses, it looses its degeneracy and very quickly the star converts into a "Horizontal Branch" star; at the centre it burns helium into carbon in the triple–$\alpha$ process; the luminosity is between 50 and 100 $L_{\odot}$. Later, when most of the helium core has been transformed into a core of carbon and oxygen, another contraction of the core follows, again leading to degeneracy of the free electron gas, while the mantle expands and the star becomes a red giant for the second time. Most of the time the energy production is by hydrogen burning into helium, this time at the outside of a thin helium layer surrounding the carbon/oxygen core. Again the red giant increases quickly in luminosity; it is now called an Asymptotic Giant Branch or AGB star and has a luminosity well above the RGB limit of 2500 $L_{\odot}$.

In stars of intermediate mass the helium core never degenerates. Consequentely intermediate–mass stars never pass through the RGB phase. However, after helium burning has converted their core into one of carbon and oxygen, core–contraction, electron–degeneracy and expansion of the mantle occur and the star becomes an AGB star. It reaches a higher luminosity than low–mass AGB stars; probably the luminosity of an AGB star in its final stage is an indication of its main–sequence mass.

High–mass stars never develop a degenerate core and their luminosity does not change much; in fact, in the HR–diagram they move almost horizontally at constant luminosity between a small hot, blue surface and a large, cool, red surface. These high–mass stars are supergiants and I will not discuss them here.

The assumption that the stars are single, or that they evolve as if they were single is a bold simplification: about half of all stars have a companion and when this companion is sufficiently close the loss of mass from the primary to the secondary may cause the secondary to take a different evolutionary path. Later, when the primary has become an inconspicuous white dwarf, the erewhile secondary is still a main–sequence star but it has element "enrichments" typical for the AGB. See the example of the S–type stars discussed farther below.

## 3. Large–amplitude, long–period pulsations, mass loss and thermal pulses

Our knowledge of the interior structure of an AGB star is derived from the calculation of sequences in time of the structure of a star of given mass. That such calculations can be trusted follows from the succesful comparison of calculated and observed HR diagrams of stellar clusters. The calculations show that AGB stars consist of a very small core (of the order of the size of the Earth) with a mass of about 0.5 $M_\odot$. The core is surrounded by a "mantle" of hydrogen with a diameter comparable to the radius of the orbit of the Earth around the Sun and containing the remainder of the stellar mass. The density contrast between core and mantle is therefore of the order of the third power of the radius of the Earth' orbit divided by the radius of the Earth: in total a factor of $1.3 \times 10^{13}$! This result is the outcome of long calculations that have been checked in an indirect manner. A more direct check will be the analysis of the lightcurve of these variable stars: the light curve contains "seismological" information. At present similar seismological evidence is being tapped from various types of variables, but not yet from the stars that interest us here.

Three properties characteristic of the AGB will have our attention. The first are the large–amplitude ($\Delta m_{\rm bol} \approx 1^m$) pulsations with long periods (between 50 and 1000 days). Such pulsations occur only in red giants of a luminosity higher than the RGB limit of 2500 $L_\odot$; hence these must be AGB stars. Mira variables form an important subgroup among the LALPVs. (I note in passing that the study of the properties of Miras will probably make rapid progress in the coming years: HIPPARCOS has measured the distance of several Miras so that we will soon have accurate luminosities. And for the first time interferometers measure the diameters of Mira variables succesfully and at important wavelengths.)

Why AGB stars develop large–amplitude pulsations and RGB stars do not, is not well understood, but there are rather convincing suggestions that the pulsations have their origin in the very large extent and the high luminosity of AGB stars, even when compared to the RGB stars: the stars are close to the point where radiation pressure blows them up; see e.g. Wagenhuber and Weiss (1994).

The second characteristic property of an AGB star is its mass loss at a rate exceeding $10^{-7} M_\odot/\rm{yr}$ at least; this happens only for stars with a pulsation period longer than 300 days. The cause of the rapid mass loss probably lies in a combination of two factors: (i) the pulsations produce a dynamic and very extended atmosphere; in the top of this atmosphere the temperature is sufficiently low and the density sufficiently high that dust particles form; (ii) these dust particles are blown out of the star by radiation pressure; the grains transfer momentum to the gas, and enough grains are formed to carry all the gas along. Full dynamical models of these pulsating stars do not yet exist: we lack an exact theory of time dependent convection. Approximations are being made: see in the next subsection.

There is a small number of stars with a thick dusty envelope and no periodic pulsations. Again there are two groups. Group one are stars that have stopped LALPV pulsations recently and are now in a transition phase where the envelope is still close to the star but will disperse quickly; this category of objects is called "post- AGB stars". Group two are the supergiants like NML Cyg and VY CMa; very few members of this group are known.

The high mass–loss rate of an AGB star implies that in a short time the star ejects its mantle into the interstellar medium so that only the core remains. Thus mass loss terminates the life of the star. When almost all of the mantle material has been lost the core becomes visible; its high temperature leads to ionization of the dispersed mantle material and the object has become a planetary nebula. A short while later the nebular gas has also been lost to the interstellar medium and the core remains as a white dwarf.

The third characteristic of AGB stars is the prediction of "thermal pulses" and the subsequent "dredge–up" of newly produced nuclei: after a thermal pulse the layers surrounding the core are convective and the new atoms are transported from the core to the atmosphere of the star. We observe this as a "peculiar chemistry". Below I will discuss these abundance anomalies in a bit more detail.

When the short AGB episode is over the star becomes first a Post–AGB star, then a planetary nebula (PN), and, finally, a white dwarf (WD). This development is so rapid that AGB–stars, Post–AGB stars, PNe and WDs emerging from stars of the same main–sequence mass must all have the same spatial and kinematical distribution in our Galaxy.

## 4. Properties of the outflowing dust–rich gas

Stars are usually treated as spherical objects. Spherical symmetry also appears to apply to AGB stars. The best evidence is the high degree of circularity found in maps of maser spots and in maps of the mid–infrared emission. When all 2D maps are circular it follows that the 3D symmetry is spherical. As a contrast, objects that have left the AGB (post–AGB stars, PNe) usually do not show circular but axial symmetry in the form of a stationary disk and a high velocity wind along both directions of the polar axis. Here I discuss the AGB stars and I assume spherical symmetry.

Hydrodynamical models of pulsating and mass–losing AGB stars have been a major topic of research, first in Sedlmayr's group in Berlin and recently also in Dorfi's group in Vienna; for both groups the focus has been on carbon–stars, because the formation of carbonaceous grains is much better understood than that of silicate dust. For a good summary on the work of both groups see Höfner et al. (1996). I still have confidence in the simple stationary hydrodynamic models with instantaneous dust formation

(Habing et al., 1994; Netzer and Elitzur, 1993) and find them instructive. These stationary models assume a spherically symmetric outflow of gas and dust; the absorption of stellar photons by the dust transfers energy and momentum from the photons to the dust and forces the dust to move outward; friction couples dust and gas efficiently and the gas follows the dust closely. An analysis of the full spectrum of a given object tells us the value of the optical depth, $\tau_c$, of the gas/dust envelope; together with a determination of the luminosity, $L_*$, we then can calculate how much momentum is transfered to the gas $(\tau_c L_*/c)$. The outflow velocity far away from where the acceleration takes place, $v_{\rm exp}$ can be measured simply and accurately from the form of the spectral lines. We then derive the *total* mass–loss rate as $\dot{M} = \tau_c L_*/v_{\rm exp}$. Because the knowledge of $\tau_c$ and $v_{\rm exp}$ also allows an independent determination of $\dot{M}_{\rm dust}$ the dust–to–gas ratio is found. The analysis of the envelope of OH26.5+0.6 by Justtanont et al. (1992, 1994, 1996) shows how much has already been learned; for example that the dust–to–gas ratio is 0.0036. Another and surprising result is the weakness of the (CO $J = 1 \rightarrow 0$)–line when compared with the ($J = 2 \rightarrow 1$)–line; the ($1 \rightarrow 0$)–line is formed at much larger distance from the star than the ($2 \rightarrow 1$)–line and the relative weakness of the second line has been interpreted as the consequence of a recent (200 yr ago), and sudden decrease of the mass loss rate by a factor 55.

The opposite change in the mass–loss rate, a sudden decrease or interruption of $\dot{M}$, has also been observed. The first recorded case concerned a carbon star with a "detached envelope" that is: the star is surrounded by an envelope, but there is an empty region between the star and the envelope. Several other carbon stars are now known that have a similar void between star and envelope; see e.g. Olofsson et al. (1993a,b) and Waters et al. (1994). The same phenomenon, a star inside a detached envelope, has also been found among M–type AGB stars. The conclusion is that the mass–loss rate of AGB stars can jump up or down in a short time; the suggestion that such a jump coincides with a thermal pulse is obvious, and has made by many.

## 5. Carbon and other chemically peculiar stars

Stellar spectroscopy started over 100 years ago. In the early years Secchi discovered that there are two kinds of red giants with very different, but always very cool spectra: M–type stars and carbon–stars. Analysis has shown that in carbon stars there are more carbon atoms than oxygen atoms ([C/O]$>$ 1) and that the reverse ([C/O]$<$ 1) is true for the M–type stars. The latter condition is the "natural" or "default" situation: interstellar matter in our Galaxy (but perhaps not in the Small Magellanic Cloud!) contains (and has always contained) more oxygen than carbon atoms and all stars are always born as "oxygen–rich". That some stars change the atomic composition of their photosphere during their evolution is unexpected and is explained as the consequence of a "dredge–up" during a thermal pulse: freshly produced carbon is transported upward when the convection reaches into deeper layers than before. Although the calculation of how much is dredged up is still very difficult and although the quantitative results are still unreliable, the observations so far have shown that single, chemically peculiar stars are always AGB stars. This reinforces the assumption that chemical peculiarity is brought about by thermal pulses.

An interesting case of chemical peculiarity is the appearance of the element Technetium in Mira stars with pulsations periods longer than $300^d$. Tc is a radio active element with a half–life time of the order of $10^5$ yr. Its presence in a stellar atmosphere must mean that the star has experienced a dredge–up less than $10^5$ yr ago. Thus one concludes that Mira variables with a pulsation period over $300^d$ have experienced one or (probably) more thermal pulses.

To confuse us, Mother Nature creates other types of chemically peculiar stars, e.g. "barium–stars", "CH–stars". In these cases it appears that each star started its life as the minor companion in a binary system. When the bigger brother became a mass–loosing AGB star the smaller brother received, via Roche–lobe overflow, "enriched" material. Ultimately big brother becomes an inconspicuous white dwarf and we now see small brother as a main–sequence star with a pattern of element abundances somewhat reminiscent of that of a red giant. Jorisse et al. (1993) have analysed very carefully a large set of S–type stars (stars in which there are as many O–atoms as C–atoms) and conclude that about 90% of the S–type stars have been "smaller brother".

Observations of carbon stars in the Magellanic Clouds show that their luminosities fall into a narrow interval. Why this is so is still not completely understood but the following (qualitative) explanation probably goes a long way: during a thermal pulse a massive AGB–star has such a high temperature in the layer between core and mantle that all the carbon atoms will be transformed into nitrogen atoms and these N–atoms are brought to the surface by convection. In such stars one expects an excess of nitrogen. Indeed, there exists a class of planetary nebulae, called "Type I", that are thought to have been formed out of the most massive AGB stars, and that have an overabundance of N. This argument thus provides an upper limit to the mass of (single) stars that become carbon rich. A lower limit to that mass is probably the consequence of lower dredge–up efficiences in stars of low $M_{\mathrm{ms}}$.

The existence of chemically peculiar stars are an indirect proof of the real occurence of thermal pulses. The first, direct detection of a star actually undergoing a thermal pulse is only very recent: Sakurai's object (e.g. Duerbeck et al., 1997).

## 6. Mass–loss as a function of time for stars of different main–sequence mass

It is now a generally accepted hypothesis that mass–loss on the AGB terminates the evolution of a star: mass–loss reduces the mantle mass to a very small value and starts the evolution of the AGB star into a "Post–AGB" star, a planetary nebula and a white dwarf. To describe the evolution during the thermal–pulsing AGB phase we need an equation that describes the mass–loss rate as a function of time and of other possible parameters ($M_{ms}$, $Z$, rotation, ...), i.e. we ask for the calculation of $\dot{M}(t; M_{ms}, Z, ...)$. An *ab initio* calculation of this function has not yet been made. This may seem odd considering the ease by which one calculates numerical stellar models and evolution tracks. The problem that one faces during the calculation of $\dot{M}(t; M_{ms})$, is that convection plays an important role and the numerical calculation of time–dependent convection is still too difficult.

The best and most extensive calculations of the evolution of an AGB star are those on carbon stars by Sedlmayr and his associates in Berlin, and by Dorfi and his colleagues in Vienna. The results of these two groups have been compared with satisfactory results; see Höfner, Fleischer et al. (1996); the only, but important shortcoming is that the pulsations are being brought about by a piston. Several models have been calculated, but a broad overview, mass loss as a function of main–sequence mass and age, is missing. More elaborate evolutionary paths have been calculated with some surrogate, phaenomenological equation to describe mass–loss: e.g. Vassiliadis and Wood (1993), Blöcker (1995) and Groenewegen and de Jong (1994). In these models a simple dependence of the mass–loss rate on stellar parameters is assumed and then the undetermined coefficients in these mass–loss rate equations are determined by comparison with known properties, e.g. the distribution of white dwarf masses (these are almost all between 0.57 and 0.62 $M_{\odot}$). For a thorough discussion of this approach see Blöcker (1995).

## 7. The distribution of dust producing stars in our Galaxy and in its nearby neighbours

Distinguish four different components of our Galaxy: halo, bulge, disk, and nucleus; the nucleus being the volume within about 200 pc from Sgr A*. I do not know of the presence of dust producing AGB stars in the halo, but it contains a few planetary nebulae and perhaps these have a dusty AGB star as progenitor. Thus the detection of halo PNe suggests the presence of dusty AGB stars. Unfortunately, so far very few dusty AGB stars have been detected in the halo; for new data on two carbon stars at high latitudes see Groenewegen et al. (1997).

The galactic disk contains most of the known AGB stars. The distribution *perpendicular* to the disk (the $z$–distribution) is more extended for AGB stars of lower main–sequence mass, a behaviour comparable to what is found for other stars, e.g. for the Mira variables. The *radial* distribution in the disk has an interesting property: almost all AGB stars with an oxygen–rich mantle (and an envelope containing silicate dust) are located within 90° from the galactic center; only a handful are known outside the first and fourth quadrant of the Milky Way, in spite of several searches directed towards the second and third quadrant. In contrast there are plenty of carbon–rich dusty AGB stars outside the solar circle; for an early, and convincing demonstration that carbon stars have a distribution in our Galaxy that differs from that of the M-giants look at figures 5 and 6 in Blanco (1965). The scarcity of M-type LPVs in the outer Galaxy may point to a lower metallicity there. I expect that the dust in the outer Galaxy is mainly carbonaceous.

The bulge of our Galaxy contains many AGB stars. Already about 50 years ago Minkowski discovered many bulge planetary nebulae and Plaut discovered hundreds of bulge Mira variables. Since then the corresponding AGB stars have been discovered in the near infrared, by IRAS or at $\lambda =$ 18cm as OH/IR stars. Several hundreds of cases are known. They allow two interesting conclusions about the bulge:

(i) there are no carbon stars among these AGB stars, except for a small number of intrinsically very faint objects that are not representative for the whole class of carbon stars;

(ii) the dusty AGB stars are indeed all LALPVs; their longest period is $800^d$; OH/IR stars of much longer periods (up to $2000^d$) exist in the galactic disk but not in the bulge.

These two observations can be explained by assuming that the bulge population contains only stars of modest main–sequence mass, stars with a life time of, say, a few Gyr.

Close to the galactic center, within 150 pc, about 200 OH/IR stars have been identified and there should be many more below the present–day detection limit. The stars occupy a flat volume around the center and they are in rapid rotation around it. The distribution of their expansion velocities, $v_{\rm exp}$, has a strong maximum around 19 km/s, in contrast with the corresponding distribution from AGB stars in the bulge or in the disk. Probably these are all of the same age. Is this sample what remains of cannibalism by our own Galaxy?

An obvious question is how many dusty AGB stars there are in our Galaxy, and how much matter they return and how much dust. From a survey for OH/IR stars, all AGB stars with silicate dust envelopes, M. Sevenster estimates about 20,000 such stars in our Galaxy, an estimate with an uncertainty of a factor of two. If we assume that there will be about 10,000 dusty carbon AGB stars, and if we take as an average mass–loss rate $3 \times 10^{-5} M_\odot/\mathrm{yr}$, one guestimates that a mass of gas of 0.9 $M_\odot$ is yearly returned to the interstellar medium. This is 1.6 times the rate estimated by Sedlmayr (see Dorschner and Henning, 1995), and that, I think, is good agreement. As the next step I guess that a fraction 0.003 of the gas is turned into dust (see the example of OH26.5+0.6, Justtanont et al., 1996) so that all AGB stars in our Galaxy together produce an yearly amount of dust of $2.7\ 10^{-3}\ M_\odot$. This is more than any other source of dust production, unless supernovae are very efficient dust producers. Assume that one supernova of type II of 30 $M_\odot$ explodes once every 50 years and that it converts 1% of its mass into dust, then the production rate is $6\ 10^{-3}\ M_\odot/\mathrm{yr}$, a factor 2.2 larger than that of the AGB stars. It seems to me that AGB stars offer a much gentler and more efficient conditions for the forming of dust particles than supernovae but this is only an intuitive argument.

Dust–producing AGB stars have also been found in the Magellanic Clouds. The initial searches have been made largely on the basis of $I$–band photographs; IRAS detections played a minor rôle because of the limited sensitivity and because of confusion. A good census of all "dust–enshrouded" AGB stars remains a very desirable goal; a large and systematic survey for these stars has been started by a French-Dutch consortium; a recent paper by van Loon et al. (1997) contains 20 to 30 dusty AGB stars. Finally I noticed the discovery of two planetary nebulae in the Sagittarius dwarf galaxy that collides with ours at a distance of 16 kpc at the other side of the galactic centre (Zijlstra and Walsh, 1997). The presence of these PNe tells us that this galaxy contains AGB stars as well.

## 8. Discussion and conclusions

The first conclusion is that the large majority of dust producing stars are LALPVs: long–period, large–amplitude variables, also called Mira variables, OH/IR stars and "stars like IRC+10216". The small percentage of other dust–surrounded stars are post–AGB stars, symbiotic stars, young planetary nebulae and a handful of peculiar stars like RCorBor and RV Tauri: in short, all stars that recently ended their AGB phase. Finally there is a small group of red supergiants and of Wolf–Rayet stars that also produce much dust; however because the stars are very rare their total dust production is probably small.

The LALPVs are all AGB stars: this follows from their red colours and their large luminosities. They probably pulsate because of the high luminosity and their very large diameter that makes the star close to break-up due to radiation pressure. A consequence of the pulsations is the loss of matter, a process that ultimately ends the stellar life. The mass–loss rate is not continuous; there is strong observational evidence that mass loss can be interrupted suddenly or can be greatly enhanced in a very short time interval.

Thermal pulses have been predicted to occur during the high–luminosity AGB phase. The pulse leads to a strong and sudden increase in luminosity. This event is followed by the injection of newly produced atoms into the mantle where these new atoms are convected to the atmosphere. An extreme consequence is the transition that a star can make from oxygen–rich (M–type) to carbon–rich (C–star). Although the basic process has now been identified for the production of "chemically peculiar" stars, there are still large discrepancies with the detailed observations. The thermal pulse calculations predict that after a short time ($10^3$ yr) the stellar luminosity returns to its pre-pulse value. Do observations confirm this sudden and temporary increase in luminosity? Until recently there existed no direct observations but that has now changed via the discovery of "Sakurai's object".

Based on the distribution of dust producing AGB stars I predict that all the dust in the bulge of our Galaxy is made of silicate: the bulge does not contain carbon–dust producing stars. In the galactic disk both C– and M–type stars are found and both types of dust are present there. In the outer Galaxy, however, there are more carbon stars than M–type stars and the dust may be predomintly carbonaceous.

I started this review in the conviction that dusty AGB stars produce all the dust but I learned that there is a black horse in the race: supernovae of type II. These supernovae will outclass the AGB stars when their efficiency in producing dust is higher than roughly 1 %. We do not know at all how high this efficiency is. I thank John Kerridge for the suggestion made during this school that a study of the properties of interstellar particles detected in meteorites may shed some light on this question because (see his contribution in this volume) the origin of these interstellar particles (AGB or SNII?) can be deduced.

## References


Blanco, V. 1965, in "Stars and Stellar Systems", volume V, "Galactic Structure", eds. A. Blaauw and M. Schmidt, University of Chicago Press, p. 241

Bressan, A., Fagotto, F., Bertelli, G., Chiosi, C. 1993, A&A Sup. 100, 647

Blöcker, T. 1995, A&A 297, 727

Dorschner, J., Henning, T., 1995, TAAR 6, 271

Duerbeck, H.W., Benetti, S. 1996, ApJL 468, L111

Gilman, R.C. 1969, ApJL 155, L185
Groenewegen, M.A.T., de Jong, T. 1994, A&A 283, 463
Groenewegen, M.A.T., Oudmaijer, R.D., Ludwig, H.-G., 1997, MNRAS 292, 686
Habing, H.J. 1996, The Astron. Astrophys. Rev. 7, 97
Habing, H.J., Tignon, J., Tielens, A.G.G.M. 1994, A&A 286, 523
Höfner, S., Fleischer, A.J., Gauger, A., Feuchtinger, M.U., Dorfi, E., Winters, J.M., Sedlmayr, E. 1996, A&A 314, 204
Jorisse, A., Frayer, D.T., Johnson, H.R., Mayor, M., Smith, V.V. 1993, A&A 271, 463
Justtanont, K., Tielens, A.G.G.M., 1992, ApJ 389, 400
Justtanont, K., Skinner, C.J., Tielens, A.G.G.M., 1994, ApJ 435, 852
Justtanont, K., Skinner, C.J., Tielens, A.G.G.M., Meixner, M., Baas, F. 1996, ApJ 456, 337
Netzer, N., Elitzur, M. 1993, apj 410, 701
Olofsson, H., Eriksson, K., Gustafsson, B., Carlström, U. 1993a, ApJS, 87, 267
Olofsson, H., Eriksson, K., Gustafsson, B., Carlström, U. 1993b, ApJS, 87, 305
Van Loon, J.Th., Zijlstra, A.A., Whitelock, P.A., Waters, L.F.B.M., Loup, C., Trams, N. 1998, A& A (in press)
Vassiliadis, E., Wood, P.R. 1993, ApJ 413, 641
Wagenhuber, J., Weiss, A. 1994, A&A 290, 807
Waters, L.B.F.M., Loup, C., Kester, D.J.M., Bontekoe, Tj. R., de Jong, T. 1994 A&A 281, L1
Zijlstra, A.A., Walsh, J.R. 1996, A&A 312, L21

# DUST FORMATION IN SUPERNOVAE

P.-O. LAGAGE
*CEA/DSM/DAPNIA/Service d'Astrophysique*
*Centre d'Etudes de Saclay,*
*F-91191 Gif-sur-Yvette Cedex*

## 1. Introduction

Dust grains are a key component of the Universe. They play a crucial role in various astrophysical processes, such as the formation of stars and planets, the catalysis of molecular hydrogen... Several important questions about dust are still under debate. One of them deals with the site of dust production.

The core of a grain is probably formed around stars and then ejected in the interstellar medium (ISM), where it can be covered with ice and organic mantles, (see M. Greenberg this Volume and references there-in). Evolved stars on the Asymptotic Giant Branch (AGB's) are generally considered as the main dust factories. Indeed, dust is kwown to form efficiently in the stellar atmosphere of these objects and to be ejected in the ISM under the radiation pressure from the central star, (see H. Habing this Volume and references there-in). However, given the amount of condensable elements ejected during a supernova (SN) explosion, supernovae have the *potentiality* of being the main dust factories, especially of silicate dust.

The idea that *part* of the dust was made from SN material, introduced 30 years ago (Cernuchi et al. 1967; Hoyle and Wickramasinghe 1970), is now strongly supported by the studies of meteorites. Indeed, SN are the only place where the nucleosynthesis can generate isotopic anomalies observed in meteorites, (see J. Kerridge this Volume and references there-in). The astrophysical evidences in favour of dust formation from SN ejecta are scarce. Several indirect evidences have been deduced from the observations of SN1987a, the supernova which exploded in the large magellanic cloud 10 years ago, (Lucy et al. 1989, Moseley et al. 1989, Suntzeff and Bouchet 1990, Bouchet et al. 1991, Wooden et al. 1993, Colgan et al. 1994). Recent infrared (IR) observations of Cassiopeia A, the youngest (300 yr old )



*J. M. Greenberg and A. Li (eds.), Formulation and Evolution of Solids in Space,* 325–329.

supernova remnant (SNR) known in our galaxy, have provided good hints that ejecta dust did also form in Cas A (Lagage et al. 1996).

This paper is devoted to a discussion of the IR searches for SN condensates in Cas A. In section 2, the reasons underlying searches for the dust in the IR radiation are discussed. Section 3 deals with the IR observations made with the IRAS satellite and Sect. 4 with the observations made with the ISO satellite. Conclusions and perspectives are drawn in Sect. 5.

## 2. Why searching for the dust in the infrared?

Dust is usually revealed thanks to interactions with UV or visible photons from an external source, generally a star. The photons are either scattered or absorbed by the dust. The absorbed photons heat the dust grains, which then radiate thermally. The thermal emission of a blackbody at a temperature between 30 and 300 K has its peak of emission in the IR, at a wavelength between 100 and 10 $\mu$m.

In the case of Cas A, the main source of heating of the dust is not photons, but ... electrons. Indeed a SNR is filled with a hot plasma, as revealed by the X-ray emission (Murray 1979, Fabian 1980). Collisions between the electrons of the hot plasma and dust are an efficient way to heat dust in SNR (Silk and Burke 1973). The equilibrium temperature of dust grains embedded in a hot plasma at a temperature of $10^7$ K and with a number density of 10 $cm^{-3}$ is in the range 50-110 K according to the particle size (Dwek 1986). Dust grains at such a temperature radiate mainly in the IR.

## 3. IRAS observation of Cas A

The first attempts to detect dust in Cas A, made either from the Kuiper Airborn Observatory (Wright et al. 1980), or from ground-based telescopes (Dinerstein et al. 1982), failed. It is only with IRAS, the first satellite devoted to IR observations (Neugebauer et al. 1984), that IR emission from Cas A was detected (Dwek et al. 1987, Braun 1987). The flux was observed in the four IRAS bands, (12, 25, 60 and 100 $\mu$m). A dust temperature around 90 K was deduced from the observations.

Two models were able to account for the IRAS flux (Dwek et al. 1987);

- in one model, the dust at the origin of the IR emission is just interstellar dust swept-up by the supernova blast wave;
- in an alternative model, the dust is made of supernova condensates heated by the reverse shock propagating through the ejecta.

The angular resolution of IRAS was not high enough to make possible a choice between the two models.

## 4. ISOCAM observation of Cas A

The ISOCAM camera (Cesarsky et al. 1996) on board on the Infrared Space Observatory (Kessler et al. 1996) was well suited for follow-up studies of the IR emission from Cas A. Indeed, the angular resolution ranges from 1.5 arcsec to 6 arcsec according to the configuration, to be compared to 30 arcsec, at best, for IRAS. A first set of observations has revealed that the IR emission at the angular resolution of ISOCAM was very patchy, (see Fig. 1 of Lagage et al. 1996).

Most of the emission observed in the IR is spatially well correlated with the so-called fast moving knots (FMK) observed in the visible radiation. These gaseous knots, discovered more than 50 years ago, are known to be made of SN nuclear burning products; when oxygen, sulfur and argon ionic lines are observed in abundance from these knots, no hydrogen, no helium lines are detected (Baade and Minkowsky 1954, van den Bergh 1971, Kirsher and Chevalier 1977, Chevalier and Kirsher 1978). The velocity of the knots is in the range of several thousands of km/s (Reed et al. 1995). The gaseous component of the knots was also observed with ISOCAM, via the forbidden ionic lines of NeII at 12.8 $\mu$m and SIV at 10.52 $\mu$m. (One knot had already been detected in the SIV line from ground-based observations (Dinerstein et al. 1987)). The good spatial correlation between the gaseous knots and the dusty knots detected in IR, led us to claim that the dust in the knots is made of ejecta material. Observing at the same time both the gas and the dust in the knots is essential to draw firm conclusions. It was not the case previously (Greidanus and Strom 1991).

The color temperature deduced from the IR emission, 220 K, seems high. How to heat a dust grain at such a temperature? The gaseous knots, which are cool and dense (electronic density $\sim 10^4$ cm$^{-3}$), are immersed in a hot plasma. Then they are expected to evaporate. The expected conditions in the evaporating flow (Dwek and Werner 1981) are such that dust grains of size smaller than 50 Å can be heated up to 220 K (Dwek 1986). The size of 50 Å is the size expected from the models of dust formation developped in the context of SN1987a (Kozasa et al. 1991). (However we cannot rule out a heating of the dust by photons, if the reverse shock propagating in the knots is rapidly radiative). A dust mass of $\sim 10^{-6}$-$10^{-7}$ solar mass is required to account for the IR emissivity from the brightest knot. Much more dust is probably inside the knots and cold. Interferometric measurements in the millimetric range are planned at IRAM to try to probe the presence of cold dust (proposal by J. Lequeux et al.).

Part of the dust, especially the dust located at the northest part of the remnant, is at a lower temperature, around 100K. This dust is not correlated with knots and is probably just swept-up interstellar dust.

## 5. Conclusions and perspectives

The first ISOCAM observations of SNR are very promissing. More ISOCAM observations of Cas A, especially in the spectro-imaging mode (spectral resolution of 40, thanks to a Circular Variable Filter), will tell us if the 10 $\mu$m feature caracteristic of silicate dust is there. The three other young galactic SN remnants, the Crab, Tycho and Kepler, will be observed. It is sure that ISO will bring an important contribution to the understanding of the role of supernovae in dust formation.

ISO will also probably help in assessing the role of SN remnant in dust destruction. According to current models, dust in the interstellar medium should be destroyed by the SN shock waves propagating in the ISM, (see Tielens this Volume and references there-in). Several ISO programmes are devoted to test this prediction. It is too early to report on the results. More at the next school!

## References


1. Baade, W.D., and Minkowski, R. (1954), *ApJ* **119**, 206.
2. Bouchet, P., Danziger, I.J., and Lucy, L.B. (1991) The bolometric light curve of SN 1987A: results from day 6161 to 1316 after outburst, *AJ* **102**, 1135-1146.
3. Braun, R. (1987) The structure and dynamics of young supernova remnants: new constraints from observations of shock-heated dust, *AA* **171**, 233-251.
4. Cesarsky, C.J., Albergel, A., Agnese, P. et al. (1996) ISOCAM in flight, *AA* **315**, L32-L37.
5. Cernuchi, F., Marsicano, F., and Codina, S. (1967), *Ann. d'Ap.* **30**, 1039.
6. Chevalier, R.A., and Kirshner, R.P. (1978) Spectra of Cassiopeia A. II Interpretation, *ApJ* **219**, 931-941.
7. Colgan, S.W.J., Haas, M.R., Erickson, E.F., et al. (1994) Day 640 infrared line and continuum measurements: dust formation in SN 1987a, *ApJ* **427**, 874-888.
8. Dinerstein, H.L., Werner, M.W., Capps, R.W., and Dwek, E. (1982) A search for hot dust in the fast moving knots in Cassiopeia A, *ApJ* **255**, 552-556.
9. Dinerstein, H.L., Lester, D.F., Rank, D.M., et al. (1987) Observations of infrared emission from a fast moving knot in Cassiopeia A, *ApJ* **312**, 314-320.
10. Dwek, E., and Werner, M.W. (1981) The infrared emission from supernova condensates, *ApJ* **248**, 138-151.
11. Dwek, E. (1986) Temperature fluctuations and infrared emission from dust particles in a hot gas, *ApJ* **302**, 363-370.
12. Dwek, E., Dinerstein, H.L., Gillett, F.C., et al. (1987) Physical processes and infrared emission from the cassiopeia A supernova, *ApJ* **315**, 571-579.
13. Fabian, A.C., Willingale, R., Pye, J.P., et al. (1980) The X-ray structure and mass of the Cassiopeia A supernova remnant, *MNRAS* **193**, 175-188.
14. Greidanus, H., and Strom ,R.G. (1991) 20 $\mu$m observations of Cassiopeia A, *AA* **249**, 521-527.
15. Hoyle, F., and Wickramasinghe, N.C. (1970) Dust in supernova explosion, *Nature* **226**, 62-63.
16. Kessler, M.F., Steinz, J.A., Anderegg, M.E., et al. (1996) The Infrared Space Observatory (ISO) mission, *AA* **315**, L27-L31.
17. Kirshner, R.P., and Chevalier, R.A. (1977) Spectra of Cassiopeia A. I. Observations, *ApJ* **218**, 142-147.

18. Kozasa, T., Hasegawa, H., and Nomoto, K. (1991) Formation of dust grains in the ejecta of SN 1987A. II, *AA* **249**, 474-482.
19. Lagage, P.O., Claret, A., Ballet, J., et al. (1996) Dust formation in the Cassiopeia A supernova, *AA* **315**, L273-L276.
20. Lucy, L.B., Danziger, I.J., Gouiffes, C., and Bouchet, P. (1989) Dust condensation in the ejecta of SN 1987a, ***Structure and Dynamics of Interstellar Medium***, IAU Col. **120**, ed. G. Tenorio-Tagle, M. Moles, & J. Melnick (Berlin: Spinger-Verlag), 164-178.
21. Murray, S.S., Fabbiano, G., Fabien, A.C., et al. (1979) High-resolution X-ray observations of Cassiopeia A supernova remnant with the Einstein observatory, *ApJ* **234**, L69-L72.
22. Moseley, S. H., Dwek, E., Glaccum, W., et al. (1989) Far-infrared observations of thermal dust emission from supernova 1987a, *Nature* **340**, 697-699.
23. Neugebauer, G., Habing, H.J., Van Duinen, R., et al. (1984) The Infrared Astronomical Satellite (IRAS) mission, *ApJ* **278**, L1-L6.
24. Reed, J.E., Hester, J.J., Fabian, A.C., and Winkler, P.F. (1995) The three-dimentional structure of the Cassiopeia A supernova remnant. I. The spherical shell, *ApJ* **440**, 706-721.
25. Silk, J., and Burke, J.R. (1974) Dust grains in a hot gas. II. Astrophysical applications, *ApJ* **190** 11-17.
26. Suntzeff, N.B., and Bouchet, P. (1990) The bolometric light curve of SN 1987a. Results from ESO and CTIO U and Q0 photometry, *AJ* **102**, 650-663.
27. van den Bergh, S. (1971) Optical studies of Cassiopeia A. III. Spectra of the supernova remnant, *ApJ* **219**, 931-941.
28. Wooden, D.H., Rank, D.M., Bregman, J.D., et al. (1993) Airborne Spectrophotometry of SN 1987a from 1.7 to 12.6 microns: time history of the dust continuum and line emission, *ApJS* **88**, 477-507.
29. Wright, E.L., Harper, D.A., Loewenstein, R.F., et al. (1980) Search for far-infrared emission from young supernova remnants, *ApJ* **240**, L157-L158.

# THE DESTRUCTION OF INTERSTELLAR DUST

A.G.G.M. TIELENS
*Nasa Ames Research Center*
*MS245-3, Moffett Field, CA 94035-1000, USA*

**Abstract.**

This paper reviews the detailed physics of the processes involved in the destruction of interstellar dust in interstellar shocks: sputtering by impacting gas atoms, and vaporization and shattering by grain–grain collisions. An analytical formalism is developed that describes these processes. The structure of interstellar shocks is discussed and the fate of grains in the cooling postshock gas is followed. Theoretical calculations show that high velocity shocks ($v > 50$ km/s) destroy interstellar dust efficiently. A simple stick model for evolution of interstellar dust in the interstellar medium (ISM), which includes dust destruction by shocks in the Warm ISM, accretion in the cloud phases, stardust injection, and mixing between the interstellar cloud and intercloud phases is described. It is concluded that fast shocks occur so frequently that the lifetime of refractory grains such as graphite and silicates is only $5 \times 10^8$ yr, much shorter than the stardust injection timescale. Consequently, growth of interstellar dust through accretion in the cloud phases of the ISM has to be rapid as well. This model is also used to derive rather directly dust lifetimes against destruction and accretion timescales from observations of depletions in the cloud and intercloud phases of the ISM. These are compared to the theoretically calculated lifetimes and the implications are discussed.

## 1. Introduction

The evolution of interstellar dust is determined by a number of complex interplaying processes. Dust is injected into the ISM by stars in various phases of their evolution. Infrared observations of giants, supergiants, novae, WC stars, and supernovae attests to the importance of nucleation and growth



*J. M. Greenberg and A. Li (eds.), Formulation and Evolution of Solids in Space,* 331–375.

of stardust in their outflows. In the ISM, dust can grow in size through accretion and through coagulation and both processes are thought to be important in dense clouds. Dust is mainly destroyed in the ISM by strong shock waves in the warm phases of the interstellar medium. Sputtering by impacting gas species and vaporization by high velocity grain–grain collisions return grain material to the gas phase. At the same time, shattering by grain–grain collisions considerably modify the grain size distribution.

This review concentrates on the latter aspect of the lifecycle of interstellar dust. It has long been known (Routly and Spitzer 1952; Siluk and Silk 1974; Barlow and Silk 1977; Cowie 1978; Jenkins et al. 1976; 1981; 1984) that gas at high velocities, presumably recently accelerated by strong shock waves, contains a larger complement of condensible species such as Ca, Fe and Si which are normally heavily depleted (ie., locked up in dust). This is generally attributed to the destruction of grains in the hot postshock gas and, over the years, detailed theoretical models for the destruction of dust in such hostile environments have been constructed with increasing degree of sophistication (Barlow 1978; Draine and Salpeter 1979; Dwek and Scalo 1980; Seab and Shull 1983; McKee et al. 1987; Jones et al. 1994; 1996). Here, I will discuss the detailed physics of the various processes involved in the destruction of interstellar dust in shock waves. This includes sputtering (§2) by impacting gas atoms either because of the inertial motion of the grains through the gas or through the thermal motion of the hot gas. Shattering and vaporization by the strong shock waves driven into grains by grain–grain collisions are described in §3. Calculations of the destruction by interstellar shock waves are reviewed in §4. These results are then used in §5 to discuss the ecology of the ISM. This section also includes a discussion of the relative importance of the various stardust sources. Theoretical and observational estimates of interstellar dust lifetimes are discussed in §6. Finally, §7 discusses estimates of dust lifetimes derived rather directly from observations of depletions in the ISM.

## 2. Sputtering

### 2.1. A "UNIVERSAL" SPUTTERING RELATION

Three different regimes can be recognized in sputtering of materials (cf., Figure 1; Sigmund 1981). At low energies, single-knock-on events dominate and, for light ions, sputtering is controlled by incoming ions reflected of deeper atoms with negligible energy loss and knocking out surface atoms. At higher energies, recoil atoms are energetic enough to generate secondary and higher generation recoils and these can be effective in knocking out surface atoms. Finally, for heavy projectiles, the density of recoil atoms is so high that the majority of the atoms are in motion. The boundaries

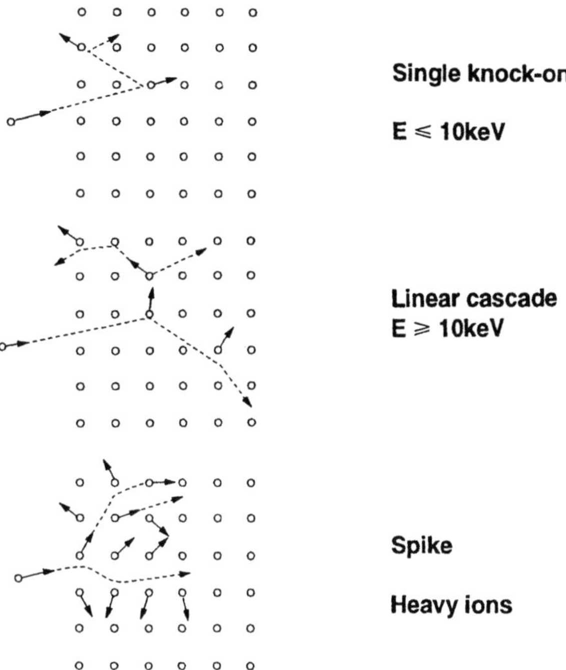


*Figure 1.* Sputtering can be divided into three regimes: 1) The single-knockon regime where the reflected projectile knocks out a near-surface atom. 2) The linear cascade regime where recoiling target atoms receive sufficient energy to generate recoil cascades. 3) The spike regime where the majority of atoms near the surface are set in motion (Adapted from Sigmund 1981).

between these different regimes are not very sharp and will depend on the energy and the nuclear charges (Andersen and Bay 1981). For sputtering in astrophysical environments, the first regime is most important.

Experimental and theoretical studies have revealed the existence of a universal sputtering law (cf., Sigmund 1981; Andersen and Bay 1981; Bohdansky 1984; Yamamura et al. 1983). Through the first and second regime, sputtering can be treated by a linear cascade process, which views the event as a series of binary collisions between atoms in the solid (Sigmund 1981). As a result of these collisions, some atoms are set in motion towards the surface with enough energy to overcome the surface binding forces. The sputtering yield can then be derived by solving the transport equation with various degrees of sophistication. Without going into the details, this results in

$$Y(E) = \frac{4.2 \times 10^{14}}{U_o} \alpha S_n(E) \quad \text{atoms/ion} \quad , \tag{1}$$

with $U_o$ the surface binding energy in eV, $\alpha$ an energy independent function of the mass ratio of target to projectile ($=0.3(M_2/M_1)^{2/3}$), and $S_n(E)$ the nuclear stopping cross section (in units of erg/cm$^2$) and the numerical factor in this equation depends on the surface interaction potential and the geometry assumed. Despite the rather complicated approximations, the physics involved in this result is simple to understand. The density of atoms set in motion scales with the deposited energy which is proportional to $\alpha S_n(E)$. The density of recoil atoms with energy $E_t$ scales with $1/E_t^2$. This reflects the energy dependence of the cross section on the transfered energy, $E_t$ (Rutherford scattering); ie., Coulomb repulsion strongly prefers collisions with low energy transfers. Hence, the density of recoil atoms with energy exceeding a threshold, $U_o$, is proportional to $1/U_o$.

The nuclear stopping cross section can be described by,

$$S_n(E) = 4\pi a Z_1 Z_2 e^2 \frac{M_1}{M_1 + M_2} s_n(\epsilon) \quad , \tag{2}$$

With $Z_i$ and $M_i$ the nuclear charges and masses of the atoms involved and $a$ the screening length ($\equiv 0.468/\left(Z_1^{2/3} + Z_2^{2/3}\right)^{1/2}$ Å)[1]. Here, $s_n$ is a "universal" function which describes the screened Coulomb interaction as a function of the dimensionless energy, $\epsilon$,

$$s_n(\epsilon) = \frac{3.441\sqrt{\epsilon}\ln(\epsilon + 2.718)}{1 + 6.35\sqrt{\epsilon} + \epsilon(-1.708 + 6.882\sqrt{\epsilon})} \quad , \tag{3}$$

and

$$\epsilon = \frac{M_2}{M_1 + M_2} \frac{a}{Z_1 Z_2 e^2} E \quad . \tag{4}$$

Hence, these set of equations show that, by appropriate scaling of target and projectile parameters, a universal relationship exists which relates the sputtering yield to the ion energy. Experimental data shows reasonable agreement with these theoretical expectations, particular at moderate energies (10 keV to MeV range; cf., Andersen and Bay 1981). At high energies and for heavy ions and or targets, discrepancies occur because of non-linear effects in the collision cascade. This range is of little interest to astrophysical sputtering and we will not further discuss it.

For astrophysics, low projectile masses and low ion energies are of particular importance and, unfortunately, this "simple" theory and experiment

[1]Note the small error in this equation (eqn. 4.2) in Tielens et al. (1994)

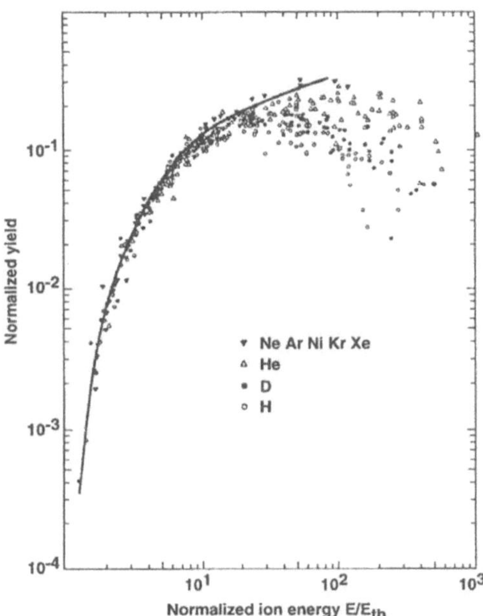


*Figure 2.* Energy dependence of the sputtering yield in the threshold regime. Comparison of experimental data and theoretical relation (taken from Bohdansky 1984).

do not agree well (theory overestimates by up to a factor of ten for proton sputtering; cf., Andersen and Bay 1981). Near threshold, two modifications are required (Bohdansky 1984; Yamamura et al. 1983). First, a large collision cascade does not develop and a considerable number of the recoiling atoms receive insufficient energy to overcome the surface potential. This effect can be corrected for by replacing the deposited energy by an effective deposited energy, $S_{n,1}$, given by

$$S_{n,1} = S_n \left(1 - \left(\frac{E_{th}}{E}\right)^{2/3}\right) \quad , \tag{5}$$

where $E_{th}$ is a threshold energy (see below). Second, the assumption of an isotropic momentum distribution also breaks down at these low energies and this leads to a similar correction factor; viz., replace the deposited energy by an effective deposited energy, $S_{n,2}$, given by

$$S_{n,2} = S_{n,1} \left(1 - \frac{E_{th}}{E}\right)^2 \quad . \tag{6}$$

For small mass ratios ($M_1/M_2 < 0.3$), sputtering near threshold is caused by reflected projectiles on their way out and this leads to a threshold energy given by

$$E_{th} = \frac{U_o}{\gamma(\gamma - 1)} \quad , \tag{7}$$

with $\gamma$ the maximum fractional energy transfer possible ($4M_1M_2/(M_1 + M_2)^2$). For large mass ratios ($M_1/M_2 > 0.3$), the threshold energy is given by

$$E_{th} = 8U_o \left(M_1/M_2\right)^{1/3} \quad . \tag{8}$$

In the threshold regime, the energy dependence of the yield is dominated by the $S_{n,1}$ and $S_{n,2}$ factors. These factors provide a very good fit to a wide variety of experimental data down to energies as low as 1.5 times threshold and yields as low as $10^{-4}$ (Fig. 2; Bohdanski 1984).

Finally, for small mass ratios, the expression for $\alpha$ yields unreasonably large values. Partly, this reflects the importance of electronic stopping which is of more importance for light projectiles. Partly, this reflects the assumption of an infinite target which allows the projectile to be reflected many times across the (imaginary) target surface. These two factors can be corrected for by multiplying the derived yield function by the ratio of the mean projected range to the average pathlength which is given by

$$\frac{R_p}{R} = \left(K\frac{M_2}{M_1} + 1\right) \quad , \tag{9}$$

where $K$ depends on the details of the nuclear interaction potential, but for practical purposes, we can adopt a target–dependent constant.

These sets of equations describe a "universal" relationship for sputtering by high energy ion–impact. The sputtering yield will depend on the material and ion characteristics, $M_1$, $M_2$, $Z_1$, and $Z_2$. Of course, it will also depend on the energy of the ion, $E$. For each target, there are two free parameters: the surface binding energy, $U_o$, and the fitting parameter, $K$ in the formula for $R_p/R$ (eqn. 9). The cohesive binding energy, $U_b$, is the energy per atom necessary to disperse a solid into its constituent atoms; ie., this corresponds to breaking all bonds between the atom pairs. Thus, bond–breaking corresponds to $2U_b$ per bond. Ignoring relaxation, an atom at the surface has half the number of normal bonds and hence we expect $U_o$ to be equal to $U_b$. This value has often been used in sputtering studies and has been shown to give good agreement with experimental data (Andersen and Bay 1981). Here, we will folow that lead with the proviso that sputtering under astrophysical conditions corresponds to relatively high doses (but at low fluence). Particularly for graphite, such high doses are known to lead to structural modification of the surface; ie., the transformation of graphite to amorphous carbon, or hydrogenated amorphous carbon if H is the ion (cf., Tielens et al. 1994 and references therein). In this exceptional case, this structural modification leads to a substantial reduction in the (surface) binding energy from 7.7 eV for graphite to 4eV for amorphous carbon. Carbon has a variety of valence structures which lead to distinct solids with different binding energies. This reduction in $U_o$ parallels the reduction in binding energy between $sp^2$ bonded carbon (aromatics) and $sp^3$ bonded carbon (aliphatics) (cf., Tielens et al. 1994). Although other materials may amorphitize under ion–bombardment, no such large reduction in binding energy is expected (or observed). The other free parameter, $K$, has a typical value of 0.4 if only elastic collisions are considered. If inelastic losses are important, this parameter reduces to $\simeq 0.1$, but other factors may influence this parameter as well (Sigmund 1981). Here we will treat it as a free parameter.

## 2.2. ASTROPHYSICAL SPUTTERING YIELDS

These formulae have been compared to a large variety of target–projectile combinations and good agreement is achieved (Andersen and Bay 1981; Bohdansky 1984; Yamamura et al. 1983), as exemplified by the comparison for sputtering near threshold shown in Figure 2. Such a good correspondence is not unexpected for what in many ways is a semi–empirical method (albeit heavily guided by theoretical studies). Here, I will concentrate further on sputtering of astrophysically relevant materials. The theoretical results are compared to available laboratory data for graphite, SiC, Fe, $SiO_2$, and $H_2O$–ice in Figure 3. These fits and their parameters are discussed in extensio in Tielens et al. (1994) and that will not be repeated here. Here, I note the overall good comparison with the available data. However, there is a general lack of data near threshold for many relevant materials and the success of the universal sputtering relation has to be judged from the general good agreement with sputtering data in this energy range displayed in Figure 2. For silicates, the situation is even worse since the little data available applies to quartz rather than Mg and/or Fe silicates. For $H_2O$ ice, the sharp rise in the sputtering yield above 1 keV results from electronic stopping losses. The energy deposition by this process has not been taken into account in the theoretical formulae summarized above and the large discrepancies are no suprise, therefore. For astrophysical purposes, where $H_2O$ ice sputtering is dominated by lower energies, this discrepancy is of no consequence, but it is of much larger importance for planetary system studies. Overall, I would summarize the state of the field by extoiling the good agreement and conclude that, if the free parameters, $K$ and $U_o$, can be constrained, sputtering yields are uncertain by less than 30% over most of the energy range.

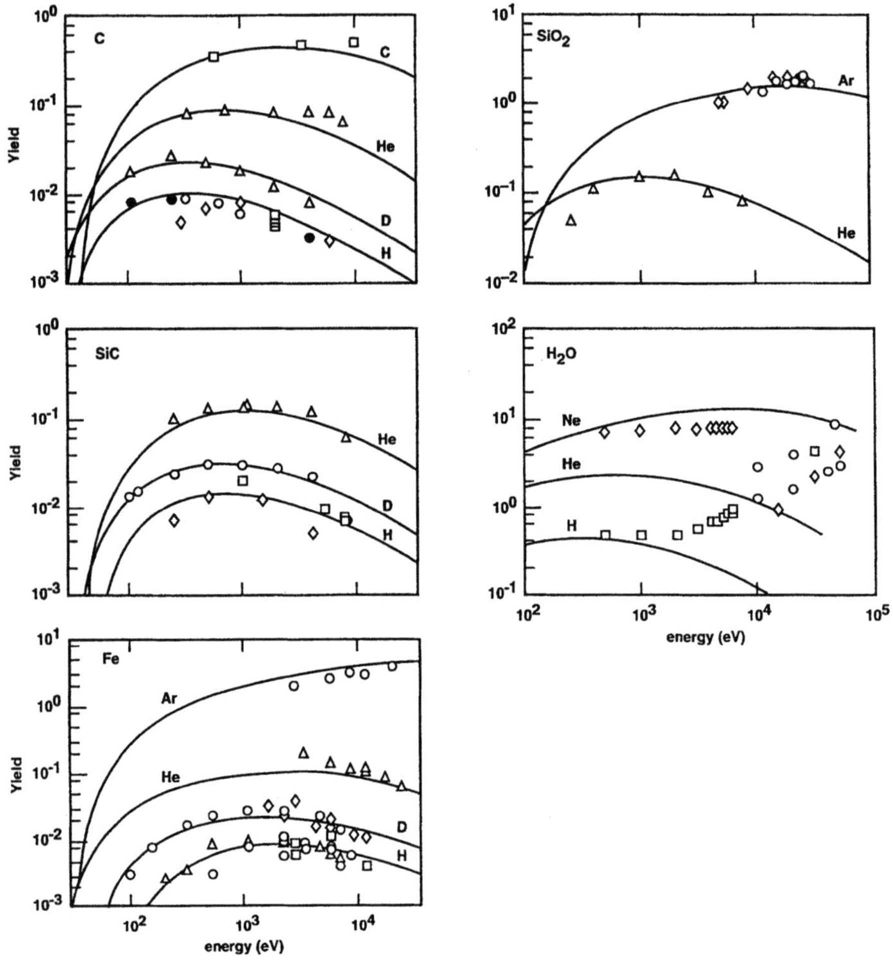


*Figure 3.* Comparison of experimental sputtering yields with the universal sputtering relation for astrophysically relevant materials. Details and references to the experimental data are given in Tielens et al. (1994).

## 2.3. OTHER SPUTTERING YIELD FORMALISMS

It is of some interest to compare these sputtering yields with previously used yields. This has recently been done by Draine (1995). Sputtering of interstellar dust has been studied by Barlow (1978), Draine and Salpeter (1979), Seab and Shull (1983), and Tielens et al. (1994). The formalism described above was taken from Tielens et al. (1994) and this is presently the better description, because of its "universal" character, because of its validation by a large body of experimental data on a large variety of projectile–target combinations, because of its close guidance by theory and, last, because of its coverage of the full energy range in one handy-dandy set of formulae. Having stated this (perhaps biased) view, I should hasten to add that

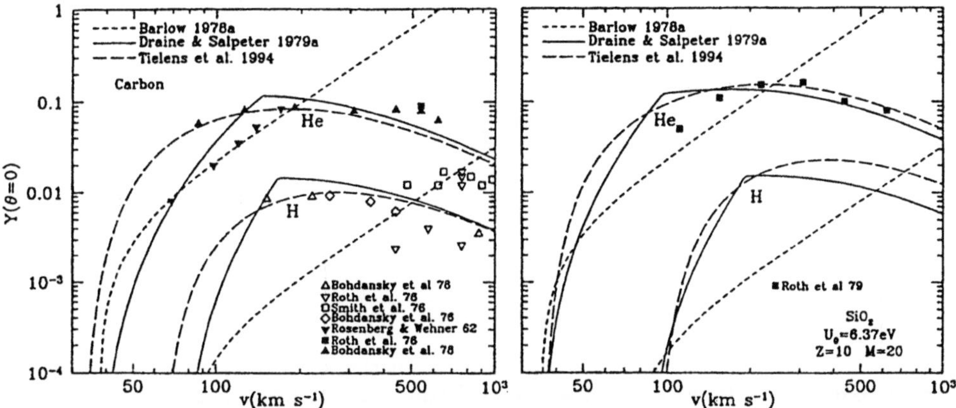


*Figure 4.* Comparison of various sputtering yield formalisms used in astrophysics (taken from Draine 1995). Left, Carbon. Right, silicates. The symbols indicate experimental data. The inverted solid triangles are based upon plasma discharge techniques and therefore suspect.

the sputtering yields derived by Draine and Salpeter (1979) are in very good agreement with these results (see below). The Barlow results for the yield, which were guided by somewhat older data using less reliable, plasma discharge techniques, can differ by an order of magnitude from the above formulation. These should not be used anymore. Sputtering yields using these different formulations are compared in Figure 4 (Draine 1995). For $SiO_2$ and $MgFeSiO_4$, the points mentioned above are obvious. For carbon, the difference between the Draine and Salpeter (1979) and Tielens et al. (1994) results are quite large near threshold. This solely results from a difference in the adopted surface binding energy. Draine and Salpeter adopted 7.7 eV, the binding energy of graphite. Tielens et al. (1994) recognized that the ion bombardment will lead to a reduction in the surface binding energy (see above) and adopted the value of 4 eV, appropriate for amorphous carbon materials (see also Strazulla et al. 1985).

Recently, pure theoretical sputtering yields calculated using the TRIM code (Eckstein 1991) have been published for silicates and graphite (Flower et al. 1996). TRIM is a numerical code which follows the pathway of ions as they penetrate and collide in a solid. It uses an approximate but accurate treatment of the individual collisions; ie., in the binary collision approximation and considering one collision at a time. The collision cascade that is set up by the penetrating ion is followed using a Monte Carlo simulation technique and that way the sputtering yield is calculated. These types of numerical studies have reached a high degree of sophistication where besides the sputtering yield also questions such as implantation, mixing,

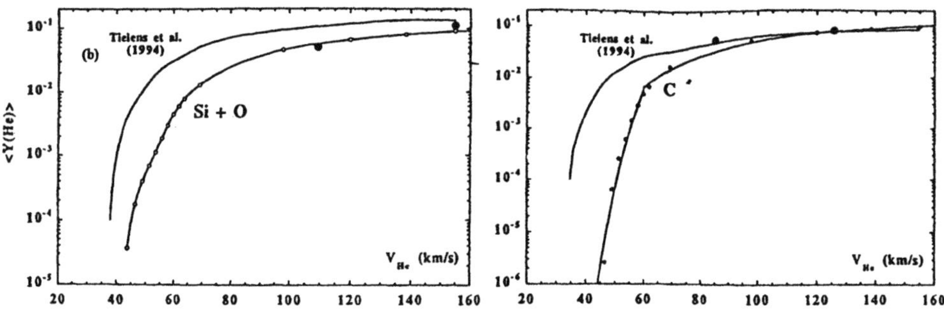


*Figure 5.* Sputtering yields for He impacts as a function of velocity computed using the TRIM code (taken from Flower et al. 1996). Right, carbon. Left silicates.

damage, and recoil accompagning ion penetration can be addressed. Impressive fits to available experimental data on sputtering of, for example, Ni have been published over a wide range of energies and penetrating ions (cf., Eckstein 1991). However, other materials are not as well fit. Generally, the surface binding energy is treated as a free parameter in these studies which is adjusted until a good agreement is achieved. Figure 5 compares the calculated sputtering yields for He bombardment of $SiO_2$ and carbon with those derived from the "universal" relation (Flower et al. 1996). The calculations adopted a binding energy of 6.4 and 7.4 eV for quartz and graphite, respectively. The latter is much higher than adopted by Tielens et al (1994) and that is the largest source of difference between the graphite curves. As for the semi–empirical formalism (cf., §2.1), TRIM calculations cannot produce a good fit to the multitude of experimental data for graphite sputtering with an adopted binding energy of 7.4 eV (Eckstein et al. 1987) and the agreement with the data for He sputtering in Figure 5 is somewhat fortuitous. The fit improves considerably when the binding energy is decreased to about 4 eV. While similar binding energy were adopted, the TRIM and "universal" $SiO_2$ sputtering yields differ considerably as well, particularly near threshold. The TRIM code is somewhat suspect for energies near threshold ($<$ 0.1 keV or 70 km/s) where electronic stopping is important and many body potentials have to be used. Also, collisions become frequent and the mean free path becomes smaller than the internuclear distance. These same points lead to the introduction of semi–empirical correction factors near threshold for the "universal" relation discussed above (cf., eqn 5 and 6) and so it is not unexpected that the TRIM code has difficulties in this same regime [although the direction of the discrepancy is somewhat puzzling]. While these semi–empirical correction factors have been "fitted" over a wide range of materials and ion combinations and have been tested to energies as low as 1.5 times the threshold energy (Fig. 2),

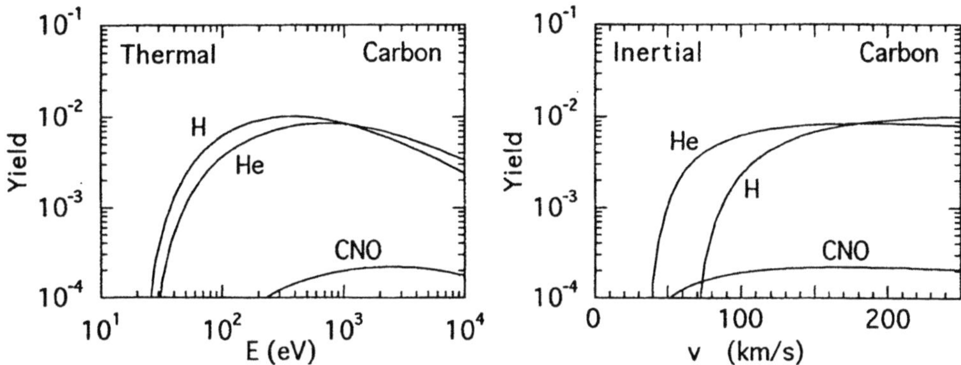


*Figure 6.* Inertial (right) and thermal (left) sputtering yields for for carbon dust for different impacting ions. These yields are displayed per H–atom and hence include a correction for the respective cosmic abundances.

how well they can be extrapolated to energies not explored, or target/ion combinations not measured is of course always the question. Nevertheless, at this point, the TRIM curves should be treated with some caution near threshold.

### 2.4. SPUTTERING IN ASTROPHYSICAL SHOCKS

These sputtering yields have to be convolved with gas phase abundances to derive relative sputtering yields. In astrophysical shocks, sputtering can be relevant in two distinct situations (cf., §4). First, for low shock velocities ($<$ 200 km/s; radiative shocks), grains are moving with respect to the gas due to their own inertia until they are dragged to a halt. Because of the mass difference, an impacting He ion will have a higher energy than an H atom and hence will have a higher yield. This difference in yield more than compensates for the difference in abundance and He sputtering dominates at low velocities ($<$ 100 km/s; Fig. 6). At higher velocities (ie., well above threshold), H sputtering becomes as important as He. Second, at high shock velocities, grains are quickly dragged to a halt while the gas stays hot (ie., adiabatic shocks). In sputtering by thermal gas, impacting ions will have the same energy distribution and, then, H and He sputtering are quite comparable over the full range. Sputtering by C,N, or O atoms is never very important in astrophysics.

## 3. Grain–Grain Collisions

When a projectile impacts a target at high velocity, shock waves will propagate from the impact area into the target and the projectile. When the shock wave reaches the back of the projectile and/or the "free" surfaces of the target, rarefaction waves will travel into the shocked regions relieving the high pressures. At the end of this (early–time) phase, the coupling of energy and momentum to the target is essentially complete. In the second phase, the excavation phase, the shock has departed from the immediate vicinity of the projectile and the transient crater. There are now two separate flows in the target: 1) The detached shock penetrating deeper and deeper into the target while compressing the material and decaying in strength. 2) The excavation flow which will eventually carve out the final crater. Figure 7 shows a schematic of the impact of a projectile on a slab. In this section, I will first discuss the shock propagation into the target, followed by a discussion of volume vaporized by the shock. Finally, I will discuss the process of shattering occuring during the excavation flow.

### 3.1. SHOCK PROPAGATION IN SOLIDS

When two solid particles collide at high velocity, strong shock waves are driven into each particle, compressing them to high pressures. Upon release, the shocked material may pass from the solid phase into the vapor phase or liquid phase, or it may shatter. The amount of vapor, melt or shattered material depend on the pressure generated by the shock. Now, define $\mathcal{M}_1$ as the Mach number of the shock which corresponds to a normalized pressure $\phi_1 \equiv P_1/\rho_o c_o^2$ with $\rho_o$ and $c_o$ the preshock pressure and sound speed, respectively. Guided by analytical, numerical and experimental studies, and making some simple approximations, the mass of target material, $M$, shocked to a pressure exceeding $\phi_1$ can then be written as (Tielens et al. 1994),

$$\frac{M}{M_P} = \frac{1+2\mathcal{R}}{2(1+\mathcal{R})^{16/9}} \frac{1}{\sigma_{1i}^{1/9}\sigma_1^{8/9}} \left(\frac{\mathcal{M}_r^2}{\mathcal{M}_1^2}\right)^{8/9} , \tag{10}$$

where $M_P$ is the projectile mass, $v_r$ (or Mach number $\mathcal{M}_r = v_r/c_o$) is the impact velocity, the $\sigma$'s are constants of order unity (Ostriker and McKee 1988), and $\mathcal{R}$ is a parameter related to the equation of state. Experimentally, it has been found that, for a large variety of materials and a wide range of shock pressures, there is a simple expression for the equation of state, generally expressed as a simple linear relationship between the shock velocity, $v_s$, and the velocity of the shocked matter, $v_1$ (McQueen et al 1970):

$$v_s = c_o + s v_1 \tag{11}$$

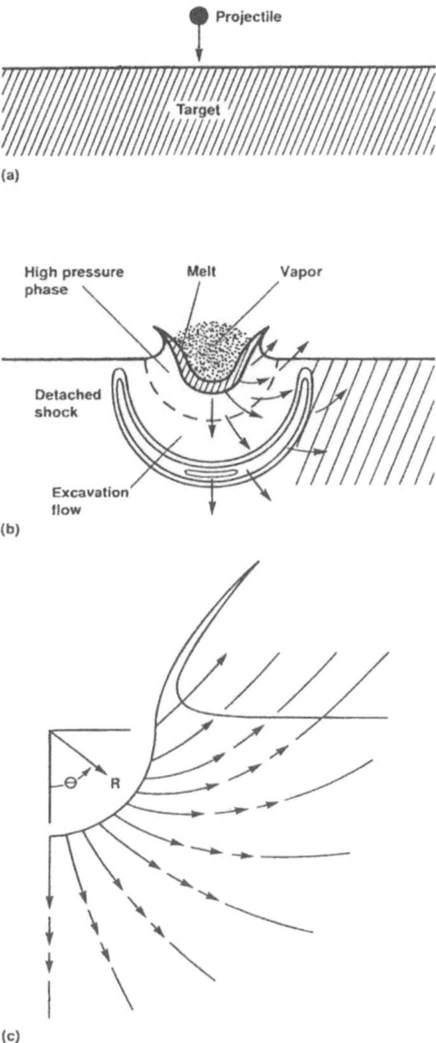


*Figure 7.* A schematic of the impact of a projectile on a semi-infinite slab (a). The shock wave driven into target and projectile will lead to vaporization and melting (b). It will also set up an excavation flow which carves out the crater. Panel (c) shows a schematic of the stream lines in the z-model for the excavation flow (see text).

The constant $c_o$ is approximately equal to the sound speed[2]. Experimental values of $c_o$ and $s$ for several astrophysically relevant materials are discussed in Tielens et al. (1994). Using this equation and the jump condition $v_1/v_s =$

[2]Unlike for an ideal gas, this implies that the ratio of the specific heats, $\gamma$, is not constant but depends on the shock velocity. Only at high Mach numbers, does $\gamma$ approach a constant ($2s$).

$P_1/\rho_o v_s^2$, we can write $\mathcal{M}_1$ in terms of $\phi_1$

$$\mathcal{M}_1 = \frac{2\phi_1}{1+(1+4s\phi_1)^{1/2}} \tag{12}$$

where $\mathcal{M}_1 \equiv v_1/c_o$ is the Mach number of the shocked material. The parameter $\sigma$ can be adequately approximated by

$$\sigma \simeq \frac{0.3(2+\mathcal{M}^{-1}-0.11)^{1.3}}{s+\mathcal{M}^{-1}-1} \quad , \tag{13}$$

with $\mathcal{M}$ the appropriate Mach number. The parameter $\mathcal{R}$ is defined as

$$\mathcal{R} \equiv \left(\frac{s\rho_o}{s_P\rho_{oP}}\right)^{1/2} \quad , \tag{14}$$

where the subscript $P$ refers to the projectile material parameters. For like–materials, $\mathcal{R}$ is equal to 1 while for very dense materials impacting low density materials, $\mathcal{R} << 1$.

Since $\sigma$ is nearly constant, we find that the pressure of the blast wave in the solid scales with $M^{-9/8}$, or slightly steeper than $V^{-1}$. Essentially, the hot shocked material is exposed to vacuum beyond the impact surface and flows outward draining energy and imparting momentum. Thus, while the flow is still self–similar, the exponent, $\eta$, describing its time evolution, $R \sim t^\eta$, lies between the values for an energy conserving (2/5) and a momentum conserving (1/4) blast wave.

The results for this simple analytical model can be compared to detailed numerical as well as experimental studies. The total damage resulting from an impact depends on the total energy coupled to the target as well as the decay of pressure with distance from the impact site. The energy coupled to the target is given by

$$E_i = \frac{1}{2} M_P v_r^2 \frac{1+2\mathcal{R}}{(1+\mathcal{R})^2} \quad , \tag{15}$$

independent of impact velocity. Thus for like–materials, the energy coupled to the blast wave in the target is equal to $E_i = 3/4E_o$. The rest is carried away by the projectile in the form of thermal energy. For a relatively dense material all the energy is transmitted to the target, whereas for a low density material only a small fraction is. Comparison of this analytical estimate with numerical calculations show reasonable agreement (Tielens et al. 1994). For self–similar blast waves with energy loss, the energy scales with radius, $E \sim R^{k_E}$. The decrease in energy and increase in momentum, $p$, leaves the quantity $Mv_s^{3\alpha} \sim E^{3\alpha-1}p^{2-3\alpha}$ conserved, where $\alpha = 2/(3+$

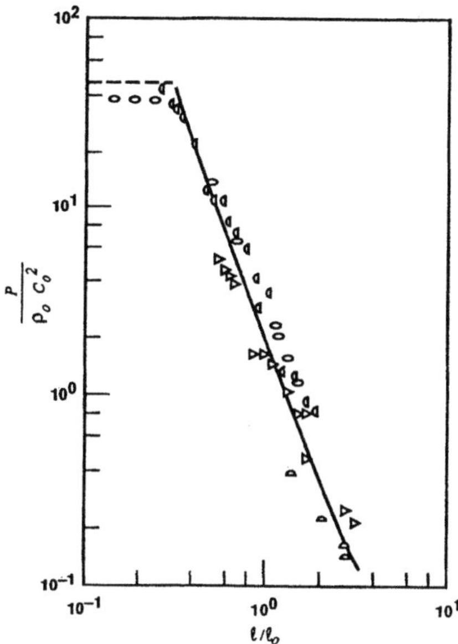


*Figure 8.* Peak pressure on the symmetry axis in units of $\rho_o c_o^2$ as a function of distance from the impact point, $l$, in units of the projectile size, $l_o$. Symbols correspond to numerical results. The solid line is the anaytical relation described in the text.

$k_E$). The value of $\alpha$ depends (very weakly) on $\gamma$ and is constrained to lie between $\frac{1}{3}$ and $\frac{2}{3}$. Extensive numerical calculations show that $\alpha$ is equal to 0.58–0.60 (Dienes and Walsh (1970) and our adopted value for $k_E$ (3/8) is consistent with this. This is illustrated in Figure 8 which compares the simple analytical results with detailed numerical calculations for the decay of the shock strength with distance in the target. The good agreement is obvious. This is really the core of the analytical model for the shock wave damage of solids. It is centered on the concept of late–stage–equivalence, demonstrated by a multitude of numerical calculations: ie., Two collisions with the same $M_P v_r^{3\alpha}$ will produce identical craters (Dienes and Walsh 1970).

### 3.2. VAPORIZATION

The high pressures produced upon shock compression can lead to a multitude of phase transitions, including vaporization, melting, and formation of various high pressure phases. Figure 9 shows a schematic phase diagram. A low–pressure and high–pressure solid, a liquid, and vapor phases as well as their mixed phases are indicated. A shock will compress the material along the Hugoniot to a maximum pressure corresponding to the strength of the shock. Upon release by the rarefaction wave, the material will expand and

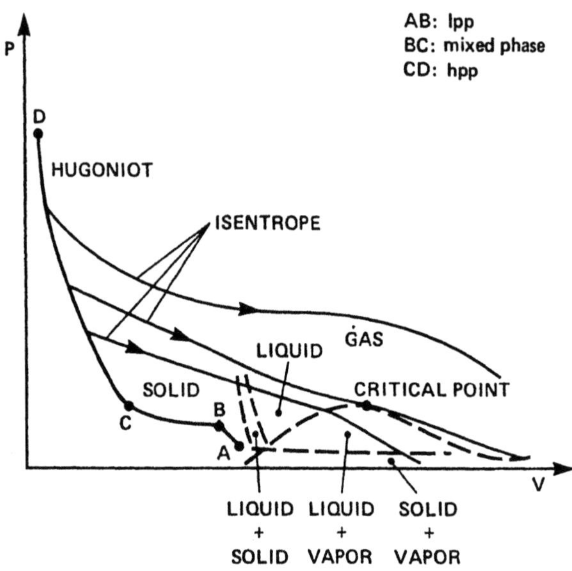


*Figure 9.* Schematic $P$, $V$ diagram of a realistic material. With increasing strength of the shock wave, the material is shocked to higher and higher pressure along the Hugoniot (A-D). Isentropic expansion behind the shock wave can take the material through various (mixed) phases, including solid, liquid, and gas, as well as low pressure and high pressure solid phases (lpp and hpp). Complete vaporization occurs on isentropes that pass through or above the critical point.

the density and temperature will decrease. Several release isentropes are indicated schematically in Figure 9. During this expansion, the material will pass through several different phase regimes. Whether a transformation will actually occur will depend on kinetic rather than thermodynamic considerations (Zel'dovitch and Raizer 1966).

There is a paucity of data on release isentropes and the minimum pressure required for complete vaporization has to be determined from theoretical arguments. Above the critical point, liquid and gas are indistinguishable and hence any isentrope passing through or above this point will lead to complete vaporization. Conversely, any lower isentrope will lead to only partial vaporization. Because entropy is conserved along an isentrope, the shock pressure required for complete vaporization can be calculated from the known entropy of the critical point. Calculations show that complete vaporization typically corresponds to an internal energy behind the shock wave of about 5 times the binding energy of the material (Zel'dovitch and Raizer 1966; Ahrens and O'Keefe 1971). Partial vaporization may start at much lower shock strength; ie., when the internal energy is equal to the binding energy. However, appreciable vaporization ($\simeq 25\%$ ) requires about

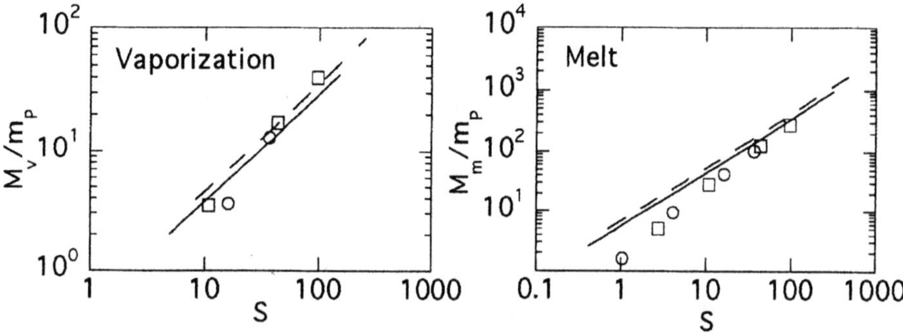


*Figure 10.* The vaporized (left) and molten (right) target mass in units of the projectile mass as a function of the shock strength parameter, $S$. Solid and dashed lines are analytical results for anorthosite and iron projectiles on an anorthosite target. Circles and squares are numerical results for anorthosite and iron projectiles on an anorthosite target (O'Keefe and Ahrens 1977).

twice the binding energy. Here is should be noted that the shock energy is partly converted into elastic energy (ie., compression against the repulsive forces of neighboring atoms) and partly into thermal energy. The former will be transformed into kinetic energy of the material upon expansion. The latter is also partly expended in expansion and partly in vaporization and, for very high shock velocities, into further ionizations etc. In other words, only the part of the shock energy remaining in the form of internal energy after the expansion to preshock densities can be used to vaporize the target.

Using a detailed numerical hydrodynamic code, O'Keefe and Ahrens (1977) have calculated the vaporized mass produced by the impact at various velocities of silicate and iron projectiles on a semi-infinite silicate slab (ie., gabbroic anorthosite, a well studied lunar highland silicate). These calculations employed an accurate equation of state, including the effects of partial and complete vaporization. The results of these calculations are compared to the semi-analytical formalism in Figure 10. Following O'Keefe and Ahrens (1977), the data are plotted as a function of the parameter $S$

$$S = \frac{\rho_0}{\rho_{0P}} \left( \frac{v}{7.44\,\mathrm{km/s}} \right)^2 \quad . \tag{16}$$

A reasonable fit to the numerical data is obtained for a threshold energy density of twice the binding energy of this silicate (i.e., $\epsilon_v = 2\epsilon_b = 3.4 \times 10^{11}$ erg/g). The corresponding threshold pressure is about 3.5 Mbar, the average of the incipient ($\simeq$1Mbar) and complete ($\simeq$6 Mbar) vaporization thresholds (O'Keefe and Ahrens 1977). Both the threshold impact velocity and the absolute amount of vaporized target mass are well predicted by the analytic theory. In essence, both depend directly on the assumed value of the threshold pressure. We can also compare the predictions for the analytic theory with numerical calculations for the melt mass generated by an impact (O'Keefe and Ahrens 1977). In this case, the threshold pressure for incipient and complete melting, 430 and 520 kbar respectively, are very similar and this free parameter is eliminated. Assuming a threshold pressure of 520 kbar, a good fit to the data is obtained at high impact pressures, but near threshold the analytical theory overestimates the molten mass considerably (Figure 10). Intuitively, it is expected that at threshold a target mass equal to the projectile mass is shocked to the threshold pressure and the numerical results seem to bear this out. The reason for this is that the pressure is relatively uniform during the first stage of the collision. However, the analytic theory was derived for the late stage, in which the blast wave is decelerating and the energy is concentrated near the surface; thus overestimating the effects of the shock. For grain–grain collisions in interstellar shocks, we are mainly interested in high velocity impacts of small projectiles on large targets and this mismatch near threshold is of little concern as long as the threshold pressure has been chosen appropriately.

### 3.3. SHATTERING

In the relevant pressure regime ($10^{11} - 10^{13}$ dyn/cm$^2$), the weak shock wave will split into two separate waves (Zel'dovitch and Raizer 1966). First, an elastic wave will propagate into the target, shocking it to the Hugoniot elastic limit ($\simeq 10^{10}$ dyn/cm$^2$). This shock wave is followed by a plastic wave which brings the material into the plastic state, thereby facilitating shearing motions. Rarefaction waves propagating into the target from the nearby "free" surfaces will unload the material and set it into motion in the direction of the free surface. This shearing flow will lead to the excavation of the crater and some of the material transported to the surface by this flow will be ejected in the form of small fragments. The crater boundary is then set by the conditions that the stresses in this flow exceed the shear strength of the material. Likewise, the fragment sizes are set by the detailed stress distribution in the cratering flow and the shear strength of the material. Extensive numerical calculations have revealed surprisingly simple cratering flows (Maxwell 1977; Orphal 1977). The flow is incompressible and in steady state and the stream lines are well described by simple power laws in spherical coordinates, characterized by the exponent $z$ (The so-called z–model; cf., Fig. 7). The parameter, $z$, has to exceed 2, corresponding to pure radial flow; ie., an explosion in an infinite medium. The numerical studies typically yield a $z$ of about 3. Late stage equivalence (cf., §3.1) requires $z \simeq 3.4$ (Jones et al. 1996). This flow field can then be used to evaluate the crater size and fragment size spectrum as a function of the impact kinetic energy and the shear strength of the material (cf., Jones et al. 1996). The fragment size distribution, $n(\tilde{a})$ is given by

$$n(\tilde{a})d\tilde{a} = \frac{12z^3(z-2)}{(z+1)^2}\tilde{a}^{\alpha}d\tilde{a} \tag{17}$$

where $\alpha = (4z+1)/(z+1)$ and $z$ is the exponent of the flow field. Here, the normalized fragment size is given by $\tilde{a} = a/a_+$ with $a$ the fragment size. The largest fragments are formed near the crater boundary and, hence, the maximum fragment size, $a_+$, is a fraction of the crater radius, $R_c$; viz.,

$$a_+ = \frac{R_c}{2z} \quad . \tag{18}$$

The exponent of this grain size distribution is not very sensitive to $z$; ie., $\alpha$ only varies from 3.0 to 3.4 for $z$ in the range 2–4. For $z$=3.4, we find $\alpha = 3.3$. This is a very general conclusion. Any comminution process, that leads to smaller sizes when the stresses involved increase, will yield an exponent of order 3. Essentially, this reflects the decrease of the average stress experienced when the volume increases (ie., energy conservation). Finally, the minimum grain size in these models is not well constrained.

For very high velocity impacts, the detached shock may break through the target before the (slow) cratering process is complete. The rarefaction wave propagating backwards into the target will then catastrophically disrupt the whole target. The grain size distribution function has not been evaluated for this case. Instead, for simplicity and consistency, it has been set equal to that derived from the z–model for the cratering flow. However, the maximum fragment size produced by the impact will depend on the tensile stresses associated with the reflected tensile wave and hence the kinetic energy of the impact,

$$a_+ = \frac{\pi^{1/3}}{2z}\frac{v_{cat}}{v_r}a_T \quad , \tag{19}$$

where $v_{cat}$ is the minimum impact velocity required for the shock to break through the backside of the target with a strength equal to the shear strength of the material, $v_r$, the impact velocity, and $a_T$ the size of the target grain (Jones et al. 1996).

These theoretical results show that, apart from known material properties and impact parameters, the crater volume depends only on the adopted value for the critical pressure. For macroscopic (ductile) bodies, measured tensile strengths of materials are typically 1 kbar ($10^9$ dyne/cm$^2$). This is much less than the ultimate tensile strength of a material ($\simeq G/6 \simeq 100$ kbar $= 10^{11}$ dyne cm$^2$, with $G$ the shear modulus). This reflects the presence of imperfections, particularly line dislocations, which act as sources of

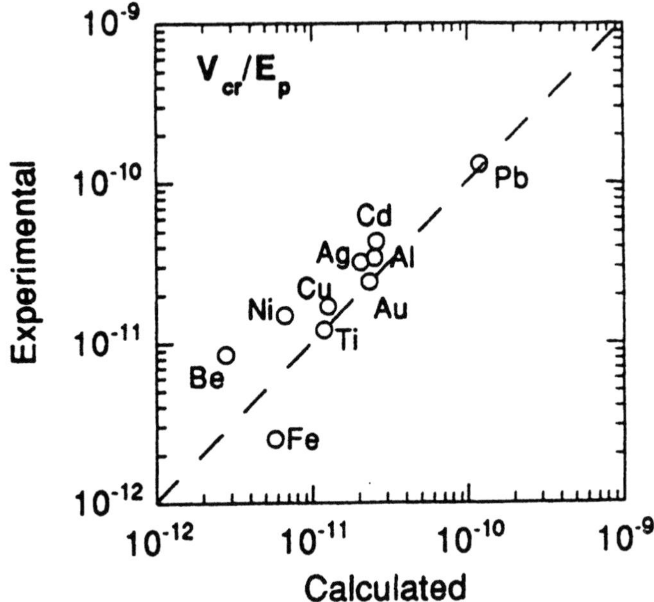


*Figure 11.* The calculated ratio of cater volume to projectile kinetic energy is compared to the measured ratio for micron–sized iron projectiles impacting various surfaces at $\simeq 5$ km/s. Experimental data taken from Rudolph (1969) and Fechtig et al. (1980).

mechanical weakness (Kittel 1976). Essentially, a crystal is highly strained near a dislocation and this promotes slip[3]. However, measured densities of dislocations are typically $10^{11}$ cm$^{-2}$ corresponding to about one dislocation through a region with a diameter of 400Å. Clearly, thus, movement of dislocations plays little role in crater formation of micron–sized and smaller bodies and the critical pressure involved should approach the ultimate tensile strength of a material. As a corrollary to this, macroscopic experiments on cratering by impacts will not provide a good bench mark for theories of shattering of interstellar grains.

Over the years, extensive experimental studies of the impact of micron-sized particles on slabs of various materials have been performed because of the micrometeroid hazards of interplanetary space, the interpretation of microcraters on lunar rocks, and the erosion of atmosphere–free planetary bodies (McDonnell et al. 1976; Fechtig et al. 1977; Mandeville 1972; Vedder 1971; Rudolph 1969). Often this data is presented in terms of the ratio of crater volume to projectile energy averaged over a large number of impacts with different velocities. In a few cases, crater volumes have been presented for individual impacts. Figure 11 compares calculated ratios of the crater

[3]Macroscopic brittle bodies, such as glasses, fracture at low pressures (1 kbar) due to the presence of minute (submm–sized) cracks which also act as stress enhancers.

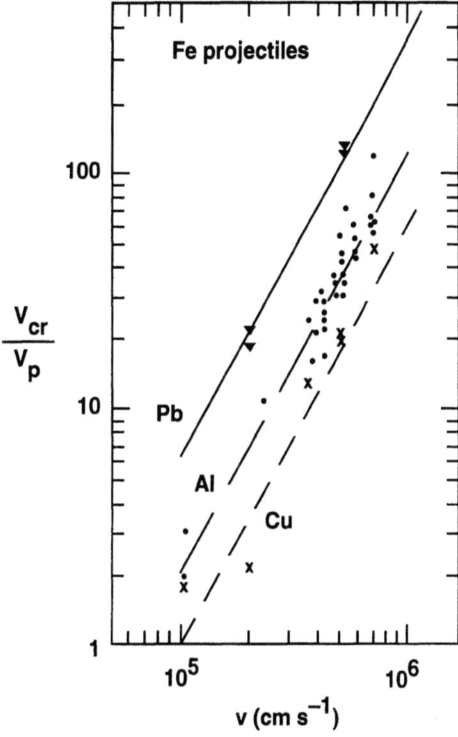


*Figure 12.* The ratio of crater volume to projectile volume for iron projectiles impacting Pb, Al, and Cu surfaces at various velocities. Data taken from Rudolph (1969). See text for details.

volume to the projectile kinetic energy to experimental data for impacts of iron projectiles on various metallic surfaces (Rudolph 1969; Fechtig, Nagel and Pailer 1980). Shock wave data has been taken from Marsh (1980). In all cases, the shear modulus was adopted for the critical pressure. The model results agree reasonably well with the data.

In Figure 12, we compare the model calculations for the ratio of crater to projectile volume as a function of impact velocity to experimental data for low velocity impacts of iron projectiles on lead, aluminium, and copper surfaces (Rudolph 1969). For lead, good agreement is obtained for a critical pressure equal to the shear modulus. For aluminium and copper, the critical pressure was lowered slightly ($\simeq$ 25 % ) from the shear modulus in order to improve the fit. This shifts the calculated curves upwards without affecting the slope noticably. Our theory reproduces the measured dependence on impact velocity well.

Finally, very little information is available on the size distribution of the fragments produced by submicron impacts. McDonnell et al. (1976) has measured experimentally the size distribution resulting from micron-sized impacts at $\simeq 5$ km s$^{-1}$ on a lunar sample. The exponent of his measured power law ($\simeq 3.5$) is in reasonable agreement with our theoretical estimate (3.3). However, SEM studies showed that the largest of his fragments, with sizes comparable to the crater radius, failed to get dislodged.

## 3.4. ASTROPHYSICAL IMPLICATIONS

Material parameters relevant for vaporization and shattering of interstellar grains have been summarized in Tielens et al (1994) and Jones et al. (1996). Typically, the critical velocity at which cratering ensues, $v_{crit}$, is 1–10 km/s. Consistent with the relative ease with which graphite flakes, its critical velocity (1.2 km/s) is at the low end of this range. Likewise, ice shatters readily (1.8 km/s). However, silicon carbide (8.8 km/s) and (single crystal) diamond (16.1 km/s) are rather robust materials which withstand shattering better than most other materials. The ratio of the vaporized mass to total mass affected by the collision (vaporized plus shattered) is well represented by $M_v/M_T = 0.76\,(G/P_v)^{1.07}$ where $G$ is the shear strength of the material and $P_v$ is the critical pressure for vaporization. For graphite and silicates, this ratio is 0.005 and 0.023, respectively. Obviously, shattering will be the dominant process affecting the interstellar grain size distribution.

Most of the shattered volume in interstellar shock waves results from high velocity impacts of small projectiles on large targets. As an example, we consider a 100 km/s impact of a 50 Å grain on a 1000 Å target. The velocity at which catastrophic destruction of the target occurs, $v_{cat}$, is 75–600 km/s depending on the material parameters. Graphite ($v_{cat}$=75 km/s) and ice (115 km/s) are the most easily destroyed this way, somewhat more easily than silicates (175 km/s). SiC (560 km/s) and diamond (1000 km/s) are the sturdiest. For a 100 Å impact, these catastrophic velocities are reduced to 23–175 km/s. Of course, a 1000Å impact only needs the critical shattering velocity for complete disruption of the target.

Figure 13 illustrates the size distribution of shattered fragments that can be expected for the impact of 50 Å graphite projectiles on 1000 Å graphite grains at various velocities. At the critical velocity, the largest fragment size is somewhat smaller than the projectile size. As the impact velocity increases, the size of the largest fragment increases as does the total crater volume. At the catastrophic velocity, the whole target grain is destroyed and the maximum fragment size is about 1/5 the target radius (independent of projectile size). Above the catastrophic velocity, the maximum size decreases with increasing velocity but the total shattered volume is constant.

The calculated fragment size distributions for different materials are

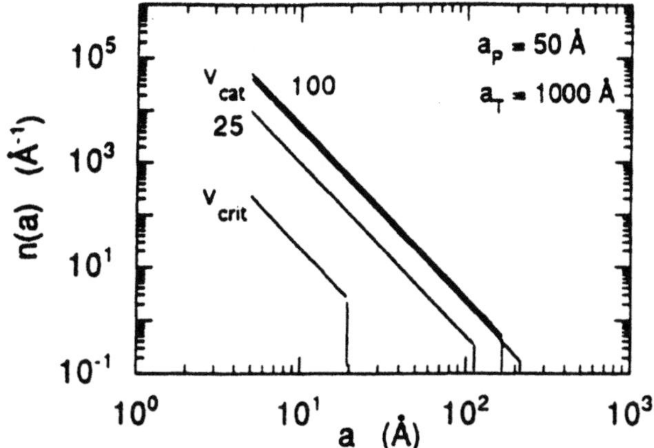


*Figure 13.* **Calculated fragment size distributions for the impact of a 50 Å projectile on a 1000 Å target at different velocities, including the critical velocity, $v_{crit}$ (=1.2 km/s), at which cratering ensues and $v_{cat}$ (75 km/s for these collision parameters) at which the grain is completely disrupted. See text for details.**

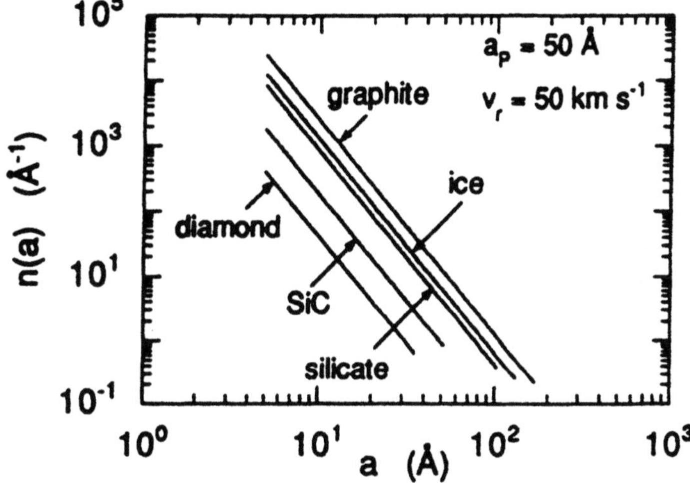


*Figure 14.* **Calculated fragment size distributions for different materials resulting from the impact of a 50 Å projectile on a 1000 Å target.**

compared in Figure 14. Like materials are assumed in these calculations. As expected from our discussion above, the total shattered volume is the largest for graphite and ice grains and smallest for SiC and diamond grains. Of course, the colliding ice grains have the smallest amount of kinetic energy and this should be kept in mind when interpreting these results. For all of these collisions, the collision velocity is below the catastrophic velocity and hence the maximum fragment size increases with the total crater volume.

## 4. Grains and Interstellar Shocks

The interstellar medium consists of various phases which differ in their physical conditions (cf., Kulkarni and Heiles 1987; Reynolds 1993; Shull 1987). Much of the mass of the ISM is localized in dense ($> 10$ cm$^{-3}$), cold ($< 100$ K) atomic or molecular clouds which have a very small filling factor (3%). Much of the volume of the interstellar medium is filled by much warmer and more tenuous intercloud phases, including the Hot Intercloud Medium (0.001 cm$^{-3}$, $10^6$ K), the Warm Neutral Medium (0.4 cm$^{-3}$, 8000 K), and the Warm Ionized Medium (0.2 cm$^{-3}$, 8000 K). The latter two differ mainly in their degree of ionization (0.1 versus 0.7). Supernova remnants expanding into the ISM will drive shock waves into these different media. The shock velocity in medium, $i$, is given by $v_i = (n_a/n_i)^{0.5}\, v_a$ where $v_a$ and $n_a$ are the shock velocity and preshock density in the ambient medium which dominates the SNR expansion (ie., the Warm Media in a two phase ISM, the HIM in a three phase ISM). The relative importance of shock destruction in these different phases for the overall life cycle of interstellar dust is controlled by the relevant volume filling factors and densities (ie., shock velocities). Thus, destruction of interstellar dust is dominated by shocks in the Warm Media because of their large volume filling factor and low densities. Shocks in the HIM are unimportant because so little total mass is involved. Shocks in the cloud phases are unimportant for the overall balance because of the high density and consequently low shock velocities (cf., Seab 1987). In the remainder of this section, we will therefore concentrate on the structure of shocks in the Warm Medium and their effects on destruction of interstellar dust.

In a general way, shocks driven by SNR can be separated into two different classes: Low velocity, radiative shocks and high velocity, adiabatic shocks. The shock cooling timescale, $\tau_{cool}$, is approximately given by

$$\tau_{cool} = \frac{2\times 10^{10}}{n}\left(\frac{v}{100\,\mathrm{km/s}}\right)^{3.2}\ \mathrm{s} \qquad (20)$$

(Hollenbach and McKee 1979). The expansion timescale, $\tau_{exp}$, of a SNR in a homogenous medium is given by

$$\tau_{exp} = 4.5\times 10^{10}\left(\frac{E}{10^{51}\,\mathrm{erg}}\right)^{1/3}\left(\frac{1}{n}\right)^{1/3}\left(\frac{100\,\mathrm{km/s}}{v}\right)^{5/3}\ \mathrm{s} \qquad (21)$$

where $E$ is the supernova explosion energy. Hence, the transition from radiative to expansion dominated regimes occurs around 150 km/s. For low velocities, gas is shocked at the shock front to a high temperature and it cools radiatively downstream. While material is advected through, the shock structure will be in a "steady" state. For high velocities, the gas is shocked to a high temperature and cools due to the expansion of the SNR. For intermediate velocities, the shock structure will be weakly time–dependent because the pressure driving the shock will decrease on a timescale comparable to the cooling timescale. The effects of this are however limited (McKee et al. 1987) and we can approximate such shocks with their steady state structure.

## 4.1. LOW VELOCITY STEADY STATE SHOCKS

### 4.1.1. *Shock structure*

Figure 15 shows the density, temperature, and electron fraction for a standard shock ($v_s$=100 km/s, $n$=0.25 cm$^{-3}$, B=3$\mu$G; Jones et al. 1994) as a function of the column density, $N_H$, behind the shock front. The time, $t$, is related to $N_H$ through $log(t(yr)) = log(N_H) - 13.9$. At the shock front, the gas motion is randomized; ie., converted into heat. Typically, behind the shock the gas will be flowing at $\frac{1}{4}$ of the shock velocity. As the gas cools behind the shockfront, it compresses. The gas is ionized around a column of $10^{14}$ cm$^{-2}$. Recombination occurs around a column of $10^{18}$ cm$^{-2}$. Further compression at that time is limited because of magnetic cushioning.

Also shown in this figure are the velocities of various-sized graphite grains. The results for silicate grains are very similar. Because of their inertia, grains will keep moving with respect to the gas at approximately $\frac{3}{4}$ of the shock speed. Because grains are also charged, they will gyrate around the magnetic field lines. The Larmor radius, $R_L$, is approximately given by

$$R_L = \frac{m_d V_d c}{Z_d e B} \simeq \frac{1.3 \times 10^{16}}{\phi} \left(\frac{a}{1000\text{Å}}\right)^2 \frac{v}{100\,\text{km/s}} \frac{10^5\,\text{K}}{T} \quad \text{cm} \qquad (22)$$

where $\phi$ is the grain potential parameter ($eU/kT$) which is typically of order unity. Thus, for reasonable sizes, grains are position coupled to the hot gas behind the shock. The magnetic moment, $\mu_d$, of the grains,

$$\mu_d = \frac{m_d v_d^2}{2B} \qquad (23)$$

is conserved and thus, as the gas compresses and the magnetic field strength increases, the grains will spin up. This process, called betatron acceleration, is counteracted by gas drag. As a result, grains will initially speed up before being dragged to a halt (Figure 15). The effects of betatron acceleration depend on the grain size as well as the shock speed. In the standard shock, all grains are betatron accelerated during the initial compression around $3 \times 10^{14}$ cm$^{-2}$. The larger grains experience a second burst around $10^{17}$ cm$^{-2}$, while the smaller grains have been stopped at that point already. Betatron acceleration is unimportant for low shock velocities ($\lesssim$50 km/s)

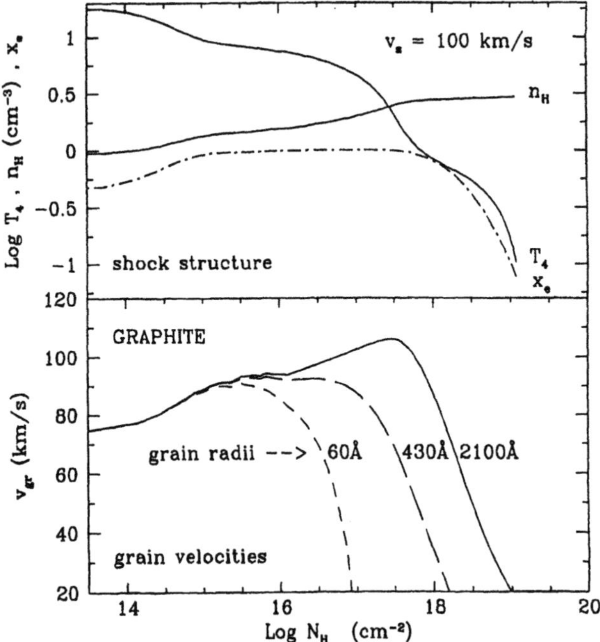


*Figure 15.* The structure of the standard shock. (Top) The postshock density, $n_H$, temperature $T_4$ in units of $10^4$ K, and electron fraction, $x_e$, as a function of column density behind the shock front. (Bottom) Velocities of different size, graphite grains as a function of the column density behind the shock front. See text for details.

and becomes progressively more important when the shock speed increases. Of course, betatron acceleration is unimportant for very fast, adiabatic shocks.

### 4.1.2. *Grain destruction and disruption*

Because of the relative motion of the gas, grains are sputtered by impacting gas atoms and ions (inertial sputtering), and vaporized by grain–grain collisions. Thermal motion of the hot postshock gas will also lead to sputtering. These processes lead to grain destruction; ie., return of solid material to the gas phase. Grain–grain collisions can also shatter grains which will lead to redistribution of the grain mass over the grain size distribution. This latter process we will call disruption to distinguish it from destruction. For the standard shock, grain destruction is dominated by inertial sputtering of large grains (Fig. 16), because they are betatron accelerated the most (cf., Fig. 15). Thermal sputtering is unimportant at these relatively low post-shock temperatures ($\lesssim 10^5$ K; cf., §2). Vaporization by grain–grain collisions is also relatively unimportant. Shattering affects the grain size distribution the most. That shattering dominates so much over vaporization reflects the

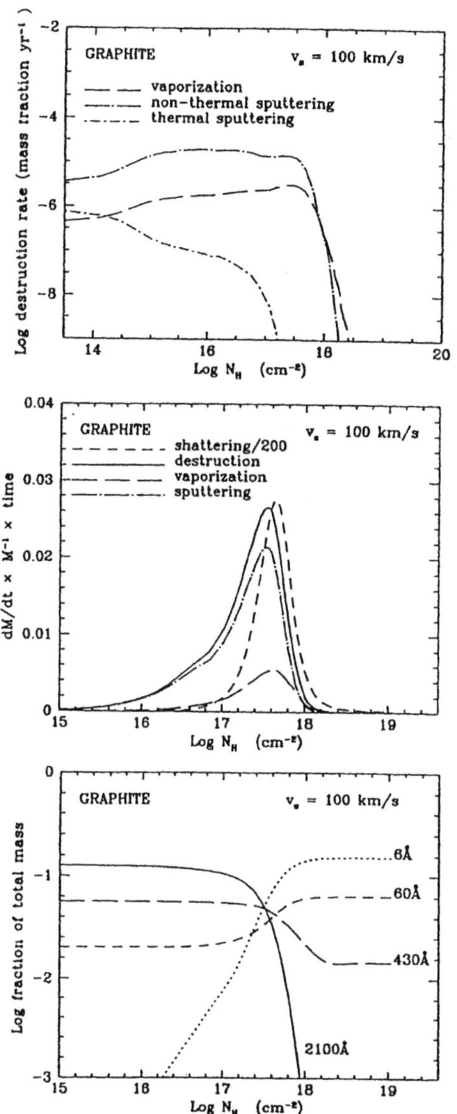


*Figure 16.* Dust destruction in the standard shock (Jones et al. 1996). (Top) Total destruction rates due to vaporization, non-thermal sputtering, and thermal sputtering, as a function of column density behind the shock front. (Middle) Fractional destruction and disruption as a function of column density behind the shock. Equal areas under the curve indicate equal contributions to the total destruction and disruption. (Bottom) The fraction of the total mass in different grain size bins as a function of the column density behind the shock front. See text for details.

much lower threshold (ie., impact at $\simeq 1$ km/s rather than 20 km/s; §3). For the standard shock, shattering rapidly removes the large grains from the grain size distribution and puts it into the small size bins ($\lesssim 300$ Å; Fig. 16).

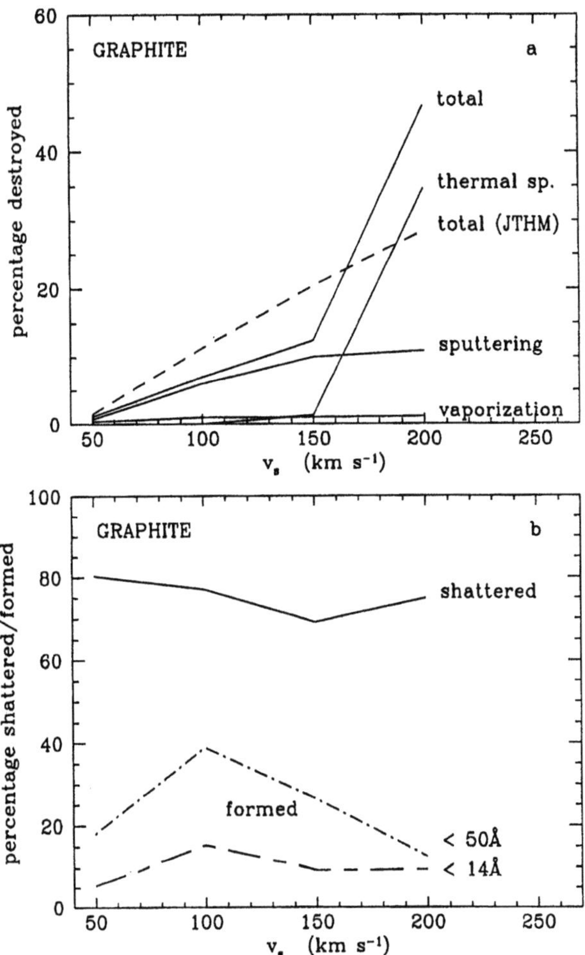


*Figure 17.* Graphite grain processing as a function of shock velocity (Jones et al. 1996). (Top) The percentage of the total mass destroyed (ie., returned to the gas phase) as a function of the shock velocity. Separate curves for inertial sputtering, thermal sputtering, and vaporization in grain-grain collisions are also shown. (Bottom) The percentage of the total graphite grain mass shattered as a function of shock velocity (solid line). The dashed and dash-dotted lines give the percentage of the mass entered into bins smaller than 14 and 50 Å, respectively. See text for details.

The fraction of the total grain mass destroyed increases with increasing shock velocity (cf., Fig. 17). Inertial sputtering dominates at velocities below $\simeq$ 200 km/s. Because of the high postshock temperatures, thermal sputtering takes over at high velocites. Vaporization is never important. At all velocities, shattering affects the grain size distribution the most. Typically, about 80% of the mass is shattered and a large fraction of that mass is turned into very small grains ($\lesssim$50Å). Processing of interstellar grains also depends on the material properties. This includes the specific density of the grain, which affects the grain motion behind the shock, as well as the sputtering, vaporization, and shattering thresholds (cf., §2 & 3). For equal sized grains, diamonds are more readily sputtered than graphite grains because of their larger mass (Fig. 18). This is also the main cause for the larger destruction of silicate, SiC and Fe grains. Ice grains, on the other hand, have very low destruction thresholds and are readily destroyed at all velocities. Because of their much lower shear strength, graphite grains are more readily shattered than SiC and diamond grains. Iron grains are also readily shattered but that mainly reflects their high specific weight (ie., long stopping length). Ice grains are destroyed before shattering collisions can take effect.

#### 4.1.3. *The grain size distribution*

As the results discussed above demonstrate, shattering dominates grain processing by interstellar shocks. The grain size distributions after processing by different shocks are compared in Figure 19 (Jones et al. 1996). Obviously, most of the large grains are shattered into very small grains. The effect of shattering increases with increasing shock velocity because of the increased relative grain velocities. At high velocities ($\gtrsim$200 km/s), betatron acceleration becomes less important, and the effects of shattering decrease somewhat. The size distribution of the fragments produced by individual grain–grain collisions is somewhat less steep (–3.3; §3) than the MRN grain size distribution exponent (–3.5). Nevertheless, the size distribution produced by the shocks is slightly steeper than MRN. Essentially, the velocity of the (target) grains increases with size due to betatron acceleration. Hence, the largest grains, which contain most of the mass, are relatively speaking shattered in smaller fragments than somewhat smaller grains.

### 4.2. ADIABATIC SHOCKS

Evaluation of the effects of adiabatic shocks requires a model for the expansion of supernova remnants (SNR) in the ISM. Here, we will follow McKee et al. (1987) for the pressure drop associated with the expansion of SNR. The expansion of a SNR in a 2-Phase as well as in a 3-Phase medium is self-similar with $R \sim t^{\eta}$. Consider an intercloud medium hit by a shock at time $t_o$ of the expansion of the SNR. The grains in the shocked gas are thermally sputtered while the SNR expands at a rate

$$\frac{1}{n}\frac{da}{dt} = f(T) \tag{24}$$

where $f(T)$ is function (essentiall the averaged sputtering yield) which depends only on the temperature. Polynomial fits to this function are given

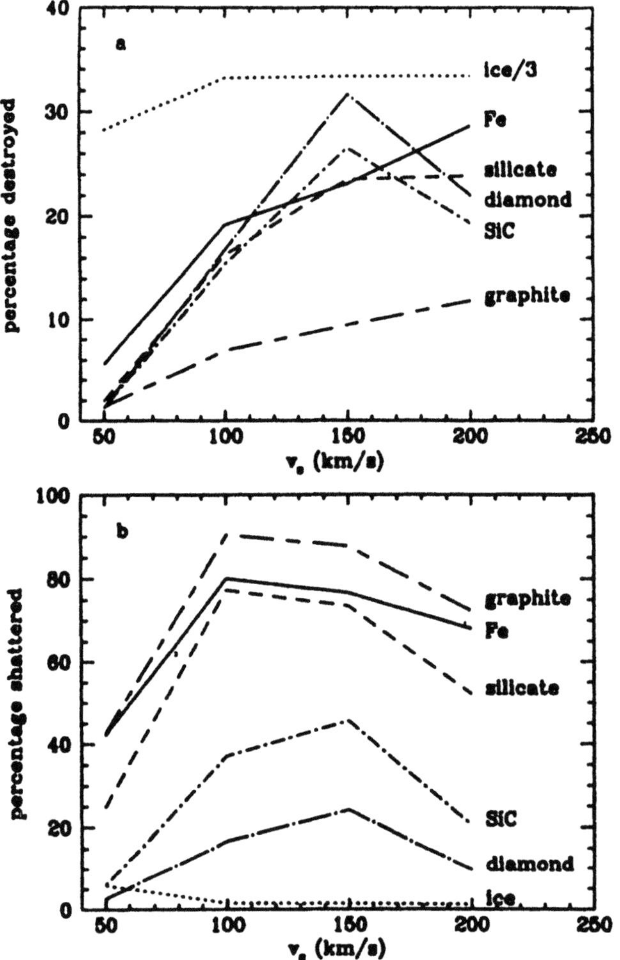


*Figure 18.* Processing of 1000 Å grains as a function of shock velocity for different grain materials (Jones et al. 1996). (Top) The percentage of the total mas destroyed (ie., returned to the gas phase) as a function of the shock velocity. Note that the curve for ice grains has been reduced by a factor 3. See text for details. (Bottom) The percentage of the grain mass shattered as a function of shock velocity.

in Tielens et al. (1994). The thickness of the grain layer removed is independent of grain size and is given by

$$\Delta a = \int_{t_0}^{t_1} n f(T) dt = \frac{20 n_o t_o}{6 \eta T_s^{\alpha}} \int_0^{T_s} T^{\alpha-1} f(T) dT \tag{25}$$

where $n_o$ is the preshock density of the Warm Medium (0.25 cm$^{-3}$), $T_s$ is the initial shock temperature (ie., at $t_o$), $\eta$ is the self-similarity exponent, and $\alpha = (9\eta - 5)/6\eta$. Now, for a two phase medium (clouds with a small filling factor embedded in the intercloud medium), $\eta = 2/5$, $t_o \simeq 2.2 \times$

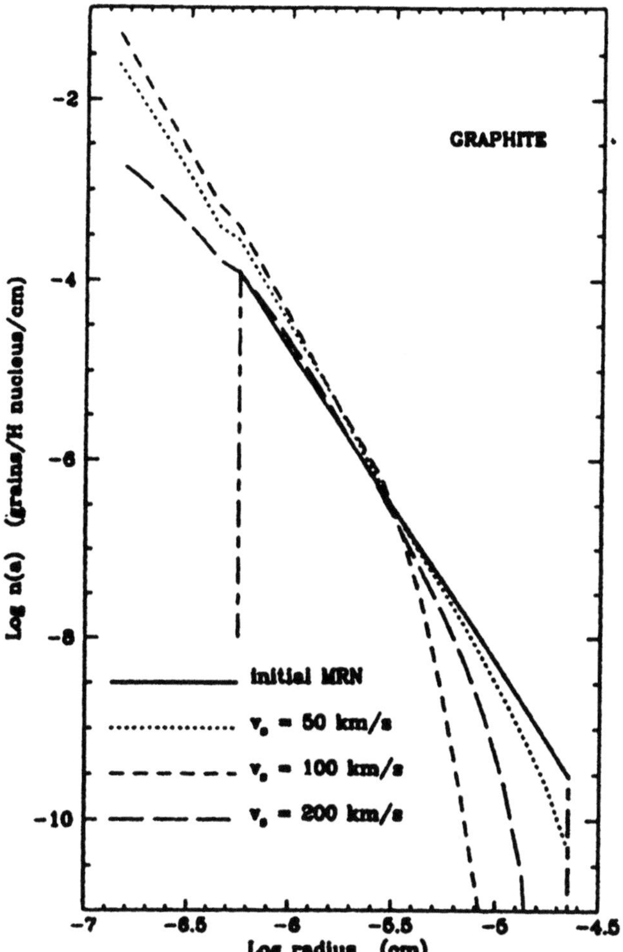


*Figure 19.* The grain size distribution for different shock velocity (Jones et al. 1996). The initial MRN grain size distribution ($n(a)$ $a^{-3.5}$, for $50 < a < 3000$ Å) is shown for comparison. See text for details.

$10^5(100\text{km/s}/v_s)^{5/3}$ yr, and

$$\Delta a = 4.6 \times 10^5 \left(\frac{100\,\text{km/s}}{v_s}\right)^{5/3} T_s^{7/12} \int_0^{T_s} T^{-19/12} f(T) dT \qquad (26)$$

For a three phase medium, the SNR expands in the HIM, and hence the shock velocity in the warm medium is reduced from the expansion velocity. In this case, $\eta = 3/5$, $t_o \simeq 1.3 \times 10^5(100\text{km/s}/v_s)^{10/9}$ yr, and

$$\Delta a = 1.8 \times 10^5 \left(\frac{100\,\text{km/s}}{v_s}\right)^{10/9} T_s^{-1/9} \int_0^{T_s} T^{-8/9} f(T) dT \qquad (27)$$

The results of these integrations are shown in Figure 20. At high velocities, silicates lose a thicker surface layer than graphite grains, reflecting the

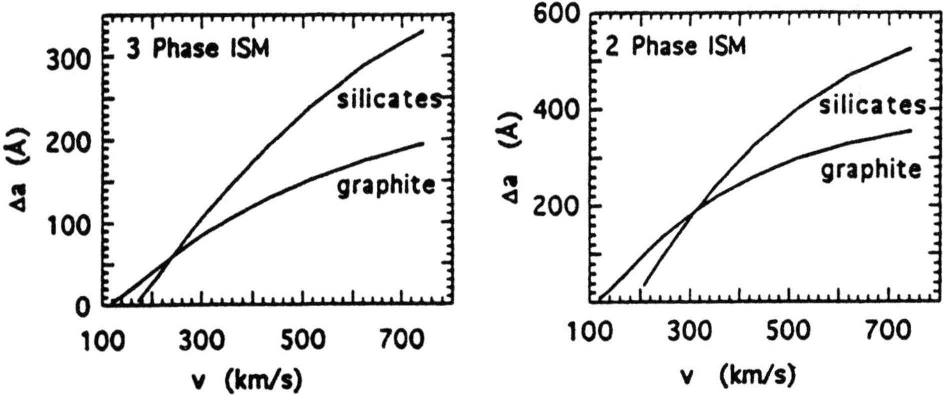


*Figure 20.* The thickness of the thermally sputtered layer as a function of the (adiabatic) shock velocity for a SNR expanding into a two phase and three phase ISM.

difference in yields at high energies. At low velocities, graphite suffers more of a loss because of its lower threshold energy (cf., §2). The results also show that, for a given speed, adiabatic shocks in a two phase medium are more destructive than in a three phase medium, because SNR expand faster in a 3 phase medium than in a 2 phase one.

## 5. Ecology of the Interstellar Medium

Dust is injected into the ISM by stars and destroyed by stardust as well as by incorporation into newly formed stars (and planets). Over the life time of the Galaxy, a steady state has ensuid and injection and destruction (approximately) balance. Here, we will develop a stick model for the evolution of dust in the ISM and the fraction of an element locked up in dust (cf., Dwek and Scalo 1980; McKee 1989). We can write for the dust fraction, $\delta$,

$$\frac{d\delta}{dt} = -(\alpha - 1)\, k_* \delta + k_{in}(\delta_o - \delta) - k_{des}\delta \tag{28}$$

where $\alpha\delta$ is the fraction of dust destroyed during star formation. In principle, $\alpha$ can be less than 1, if some dust formed in the circumstellar disk is entrained in the protostellar outflow and ejected into the ISM, and greater than 1, if the shocks driven by the outflow destroy more dust than the outflow injects. Since neither of these processes is well known and they are likely of little importance compared to the stellar injection rates and supernova destruction rates, we will adopt $\alpha = 1$. In that case, the steady state fraction of an element locked up in dust is given by,

$$\delta = \delta_o \left(1 + \frac{k_{des}}{k_{in}}\right)^{-1} \tag{29}$$

Thus, in order to determine the fraction of an element tied up in dust, we have to evaluate the rate at which that element is injected into the ISM, the fraction of that element injected in the form of dust, and the destruction rate of dust by SNR.

## 5.1. INJECTION RATES

Stars in various stages of their lives indiscriminantly pollute their environment with gas, dust, and metals. The relative importance of these different types of objects is not well known. Table 1 summarizes various gas and dust (carbonaceous and silicate/metalic) mass injection rates into the ISM (Tielens 1989). The quoted values are uncertain – some more than others – and are often based upon various assumptions. Even estimates of the relative importance of high mass stars versus low mass stars vary continously due to new developments in, for example, the determination of mass-loss rates (eg., IR studies) or nucleosynthetic reaction rates (eg., the $^{12}C(\alpha,\gamma)^{16}O$ rate). Thus, while 20 years ago high mass stars ($M>8\ M_\odot$) where thought to be mainly responsible for the carbon in the interstellar medium (Audouze and Tinsly 1976), in more recent studies carbon-rich red giants seem to dominate (Jura 1987). These injection rates also vary across the Galaxy. For example, because of the general increase in metalicity towards the inner galaxy, the ratio of the O–rich to C–rich giants increases towards the galactic center. For the same reason, Wolf Rayet stars are limited to the inner galaxy. Keeping these caveats in mind, the remainder of this section reviews the contribution of different objects to the dust and gas budget of the ISM.

*Red Giants*: Low mass stars ($<8M_\odot$) predominantly contribute during the Asymptotic Giant Branch (AGB) phase. In this short–lived phase ($\simeq 10^5$ yr), $\simeq$ 0.5-7 $M_\odot$ is ejected in the form of a cool dusty wind. Because of their long lifetimes and because they lock up much of the (stellar) mass of the Galaxy, low–mass stars have to dominate the (hydrogen) mass budget. The total mass injection rate can be estimated from the white dwarf birth rates ($\simeq 1\ yr^{-1}$), assuming that each has lost about 1 $M_\odot$ on the AGB, to be equal to 1 $M_\odot\ yr^{-1}$, corresponding to about $10^{-3}\ M_\odot\ kpc^{-2}\ yr^{-1}$ (Jura 1987). This is supported by IR and CO studies of the dust and gas mass-loss rate from these objects, which yield values between $5-9\times10^{-4}$ $M_\odot\ yr^{-1}\ kpc^{-2}$ (Thronson et al. 1987; Clausen et al. 1987; Loup et al. 1993). Estimates on the return rate by O-rich giants agree on approximately

TABLE 1. Interstellar gas and dust budgets[a]

| Source | $\dot{M}^b_H$ | $\dot{M}^c_C$ | $\dot{M}^d_{sil}$ |
|---|---|---|---|
| C–rich giants | 5 (-4) | 2 (-6) | – |
| O–rich giants | 5 (-4) | – | 3 (-6) |
| Novae | 6 (-6) | 3 (-7) | 3 (-8) |
| SN type Ia | – | 3 (-7)[e] | 2 (-6)[e] |
| OB stars | 3 (-5) | – | – |
| Red supergiants | 2 (-5) | – | 2 (-7) |
| WC 8-10 | 1 (-4)[f] | 6 (-8) | – |
| SN type II | 1 (-4) | 2 (-6)[e] | 1 (-5)[e] |
| Recycle [g] | 5 (-3) | - | - |
| Magellanic stream | 5 (-4) | 1 (-6)[h] | 1 (-6)[h] |
| Star formation | -1 (-3) | - | - |

[a]Mass injection rates in units of $M_\odot$ kpc$^{-2}$ yr$^{-1}$. Values differ slightly from previous versions of this table (Tielens 1989; 1994), largely because here dust injection rates are given as the maximum possible. [b] Total gas mass injection rate.
[c] Carbon dust injection rate.
[d] Silicate and metal dust injection rate.
[e] Fraction and composition of of dust formed in SN is presently unknown. These values correspond to upper limits.
[f] H–mass injection by all Wolf Rayet stars. WC stars are actually H–defficient and do not contribute at all.
[g] Mixing through the halo. Some dust ($\simeq 1/3$) is destroyed in the shocks that bring the material up (Savage and Sembach 1996).
[h] A dust-to-gas ratio of 0.002 is assumed (0.2 of the interstellar value; Koornneef 1982).

$5 \times 10^{-4}$ $M_\odot$ yr$^{-1}$ kpc$^{-2}$, but those for C-rich giants range from $5 \times 10^{-5}$ to $5 \times 10^{-4}$ $M_\odot$ yr$^{-1}$ kpc$^{-2}$ (Clausen et al. 1987; Thronson et al. 1987). The giants with a high mass-loss rate ($> 10^{-5}$ $M_\odot$ yr$^{-1}$) dominate the total gas return to the galaxy and this discrepancy centers on sample incompleteness correction factor required to derive the total number of such C-rich objects present in the galaxy from the few actually observed. We will adopt a return rate by C-rich giants of $5 \times 10^{-4}$ $M_\odot$ yr$^{-1}$ kpc$^{-2}$.

O-rich and C-rich giants inject silicates and carbonaceous dust, respectively. For the former, we adopt that all the elemental Si condensed in the form of silicates. The value in Table 1 reflects then solar abundances and a mass per silicon atom of 172 amu. For the latter, we have to account for C enrichment due to dredge up - we adopt a carbon abundance of 0.014 by mass - and subtract the fraction of the elemental C locked up in CO (0.01). The uncertainties in these values are obvious.

*Novae*: Nova events are generally interpreted as a thermonuclear runaway in an accreting hydrogen shell on top of a degenerate C-O-white dwarf in a binary system (Truran 1986). With a nova rate of 40 yr$^{-1}$ and an ejection of $10^{-4}$ $M_{\odot}$ per event, the total mass injection rate is about $6 \times 10^{-6}$ $M_{\odot}$ $yr^{-1}$ $kpc^{-2}$. Typically H is depleted by a factor of about 3 with respect to He and little additional heavier elements like $^{12}C$ have been synthesized (Ferland and Shields 1978; Gallagher et al. 1980). Novae show the spectral signature of PAHs and silicates, while a featureless dust component is often taken to be amorphous carbon (Gehrz et al. 1994). Assuming solar abundances for C and Si (wrt total mass) leads to $3 \times 10^{-8}$ and $4 \times 10^{-8}$ $M_{\odot}$ $yr^{-1}$ $kpc^{-2}$ for the carbon dust and silicate dust injection rates, respectively (cf., Table 1).

*Supernovae*: There are two types of SN, whose only connection is the large amount of energy released ($\simeq 10^{51}$ erg) and the explosive nucleosynthesis of heavy elements. Type II are due to the gravitational collapse in massive stars ($M > 8$ $M_{\odot}$ ). Type Ia's result from a thermonuclear instability of an accreting white dwarf due to the accumulation of a critical mass (Woosley and Weaver 1986). The total supernova rate has been estimated to be about $2 \times 10^{-5}$ $yr^{-1}$ $kpc^{-2}$ with about equal rates for type I and II (Tammann 1982). The typical type II supernova has a main sequence mass of 25 $M_{\odot}$ of which about 2 $M_{\odot}$ remains behind in a neutron star (Woosley and Weaver 1986), this yields a total H mass injection rate of about $2 \times 10^{-4}$ $M_{\odot}$ $yr^{-1}$ $kpc^{-2}$.

Studies of SN1987A have conclusively demonstrated that dust does condens in type II SN ejecta (Wooden 1997; McCray 1993). However, the amount and type of dust are controversial. The IR spectra are featureless indicating either amorphous carbon or metalic grains or optically thick (silicate ?) emission (ie., large grains or clumps). Theoretically, SN dust chemistry has been evaluated based upon the onion-skin layered structure of theoretical nucleosynthesis stellar models, assuming no mixing (cf., Lattimer et al. 1978). In that case, metalic, amorphous carbon, and silicate grains are all expected to condens in different zones of the ejecta. However, isotopic analysis of graphite and SiC stardust isolated from meteorites indicates that there was some (but incomplete) microscopic mixing of the different nucleosynthesis zones (Zinner 1997). Here, we will assume that the carbon-rich zones condensed carbonaceous dust. While the FeNi core likely condensed metalic grains, we have lumped those for simplicity with the silicate grains in Table 1. A type II supernova ejects about 4 $M_{\odot}$ in the form of freshly synthesized heavy elements, of which 0.25 $M_{\odot}$ is in the form of C (Woosley and Weaver 1986). The C injection rate from type II supernovae is then about $2 \times 10^{-6}$ $M_{\odot}$ $yr^{-1}$ $kpc^{-2}$ (cf., table 1). For silicates, we assume a silicon overproduction factor of 15 with respect to solar, and a atomic weight of 172 per Si atom resulting in $2 \times 10^{-5}$ $M_{\odot}$ $yr^{-1}$ $kpc^{-2}$. This corresponds to an assumed total dust mass of $\simeq 2$ $M_{\odot}$ per SN, about half of the total mass in heavy elements ejected.

Type I supernovae will contribute much less to the galactic enrichment of intermediate-mass elements such as C and Si. Calculations typically yield a C/Fe and Si/Fe mass fraction ratio of about $3 \times 10^{-2}$ and $4 \times 10^{-1}$ of the solar ratio (Thielemann et al. 1986). Since about 0.5 $M_\odot$ of Fe is required to explain the lightcurves of type I supernovae (Woosley and Weaver 1986), we adopt the production of about $2.5 \times 10^{-2}$ $M_\odot$ of carbon dust per type I SN explosion, corresponding to $3 \times 10^{-7}$ $M_\odot$ yr$^{-1}$ kpc$^{-2}$. For the mineral grains we will add 0.5 $M_\odot$ of silicates to the Fe mass and lump it all under silicates in Table 1. The total injection rate is then $10^{-5}$ $M_\odot$ yr$^{-1}$ kpc$^{-2}$.

*OB stars*: Estimates of the total mass-loss rate of OB stars range from 1 to $3 \times 10^{-4}$ $M_\odot$ yr$^{-1}$ kpc$^{-2}$ (Abbott 1982; van Buren 1985). These stars do not produce any dust. Adopting solar abundance, this translates into about $10^{-6}$ and $2 \times 10^{-7}$ $M_\odot$ yr$^{-1}$ kpc$^{-2}$ of gaseous C and Si injected into the ISM (cf., Jura 1987).

*Red supergiants* During their red supergiant phase, massive stars inject gas at low velocities in a dust-driven wind into the ISM. The total mass injection rate is about $2 \times 10^{-5}$ $M_\odot$ yr$^{-1}$ kpc$^{-2}$ (Jura and Kleinmann 1990). About 1% of that is in the form of silicates.

*Wolf-Rayet stars*: Population I Wolf Rayet stars are the descendants of massive ($M$>25-40$M_\odot$ ) OB stars where substantial mass-loss during the H burning phase as well as mixing processes have "uncovered" enriched stellar core material on the surfaces of these objects (Chiosi and Maeder 1986). The WC stars contain He, C and O and no evidence for H or N, while the WN have greatly enhanced He and N and no H (Willis 1982). The total mass-loss rate of Wolf Rayet stars is estimated to be about $10^{-4}$ $M_\odot$ yr$^{-1}$ kpc$^{-2}$ where about 5% is due to WC 8-10, which are associated with carbon dust formation with a total dust injection rate of $5 \times 10^{-8}$ $M_\odot$ yr$^{-1}$ kpc$^{-2}$ (Abbott and Conti 1987; van der Hucht et al. 1987).

## 5.2. DUST AND GAS BUDGET OF THE GALAXY

From this discussion it is obvious that, although the mass return rate is dominated by red giants, many different object contribute substantially to the dust budget of the galaxy (Table 1). With a total interstellar gas mass of $4.5 \times 10^9$ $M_\odot$ (Scoville and Sanders 1987), the injection timescale is $\simeq 3 \times 10^9$ yr. The total gas injection rate approximately balances the loss of gas to form stars and the ISM is in approximate steady state. The total dust injection rate corresponds to a dust-to-gas ratio in the ejecta of 1.5%, largely due to the synthesis of heavy elements in type II supernova. This is somewhat more than current estimates of interstellar dust-to-gas ratio (1%). This enrichment parallels the general elemental enrichment of the ISM. Of course, besides the various uncertainties noted in the previous section, this enrichment might also be counteracted by infall of metal-deficient gas stripped from dwarf systems such as the Magellanic clouds, from intergalactic space, or left-overs from the formation of the local group (Audouze and Tinsley 1976, Rana 1991). The total infall rate is about $5 \times 10^{-3}$ $M_\odot$ $yr^{-1}$ $kpc^{-2}$ but much of that is merely circulation between the plane and the lower halo in a galactic fountain (Bregman 1980; Wakker 1990). The Magellanic stream is an order of magnitude smaller, which is actually still an appreciable fraction of the total gas injection rate by stars. This may have profound impact on the metalicity of the galaxy. For example, the low metal abundances in the Orion region and the origin of Gould Belt have been attributed to the infall of metal–poor material from the Magellanic stream (Meyer et al. 1994).

If supernovae are efficient dust factories, the dust injection rate is $4.7 \times 10^{-6}$ and $1.5 \times 10^{-5}$ $M_\odot$ $yr^{-1}$ $kpc^{-2}$ for carbonaceous and mineral stardust, respectively (Table 1). However, if not, these numbers drop to $2.4 \times 10^{-6}$ and $3.2 \times 10^{-6}$, respectively. Even if all the C produced by SN condenses, the total fraction of the elemental C injected into the ISM in dust form is only 0.25 (eg., $\delta_o$(C)=0.25). The remainder is injected in gaseous form by OB stars, WC stars, O-rich giants, and WC stars. In contrast, if all the Si in SN ejecta condenses as silicates and as SiC in C-rich giants, $\delta_o$(Si)=0.95 (cf., Jura 1987). If SN do not contribute to the silicate budget, then $\delta_o$(Si) drops to 0.4.

## 6. Dust lifetimes

The destruction rate, $k_{des}$, of interstellar dust by supernova shock waves is given by (McKee 1989),

$$k_{des} M_{ISM} = \frac{1}{\tau_{SN}} \int \epsilon(v_s) dM_s(v_s) \tag{30}$$

where $M_{ISM}$ is the total mass of the ISM ($8 \times 10^9$ $M_\odot$, $\tau_{SN}$ is the effective interval between supernova explosions, $\epsilon(v_s)$ is the fraction of dust destroyed by a shock of velocity $v_s$, and $M_s(v_s)$ is the mass shocked to a velocity of at least $v_s$. The destruction efficiency of interstellar shocks has been discussed in §4.1.2 and results are shown in Figures 17 and 18. The other two factors that enter into this expression are discussed here.

### 6.1. THE SHOCKED MASS PER SUPERNOVA

The mass processed by a SNR depends on the structure of the ISM (McKee 1989). In a two phase (homogenous) ISM, the mass, $M_s(v_s)$, shocked to a velocity of at least $v_s$ is given by

$$M_s(v_s) \simeq 6800 \left(\frac{100\,\mathrm{km/s}}{v_s}\right)^2 \frac{E}{10^{51}\,\mathrm{erg}} \quad \mathrm{M}_\odot\,, \tag{31}$$

in the Sedov–Taylor (energy conserving) phase. In the radiative phase ($v_s <$ 150 km/s), the SNR energy is no longer conserved and the mass processed is given by

$$M_s(v_s) \simeq 2460 \left(\frac{100\,\mathrm{km/s}}{v_s}\right)^{9/7} \left(\frac{E}{10^{51}\,\mathrm{erg}}\right)^{0.95} n^{-0.1} \quad \mathrm{M}_\odot\,, \tag{32}$$

where $n$ is the ambient, preshock density (Cioffi et al. 1988).

In a three phase medium, the SNR will expand in the hot phase which has the largest volume filling factor ($f_h \simeq 0.7$). Dust destruction still occur in the Warm ISM and, while the shock in that medium might be radiative, the SNR expands adiabatically in the hot phase. Hence, the mass of warm medium shocked to a velocity exceeding $v_s$, is given by

$$M_s(v_s) \simeq 6800 \left(\frac{f_W}{f_h}\right) \left(\frac{100\,\mathrm{km/s}}{v_s}\right)^2 \frac{E}{10^{51}\,\mathrm{erg}} \quad \mathrm{M}_\odot\,, \tag{33}$$

where $f_W$ is the filling factor of the warm medium ($\simeq$0.3).

## 6.2. THE EFFECTIVE SUPERNOVA RATE

The effective supernova rate has been extensively discussed by McKee (1989). The supernova rate in our Galaxy is $2 \times 10^{-2}$ yr$^{-1}$. However, not every supernova interacts effectively with the ISM (cf., McKee 1989). About half of the SN are type Ia which have low mass progenitors with a large galactic scale height. The correction factor for this is about 1/3 (Heiles 1987). Half of the supernovae type II occur in associations. The first star to explode in an association will be very effective in destroying dust while sweeping up matter in a shell. However, subsequent SNe will interact predominantly with the gas inside the shell and the shell itself and do not contribute much to the dust destruction therefore. The expansion of the supershell is too slow to destroy much dust (McKee 1989). Given that a typical OB association yields 40 SNe (Heiles 1987), the destruction per supernova in an OB association is only $\simeq$0.1 of that for field SNe. The other half of type II SNe are isolated – due to field stars and run-aways. Here again a correction factor ($\simeq$0.6) for the increased scale-height as compared to the Warm ISM is needed (McKee 1989). Hence, in summary, we find for the effective supernova timescale, $\tau_{SN}$,

$$\frac{1}{\tau_{SN}} = \left(0.4 \times \frac{1}{2} + (0.5 \times 0.1 + 0.5 \times 0.6) \times \frac{1}{2}\right) \times 2 \times 10^{-2} \simeq 8 \times 10^{-3} \quad \mathrm{yr}^{-1}\,. \tag{34}$$

Using the results for dust destruction as a function of shock velocity, the interstellar mass shocked to different velocities, and the effective supernova rate, the rate for dust destruction can be evaluated from equation 30. This has been done with increasing degree of sophistication by Draine and Salpeter (1979), McKee et al. (1987), McKee (1989), and Jones et al. (1994; 1996). Adopting an MRN grain size distribution, graphite dust is destroyed in the equivalent of about 8 $M_\odot$ per year of interstellar gas. For comparison to Table 1, this corresponds to $1.3 \times 10^{-2}$ $M_\odot$ $yr^{-1}$ $kpc^{-2}$ for the rate at which the (gas) mass of the ISM is processed (ie., cleaned of dust). For clarity, with a dust-to-gas ratio of 1% by mass, the rate at which dust mass is destroyed is a factor 100 smaller. For silicates, the values are slightly larger, 12 $M_\odot$ $yr^{-1}$ and $2 \times 10^{-2}$ $M_\odot$ $yr^{-1}$ $kpc^{-2}$. This results in calculated grain lifetimes against destruction (return to the gas phase) of $6 \times 10^8$ and $4 \times 10^8$ yr for graphite and silicate grains, respectively.

Besides grain material, the dust destruction rate depends also on the grain size (Seab and Shull 1983; Jones et al. 1994; 1996). The above quoted values are averaged over the MRN grain size distribution and refer to the grain mass which is mainly contained in large grains ($\gtrsim$1000 Å). All velocity shocks contribute about equally to the destruction of large grains. In contrast, the destruction of small grains is dominated by thermal sputtering in adiabatic shocks. In a two phase ISM, a 200 Å or less graphite of silicate grain is completely sputtered away in any shock with $v_s > 300$ km/s (cf., Fig. 20). So, all the small grains in the equivalent of 750 $M_\odot$ of warm interstellar gas are destroyed by a single SNR, equivalent to 6 $M_\odot$ $yr^{-1}$ or, averaged over the galaxy, $10^{-2}$ $M_\odot$ $yr^{-1}$ $kpc^{-2}$. In a 3 phase medium, a 200 Å or less silicate grain will be destroyed by a shock faster than 450 km/s. For a graphite grain a shock velocity in excess of 700 km/s is required and this can be ignored. Taking $f_H = 0.7$ and $f_W = 0.3$, in this case, about 140 $M_\odot$ are processed by a single SNR (1 $M_\odot$ $yr^{-1}$ and $1.6 \times 10^{-3}$ $M_\odot$ $yr^{-1}$ $kpc^{-2}$). Thus, 200 Å grains are much better protected in a 3 phase medium than in a 2 phase medium because a SNR expends much of its energy expanding in the hot phase rather than the warm phase and because the expansion is faster (cf., §4.2). The lifetimes of small grains are then $7.5 \times 10^8$ and $4.5 \times 10^9$ yr in a 2 phase and 3 phase ISM, respectively. These are lifetimes against complete destruction of a 200 Å grain but even removal of only 100 Å (which requires $v_s > 300$ km/s in a 3 phase medium) would still return 7/8 of the grain mass to the gas and such a shock occurs about 2 as often.

## 7. Depletion

With a dust destruction timescale of $\simeq 5 \times 10^8$ yr and an injection timescale of $\simeq 3 \times 10^9$ yr, the fraction of an element locked up in stardust is about 0.14 of the fraction initially injected as dust (cf., eqn 29; $\delta/\delta_o \simeq 0.14$). Observed depletions of Si, Fe, and Mg are much larger, typically 0.9 in the cloud medium. Clearly there has to be an efficient way to return gas to the solid phase in the ISM itself. This discrepancy between dust lifetimes and depletions becomes even more poignant if SNe do not inject much dust and $\delta_o$ is small. This problem has been the subject of much research but has not been satisfactoraly resolved (Dwek and Scalo 1980; Seab and Shull 1983; McKee et al. 1987; Jones et al. 1994). It seems that reaccretion of various elements in the ISM has to be an efficient process. Moreover, since IR observations show that the elemental Si missing from the gas phase is locked up in silicate dust (cf., Tielens and Allamandola 1987), this accretion process, perhaps assisted by FUV photolysis, has to lead to the formation of silicate material rather than incorporation into a carbonaceous-based grain mantle. While there is some laboratory support for this scenario (Nuth and Moore 1988), further studies are required to evaluate the actual merits of this suggestion for the ISM.

Up to now, we have concentrated on the lifetime against destruction (grain $\longrightarrow$ gas). The lifetime of large grains against shattering into small fragments is much less, about $10^8$ yr for both graphite and silicate grains (Jones et al. 1996). These fragments are typically less than 200 Å in size. Such small, 200 Å grains survive processing by interstellar shocks much better. For example, in a 3 phase medium, an average batch of interstellar material still contains about 60% of the small silicate grains (albeit reduced in size) and essentially all the graphite grains initially injected. Actually, this suggests that rather than being destroyed (ie., returned to the gas), large grains will predominantly be shattered into small fragments by the first shock encountered. For the rest of this sojourn into the Warm ISM, they would be well protected against destruction in shocks. However, while this seemingly solves the depletion problem discussed above, it replaces it with a grain-size crisis. Interstellar visual extinction testifies to the dominance of large grains in the dust mass distribution. Thus, the small grains produced by shocks in the Warm ISM have to be reassembled rapidly into large grains in diffuse clouds in order to produce the total visual extinction observed. Coagulation may indeed be a very important process in interstellar clouds (Chokshi et al. 1993; Dominik and Tielens 1997). Taken together this seems to point towards a model where interstellar dust cycles back and forth between large agglomorates formed by coagulation of very small grains in diffuse and molecular clouds and small fragments produced by

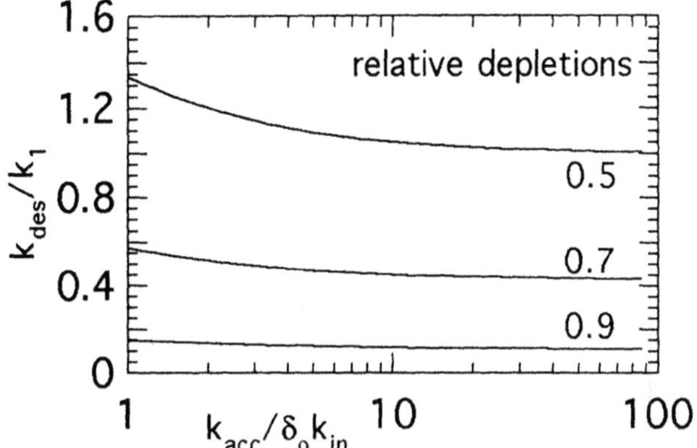


*Figure 21.* The ratio of the intercloud destruction rate to the intercloud-to-cloud mixing rate ($k_{des}/k_1$) is shown as a function of the ratio of the cloud accretion rate to the dust injection rate ($k_{acc}/\delta_o k_{in}$) for various intercloud-to-cloud depletion ratios ($\delta_i/\delta_c$).

shattering in the Warm ISM. Because of their low strength, these agglomorates will fall apart in the first, gentle shock they encounter in the Warm ISM without undergoing much actual destruction in that shock or in any further shocks encountered. Further studies wil be required to put these ideas on a firm footing.

Depletion variations between the different phases of the ISM can be used to get an "observational" handle on the rate at which dust is destroyed in the Warm ISM and reformed through accretion in the cloud phase (Tielens 1997). The simple model for the depletion described by equation (29) can be extended to include the presence of different phases. Consider a Warm phase with depletion, $\delta_i$, in which dust is destroyed with rate, $k_{des}$, and a cloud phase with depletion, $\delta_c$, in which dust is reformed at rate, $k_{acc}$. The two phases mix at rates $k_1$ ($i \rightarrow c$) and $k_2$ ($c \rightarrow i$). Dust injection occurs at a rate $\delta_o k_{in}$. Making a few simplifying assumption leads then to the following two equations (Tielens 1997),

$$\frac{\delta_c}{\delta_i} = \left(1 + \frac{k_{des}}{k_1}\left(\frac{\frac{1}{2} + \frac{k_{acc}}{\delta_o k_{in}}}{1 + \frac{k_{acc}}{\delta_o k_{in}}}\right)\right) \quad , \tag{35}$$

and

$$\frac{\delta_c}{1-\delta_c} = \frac{k_{acc}}{k_2}\left(1 + \frac{k_1}{k_{des}}\right) \quad . \tag{36}$$

These two equations link relative and absolute cloud and intercloud depletions to ratios of rates for the various processes involved: destruction, accretion, injection, and mixing.

The relation between $k_{des}/k_1$ and $k_{acc}/\delta_o k_{in}$ given by equation (30) is shown in Figure 21 for different values of the depletion ratios, $\delta_i/\delta_c$. Note that the y–axis is linear while the x–axis is logarithmic. Clearly, this depletion ratio is very sensitive to $k_{des}/k_1$ and very insensitive to $k_{acc}/\delta_o k_{in}$. Whether accretion or injection dominates dust formation is not terribly important for the relative depletion ratios. However, if there were no destruction, this depletion ratio would be unity since any accretion in the cloud phase would be immediately passed on to the intercloud phase. The presence of differences in depletion between the intercloud and cloud phases of the ISM have long been surmized from Copernicus data (cf., Spitzer 1985; Jenkins 1989). The Goddard High Resolution Spectrograph can measure directly absorption features due to velocity components along various sight–lines corresponding to intercloud and cloud phases (cf., Savage and Sembach 1996). The observed $\delta_i/\delta_c$ of 0.65 and 0.95 for Si and Fe correspond then to $k_{des}/k_1$ of 0.5 and 0.06. Obviously, substantial differences in depletion between the intercloud and cloud phases of the ISM imply directly that the destruction rate is comparable to the mixing timescale between the phases. The latter is much shorter than the stardust injection rate. In the plane, the cloud phase contains about 10 times more mass than the intercloud phase, so that in steady state $k_1/k_2$=10. Cloud lifetimes are $\simeq 3 \times 10^7$ yr (McKee 1989), which then implies a lifetime against destruction averaged over the ISM phases of $\simeq 6 \times 10^7$ and $\simeq 5 \times 10^8$ yr for silicon bearing and iron bearing interstellar dust (Tielens 1997). Thus, Si and Fe are in two distinct interstellar dust components with quite different destruction lifetimes. Comparison with the theoretical calculations discussed above imply binding energies of 1–2 and 5 eV for these two grain components. Hence, much of the Si may be in a rather volatile grain mantle accreted on a refractory Fe–bearing core (cf., Joseph 1988; Sofia et al. 1994). The depletion of C in the intercloud medium is difficult to measure and hence whether this grain mantle is carbon-based or mineral-based is presently unknown.

While relative depletions are largely controlled by the ratio of the destruction rate to the mixing rate, absolute depletion levels are set by the balance between accretion in the cloud phase and destruction in the intercloud phase. For given absolute depletion levels, Figure 2 shows the relation between the destruction-to-accretion rate ratio ($k_{des}/k_{acc}$) and the destruction-to-mixing rate ratio ($k_{des}/k_1$), illustrating the insensitivity to $k_{des}/k_1$. The measured absolute cloud depletion levels for Si and Fe, 0.95 and 0.993, imply then $k_{acc}/k_1$ is 0.6 and 0.8, respectively. In contrast to the destruction timescales, the accretion timescales are very similar for these elements. Likely, accretion is limited by the collision rate with grains and not by the specific chemical interaction with the surface of an element.

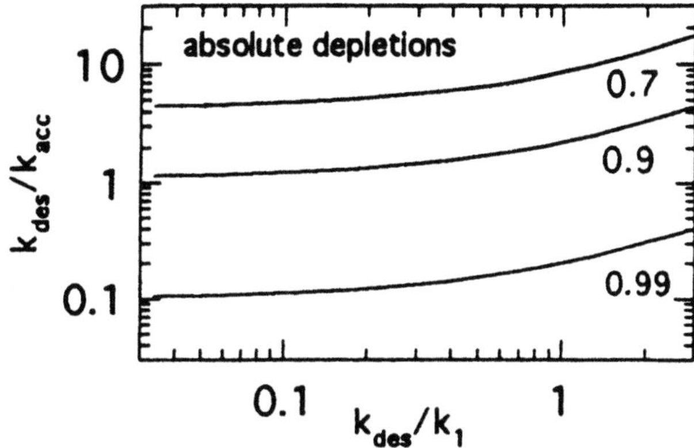


*Figure 22.* The ratio of the intercloud destruction rate to the cloud accretion rate ($k_{des}/k_{acc}$) is shown as a function of the ratio of the intercloud destruction rate to the intercloud-to-cloud mixing rate ($k_{des}/k_1$) for various absolute cloud depletion ratios ($\delta_c$).

These observed cloud depletions refer mainly to $\zeta$ Oph, which is known to be unusually dense ($\simeq 300$ cm$^{-3}$; van Dishoeck and Black 1986), and this is reflected in the high derived accretion rates ($\simeq 2 \times 10^{-7}$ yr$^{-1}$).

## Acknowledgements

Research on interstellar dust at NASA Ames is funded through the Theory, SOFIA, and IR astronomy programs (RTOPs 399-28-10-08, 263-10-97-03, 188-40-01).

## References

Abbott, D.C., 1982, ApJ, 263, 723
Abbott, D.C., Conti, P.S., 1987, Ann Rev Astr Ap, 25, 113
Ahrens J.T., O'Keefe, J.D., 1971, Moon, 4, 214
Andersen, H.H., Bay, H.L., 1981, in Sputtering by Particle Bombardment, ed. R. Behrisch, (Berlin: Springer Verlag), 145
Audouze, J. Tinsley, B.M., 1976, Ann Rev Astr Ap, 14, 43
Barlow, M.J., 1978, MNRAS, 183, 367
Barlow, M.J., Silk, J., 1977, ApJL, 211, L83
Bohdansky, J., 1984, J Nucl Instr Meth Phys Res B, 2, 587
Bregman, J.N., 1980, ApJ, 237, 280
Chiosi, C., Maeder, A., 1986, Ann Rev Astr Ap, 24, 329
Chokshi, A., Tielens, A.G.G.M., Hollenbach, D.J., 1993, ApJ, 407, 806
Cioffi, D.F., McKee, C.F., Bertschinger, E., 1988, ApJ, 334, 252
Clausen M.J., Kleinmann, S.G., Joyce, R.R., Jura, M., 1987, ApJS, 65, 385
Cowie, L.L., 1978, ApJ, 225, 887
Dienes, J.K., Walsh, J.M., 1970, in High-Velocity Impact Phenomenae, ed. R. Kinslow, (New York: Academic Press),
Dominik, C., Tielens, A.G.G.M., 1997, ApJ, 480, 647
Draine, B.T., 1995, Ap Space Sci, 233, 111
Draine, B.T., Salpeter, E.E., 1979, ApJ, 231, 438
Dwek, E., Scalo, J.M., 1980, ApJ, 239, 193

Eckstein, W., 1991, Computer Simulations of Ion-Sold Interactions, (Berlin: Springer Verlag)
Fechtig, H., Nagel, K., Pail, N., 1980, in Solid Particles in the Solar System, eds., I. Halliday, B.A. McIntosh, (Dordrecht: Reidel), 357
Fechtig, H., Nagel, K., Stahle, V., Grögler, N., Schneider, E., Neukum, G., 1977, Proc Lunar Sci Conf, 7, 889
Ferland, G.J., Shields, G.A., 1978, ApJ, 226, 172
Flower, D.R., Pineau des Forêts, G., Field, D., May, P.W., 1996, MNRAS, 280, 447
Gallagher, J.S., Hege, E.K., Kopriva, D.A., Williams, R.E., Butcher, H.R., 1980, ApJ, 237, 55
Gehrz, R.D., Greenhouse, M.A., Hayward, T.L., Houck, J.R., Mason, C.G., Woodward, C.E., 1994, ApJL, 448, L119
Heiles, C., 1987, ApJ, 315, 555
Hollenbach, D.J. McKee, C.F., 1989, ApJ, 342, 306
Jenkins, E.B., 1989, in Interstellar Dust, eds. L.J. Allamandola, A.G.G.K. Tielens, (Dordrecht: Reidel), 56
Jenkins, E.B., Silk, J., Wallerstein, G., 1976, ApJL, 209, L87
Jenkins, E.B., Silk, J., Wallerstein, G., Leep, E.M., 1981, ApJ, 248, 977
Jenkins, E.B., Silk, J., Wallerstein, G., 1984, ApJ, 278, 649
Jones, A.P., Tielens, A.G.G.M., Hollenbach, D.J., McKee, C.F., 1994, ApJ, 433, 797
Jones, A.P., Tielens, A.G.G.M., Hollenbach, D.J., 1996, ApJ, 469, 740
Joseph, C.L., 1988, ApJ, 335, 157
Jura, M, 1987, in Interstellar Processes, D.J. Hollenbach, H. Thronson, (Dordrecht: Reidel), 3
Jura, M., Kleinmann, S.G., 1990, ApJS, 73, 769
Kittel, P., 1976, Introduction to Solid State Physics, (New York, Wiley)
Kulkarni, S.R. Heiles 1987, in Interstellar Processes, D.J. Hollenbach, H. Thronson, (Dordrecht: Reidel), 87
Lattimer, J.M., Schramm, D.N., Grossman, L., 1978, ApJ, 219, 230
Loup, C., Forveille, T., Omont, A., Paul, J.F., 1993, AAS, 99, 291
Mandeville, J.C., 1972, Earth Planet Sci Lett, 15, 110
Marsh, S.P., 1980, LASL Shock Hugoniot Data (Berkeley: Univ California Press)
Maxwell,D.E., 1977, in Impact and Explosion Cratering, ed. D.J. Roddy, R.O. Pepin, R.B. Merrill, (New York: Pergamon Press), 1003
McCray, R., 1993, Ann Rev Astr Ap, 31, 175
McDonnell, J.A.M., Flavill, R.P., Carey, W.C., 1976, Proc Lunar Sci Conf, 7, 1055
McKee, C.F., 1989, in Interstellar Dust, eds. L.J. Allamandola, A.G.G.M. Tielens, (Dordrecht: Reidel), 431
McKee, C.F., Hollenbach, D.J., Seab, C.G., Tielens, A.G.G.M., 1987, ApJ, 318, 674
McQueen, R.G., Marsh, S.P., Taylor, J.W., Fritz, J.N., Carter, W.J., 1970, in High Velocity Impact Phenomenae, ed. R. Kinslow, (New York: Academic Press), 294
Meyer, D.M., Jura, M., Hawkins, I., Cardelli, J.A., 1994, ApJL, 437, L59
Nuth, J.A. Moore, M.H., 1988, ApJL, 329, 113
O'Keefe J.D., Ahrens, J.T., 1977, Proc Lunar Sci Conf, 8, 3357
Orphal, D.L., 1977, in Impact and Explosion Cratering, ed. D.J. Roddy, R.O. Pepin, R.B. Merrill, (New York: Pergamon Press), 907
Ostriker, J. McKee, C.F., 1988, Rev Mod Phys, 60, 1
Rana, N.C., 1991, Ann Rev Astr Ap, 29, 129
Reynolds, R.J. 1993, ApJ, 392, L35
Routly, P.M., Spitzer, L.,1952, ApJ, 115, 227
Rudolph, V., 1969, Z. Nat, 24A, 326
Savage, B.D., Sembach, K.R., 1996, Ann Rev Astr Ap, 34, 279
Scoville, N.Z., Sanders, D.B., 1987, in Interstellar Processes, eds D.J. Hollenbach and H.Thronson, (Dordrecht: Reidel), 21
Seab, C.G., 1987, in Interstellar Processes, D.J. Hollenbach, H. Thronson, (Dordrecht:

Reidel), 491
Seab, C.G. Shull, J.M., 1983, ApJ, 275, 652
Shull, J.M. 1987, in Interstellar Processes, D.J. Hollenbach, H. Thronson, (Dordrecht: Reidel), 225
Sigmund, P., 1981, in Sputtering by Particle Bombardment, ed. R. Behrisch, (Berlin: Springer Verlag), 9
Siluk, R.S., Silk, J., 1974, ApJ, 192, 51
Sofia, U.J., Cardelli, J.A., Savage, B.D., 1994, ApJ, 430, 650
Spitzer, L., 1985, ApJL, 290, L21
Strazulla, G., Baratta, G.A., Magazzu, A., 1985, Astron Express, 1, 143
Tamman, G.A., 1982, in Supernovae: A survey of Current Research, eds. M.J. Rees, R.J. Stoneham, (Dordrecht: Reidel), 371
Thielemann, F.K., Nomoto, K., Yokoi, K., 1986, in Nucleosynthesis and its Implications on Nuclear and Particle Physics, eds. J. Audouze, N. Mathieu, (Dordrecht: Reidel), 131
Thronson, H.A., Latter, W.B., Black, J.H., Bally, J. Hacking, P., 1987, ApJ, 322, 770
Tielens, A.G.G.M., in Carbon in the Galaxy, eds. J.Tarter, S. Chang, D. DeFrees, NASA-CP 3061, 59
Tielens, A.G.G.M., 1997, ApJ, submitted
Tielens, A.G.G.M., Allamandola, L.J., 1987, in Interstellar Processes, eds. D. Hollenbach, H. Thronson, (Reidel, Dordrecht), 397
Tielens, A.G.G.M., McKee, C.F., Seab, C.G., Hollenbach, D.J., 1994, ApJ, 431, 321
Truran, 1986, in Nucleosynthesis and its Implications on Nuclear and Particle Physics, eds. J. Audouze, N. Mathieu, (Dordrecht: Reidel), 97
van Buren, D., 1985, ApJ, 294, 567
van der Hucht, K.A., Williams, P.M., Thé, P.S., 1987, QJRAS, 28, 254
van Dishoeck, E.F., Black, J.H., 1986, ApJS, 64, 109
Vedder, J.F., 1971, Earth Planet Sci Lett, 11, 291
Wakker, B, 1990, Ph. D. Thesis, Groningen
Wooden, D., 1997, in Laboratory Studies of Presolar Grains: Implications for Astrophysics, eds. T. Bernatowitz, E. Zinner, (AIP), in press
Woosley, S.E., Weaver, T.A., 1986, Ann Rev Astr Ap, 24, 205
Yamamura, Y., Matsunami, N., Itoh, N., 1983, Rad Eff, 71, 65
Zel'dovich, Y.B. Raizer, Y.P., 1966, Physics of Shock Waves and High Temperature Phenomenae, (New York: Academic Press)
Zinner, E., 1997, in Laboratory Studies of Presolar Grains: Implications for Astrophysics, eds. T. Bernatowitz, E. Zinner, (AIP), in press

# DUST COAGULATION AND THE STRUCTURE OF DUST AGGREGATES IN SPACE

C. DOMINIK
*Leiden Observatory, P.O. Box 9513, 2300 RA Leiden, The Netherlands*

**Abstract.** Coagulation is an important grain processing mechanism in the interstellar medium as well as in the protoplanetary disk. We discuss the basic physics involved when two grains make contact and stick together because of adhesive forces and dissipative processes. We also discuss the consequences which arise for the stability of dust aggregates formed in space.

## 1. Introduction

In a course on the formation and evolution of interstellar dust grains it is important to realize that dust grains are in fact heavily processed in interstellar space. John Kerridge has reported in this volume on the "interstellar" grains found in meteorites which still carry strong signatures of their origin. Some, like the large mono-crystalline silicon carbide grains have seen very little if any processing. However, it is quite likely that these are the "lucky guys" because we have ample evidence of grain processing. Most drastically this is reflected in the apparently high destruction rates of grains in shocks (see the chapter by Tielens) which result both from theoretical models and from observations of high gas-phase abundances of refractory species in post-shock medium. It is therefore required that the destroyed dust is put together again in some fashion, either chemically by gas accretion or physically by dust coagulation. Even though it is not clear what contribution coagulation can make to solve this problem, there is evidence for it in the denser interstellar medium. This can be observed by changes in the interstellar extinction. The extinction at different wavelengths is due to particles of different sizes. While the visual extinction is largely due to particles with radii around 1e-5 cm (Mathis *et al.* 1977), the UV extinction



*J. M. Greenberg and A. Li (eds.), Formulation and Evolution of Solids in Space,* 377–387.

is dominated by much smaller grains. The fact that the ratio between visual and UV extinction is much larger in dense regions finds its most probable explanation in the depletion of small grains by coagulation onto the surface of large grains (Cardelli *et al.* 1989, Kim *et al.* 1994).

The second location where dust coagulation is important are the accretion and protoplanetary disks around (proto) stars. In the earlier phases, coagulation increases the grains sizes (and consequently decreases the visual absorption coefficient per unit mass, with profound influences on the thermodynamic structure of the disk (Schmitt *et al.* 1997)). In the later phases, coagulation leads to the accumulation of bodies up to km sizes (Weidenschilling 1997).

The properties of the aggregates formed in these environments are of great interest. Aggregate structure and composition are the basic factors, governing absorption and emission cross sections, collision cross sections and stability alike. In calculations of grain coagulation, the structure of the aggregates is usually ignored. The few exceptions to this rule include the treatment of aggregation in interstellar clouds by Ossenkopf (1993) and some calculation by Weidenschilling *et al.* (1989). However, even in these calculations, the structure of the aggregates is determined by assuming that the aggregates grow without restructuring. In the last few years, we have studied in detail the forces acting at a contact between grains in order to provide better means to support or falsify such an assumption. We will discuss the results in this contribution.

## 2. Basic Processes Influencing Aggregate Structure

Let us first examine the basic process which are important factors for the structure of aggregates formed in the interstellar medium. If we assume for a moment that grains stick when they hit and do not allow for later rearrangement of contacts, there are two factors: The growth mode and the approach trajectory.

The approach trajectory is the path on which a new particle approaches an already existing cluster. A particle in a gas always make a random walk in which it is moved around by collisions with the gas atoms. This motion is called Brownian motion. It is possible to define a mean path length between two direction changes in such a motion (Einstein 1905). If this path length is small compared to the size of the aggregate, the particle spends a long time near the outer boundary of the aggregate and consequently will likely stick to one of the outermost grains in the aggregate. This case, well realized in liquid suspension, produces very porous structures (Weitz *et al.* 1985).

If on the other hand the average path length is much larger than the size of the aggregate, it is valid to assume that the particle approaches on

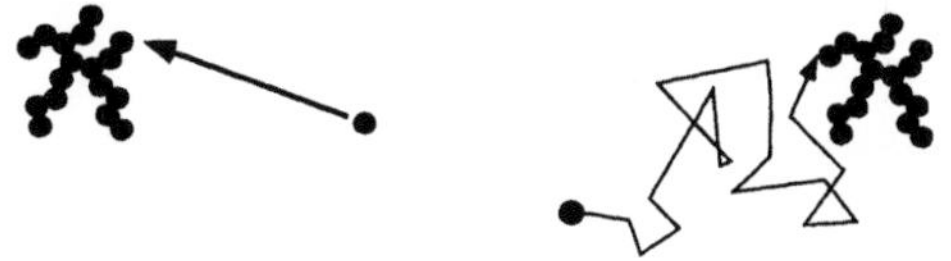


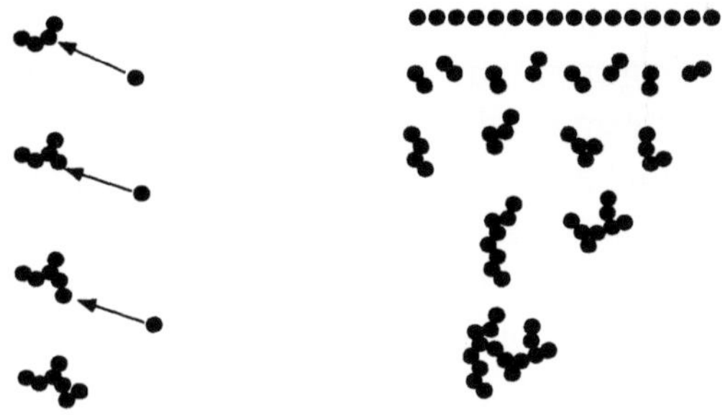


*Figure 1.* Approach trajectory and growth mode both influence the structure of an aggregate formed in interstellar space. See text.

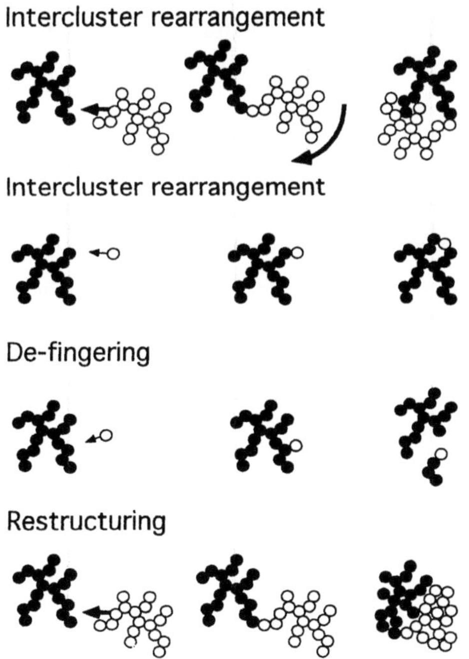


*Figure 2.* Different mechanisms can be responsible for the compaction of aggregates.

a straight line, with a larger probability to hit the denser central part of an aggregate. Consequently, these processes tend to produce more compact aggregates. In astrophysical situations, we always have the case of ballistic trajectories. The surrounding gases are so rarefied that the mean free path of a (sub)-micron dust grain is huge compared to any aggregates (Blum *et al.* 1996).

The second basic property of the growth process is the distinction between cluster-cluster and particle-cluster aggregation. In the latter case, only single particles are added in each growth step, while in the first case, single particles first make dimers, then dimers aggregate further etc. Particle-cluster aggregation can be shown to produce relatively compact aggregates, with a "fractal" dimension of 3 (Ball & Witten 1984). Thus, large aggregates would have a similar density as small ones, even though the structure may be porous. Cluster-Cluster aggregation produces much more open structures, with fractal dimensions less than 2 (Witten & Cates 1986). In these cases, the density in the aggregate decreases as a function of size. If we are looking at a system with a monodisperse size distribution, cluster-cluster aggregation will always dominate. All particles have the same collision timescale $\tau_c$, and after very few $\tau_c$, no single grains will be left which could drive a particle-cluster aggregation process. The size distribution will remain highly peaked throughout the growth process. For particle-cluster aggregation to be dominant, it is necessary to continuously destroy some of the aggregates in order to replenish the monomer density. Alternatively, if the collision frequency between large aggregates and small grains is much higher than between individual grains, some large aggregates may consume all particles in a particle-cluster aggregation process. This is conceivable in turbulent velocity fields where the small grains are coupled well to the gas while large compact aggregates may decouple from the gas and introduce such a velocity gradient with size.

If aggregation occurs with fractal dimension about 2, it is easy to see that at some point aggregates need to be compacted. With fractal dimension $d = 2$, the density varies as $\rho \propto r^{d-3} \approx r^{-1}$. If the aggregate consists of $1\mu$m grains, a one meter sized aggregate will have only $10^{-6}$ of the bulk density. However, in the solar system we will eventually have to form compact objects like comets and planetesimals. To create these, compaction of aggregates will be required. Figure 2 shows different mechanisms which can contribute to making more compact structures. The first two are called inter–cluster restructuring. Upon collision, the two clusters first make on contact, then rotate around this contact point to make a second and finally a third contact which completely defines their relative position. This process has been studied by Richardson (1995). Its critical assumption is that the new contacts between the two aggregates are much weaker than the contacts inside each aggregate. It allows to increase the fractal dimension of cluster-cluster aggregates by approximately 0.25 (Meakin & Jullien 1988). Breaking off long fingers of an aggregate also makes the average porosity smaller. The last and possibly most efficient mechanism is restructuring, in which all contacts inside an aggregate are rearranged to reach more compact structures. This has been studied recently by Dominik & Tielens (1997).

## 3. Collision and Sticking of Grains

The first question to address is whether two grains colliding in interstellar space will actually stick, or simply bounce. Two grains which move relatively to each other through space are clearly not a bound system to begin with - the total energy (kinetic energy minus binding energy) is positive because at larger distances the binding energy is very low. That means, if the two grains collide, they can only stick of they have some way of removing their kinetic energy - i.e., the collision must be inelastic. This has been discussed in astronomical context by Chokshi *et al.* (1993). The energy may be lost in a variety of ways. Plastic work is a possibility. If the grains materials were very soft, like wet clay for example, they could simply loose the collision energy by irreversibly deforming their rounded surface into flat surfaces in contact. However, at least in the laboratory small particles are normally harder than bulk material. Therefore plastic work is not considered to be a main contributor to the energy dissipation. Another possibility is to transform some of the energy into *elastic* energy by straining the material. This is different from plastic work as the deformation of the material is reversible. The elastic deformation of materials in contact has been studied extensively in the field of adhesion and contact physics. The basis is the classical work by Hertz (1896) which describes the deformation of two parabolic surfaces which are pressed together by an external force. Within the frame of this description there is no room for energy dissipation as the deformation is static, ideally elastic and reversible. In order to find the cause for sticking of small particles we need to take into account the adhesive forces between grains. The weakest and always present force is the van der Waals attraction which is of the order of 10–30 ergs/cm$^2$. These forces are dominant for apolar, non-metallic surfaces, like graphite or silicate. Much larger forces can act between polar materials (like water ice) or metallic grains (iron) where the typical surface energies reach from several 100 to several 1000 ergs/cm$^2$. These forces now lead to a different form of contact. While in the Hertz case the equilibrium state without external forces is a contact area with size zero, the adhesion case leads to a competition between the attractive forces and the elastic repulsion. The equilibrium state of such a contact between spherical grains is a finite contact circle with radius (Johnson *et al.* 1971)

$$a_0 = \left(\frac{9\pi\gamma R^2}{\mathcal{E}^\star}\right)^{1/3} \tag{1}$$

where $R = 1/(R_1^{-1} + R_2^{-1})$ is the reduced radius of the two grains in contact, $\gamma$ is the interface energy and $\mathcal{E}^\star = 1/((1-\nu_1^2)/\mathcal{E}_1 + (1-\nu_2^2)/\mathcal{E}_2)$ is an elastic material constant depending upon Young's modulus $\mathcal{E}$ and Poisson's ratio $\nu$ of both materials in contact. The way in which such a contact stores energy in elastic strain is shown in Fig. 3. The main mechanism is a hysteresis in the contact making and breaking process. The two surfaces jump into contact when the separation of the spherical surfaces first becomes zero. However, the contact is lost at a larger distance, when the distance between the undeformed surfaces would be

$$\delta_c = \frac{1}{2}\frac{a_0^2}{6^{1/3}R} > 0 \tag{2}$$

and the required force is

$$F_c = 3\pi\gamma R \quad . \tag{3}$$

In order to form the neck of material indicated in Fig. 3f, energy is required. This energy is given by $E_{\mathrm{JKRS}} \approx 0.09 F_c\delta_c$ (Chokshi *et al.* 1993).

When the initial kinetic energy of the impact is less than this energy, neck formation alone is enough to cause sticking of the grains. Chokshi *et al.* (1993) also studied the excitation of elastic waves by the impact which is another way of storing away kinetic impact energy. They showed that the total energy stored in a collision in elastic waves is given by $E_{\mathsf{stick}} = 0.4 \times F_c\delta_c$.

Grains will stick when they collide with impact energies below this limit. For $10^{-5}$cm radius silicate grains, these energies correspond to approximately 60cm/s. For icy grains, one finds approximately 2000 cm/s. From this we can already see that sticking of grains is highly material dependent. Ice is softer than silicate and exhibits stronger attractive forces (hydrogen bonds), both of which work for better sticking properties. In the dark and dense interstellar medium, grains are known to have ice coatings. In these regions, coagulation will be enhanced.

## 4. Tangential Forces

For the restructuring of aggregates, tangential forces in the contact area are very important, because these are the forces which can move contacts around. In fig. 4 we show the six different degrees of freedom a contact has:

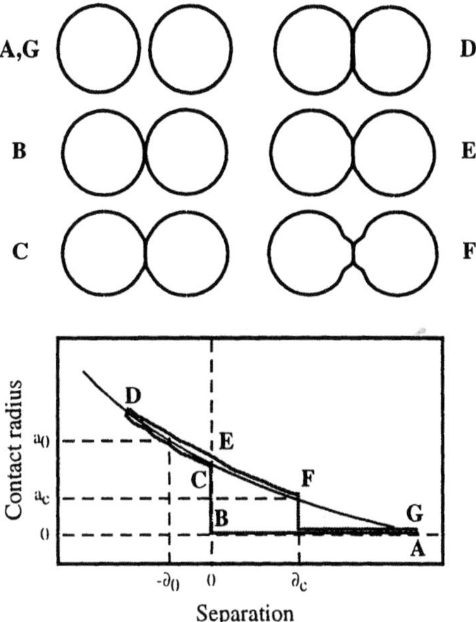


*Figure 3.* The hysteresis of an adhesive contact. Two grains make first contact when the surfaces touch (A,B). The radius of the contact area jumps from zero to $a_0$. As the grains move further together, the contact area increases, beyond the zero–external–force equilibrium (C) to a maximum (D). Here the motion turns around and the separation–contact relation follows the same track backwards. However, the grains do not separate at (E), but stay in contact. A neck of material is pulled out until finally at (F) the surfaces separate.

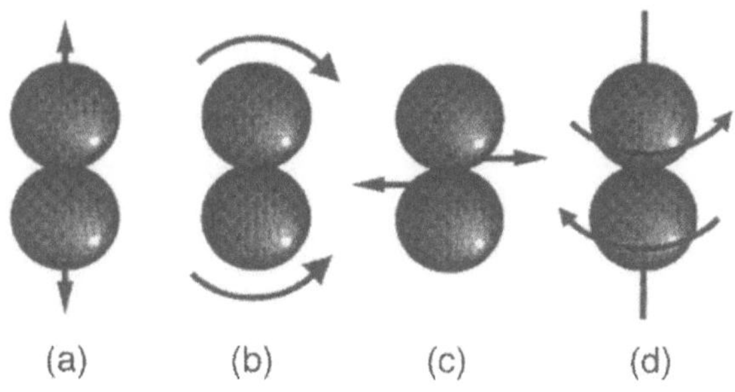


*Figure 4.* The different degrees of freedom of a contact.

one pulling the grains apart, two rolling, two sliding, and one twisting them around an axis through the centers of the spheres. Each relative motion of two grains in contact can be broken down into motions in these 6 directions. For small relative motions, the materials must respond elastically, so in each direction a small motion is possible which is only straining the material

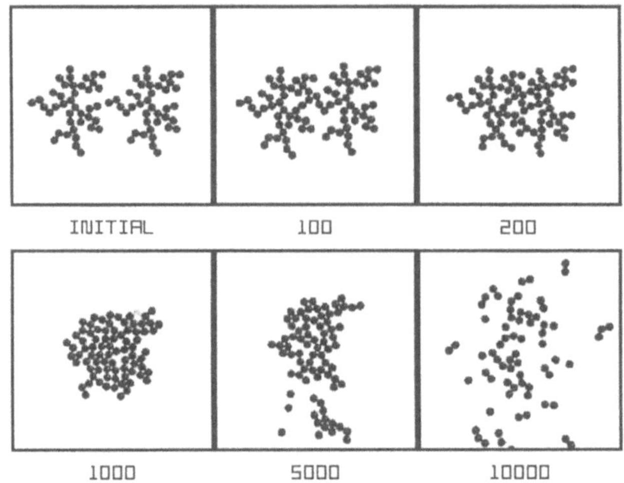


*Figure 5.* Final states of the collision of two clusters of 40 ice grains each. The first panel shows the clusters just before collision, the other panels after a collision with the specified velocity.

but not actually moving the contact around. When some threshold in the displacement is reached, the contact starts to slip and moves irreversibly. An investigation of the forces and friction processes involved shows that the most important channel for energy dissipation in such a contact is rolling (Dominik & Tielens 1995). The forces resisting sliding and twisting are so large, that the minimum energy required to move a contact is at least as large as the breakup energy (Dominik & Tielens 1996).

Looking at a cluster in this way means that we do not view the aggregate as one rigid structure, but rather as an N-body system. The N grains move under the forces transmitted through the contacts. We have developed a numerical code which allow to model such a system. Technically very similar to molecular dynamics, it requires expensive calculations due to the steep force gradients and high eigenfrequencies involved. This code has been used to simulate collisions of clusters in 2 dimensions. An example is shown in fig. 5. The figure shows the collision between two identical clusters made of 1000Å grains. It is easy to see that at small velocities, the clusters just stick with hardly any restructuring. At intermediate velocities, compaction can be very efficient, while high collision velocities break up the aggregates into single grains.

The clusters don't act as rigid structures in the collisions. At the intermediate velocities, is seems to be soft and absorb the collision energy completely. Rigid aggregates would not stick at the same collision velocities, but just bounce. Only the internal dissipation channels allow sticking

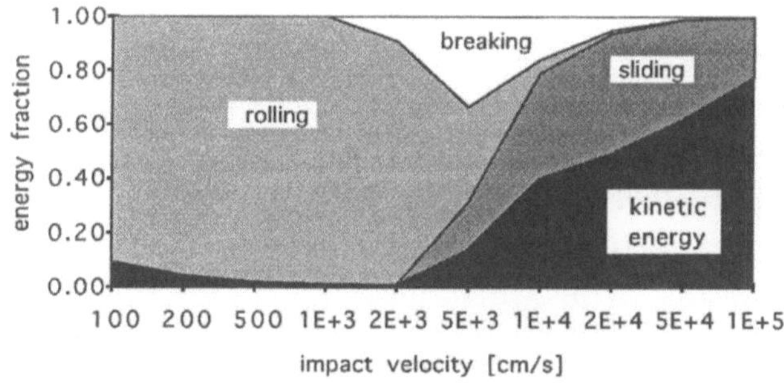


*Figure 6.* The different energy sinks in the collision shown in fig. 5 as a function of collision velocity. In the non-destructive collisions, almost all energy is dissipated in rolling of contacts.

to occur in the shown cases. In the sticking collisions, rolling is clearly the dominant channel through which the collision energy is dissipated. This can be seen from fig. 6. Other channels like sliding become important only in the high velocity collisions ($v_{\rm coll} > 2000$km/s), which destroy the aggregates at least partially.

In order to considerably deform an aggregate, it can be shown that enough energy is needed to move (roll) each contact over a distance comparable to the radius of a constituent grain. For silicate grains with $0.1\mu$m radii, this energy is typically $10^{-10}$ergs/contact. Since tenuous clusters have typically one contact per grain, this means we need approximately the same energy per grain, corresponding to a relative velocity of 135cm/s. Comparing this to the typical relative velocities of small grains in the early solar nebula, which are of the order of 0.01...1cm/s, we see immediately that the initial growth of aggregates in the solar nebula will produce uncompacted cluster-cluster aggregates with a dimension less than 2. However, once large

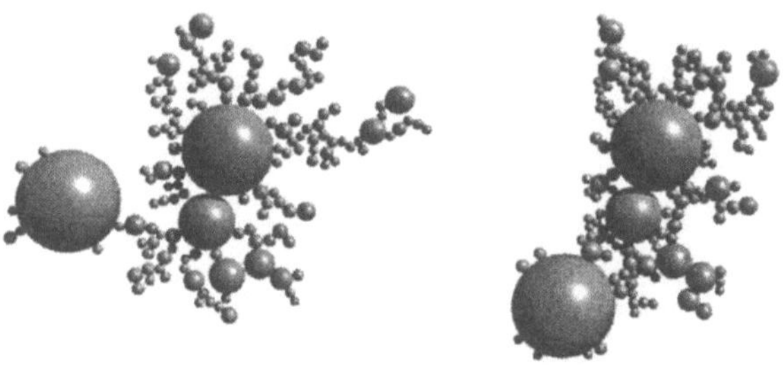

*Figure 7.* Comparison between the structures of an aggregate formed in a molecular cloud with low collision velocities (left) and large (turbulent) velocities (right).

particles become involved, the relative velocities increase and compaction should be an important factor. At what size exactly this will happen is still under investigation.

In the interstellar medium, relative motions of dust grains can also be faster due to the turbulence of the medium (Draine 1985). Under such conditions, compaction may play a role also during early stages of the growth process. Fig. 7 shows a comparison between the results of two growth processes, one with small and one with large relative velocities, but otherwise set up identically. It can easily be seen that the higher velocities result in a much compacted structure. The 2D porosities of both clusters are 66% and 53%, respectively. Much larger differences can be expected in 3D calculations, which are now under way.

## References


Ball, R. C. and Witten, T. A., 1984. Causality Bond on the Density of Aggregates. *Phys. Rev. A*, **29**, 2966–2967.

Blum, J., Wurm, G., Kempf, S., and Henning, T., December 1996. The brownian motion of dust particles in the solar nebula: An experimental approach to the problem of pre-planetary dust aggregation. *Icarus*, **124**, 441–451.

Cardelli, J. A., Clayton, G. C., and Mathis, J. S., 1989. The relationship between infrared, optical, and ultraviolet extinction. ApJ, **345**, 245–256.

Chokshi, A., Tielens, A. G. G. M., and Hollenbach, D., 1993. Dust Coagulation. ApJ, **407**, 806–819.

Dominik, C. and Tielens, A., 1995. Resistance to rolling in the adhesive contact of two elastic spheres. *Phil. Mag. A*, **72**, 783–803.

Dominik, C. and Tielens, A., 1996. Resistance to sliding on atomic scales in the adhesive contact of two elastic spheres. *Phil. Mag. A*, **73**, 1279–1302.

Dominik, C. and Tielens, A., 1997. Coagulation of dust grains and the structure of dust aggregates in space. ApJ, **480**, 647–673.

Draine, B. T., 1985. Grain Evolution in Dark Clouds. In Black, D. C. and Matthews, M. S., editors, *Protostars and Planets II*, page 621, University of Arizona Press, Tuscon.

Einstein, A., 1905. Über die von der molekularkinetischen Theorie der Wärme geforderte Bewegung von in Flüssigkeiten suspendierten Teilchen. *Ann. d. Phys.*, **XVII**, 549–560.

Hertz, H., 1896. *Miscellaneous Papers.* Macmillan, New York.

Johnson, K. L., Kendall, K., and Roberts, A. D., 1971. Surface energy and the contact of elastic solids. *Proc. R. Soc. Lond.*, **324**, 301–313.

Kim, S.-H., Martin, P. G., and Hendry, P. D., 1994. The Size Distribution of Interstellar Dust Particles as Determined from Extinction. ApJ, **422**, 164–175.

Mathis, J. S., Rumpl, W., and Nordsieck, K. H., 1977. The Size Distribution of Interstellar Dust Grains. ApJ, **217**, 425–433.

Meakin, P. and Jullien, R., 1988. Structural Readjustment Effects in Cluster-CLuster Aggregation. *J. Chem. Phys.*, **89**, 246–250.

Ossenkopf, V., 1993. Dust coagulation in dense molecular clouds: the formation of fluffy aggregates. A&A, **280**, 617–646.

Richardson, D. C., 1995. A self-consistent numerical treatment of fractal aggregate dynamics. *Icarus*, **115**, 320–335.

Schmitt, W., Henning, T., and Mucha, R., September 1997. Dust evolution in protoplanetary accretion disks. A&A, **325**, 569–584.

Weidenschilling, S., 1997. The origin of comets in the solar nebula: A unified model. *Icarus*, **127**, 290–306.

Weidenschilling, S. J., Donn, B. D., and Meakin, P., 1989. *The Formation and Evolution of Planetary Systems*, chapter The Physics of Planetesimal Formation, page 131. Cambridge University Press.

Weitz, D. A., Hunag, J., Lin, M. Y., and Sung, J., 1985. Limits of the fractal dimension for irreversible kinetic aggregation of gold colloids. *Phys. Rev. Lett.*, **54**, 1416.

Witten, T. A. and Cates, M. E., 1986. Tenuous Structures from Disorderly Growth Process. *Science*, **232**, 1607–1611.

# SOLIDS AND VOLATILES IN COMETS

*From Cometary Nuclei to Cometary Atmospheres*

J. CROVISIER
*Observatoire de Paris-Meudon*
*F-92195 Meudon, France*
crovisie@obspm.fr

**Abstract.** Comets occupy a key position in the cycles of ices and dust in the Universe. Their nuclei have mostly kept intact the composition and morphology they had at their formation. The return of a comet close to the Earth and Sun is thus an appreciated opportunity to investigate the original state of the material which built up our Solar System. The physical processes occurring in cometary nuclei and cometary atmospheres, as well as the interface between the two media, are reviewed. The status of our present knowledge of the chemical composition and physical state of comets is presented. The recent observations of comets C/1996 B2 (Hyakutake) and C/1995 O1 (Hale-Bopp), which provided a bonanza of new cometary molecules, are outlined. There are strong analogies between interstellar and cometary ices. In contrast, the silicates of cometary dust (for comet Hale-Bopp) are akin to circumstellar, rather than interstellar dust. Different families of comets may have different composition and history, however. These clues will have to be taken into consideration to achieve a consistent scenario of cometary formation.

## 1. Introduction: comets, a historical perspective

Comet nuclei are conglomerates of ices and dust particles. They spend most of their time at large heliocentric distances as bare solid bodies. When one of them enters the inner Solar System, heating by the Sun causes sublimation of the ice, release of dust, and formation of an expanding atmosphere and of a tail: this is the popular *comet* phenomenon, which may be very spectacular (Fig. 1). Thus, if we unfortunately cannot witness the formation of cometary nuclei — at least in our own Solar System — we have the priv-



*J. M. Greenberg and A. Li (eds.), Formulation and Evolution of Solids in Space,* 389–426.

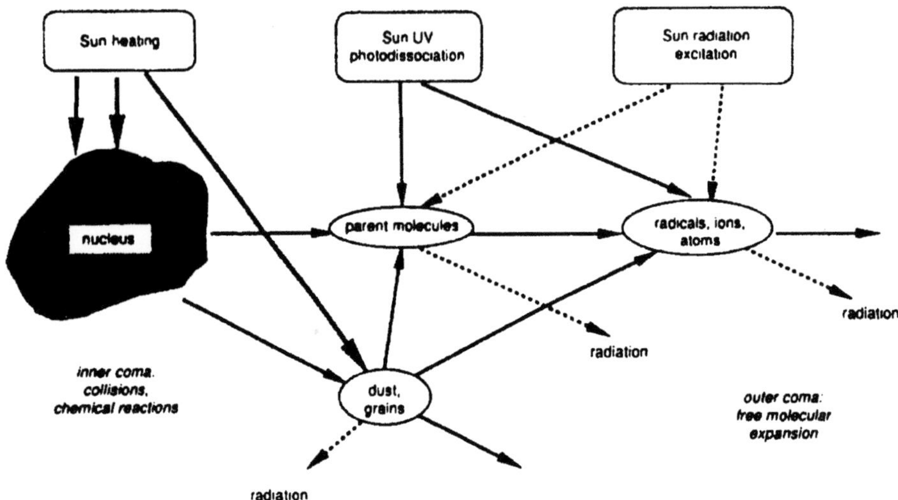


*Figure 1.* Physical and molecular processes initiated by solar radiation in comets.

ilege to see in many occasions the progressive (or catastrophic) evolution and destruction of these bodies.

Besides the show, the interest for studying comets is at least twofold. First, they are pieces of cosmic matter testifying for the formation of our Solar System. Second, the variety of physical conditions and processes that can be encountered in these objects makes them an appreciated natural laboratory.

Table 1 lists some of the most famous comets as well as several comets which were recently well studied. Since the rule for naming comets changed recently, the correspondence between the old designations (*provisional* and *definitive*) and the new ones is also given, so that the reader could find his way in the literature.

Spectroscopy and the long-term study of the evolution of comets enabled us to understand their nature and composition. It was already known in the 50's that their activity was governed by water sublimation and at that time emerged the *dirty snowball* concept. However, the detailed composition of cometary volatiles was then very uncertain. One had to rely on the presence of secondary species — radicals, ions and atoms — in the visible cometary spectra and to *guess* the nature of their parents. One had to wait for the emergence of astronomical spectroscopy in other spectral domains to see the first direct detections of cometary parent molecules: carbon monoxide in the ultraviolet in 1976 in comet West [49], hydrogen cyanide at radio wavelengths [44] and water in the infrared [86] in 1985 in comet 1P/Halley.

TABLE 1. A selection of famous comets. Comets are listed by order of perihelion date for non-periodic or long-period comets, and by order of periodic comet number for short-period comets. For comets that appeared before 1995, the old-style provisional and definitive numberings are also given.

| Comet | | | perihelion date | $q$ [AU] |
|---|---|---|---|---|
| C/1882 R1 Great Comet | 1882 II | 1882b | 17 Sep. 1882 | 0.00775 |
| C/1965 S1 Ikeya-Seki | 1965 VIII | 1965f | 21 Oct. 1965 | 0.00779 |
| C/1969 Y1 Bennett | 1970 II | 1969i | 20 Mar. 1970 | 0.538 |
| C/1973 E1 Kohoutek | 1973 XII | 1973f | 28 Dec. 1973 | 0.142 |
| C/1975 V1 West | 1976 VI | 1975n | 25 Feb. 1976 | 0.197 |
| C/1980 E1 Bowell | 1982 I | 1980b | 12 Mar. 1982 | 3.364 |
| C/1983 H1 IRAS-Araki-Alcock | 1983 VII | 1983d | 21 May 1983 | 0.991 |
| C/1986 P1 Wilson | 1987 VII | 1986l | 20 Apr. 1987 | 1.200 |
| C/1989 X1 Austin | 1990 V | 1989$c_1$ | 10 Apr. 1990 | 0.350 |
| C/1990 K1 Levy | 1990 XX | 1990c | 24 Oct. 1990 | 0.939 |
| C/1996 B2 Hyakutake | | | 1 May 1996 | 0.230 |
| C/1995 O1 Hale-Bopp | | | 1 Apr. 1997 | 0.914 |
| D/1993 F2 Shoemaker-Levy 9 | | 1993e | | |
| 1P/Halley | 1986 III | 1982i | 9 Feb. 1986 | 0.587 |
| 2P/Encke | | | 23 May. 1997 | 0.331 |
| 21P/Giacobini-Zinner | 1985 XIII | 1984e | 5 Sep. 1985 | 1.028 |
| 22P/Kopff | | | 2 Jul. 1996 | 1.585 |
| 23P/Brorsen-Metcalf | 1989 X | 1989o | 11 Sep. 1989 | 0.479 |
| 29P/Schwassmann-Wachmann 1 | | | | 5.772 |
| 46P/Wirtanen | | | 14 Mar. 1997 | 1.083 |
| 73P/Schwassmann-Wachmann 3 | | | 22 Sep. 1995 | 0.936 |
| 95P/Chiron | | | 14 Feb. 1996 | 8.454 |
| 109P/Swift-Tuttle | 1992 XXVIII | 1992t | 12 Dec. 1992 | 0.958 |

The space exploration of P/Halley by several space probes [88] permitted for the first time the imaging of a comet nucleus and the *in situ* observation of coma material by mass spectroscopy. Then a rapid sequence of new discoveries took place as new technologies were available or improved. These two last years (1996 and 1997) we were granted with the apparition of two exceptional comets: C/1996 B2 (Hyakutake), which passed at only 0.10 AU from the Earth, and C/1995 O1 (Hale-Bopp), the *Great Comet* of this century, which passed at perihelion on 1 April 1997.

TABLE 2. Characteristics of cometary nuclei and other Solar System small bodies (compiled from various sources).

| Object | revolution period [y] | radius [km] | rotation period [h] | albedo |
|---|---|---|---|---|
| *Comets:* | | | | |
| IRAS-Araki-Alcock 1983 VII | long | $8 \times 3$ | ? | 0.03 |
| Sugano-Saigusa-Fujikawa 1983 V | long | 0.4 | | 0.03? |
| C/1995 O1 (Hale-Bopp) | long | 22 | 11.5 | ? |
| 109P/Swift-Tuttle | 135 | 12 | 72. | 12 |
| 1P/Halley | 76.0 | $8 \times 4$ | complex | 0.04 |
| 28P/Neujmin 1 | 18.2 | $11 \times 9$ | 12.7 | 0.02 |
| 29P/Schwassmann-Wachmann 1 | 14.9 | 20 | complex? | 0.10 |
| 49P/Arend-Rigaux | 6.8 | $7 \times 4$ | 13.6 | 0.03 |
| 31P/Schwassmann-Wachmann 2 | 6.4 | 3 | 5.6 | 0.04? |
| 10P/Tempel 2 | 5.5 | $8 \times 4$ | 8.9 | 0.02 |
| 2P/Encke | 3.3 | $\sim 2$ | 15.1 | $\sim 0.03$ |
| C/1996 B2 (Hyakutake) | long | $\sim 1$ | $\sim 6$ | |
| 46P/Wirtanen | 5.5 | $\sim 0.6$ | 6? | |
| *Asteroids:* | | | | |
| (4179) Toutatis | 4.0 | 2.3 | complex | |
| (1) Ceres | 4.6 | 456 | 9.1 | 0.10 |
| (243) Ida | 4.8 | 28 | 4.5 | |
| *Centaurs:* | | | | |
| (2060) Chiron = 95P/Chiron | 50.9 | 90 | 5.9 | 0.13 |
| (5145) Pholus | 92.3 | $\sim 100$ | ? | ? |
| *Trans-Neptunian objects:* | | | | |
| Pluto | 247.7 | 1150 | 153 | 0.61 |
| 1992 $QB_1$ | $\sim 285$ | $\sim 100$ | ? | ? |

## 2. Physical processes in comets.

### 2.1. PHYSICAL PROCESSES IN COMETARY NUCLEI

#### 2.1.1. *The physical nature of cometary nuclei*

Little is know on the actual physical properties of comet nuclei. Only 1P/Halley's nucleus has been explored and imaged by a space probe. A bare cometary nucleus can in principle be observed in the same way as an asteroid, providing estimation of its size and albedo from visible and infrared photometry, and of its rotation period from the light curve. However, this is only possible when the comet is inactive, i.e. far from the Sun when the object is very faint. Such informations are summarized in Table 2.

The density of cometary nuclei is still very poorly known. Their masses have never been measured directly (this could be done in the future by orbiting a spacecraft like *ROSETTA* around a comet nucleus). They have been inferred, quite indirectly, by comparing the forces of the rocket effect, due to anisotropic outgassing, to the non-gravitational forces estimated from the perturbations of their orbits [93]. The inferred densities are poorly constrained: from 0.1 to 0.9.

The tensile strength of cometary nuclei is very low. This is testified by the high fragility of these bodies which are very easily disrupted by tidal forces when passing close to Jupiter (e.g. D/1993 F2 (Shoemaker-Levy 9); [89, 114]) or to the Sun (Sun-grazing comets such as C/1965 S1 (Ikeya-Seki)). But there are also many examples of comets which split *spontaneously*, merely under the stresses caused by their activity [25]. Tensile strengths as low as $10^{-3}$ bar are estimated [59]. This further suggests that cometary nuclei have a low density and are probably a very porous medium.

#### 2.1.2. *Physical processes in cometary nuclei and the activity of distant comets*

The activity of comets is governed by the sublimation of the nucleus ices caused by solar insolation. Close to the Sun, water is believed to be the major sublimating species. At heliocentric distances $r$ larger than $\sim 4$ AU, however, the equilibrium temperature of the nucleus surface is too low to permit efficient water sublimation. Many examples of cometary activity at larger distances have however been observed: 95P/Chiron at $r \sim 11$ AU, 29P/Schwassmann-Wachmann 1 at $r \sim 6$ AU, the outburst of 1P/Halley observed at $r \sim 14$ AU, C/1995 O1 (Hale-Bopp) already quite active when discovered at $r = 7$ AU. Thus, species more volatile than water should be responsible for these distant activities.

Table 3 lists the sublimation temperatures for a selection of molecular species identified, or likely to be present, in cometary ices. These temperatures, which correspond to pure ice species in equilibrium with their own vapours, must be taken with a pinch of salt, because real cometary ices are mixtures which behave in a much more complicated way. For instance, a significant fraction of CO or $CO_2$ can remain trapped in water ice at temperatures higher than 100 K [96].

A significant step was made in the understanding of distant activity of comets when important quantities of CO were found to be outgassed from comets 29P/Schwassmann-Wachmann 1 at $r = 6$ AU ([34, 98]; Fig. 3) and C/1995 O1 (Hale-Bopp) at $r = 6.7$ AU [10, 69]. For this latter comet, it was possible to watch for the apparition of every molecular species when it approached from 7 to 1 AU to the Sun (Table 4).

Radio observations of comet C/1995 O1 (Hale-Bopp) also allowed us to monitor the production rates of various molecules from $r = 7$ to 1 AU

TABLE 3. Sublimation temperatures of selected cometary volatiles. Sublimation temperatures are given for a pure species in equilibrium with its own vapour, the total gas density being $10^{13}$ $cm^{-3}$ with the listed relative abundance [X]/[H] of the given species. These temperatures are either taken from [110] (Y) or extrapolated from the Clausius-Clapeyon equation, using the data of the *Handbook of Chemistry and Physics* for species in the solid (S) or liquid (L) phase.

| Species | | log [X]/[H] | $T_{subl}$ | Reference |
|---|---|---|---|---|
| anthracene | $C_{14}H_{10}$ | −8 | 200 | HCP(S) |
| phenanthrene | $C_{14}H_{10}$ | −8 | 160 | HCP(S) |
| naphtalene | $C_{10}H_8$ | −8 | 160 | HCP(S) |
| water | $H_2O$ | −4 | 152 | Y |
| hydrogen peroxide | $H_2O_2$ | −7 | 128 | HCP(S) |
| formic acid | HCOOH | −7 | 112 | Y |
| acetic acid | $CH_3COOH$ | −8 | 106 | HCP(S) |
| ethanol | $C_2H_5OH$ | −8 | 105 | HCP(L) |
| benzene | $C_6H_6$ | −8 | 104 | HCP(S) |
| methanol | $CH_3OH$ | −8 | 99 | Y |
| hydrogen cyanide | HCN | −7 | 95 | Y |
| methyl formate | $CH_3OHCO$ | −8 | 95 | HCP(L) |
| methyl cyanide | $CH_3CN$ | −8 | 93 | HCP |
| sulphur dioxide | $SO_2$ | −8 | 83 | Y |
| ammonia | $NH_3$ | −7 | 78 | Y |
| carbon disulphide | $CS_2$ | −8 | 78 | HCP(L) |
| cyanoacetylene | $HC_3N$ | −8 | 74 | Y |
| carbon suboxide | $C_3O_2$ | −8 | 73 | HCP(L) |
| acetaldehyde | $CH_3CHO$ | −8 | 73 | HCP(L) |
| carbon dioxide | $CO_2$ | −5 | 72 | Y |
| propyne | $CH_3CCH$ | −8 | 65 | Y |
| formaldehyde | $H_2CO$ | −8 | 64 | Y |
| dimethyl ether | $CH_3OCH_3$ | −8 | 60 | HCP(L) |
| nitrous oxide | $N_2O$ | −7 | 59 | HCP(S) |
| hydrogen sulphide | $H_2S$ | −7 | 57 | Y |
| carbonyl sulphide | OCS | −8 | 57 | HCP(L) |
| acetylene | $C_2H_2$ | −8 | 54 | HCP(S) |
| ethane | $C_2H_6$ | −8 | 44 | HCP |
| ethylene | $C_2H_4$ | −8 | 38 | HCP(L) |
| methane | $CH_4$ | −4.5 | 31 | Y |
| carbon monoxide | CO | −5 | 24 | Y |
| dinitrogen | $N_2$ | −4 | 22 | Y |
| dihydrogen | $H_2$ | 0 | 5 | Y |

([9]; Fig. 2). The evolution of the production rates differs from one species to another. For CO and for OH (which traces $H_2O$), slope variations are observed, which may be related to different regimes of sublimation. For instance, water was detected as far as $r = 4.7$ AU and showed a steep increase up to $r \sim 3$ AU: this may be due to water sublimated from icy grains at these large distances (such icy grains were indeed observed at $r \sim 7$ AU: [41]; Fig. 7).

### 2.1.3. *Models of cometary nuclei*

A series of thermal models of cometary nuclei have been achieved by several groups (e.g. [24, 72, 96, 112, 115]). They have to take into account:

- the sublimation/recondensation of volatiles in a porous matrix;
- the amorphous-crystalline water ice phase transition, which is exothermal and may induce sublimation of volatiles;
- gas transport in a porous medium;
- heat transport;
- dust transport and mantle formation/destruction, controlling solar heating and ice sublimation;
- the role of nucleus rotation and of the spin axis direction in day-night and seasonal effects;
- the secular variation over several orbits and the possibility of orbit variations.

These models are solving through numerical integration (now in several dimensions) for the gas production evolution of several species as a function of a set of initial parameters (nucleus composition and physical parameters, comet orbital parameters). The main result is that the abundances of volatiles in the coma are a complex function of the abundances of the ices in the nucleus, of the heliocentric distance and of the comet history. In other terms, sublimation fractionation is an important effect, and it would be an error to consider the volatile abundances measured in the coma as directly representative of the cometary ice composition.

## 2.2. PHYSICAL PROCESSES IN COMETARY ATMOSPHERES

Although the physical processes in cometary atmospheres are not directly relevant to the knowledge of cometary nuclei, they must be fully understood in order to relate the secondary products — which are those observed by remote sensing in cometary atmospheres — to the original material of cometary nuclei.

TABLE 4. Molecules observed in distant comets. Each entry gives the observation at the farthest distance $r$ from the Sun at which the given molecule has been observed.

| Comet | $r$ [AU] | molecule | $Q$ $10^{28}$ s$^{-1}$ | reference |
|---|---|---|---|---|
| C/1980 E1 (Bowell) | 5.2 | OH | 7. | [4] |
| | 4.6 | CN | 0.00002 | [4] |
| 1P/Halley | 4.8 | CN | 0.0003 | [107] |
| | 2.8 | OH (radio) | | [55] |
| | 2.6 | OH (UV) | | [50] |
| 95P/Chiron | 11.3 | CN$^{a)}$ | 0.14 | [23] |
| | | CO (radio)$^{a)}$ | | *IAUC* 6193 |
| 29P/Schwassmann-Wach. 1 | 5.8 | CN | 0.0008 | [26] |
| | 6.1 | CO (radio) | 4. | [34, 98] |
| | 6. | $CO^+$ | | *several* |
| C/1995 O1 (Hale-Bopp) | 6.7 | CO (radio) | 2. | [10, 69] |
| | 6.8 | CN | 0.006 | [54, 100] |
| | 4.9 | $CH_3OH$ (radio) | 0.09 | *IAUC* 6382, [9] |
| | 4.8 | OH (UV) | 1.5 | [101] |
| | 4.8 | CS (UV) | 0.012 | [101] |
| | 4.8 | HCN (radio) | 0.02 | *IAUC* 6377 |
| | 4.7 | OH (radio) | 5. | [9] |
| | 4.7 | $CO_2$ (IR) | 1.3 | [36] |
| | 4.6 | $C_2$ | 0.05 | [91] |
| | 4.6 | $C_3$ | | [91] |
| | 4.3 | $H_2S$ (radio) | 0.03 | *IAUC* 6408, [9] |
| | 4.3 | $NH_2$ | | [91] |
| | 4.1 | CS (radio) | 0.014 | [9] |
| | 4.1 | $H_2CO$ (radio) | 0.01 | [9] |
| | 3.4 | $CH_3CN$ (radio) | 0.012 | [9] |
| | 3.4 | $CO^+$ | | *IAUC* 6468 |
| | 2.8 | $H_2O^+$ | | [91] |
| | 2.4 | HNC | | [9] |

a) unconfirmed observation; see [92].

### 2.2.1. *Chemistry in cometary atmospheres*

The chemistry of cometary atmospheres and tails is dominated by photolysis due to solar irradiation and interaction with the solar wind. Photolysis is mainly destructive, causing the disruption of the *parent molecules* directly sublimated from the nucleus ices into radicals, molecular ions, atoms and atomic ions — all these species being usually designated as *daughter molecules* (which is a misnomer). Some of the photolysis rates $\beta$ of cometary

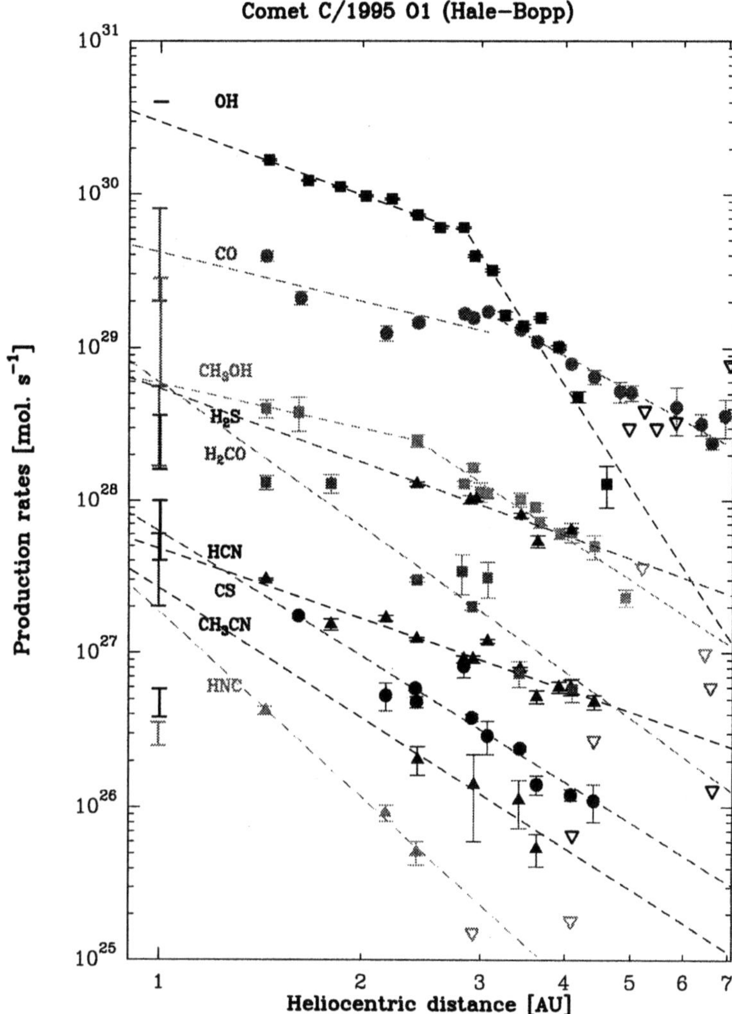


*Figure 2.* The evolution of molecular production rates in comet C/1995 O1 (Hale-Bopp) as a function of heliocentric distance, derived from pre-perihelion radio observations (from [9]).

parent molecules, from which the lifetimes $1/\beta$ and scalelengths $v_{exp}/\beta$ may be inferred, are listed in Table 5. At $r = 1$ AU, molecular lifetimes range from hours to days .

Solar photolysis leads to partial ionisation of the cometary atmosphere. Interaction with the energetic ions of the solar wind leads ultimately to the formation of the ionic tail (its main constituents being $H_2O^+$, $CO^+$,

TABLE 5. Photo-destruction rates of cometary parent molecules for quiet Sun at 1 AU. Compiled from [32, 65], and references therein.

| | Molecule | | rate [$s^{-1}$] |
|---|---|---|---|
| *water group:* | water | $H_2O$ | $1.3 \times 10^{-5}$ |
| | hydrogen peroxide | $H_2O_2$ | $1.3 \times 10^{-4}$ |
| *hydrocarbons:* | methane | $CH_4$ | $8.0 \times 10^{-6}$ |
| | acetylene | $C_2H_2$ | $1.4 \times 10^{-5}$ |
| | ethylene | $C_2H_4$ | $4.8 \times 10^{-5}$ |
| | ethane | $C_2H_6$ | $1.1 \times 10^{-5}$ |
| | benzene | $C_6H_6$ | $1.1 \times 10^{-3}$ |
| *CO group:* | carbon monoxide | $CO$ | $7.5 \times 10^{-7}$ |
| | carbon dioxide | $CO_2$ | $2.0 \times 10^{-6}$ |
| | carbon suboxide | $C_3O_2$ | $1.1 \times 10^{-4}$ |
| *CHO species:* | formaldehyde | $H_2CO$ | $2.0 \times 10^{-4}$ |
| | methanol | $CH_3OH$ | $1.3 \times 10^{-5}$ |
| | formic acid | $HCOOH$ | $3.2 \times 10^{-5}$ |
| | ethanol | $C_2H_5OH$ | $1.8 \times 10^{-5}$ |
| | dimethyl ether | $(CH_3)_2O$ | $3.1 \times 10^{-5}$ |
| | acetic acid | $CH_3COOH$ | $5.1 \times 10^{-5}$ |
| *nitrogen compounds:* | ammonia | $NH_3$ | $1.5 \times 10^{-4}$ |
| | hydrogen cyanide | $HCN$ | $1.5 \times 10^{-5}$ |
| | cyanoacetylene | $HC_3N$ | $6.6 \times 10^{-5}$ |
| | methyl cyanide | $CH_3CN$ | $6.7 \times 10^{-6}$ |
| | isocyanic acid | $HNCO$ | $2.9 \times 10^{-5}$ |
| *sulphur compounds:* | hydrogen sulphide | $H_2S$ | $2.5 \times 10^{-4}$ |
| | carbon disulphide | $CS_2$ | $1.7 \times 10^{-3}$ |
| | sulphur monoxide | $SO$ | $1.5 \times 10^{-4}$ |
| | sulphur dioxide | $SO_2$ | $2.1 \times 10^{-4}$ |
| | carbonyl sulphide | $OCS$ | $9.4 \times 10^{-5}$ |

$CO_2^+$). In the collisional region of the coma, ion-neutral reactions are much more efficient that neutral-neutral reactions, and lead to the formation of protonated molecules. Of these, $HCO^+$ and $H_3O^+$ have been recently identified by radio spectroscopy in C/1995 O1 (Hale-Bopp). But several others such as $NH_4^+$ and $H_3S^+$ have been inferred from the mass spectra observed *in situ* in 1P/Halley.

An important ingredient to cometary chemistry is the role of grains, which are now recognized to be a significant source of release of molecules in the coma. This will be detailed in Section 3.3.

### 2.2.2. *Physics of cometary atmospheres*

Due to the very low gravity of comets, cometary atmospheres are not captive: they expand. This expansion, which is almost adiabatic, is similar to those of supersonic jets. The kinetic temperature is dropping rapidly with distance to nucleus. After several thousands kilometres from the nucleus, collisions are rare and the expanding atmosphere is similar to a free molecular jet.

The photodissociation of parent molecules by solar UV radiation produces high-velocity atoms and radicals which, if they can get thermalized by collisions, will heat the coma and increase its expansion velocity. The cometary atmosphere can also be radiatively cooled; the most efficient process is the emission of water rotational lines, but the large optical depth of these lines limits the efficiency of this process.

The temperatures and expansion velocities of the coma were measured *in situ* in comet Halley by the *Giotto* probe. They can also be measured by remote sensing. The analysis of radio line shapes (Fig. 3) provides the expansion velocity. The rotational population distribution of molecules, which can be determined from their rotational radio lines (Fig. 4) or their rovibrational IR lines (Fig. 12), provides the coma temperature.

Observed expansion velocities $v_{exp}$ range from less than 0.5 to more than 1 km $s^{-1}$, and temperatures $T_{kin}$, from about 10 K to about 100 K, when comets are approaching from $r \sim 6$ to $\sim 1$ AU [9]. $v_{exp}$ and $T_{kin}$ are controlled not only by $r$, but also by the total gas production rate $Q$ of the comet, which governs the collision rates and thus the efficiency of the photolytic heating. Except for comet Halley for which they were measured *in situ*, the velocity and temperature profiles (i.e. their variation with distance to nucleus) are still poorly known.

Many detailed models of the hydrodynamical processes within cometary atmospheres have been achieved [29, 30]. They are qualitatively in good agreement with the observed $v_{exp}$ and $T_{kin}$, and their variations with $r$, $Q$ and distance to nucleus.

The formation and evolution of cometary tails, their relation with interstellar dust, will not be reviewed here (see [42]).

## 3. The composition of comets from remote-sensing observations

Our knowledge of the composition of cometary ices built up progressively. First, by guessing which molecules could be the parents of the radicals identified in the visible spectra. Then, with the help of the newly developed techniques: ultraviolet, radio and infrared spectroscopy, which enabled us to look for the signatures of parent molecules, not present in the visible domain. The space exploration of Halley's comet in 1986 and the ground-

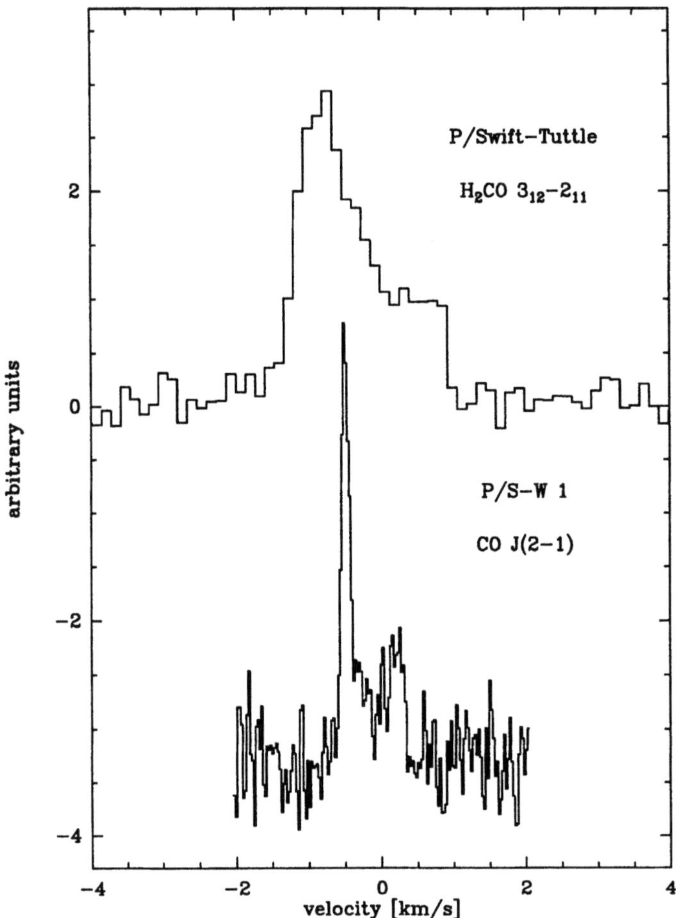


*Figure 3.* Two extreme examples of radio line profiles probing the kinematics of the coma. Top: the $3_{12} - 2_{11}$ formaldehyde line in 109P/Swift-Tuttle at 1 AU from the Sun; at this distance, strong sunward jets were present and the coma had a high expansion velocity (from [43]). Bottom: the $J(2 - 1)$ carbon monoxide line in 29P/Schwassmann-Wachmann 1 at 6 AU from the Sun; the coma was very cold (probably less than 10 K to explain the very narrow feature at negative velocities) and the expansion velocity was small (from [34]).

based International Halley Watch were an important milestone. Then several bright comets contributed to improve our knowledge: C/1986 P1 (Wilson), C/1989 X1 (Austin), C/1990 K1 (Levy), 109P/Swift-Tuttle...But

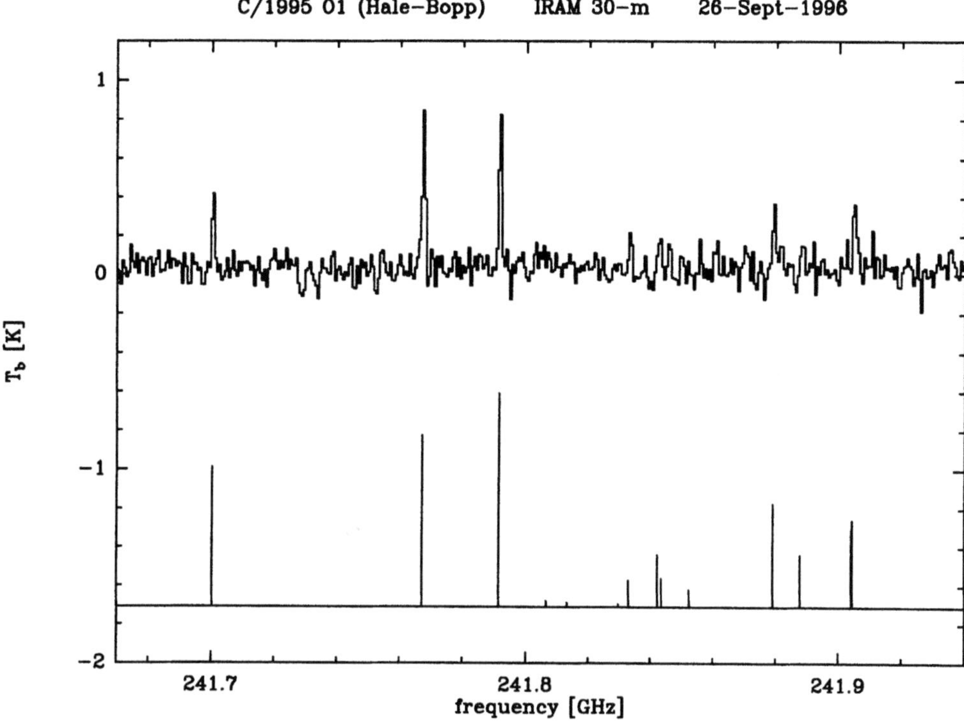


*Figure 4.* The lines of methanol around 242 GHz observed at IRAM in comet C/1995 O1 (Hale-Bopp), from which an accurate rotational temperature can be determined (from [11].)

a dramatic new step was made recently with the two exceptional comets C/1996 B2 (Hyakutake) and C/1996 B2 (Hyakutake).

### 3.1. THE COMPOSITION OF COMETARY ICES

Spectroscopy is a very powerful tool for investigating the composition of cometary gases. Cometary molecules are emitting through their electronic systems and vibrational bands following fluorescence excited by the Sun's radiation. Emission rates (the so-called $g$-factors, i.e. the numbers of photons emitted per second by one molecule) are typically $10^{-4}$ $s^{-1}$ for fundamental bands of vibrations, $10^{-3}$ to $10^{-2}$ $s^{-1}$ for electronic systems at 1 AU from the Sun. Unfortunately, for *parent* molecules, the electronic systems are not in the visible domain, but rather in the UV where their excitation by the Sun leads to destruction of the molecule rather than to fluorescence. Important exceptions are for CO, SO, $SO_2$ and $S_2$ which have strong non-dissociative electronic bands in the UV or near-UV. Vibrational fundamental bands are all in the IR, longward of 2.5 $\mu$m. Rotational lines are in the radio domain. They are excited both radiatively, at the final step of

TABLE 6. Parent molecules in comets: the chronology of first detections.

| Year | comet/technique | molecules |
|---|---|---|
| 1973 | C/1973 E1 (Kohoutek) | |
| | radio | HCN? |
| 1976 | C/1975 V1 (West) | |
| | UV | CO |
| 1983 | C/1983 H1 (IRAS-Araki-Alcock) | |
| | radio | $NH_3$? |
| | UV | $S_2$ |
| 1985–86 | 1P/Halley | |
| | radio | HCN, $H_2CO$? |
| | IR | $H_2O$, $CO_2$ |
| 1990 | C/1989 X1 (Austin), C/1990 K1 (Levy) | |
| | radio | $H_2CO$, $H_2S$, $CH_3OH$ |
| 1996 | C/1996 B2 (Hyakutake) | |
| | radio | $NH_3$, HDO, HNC, $CH_3CN$, OCS? HNCO? |
| | IR | $CH_4$, $C_2H_2$, $C_2H_6$ |
| 1997 | C/1995 O1 (Hale-Bopp) | |
| | radio | OCS, HNCO, $HC_3N$, $SO_2$, HCOOH, $H_2CS$, $NH_2CHO$, $HCOOCH_3$ |

the fluorescence cascades, and thermally by collisions. Radio spectroscopy proved to be a very sensitive way to probe even minor constituents. However, a whole class of molecules escape radio observations: the non-polar symmetric molecules, among which $CO_2$ and most of the simplest hydrocarbons. On the other hand, water and light hydrides have their stronger rotational lines in a submillimetric domain inaccessible from the ground.

It is usual to evaluate the abundances of cometary volatiles with respect to water, which appears to be the main constituent of cometary ices. Water, however, is very difficult to observe directly and its production rate has to be evaluated from the observations of its secondary products, such as the OH radical or the O and H atoms. Water has been directly observed from space in a very few number of comets, using the Kuiper Airborne Observatory [76, 86, 103], the *Vega* probes [28] or *ISO* [36, 38]. However, it has been recently demonstrated that the lines of a hot vibrational band of water can be observed from the ground using high-resolution infrared spectroscopy [84].

Table 6 lists the successive discoveries of cometary parent molecules from remote sensing and Table 7 summarizes our current knowledge of the abundances of cometary ices. It must be emphasized that these abundances may be still ill-known due to shortcomings in our knowledge of the molec-

TABLE 7. Abundances of cometary volatiles. This table has been established from [12,31] and from the preliminary reports given in the *IAU Circ.* for the results on comets C/1996 B2 (Hyakutake) and C/1995 O1 (Hale-Bopp).

| molecule | abundance (a) | method (b) | Nb (c) | comments |
|---|---|---|---|---|
| $H_2O$ | =100. | IR | 6 | also indirect (from OH, O, H) |
| HDO | 0.05 | radio | 2 | |
| CO | 2–20. | UV, radio, IR | > 5 | extended source? |
| $CO_2$ | 2–6. | IR | 2 | |
| $H_2CO$ | 0.05–4. | radio | > 5 | extended source |
| $CH_3OH$ | 1–7. | radio, IR | > 5 | |
| HCOOH | ~ 0.1 | radio | 1 | |
| HNCO | 0.07 | radio | 2 | |
| $NH_2CHO$ | ~ 0.1 | radio | 1 | |
| $HCOOCH_3$ | 0.05 | radio | 1 | |
| $CH_4$ | 0.7 | IR | 2 | |
| $C_2H_2$ | 0.3–0.9 | IR | 2 | |
| $C_2H_6$ | 0.4 | IR | 2 | |
| $NH_3$ | 0.5 | radio | 3 | also indirect (from NH, $NH_2$) |
| HCN | 0.1–0.2 | radio, IR | > 5 | |
| HNC | 0.01 | radio | 2 | |
| $CH_3CN$ | 0.01 | radio | 2 | |
| $HC_3N$ | 0.01 | radio | 1 | |
| $H_2S$ | 0.3 | radio | > 5 | |
| $H_2CS$ | 0.01 | radio | 1 | |
| CS | 0.1 | UV, radio | > 5 | from $CS_2$? |
| OCS | 0.3 | radio | 2 | |
| SO | ~ 0.5 | radio | 1 | from $SO_2$? |
| $SO_2$ | ~ 0.1 | radio | 1 | |
| $S_2$ | 0.005–0.2 | UV | 2 | |

(a) abundance relative to water (100); listed abundances may be uncertain by a factor of 2 or more for some species.
(b) method of observation.
(c) number of comets in which this species was reliably and directly observed.

ular excitation conditions or space distributions. It must also be noted that many molecules with significant abundances may still have escaped detection, just because they do not have spectral signals appropriate for observation with present techniques (this may well be the case for the $H_2$, $N_2$ and $O_2$ molecules).

Detailed reviews on the composition of cometary volatiles may be found elsewhere [12, 16, 31, 52, 87]. They will not be repeated here, but a few

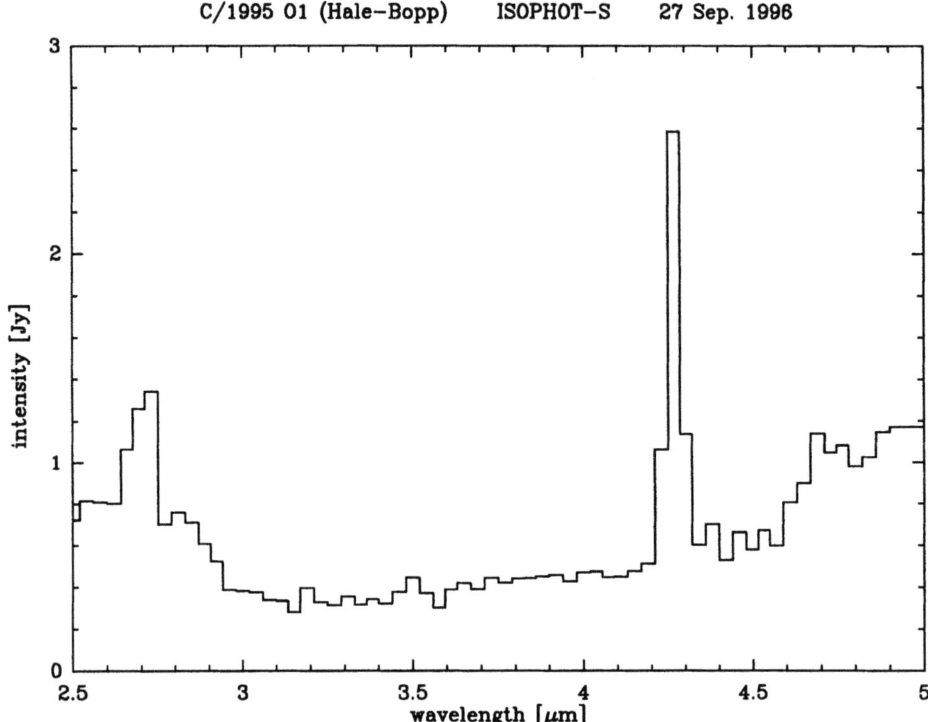


*Figure 5.* The 2.5–5 $\mu$m spectrum of comet C/1995 O1 (Hale-Bopp) observed at 2.9 AU from the Sun with the PHT-S instrument of *ISO*. The aperture is 24" × 24" and the spectral resolution is $\lambda/\delta\lambda \sim 90$. One can see the $\nu_3$ band of $H_2O$ at 2.9 $\mu$m, the $\nu_2$ band of $CO_2$ at 4.25 $\mu$m, and the $v(1-0)$ band of CO at 4.65 $\mu$m (from [38]).

points raised by recent observations will be stressed below.

- Carbon dioxide is an important constituent of cometary ices, but it is very difficult to observe. It could only be observed directly from space in comets 1P/Halley from the *Vega* probes [28] and in C/1995 O1 (Hale-Bopp) with *ISO* (Fig. 5; [36, 38]). However, the *Cameron* bands of CO in the UV observed in comets are believed to come from prompt emission following photodissociation of $CO_2$ and could trace carbon dioxide [102].
- The $C_2$ and $C_3$ radicals have been for a long time *orphan molecules* for which no parent could be identified. This is no longer the case with the identification of several species with a $C_2$ skeleton ($C_2H_2$, $C_2H_6$, $CH_3CN$) and one with a $C_3$ skeleton ($HC_3N$).
- Many nitrogen-bearing molecules are now known; several nitriles (HCN, HNC, $HC_3N$, $CH_3CN$), HNCO [77], $NH_2CHO$ and ammonia which is now firmly identified from radio observations [106]. The nitrogen depletion formerly found in comets is no longer a problem [51,

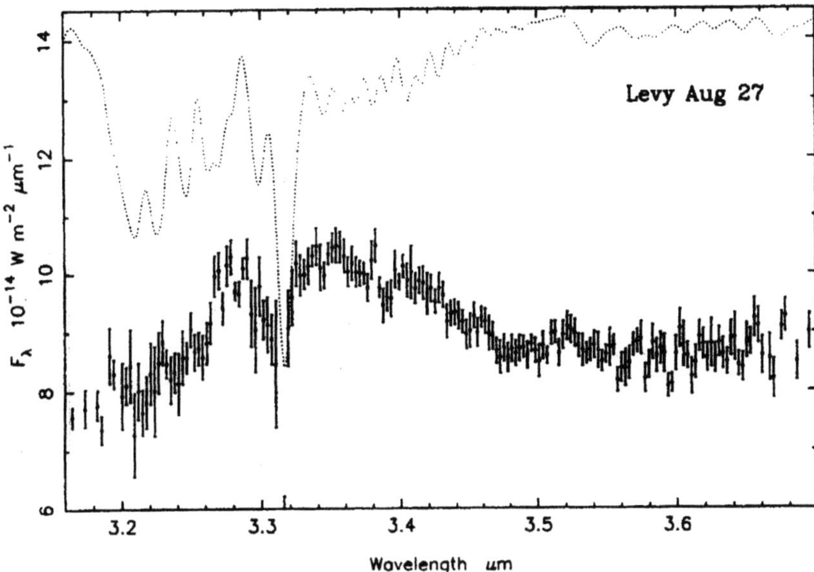


*Figure 6.* The spectrum of comet C/1990 K1 (Levy) observed on August 27 at the UKIRT telescope (from [39]). The top curve shows the transmission of the atmosphere. The feature at 3.52 $\mu$m is attributed to methanol ($\nu_3$ band), the feature at 3.28 $\mu$m is possibly due to aromatics, and the broader feature at 3.4 $\mu$m is due to methanol and other species.

108, 109]. $N_2$, however, is still an elusive molecule (only inferred from difficult observations of the $N_2^+$ ion; [109]).

- Sulphur is now well represented with $H_2S$, OCS, CS (which could trace $CS_2$), SO, $SO_2$ and $H_2CS$ [37, 81]. The $S_2$ molecule has now been observed in two comets which passed close to the Earth — C/1983 H1 (IRAS-Araki-Alcock) [1] and C/1996 B2 (Hyakutake) — but its origin is still mysterious. However, it has been suggested that it could have been produced by protoprocessing of interstellar dust mantles [111].
- The 3.2–3.5 $\mu$m emission band in comets has had for a long time a mysterious origin. Methanol is known to be an important contributor [14, 18, 19, 40, 63, 64]. Other contributors among species recently identified are ethane ($C_2H_6$) and formic acid (HCOOH). But other contributors, possibly in the dust phase, may still be present. A discrete band at 3.28 $\mu$m seen in several comets (Fig. 6; [14, 39]) is indicative of the presence of aromatics.
- There are still many unidentified lines in visible and IR cometary spectra, presumably due to ions, radicals or even parent molecules [6, 21, 22, 83]. Their identification is pending to laboratory work.

TABLE 8. The elemental composition of comets. Average elemental abundances measured in comet C/1965 S1 (Ikeya-Seki) (from [7]), in Halley's dust grains and in the whole comet dust and volatiles (from [67]), and in other solar system objects (from [5]). The abundances are normalized to Mg (to the solar system abundance of Fe for Ikeya-Seki, in which Mg was not observed).

| element | Ikeya-Seki[a)] | P/Halley | | Solar system | CI-chondrite |
|---|---|---|---|---|---|
| | | dust | dust + ice | | |
| H | | 2 025. | 4 062. | $2.6\times10^6$ | 492. |
| Li | † | | | 0.0053 | 0.0053 |
| C | | 814. | 1 010. | 0. | 70.5 |
| N | | 42. | 95. | 291. | 5.6 |
| O | | 890. | 2 040. | 2 216. | 712. |
| Na | † | 10. | 10. | 5.3 | 5.3 |
| Mg | (=100.0) | =100.0 | =100.0 | =100.0 | =100.0 |
| Al | | 6.8 | 6.8 | 7.9 | 7.9 |
| Si | | 185. | 185. | 93. | 93. |
| P | | | | 1.0 | 1.0 |
| S | | 72. | 72. | 48. | 48. |
| K | † | 0.2 | 0.2 | 0.35 | 0.35 |
| Ca | | 6.3 | 6.3 | 5.7 | 5.7 |
| Ti | <0.02 | 0.4 | 0.4 | 0.22 | 0.22 |
| V | 0.01 | | | 0.027 | 0.027 |
| Cr | 0.08 | 0.9 | 0.9 | 1.3 | 1.3 |
| Mn | 0.5 | 0.5 | 0.5 | 0.89 | 0.89 |
| Fe | 84. | 52. | 52. | 84. | 84. |
| Co | 0.4 | 0.3 | 0.3 | 0.21 | 0.21 |
| Ni | 7.2 | 4.1 | 4.1 | 4.6 | 4.6 |
| Cu | 0.2 | | | 0.049 | 0.049 |

a) In addition, abundances relative to Na were determined to be $1.6 \times 10^{-3}$ for K and $< 2.5 \times 10^{-5}$ for Li [90].

## 3.2. THE COMPOSITION OF COMETARY DUST

The *elemental* composition of cometary material could be determined (or constrained) in specific occasions.

- Comets passing very close to the Sun — the so-called *sun-grazers* — are disrupted and their refractories are volatilised. Their spectra show many atomic (mainly metallic) lines. A well documented case was that of comet C/1965 S1 (Ikeya-Seki) ([6, 90]; Table 8).
- The *Giotto* and *VEGA* mass spectrometers determined the elemental composition of 1P/Halley's dust particles ([67, 70]; Table 8).
- Recently, comet D/1993 F2 (Shoemaker-Levy 9) crashed on Jupiter. Its material was totally volatilised. The spectra lines of many atomic

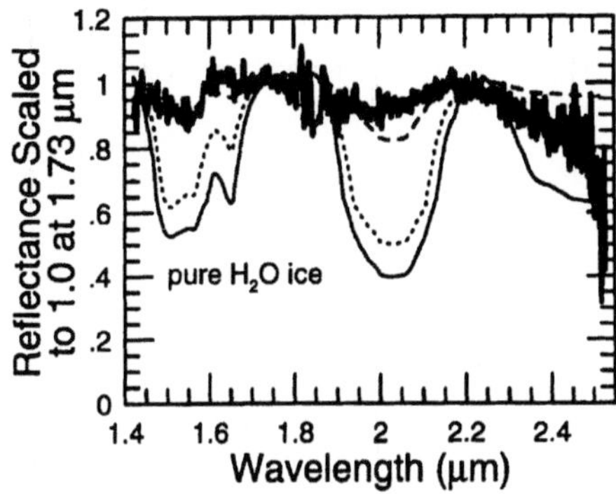


*Figure 7.* The 1.4–2.5 $\mu$m spectrum of comet C/1995 O1 (Hale-Bopp) observed at 7 AU from the Sun at UKIRT. The broad absorptions at 1.5 and 2.05 $\mu$m are attributed to water ice. The plain and dotted lines show a model spectrum of water ice (from [41]).

and molecular species ($H_2O$, CO, $NH_3$, HCN, $S_2$, CS, OCS, silicates), normally not present in the Jupiter spectrum, appeared during or just after the impact. It is, however, not easy to distinguish between original cometary material and processed Jupiter atmosphere. Atomic lines (Li, Na, Ca, Si, Fe, Mg, Mn, Cr) however, seen in the plumes following the impacts, could be attributed unambiguously to cometary matter (see review in [33] and references therein).

The study of cometary dust by mass spectroscopy, which is the subject of another lecture [70], will not be reviewed here. The composition of cometary dust can be also investigated by infrared spectroscopy. Infrared spectroscopy at low resolution revealed excess emissions at 10 and 20 $\mu$m, which correspond to the stretching and bending modes of the Si–O bond, respectively. Hence the hypothesis that silicates are an important constituent of cometary dust.

Two main kinds of silicates are olivines and pyroxenes. Olivines are solid solutions of fayalite ($Fe_2SiO_4$) and forsterite ($Mg_2SiO_4$). Pyroxenes are solid solutions of enstatite ($MgSiO_3$), ferrosilite ($FeSiO_3$) and wollastonite ($CaSiO_3$). The IR spectrum of all these species, when amorphous, presents only broad, featureless bands at 10 and 20 $\mu$m. In the crystalline form, these species have a structured 10 $\mu$m band and present a series of additional features at wavelengths depending upon their composition.

Detailed ground-based observations of the 10 $\mu$m band in several comets showed revealed that cometary dust could contain amorphous or crys-

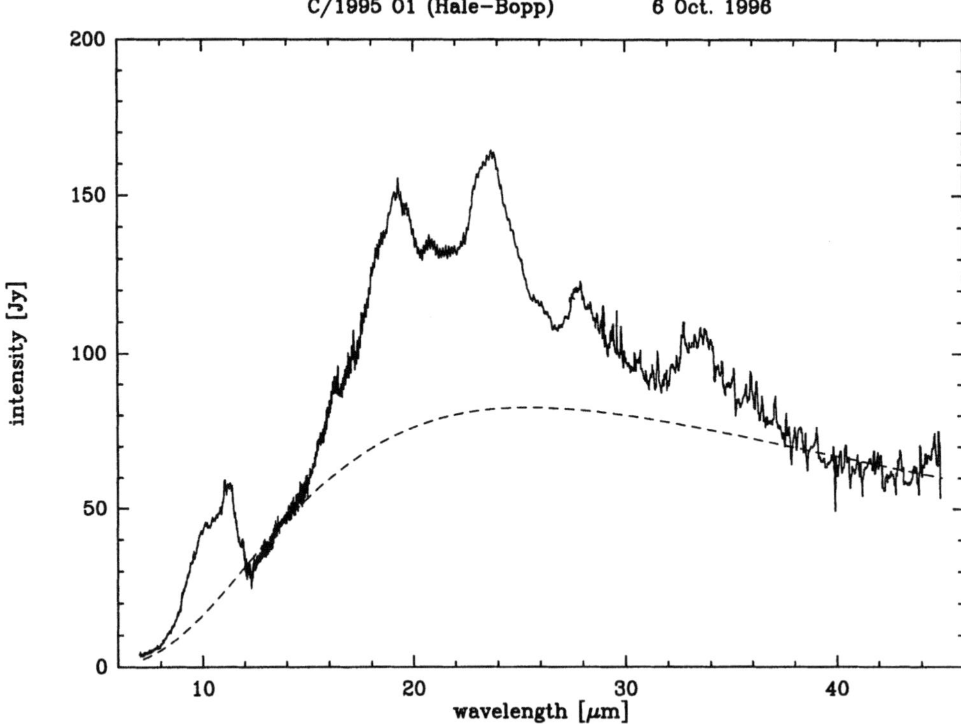


*Figure 8.* The 7–45 $\mu$m spectrum of comet C/1995 O1 (Hale-Bopp) at 2.9 AU from the Sun observed by the short-wavelength spectrometer of *ISO*. The spectral resolution has been degraded to $\lambda/\delta\lambda \sim 500$. The dotted line shows a 200 K black body emission. All spectral features in this region are attributed to Mg-rich crystalline olivine (from [38]).

talline silicates, depending upon the object (e.g. [60, 61]). Only for comet 1P/Halley could the dust spectrum be investigated at longer wavelengths, revealing features indicative of crystalline silicates [56, 62].

*ISO* observations of the complete 7 to 45 $\mu$m spectrum of comet Hale-Bopp at high-resolution (Fig. 8; [38]) showed a spectrum which is best matched by crystalline Mg-rich olivine (i.e. forsterite $Mg_2SiO_4$).

### 3.3. GRAINS AS SOURCES OF VOLATILES

It has been inferred in several occasions that grains could be important sources of volatiles in the coma. This is supported by several pieces of evidence:

- The necessity to include *extended* sources to explain the space distribution of CO [47] and $H_2CO$ [16, 27].
- The space distribution of the CN radical [2], showing jets which could be related to dust jets.

- The production of water at large heliocentric distances, which can only be explained by superheated icy grains [9].
- The water ice absorption feature observed in comet Hale-Bopp at $r \sim$ 7 AU (Fig. 7; [41]), pointing to the presence of icy grains
- The high CHON content of some grains observed by mass spectroscopy [67, 70], attributed to an organic mantle.

Icy grains are undoubtedly subject to a high sublimation rate and important contributors to the gas production (see Section 2.1.2). Their lifetime, however, is very short at small $r$'s. They thus can difficultly account for the distributed sources of CO and $H_2CO$ at these heliocentric distances. The sublimation of *semi-refractory* species can also occur (e.g. PAH's).

The release of molecules or radicals from the refractory organic mantle of grains requires high energy processes. Sputtering by solar UV photons or high-velocity ions have been invoked. These processes, however, are highly inefficient, because of the small cross-section of molecules condensed on grains, and cannot account for significant molecular production rates.

### 3.4. THE DIVERSITY OF COMETS

Are comets all alike, or do they present diversity of composition? If existing, is this diversity related to a diversity of origin or history?

There are extreme cases of diversity: comet C/1988 Y1 (Yanaka) was dramatically depleted in $C_2$ and CN, but showed prominent bands of $NH_2$ [53]; comet 43P/Wolf-Harrington was depleted in $C_2$ and $C_3$ relative to CN and OH [95]. A systematic study of cometary composition, using narrow-band photometric observations of the OH, CN, $C_2$, $C_3$ and NH, has been undertaken on 85 comets [3]. A class of *carbon-depleted comets* which exhibits a lower $Q[C_2]/Q[CN]$ was identified. This class is significantly correlated with Jupiter-family comets

It would be interesting to investigate the meaning of such a correlation, by studying the variations of relative abundances of parent molecules rather than the daughter products. Unfortunately, such data, which require observations at radio or IR wavelengths, are available for less than ten comets, among which Jupiter-family comets (which are faint objects) are poorly represented. If important variations on molecular abundances, such as those of $CH_3OH$ or $H_2CO$, do exist, it is not yet possible to relate them to cometary families.

Concerning cometary dust, the 10-$\mu$m feature shape points to either crystalline or amorphous silicates. Both classes are represented for all families of comets [60, 61].

TABLE 9. Comparison of the abundances of ices in the interstellar medium (towards IRS9) and of cometary volatiles (at $\sim$ 1 AU). Cometary abundances are from Table 7. ISM ice abundances are taken from a compilation of [97] from various sources as well as from recent *ISO* results [45, 105].

| species | interstellar ices | | cometary volatiles | |
|---|---|---|---|---|
| $H_2O$ | =100 | | =100 | |
| CO | 10–40 | | 2–20 | |
| $CH_3OH$ | 5 | | 1–7 | |
| $CO_2$ | 10 | | 2–6 | |
| $H_2CO$ | 2–6 | tentative | 0.05–4 | |
| HCOOH | 3 | tentative | $\sim 0.1$ | |
| $CH_4$ | 1–2 | | 0.7 | |
| other hydrocarbons | ?? | | $\sim 1$ | $C_2H_2$, $C_2H_6$ |
| $NH_3$ | $< 10$ | | 0.5 | |
| $O_3$ | $\leq 2$ | | ?? | |
| XCN | $\leq 0.5 - 2$ | | 0.2 | nitriles + HNCO |
| OCS, XCS | 0.2 | | 0.4 | OCS + CS |
| $SO_2$ | ?? | | $\sim 0.1$ | |
| $H_2$ | $>\sim 1$ | | ?? | |
| $N_2$ | ?? | | ?? | |
| $O_2$ | ?? | | ?? | |

## 4. The origin of comets

The composition and physical state of cometary material are important clues to the origin and formation of comets.

### 4.1. COMETARY MATTER AND INTERSTELLAR MATTER

#### 4.1.1. *The gross composition of cometary ices.*

Table 9 compares the composition of interstellar ices with that of the cometary volatiles. One must have in mind the importance of sublimation fractionation already stressed out in Sections 2.1.2 and 2.1.3. However, it is striking that cometary and interstellar ices have quite comparable relative abundances.

It has been noted [21, 83] that the important abundance of $C_2$ hydrocarbons ($C_2H_2$ and $C_2H_6$) relative to methane excludes a formation chemical reactions at thermal equilibrium.

The HNCO molecule recently identified might be the sublimation product of the $OCN^-$ ion supposed to be present in ISM ices by [97].

4.1.2. *Cometary dust.*

C/1995 O1 (Hale-Bopp) is the first comet for which, thanks to *ISO* observations, a constituent of dust could be clearly identified: Mg-rich crystalline silicates, from the 6–45 $\mu$m high-resolution spectrum (Fig. 8; [38]).

There is no evidence of crystalline silicates in ISM dust. This is attested by the featureless 10 and 20 $\mu$m absorption bands and the absence of other dust features (see e.g. the IR absorption spectrum observed by *ISO* towards the Galactic Centre; [79]).

The dust in discs or envelopes of some stars do contain crystalline silicates. The $\beta$ Pic disc presents a 10 $\mu$m feature with the characteristic structure of crystalline silicates [73]. Another case is HD 100546 is an object intermediate between Herbig Ae/Be stars and Vega-type objects for which *ISO* observations are available [99]. Its 6–45 $\mu$m spectrum shows the same features as that of comet Hale-Bopp, suggesting that dust in its disc also contains crystalline forsterite.

It may be surprising that cometary dust — at least that analyzed in comet Hale-Bopp — only shows the features of crystalline forsterite. The silicates of interplanetary dust particles (*IDPs*) contain a large fraction of pyroxenes and are mostly amorphous [20].

It has been argued that crystalline silicates could form in the coma itself through annealing of amorphous silicates in dust grains heated by the Sun. However, this hypothesis is ruined by the observation that olivine could be crystalline also far from the Sun (at $r = 4.6$ AU in comet Hale-Bopp; [36]) where the temperature is obviously too small for this process to occur.

4.1.3. *Isotopic ratios.*

Besides deuterium, isotopic species abundances are available for $^{13}$C, $^{15}$N, $^{18}$O, and $^{34}$S (Table 10). Measurements come either from *in situ* mass spectroscopy of dust particle in 1P/Halley, or high-resolution spectroscopy of recent comets (radio and visible). If one considers only the results of the highest accuracy, the isotopic ratios are *normal* (solar). One should note that the isotopic ratios for these species in the interstellar medium are also *normal.* However, *Giotto* measurements of individual grains revealed a high dispersion of the [$^{13}$C]/[$^{12}$C] ratio [67, 70].

The situation is quite different for deuterium; The [D]/[H] ratio has been measured in 1P/Halley from $H_3O^+$ observed by mass spectroscopy, and converted by modelling ion-neutral reactions to [D]/[H] in cometary water ice ([8, 48]; Table 10). A radio line of HDO was also observed in comet C/1996 B2 (Hyakutake) (Fig. 9). For both comets [D]/[H] $\sim 3\times10^{-4}$. This shows an important deuterium enrichment over the [D]/[H] values for the local interstellar medium ($\sim 1.5 \times 10^{-5}$) and the protosolar nebula ($\sim 2 \times 10^{-5}$). It is even larger than the Earth oceans value ($1.5 \times 10^{-4}$),

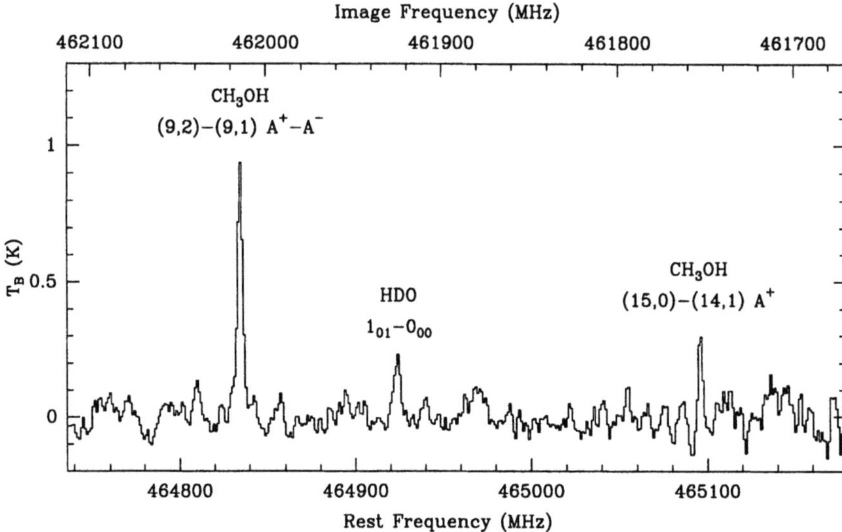


*Figure 9.* The $1_{10} - 0_{00}$ HDO line at 465 GHz observed in comet C/1996 B2 (Hyakutake) at the Caltech Submillimeter Observatory. Two lines of methanol were observed in the same spectrum (from [13]).

suggesting that the Earth oceans were not due to infalling comets. On the other hand, important deuterium enrichments are also found for several interstellar molecules, including interstellar water.

### 4.1.4. *The [HNC]/[HCN] ratio.*

The HNC molecule has been detected in comets C/1996 B2 (Hyakutake) and C/1995 O1 (Hale-Bopp) (Fig. 10; [66]). HNC is an unstable isomer of hydrogen cyanide HCN: at room temperature, [HNC]/[HCN] $< 10^{-15}$. In the interstellar medium, HNC is an abundant molecule with [HNC]/[HCN] as large as 1 in some clouds. In comets, [HNC]/[HCN] can be readily derived, because the same rotational transitions of the two molecules are observed in the same conditions and may be supposed to be excited in a similar way (a correction for possible saturation of the HCN lines is however necessary). A ratio [HNC]/[HCN] $\sim 0.06$ was observed for comet Hyakutake at $r \sim 1.1$ AU [66], which is typical of interstellar abundance ratios. This shows that cometary volatiles were not (all) formed by thermal equilibrium processes. In the interstellar medium, HNC and HCN are believed to be formed as the result of ion-neutral reactions, one of the final processes being

TABLE 10. Isotopic ratios in comets.

| Isotopes | | comet | method | cosmic | comet value | ref. |
|---|---|---|---|---|---|---|
| D/H | $H_3O^+$ | 1P/Halley | mass | $1.5 \times 10^{-5}$ | $3.08 \pm 0.53 \times 10^{-4}$ | [8] |
| | | | | | $3.02 \pm 0.22 \times 10^{-4}$ | [48] |
| | OH | 1P/Halley | UV | | $< 5 \times 10^{-4}$ | [94] |
| | HDO | C/1990 K1 | radio | | $< 7 \times 10^{-3}$ | [35] |
| | | C/1996 B2 | radio | | $2.9 \pm 1.0 \times 10^{-4}$ | [12] |
| | | C/1995 O1 | radio | | $3.3 \pm 0.8 \times 10^{-4}$ | [113] |
| $^{18}O/^{16}O$ | $H_3O^+$ | 1P/Halley | mass | $2.0 \times 10^{-3}$ | $1.93 \pm 0.12 \times 10^{-3}$ | [8] |
| | | | | | $2.13 \pm 0.18 \times 10^{-3}$ | [48] |
| $^{13}C/^{12}C$ | CN | 1P/Halley | visible | $1.1 \times 10^{-2}$ | $1.05 \pm 0.13 \times 10^{-2}$ | [71] |
| | HCN | C/1996 B2 | radio | | $2.9 \pm 1.0 \times 10^{-2}$ | [77] |
| | | C/1995 O1 | radio | | $1.11 \pm 0.18 \times 10^{-2}$ | [78] |
| | | C/1995 O1 | radio | | $0.90 \pm 0.09 \times 10^{-2}$ | [80] |
| $^{15}N/^{14}N$ | CN | 1P/Halley | visible | $3.6 \times 10^{-3}$ | $< 3.6 \times 10^{-3}$ | [71] |
| | HCN | C/1995 O1 | radio | | $3.1 \pm 0.4 \times 10^{-3}$ | [80] |
| $^{34}S/^{32}S$ | | 1P/Halley | mass | $4.2 \times 10^{-2}$ | $4.5 \pm 1.0 \times 10^{-2}$ | [74] |
| | CS | C/1995 O1 | radio | | $3.7 \pm 0.4 \times 10^{-2}$ | [80] |

the dissociative recombination of $HCNH^+$ which leads to the two molecules with equal branching ratios.

In comet Hale-Bopp, [HNC]/[HCN] was found to increase from less than 0.03 to $\sim 0.50$ when the comet approached from 2.4 to 1 AU. The reason for such a variation is still ill-understood; the formation of HNC in the coma by chemical reactions which at first look seemed to be unlikely, has to be considered.

### 4.1.5. *The ortho-to-para ratio (OPR) of water.*

Molecules with hydrogen atoms at symmetrical positions may exist in different nuclear-spin species according to the relative orientations of the spins of their H-atoms. These species are called *ortho* and *para* for molecules with two H-atoms like $H_2$, $H_2O$ or $H_2CO$, *A* and *E* for molecules with three H-atoms like $NH_3$ or $CH_3OH$, *A*, *E* and *F* for $CH_4$, and so forth. Conversions between different species, either by radiative transition or by collisions, are strictly forbidden, at least in the gas phase, and presumably very slow (by proton-exchange mechanisms) in the solid phase [85]. Therefore, the *ortho*-to-*para* ratio (*OPR*) (or *E/A* ratio) of the molecular rotational distribution is characteristic of the formation conditions of the molecules. Fig. 11 shows the *OPR* (or *E/A*) as a function of the equilibrium temperature for various

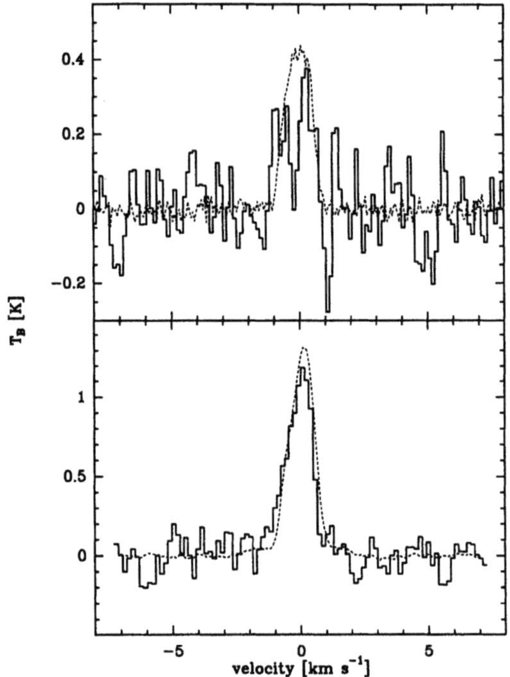


*Figure 10.* The $J(4-3)$ lines of the HCN (solid lines) and HNC (dotted lines) molecules observed in comet C/1996 B2 (Hyakutake) at the James Clerk Maxwell Telescope (top) and at the Caltech Submillimeter Observatory (bottom). The HCN lines have been scaled down by a factor of 10 (from [66]).

molecules. For increasing temperatures, the *OPR* rapidly tends to the statistical weight ratio and is only useful for probing very cold temperatures, except for the $H_2$ molecule (which is of little interest for comets because it is not expected to be incorporated in the nucleus in significant abundances and because its transitions are excessively weak and not observed). Temperatures of the order of 30 K may be probed by $H_2O$, $NH_3$ or $CH_4$. The molecules $H_2S$, $H_2CO$ or $CH_3OH$ would only be sensitive to cooler temperatures.

First attempts have been made to evaluate the *OPR* of water from the *KAO* observations of comets 1P/Halley and C/1986 P1 (Wilson) [82]. For comet Halley, an *OPR* of 2.3 was claimed, smaller than the limit value of 3 and corresponding to an equilibrium temperature of 23 K; a value close

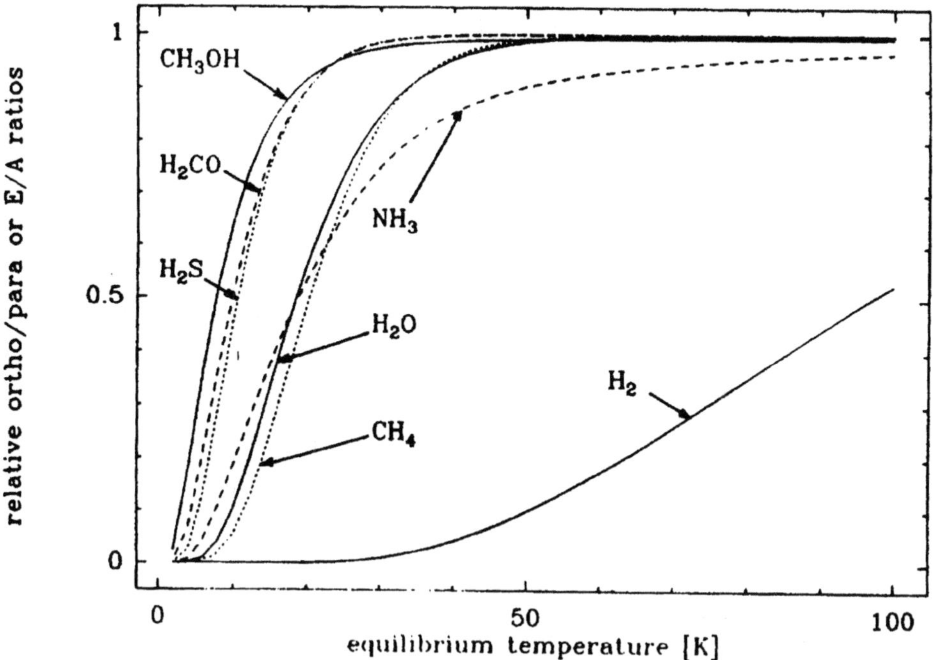


*Figure 11.* The *ortho*-to-*para* and $E/A$ population ratios as a function of equilibrium temperature for several molecules with identical H atoms. The ratios have been normalized to their limit values for high temperatures.

to 3 was found for comet Wilson. These results are controversial, however, because retrieving the rotational population distributions from the *KAO* is model-dependent [15].

A high-resolution spectrum of the whole $\nu_3$ band of water was observed in comet C/1995 O1 (Hale-Bopp) with *ISO* (Fig. 12). The retrieved *OPR*, taking into account opacity effects, is $2.45 \pm 0.1$ and corresponds to a temperature of 25 K [38]. This could be the formation temperature of water, thus excluding formation in a warm Solar Nebula. The possibility of equilibration in the nucleus during the long storage time in the Oort's cloud, however, should also be investigated.

## 4.2. SCENARIOS FOR COMETARY FORMATION AND EVOLUTION

Our present understanding is that there are at least two different classes of comets with different origins (Fig. 13; [75]).

*New comets* have orbits of any possible inclination. They were shown by Oort to come from a very distant isotropic reservoir, now known as the *Oort cloud.* Following gravitational perturbations, they could either be ejected from the Solar System or migrate as short-period comets ($P < 200$ years)

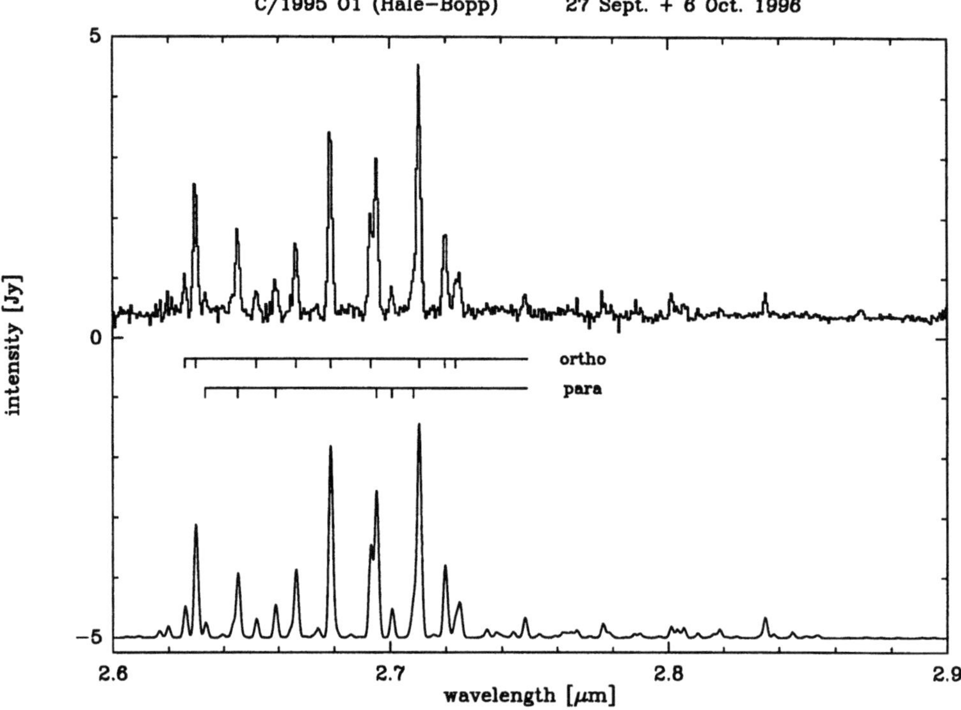


*Figure 12.* The spectrum of the 2.7 $\mu$m water band in comet C/1995 O1 (Hale-Bopp) observed by the short-wavelength spectrometer of *ISO* at 2.9 AU from the Sun. The resolution is $\lambda/\delta\lambda \sim 1500$ and the aperture is 14" $\times$ 20". The curve in the bottom shows a synthetic fluorescence spectrum of water computed for a rotational temperature of 28.5 K and an ortho-to-para ratio of 2.45. All the observed lines are due to water, except the line at 2.869 $\mu$m which is from OH (from [38]).

such as those of the *Halley family.*

*Jupiter family* comets are short-period comets ($P < 20$ years) with low-inclination orbits. They cannot come from the isotropic Oort cloud. It was postulated by Edgeworth and Kuiper that they should come from a reservoir beyond Neptune: the so-called *Edgeworth-Kuiper Belt* [104]. Confirmation of the existence of such a reservoir came recently with the detection of several tens of Kuiper Belt objects [68]. *Centaurs*, which are asteroids with highly eccentric, Saturn-crossing orbits (such as the enigmatic Chiron) (Table 2), could be intermediate objects, migrating from the Kuiper Belt to the Jupiter family.

These two reservoirs have quite different origins. Kuiper Belt objects are believed to be formed where they are now, in the trans-Neptunian outskirts of the Solar System. In contrast, the Oort cloud consists of objects formed in the inner Solar System and ejected by gravitational perturbation after formation of the giant planets. Hence, the two classes of comets, formed in different regions and having experienced different evolution, could be of

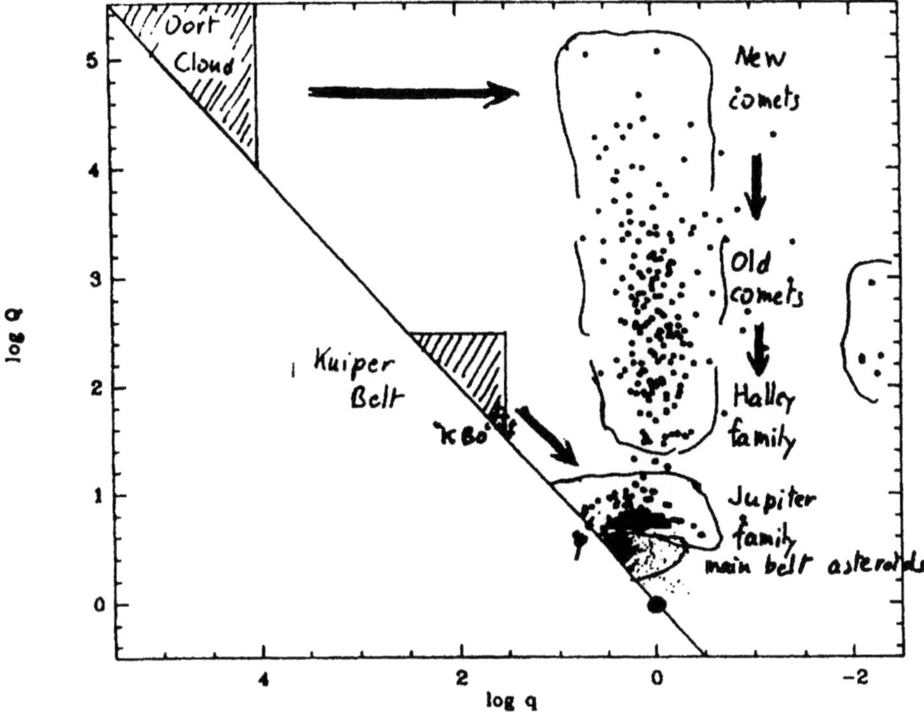


*Figure 13.* The different families of comets and their dynamical evolution. They are represented in a $Q$ (aphelion distance) – $q$ (perihelion distance) diagram. Adapted from [75].

quite different composition.

## 5. Conclusion

Our knowledge of comets has just been completely renewed with the exceptional observations of comets C/1996 B2 (Hyakutake) and C/1995 O1 (Hale-Bopp). This course was given and written during the rush hours of the comet Hale-Bopp observing campaign. Thus it only reports on the highlights of the observations which were known at that time.

There is now a long series of pieces of evidence which show that cometary material is akin to interstellar matter: the composition of cometary ices compared to interstellar ices, the isotopic ratios and especially the high deuterium abundance, the high abundance of cometary HNC, the *ortho*-to-*para* ratio of cometary water. This seems to comfort the pioneering ideas of Greenberg who suggested that comets formed from unprocessed interstellar grains (e.g. [58, 59]). One cannot exclude, however, that part of the interstellar ices have been processed in the Solar Nebula, but this should have been done by processes mimicking interstellar processes.

The silicates of cometary dust are found to be in the crystalline form in about one comet out of two, and in the case of comet Hale-Bopp, crystalline forsterite has been more specifically identified. This is also the case for the dust contained in the discs surrounding Vega-type stars, which are believed to be akin to protoplanetary discs. Bona fide interstellar silicates, however, seem to be amorphous. We thus have to understand how cometary nuclei could have incorporated ices so similar to primitive interstellar ices together with processed silicates.

The next quantum leaps in our understanding of comets will probably be made with the expected results from the space missions which are at various steps of preparation.

- *NEAR*, launched in February 1996, is in fact aimed at asteroid (433) Eros. It will investigate how asteroidal material is related to cometary material.
- *Stardust*, a JPL/NASA mission, is to be launched on June 1999, to fly-by comet 81P/Wild 2 on March 2004 with a relative velocity of $\sim 5$ km $s^{-1}$, to collect samples of cometary dust and to return them on Earth on January 2006.
- *ROSETTA* is an ESA mission to be launched towards comet 46P/Wirtanen in 2003. It will meet the comet at 3.8 AU from the Sun in May 2012 and accompany it until perihelion at 1.05 AU in July 2013. This very elaborated mission will include remote sensing instruments at all wavelengths, *in situ* analysers for cometary atmosphere material, and a lander which will probe the composition and physical properties of the nucleus of the comet.
- *Champollion (DS 4)* is a mission under study at JPL, to be launched in 2003 on a solar-electric power carrier spacecraft. It would meet comet 9P/Tempel 1 on 2006, send a lander on its nucleus, for *in situ* analysis and returning samples to Earth on 2010.

However, one should not forget that cometary science cannot focus only on exceptional comets and on space mission targets. A thorough study of the whole population of comets requires observations of a large sample of these objects, for which long-term ground-based efforts are needed.

## 6. References

1. A'Hearn, M.F., Feldman, P.D. and Schleicher, D.J. (1983) The discovery of $S_2$ in comet IRAS-Araki-Alcock 1983d. *Astrophys. J.* **274**, L99–L103.
2. A'Hearn, M.F., Hoban, S., Birch, P.V, *et al.* (1986) Cyanogen jets in comet Halley. *Nature* **324**, 649–651.

3. A'Hearn, M.F., Millis, R.L., Schleicher, D.G., Osip, D.J. and Birch, P.V. (1995) The ensemble properties of comets: results from narrowband photometry of 85 comet., 1976–1992, *Icarus* **118**, 223–270.
4. A'Hearn, M.F., Schleicher, D.G., Feldman, P., Millis, R.L. and Thompson, D.T. (1984) Comet Bowell 1980b. *Astron. J.* **89**, 579–591.
5. Anders, R. and Grevesse, E. (1989). Abundances of the elements: meteoritic and solar. *Geochim. Cosmochim. Acta* **53**, 197–214.
6. Arpigny, C. (1995) Spectra of comets: ultraviolet and optical regions. In *Laboratory and Astronomical High Resolution Spectra*, A.J. Sauval et al. (edts), ASP Conference Series **81**, pp. 362–382.
7. Arpigny, C., Dossin, F., Woszczyk, A. *et al.* (1995) *Atlas of cometary spectra.* Kluwer, Dordrecht (*in press*).
8. Balsiger, H., Altwegg, K. and Geiss, J. (1995) D/H and $^{18}O/^{16}O$ ratio in the hydronium ion and in neutral water from in situ ion measurements in comet Halley. *J. Geophys. Res.* **100**, 5827–5834.
9. Biver, N., Bockelée-Morvan, D., Colom, P., *et al.* (1997) Evolution of the outgassing of comet Hale-Bopp (C/1995 O1) from radio observations. *Science* **275**, 1915–1918.
10. Biver, N., Rauer, H., Despois, D., *et al.* (1996) Substantial outgassing of carbon monoxide from comet Hale-Bopp at large heliocentric distance. *Nature* **380**, 137–139.
11. Biver, N, *et al.* (1997) *in preparation.*
12. Bockelée-Morvan, D. (1997) Cometary volatiles: the status after comet C/1996 B2 Hyakutake. In IAU Symposium 178, *Molecules in Astrophysics: Probes & Processes*, edt. E. van Dishoeck, Kluwer, Dordrecht, pp. 219–235.
13. Bockelée-Morvan, D., Gautier, D., Lis, D.C., *et al.* (1998) Deuterated water in comet C/1996 B2 (Hyakutake) and its implication for the origins of comets. *Icarus* (*in press*).
14. Bockelée-Morvan, D., Brooke, T.Y. and Crovisier, J. (1995) On the origin of the 3.2–3.6 $\mu$m emission features in comets. *Icarus* **116**, 18–39.
15. Bockelée-Morvan, D. and Crovisier, J. (1990) The ortho-para ratio of water in comets: can reliable values be derived from the observations of the 2.7 $\mu$m band by the KAO? In *Asteroids, Comets, Meteors III*, edts C.I. Lagerkvist *et al.*, Uppsala, 263–265.
16. Bockelée-Morvan, D. and Crovisier, J. (1992) The formation and composition of the atmosphere of small solar system bodies. In *Observations and Physical Properties of Small Solar System Bodies* A. Brahic, J.-C. Gérard, J. Surdej edts, Institut d'Astrophysique de Liège, pp. 65–82.
17. Bockelée-Morvan, D. and Crovisier, J. (1992) Formaldehyde in comets: II. Excitation of the rotational lines. *Astron. Astrophys.* **264**, 282–291.

18. Bockelée-Morvan, D., Crovisier, J., Colom, P. and Despois, D. (1994) The rotational lines of methanol in comets Austin 1990 V and Levy 1990 XX. *Astron. Astrophys.* **287**, 647–665.
19. Bockelée-Morvan, D., Padman, R., Davies, J.K. and Crovisier, J. (1994) Observations of submillimetre lines of $CH_3OH$, HCN, and $H_2CO$ in comet P/Swift-Tuttle with the James Clerk Maxwell Telescope. *Planet. Space Sci.* **42**, 655–662.
20. Bradley, J. (1997) Interstellar dust: evidence from interplanetary dust particles. In *Formation and Evolution of Solids in Space*, this volume.
21. Brooke, T.Y., Tokunaga, A.T., Weaver, H.A., *et al.* (1996) Detection of acetylene in the infrared spectrum of comet Hyakutake. *Nature* **383**, 606–608.
22. Brown, M.E., Bouchez, A.H., Spinrad, H. and Johns-Krull, C.M. (1996) A high-resolution catalog of cometary emission lines. *Astron. J.* **112**, 1197–1202.
23. Bus, S.J., A'Hearn, M.F., Schleicher, D.G. and Bowell, E. (1991) Detection of CN emission from (2060) Chiron. *Science* **251**, 774–777.
24. Capria, M.T., Capaccioni, F., Coradini, A., *et al.* (1996) A P/Wirtanen evolution model. *Planet. Space Sci.* **44**, 987–1000.
25. Chen J. and Jewitt D. (1994) On the rate at which comets split. *Icarus* **108**, 265–271.
26. Cochran, A.L. and Cochran, W.D. (1991) The first detection of CN and the distribution of $CO^+$ gas in the coma of comet P/Schwassmann-Wachmann 1. *Icarus* **90**, 172–175.
27. Colom, P., Crovisier, J., Bockelée-Morvan, D., Despois, D. and Paubert, G. (1992) Formaldehyde in comets: I. Microwave observations of P/Brorsen-Metcalf (1989 X), Austin (1990 V) and Levy (1990 XX). *Astron. Astrophys.* **264**, 270–281.
28. Combes, M., Moroz, V.I., Crovisier, J., *et al.* (1988) The 2.5 to 12 $\mu$m spectrum of comet Halley from the IKS-VEGA experiment. *Icarus* **76**, 404–436.
29. Crifo, J.F. (1991) Hydrodynamic models of the collisional coma. In *Comets in the Post-Halley Era*, edts R.L. Newburn *et al.*, Kluwer, Dordrecht, pp. 937–989.
30. Crifo, J.F. (1995). A general physicochemical model of the inner coma of active comets. 1: Implications of spatially distributed gas and dust production. *Astrophys. J.* **445**, 470–488.
31. Crovisier, J. (1994) Molecular abundances in comets. In *Asteroids, Comets, Meteors 1993*, edts A. Milani *et al.*, Kluwer, Dordrecht, pp. 313–326.
32. Crovisier, J. (1994) Photodestruction rates for cometary parent molecules. *J. Geophys. Res. Planets* **99**, E4, 3777–3781.

33. Crovisier, J. (1996) Observational constraints on the composition and nature of comet D/Shoemaker-Levy 9. In *The Collision of Comet Shoemaker-Levy 9 and Jupiter*, K.S. Noll, H.A. Weaver and P.D. Feldman edts, Cambridge University Press, pp. 31–54.
34. Crovisier, J., Biver, N., Bockelée-Morvan, D., *et al.* (1995) Carbon monoxide outgassing from comet P/Schwassmann-Wachmann 1. *Icarus* **115**, 213–216.
35. Crovisier, J., Bockelée-Morvan, D., Colom, P., Despois, D. and Paubert, G. (1993) A search for parent molecules at millimetre wavelengths in comets Austin 1990 V and Levy 1990 XX: upper limits for undetected species. *Astron. Astrophys.* **269**, 527–540.
36. Crovisier, J., Brooke, T.Y., Hanner, M.S., *et al.* (1996) The infrared spectrum of comet C/1995 O1 (Hale-Bopp) at 4.6 AU from the Sun. *Astron. Astrophys.* **315**, L385–L388.
37. Crovisier, J., Despois, D., Bockelée-Morvan, D., Colom, P. and Paubert, G. (1991) Microwave observations of hydrogen sulfide and searches for other sulfur compounds in comets Austin (1989c1) and Levy (1990c). *Icarus* **93**, 246–258.
38. Crovisier, J., Leech, K., Bockelée-Morvan, D., *et al.* (1997) The spectrum of comet Hale-Bopp (C/1995 O1) observed with the Infrared Space Observatory at 2.9 AU from the Sun. *Science* **275**, 1904–1907.
39. Davies, J.K., Green, S.F. and Geballe, T.R. (1991) The detection of a strong 3.28 micron emission feature in comet Levy. *Month. Not. Roy. Astron. Soc.* **251**, 148–151.
40. Davies, J.K., Mumma, M.J., Reuter, D.C., *et al.* (1993) The infrared (3.2-3.6 $\mu$m) spectrum of comet P/Swift-Tuttle: detection of methanol and other organic. *Month. Not. Roy. Astron. Soc.* **265**, 1022–1026.
41. Davies, J.K., Roush, T.L., Cruikshank, D.P., *et al.* (1997) The detection of water ice in comet Hale-Bopp. *Icarus* **127**, 238–245.
42. Dermott, S.F. (1997) New problems in the zodiacal light. In *Formation and Evolution of Solids in Space*, this volume.
43. Despois, D., Biver, N., Bockelée-Morvan, D., *et al.* (1996) Radio line observations of comet 109P/Swift-Tuttle at IRAM. *Planet. Space Sci.* **44**, 529–539.
44. Despois D., Crovisier J., Bockelée-Morvan, D., *et al.* (1986) Observations of hydrogen cyanide in comet Halley. *Astron. Astrophys.* **160**, L11–L12.
45. d'Hendecourt, L., Jourdain de Muizon, M., Dartois, E., *et al.* (1996) ISO-SWS observations of solid state features towards RAFGL 7009S. *Astron. Astrophys.* **315**, L365–L369.
46. Eberhardt, P. (1997) Volatiles, isotopes and origin of comets. In *Asteroids, Comets, Meteors 1996*, COSPAR Colloquium **10** (*in press*).

47. Eberhardt, P., Krankowsky, D., Schulte, W., *et al.* (1987) The CO and $N_2$ abundance in comet P/Halley. *Astron. Astrophys.* **187**, 481–484.
48. Eberhardt, P., Reber, M., Krankowsky, D. and Hodges, R.R. (1995) The D/H and $^{18}O/^{16}O$ ratios in water from comet P/Halley. *Astron. Astrophys.* **302**, 301–316.
49. Feldman, P.D. and Brune, W.H. (1976) Carbon production in comet West (1975n). *Astrophys. J.* **209**, L45–L48.
50. Feldman, P.D., Festou, M.C., A'Hearn, M.F., *et al.* (1987) IUE observations of comet P/Halley: evolution of the ultraviolet spectrum between September 1985 and July 1986. *Astron. Astrophys.* **187**, 325–328.
51. Feldman, P.D., Fournier, K.B., Grini, V.P. and Zvereva A.M. (1993) The abundance of ammonia in comet P/Halley derived from ultraviolet spectrophotometry of NH by ASTRON and IUE. *Astrophys. J.* **404**, 348–355.
52. Festou, M.C., Rickman, H. and West, R.M. (1993) Comets. *Astron. Astrophys. Rev.* **4**, 363–447 and **5**, 37–163.
53. Fink, U. (1992) Comet Yanaka (1988r): A new class of carbon-poor comet. *Science* **257**, 1926–1929.
54. Fitzsimmons, A. and Cartwright, I.M. (1996) Optical spectroscopy of comet C/1995 O1 Hale-Bopp. *Month. Not. Roy. Astron. Soc.* **278**, L37–L40.
55. Gérard, E., Bockelée-Morvan, D., Bourgois, G., Colom, P. and Crovisier, J. (1987) 18-cm wavelength radio monitoring of the OH radical in comet P/Halley 1982i. *Astron. Astrophys.* **187**, 455–461.
56. Glaccum, W., Moseley, S.H., Campins, H. and Loewenstein, R.F. (1987) Airborne spectrophotometry of P/Halley from 20 to 65 microns. *Astron. Astrophys.* **187**, 635–638.
57. Greenberg, J.M. (1982) What are comets made of? A model based on interstellar grains. In *Comets*, L.L. Wilkening edt., Univ. Arizona Press, Tucson, pp. 383–395.
58. Greenberg, J.M. and Hage, J.I. (1990) From interstellar dust to comets: a unification of observational constraints. *Astrophys. J.* **361**, 260–274.
59. Greenberg, J.M., Mizutani, H. and Yamamoto, T. (1995) A new derivation of the tensile strength of cometary nuclei: application to comet Shoemaker-Levy 9. *Astron. Astrophys.* **295**, L35–L38.
60. Hanner, M.S., Hackwell J.A., Russell, R.W. and Lynch, D.K. (1994) Silicate emission feature in the spectrum of comet Mueller 1993a. *Icarus* **112**, 490–495.
61. Hanner, M.S., Lynch, D.K. and Russell, R.W. (1994) The 8–13 micron spectra of comets and the composition of silicate grains. *Astrophys. J.* **425**, 274–285.

62. Herter, T., Campins, H. and Gull, G.E. (1987) Airborne spectrophotometry of P/Halley from 16 to 30 microns. *Astron. Astrophys.* **187**, 629–631.

63. Hoban, S., Mumma, M., Reuter, D., *et al.* (1991) A tentative identification of methanol as the progenitor of the 3.52 $\mu$m feature in several comets. *Icarus* **93**, 122–134.

64. Hoban, S., Reuter, D.C., DiSanti, M.A., Mumma, M.J. and Elston, R. (1993) Infrared observations of methanol in comet P/Swift-Tuttle. *Icarus* **105**, 548–556.

65. Huebner, W.F., Keady, J.J and Lyon, S.P. (1992) Solar photorates for planetary atmospheres and atmospheric pollutants. *Astrophys. Space Science* **195**, 1–294.

66. Irvine, W.M., Bockelée-Morvan, D., Lis, D.C., *et al.* (1996) Spectroscopic evidence for interstellar ices in comet Hyakutake. *Nature* **383**, 418–420.

67. Jessberger, E.K. and Kissel, J. (1991) Chemical properties of cometary dust and a note on carbon isotopes. In *Comets in the Post-Halley Era*, edts R.L. Newburn *et al.*, Kluwer, Dordrecht, pp. 1075–1092.

68. Jewitt, D.C. and Luu, J.X. (1995) The solar system beyond Neptune. *Astron. J.* **109**, 1867–1876.

69. Jewitt, D., Senay, M. and Matthews, H. (1996) Observations of carbon monoxide in comet Hale-Bopp. *Science* **271**, 1110–1113.

70. Kissel, J. (1997) In situ measurements of evolved solids in space, with emphasis on cometary particles. In *Formation and Evolution of Solids in Space*, this volume.

71. Kleine, M., Wyckoff, S., Wenhinger, P.A. and Peterson, B.A. (1995) The carbon isotope abundance ratio in comet Halley. *Astrophys. J.* **439**, 1021–1033.

72. Klinger, J., Levasseur-Regourd, A.-C., Bouziani, N. and Enzian, A. (1996) Towards a model of cometary nuclei for engineering studies for future space missions to comets. *Planet. Space Sci.* **44**, 637–653.

73. Knacke, R.F., Fajardo-Acosta, S.B., Telesco, C.M., *et al.* (1993) The silicates in the dust of $\beta$ Pictoris. *Astrophys. J.* **418**, 440–450.

74. Krankowsky, D., Lämmerzahl, P., Herrwerth, I., *et al.* (1986) In situ gas and ion composition measurements at Comet Halley, *Nature* **321**, 326–329.

75. Kresàk, L. (1994) Comets, existing populations. In *Asteroids, Comets, Meteors 1993*, edts A. Milani *et al.*, Kluwer, Dordrecht, pp. 77–94.

76. Larson, L., Weaver, H.A., Mumma, M.J. and Drapatz, S. (1988) Airborne infrared spectroscopy of comet Wilson (1986l) and comparisons with comet Halley, *Astrophys. J.* **338**, 1106–1114.

77. Lis, D.C., Keene, J., Young, K., *et al.* (1997) CSO observations of comet C/1996 B2 (Hyakutake). *Icarus* **130**, 355–372.

78. Lis, D.C., Gardner, M., Phillips, T.G., *et al.* (1997) Comet C/1995 O1 (Hale-Bopp., *IAU Circ.* 6566.

79. Lutz, D., Feuchtgruber, H., Genzel, R., *et al.* (1996) SWS observations of the Galactic center. *Astron. Astrophys.* **315**, L269–L272.

80. Jewitt, D.C., Matthews, H.E., Owen, T. and Meier, R. (1997) Measurements of $^{12}C/^{13}C$, $^{14}N/^{15}N$, and $^{32}S/^{34}S$ ratios in comet Hale-Bopp (C/1995 O1). *Science* **278**, 90–93

81. Meier, R. and A'Hearn, M.F. (1997) Atomic sulfur in cometary coma based on UV spectra of the SI triplet near 1814 A. *Icarus* **125**, 164–194.

82. Mumma, M.J., Blass, W.E., Weaver, H.A. and Larson, H.P. (1988) Measurements of the ortho-para ratio and nuclear spin temperature of water vapor in comets Halley and Wilson (1986l) and implications for their origin and evolution. In *Proc. STScI Conference on Origins and Evolution of Planetary Systems*, STScI, Baltimore.

83. Mumma, M.J., DiSanti, M.A., Dello Russo, N., *et al.* (1996) Detection of abundant ethane and methane, along with carbon monoxide and water, in comet C/1996 B2 Hyakutake): evidence for interstellar origin. *Science* **272**, 1310–1314.

84. Mumma, M.J., DiSanti, M.A., Tokunaga, A.T. and Roettger, E.E. (1995) Ground-based detection of water in comet Shoemaker-Levy 1992 XIX; probing cometary parent molecules by hot-band fluorescence. *Bull. Amer. Astron. Soc.* **27**, 1144.

85. Mumma, M.J., Weaver, H.A. and Larson, H.P. (1987) The ortho-para ratio of water in comet P/Halley. *Astron. Astrophys. 187*, 419–424.

86. Mumma, M.J., Weaver, H.A., Larson, H.P., Davis, D.S. and Williams, M. (1986) Detection of water vapor in Halley's come., *Science* **232**, 1523–1528.

87. Mumma, M.J., Weissman, P.R. and Stern, S.A. (1993) Comets and the origin of the solar system: Reading the Rosetta stone. In *Protostars and Planets III*, E.H. Levy & J.I. Lunine (edts), University of Arizona Press, Tucson, pp. 1177–1252.

88. Newburn, R.L., Neugebauer, M. and Rahe, J. (edts) (1991) *Comets in the Post-Halley Era.* Kluwer, Dordrecht.

89. Noll, K.S., Weaver, H.A. and Feldman, P.D. (edts) (1996) *The Collision of Comet Shoemaker-Levy 9 and Jupiter*. Cambridge University Press, Cambridge.

90. Preston, G.W. (1967) The spectrum of comet Ikeya-Seki (1965f). *Astron. J.* **147**, 718–742.

91. Rauer, H., Arpigny, C., Boehnhardt, H., *et al.* (1997) Optical observations of comet Hale-Bopp (C/1995 O1) at large heliocentric distances before perihelion. *Science* **275**, 1909–1912.

92. Rauer, H., Biver, N., Crovisier, J., *et al.* (1997) Millimetric and optical observations of Chiron. *Planet. Space Sci.* **45**, 799–805.

93. Rickman, H. (1994) Cometary nuclei. In *Asteroids, Comets, Meteors 1993*, edts A. Milani *et al.*, Kluwer, Dordrecht, pp. 297–312.

94. Schleicher, D.G. and A'Hearn, M.F. (1986) Comets P/Giacobini-Zinner and P/Halley at high dispersion. In *New Insights in Astrophysics*, ESA **SP-263**, 31–33.

95. Schleicher, D.G., Bus, S.J. and Osip, D.J. (1993) The anomalous molecular abundances of comet P/Wolf-Harrington. *Icarus* **104**, 157–166.

96. Schmitt, B. (1992) Thermal and physico-chemical processes in cometary nuclei. In *Interrelations between Physics and Dynamics for Minor Bodies in the Solar System*, edts D. Benest & C. Froeschlé, Éditions Frontières, pp. 265–307.

97. Schutte W. (1997) Laboratory simulation of processes in interstellar ices. In *Formation and Evolution of Solids in Space*, this volume.

98. Senay, M.C. and Jewitt, D. (1994) Coma formation driven by carbon monoxide release from comet Schwassmann-Wachmann 1. *Nature* **371**, 229–231.

99. Waelkens, C., Waters, L.B.F.M., de Graauw, M.S., *et al.* (1996) SWS observations of young main-sequence stars with dusty circumstellar disks. *Astron. Astrophys.* **315**, L245–L248.

100. Wagner, R.M. and Schleicher, D.G. (1997) The spectrum and spatial distribution of cyanogen in comet Hale-Bopp (C/1995 O1) at large heliocentric distance. *Science* **275**, 1918–1920.

101. Weaver, H.A., Feldman, P.D., A'Hearn, M.F., *et al.* (1997) The activity and size of the nucleus of comet Hale-Bopp (C/1995 O1). *Science* **275**, 1900–1904.

102. Weaver, H.A., Feldman, P.D., McPhate, J.B., *et al.* (1994) Detection of CO Cameron band emission in comet P/Hartley 2 (1991 XV) with the Hubble Space Telescope. *Astrophys. J.* **422**, 374–380.

103. Weaver, H.A., Mumma, M.J., Larson, H.P. and Davis, D.S. (1986) Post-perihelion observations of water in comet Halley. *Nature* **324**, 441–444.

104. Weissman, P.R. (1995) The Kuiper Belt. *Ann. Rev. Astron. Astrophys.* **33**, 327–357.

105. Whittet, D.C.B., Schutte, W.A., Tielens, A.G.G.M., *et al.* (1996) An ISO SWS view of interstellar ices — first results. *Astron. Astrophys.* **315**, L357–L360.

106. Wootten, A., Butler, B., Bockelée-Morvan, D., *et al.* (1996) Detection of ammonia in comet C/1996 B2 (Hyakutake). In *Asteroids, Comets, Meteors 1996*, 8–12 July 1996, Versailles, France.
107. Wyckoff, S.W., Wagner, R.M., Wehinger, P.A., Schleicher, D.G. and Festou, M.C. (1985) Onset of sublimation in comet Halley (1982i). *Nature* **316**, 241–242.
108. Wyckoff, S.W., Tegler, S.C. and Engel, L. (1991) Nitrogen abundance in comet Halley. *Astrophys. J.* **367**, 641–648.
109. Wyckoff, S.W., Tegler, S.C. and Engel, L. (1991) Ammonia abundances in four comets. *Astrophys. J.* **368**, 279–286.
110. Yamamoto, T. (1985) Formation environment of cometary nuclei in the primordial solar nebula. *Astron. Astrophys.* **142**, 31–36.
111. Grim, R.J.A., Greenberg, J.M. (1987) Photoprocessing of $H_2S$ in interstellar grain mantles as an explanation for $S_2$ in comets. *Astron. Astrophys.* **181**, 155–168.
112. Haruyama, J., Yamamoto, T., Mizutani, H. and Greenberg, J.M. (1993) Thermal history of comets during residence in the Oort cloud: effect of radiogenic heating in combination with the very low thermal conductivity of amorphous ice. *JĠeophys. Res.* **98 E8**, 15 079–15 090.
113. Meier, R., Owen, T.C., Matthews, H.E., *et al.* (1998) A determination of the HDO/$H_2O$ ratio in comet C/1995 O1 (Hale-Bopp). *Science* **279**, 842–844.
114. Sekanina, Z. (1996) Tidal breakup of the nucleus of comet Shoemaker-Levy 9. In *The Collision of Comet Shoemaker-Levy 9 and Jupiter*, K.S. Noll, H.A. Weaver and P.D. Feldman (edts) (1996) . Cambridge University Press, Cambridge, pp. 55–80.
115. Tancredi, G., Rickman, H., Greenberg, J.M. (1994) Thermochemistry of cometary nuclei: I. The Jupiter family case. *Astron. Astrophs.* **286**, 659–682.

# IN SITU MEASUREMENTS OF EVOLVED SOLIDS IN SPACE WITH EMPHASIS ON COMETARY PARTICLES

J. KISSEL
*Max-Planck-Institut für Kernphysik, P/O Box 103980, 69029 Heidelberg*

**Abstract.** In March of 1986 several spacecrafts went by Halley's comet. Time-of-flight mass spectrometers carried aboard provided in situ measurements of the chemical composition of the dust particles emitted from the nucleus of the comet. The variation in the composition from one grain to another was found to be surprisingly large. Although the ratio of rock forming elements to those forming organic compounds varied enormously, there was no hint of a separate occurrence of one component without the other. This intimate mixture of rocky and organic solids as well as the wide variation of chemical composition points to the fact that this material is indeed left over from the formation of the solar system as aggregates of core-mantle particles evolved from those seen in interstellar space. Infrared observations of interstellar space show the presence of large quantities of such small dust particles.

## 1. Introduction

It is now generally accepted that new stars are born as a result of the gravitational collapse of a large cloud of gas and dust. Such a cloud can become unstable by some triggering event, such as collision with another cloud, a supernova explosion, magnetic disturbances, or any other event that removes the balancing feature, be it electromagnetic fields within the cloud or be it small relative velocities of its constituents. The conservation of momentum and angular momentum cause the material to settle into a rotating disk, with the major part accumulating in its center, thus providing the material for the new star. During this process, most of the gravitational energy is converted into thermal energy, into heat. The inner parts of this system become so hot, that essentially all material will be gaseous or at least molten, and in such a way homogenized with respect to its elemental



*J. M. Greenberg and A. Li (eds.), Formulation and Evolution of Solids in Space,* 427–445.

and isotopic inventory. Separation of material occurs during the cooling phase, when various minerals form at different temperatures and pressures. Once there is enough material in the central body, nuclear fusion starts and after another million years the star is heated up and its stellar winds sweep out the material around it that has not accreted into larger bodies like planets, moons, or at least meteoroids.

Only in regions far away from the hot center, the material never reaches temperatures required for melting/sublimation of even ices. It is in these cold zones where the original material from the collapsing cloud has a chance to survive. Material at these large distances from the sun was influenced by the big planets that formed there. Some of these small bodies are now discovered as the 'Kuiper belt family', while most other small bodies were ejected into the 'Oort cloud'. The latter reservoir of small bodies is established from the fact that the majority of small cold bodies, 'fresh comets', that visit the inner solar system for the first time have a semi-major axis of greater than several thousands of astronomical units.

Comets have long been a mystery for mankind, as they appeared unexpectedly out of the skies, displaying a sometimes spectacular view, then apparently disappearing again into the dark skies. It was not until 1705, when Sir Edmund Halley predicted the return of a comet (later named after him), that comets were accepted as orderly members of the solar system. As late as 1950 Whipple (1950) proposed a comet model, that was able to explain part of the features that were observed. Finally in 1986 the fact was established that comets (*comet* is a generic name for the entire phenomenon) have a nucleus, as images were returned by the international fleet of comet missions. It is perhaps also interesting to note that the nucleus of comet Halley turned out to be darker than most scientists had expected even though, based on the observed optical properties of interstellar dust as aggregated into a fluffy model of comets (Greenberg 1979, 1982), it had been predicted to have a reflectivity no higher than 6% (Greenberg 1986a, b). Cometary nuclei were also identified by long range radar measurements (see e.g. Kamoun et al. 1982). It is only recently that better optics and light detectors have led to the detection of fairly large bodies in the Neptune range, members of the Kuiper belt.

Meteorites are samples of extraterrestrial material fallen on the earth. When found and collected they are available for detailed chemical analysis. Meteorites have been studied for many years and a scheme for classification has been established, largely based on the degree of processing, modification the meteorite has experienced. Most analyses have shown that meteoritic material with respect to its isotopic composition is thoroughly mixed. It had been found early on, however, that the noble gases Neon and Xenon sometimes showed an unusual isotopy. Even though noble gases do not occur in great quantities in meteorites, they are more easily accessible, as they can be degassed from the whole body into a very sensitive instrument, while a solid of the same mass would still be very small and tricky to find, and even more difficult to isolate for analysis.

Consequently, it took until recently, about the time of the Halley fly-bys, that Lewis & Anders (1983) were able to extract from pieces of primitive Meteorites the 10 ppm phase that holds those tiny, often only a few nm sized stardust grains, that have conserved the isotopic signature of their mother-star. The details of the isotopy of Carbon, Nitrogen, Oxygen, Magnesium, Aluminium, Silicon, and Titanium not only provide information on the number of types of stars that must have provided material to the solar system, but now, the data also put limits on the models for the nuclear processes active in stars of different age and size (see e.g. Anders & Zinner 1993, 1994).

An other source of extraterrestrial material are the Interplanetary Dust Particles (IDP) collected in the earth's atmosphere. Despite the fact, that extensive laboratory analyses have been conducted on them, it was not possible to prove definitively that any one of these meteoritic particles had a cometary origin. It is nearly certain that there were cometary particles among those that have been studied but it has not been possible to tell which ones were asteroidal and which ones were cometary, even though there is strong evidence in some cases (see J. Bradley, this issue). All the information on the composition of cometary particulates came from astronomical or other indirect measurements, like e.g. the spectral analysis of cometary meteors (Millman 1982), or rocket borne measurement of mesospheric ion enhancements during meteor showers (Goldberg & Aiken 1973).

## 2. The Halley Fly-By Missions

In March 1986, several spacecrafts encountered comet p/Halley delivering a set of instruments for various investigations in its coma. Among the main objectives was to verify whether comets were formed in the early phase of our planetary system and did indeed maintain the interstellar material largely unaltered. They were pictured as a fluffy accumulation of interstellar dust (Greenberg 1982). As comets close to the sun emit large amounts of water gas, and obviously, mineral dust, too, comets were earlier seen as "dirty snowballs" (Whipple 1950) which were activated under the heat of the solar radiation. Gases would be ionized, and dragged along by the solar wind and the interplanetary magnetic field, while the dust particles were influenced by the momentum picked up from the sun's light. By this process, the well-known tails would form, which can be distinguished by their shape as the dust tail, and the plasma tail.

It came as a big surprise to many, when the cameras carried onboard the two Soviet space probes Vega-1 and Vega-2 and the European craft Giotto located a rather dark (Albedo <4%) and large, irregular nucleus of p/Halley. The mean density of the nucleus is still not known, but seems to be well below $1\,g\,cm^{-3}$. This nucleus showed jet-like activities over some 10% of its surface, which seems to go on partly also on the comet's night side. There was a rather bright background of scattered light against which the dark nucleus was seen. This phenomenon might be indicative of exothermal processes activated/initiated by the solar heat input. All the more important was therefore the chemical analysis of the cometary dust. This was performed on all 3 spaceprobes with almost identical dust impact mass spectrometers (PUMA on the VEGA probes and PIA on Giotto). These time-of-flight mass spectrometers which differed in details of the Target and the Reflector layout, had an added aim of studying as well the physics of the dust impact process itself. This knowledge is necessary for the interpretation of the data and will be discussed next.

The reason that p/Halley was explored by fly-by missions comes from the fact that the comet is on a retrograde orbit. The relative speeds of the spacecrafts were 79, 77, $69\,km\,s^{-1}$ for VEGA 1, VEGA 2, and GIOTTO, respectively. A significant reduction of this speeds was out of reach at that time. The impact of particles at those speeds result in highly compressed states for at least a short ($\sim$nsec) time. This regime of the equations of states is difficult to access experimentally. (Most data come from tests of nuclear devices, and are not openly accessible). The so called 'ab initio' calculations are successfully used to describe what happens mechanically, i.e., the cratering process. For the representation of the ionized states of the material after the impact the situation is very difficult, and produces credible results for simple projectile- and target materials only. In this respect the Halley measurements also help in the understanding of the ion formation process as it occurs upon impact (Hornung et al. 1996).

## 3. The Time-of-Flight Dust-Impact Mass-Spectrometers

Since all instruments were very similar, we will only describe the PIA instrument for GIOTTO. A detailed description is given in Kissel (1986). Fig. 1 shows a schematic of the impact analyzer. Dust particles enter the instrument through the optical baffle, and hit the target. For PIA it was a Pt foil doped with 4% Ag, solid Ag for PUMA 1 and a silver foil for PUMA 2. Instantaneously positive ions, negative ions, electrons secondary ejecta, and a light flash appear. An electric field of $-1\,kV\,cm^{-1}$ in front of the target pulls the positive charge into the instrument for the use of mass-spectrometry. The negative charge collected at the target, and the light flash were used for coincidences, eliminating triggers from sporadic noise pulses. (The optical baffle was used to reduce the background for the light flash detector.) A separate target was used to collect charge from the secondary particulates, and at an electrode in the first flight tube the total charge that entered the instrument was monitored. The ions would arrive at the end of the instrument separated by their flight time that depends in a well known way on their mass as well on any initial energy they acquire in the formation process. The ion reflector, however, first described by Mamyrin et al. (1973), removed all first order effects initial energy would have on the flight times. At the same time the potential of the end plate of the reflector is limiting the maximum energy ions can have, which will pass. This feature was used in the VEGA instruments. After detection of the ions with the secondary electron multiplier, the signal was digitized at 15 MegaSamples/s and prepared for telemetry. For more details and other modes of signal processing see Kissel (1986).

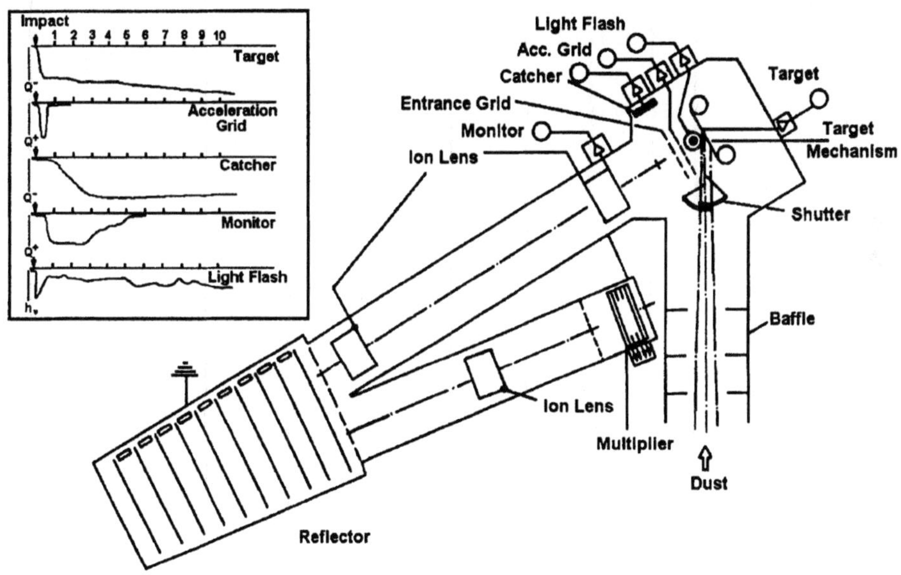


*Figure 1.* Cross Section of an Impact Analyser such as PIA or PUMA. Dust particles enter from bottom to top, practically as a parallel beam. The baffle structure keeps light away from the target area to a degree compatible with the angle of dust acceptance of the Analyser. An electric field in front of the target separates positive ions from other charged species and accelerates them into the analyser. The impact is detected by a photomultiplier for the 'impact light flash' and the charges on either target- and acceleration grid- amplifiers. After travelling through the drift tube, the reflector and the outleg drift tube, ions separated in time by their mass are detected by the ion detector.

During the fly-bys the three instruments on the three different spacecrafts performed differently. PUMA 1 on VEGA 1 apparently showed the fewest problems. It behaved flawlessly, although it exposed a deficiency known before launch that the time-squared generator started with a negative slope. PUMA 2 on VEGA 2 suffered from an operational problem, which apparently drew heavily on the main bus voltage, which in turn set the multiplier high voltage to its lower default value and consequently reduced the instruments sensitivity in the spectrum channel and at the same time lowered its dynamic range to 3.5 instead of 5 decades. PIA on GIOTTO, last but not least, already early in the encounter had lost 2 out of 6 subunits of the amplifier attached to its multiplier therefore losing the range for the smallest mass lines as well as the subrange two levels above where normally the lower isotopes of the major elements would show up.

## 4. The Physics of Ion Formation

While a mass spectrometer can only be used with ions, it is the amount of atoms and molecules that make up the dust particle along with its physical parameters which are of interest. In order to derive the chemical composition and physical parameters such as size and density of the dust particles from the mass spectra it is necessary to know the physics of the underlying ion formation process. Laboratory tests are of little help, since only small (few 100 nm diameter), conductive particles can be accelerated to the speeds needed here. Tests were performed (Knabe & Krueger 1982), however, and showed that the types of ions and their quantity differs significantly if the impact speeds are below $15\,\mathrm{km\,s^{-1}}$ or above $20\,\mathrm{km\,s^{-1}}$.

At the speeds reached during the fly-bys the material of the particle and the target are compressed up to 3-fold densities. At these conditions an atom's energy levels are highly degenerated. The calculation of the residual ionization depends highly on assumptions for the conditions in the compressed state and during expansion. A clue can be the observation of doubly charged atoms, especially Calcium which has a very low second ionization energy. The types of ions observed gave motivation to compare ion generation by particle impact to other methods of ion formation by sudden dissipation of energy, as is known to occur with ions as primary particles (SIMS), or fission fragments or even laser light. It looks as if the energy of the ionic grid does not couple to the electronic system fast enough to generate multiply charged ions. The results are summarized in Kissel & Krueger (1987a). From the analysis of laboratory data we also found that ions from the target and from the particle follow different yield functions. This allowed us to also get an estimate on mass and density of the impacting particle (Maas et al. 1989). The modeling of the ion formation process was so successful, that the elemental composition of the cometary particles was derived as meteoritic or solar for all but the most volatile elements (H, C, N, O), see e.g. Jessberger et al. (1986).

More recently, years after the fly-bys, we tried another approach to the ion formation issue, by using standard 'ab initio' calculations. This time mass and density of individual particles as determined from the mass spectra were used as starting values. From these the ionization degree was calculated and the major mass lines of the actual mass spectra taken as target values. The result was a new set of data for mass and density. The procedure was repeated until it converged. It turned out that the values obtained that way covered a variation by a factor of two smaller than did the starting values (Hornung & Kissel 1994). The analysis of the results showed that the oxygen in the particles acted as a very efficient electron catcher, reducing the ionization degree after the impact.

## 5. Results

### 5.1. ELEMENTAL COMPOSITION

The first result that became available after the fly-bys was the elemental analysis of cometary dust particles (see Fig. 2). Due to the high impact speeds, singly charged atomic ions formed 99.9% of the mass spectra. Within the accuracy of the ion yields (factor of 2) as stated above the major rock forming elements were found to occur at relative abundances similar to CI or solar composition. The very volatile elements H, C, N, and O, however, were more abundant in the cometary dust relative to the CI chondrites, yet still depleted relative to the solar photosphere. The initial analyses have been performed by different authors independently, emphasizing slightly different goals in their work. The results, however, agreed well within the usual factor of two [see Langevin et al. (1987), Krueger & Kissel (1987)]. In total 30 elements could be detected in the cometary dust, the abundance of the major elements C, N, O, Mg, Al, Si, S, Ca, Fe, was determined within a factor 2, while for the more difficult to measure; i.e., less abundant elements H, Li, B, Na, P, Cl, K, Ti, Cr, Mn, Co, Ni, Cu only a factor of 10 was possible.

Analysis of the data showed that the organic (H, C, O, N) ions had on the average a higher initial energy than the rocky ions which led Krueger & Kissel (1987) to infer a core mantle structure of the dust particles. Thus according to Jessberger and Kissel(1991) "The existence of the previously postulated (Greenberg 1982) core-mantle grains seem to be substantiated by the data". For the least abundant elements V, Zn, Ga, Ge, Se, Br, Sr, and Y only their presence could be established.

In general it was not possible to directly measure the isotopy of the el-

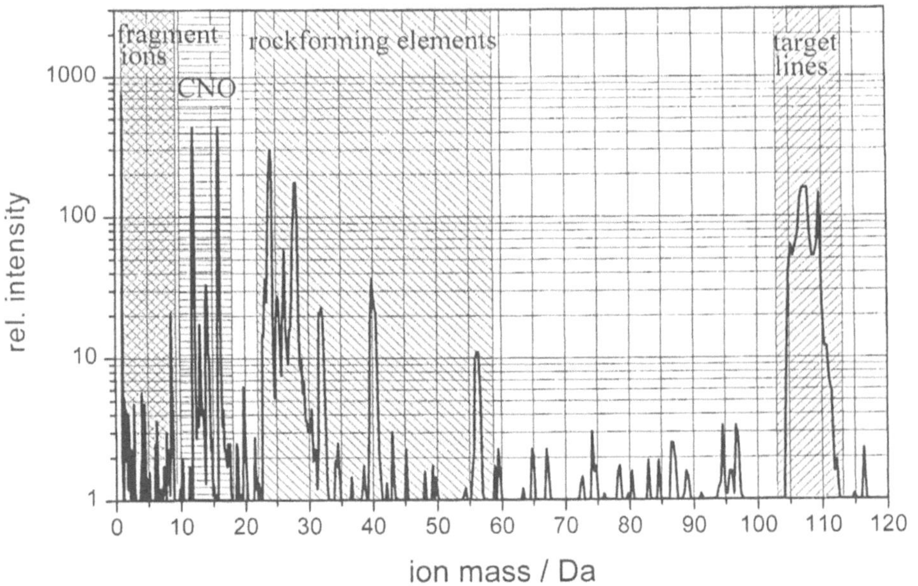


*Figure 2.* Typical mass spectra from impacts of cometary dust particles. The three subpanels show: a) a particle dominated by rock-forming elements ('mineral composition'); b) a particle dominated by light elements H, C, N, and O ('CHON' or 'organic'); c) all data from PUMA-1 on board VEGA-1 raw data, with time-of-flight scale converted to Daltons.

ements, rather their well established terrestrial averages have been used to discriminate atomic ions from molecular interferences like e.g. in the case of mass 24 which may be assigned to Mg or $C_2$. In a few cases, however, it seems that particles with a particular carbon isotopy may have been measured: The mass 12 was very much higher than the corresponding m/z=13 peak would suggest [see Jessberger & Kissel (1991) for details]. This might well be evidence for the presence of the so called "light carbon" which had been detected in fractions of primitive meteorite (Anders & Zinner 1994).

As described above, values of mass and density for individual cometary particles could be determined from the atomic ions in the mass spectra. Maas et al. (1989) showed that the particles dominated by light elements have a lower density and larger masses than particles that are dominated by the rock-forming elements. This is consistent with the picture of a fluffy nature of cometary dust, the exact structure of the particles can, however, not be inferred from the measurements.

The data on elemental composition has also be used to search for clues to the mineral composition of the cometary dust particles by Schulze & Kissel (1992) and more recently by Schulze et al. (1997). The variation of element ratios did not allow to identify individual minerals unambiguously, but some peculiarities could be found. It may well be an indication that the cometary dust particles of typically 100 nm in size are still mixtures of several minerals of different composition. While Mg/Fe ratios vary over almost six orders of magnitude, the variation of the ratio Mg/Si is limited to a factor of two. The oxygen in the inorganic phase was not easy to trace, since the ubiquitous organic phase made it difficult to correlate it with the rock-forming elements. It seems, however, that oxygen is more tied up in the silicates rather than in Fe-oxides, since the spectra with the biggest Fe-lines show almost no oxygen, but rather sulfur. Again it was also difficult to discriminate between FeS or $FeS_2$, perhaps because sulfur may be a constituent of the organic phase as well. Overall the data seem to point to a high temperature origin of the minerals with essentially no low temperature or even aqueous alteration.

A particular behavior of all three instruments has been seen during their flybys, which has not yet been unambiguously interpreted: Data frames that were triggered by stray light from the coma showed pulses in the spectrum channel even at distances from the comet nucleus of 1 million kilometers. Two groups [Sagdeev et al. (1989) and Utterback & Kissel (1990)] have worked on the data and have come to the conclusion, that most likely very small dust particles with masses in the order of $10^{-17}$ g caused the signals seen. Such particles are very abundant in the size spectrum of diffuse cloud interstellar dust (Li & Greenberg 1997) and would be accreted along with the volatiles (ices) in the final collapse phase of the protosolar nebula (Greenberg 1984, 1988). From Utterback & Kissel (1990) we quote the main features of the phenomenon: event rates and amplitude distribution of the signals in all three instruments are strikingly similar. The maximum signal falls off with larger distance to the nucleus, and the corresponding number densities fall off with 1/R, suggesting a disintegration of larger particles into smaller ones as they get away from the nucleus. The biggest question mark in the interpretation of the signals is, however, the fact that despite a lot of statistical work a composition of such particles could not be obtained. The particles would be so small that they would not be observed in the optical, but rather in the extreme ultraviolet or in the X-ray range. Soft X-rays have been recently observed, indeed, by Dennerl et al. (1996). The discussion about the source of the X-rays is, however, still ongoing, with reactions of comet gas with solar wind constituents or scattering by 'attodust' being the possible sources (see Kransopolsky 1996, Kransopolsky et al. 1997).

## 5.2. MOLECULAR COMPOSITION

After evaluation of the major lines in the mass spectra a detailed analysis showed the presence of mass peaks associated with molecular ions. A complicated analysis described by Kissel & Krueger (1987b) identified a lot of fragment ions, that made it possible to characterize the chemical environment of the cometary organics. The presence of unsaturated nitrogen-containing species like nitriles, aldimines, enamines and possibly their polymerization products is assured. The latter ones are the main types of species which are able to explain some of the features found in the analysis; i.e., pyrrol, pyrazol/imidazol, pyridine, pyrimidine and its derivatives are very likely to be present in the dust. Due to the low fraction of nitrogen relative to carbon found in the elemental composition, it may be surprising at first sight not to find the unsaturated carbohydrides to a larger extent. However, their probability of forming ions in desorption is a factor of $\sim$ 100 lower than the related nitrogen-containing ones, because they lack a charge center and thus have only a low basicity. Furthermore, they are consequently more volatile, so they may be evaporated out of the dust before impact. Nevertheless, we conclude that a considerable fraction of alkenes, alkynes, or cyclic, especially aromatic, molecules (like benzene, toluene) may well be present. For heavier ions the interpretation became ambiguous.

The presence of oxygen atoms in molecules may well occur more frequently with larger molecules, but could not be proven. No indications for alcohol or sugars as N-free carbohydrates have been found, nor alpha-amino acids, although these would produce large yields of ions. They are, if formed at all, at least a factor of 30 less abundant than the equally biologically important pyrimidines and purines. It is puzzling that the presence of oxygen in the organic molecules was not found to a larger extent, although there is much oxygen within the dust, as seen from the elemental composition. With a possibly strong acidic behaviour they would rather show up as negative ions, which we did not measure. The same would hold for S-containing species, which rather would form negative ions from thiols, thials and so on. The molecular composition of the organic material is of the highest interest, since comets are supposed to have brought a large share of the organic inventory to the early earth. It is also speculated that they may have played an important role in the origin of life on earth (Greenberg 1981, 1986c). The hypothesis that the organic part of the dust of a comet contains a large fraction of biological material has been disproved at least for that of comet Halley. Firstly, dried cells should contain at least some fraction of Na and K. Less than 1 ppm in the organic material of the dust particles has been found, however. Secondly, no indications have been found for a significant presence of oxygen-rich compounds such as sugars and peptides, although the latter molecules should show up in the mass spectra.

Nevertheless the question of the origin of life in the context of primordial matter as found in comets has become even more exciting. Krueger & Kissel (1989) stress that the substance classes, which seem to be present in the cometary dust are highly reactive especially in warm water. Imagine that some mechanism brings such substances into contact with liquid water:

-The unsaturated carbohydrides may add water molecules thus reacting to give carbohydrates such as sugars.

-The nitrogen-containing species may react to give bases of nucleosides, if not already present.

-The mineral core (Greenberg 1982) may serve as the phase boundary necessary to give a local concentration gradient, and also as a source for dissolved phosphoric acid.

With such a model, the chemical prerequisites and also the requirements of nonlinear non-equilibrium thermodynamics are fulfilled to trigger, for example, hyper cyclic behavior. High affinity, high concentration gradient and localization of the reactions have been considered to trigger the self-organization of nucleic acids. This is known to be possible even without amino acids at this early stage because the replication of RNA is not necessarily associated with transcription, when $Zn^{2+}$-catalysts are present, as is the case with the cometary dust.

### 5.3. DUST DENSITY

The values of the mass density of particles (see Fig. 3) have to be taken with some caution before they are compared with model data. From the laboratory results that Kissel & Krueger (1987a) have reported the amount and type of ions depend on their mass and density. This has to be interpreted, however, as a homogeneous spherical projectile of that mass and density would impact on the target. No extrapolation to fluffy particles of irregular shape is possible without further laboratory work, which in turn is difficult as such particles cannot be accelerated softly enough to the high velocities needed for comparison. In addition, it should be recollected that the mass-density determinations come from the PUMA-1 instrument data, which were taken outside 10,000 km from the comet nucleus. It is therefore conceivable that the mass-densities are significantly higher than they are near the nucleus, as they could increase as a result of fractionation and as some of the volatiles are released to the gas phase. Along these lines Krueger et al. (1991) report an average molecular mass of 250-350 u based on their GIOTTO observations, a value too high than what one would expect for the volatile phase of the dust.

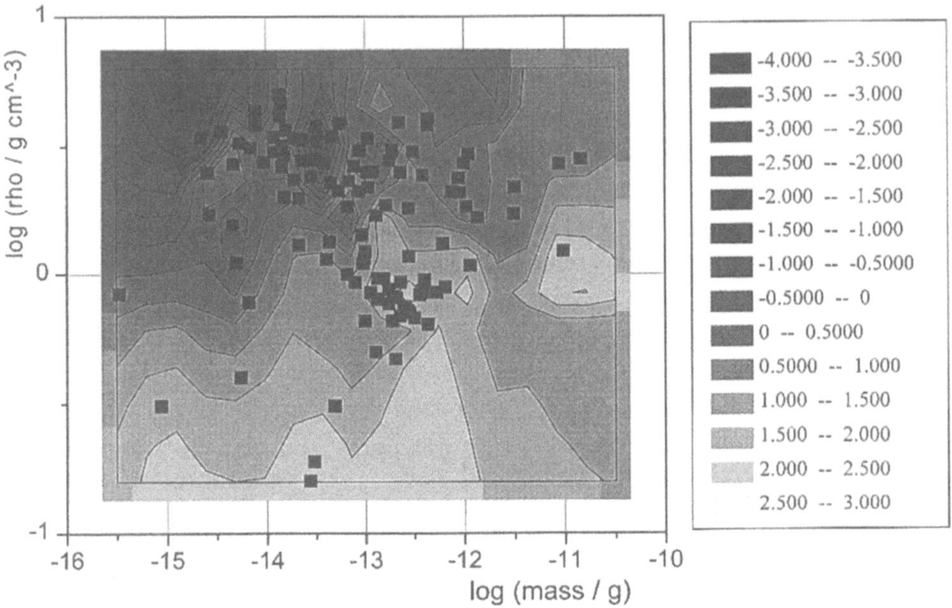


*Figure 3.* Mass and density of 112 PUMA-1 particles are derived after Mass et al. (1989).

## 6. Open Questions

Even before space missions to comets have been seriously considered, many features of comets have been derived from astronomical observations of their dynamics. The elemental, molecular, and particulate composition has been inferred. Most of these features could, however, only be observed as components of the coma (cometary atmosphere), since the cometary nucleus would be too small to be resolved. This consequently meant that the material had to be thermally altered, or at least be transported into the coma by some mechanism, that apparently depends on the solar radiation input. Some examples of relevant papers should be given here, which, however, is by no means meant to be either complete or to list the most important contributions.

In a recent paper Sekanina (1996) gives an overview on the scientific research into comets from the 17th century. It is amazing to see, how early the main elements for comet models, still discussed today, have been developed. All major models assume a solid nucleus within the cometary coma. This nucleus in turn is thought to be made up by many individual pieces, be it in loose contact as in the so called 'sandbank' model, be it as meteoritic type solids grains 'glued' together by water ice. The latter, described as 'dirty ice ball' by Whipple (1950) has been widely accepted and also was used for the design of the missions to Halley's comet and the instruments flown on them. While some of the physical properties were predicted reasonably well, it completely failed with respect to cometary chemistry.

Another model by Greenberg (1982), however, provides reasonably good qualitative and to some extent even quantitative agreement with respect to most observational facts on comets, including the phenomenon of nucleus splitting (Greenberg, Mizutani & Yamamoto 1995). It is based on the commonly accepted model, that comet nuclei are formed from interstellar material along with the whole solar system, and have been kept at temperatures below 30-50 K since then. More recently, Greenberg & Li (1997) and Raulin & Greenberg (1997) expand and improve the quantitative features of this model by applying boundary conditions such as the elemental inventory, the interstellar particle inventory, and the organic material available for the formation of the solar system. These efforts of improving the model on how to make a comet nucleus need an equally intense effort to measure the chemical and physical properties of cometary nuclei, as well as of processes that control the evolution of these nuclei from formation to the present time when they become observable in the inner solar system.

Sekanina & Farrell (1982) have used astronomical observations and analyzed the size distribution and motion of comet released dust particles in a most impressive way. Sekanina (1997) has also summarized observational facts on the splitting of cometary nuclei to derive boundaries for their physical properties. Hanner et al. (1993) (and others) report observational facts with respect to the elemental and some molecular components.

The space missions to Halley's comet were the first to undertake in situ measurements. The instruments have helped to put boundaries to some of the questions. The camera showed for the first time the real comet nucleus: a dark body of irregular shape, from which material is emitted in some isolated areas. It is still open, however, what the density and the internal structure of such a nucleus is. The infrared spectrometer onboard VEGA 2 showed the presence of solid water ice, yet also surface temperatures of 320 K to 400 K (Emerich et al. 1987).

In an attempt to simulate cometary activity Lammerzahl et al. (1995) have found that even surfaces of pure water ice if irradiated do not produce the sublimation rates assumed for the so-called active areas on the nucleus. They show, however, the formation of an inert crust. It might well be that the gas is produced below the surface, flows to, and exits at these areas. Even though silicate features are identified in astronomical observations, the dust mass spectrometers clearly showed that all dust particles are an intimate mixture of mineral grains with an organic component. Indications are also that dust cometary particles are not made from single minerals but are again a mixture of several minerals.

As to the organic component the abundance ratios of atomic ions of C, N, and O seem to be fairly constant. In the analysis of the molecular, organic ions, however, Kissel & Krueger (1987b) found hints to a lot of different substance classes. It appears as if all stable molecules that can be made of C, N, and O (and of course containing also other elements) will occur. So it is yet to be determined, whether the concept of a few 'mother'-molecules decaying into smaller 'daughter'-molecules can really be applied.

Most models for cometary dust consider size ranges down to about $0.1\,\mu$m, representing what has been observed and surely identified. In the Halley missions the mass spectrometers showed signals, however, that appeared on the respective days of their flyby that were associated with the comet. The signals were too small to trigger the mass-spectrometer function of the instrument, but also too big to be caused by individual gas molecules. Sagdeev et al. (1989) and in more detail Utterback & Kissel (1990) have discussed these features and come to the conclusion that the presence of very small dust like $10^{-18}$ g 'attodust' is very likely, but could not be proven beyond any doubt.

## 7. Next Opportunities

The space missions to Halley's comet have helped to put closer limits on comet models, but have certainly not answered all open questions about the nature of comets. This has been recognized and up to now there are three more missions set to go to other comets.

The next one will be in NASA's 'New Millennium' technology program: Deep Space One. Its launch is planned for July 1st 1998. After a flyby at the asteroid McAuliffe and a gravity assist at Mars it will flyby comet West-Kohoutek-Ikemura on June 8th 2000 (see Fig. 4). DS-1 will provide nucleus images in various wavelength ranges at a resolution of about 5-10 m/pixel. The purpose of this mission is to demonstrate flight readiness of new technologies, among which the implementation of a Xenon operated ion drive and autonomous navigation are the most demanding ones. This will provide images of a comet nucleus for only the second time.

The next mission after that will be in NASA's DISCOVERY-program: STARDUST (Fig. 5). Launch is scheduled for January/February 1999. After a first orbit in space, a delta-v maneuver, an earth gravity assist maneuver it will flyby at comet Wild-2 in December of 2003 at about $6\,\mathrm{km\,s^{-1}}$ only. Another two years later it will return a container, which was exposed at the comet and bring samples of cometary dust (and gas) back to earth. On its way it will also expose some surface to the flux of interstellar dust, and at the comet also monitor the dust impact rate, and the return samples of the

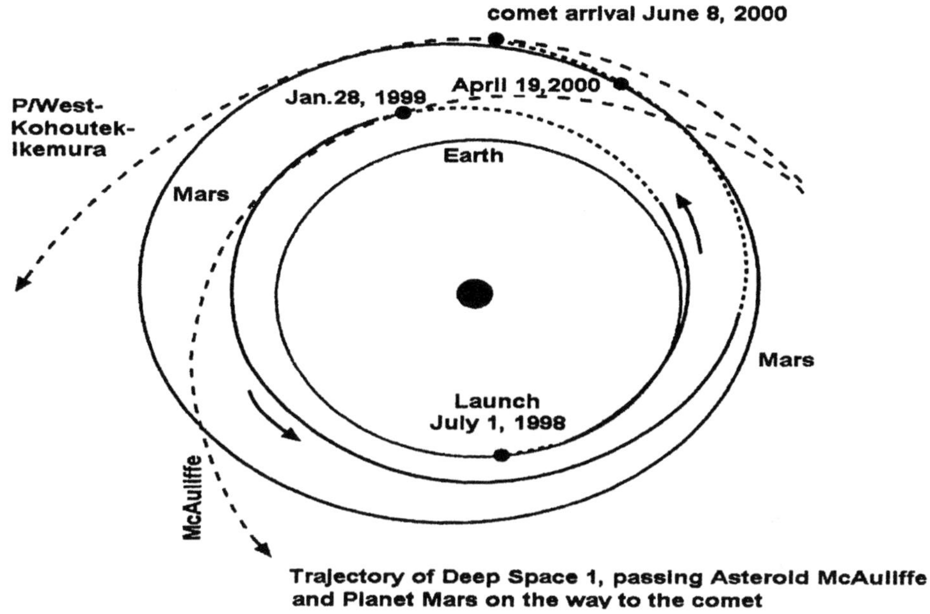


*Figure 4.* Trajectory of NASA's Deep Space-1 mission. It will be the first solar electric powered, autonomously navigating spacecraft. It will be an route to comet P/West-Kohoutek-Ikemura.

gaseous constituents of the coma of the comet. Another instrument will be CIDA, a dust impact mass spectrometer, equipped with a $100\,\mathrm{cm}^2$ target for the measurement of interstellar dust on the way to the comet and also provide the possibility to measure negative ions upon dust impacts.

After these two mission, the ESA-cornerstone mission ROSETTA is set for launch in January 2003 with an ARIANE-5 launcher. Since this launcher is not designed for missions into interplanetary space, it takes 1 Mars- and 2 Earth- gravity assist manoeuvers to get ROSETTA on track to comet Wirtanen-2, where it will -after deceleration- rendezvous (relative velocity $= 0$) in 2011 (Fig. 6). On the way it will also visit two asteroids Rodan and Mimistrobell. After achieving rendezvous at Wirtanen-2 ROSETTA will drop a surface probe onto its nucleus. ROSETTA and its lander system carry by far the most complex payload of these missions. Mass spectrometers for gas and dust are part of both crafts, also gas chromatographs for special purpose analysis, cameras and microwave, IR - UV spectrometers of course. The orbiter even has an atomic force microscope for dust morphology. The lander adds instrumentation for sub-surface sampling and acoustic and electric probing of the nucleus. It will be the first rendezvous with a comet and the first probe on a cometary nucleus.

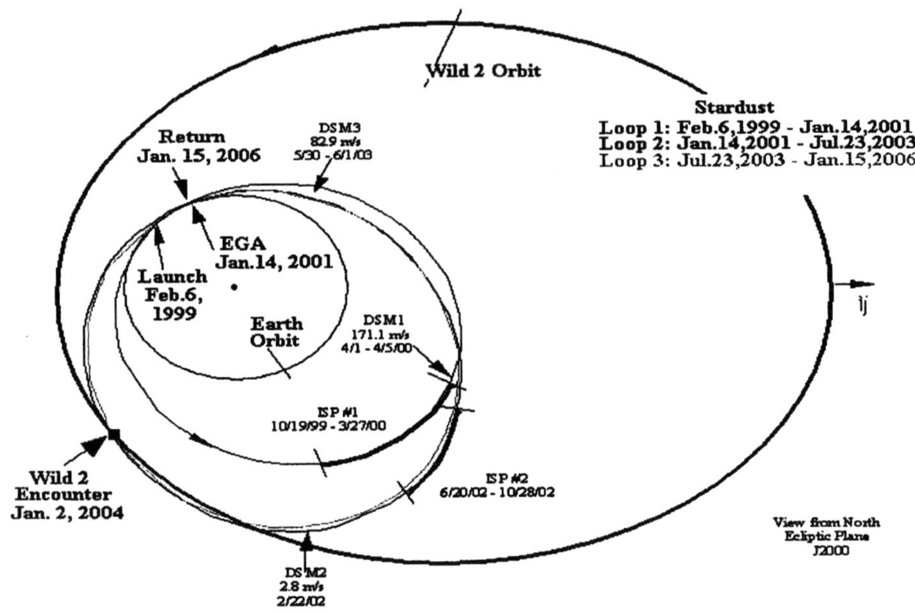


*Figure 5.* Trajectory of NASA's Stardust mission which will carry an impact analyser for interstellar and cometary dust (CIDA). The mission will ultimately return a sample container with both interstellar and cometary dust to Earth.

There are plans, proposals, however, for further comet missions. In NASA's New Millennium program they might implement as DS-4 to drop a probe on a cometary nucleus, now that ESA has eliminated the second lander from ROSETTA, which would give to NASA the 'first' probe on a comet nucleus.

In the DISCOVERY program, CONTOUR would be a mission that, if selected, flies by three different comets. With the concept of using earth returns for gravity assist maneuvers to change the orbit, it even has the option to intercept a fresh new comet.

## 8. Conclusion

Comets are indeed the most fascinating small objects in the solar system. Their unique nature makes them display a variety of features, as they approach the sun on their orbit and react to the increased heat influx. It is certainly safe to assume that all comets are created equal with respect to their major constituents, although differences will be found in nuclei having different masses and a different orbital history. Comets are also the 'cheapest supplier' of real interstellar material in the state as it was present when

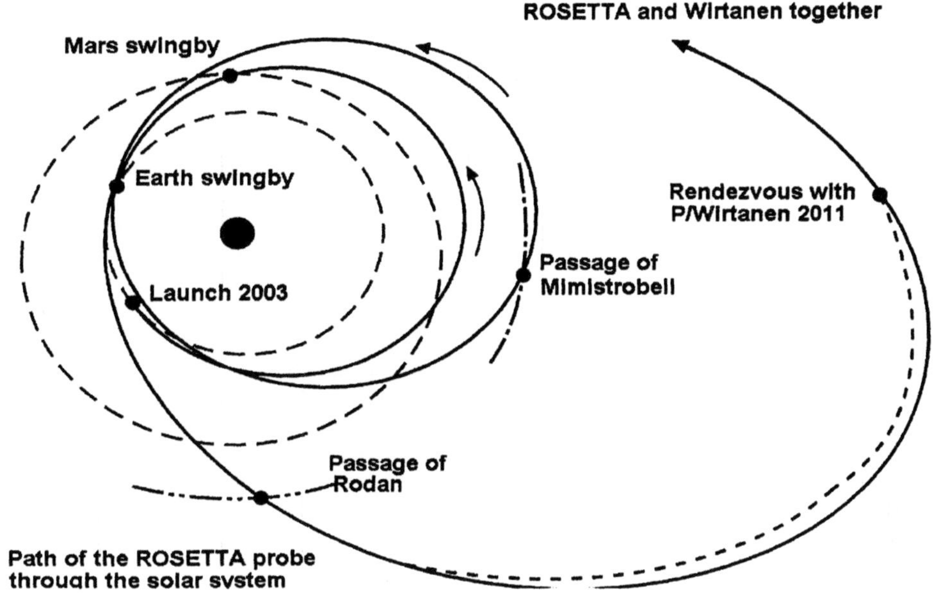


*Figure 6.* Trajectory of ESA's ROSETTA mission. The spacecraft will carry an analyser for collected cometary dust operating as time-of-flight Secondary-Ion-Mass-Spectrometer (COSIMA).

the solar system formed. Such interstellar material is the clue to our understanding of the formation of the solar system, and possibly also of the generation of the biosphere, since all elements beyond helium have been brought to the solar system by the initial molecular cloud and the dust contained in it.

While the first space missions to a comet have revealed a lot of new details, allowing for better modeling, it will take more missions in the future to fully understand the formation, alteration and decay of cometary nuclei.

## References

Anders, E., and Zinner, E. (1993) Interstellar grains in primitive meteorites: diamond, silicon carbide, and graphite, Meteoritics 28, 490-514.

Anders, E., and Zinner, E. (1994) Kuiper Prize Lecture: sources of solar-system carbon, Icarus 112, 303-309.

Dennerl, K., Englhauser, J., and Trümper, J. (1996) X-Ray detections of comets, IAU Circular 6413, 2.

Emerich, C., Lamarre, J.M., Moroz, V.I., Combes, M., Sanko, N.F., Nikolsky, Y.V., Rocard, F., Gispert, R., Coron, N., Bibring, J.P., Encrenaz, T., and Crovisier, J. (1987) Temperature and size of the nucleus of comet P/Halley deduced from IKS infrared VEGA 1 measurements, Astron. Astrophys. 187, 839-842.

Goldberg, R.A., and Aikin, A.C. (1973) Comet Encke meteor metallic ion identification by mass spectrometer, Science 180, 292-296.
Greenberg, J.M. (1979) From interstellar dust to comets to the zodiacal light, in Cometary Missions (eds. W.I. Axford, H. Fechtig, J. Rahe) Veroeff. Remeis Sternwarte, Bamberg bd. XII nr. 132, p. 119-126.
Greenberg, J.M. (1981) Chemical evolution of interstellar dust – a source of prebiotic material? in Comets and the Origin of Life (ed. C. Ponnamperuma) Reidel, Dordrecht, p. 111-127.
Greenberg, J.M. (1982) What are comets made of? A model based on interstellar dust, in Comets (ed. L.L. Wilkening), University of Arizona Press, p. 131-163.
Greenberg, J.M. (1984) A fine mist of very small comet dust particles, Adv. Space. Res. 4, 211-212.
Greenberg, J.M. (1986a) Fluffy comets, in Asteroids, Comets, Meteors II (eds. C.I. Lagerquist, et al.) Rprocentrum HSC, Uppsala, p. 221-223.
Greenberg, J.M. (1986b) Predicting that comet Halley is dark, Nature 321, 385
Greenberg, J.M. (1986c) The chemical composition of comets: an possible contribution to planet composition and evolution, in The Galaxy and the Solar System (eds. R. Smoluchowski, J.N. Bahcall, M.S. Mathews), University of Arizona Press, p. 103-115.
Greenberg, J.M. (1988) The interstellar dust model of comets: post Halley, in Dust in the Universe (eds. M.E. Bailey, D.A. Williams), Cambridge University Press, p. 121-143.
Greenberg, J.M. (1998) Making a comet nucleus, Astron. Astrophys. 330, 375-380.
Greenberg, J.M., and Li, A. (1997) Silicate core-organic refractory mantle particles as interstellar dust and as aggregated in comets and stellar disks, Adv. Space. Res. 19, 981-990.
Hanner, M.S., Russell, R.W., and Lynch, D.K. (1993) Infrared spectroscopy and photometry of comet Austin 1990 V, Icarus 101, 64-70.
Hornung, K., and Kissel, J. (1994) On shock wave impact ionization of dust particles, Astron. Astrophys. 291, 324-336.
Hornung, K., Malama, Yu., and Thoma, K. (1996) Modeling of the very high velocity impact process with respect to in-situ ionization measurements, Adv. Space Res. 17, 77-86.
Jessberger, E.K., Kissel, J., Fechtig, H., Krueger, F.R. (1986) On the average chemical composition of cometary dust, ESA SP-249, p. 27-30.
Jessberger, E.K., and Kissel, J. (1991) Chemical properties of cometary dust and a note on carbon isotopes, in Comets in the Post-Halley Era (eds. R. Newburn, M. Neugebauer, J. Rahe), Kluwer Academic Publishers, Dordrecht, p. 1075-1092.
Kamoun, P.G., Campbell, D.B., Ostro, S.J., Pettengill, G.H., and Shapiro, I.I. (1982) Comet Encke - radar detection of nucleus, Science 216, 293-295.
Kissel, J. (1986) The GIOTTO particulate impact analyser, ESA SP-1077, p. 67-83.
Kissel, J., and Krueger, F.R. (1987a) Ion formation by impact of fast dust particles and comparison with related techniques, Appl. Phys. A 42, 69-85.
Kissel, J., and Krueger, F.R. (1987b) The organic component in dust from comet Halley as measured by the PUMA mass-spectrometer on board VEGA 1, Nature 326, 755-760.
Knabe, W., and Krueger, F.R. (1982) Ion formation from Alkali iodide solids by swift dust particle impact, Z. Naturforsch. 37a, 1335-1340.
Kransopolsky, V.A. (1996) On the nature of soft X-ray radiation in comet Hyakutake 1996 B2, Bull. American Astron. Soc., DPS meeting #28, #09.37.
Kransopolsky, V.A., Mumma, M.J., and Abbott, M.J. (1997) EUVE observations of soft X-ray emissions from comets 6P/d'Arrest, Bradfield (C/1995 Q1), Hyakutake (C/1996 B2), and Hale-Bopp (C/1995 O1), Bull. American Astron. Soc., DPS meeting, #29, #32.09.
Krueger, F.R., and Kissel, J. (1987) The chemical composition of the dust of comet P/Halley as measured by PUMA' on board VEGA-1, Naturwissenschaften 74, 312-316.

Krueger, F.R., Korth, A., Kissel, J. (1991) The organic matter of comet Halley as inferred by joint gas phase and solid phase analyses, Space Science Rev. 56, 167-175.

Krueger, F.R., and Kissel, J. (1989) Biogenesis by cometary grains – thermodynamic aspects of self-organisation, Origin of Life 19, 87-94.

Lämmerzahl, P., Gebhard, J., Grün, E., and Klee, G. (1995) Gas release from ice/dust mixtures: results from eleven KOSI experiments, Planetary and Space Science 43, 363-373.

Langevin, Y., Kissel, J., Bertaux, J.-L., and Chassefire, L. (1987) First statistical analysis of 5000 mass spectra of cometary grains obtained by PUMA-1 (VEGA-1) and PIA (GIOTTO) impact ionisation mass spectrometers in the compressed modes, Astron. Astrophys. 187, 761-766.

Lewis, R.S., and Anders, E. (1983) Interstellar matter in meteorites, Scientific American 249, 66-77.

Li, A., Greenberg, J.M. (1997) A unified model of interstellar dust, Astron. Astrophys. 323, 566-584.

Mamyrin, B.A., Karatyev, V.I., Shmikk, D.V., and Zagulin, V.A. (1973) Mass-Reflection – A new high-resolution nonmagnetic time-of-flight mass-spectrometer, Zh Eksp. i. Teor. Fiz. 64, 82 (Sov. Phys. JETP 37, No.1, July 1973).

Maas, D., Krueger, F.R., and Kissel, J. (1989) Mass and density of silicate- and CHON-type dust particles released by comet P/Halley, in Asteroids, Comets, Meteors III, p. 389-392

Millman, P.M., (1972) Cometary meteoroids, in From Plasma to Planet, Nobel Symp. No. 21 (ed. A. Elvius), p. 157-168.

Raulin, F., and Greenberg, J.M. (1997) From the interstellar clouds, through the inner to the outer solar system: a universally distributed complex organic chemistry, Adv. Space Res. 19, 975-978.

Sagdeev, R.Z., Evlanov, E.N., Fomenkova, M.N., Prilutskii, O.F., Zubkov, B.V. (1989) Small size dust particles near Halley's Comet, Adv. Space Res. 9, 263-267.

Schulze, H., and Kissel, J. (1992) Chemical heterogeneity and mineralogy of Halley's Dust, Meteoritics 27, 286-287.

Schulze, H., Kissel, J., Jessberger, E.K. (1997) Chemistry and mineralogy of comet Halley's dust, in From Stardust to Planetesimals (eds. Y.J. Pendleton, and A.G.G.M. Tielens), ASP Conference Series, Vol. 122, p. 397.

Sekanina, Z., and Farrell, J.A. (1982) Two dust populations of particle fragments in the striated tail of Comet MRKOS 1957V, Astron. J. 87, 1836-1853.

Sekanina, Z. (1996) Comet D/1993 F2 (Shoemaker-Levy 9) as culprit of a revived controversy? in International Comet Quaterly 97, p. 1-11.

Utterback, N.G., and Kissel, J. (1990) Attogram dust cloud a million kilometers from comet Halley, Astron. J. 100, 1315-1322.

Whipple, F. L. (1950) A Comet Model, Astrophys. J. 111, 375-394.

# INTERSTELLAR MATERIAL IN METEORITES

JOHN F. KERRIDGE
*Department of Chemistry, UCSD, La Jolla, CA 92093*

**Abstract.** Many primitive meteorites contain traces of presolar material in the form of mineral grains which were formed in a variety of stellar objects before formation of the Solar System. Most such grains identified to date are carbonaceous: diamond, SiC, and graphite; but some are oxides such as corundum. Chemical, isotopic and physical properties of the presolar grains have yielded information about nucleosynthesis and grain formation in certain stellar objects, such as AGB stars, supernovae and novae, and about chemical evolution in the Galaxy. Some primitive meteorites also contain organic matter apparently synthesised abiotically in an asteroidal environment from precursor molecules derived from interstellar clouds.

## 1. Introduction

Although the meteorites that fall on Earth are fragments of asteroids, bodies large enough to have experienced a certain amount of secondary alteration analogous to geological processes on Earth, a few retain within their mineral fabric primitive material which has experienced little or no such alteration. Some of that material was formed in the primordial solar nebula, before accumulation of the planets. But mixed in with this mundane, locally produced material is a small fraction consisting of presolar material: mineral grains or chemical compounds whose formation predated the existence of the Solar System. Some of that material owes its origin to chemical reactions taking place in interstellar molecular clouds, such as the one from which the Solar System formed by fragmentation and collapse into a gravitationally bound protosolar nebula. The record of such interstellar material survives in primitive meteorites as organic matter, which will be described in section 3 of this chapter. The remainder of the presolar material consists of mineral grains which condensed in the outflows from a variety of stellar objects, including supernovae, novae, red giants, and stars on the Asymp-



*J. M. Greenberg and A. Li (eds.), Formulation and Evolution of Solids in Space,* 447–484.

totic Giant Branch (AGB). This material will be described in section 2. (Strictly, such material is more correctly described as "circumstellar", but the term "interstellar" is in such common use that we shall employ it here as well.)

Because tangible samples can be analysed in the laboratory to yield data that are impossible to obtain by means of astronomical techniques, the study of presolar material in meteorites can potentially generate important new insights into stellar and interstellar processes. So far, this promise has been more realised for circumstellar grains than for interstellar organic molecules, at least in part because of the greater resistance of solid mineral grains against secondary alteration.

The classification of meteorites is an arcane subject, the details of which need not concern us here. Sufficient for our purposes is the knowledge that the most primitive meteorites are known as chondrites, because they contain chondrules, mm-sized silicate spherules that almost certainly were formed in the solar nebula. A small proportion of chondrites have experienced aqueous alteration on their parent asteroids; these are known as carbonaceous chondrites. The great majority of the remainder are known as ordinary chondrites, which have been thermally metamorphosed to a greater or lesser extent. Those that have experienced minimal metamorphic heating are known as unequilibrated ordinary chondrites (UOC). Both UOC and carbonaceous chondrites can contain interstellar material, but the great bulk of studies to date have been carried out on material extracted from carbonaceous chondrites, particularly Murchison, which is probably the least contaminated of these meteorites, and is relatively plentiful.

Excellent recent reviews of the work on the interstellar grains [Ott 1993; Anders & Zinner 1993] and the interstellar organic matter [Cronin & Chang 1993] include comprehensive bibliographies. Consequently, the bibliography for this review will tend to concentrate on papers published since 1993, with earlier work represented by the 1993 reviews plus a limited number of landmark papers.

## 2. Interstellar Grains

We define grains as solid particles, as distinct from molecules, which can be extracted from meteorites in a form that permits them to be directly observed. Thus, this definition does not include the pervasive $^{16}O$-rich component present most noticeably in refractory phases found in primitive meteorites [e.g., Clayton et al. 1973], for which a carrier phase has yet to be identified. Note that the term "grain" carries no connotations of crystallinity; although most interstellar grains discovered to date reveal a greater or lesser degree of structural ordering, some amorphous material is present and such material may well be abundant in the interstellar medium.

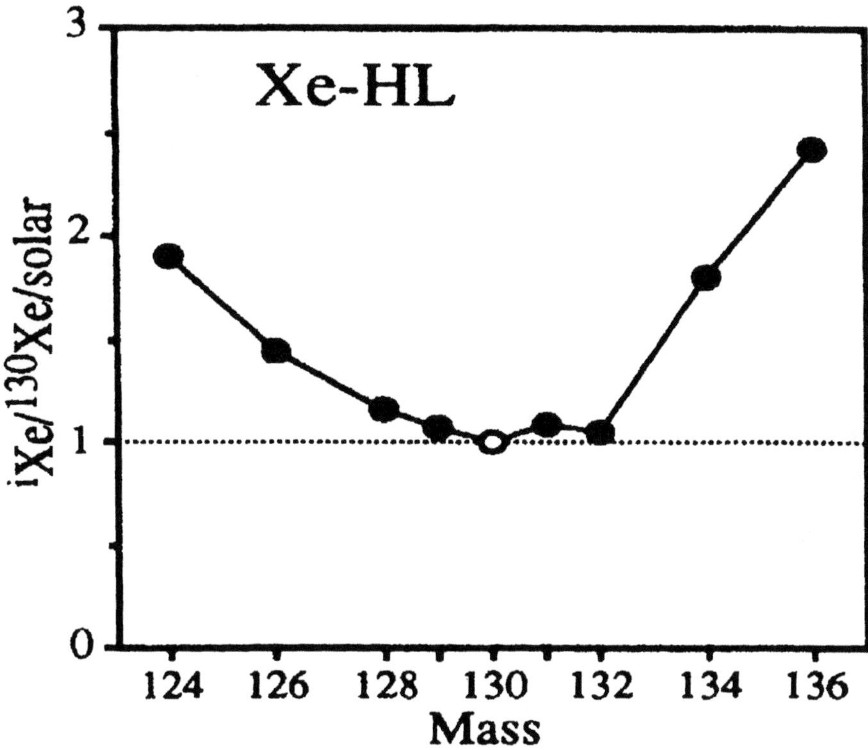


*Figure 1.* Isotopic composition of Xe-HL. This represents the most extreme composition measured so far, much greater than that originally observed by Reynolds and Turner [1964]. Note enrichment in both heavy (r-process) and light (p-process) isotopes. From Anders & Zinner [1993].

## 2.1. HOW INTERSTELLAR GRAINS ARE IDENTIFIED

The trail that led eventually to identification of individual interstellar grains in meteorites began in 1964 with an isotopic analysis of the noble gas xenon extracted from the carbonaceous chondrite, Renazzo [Reynolds and Turner 1964]. Those authors noted that Renazzo contained an anomalous Xe component in which both heavy and light isotopes were enriched relative to normal Solar-System Xe, Figure 1. Because the measured light-isotope enrichments barely exceeded their experimental uncertainties, they focused on the heavy-isotope enrichments which broadly resembled patterns produced by fission of transuranic nuclides. Although the Renazzo pattern differed from that of any known fissile nuclide, they speculated that a combination of mass-dependent fractionation of normal Solar-System Xe and addition of some fissiogenic Xe could yield the observed anomalous component. This interpretation was further developed by Anders and coworkers [e.g., Anders and Heymann 1969; Anders 1981] who argued that the fissiogenic Xe could have been supplied by an extinct superheavy element, with atomic number close to 114 where an island of (nuclear) stability was thought to exist. An alternative interpretation was put forward by Manuel et al. [1972] who proposed that both light and heavy isotopes of Xe could be synthesised in (different layers of) a Type II supernova. Reynolds et al. [1978] proposed that the carrier phase of the anomalous Xe was carbonaceous, a conclusion that was confirmed by Swart et al. [1983]. Although that carbonaceous host phase was characterised by an isotopic composition remarkably close to that of average Solar-System carbon, subsequent studies showed that it contained, besides the anomalous Xe, isotopically anomalous nitrogen [Lewis et al. 1983], thereby seriously weakening the superheavy-element hypothesis and strengthening the supernova, or interstellar, hypothesis. A few years later, the carrier phase was firmly identified as interstellar diamond, as discussed in more detail later.

While anomalous Xe was still being tracked to its host phase, another isotopically anomalous noble-gas component put in an appearance. This consisted of a neon component known as Ne-E, highly enriched in the isotope $^{22}$Ne [Black and Pepin 1969]. Successive experiments over the next 20 years increased the observed level of enrichment from a factor of a few to more than two orders of magnitude, Figure 2; in fact, in some samples the Ne consists essentially of pure $^{22}$Ne [e.g., Amari et al. 1995b]. Eventually, the carrier phase of the most highly enriched Ne was identified as graphite, while a less-highly enriched variety was shown to be present in silicon carbide, SiC, both of interstellar origin.

These tales of pursuit and successful denouement yield two important lessons. First, the trails were marked by noble gases. Second, the signature of an interstellar origin is based upon (anomalous) isotopic composition. These two themes will reccur throughout this section.

### 2.2. HOW INTERSTELLAR GRAINS ARE ISOLATED FOR STUDY

The traditional method for separating interstellar grains from the mundane mineral matrix in which they normally reside is by dissolving the latter in various acids until only acid-insoluble material remains, of which a significant proportion consists of interstellar grains, most of them carbonaceous. Unfortunately, this approach also destroys both the petrographic context within which the interstellar grains exist in the meteorites, and also any acid-soluble interstellar grains that might be present. Consequently, there have been attempts to develop a non-destructive technique for identifying interstellar grains in situ within a meteorite, some of which have been successful [e.g., Alexander et al. 1992]. This confirmed that interstellar SiC occurs as individual grains randomly dispersed throughout the meteorite matrix, but the approach has proved less effective than acid dissolution at making interstellar grains available for analysis, and was incapable, as developed, of identifying species other than SiC (or $Si_3N_4$). Thus, the findings reviewed here were obtained exclusively by analysis of grains isolated by acid dissolution.

Progress of the acid-dissolution procedure is customarily monitored by means of noble-gas analyses performed on small aliquots at each step in the procedure; concentrations of the anomalous noble-gas components increase as the proportion of interstellar grains in the acid-insoluble fractions increases. Typically, the procedure [Amari et al. 1994] begins with freeze-thaw disintegration in order to expedite reaction rate. The first acid treatment utilises an HF/HCl mixture alternating with HCl alone. This removes silicate minerals and metal and some oxides and sulfides. Elemental sulfur is then removed with potassium hydroxide, and kerogen-like organic matter (which is also of interest in its own right, as discussed in section 3) is oxidised using dichromate and hydrogen peroxide. The residue at this point consists predominantly of interstellar diamonds, graphite and SiC, with some locally produced spinel and chromite. These last are removed by means of sulfuric acid. Diamonds can be separated in a colloidal extract, which needs some treatment with perchloric acid to remove persistent traces of kerogen-like material, and then graphite and SiC can be separated from each other on the basis of density. For some studies, both graphite and SiC can be further subdivided according to grain size and/or density. The full procedure is described in detail by Amari et al. [1994].

### 2.3. ANALYTICAL TECHNIQUES USED IN THE STUDY OF INTERSTELLAR GRAINS

Methods used to analyse the isolated fractions of interstellar grains may be conveniently divided into two categories: Those utilising individual grains; and those employing samples consisting of large numbers of such grains. Historically, the latter dominated the early work, and they still represent the only way of performing certain analyses, but during the past eight years or so they have been largely supplanted by techniques, such as the ion microprobe, capable of yielding data on individual grains. Because interpretation of the meteoritic evidence in terms of stellar processes commonly requires use of a broad range of analytical data, it is frequently necessary to draw upon the results of both bulk and individual analyses, which can introduce some uncertainty into those interpretations, as will become apparent in the next section. The point was made earlier that noble gases have been key to

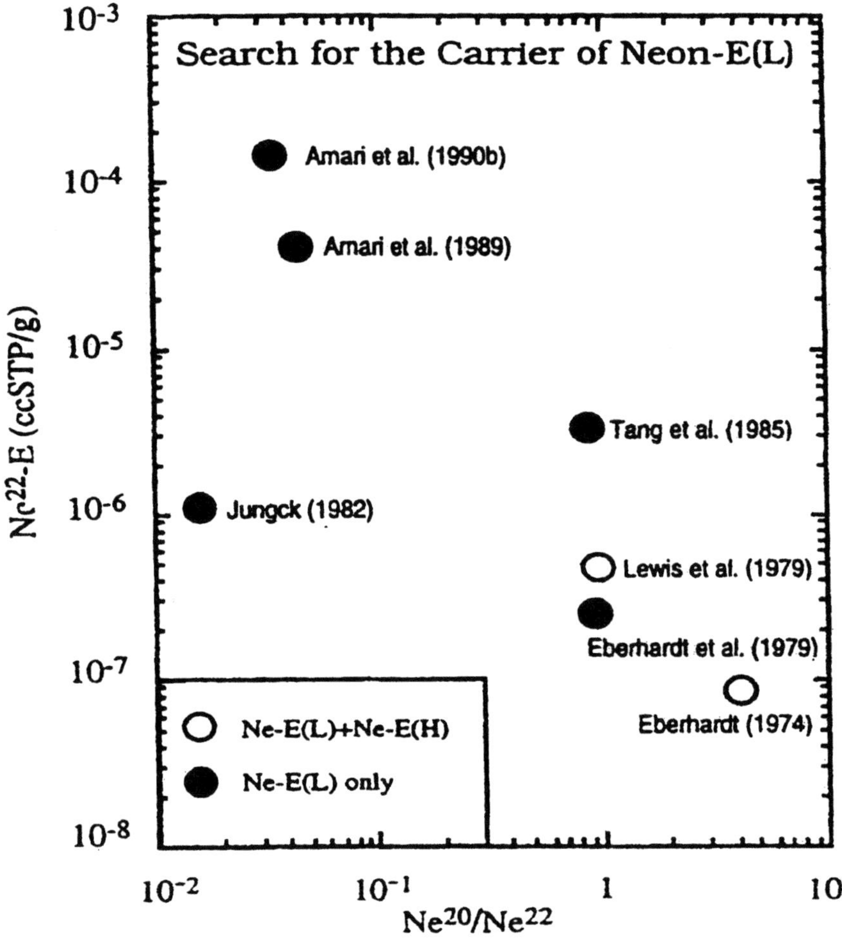


*Figure 2.* Successive stages in the isolation of the carrier of Ne-E(L). More-recent studies are characterised by isotopic ratios which are > 1000-fold enriched relative to the original discovery by Black and Pepin [1969]. From Amari et al. [1995].

identifying and locating the different kinds of interstellar grains in meteorites. Except for a few analyses of He and Ne in large individual graphite and SiC grains, all noble-gas analyses to date have been performed on bulk fractions. In a typical analysis, a sample of such a fraction, some 2 to 40 $\mu$g in size, is heated in steps to about 1800°C, with the gas evolved at each step collected, purified and analysed in the mass spectrometer [e.g., Lewis et al. 1994]. In the earliest work, a similar approach, utilising stepped combustion of bulk fractions, was adopted for isotopic analysis of C and N [e.g., Swart et al. 1983], and is still required for the nano-scale diamonds [e.g., Russell et al. 1996], but for graphite and SiC such analyses are now invariably performed using the ion microprobe.

Development of the CAMECA IMS 3f ion microprobe into a device capable of isotopic analysis of micrometer-sized particles revolutionised the study of interstellar grains in meteorites [Zinner 1989]. This approach now represents the technique of choice, not simply for C and N, but for essentially any element present at greater than trace concentrations in the grains. Such elements presently include O, H, Mg, Si, K, Ca, and Ti. Furthermore, the ion probe can be used to obtain elemental concentrations for an additional range of elements, including Al, V, Fe, Sr, Y, Zr, Nb, Ba, Ce, and Nd [Amari et al. 1995a].

Because of the pivotal role that noble gases have played in identifying interstellar grains and in elucidating their history, there is a clear need for a technique capable of yielding noble-gas analyses on single micrometer-sized grains. Such a technique has been developed, based on extraction of gas using a finely focused laser and analysis of the resulting gas in a high-sensitivity mass spectrometer. At present, this technique lacks the sensitivity to analyse the heavy noble gases, but it has been used successfully to generate He and Ne data for individual grains of SiC and graphite from Murchison [Nichols et al. 1992; 1993; 1994]. This has permitted identification of several Ne-E-rich grains of SiC and graphite. The former are relatively rare, $\sim 5\%$ of all interstellar SiC; the latter are relatively common, $\sim 30\%$ of interstellar graphites.

Finally, electron microscopy has yielded valuable morphological, textural, compositional, and crystallographic information on individual grains. Imagery on a scale of tens to hundreds of nanometers can be obtained by means of scanning electron microscopy, together with elemental analyses by means of energy-dispersive X-ray spectroscopy [e.g., Bernatowicz et al. 1996]. Even higher magnifications are available using transmission electron microscopy, capable of resolving a small fraction of a nanometer. This technique, which is typically applied to ultrathin sections made by either ion-thinning or microtoming, can also yield elemental compositions from X-ray spectra and crystal structural data from electron diffraction patterns [e.g., Daulton et al. 1996; Bernatowicz et al. 1996].

### 2.4. CLASSIFICATION OF METEORITIC INTERSTELLAR GRAINS

Table 1 lists the types of interstellar grains discovered so far in primitive meteorites. As noted previously, these consist predominantly of acid-insoluble carbonaceous material. It is now clear that these grains originated in a wide variety of stellar sources; in fact, it is likely that all the major sources of interstellar dust are represented in the meteoritic population. In this section, we summarise what is presently known about those stellar sources based on properties of the interstellar grains. It is important to note that these interpretations have resulted from the interplay between nucleosynthetic theory and isotopic (and to a less extent elemental) measurements of the grains. Models of stellar nucleosynthesis have guided the analytical work, and the results of those analyses have in turn served to refine the models.

TABLE 1. Varieties of interstellar grain observed in meteorites.

| Name | Formula | Abundance* | Noble-gas marker |
|---|---|---|---|
| Diamond | C | ~ 400 ppm | Xenon-HL |
| Silicon Carbide | SiC | ~ 7 ppm | Xenon-S; Neon-E(H) |
| Graphite | C | < 2 ppm | Neon-E(L) |
| Corundum | $Al_2O_3$ | < 1 ppm? | n/a |
| Silicon Nitride | $Si_3N_4$ | ~ 2 ppb | ? |
| Spinel | $MgAl_2O_4$ | trace | n/a |

* In Murchison.

2.4.1. *Silicon carbide*

Classification of interstellar SiC has been based mainly on the isotopic composition of C and N, Figure 3, and the abundances of certain elements, such as Sr, Ba, Y, Zr, La, Nb and W, produced by s-process nucleosynthesis (involving neutron capture at a relatively slow rate). Historically, the connection between interstellar SiC (although its identity was then still unknown) and the s-process was made by Srinivasan and Anders [1978] who recognised that a Xe component present in carbonaceous chondrites matched remarkably well, Figure 4, the calculated composition of s-process Xe [Clayton and Ward 1978]. Silicon carbide was also found to be the host of some of the $^{22}$Ne-enriched neon, Ne-E, mentioned earlier. (Even before its specific host phases were identified, it was recognised that Ne-E consists of two distinct components, one which was pure $^{22}$Ne, released at relatively low temperatures in pyrolysis, and hence known as Ne-E(L), and one with a somewhat less extreme isotopic composition, released at high temperatures, and known as Ne-E(H). The former is found to reside in interstellar graphite grains, the latter in SiC.)

Table 2 gives the stellar sources assigned to SiC, using a matrix based upon C/N isotopes and s-process abundance patterns. The latter were derived from models of Lodders and Fegley [1995] which were themselves based on data of Amari et al. [1995a]. Note that, except for the relatively rare anomalous grains X, Figure 3, all the SiC grains identified so far were derived from carbon stars lying on the asymptotic giant branch (AGB) of

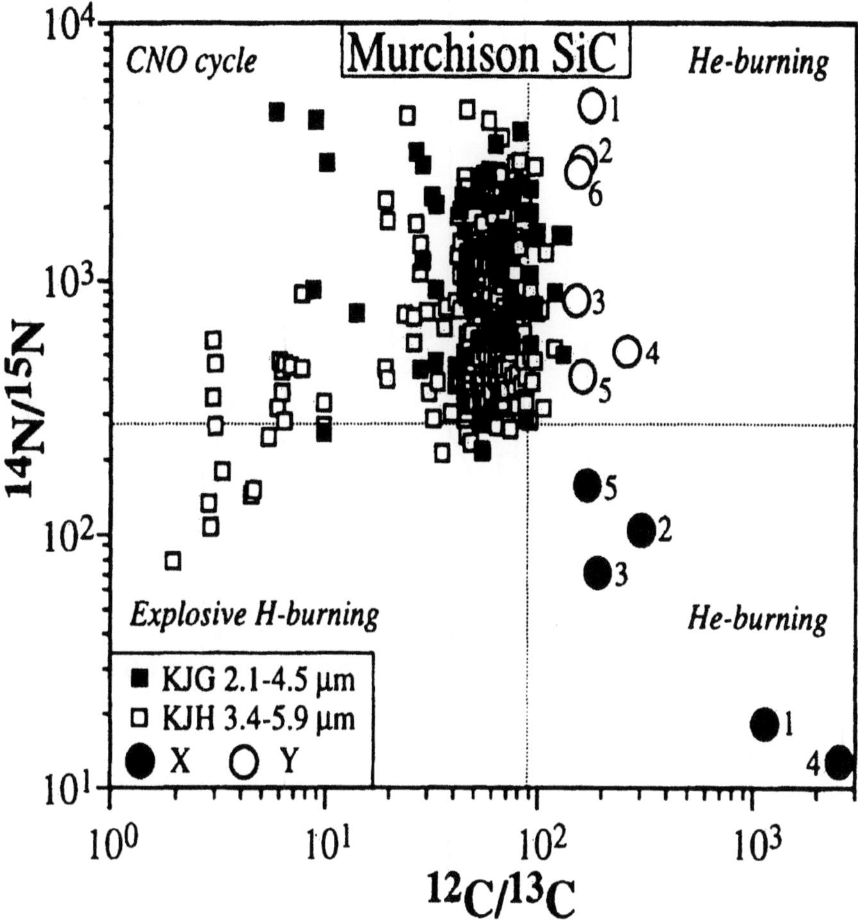


*Figure 3.* Summary of C- and N-isotopic composition of individual SiC grains extracted from the Murchison meteorite. Note that most compositions, "Mainstream", reflect nucleosynthesis by hydrostatic H-burning in the CNO cycle. The grains with extremely low $^{12}C/^{13}C$ ratios are termed "type A". Those intermediate between Mainstream and type A are termed "type B". Note anomalous grains "X" in the lower right quadrant, and "Y" in upper right quadrant. See Table 2. From Anders and Zinner [1993].

the H ertzprung-Russell stellar classification diagram. Although some nucleosynthetic details still need further refinement, production of the meteoritic SiC grains is reasonably well understood at this time.

### 2.4.2. *Graphite*

By contrast, significant uncertainty remains in the source assignments for most of the interstellar graphite. Much of that uncertainty stems from the fact that, although there is now a considerable body of data on individual

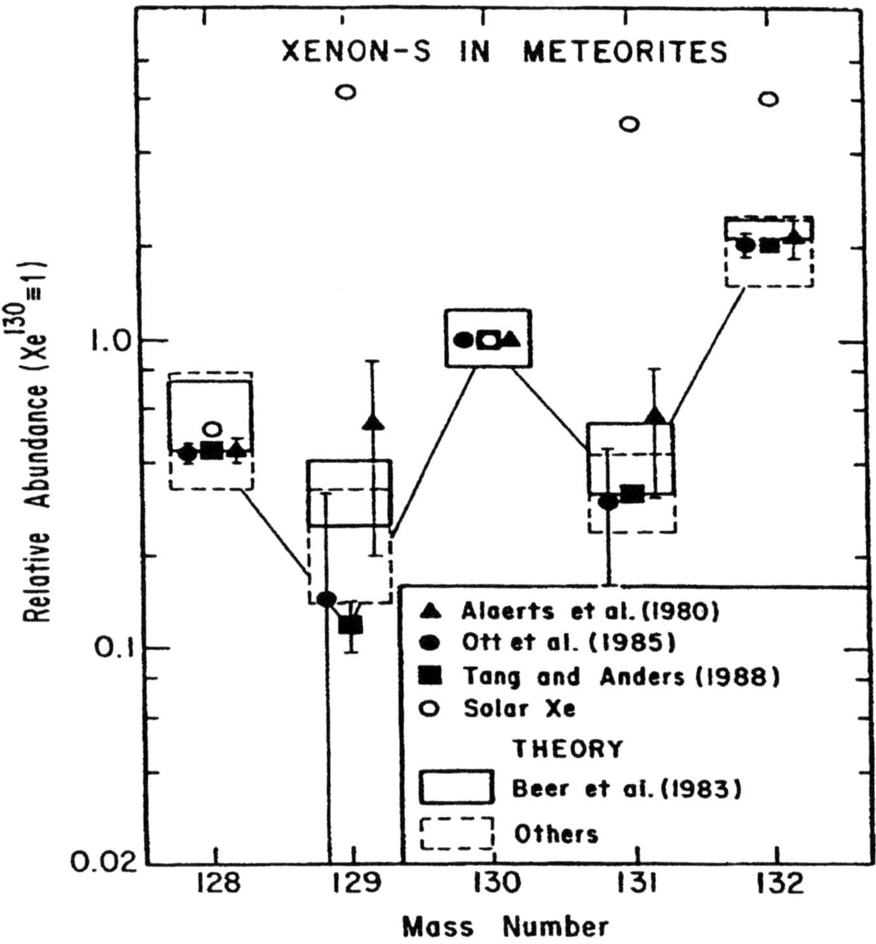


*Figure 4.* Isotopic composition of Xe-S extracted from meteorites (filled symbols). Note marked difference from solar Xe (open circles) but good agreement with s-process predictions (boxes). From Anders [1988].

graphite grains [e.g., Hoppe et al. 1995], most of the key noble-gas data have been obtained on bulk fractions, as noted earlier, and correlating such data with those for individual grains has not been straightforward. The most useful discriminants for the interstellar graphite have been C-isotopic composition, Figure 5, and density. Table 3 gives tentative source assignments based on those parameters. Except for a handful of low-density grains for which a supernova origin is reasonably secure [Amari et al. 1996; Nittler et al. 1996], most of the graphite is characterised by isotopic signatures suggestive of both massive stars ($> 10\,M_\odot$), most likely Wolf-Rayet stars ($> 25\,M_\odot$) [Hoppe et al. 1995], and relatively low-mass ($1 - 3\,M_\odot$) AGB

TABLE 2. Stellar sources inferred for meteoritic interstellar SiC grains.

| Chemical groups* (s-process enrichment) | I (x 1) | II (x 10) | III (x 3-5) | IV (x 3 [L]; x 10 [H]) |
|---|---|---|---|---|
| Isotopic Groups | | | | |
| Mainstream | N (- s) | N | N | N |
| A, B | J | - | CH (sg) | CH (g) |
| X | - | - | SN I or II | - |
| X-57 | - | - | SN II | - |
| Y | - | N($^{12}$C) | - | - |

* Groups based on abundances of s-process nuclides. Figures in brackets give enrichments relative to average solar system abundances. L = light Rare Earth Elements, H = heavy Rare Earth Elements.
Notes: N = N-type carbon stars; (-s) = no s-process; J = N, J subgroup of N-type stars CH = CH-type carbon stars; (sg) = (sub-giant); (g) = (giant); SN = supernova; X-57 = grain with extreme $^{28}$Si enrichment & inferred $^{44}$Ti abundance; ($^{12}$C) = enriched in $^{12}$C by $3\alpha$-process.
Refs: Amari et al. [1992]; Hoppe et al. [1996b]; Lodders and Fegley [1995].

stars in their thermally pulsing (TP) phase [Amari et al. 1995b]. Since the stellar conditions implied by these assignments are mutually incompatible, the different signatures presumably reside in different subpopulations of the graphite, but these remain to be identified. Note that at least two kinds of AGB star are represented in the graphite record, differing markedly in their mean neutron exposure.

The noble-gas "signpost" that led to interstellar graphite was the $^{22}$Ne-rich component that is released at relatively low temperatures, Ne-E(L). More detailed analysis [Amari et al. 1995b] has established that Ne-E(L) consists of at least two presolar components: Ne-G, produced, as Ne, in the He-shell of AGB stars; and Ne-R, produced as $^{22}$Na, most probably in a nova.

### 2.4.3. *Diamond*

Because of their extremely small size, 1-4 nm diameter, all chemical and isotopic data on meteoritic interstellar diamonds have been obtained on bulk fractions. As noted earlier, the path to interstellar diamond was marked by Xe enriched in both heavy and light isotopes, Xe-HL, for which a supernova origin seems most plausible [Manuel et al. 1972; Clayton 1989]. Compared with SiC and graphite, interstellar diamond appears to be, at least physically, quite homogeneous. Stepwise pyrolysis reveals significant isotopic variation in N and the noble gases, but all attempts at physical separation of these components have failed. However, it is reasonable to

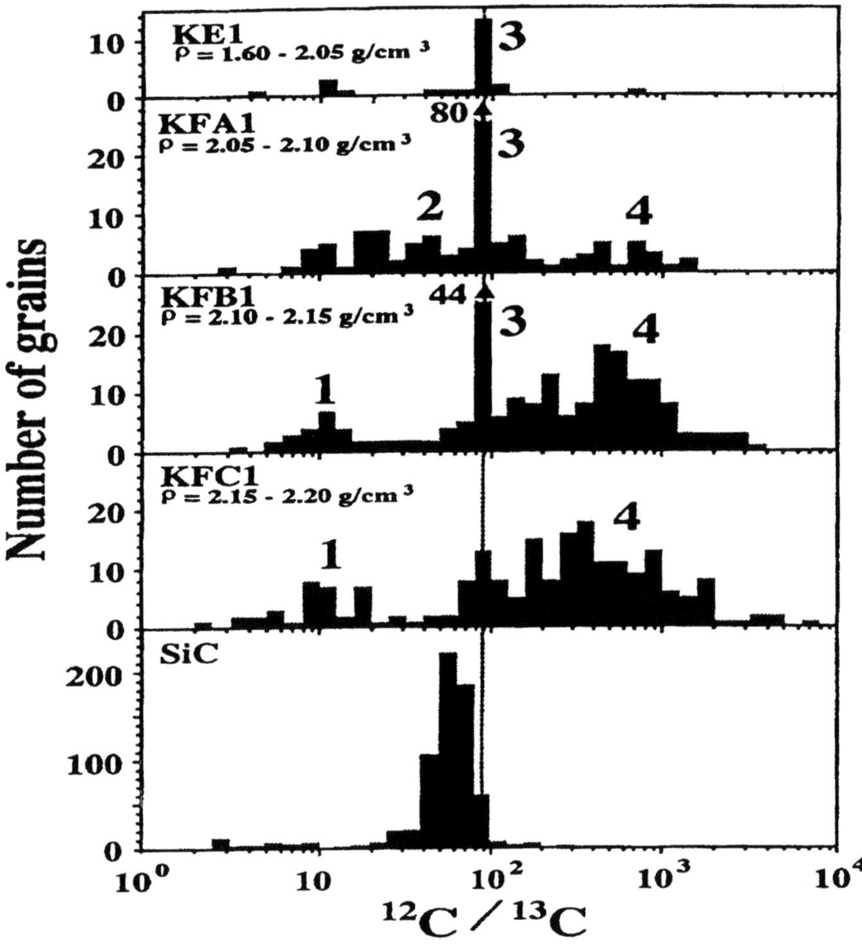


*Figure 5.* Carbon isotopic compositions of graphite grains separated from Murchison (four upper panels) compared with SiC (lowest panel). The composition of typical solar-system C is indicated by the dotted line. Four distinct isotopic groups can be identified among the graphite grains. From Hoppe et al. [1995].

conclude that more than one stellar source contributed to the meteoritic diamond. In particular, because only about one diamond in a million contains a Xe atom, it may well be that those diamonds that contain Xe-HL represent a discrete subpopulation of the interstellar diamonds.

One of the most striking features of the diamond is its C-isotopic composition, which is only some 3 to 4% lighter than average Solar-System C [e.g., Russell et al. 1996]. Clayton et al. [1995] have argued that this $^{12}C/^{13}C$ ratio is a natural consequence of mixing between He-rich and H-rich layers within a Type II supernova, of nominally 25 $M_{\odot}$, but although the observed

TABLE 3. Stellar sources inferred for meteoritic interstellar graphite grains.

| Density | Low | | | Medium | High |
|---|---|---|---|---|---|
| | Main stream | $^{44}$Ti | $^{41}$Ca | | |
| Carbon Isotopic Group | | | | | |
| 1 | ? [nova] | SNII (He/C, He/N, core) | - | ? [nova] | ? [nova] |
| 2 | WR (WN/WC) AGB (low-n exp) | - | SNII (He/C, He/N) [WR?] | WR (WN/WC) AGB (low-n exp) | - |
| 3 | WR (WN/WC) AGB (low-n exp) | SNII (He/C, He/N, Si/S) | SNII (He/C, He/N) | WR (WN/WC) AGB (low-n exp) | WR (WN/WC) AGB (high-n exp) |
| 4 | WR (WC) AGB (low-n exp) | - | SNII (He/C, He/N, O-rich) | WR (WC) AGB (low-n exp) | WR (WC) AGB (high-n exp) |

Notes: $^{44}$Ti,$^{41}$Ca = evidence for extinct radionuclides; SNII = Type II supernova (He/C, He/N, etc. = concentric zones in supernova); WR = Wolf-Rayet star (WN = H-burning products at surface; WC = He-burning products at surface); AGB = star on Asymptotic Giant Branch; [square brackets] = minor contribution.
Refs: Amari et al. [1995b, 1996]; Hoppe et al. [1995]; Nittler et al. [1996].

$^{12}C/^{13}C$ ratio is consistent with such an origin, the uncertainties in input parameters of their model preclude a unique identification with supernova C. Nonetheless, there is good reason to believe that the meteoritic interstellar diamonds formed by chemical vapor decomposition (CVD) in the expanding envelopes of (presumably a number of) supernovae [Lewis et al. 1987; Wdowiak 1987], although Tielens et al. [1987] suggested that grain-grain collisions in interstellar shocks could convert amorphous or graphitic C into diamond. This mechanism has been criticised by Anders and Zinner [1993] on the grounds that the low efficiency of such a process would leave the interstellar medium with considerably more of the pre-shock forms of

C than of diamond, whereas the meteorite record contains more diamond than graphite [Tang et al. 1989].

A recent microstructural comparison of interstellar diamonds with examples of CVD and shock-synthesised diamonds [Daulton et al. 1996] showed that the meteoritic diamonds possess features diagnostic of isotropic growth, consistent with a CVD origin but unlike the record of anisotropic growth characteristic of shock synthesis.

2.4.4. *Noncarbonaceous grains*

It has been clear from the outset that, of the interstellar grains present in the early Solar System, it would be most unlikely that only carbonaceous varieties would be incorporated into primitive meteorites. However, the search for noncarbonaceous grains of interstellar origin in such meteorites proved to be remarkably difficult, though ultimately successful with identification of nearly 100 interstellar oxide grains (corundum, $Al_2O_3$, and spinel, $MgAl_2O_4$), and nine grains of silicon nitride. It should be noted that discovery of these grains did not directly involve the use of noble gases, unlike the case for the carbonaceous grains.

*Oxides.* Data have been reported for 88 interstellar corundum grains (plus two spinels with basically the same isotopic properties as the corundum) separated from both carbonaceous chondrites and UOC [e.g., Nittler et al. 1997]. The interstellar corundum represents about 0.2% of the $Al_2O_3$ grains present in such meteorites. The crystal chemistry of corundum accepts much lower trace-element contents than either SiC or graphite, so that isotopic data for interstellar corundum are restricted to O and Mg; however these have proved quite diagnostic of grain sources. Somewhat more than half of the grains exhibit $^{26}$Mg excesses attributable to decay of extinct $^{26}$Al, yielding inferred initial $^{26}Al/^{27}Al$ ratios in the range $1.2\times10^{-4}$ to $1.6\times10^{-2}$. With one possible exception, all of the grains studied so far originated apparently in low- or intermediate-mass O-rich red giant or AGB stars. The O-isotopic composition of the exception suggests production in either a high-mass AGB star or a Wolf-Rayet star. It is noteworthy that no corundum grains reveal isotopic signatures of a supernova origin.

Figure 6 illustrates the O-isotopic data for the interstellar corundum grains and shows how they have been divided into four groups. Excesses in $^{26}$Mg occur in all groups but are predominant in groups 2 and 4, conspicuous in group 1, and modest in group 3. Although all four groups are assigned to broadly similar types of stars, the spread in isotopic compositions observed demonstrates clearly that these grains were derived from a population of Red Giant Branch (RGB) and AGB stars varying significantly in mass, metallicity, and initial O-isotopic composition.

The detailed isotopic evolution of RGB/AGB stars is a complex topic

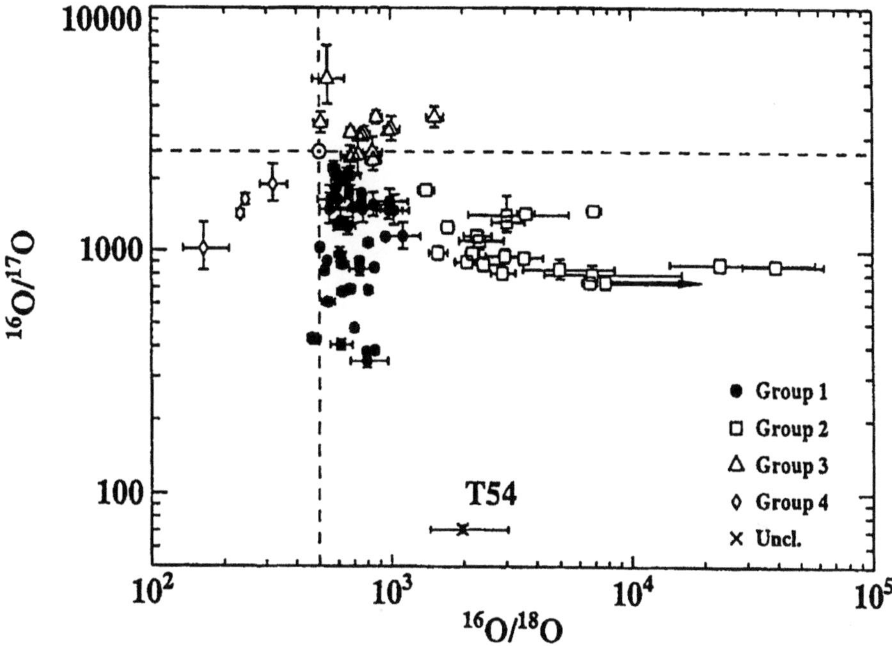


*Figure 6.* Oxygen isotopic compositions of interstellar oxide grains separated from primitive meteorites. Those compositions, except for that of grain T54, fall into four distinct groups; see text. From Nittler et al. [1997].

beyond the scope of this chapter. In brief, the stellar surface, from which outflow leads to condensation of grains from the cooling gas, is periodically "contaminated" by the products of CNO-cycle nucleosynthesis (and, later, by s-process products) which are convectively dredged up from interior regions. The temperature at which that nuclear burning took place is a strong function of stellar mass. The first dredge-up, which occurs while the star is still on the RGB lowers the $^{16}O/^{17}O$ ratio and raises the $^{16}O/^{18}O$ ratio at the surface. Following the RGB phase, He burning in the core takes place until destruction of all the core He causes the star to enter its thermally pulsing (TP) AGB stage, at which point, intermediate-mass stars ($> 3\,M_{\odot}$) experience a second convective dredge-up. This has little effect on the O isotopes of stars with metallicities around the solar value of 0.02, but for lower-metallicity stars, in which the first dredge-up would have been weak or nonexistent, the second dredge-up produces much the same isotopic effect as the first dredge-up did in higher-metallicity stars. Consequently, TP-AGB stars have initial O-isotopic compositions rather closely defined by their mass and metallicity, Figure 7. This composition apparently does not change significantly during the TP-AGB phase, despite a series of con-

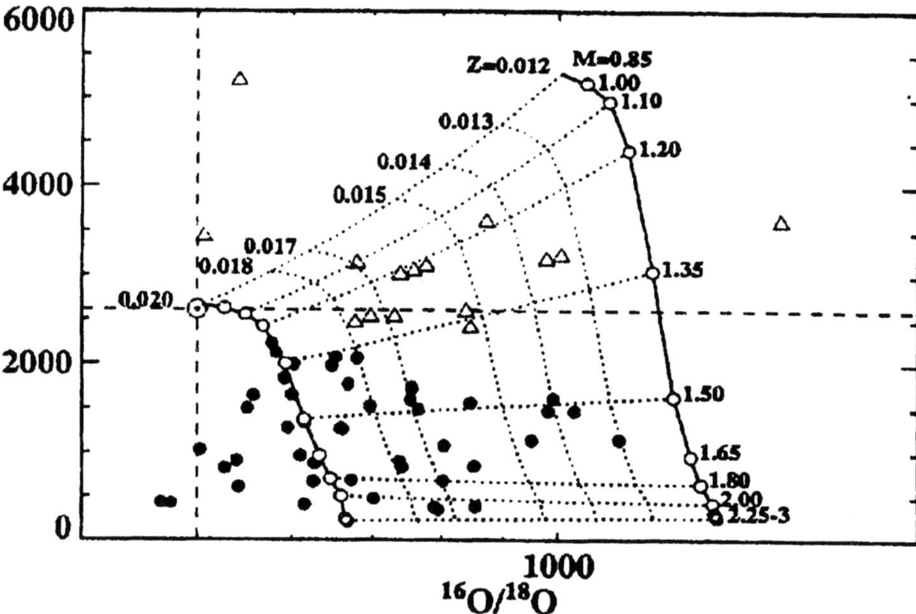


*Figure 7.* Compositions of oxide grains of groups 1 and 3 (see Figure 6 for symbol definitions) compared with predictions of isotope ratios following first dredge-up in red giant stars, for different initial mass, M (in units of solar mass) and metallicity, Z. Each open circle corresponds to predictions for an individual star. Meteoritic oxide grains apparently come from several different stars with distinct masses and initial compositions. From Nittler and Cowsik [1997].

vective episodes, known as the third dredge-up, in stars above about 1.5 $M_\odot$. However, this dredge-up does bring to the surface $^{26}$Al and the products of s-process nucleosynthesis. Another process that is believed to take place in low-mass AGB stars is known as "cool bottom processing", in which hydrodynamically induced mixing cycles material from near-surface regions to a depth where partial H burning can occur and back to the surface. This mainly has the effect of partially destroying $^{18}$O, Figure 8. We are now in a position to identify the likely sources of groups 1-4.

Group 1 grains apparently formed in TP-AGB stars with mass $> 1.4\,M_\odot$, Figure 7, though some, with low inferred $^{26}$Al/$^{27}$Al ratios, may have formed earlier, during the RGB stage, or later, during the planetary-nebula stage (after the $^{26}$Al had decayed).

Group 2 grains match the prediction of "cool bottom processing", Figure 8, so are believed to have formed in TP-AGB stars with mass $< 1.5\,M_\odot$ which underwent mixing between the convective envelope and the H-burning region.

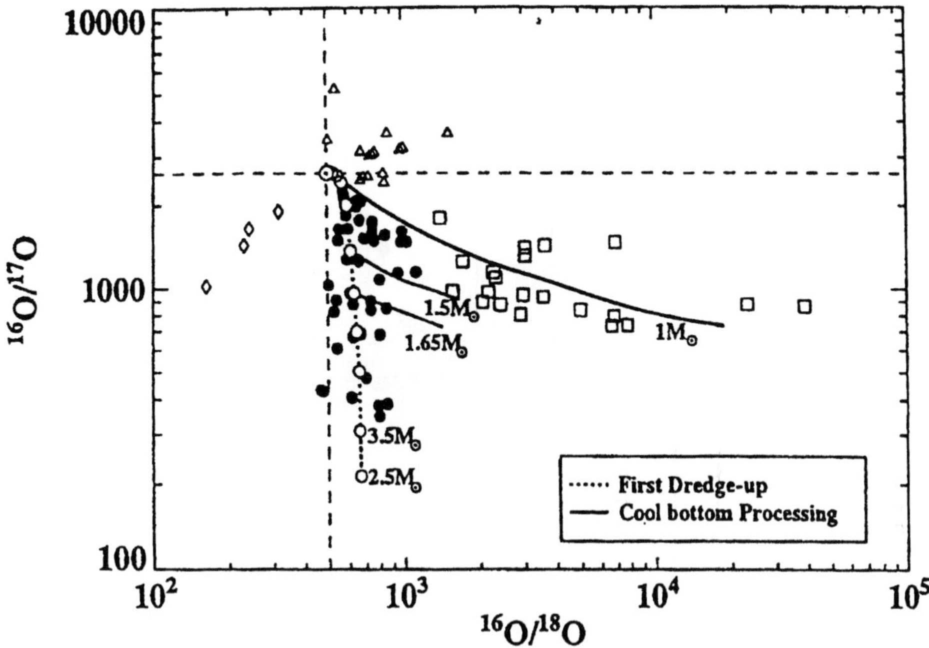


*Figure 8.* Compositions of oxide grains of Groups 1-3, compared with predictions of those produced by deep extra mixing in low-mass AGB stars, known as "cool bottom processing". See Figure 6 for symbol definitions. Note good agreement for Group 2 grains, suggesting that their $^{18}O$ depletions were caused by deep mixing. Apparently such mixing occurs only in stars below about $1.5\,M_{\odot}$ . From Nittler et al. [1997].

Group 3 grains resemble group 1 but with production in low mass stars ($< 1.4\,M_{\odot}$) with high initial $^{16}O/^{17}O$ and $^{16}O/^{18}O$ ratios indicative of formation early in Galactic history. The low incidence of $^{26}Mg$ excesses in this group suggests that grain production took place during the RGB stage.

Group 4 is the least well understood group at this time. The relatively low $^{16}O/^{18}O$ ratios are apparently inconsistent with production in low-mass AGB stars, as suggested by the $^{16}O/^{17}O$ ratios and $^{26}Mg$ excesses. Dredge-up of freshly synthesised $^{18}O$ in certain AGB stars is a possibility.

The strong dependence of $^{16}O/^{18}O$ after first (and second) dredge-up on metallicity raises the possibility of using the data for Groups 1 and 3 to track the effects of galactic chemical evolution and even to generate an estimate for the age of the galaxy, see below.

*Silicon Nitride.* Nine $Si_3N_4$ grains extracted from primitive meteorites have been identified as probably interstellar [Nittler et al. 1995]. Their abundance is much lower than that of SiC, about 2 ppb in Murchison, and

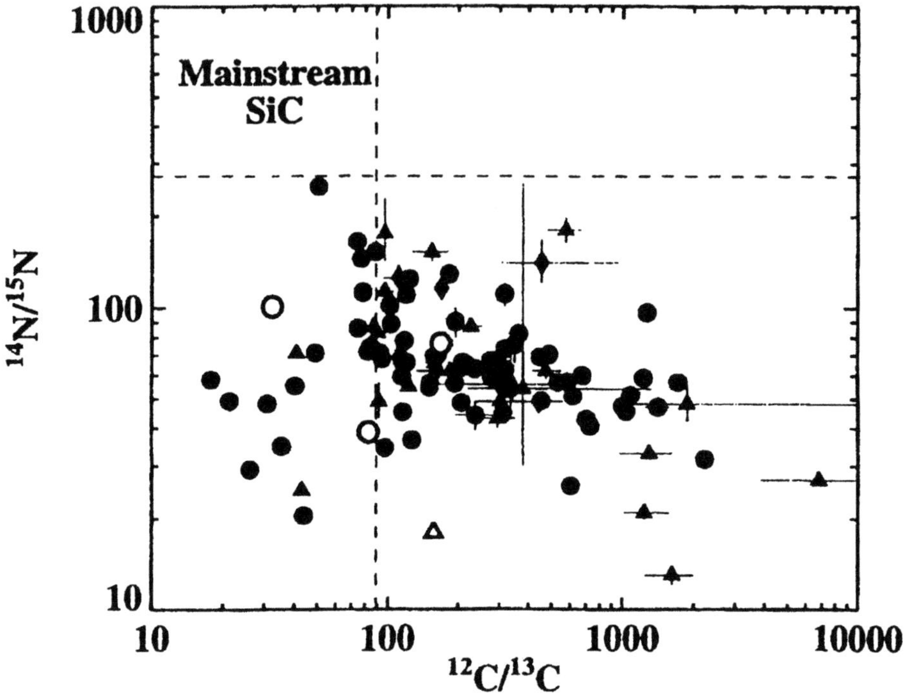


*Figure 9.* Carbon- and N-isotopic compositions of $Si_3N_4$ from the Murchison meteorite, open symbols, compared with those of SiC grains "X", solid symbols. From Nittler et al. [1995].

isotopically they differ markedly from mainstream SiC grains. However, their isotopic compositions, Figure 9, bear a strong resemblance to those of SiC grains X, for which a supernova origin has been inferred, Table 2, though for both types of grain there is a problem in accounting for their $^{26}Al/^{27}Al$ - $^{14}N/^{15}N$ systematics, Figure 10.

## 2.5. NATURE OF INFORMATION DERIVED FROM STUDY OF INTERSTELLAR GRAINS

As noted earlier, the great merit of meteoritic interstellar grains is that they can be studied with techniques that cannot be applied to astronomical objects. The small size of such grains limits the information that can be acquired from them, but in most cases it is possible to obtain elemental and isotopic analyses, and morphological and textural information. Results from such studies permit conclusions to be drawn regarding nucleosynthetic, physical, and chemical processes taking place in a variety of stellar objects. Some examples of such studies are given here.

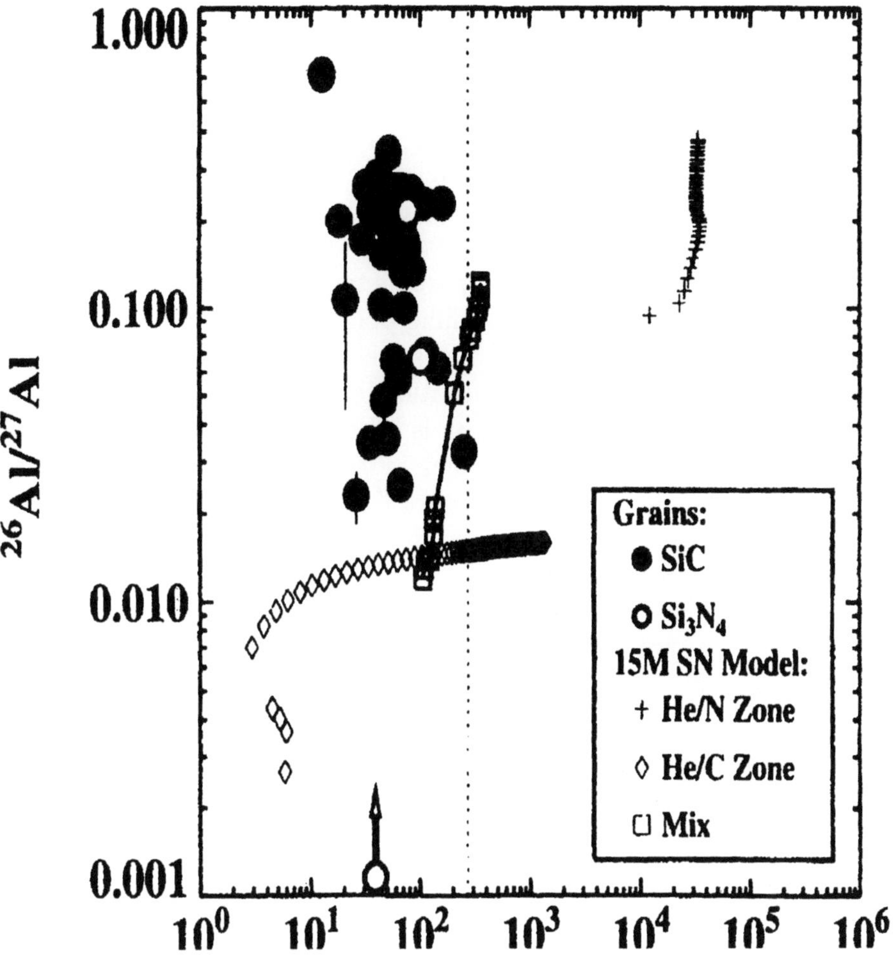


*Figure 10.* Comparison of Al- and N-isotopic compositions observed in $Si_3N_4$ and SiC from M urchison with those predicted for different zones in a $15\,M_\odot$ supernova. Most observed compositions cannot be explained by current supernova models. From Nittler et al. [1995].

### 2.5.1. *Nucleosynthesis*

The point was made earlier that application of meteoritic data to issues of nucleosynthesis is an iterative process, in which theory and observation feed back on each other permitting identification of stellar sources, refinement of nucleosynthesis modelling, and definition of future analytical tests of such models. A good example of such iteration involves the abundances of certain s-process nuclides in meteoritic SiC grains, whose compositions reflect mixing of normal Solar-System matter with pure s-process material, such as the anomalous Xe isotopes which led to discovery of these interstellar grains, see above.

As its name implies, the s-, or slow-neutron, process takes place in stellar environments in which neutron captures take place on a sufficiently long timescale that most $\beta$-emitting radionuclides decay before they can capture another neutron. This results in production of nuclides that lie along what is termed the "valley of stability". If, however, a radionuclide has an intermediate half-life, i.e., one that is comparable to neutron-capture timescales, it can create what is termed a "branching point", so that the magnitude of the local neutron density will in such a case control the path along which s-process synthesis proceeds. Such a situation is found for two of the isotopes of Kr, Figure 11. The precursor of $^{86}$Kr is $^{85}$Kr, which has a half-life of 10.8 years. Consequently, in an environment in which neutron captures occur more frequently than every 10 years or so, $^{86}$Kr is produced, but if the neutron density is too low, the product is $^{85}$Rb. A similar, but somewhat more complicated situation applies to the isotope $^{80}$Kr, whose relevant precursor is $^{79}$Se. The ground state of that nuclide is relatively long-lived (about $6.5 \times 10^4$ years), so that at low temperatures neutron capture predominates with production of $^{79}$Se. But at higher temperatures, an isomeric state, $^{79m}$Se, is populated and the rapid decay of this state (3.9 minute half-life) can lead via neutron capture on $^{79}$Br and $\beta$-decay of $^{80}$Br to production of $^{80}$Kr. Consequently, the abundance of $^{80}$Kr can act as a thermometer for an s-process environment, though if the neutron density is sufficiently high, $^{79m}$Se can capture a neutron before it decays, cutting off production of $^{80}$Kr. It follows that a combination of $^{86}$Kr and $^{80}$Kr abundances can define quite closely the conditions, i.e., temperature and neutron density, in which s-process material was produced. Those variables are themselves functions of two basic input parameters: The metallicity of the star, expressed here as log[Fe/H], where Fe/H is normalised to the solar value; and the so-called "overlap factor", r, which reflects the extent to which material in the envelope of a thermally pulsing AGB star is exposed to neutrons in successive pulses.

Applying this approach to meteoritic SiC, we find that the $^{86}$Kr/$^{82}$Kr and, to a less extent, $^{80}$Kr/$^{82}$Kr ratios are a function of grain size, Figure 12, from which it can be seen that this trend is well matched by predictions for the He-shells of low-mass ($1 - 3\,M_\odot$) AGB stars with a range of metallicities and/or overlap factors. An immediate conclusion is that the meteoritic SiC appears to be derived from a number of different AGB stars. The causal connection between s-process nuclides and grain size is unclear, because nucleation and growth of the grains would have occurred in a much cooler region than those in which the s-process operated. Presumably growth conditions for the grains are also governed in some basic

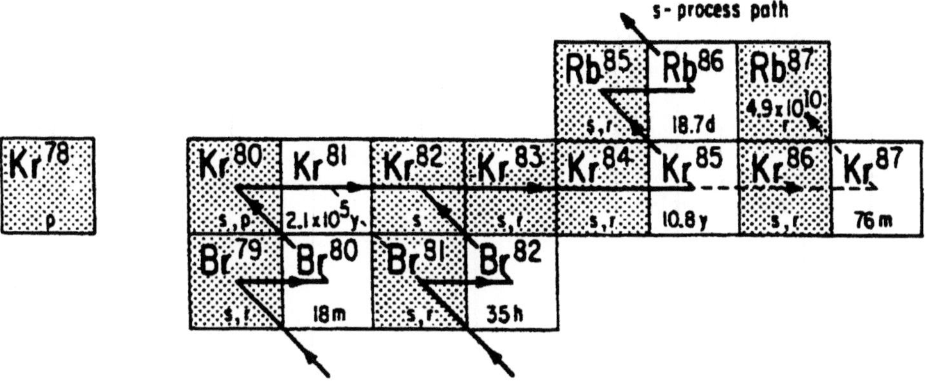


*Figure 11.* Production of the Kr isotopes in the s-process. If $^{85}$Kr has time to decay between neutron captures, little or no $^{86}$Kr is produced. A similar, but somewhat more complicated, situation applies to $^{80}$Kr, see text. From Anders [1988].

way by the mass and/or metallicity of the star. Finally, optimising the fit of the nucleosynthesis model predictions to the meteoritic SiC data implies the need for some minor adjustments to the relevant neutron cross-section data. It is worth stressing that the insights into details of s-process synthesis revealed by the SiC could not be obtained by analysis of average Solar-System material, nor by astronomical observations.

Similarly, detailed analyses of the SiC grains X and certain anomalous graphite grains have yielded new insights into the nucleosynthesis that takes place within supernovae and particularly into the way in which mixing takes place between different layers within a supernova. For example, evidence for extinct $^{44}$Ti in some low-density graphite grains not only points towards an origin within a Type II supernova for such grains, but requires mixing of core material into the near-surface He/C and He/N shells, without significant contamination by intervening shells [Nittler et al. 1996].

It should be pointed out that the above examples represent the tip of the iceberg of nucleosynthetic information derived so far from the study of interstellar grains in meteorites.

### 2.5.2. *Physical processes*

Compared with the wealth of nucleosynthetic information that has been derived from the study of interstellar grains, relatively little attention has been paid to the record within them of other physical processes taking place in stars. A major contribution in this area is a study by Bernatowicz et al. [1996] on graphite spherules from Murchison. Those spherules, which ranged in size from 1 to 10 $\mu$m, with a mean of about 2 $\mu$m, were sectioned

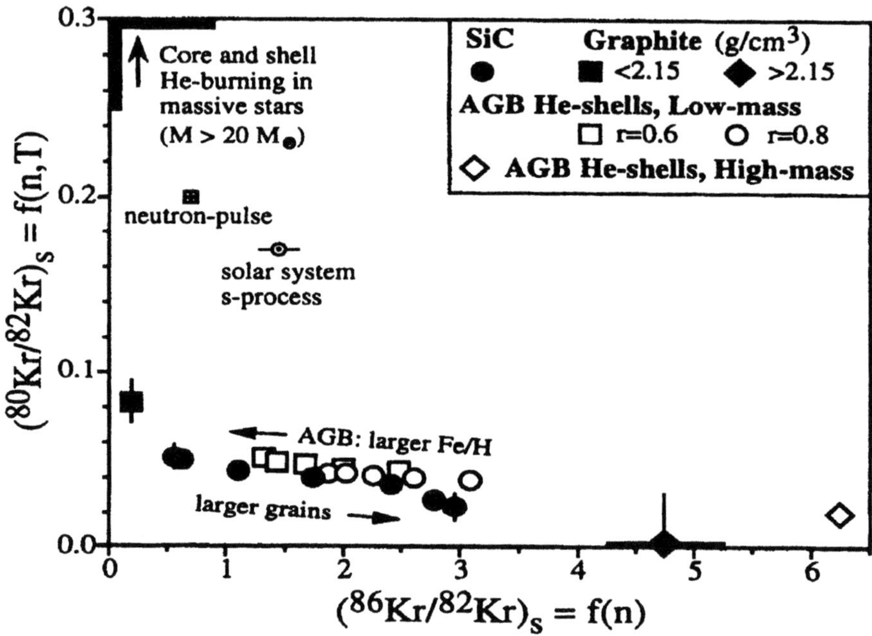


*Figure 12.* Krypton isotopic compositions of Murchison SiC and graphite (filled symbols), compared with predictions for the He-shells of low-mass AGB stars of varying metallicity, given by Fe/H. Note excellent agreement for meteoritic SiC, and striking differences between low- and high-density graphite lying on opposite extensions of the SiC trend. From Anders and Zinner [1993].

using an ultramicrotome and then examined in a transmission electron microscope equipped with an X-ray detector and an electron energy spectrometer. This study yielded important information about the crystalline structure of the graphite and the nature of refractory metal carbide inclusions that commonly occurred within the spherules. These observations set useful constraints on the stellar conditions in which the spherules formed.

More than 80% of the spherules studied consisted of mantles of well-crystallised graphite surrounding cores comprised of chaotically packed graphene sheets, i.e., two-dimensional networks of aromatic hexagonal rings. The absence of any ordered layering of these sheets suggests that they may be distorted by localised replacement of hexagons by, for example, pentagons, as in fullerenes. The broadening of electron diffraction peaks from the core region yielded a mean size for the graphene entities of about 3-4 nm, but electron energy loss spectroscopy showed that the space between those entities must be largely filled with much smaller aromatic entities, possibly polycyclic aromatic hydrocarbons (PAHs) such as dicoronene,

$C_{48}H_{20}$, which have been independently observed in samples of interstellar graphite [Messenger et al. 1994]. It seems likely that the PAHs and graphene sheets formed by homogeneous nucleation in stellar regions where condensation took place at relatively high supersaturation levels. By contrast, those graphite spherules containing refractory carbide inclusions apparently formed by heterogeneous nucleation on such previously condensed carbides.

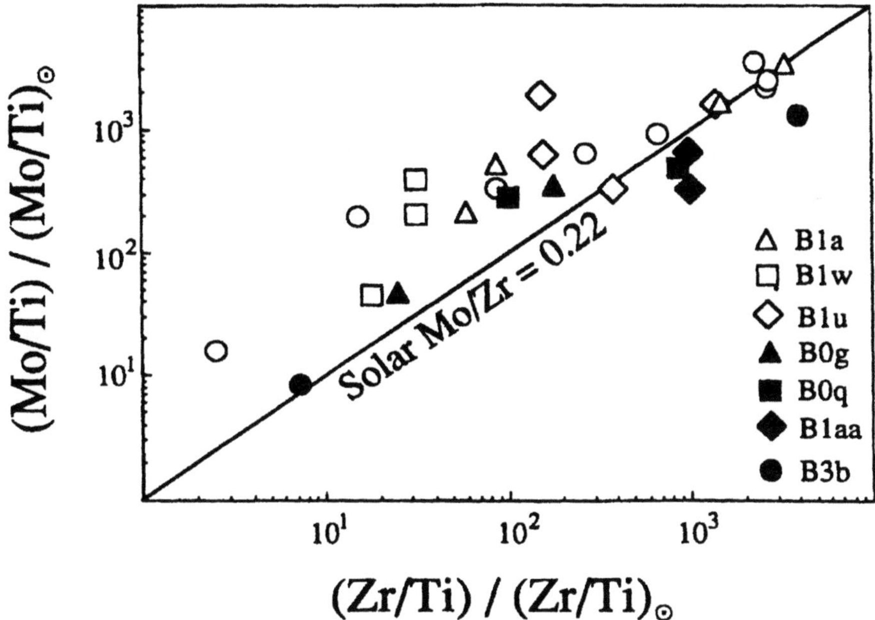


*Figure 13.* Evidence for chemical fractionation during condensation of circumstellar carbide crystals present within meteoritic graphite grains. The measured Mo/Ti and Zr/Ti ratios, normalised to solar abundances, reveal wide variations and large enrichments of both Mo and Zr. From Bernatowicz et al. [1996].

The carbide inclusions, found in at least one third of the spherules studied, range in size from 5 to 220 nm and are predominantly TiC with variable amounts of ZrC and MoC in solid solution. Only one of the spherules, out of 67 studied, contained SiC as an inclusion. The range of elemental compositions of inclusions within a single spherule precludes in situ formation and proves that they formed by condensation prior to graphite formation, i.e., at temperatures above the condensation point of graphite. Relative to Ti, both Zr and Mo are highly enriched in inclusions, when compared with solar abundances, Figure 13. Part of this enrichment can certainly be ascribed to an addition of elements produced in the s-process, but the maximum enrichment observed exceeds either s-process theory or astronomical measurement in appropriate stars. Consequently, it is necessary to appeal to elemental fractionation during the condensation process, i.e., enrichment of the earliest-formed, Zr, Mo-rich, condensate relative to the Ti-containing gas. Similarly, the presence of points to the left of the solar Zr/Mo line in Figure 13 can be attributed to carbide condensation from gas relatively depleted in Zr by prior condensation of ZrC at higher temperatures, whereas points to the right of that line probably represent such earlier-formed material. In general, condensation of refractory carbides before that of graphite points to C/O ratios no greater than about 1.2, assuming pressures in the (astrophysically reasonable) range of $10^{-7}$ to $10^{-4}$ bars.

It is worth noting that the scarcity of SiC inclusions in graphite indicates that graphite condensation is typically not kinetically inhibited in astrophysical environments, i.e., that calculations based upon equilibrium thermodynamics are apparently appropriate for such environments. Furthermore, because most SiC forms by reaction between previously condensed graphite and Si still in the gas, we would expect not only few instances of SiC embedded within graphite, but also an overall greater abundance of SiC relative to graphite, as observed in primitive meteorites [Sharp & Wasserburg 1995].

Finally, it can be shown that growth of both the graphite spherules and their refractory carbide inclusions to the sizes observed in primitive meteorites is inconsistent with plausible values of pressure and mass-loss rate for the bulk atmospheres of low-mass AGB stars. This implies that grain growth occurred predominantly within clumps, i.e., regions of enhanced density within such atmospheres. The same conclusion can be derived based on the sizes of SiC grains, of apparently AGB-star origin, observed within primitive meteorites and may also be extended to those graphites and SiC grains for which a supernova origin has been inferred.

Interstellar SiC fractions contain excess $^{21}$Ne relative to the isotopic compositions calculated for the He shells of low-mass AGB stars, suggestive of spallogenic production by galactic cosmic rays during the grains' lifetime in interstellar space [Tang & Anders 1988]. Such a cosmogenic origin is supported by the fact that interstellar graphite, which lacks an appropriate target nuclide, reveals no such $^{21}$Ne excess. The data for SiC are consistent with the meteoritic samples being a mixture of a relatively small proportion of grains containing both trapped and cosmogenic Ne diluted by a much larger number of gas-poor grains, apparently degassed at about the time that they were incorporated into the solar nebula. After correcting for such secondary processing, nominal cosmogenic ages of 900 to 1600 Myr are obtained. Whether these ages are sensibly longer than the 400 Myr lifetime calculated for refractory interstellar grains [McKee 1989] is unclear at this time.

2.5.3. *Chemical information*

The isotopes of O possess an interesting property that makes them particularly suitable for monitoring the progress of galactic chemical evolution. Oxygen-16 is what is known as a primary nuclide, i.e., it can be synthesised in a star initially made of only H and He. On the other hand, $^{17}O$ and $^{18}O$ are secondary nuclides, which can only be synthesised in stars containing previously made CNO-cycle nuclides. Because $^{16}O$ abundance increases linearly with metallicity, whereas those of $^{17}O$ and $^{18}O$ increase as the square of the metallicity, it follows that the $^{17}O/^{16}O$ and $^{18}O/^{16}O$ ratios increase approximately linearly with metallicity, which itself increases with galactic age. Consequently, the spread in initial $^{17}O/^{16}O$ and $^{18}O/^{16}O$ ratios for corundum grains of Groups 1 and 3, Figure 7, tracks the chemical evolution of the Galaxy prior to formation of the Solar System. Furthermore, with the aid of reasonable assumptions concerning the ages of the progenitor stars, it is possible to estimate from the data in Figure 7 an age for the Galaxy equal to 14.4 Gyr, in good agreement with independent studies of globular clusters and The abundances in metal-poor stars [Nittler and Cowsik 1997].

Similar information about Galactic chemical evolution can also be derived from the spread in Si-isotopic data for SiC grains of types A and B [Gallino et al. 1994; Hoppe et al. 1996a].

## 3. Interstellar Molecules

Carbon takes many different forms in primitive meteorites. In addition to minor quantities of interstellar carbonaceous grains, such meteorites contain more or less abundant organic matter, and quite commonly significant quantities of carbonate minerals. The last of these were formed during secondary mineralisation on asteroids that served as the parent objects of the meteorites. But the organic matter, although it was clearly affected by that secondary processing, apparently still bears a record of precursor organic compounds of interstellar origin.

### 3.1. HOW INTERSTELLAR ORGANIC MOLECULES ARE RECOGNISED

The role that noble gases play in signposting the way to interstellar grains is played in the case of interstellar molecules by deuterium. Fractionation of the H isotopes during reactions in interstellar clouds exceed by several orders of magnitude those characteristic of chemical reactions on Earth, and plausibly on other Solar-System objects. Consequently, it was the discovery of large D/H ratios in meteoritic organic matter that first led to the suggestion that interstellar molecules might be involved in the production of such organics [Kolodny et al. 1980], and current attempts to define the interstellar component more closely still rely upon very large D enrichments as the signature of such a component [e.g., Kerridge 1992; Krishnamurthy et al. 1992].

A key issue in the study of meteoritic organic matter is the need to discriminate between interstellar synthesis of organic molecules and production of such molecules by processes within the early Solar System, including those that took place on accreted objects, such as planetesimals, asteroids, or planets. This topic has been the subject of considerable controversy and, although a consensus has developed in the past few years that most of the organic molecules in primitive meteorites were formed when reactive interstellar molecules came into contact with liquid water in the near-surface regions of a primitive asteroid, many details of this episode are still poorly understood and a possible role for organic synthesis from simple gases in the solar nebula cannot be rigorously excluded.

### 3.2. METHODS EMPLOYED FOR ISOLATING ORGANIC FRACTIONS

As will be discussed in more detail below, meteoritic organic matter can be conveniently subdivided into an insoluble fraction, a macromolecular complex much resembling terrestrial kerogen, and a soluble fraction consisting of over 400 individual compounds, that number being almost certainly governed by analytical sensitivity rather than by natural diversity, as individual abundances exhibit a strong tendency to decrease with increasing molecular weight and/or complexity.

The kerogen-like material is generally prepared for analysis by dissolving the soluble fraction in a variety of solvents and then demineralising the meteorite by acid dissolution similarly to the manner in which interstellar grains are isolated. In fact, samples of meteoritic kerogen-like material invariably contain SiC, graphite and diamond but in amounts that are generally too small to exert much influence on analytical results. A typical sample will contain some 80 to 90% organic matter by weight, the bulk of the remainder being insoluble flourides produced during the acid-dissolution process.

The first step in preparing a sample of soluble organic matter is, not surprisingly, to extract the meteorite with one or more solvents, the most commonly employed being water, methanol, and benzene. Further separation into specific characterisable molecular fractions is invariably performed by some kind of chromatographic technique, GC/MS or HPLC in the case of molecular identification and quantitation, silica-gel or acid-column chromatography when subsequent isotopic analysis is planned.

## 3.3. ANALYTICAL TECHNIQUES USED TO STUDY METEORITIC ORGANIC MATTER

The structure of the kerogen-like fraction has been studied using a variety of techniques, including $^{13}C$ NMR and a number of chemical methods designed to break the material down into comparatively simple, identifiable moieties [Cronin et al. 1987; Hayatsu et al. 1977, 1980]. These revealed a three-dimensional complex of relatively small aromatic clusters cross linked by predominantly methylene chains and ether linkages. The actual size of the aromatic clusters is controversial, estimates ranging from 3-4 rings/moiety [Hayatsu et al. 1983] to $> 8$ [Kerridge et al. 1987].

Bulk analyses of the elemental and isotopic composition of the kerogen-like material are carried out by oxidising the material to $CO_2$, $H_2O$, and $N_2$, using either an atmosphere of pure $O_2$ or admixture with an oxide such as CuO. The $CO_2$ and $N_2$ are purified, quantitated to give C and N abundances, and then analysed in a mass spectrometer, of either static or dynamic design, to yield $^{13}C/^{12}C$ and $^{15}N/^{14}N$ ratios. The $H_2O$ is reduced, by either U or Zn metal, to give $H_2$ which is quantitated and introduced into a dynamic mass spectrometer to yield its D/H ratio. It is also possible to apply the same approach in stepwise fashion, either by combustion in $O_2$ or by pyrolysis in vacuum, to explore the possible presence within the kerogen-like material of isotopically distinct components of plausibly different origins.

Molecular characterisation of soluble fractions is performed by either GC/MS or HPLC. In addition, GC/MS can be used to search for possible chirality in meteoritic organic molecules, see below [Cronin and Pizzarello 1997]. Isotopic analysis of soluble material has generally been carried out by oxidising the fraction using CuO and processing the resulting $CO_2$, $H_2O$, and $N_2$ as described above, but a few analyses have generated $^{13}C/^{12}C$ data by means of GC-IRMS, a technique in which the output from a GC column is combusted directly to $CO_2$ [Engel et al. 1990; Gilmour and Pillinger 1994]. The more common CuO-based process has been applied both to bulk extracts, e.g., all compounds extracted by a specific solvent such as methanol [Becker and Epstein 1982], and to chromatographically separated fractions, e.g., amino acids [Pizzarello et al. 1991].

## 3.4. CHARACTERISATION OF METEORITIC ORGANIC MATTER

There is good reason to believe that at least a major part of the kerogen-like fraction had a different origin from the soluble fraction. Consequently, it is convenient to discuss the two fractions separately.

### 3.4.1. *Kerogen-like material*

Two obvious questions that may be asked about the kerogen-like fraction are: Was it produced analogously to similar material on Earth, by thermal degradation of an ensemble of simpler molecules, such as those comprising the soluble fraction? Are the aromatic moieties related to the aliphatic side chains and cross linkages, for example by thermal aromatisation of the latter? The main line of evidence bearing on these questions consists of the isotopic compositions of the relevant entities.

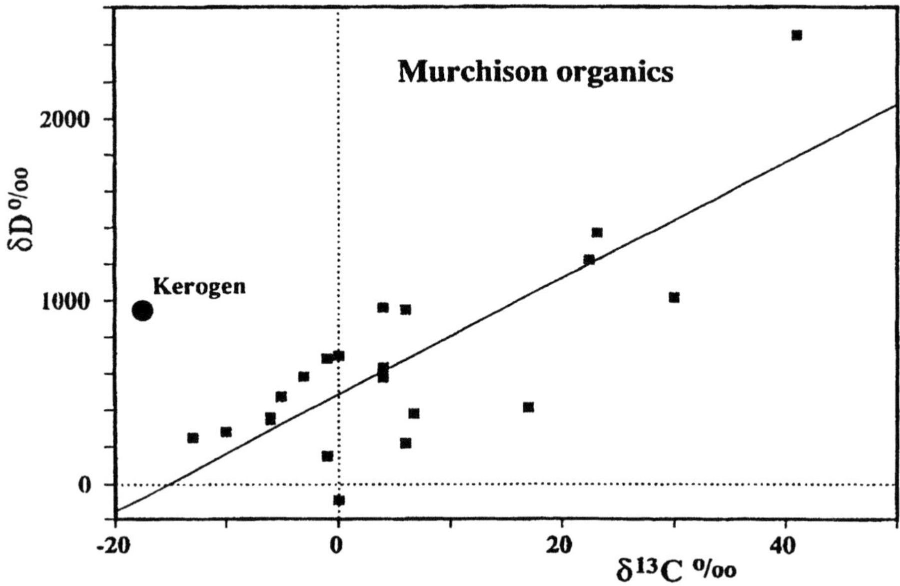


*Figure 14.* D/H ratios and C-isotopic compositions of individual organic fractions extracted from the Murchison meteorite. Square symbols represent soluble fractions; round symbol represents the insoluble kerogen-like material. Data expressed in delta notation, as parts per thousand relative to terrestrial standards. Data from Cronin et al. [1993], Epstein et al. [1987], Kerridge et al. [1987], Krishnamurthy et al. [1992], and Pizzarello et al. [1991, 1994].

A plot of the $^{13}C/^{12}C$ and D/H ratios for all the meteoritic organic fractions analysed so far, Figure 14, reveals that the kerogen-like fraction lies well off a main trend defined by the individual soluble fractions. The physical meaning of the main trend is not clear. Superficially, it resembles a mixing line between a D-, $^{13}C$-poor component and a component highly enriched in D and significantly enriched in $^{13}C$. The D-,$^{13}C$-poor component plots suspiciously close to the composition of common terrestrial organic contamination, but the great care with which most of these analyses were carried out makes such an identification unlikely. A more plausible interpretation of that component might be that it represents "normal Solar System organic matter". This issue will be addressed further below. It is tempting to ascribe the D-, $^{13}$C-rich component to interstellar organic matter, given the likely involvement of interstellar chemistry in generating the observed D enrichment, but it is not at all clear that D enrichment would necessarily be accompanied in interstellar clouds by $^{13}$C enrichment, or that, if such C isotope fractionation did occur, it would lead to a $^{13}C/^{12}C$ so close to the average Solar System value.

But whatever the explanation for the main trend in Figure 14, it is clear that the kerogen-like material cannot be simply manufactured by processing material lying on that trend. The simplest conclusion is that the soluble and insoluble fractions were derived from different isotopic reservoirs. Furthermore, the results of stepwise isotopic analysis of the kerogen-like material reveals the presence of at least two major and two minor isotopically distinct components, Figure 15. The two major components may be identified with aromatic and aliphatic moieties, respectively, and the isotopic difference between these precludes any simple genetic relationship between them. Consequently, it appears that the kerogen-like material was formed from multiple precursors under conditions which prevented isotopic equilibration between them. It is not possible to define either those conditions or the precursor compounds at this time.

Stepwise analysis of the kerogen-like fraction also reveals a range in $^{15}N/^{14}N$ ratios of at least 10% [Kerridge 1983; Kerridge et al. 1987], the details of which suggest that at least three isotopically distinct components are present, though these cannot yet be correlated with the $D/^{13}C$ components.

### 3.4.2. *Soluble molecular fractions*

Before spelling out the salient characteristics of the soluble fraction, it is important to provide a little historical perspective. The presence of organic molecules in carbonaceous meteorites has been recognised for nearly two hundred years, and throughout those years there have been periodic flurries of interest in the possibility that the meteoritic molecules were biologically produced. Each claim of biogenicity has, however, been convincingly refuted, and there is now a clear consensus that meteoritic organic matter was abiotically produced. Consequently, the rationale for the study of such material has shifted from the possibility of alien life forms to the desirability of obtaining insight into the nature of prebiotic chemical evolution, the record of which has been expunged on Earth. In other words, understanding how the population of organic molecules in meteorites was synthesised could shed light on how the Earth's "primordial soup" came into existence.

Although over a dozen possible mechanisms can be identified that were

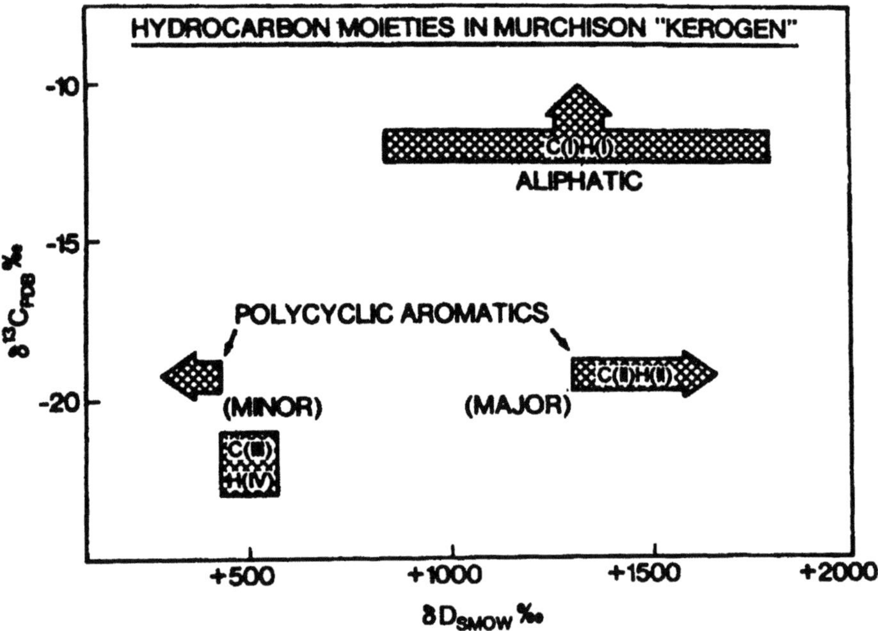


*Figure 15.* Representation of C- and H-isotopic compositions of distinct components inferred to be present in kerogen-like material from the Murchison meteorite. From Kerridge et al. [1987].

potentially capable of contributing organic matter to the early Solar System, for most of the past thirty years discussion has tended to focus on the competing merits of just two reaction schemes, with a third entering the arena about fifteen years ago. The first two are the Miller-Urey synthesis [e.g., Miller 1957] and the Fischer-Tropsch synthesis [e.g., Hayatsu and Anders 1981]; the third consists of ion-molecule reaction in interstellar clouds [e.g., Kerridge 1983]. The Miller-Urey synthesis was originally advanced as an explanation for the production of organic matter on the early Earth, and it was never clear that it was applicable in the nebular or asteroidal environment experienced by meteoritic material. Its major claim to consideration rested on the similarity between the amino-acid distribution produced by such syntheses in the laboratory and those in carbonaceous chondrites [Wolman et al. 1972]. It now seems likely, however, that the amino-acid distribution is a function, not of the Miller-Urey synthesis itself, but of the Strecker-cyanohydrin synthesis, Figure 16, acting upon the Miller-Urey products [e.g., Peltzer et al. 1984], a synthesis that could have operated equally well on precursors originating from different reaction schemes.

The case for the Fischer-Tropsch synthesis, a reaction scheme for the hydrogenation of CO visualised by Anders and coworkers [e.g., Hayatsu

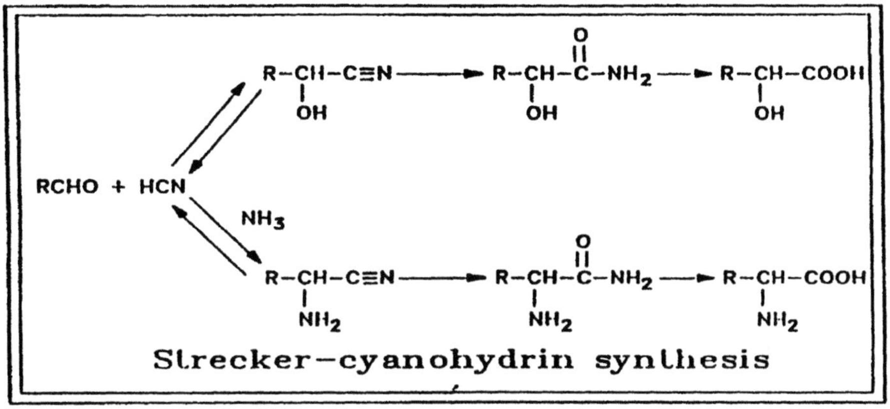

RCHO + HCN ⇌ R–CH(OH)–C≡N → R–CH(OH)–C(=O)–NH$_2$ → R–CH(OH)–COOH

RCHO + HCN ⇌ (NH$_3$) R–CH(NH$_2$)–C≡N → R–CH(NH$_2$)–C(=O)–NH$_2$ → R–CH(NH$_2$)–COOH

Strecker–cyanohydrin synthesis

*Figure 16.* Schematic representation of the pathways making up the Strecker cyanohydrin synthesis. This reaction scheme can take place in an aqueous medium containing aldehydes (and/or ketones), cyanide, and ammonia. After Peltzer et al. [1984].

and Anders 1981] as having been catalysed by condensation of magnetite and/or clay minerals in the solar nebula prior to agglomeration of planetesimals, rested heavily upon its characteristically selective production of straight-chain alkanes, such as were found in most analyses of carbonaceous chondrites. However, a pivotal study by Cronin and Pizzarello [1990] demonstrated convincingly that those straight-chain alkanes represented contamination of the meteorites by terrestrial organic matter, despite the efforts of previous groups to minimise such problems. Cronin and Pizzarello showed that the dominant aliphatics present in a pristine sample of Murchison were branched alkyl-substituted mono-, di-, and tricyclic alkanes. Similar studies have established that the great majority of organic molecules in carbonaceous chondrites are characterised by: Homologous series in which concentration decreases systematically with increasing carbon number; A predominance of branched-chain isomers over straight chains; Complete structural diversity, with all stable isomers present at any given carbon number. (It should be noted that these conclusions are strictly only valid for (a) those species abundant enough to be detected by present-day analytical techniques, and (b) the Murchison meteorite, though there is no reason to suppose that it might be anomalous in this regard.) These characteristics favor synthesis involving carbon-chain build up by means of free-radical reactions; they are inconsistent with any kind of catalysed reaction scheme, such as Fischer-Tropsch [Cronin and Chang 1993].

Several attempts have been made to determine whether those meteoritic organic molecules possessing a chiral C atom exhibit an enantiomeric excess, i.e., a "handedness", as is found in such molecules when they participate in terrestrial biochemistry. Until recently, all apparently successful attempts were shown to be due to terrestrial contamination, but Cronin and Pizzarello [1997] have demonstrated a clear L-enantiomeric excess in a Murchison amino acid, 2-amino-2, 3-dimethylpentanoic acid, which is not found naturally on Earth. The significance of this finding is discussed in the following section.

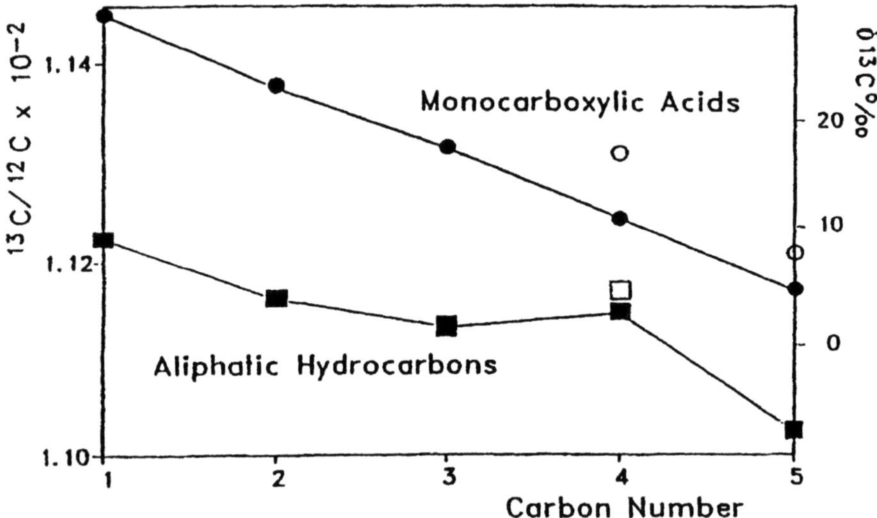


*Figure 17.* Carbon isotopic ratios for individual aliphatic hydrocarbons and carboxylic acids, extracted from the Murchison meteorite, plotted against C number. Thus, 1 = methane or $CO_2$, 2 = ethane or acetic acid, and so forth. Values for branched isomers indicated by open symbols. After Yuen et al. [1984].

Isotopic characteristics of organic molecules extracted from Murchison may be summarised as follows: All organic species that have been prepared in adequately pure form reveal D enrichments greatly in excess of any observed naturally on Earth; Similarly, all extracts containing organic N have given $^{15}N/^{14}N$ ratios well above those found on Earth, ranging up to values more than 10% greater than that of atmospheric N2; Carbon isotopic compositions of organic fractions cover a wide range, showing both depletion and enrichment of $^{13}C$ relative to the terrestrial standard, Figure 14; For light alkanes and monocarboxylic acids, $^{13}C/^{12}C$ decreases systematically with increasing C number, Figure 17. This last observation is of considerable significance, as it points strongly towards C-chain build up by isotopically competitive addition of individual C atoms [Yuen et al. 1984].

Some further comments on these observations are in order. The apparent correlation between D/H and $^{13}C/^{12}C$ for soluble species has already been noted, Figure 14. A hint of a similar correlation between D/H and $^{15}N/^{14}N$ [Kerridge 1980] is not supported by more recent studies [Pizzarello et al. 1994]. The apparent $^{13}C$ enrichment of the carboxylic acids relative to the alkanes, Figure 17, is probably due to isotopically heavy $CO_3^{2-}$; $^{13}C/^{12}C$ of the aliphatic C in the carboxylic acids is probably close to that in the alkanes, suggestive of origin from a common reservoir.

## 3.5. NATURE OF INFORMATION DERIVED FROM STUDY OF ORGANIC MATTER

It will have become apparent from the foregoing that, although the D enrichment of meteoritic organic matter points strongly towards interstellar processes in much the same way as the anomalous noble gases in the SiC, graphite, and diamond, identification, let alone isolation, of the D carrier has been much less successful than in the case of the interstellar grains. Consequently, there has been much less opportunity for feedback of meteoritic data into the broader astrophysical context.

### 3.5.1. *Chemical information*

The broad outlines of the history of organic matter in meteorites are now generally agreed upon: Synthesis of a population of organic molecules and radicals by ion-molecule reactions in dense interstellar clouds, leading to sequestration of those molecules and radicals in the icy mantles of interstellar grains, within which they underwent some molecular modification as a result of photolysis [e.g., Greenberg 1984] until such time as the grains were incorporated into the nascent Solar System, when they could experience heating and/or alteration by liquid water in a planetesimal or planetary setting [e.g., Cronin and Chang 1993]. (Note that there is abundant mineralogical and petrologic evidence that carbonaceous chondrites experienced extensive aqueous alteration on their parent asteroids [e.g., Kerridge and Bunch 1979].) However, many details in this sequence of events are still uncertain. Thus, the exact nature of the interstellar precursors that led to such meteoritic organic molecules as the amino and hydroxy acids is not clear. Candidates are: Aldehydes/ketones + HCN + ammonia (via the Strecker-cyanohydrin synthesis [Peltzer et al. 1984]); $\alpha$-amino and $\alpha$-hydroxy nitriles (subsequently hydrolysed in an asteroidal environment); Polymerised cyanide (also subsequently hydrolysed). In fact, synthesis of at least some amino and hydroxy acids directly in the interstellar medium cannot be ruled out, and may indeed be indicated by the evidence for enantiomeric excesses [Cronin and Pizzarello 1997], see below.

Also, although Figure 14 implies mixing between an interstellar organic component and "normal Solar System organic matter", the location of the putative interstellar component on the trend line cannot be defined, and it is not known whether the "normal" component consists of organic matter synthesised from purely Solar-System sources or of presolar material that has undergone extensive isotope exchange within the Solar System. Furthermore, the alkanes and carboxylic acids form part of the trend in Figure 14, raising the question of whether the relatively subtle variations in $^{13}C/^{12}C$ recorded in Figure 17 could have been established in the interstellar medium, and preserved since then, or whether they resulted from purely Solar-System processing.

In principle, when the relative contributions due to interstellar, nebular, and asteroidal processing of meteoritic organic matter are reliably identified, the detailed composition of that organic matter will afford valuable insights into the chemical processes in those environments. Unfortunately, we are presently far from such an understanding, and in fact it is problematical whether the natural scrambling of the meteoritic record will ever permit construction of a unique scenario.

### 3.5.2. *Information about physical processes*

The observation of L enantiomeric excesses in certain meteoritic amino acids has a number of interesting implications. First, although the magnitude of the observed excess in L-2-amino-2,3,-dimethylpentanoic acid is less than 10%, this could plausibly be amplified during prebiotic chemical evolution, leading to homochirality at the time of the emergence of life [e.g., Bonner 1991]. Second, although the observations to date cannot define the process responsible for creating the enantiomeric excess, a plausible explanation invokes asymmetric photolysis of interstellar molecules by circularly polarised light emanating as synchrotron radiation from neutron stars [Rubenstein et al. 1983; Greenberg 1996; Cronin and Pizzarello 1997], broadly consistent with the idea expressed above that interstellar processes played a major role in shaping the organic population in primitive meteorites. Third, existence of an L-enantiomeric excess in both stereoisomers of 2-amino-2,3,dimethylpentanoic acid, as observed, is apparently inconsistent with production of that amino acid in the Strecker synthesis. This follows because, if the handedness was carried by the putative interstellar ketone precursor, the Strecker synthesis, which is non-selective, would have produced an L excess in only one diastereomeric pair, not in both. Consequently, Cronin and Pizzarello [1997] propose that the $\alpha$-dialkyl amino acids were produced directly in the interstellar medium, unlike the $\alpha$-hydrogen amino acids, to which they ascribe a two-stage origin, i.e., interstellar synthesis of precursors followed by Strecker, or similar, reactions in an asteroidal environment. Bada [1997], however, has pointed out that such a scenario is difficult to reconcile with the fact that Strecker synthesis in the laboratory produces both $\alpha$-hydrogen and $\alpha$-dialkyl amino acids. Whatever the resolution of this issue, it is clear that measurement of an enantiomeric excess has added a powerful new constraint to theories for the origin(s) of amino acids, and other organic molecules, in primitive meteorites.

Even though we cannot specify with any certainty at this time the nature of the interstellar organic molecules that were incorporated into protometeoritic material in the early Solar System, we can be sure that at least some of them were volatile, reactive, and quite highly reduced. It follows that their survival imposes severe constraints on conditions, in particular temperature, at the time and place that they entered the Solar System. Evaporation and recondensation during entry into the solar nebula would have destroyed the record of interstellar isotopic fractionation and enantiomeric excess preserved in meteoritic organic matter.

### 3.5.3. *Implications for the origin of life*

Finally, it is worth noting that meteoritic organic matter must have fallen on the prebiotic Earth, thereby potentially contributing to the "primordial soup" from which life is believed to have arisen. The relative proportions of exogenous and possible endogenous organic matter on the early Earth are uncertain, but the possibility exists that interstellar chemistry and physics played a key role in paving the way to the emergence of life on Earth.

## References


Alexander, C.M.O'D., Swan, P.D., and Walker, R.M. (1992) Continued in situ studies of interstellar grains in primitive meteorites, Lunar Planet. Sci. XXIII, 9-10.

Amari, S., Hoppe, P., Zinner, E., and Lewis, R.S. (1992) Interstellar SiC with unusual isotopic composition: Grains from a supernova? Astrophys.J. 394, L43-L46.

Amari, S., Lewis, R.S., and Anders, E. (1994) Interstellar grains in meteorites: I. Isolation of SiC, graphite, and diamond: size distributions of SiC and graphite, Geochim. Cosmochim. Acta 58, 459-470.

Amari, S., Hoppe, P., Zinner, E., and Lewis, R.S. (1995a) Trace-element concentrations in single circumstellar silicon carbide grains from the Murchison meteorite, Meteoritics 30, 679-693.

Amari, S., Lewis, R.S., and Anders, E. (1995b) Interstellar grains in meteorites: III. Graphite and its noble gases, Geochim. Cosmochim. Acta 59, 1411-1426.

Amari, S., Zinner, E., and Lewis, R.S. (1996) $^{41}$Ca in presolar graphite of supernova origin, Astrophys. J. 470, L101-L104.

Anders, E. (1981) Noble gases in meteorites: Evidence for presolar matter and superheavy elements, Proc.Roy.Soc.London A374, 207-238.

Anders, E. (1988) Circumstellar material in meteorites: Noble gases, carbon and nitrogen, in J.F. Kerridge and M.S. Matthews, (eds.), Meteorites and the Early Solar System, University of Arizona Press, pp. 927-955.

Anders, E. and Heymann, D. (1969) Elements 112 to 119: Were they present in meteorites? Science 164, 821-823.

Anders, E. and Zinner, E. (1993) Interstellar grains in primitive meteorites: Diamond, silicon carbide, and graphite, Meteoritics 28, 490-514.

Bada, J.L. (1997) Extraterrestrial handedness? Science 275, 942-943.

Becker, R.H. and Epstein, S. (1982) Carbon, hydrogen and nitrogen isotopes in solvent-extractable organic matter from carbonaceous chondrites, Geochim. Cosmochim. Acta 46, 97-103.

Bernatowicz, T.J., Cowsik, R., Gibbons, P.C., Lodders, K., Fegley, B., Amari, S., and Lewis R.S. (1996) Constraints on stellar grain formation from presolar graphite in the Murchison meteorite, Astrophys. J. 472, 760-782.

Black, D.C. and Pepin, R.O. (1969) Trapped neon in meteorites, II. Earth Planet. Sci. Lett. 6, 395-405.

Bonner, W.A. (1991) The origins and amplification of molecular chirality, Orig. Life Evol. Biosph. 21, 59-111.

Clayton, D.D. (1989) Origin of heavy xenon in meteoritic diamonds, Astrophys.J. 340, 613-619.

Clayton, D.D. and Ward, R.A. (1978) S-process studies: Xenon and krypton isotopic abundances, Astrophys. J. 224, 1000-1006.

Clayton, D.D., Mayer, B.S., Sanderson, C.J., Russell, S.S., and Pillinger, C.T. (1995) Carbon and nitrogen in Type II supernova diamonds, Astrophys.J. 447, 894-905.

Clayton, R.N., Grossman, L., and Mayeda, T.K. (1973) A component of primitive nuclear composition in carbonaceous meteorites, Science 182, 485-488.

Cronin, J.R. and Chang, S. (1993) Organic matter in meteorites: Molecular and isotopic analyses of the Murchison meteorite, in J.M. Greenberg et al. (eds.), The Chemistry of Life's Origins, Kluwer Academic Publ., Dordrecht, pp. 209-258.

Cronin, J.R. and Pizzarello, S. (1990) Aliphatic hydrocarbons of the Murchison meteorite, Geochim. Cosmochim. Acta 54, 2859-2868.

Cronin, J.R. and Pizzarello, S. (1997) Enantiomeric excesses in meteoritic amino acids, Science 275, 951-955.

Cronin, J.R., Pizzarello, S., and Frye, J.S. (1987) $^{13}$C NMR spectroscopy of the insoluble carbon of carbonaceous chondrites, Geochim. Cosmochim. Acta 51, 299-303.

Daulton, T.L., Eisenhour, D.D., Bernatowicz, T.J., Lewis, R.S., and Buseck, P.R. (1996) Genesis of presolar diamonds: Comparative high-resolution electron microscopy study of meteoritic and terrestrial nano-diamonds, Geochim. Cosmochim. Acta 60, 4853-4872.

Engel, M.H., Macko, S.A., and Silfer, J.A. (1990) Carbon isotope composition of individual amino acids in the Murchison meteorite, Nature 348, 47-49.

Gallino, R., Raiteri, C.M., Busso, M., and Matteucci, F. (1994) The puzzle of silicon, titanium and magnesium anomalies in meteoritic silicon carbide, Astrophys. J. 430, 858-869.

Gilmour, I. and Pillinger, C.T. (1994) Isotopic compositions of individual polycyclic aromatic hydrocarbons from the Murchison meteorite, Mon. Not. Roy. Astron. Soc. 269, 235-240.

Greenberg, J.M. (1984) Structure and evolution of interstellar grains, Sci. Amer. 250, 124-135.

Greenberg, J.M. (1996) Chirality in interstellar dust and comets: Life from dead stars, in D.B. Cline (ed.), Physical Origin of Homochirality, AIP, Santa Monica, CA, pp. 185-210.

Hayatsu, R. and Anders, E. (1981) Organic compounds in meteorites and their origins, Topics Curr. Chem, 99, 1-37.

Hayatsu, R., Matsuoka, S., Scott, R.G., Studier, M., and Anders, E. (1977) Origin of organic matter in the early solar system - VII. The organic polymer in carbonaceous chondrites, Geochim. Cosmochim. Acta 41, 1325-1339.

Hayatsu, R., Winans, R.E., Scott, R.G., McBeth, R.L., Moore, L.P., and Studier, M.H. (1980) Phenolic ethers in the organic polymer of the Murchison meteorite, Science 207, 1202-1204.

Hayatsu, R., Scott, R.G., and Winans, R.E. (1983) Comparative structural study of meteoritic polymer with terrestrial geopolymers coal and kerogen, Meteoritics 18,

310.

Hoppe, P., Amari, S., Zinner, E., and Lewis, R.S. (1995) Isotopic compositions of C, N, O, Mg, and Si, trace element abundances, and morphologies of single circumstellar graphite grains in four density fractions from the Murchison meteorite, Geochim. Cosmochim. Acta 59, 4029-4056.

Hoppe, P., Kocher, T.A., Strebel, R., Eberhardt, P., Amari, S., and Lewis, R.S. (1996a) Origin of circumstellar SiC grains with low $^{12}C/^{13}C$ ratio: A multiple star scenario, Lunar Planet. Sci. XXVII, 561-562.

Hoppe, P., Strebel, R., Eberhardt, P., Amari, S., and Lewis, R.S. (1996b) Type II supernova matter in a silicon carbide grain from the Murchison meteorite, Science 272, 1314-1316.

Kerridge, J.F. (1980) Isotopic clues to organic synthesis in the early solar system, Lunar Planet. Sci. XI, 538-540.

Kerridge, J.F. (1983) Isotopic composition of carbonaceous-chondrite kerogen: Evidence for an interstellar origin of organic matter in meteorites, Earth Planet. Sci. Lett.64, 186-200.

Kerridge, J.F. (1991) A note on the prebiotic synthesis of organic acids in carbonaceous meteorites, Orig. Life Evol. Biosph. 21, 19-29.

Kerridge, J.F. and Bunch, T.E. (1979) Aqueous alteration on asteroids: Evidence from carbonaceous meteorites, in T. Gehrels, (ed.), Asteroids, University of Arizona Press, Tucson, pp. 745-764.

Kerridge, J.F., Chang, S., and Shipp, R. (1987) Isotopic characterisation of kerogen-like material in the Murchison carbonaceous chondrite, Geochim. Cosmochim. Acta 51, 2527-2540.

Kolodny, Y., Kerridge, J.F., and Kaplan, I.R. (1980) Deuterium in carbonaceous chondrites, Earth Planet. Sci. Lett. 46, 149-158.

Krishnamurthy, R.V., Epstein, S., Cronin, J.R., Pizzarello, S., and Yuen, G.U. (1992) Isotopic and molecular analyses of hydrocarbons and monocarboxylic acids of the Murchison meteorite, Geochim. Cosmochim. Acta 56, 4045-4058.

Lewis, R.S., Anders, E., Wright, I.P., Norris, S.J., and Pillinger, C.T. (1983) Isotopically anomalous nitrogen in primitive meteorites, Nature 305, 767-771.

Lewis, R.S., Tang, M., Wacker, J.F., Anders, E., and Steel, E. (1987) Interstellar diamonds in meteorites, Nature 326, 160-162.

Lewis, R.S., Amari S., and Anders, E. (1994) Interstellar grains in meteorites: II. SiC and its noble gases, Geochim. Cosmochim. Acta 58, 471-494.

Lodders, K. and Fegley, B. (1995) The origin of circumstellar silicon carbide grains found in meteorites, Meteoritics 30, 661-678.

Manuel, O.K., Hennecke, E.W., and Sabu, D.D. (1972) Xenon in carbonaceous chondrites, Nature 240, 99-101.

McKee, C. (1989) Dust destruction, in L.J. Allamandola and A.G.G.M. Tielens (eds.), Interstellar Dust; IAU Symp. 135, Reidel, Dordrecht, pp. 431-433.

Messenger, S., Alexander, C., Gao, X., Swan, P., Walker, R., Clemett, S., Maechling, C., and Zare, R. (1994) Measurement of aromatic hydrocarbons in interstellar graphite grains: I. Selection and isotopic measurement of graphite, Meteoritics 29, 502.

Miller, S.L. (1957) The mechanism of synthesis of amino acids by electric discharges, Biochim. Biophys. Acta 23, 480-489.

Nichols, R.H., Hohenberg, C.M., Hoppe, P., Amari, S., and Lewis, R.S. (1992) $^{22}$Ne-E(H) and $^{4}$He in single SiC grains and $^{22}$Ne-E(L) in single C$\alpha$ grains of known C-isotopic composition, Lunar Planet. Sci. XXIII, 989-990.

Nichols, R.H., Amari, S., Hohenberg, C.M., Hoppe, P., and Lewis, R.S. (1993) $^{20,22}$Ne-E(H) and $^{4}$He measured in single SiC grains of known C-isotopic composition, Meteoritics 28, 410-411.

Nichols, R.H., Kehm, K., Brazzle, R., Amari, S., Hohenberg, C.M., and Lewis, R.S. (1994) Ne, C, N, O, Mg and Si isotopes in single interstellar graphite grains: Multiple stellar sources for Neon-E(L), Meteoritics 29, 510-511.

Nittler, L.R. and Cowsik, R. (1997) Galactic age estimates from O-rich stardust in meteorites, Phys. Rev. Lett. 78, 175-178.

Nittler, L.R., Hoppe, P., Alexander, C.M.O'D., Amari, S., Eberhardt, P., Gao, X., Lewis, R.S., Strebel, R., Walker, R.M., and Zinner, E. (1995) Silicon nitride from supernovae, Astrophys.J. 453, L25-L28.

Nittler, L.R., Amari, S., Zinner, E., Woosley, S.E., and Lewis, R.S. (1996) Extinct $^{44}$Ti in presolar graphite and SiC: Proof of a supernova origin, Astrophys. J. 462, L31-L34.

Nittler, L.R., Alexander, C.M.O'D., Gao, X., Walker, R.M., and Zinner, E. (1997) Stellar sapphires: The properties and origins of presolar $Al_2O_3$ in meteorites, Astrophys. J., in press.

Ott, U. (1993) Interstellar grains in meteorites, Nature 364, 25-33.

Peltzer, E.T., Bada, J.L., Schlesinger, G., and Miller, S.L. (1984) The chemical conditions on the parent body of the Murchison meteorite: Some conclusions based on amino, hydroxy, and dicarboxylic acids, Adv. Space Res. 4, 69-74.

Pizzarello, S., Krishnamurthy, R., Epstein, S., and Cronin, J.R. (1991) Isotopic analysis of amino acids from the Murchison meteorite, Geochim. Cosmochim. Acta 55, 905-910.

Pizzarello, S., Feng, X., Epstein, S., and Cronin, J.R. (1994) Isotopic analyses of nitrogenous compounds from the Murchison meteorite: Ammonia, amines, amino acids, and polar hydrocarbons, Geochim. Cosmochim. Acta 58, 5579-5587.

Reynolds, J.H. and Turner, G. (1964) Rare gases in the chondrite Renazzo, J. Geophys. Res. 69, 3263-3281.

Reynolds, J.H., Frick, U., Neil, J.M., and Phinney, D.L. (1978) Rare-gas-rich separates from carbonaceous chondrites, Geochim. Cosmochim. Acta 42, 1775-1797.

Rubenstein, E., Bonner, W.A., Noyes, H.P., and Brown, G.S. (1983) Supernovae and life, Nature 306, 118

Russell, S.S., Arden, J.W., and Pillinger, C.T. (1996) A carbon and nitrogen isotope study of diamond from primitive meteorites, Meteoritics 31, 343-355.

Sharp, C.M. and Wasserburg, G.J. (1995) Molecular equilibria and condensation temperatures in carbon-rich gases, Geochim. Cosmochim. Acta 59, 1633-1652.

Srinivasan, B. and Anders, E. (1978) Noble gases in the Murchison meteorite: Possible relic of s-process nucleosynthesis, Science 201, 51-56.

Swart, P.K., Grady, M.M., Pillinger, C.T., Lewis, R.S., and Anders, E. (1983) Interstellar carbon in meteorites, Science 220, 406-410.

Tang, M. and Anders (1988) Interstellar silicon carbide: How much older than the solar system? Astrophys.J. 335, L31-L34.

Tang, M., Anders, E., Hoppe, P., and Zinner, E. (1989) Meteoritic SiC and its stellar sources: Implications for galactic chemical evolution, Nature 339, 351-354.

Tielens, A.G.G.M., Seab, C.G., Hollenbach, D.J., and McKee, C.F. (1987) Shock processing of interstellar dust: Diamonds in the sky, Astrophys.J. 319, L109-L113.

Wdowiak, T.J. (1987) Diamond formation in carbon-star atmospheres, Nature 328, 385.

Wolman, Y., Haverland, W.J., and Miller, S.L. (1972) Nonprotein amino acids from spark discharges and their comparison with the Murchison meteorite amino acids, Proc. Natl. Acad. Sci. USA 69, 809-811.

Yuen, G., Blair, N., DesMarais, D.J., and Chang. S. (1984) Carbon isotope composition of low molecular weight hydrocarbons and monocarboxylic acids from Murchison meteorite, Nature 307, 252-254.

Zinner, E. (1989) Isotopic measurements with the ion microprobe, Proc. workshop on new frontiers in stable isotope research, USGS Bulletin 1890, 145-162.

# INTERSTELLAR DUST - EVIDENCE FROM INTERPLANETAR' DUST PARTICLES

JOHN P. BRADLEY
*MVA Inc, 5500/200 Oakbrook Parkway, Norcross GA 30093*
*and*
*School of Materials Science and Engineering, Georgia Institute of Technology, Atlanta GA 30332, USA*

**Abstract.** Interplanetary dust particles (IDPs) collected from the stratosphere are derived from asteroids and comets. The chondritic porous (CP) subset of IDPs, some of which are almost certainly cometary, are physically, chemically, and isotopically primitive objects. Large D/H and $^{15}$N isotopic anomalies provide evidence for survival of presolar interstellar components. Embedded silicate grains exhibit both crystallographic and trace element evidence of formation by (nebular) gas phase condensation. The fine-grained matrices of CP IDPs contain GEMS, submicrometer spheroids whose exotic properties resemble those of interstellar "amorphous silicate" grains which are ubiquitous throughout interstellar space. Some GEMS are embedded in isotopically non-solar carbonaceous matrix (with $^{15}\mathrm{N} \sim 25 - 40\%$) (Messenger et al., 1996), providing a further hint of their presolar origins. It is possible, but as yet unproven, that some CP IDPs are composed exclusively of aggregates of interstellar and circumstellar dust grains, and that they carry a detailed record of fundamental grain formation and destruction processes in space.

## 1. Introduction

Interplanetary dust particles (IDPs) have been collected from the stratosphere for more than twenty years (Brownlee, 1985; Sandford, 1989). The particles are collected at an altitude of 20-25 km by direct impaction onto 33 cm$^2$ (and sometimes larger) flat-plate collectors (called "flags") mounted on wing pylons of NASA ER2 aircraft (Warren and Zolensky, 1994). The leading surface of each "flag" is coated with low-viscosity silicone oil which traps particles impacting from the airstream. Since the IDP collection rate



*J. M. Greenberg and A. Li (eds.), Formulation and Evolution of Solids in Space,* 485–503.

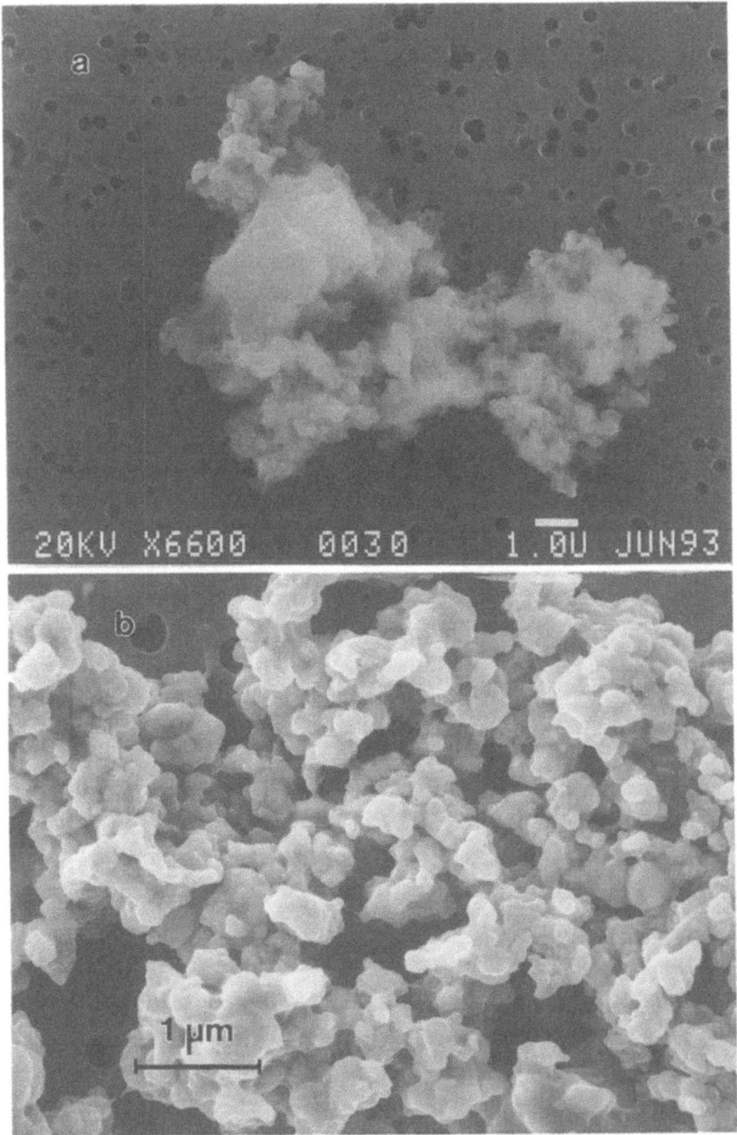


*Figure 1.* Scanning electron microscope (SEM) images of chondritic porous CP IDPs. (a) IDP U2012C-2I. The atmospheric entry speed this IDP (16.5 km $sec^{-1}$), derived from its Helium release temperature, suggests that this IDP is probably of cometary origin (see Brownlee et al., 1995). (b) High-resolution (field emission) SEM image of fine-grained matrix of CP IDP RB12A8 which consists mostly of aggregated spheroidal GEMS.

is extremely low, approximately one 10 $\mu$m diameter IDP per hour of exposure, most collectors are flown on multiple missions before they are returned to a Class 100 clean-room facility at NASA Johnson Space Center for removal and curation of the collected IDPs.

Most stratospheric IDPs particles are between 5 and 25 $\mu$m in diameter with major and trace element compositions within a factor of three chondritic (Schramm et al. 1989; Arndt et al., 1996). The extraterrestrial origins of the collected IDPs have now been firmly established by measurement of high concentrations of solar wind noble gases (Rajan et al., 1977), solar flare tracks and solar wind irradiated rims (Bradley et al., 1984a), and non-solar isotopic abundances of H and N (McKeegan et al. 1985; Staderman et al., 1989; Messenger et al., 1996). Transmission infrared spectroscopy and analytical electron microscopy studies have shown that there several (mineralogical) classes of chondritic IDPs (Sandford and Walker, 1985; Bradley et al., 1989; Germani et al., 1990). An expanding data base on the optical properties (visible and infrared), mineralogy, petrography, element abundances, isotopic compositions, and organic constituents in each of the classes of chondritic IDPs is being compiled.

Why is such an intensive effort being expended on rare, nanogram-mass IDPs when kilogram-mass conventional meteorites with similar "chondritic" compositions are widely available? There are at least four reasons for the focus on IDPs. First, even the most chemically primitive classes of meteorites, the carbonaceous chondrites, have undergone significant post-accretional processing (e.g. compaction, aqueous alteration, thermal metamorphism) on their (asteroidal) parent-body regoliths. Although refractory pre-solar grains in meteorites (e.g. graphite, diamond, and silicon carbide) have clearly survived parent body processing (Ott, 1993; Anders and Zinner, 1993), other less robust presolar grains may have been destroyed. Some chondritic IDPs, on the other hand, are extremely fragile fluffy objects which appear to have suffered little if any parent body processing and, at least in principle, **IDPs are likely to preserve a much broader range of presolar interstellar and circumstellar materials than meteorites**. Second, most meteorites are believed to be derived from <2% of the asteroids (Gaffey et al. 1993; Farinella, 1993). IDPs are derived from a much broader range of parent bodies which may include smaller bodies like outer asteroids and comets that are not represented among meteorites (Jewitt and Luu, 1990; Brownlee, 1994). Comets and outer asteroids are less likely to have undergone parent body processing and are assumed to be more primitive than the parent bodies of meteorites (Greenberg and Hage, 1990; Mumma et al., 1993).

A third reason for the focus of IDP research concerns past and future spacecraft missions. The 1986 Comet Halley fly-by missions provided the

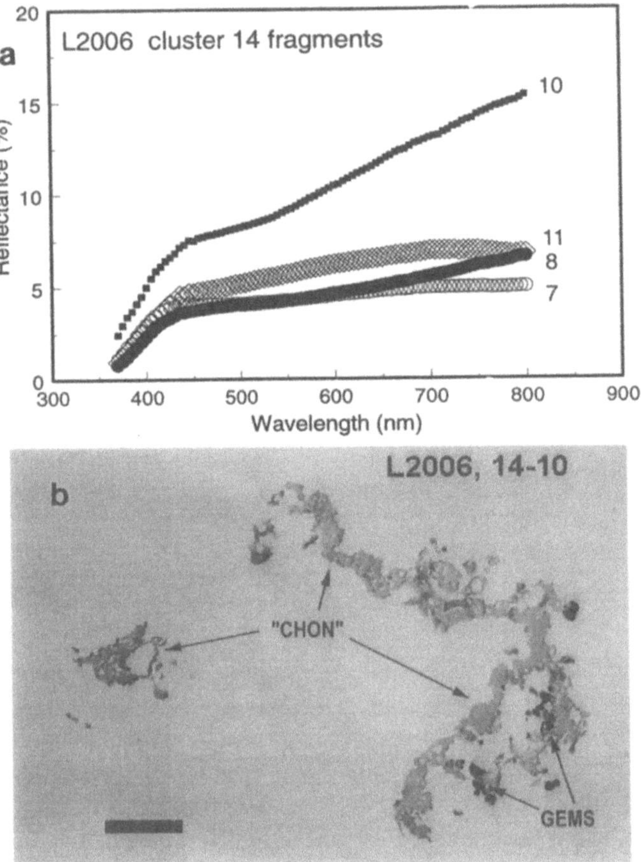


*Figure 2.* (a) Reflectance spectra of multiple fragments (labeled 7, 8, 10, & 11) from a carbon-rich (cluster) CP IDP L2006 #14. The most carbon-rich fragment (#10) below, which contains > 90% C by volume, exhibits spectral redness comparable to some comets and outer asteroids (from Bradley et al., 1996). (b) Brightfield TEM image of an ultramicrotomed thin-section (< 100 nm thick) of fragment 10 which is analogous to CHON particles in comet Halley (see Kissel and Kreuger, 1987; Fomenkova et al., 1994). The fragment consists mostly of GEMS embedded in carbonaceous material.

first analyses of comet dust, and these analyses provided the opportunity to compare the properties of the various classes of IDPs with those of *bona fide* comet dust. Two scheduled spacecraft missions STARDUST and ROSETTA will analyze and/or collect both cometary and interstellar dust grains. Finally, and perhaps most important, laboratory micro-analytical instrumentation has evolved to the point where a wide range of analyses can be performed on individual IDPs. Many of these analyses were not possible as recently as five years ago. The data base of analytical information obtained using these new instrumental techniques may be critical for fully exploiting data and samples collected during future missions.

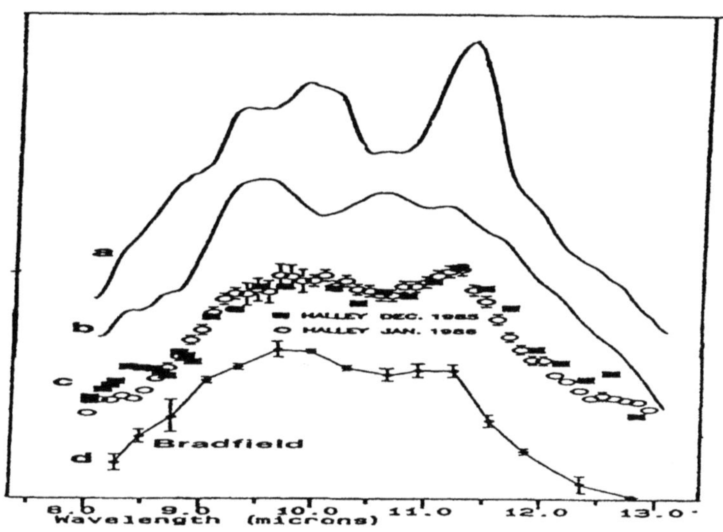


*Figure 3.* Comparison of the silicate emission features of two (glass-rich) CP IDPs and comets Halley and Bradfield. (a) CP IDP W7027A11, (b) U219C11, (c) Halley, (d) Bradfield (comet spectra courtesy of M. Hanner).

## 2. Sources of IDPs - asteroids, comets, and the interstellar dust connection

Comets and asteroids are the major sources of dust at 1 AU (Brownlee, 1985; 1994). This dust ranges from submicrometer to sub-millimeter in size with the peak of the mass flux at $\sim 200\,\mu$m diameter ($10^{-5}$ gm mass). Although the relative proportions of comet and asteroid dust probably vary, it is likely that asteroids are the dominant source (Flynn, 1990; Dermott et al., 1994; Liou et al., 1995; Liou and Zook, 1996). To date, several clay-rich "chondritic smooth" (CP) IDPs have been mineralogically linked directly to types CI and CM chondritic meteorites and hence a main-belt asteroidal origin, and it is possible that all clay-rich IDPs, about half of the collected IDPs, are from asteroids (Bradley and Brownlee, 1992; Keller et al., 1992; Rietmeijer, 1996). However, stratospheric-collected IDPs are a highly biased sampling of the dust at 1 AU because they sample predominantly the 5-25 $\mu$m diameter size range. Apart from the very real possibility that different parent-body sources generate different size ranges of particles, the collected IDPs include a delicate "chondritic porous" (CP) class of

IDPs which are simply too fragile to survive atmospheric entry as objects larger than > 50 $\mu$m diameter) (see Fig. 1). This selection effect favoring fragile materials almost certainly enhances collection of cometary IDPs because observations of the atmospheric fragmentation of cometary meteors have shown that they are predominately extremely fragile objects (Verniani, 1969).

Evidence from optical spectroscopy of IDPs suggests that some of them are from comets or "comet-like" outer asteroids. Reflectance spectra collected over the visible wavelength range 380-850 nm from chondritic IDPs show that whereas most (clay-rich) CS IDPs yield spectra similar to CI and CM meteorites and main-belt type C asteroids, most fragile CP IDPs yield spectra similar to outer P and D asteroids (Bradley et al., 1996). (There are no known spectral analogues for P and D asteroids among meteorites). The most carbon-rich CP IDPs are spectrally "red" with a degree of redness comparable with the comet-like outer asteroid Pholus (Fig. 2) (Jewitt and Luu, 1990; Binzel, 1992; Bradley et al., 1996). Transmission infrared spectra collected from some of the most fragile IDPs exhibit 10 $\mu$m "silicate features" similar to those of some comets and certain interstellar "silicate" emission features (Bradley et al., 1992) (Fig. 3).

Dust grain analyses from the 1986 Halley missions provided direct evidence that CP IDPs are consistent with comet grains. Several spacecraft encountered comet Halley in 1986 and two of them, Giotto and Vega, carried time-of-flight mass spectrometers called PIA and PUMA respectively (Kissel and Krueger, 1987; Jessberger et al., 1987). These instruments measured the compositions of several thousand submicrometer-size comet grains. Although uncertainties about (impact-ionization) ion yields, possible non-linear instrument performance, and telemetry constraints complicated interpretation of most spectra, $\sim$ 100 high quality "mode-zero" spectra were obtained (Lawler et al., 1989). The bulk composition of Halley s dust, like CP IDPs, is approximately chondritic (Jessberger et al., 1988). Carbon-rich "CHON" particles dominated by carbon and/or organic matter comprised $\sim$ 22% of the particles (Fomenkova et al., 1994). Some carbon-rich CP IDPs with non-solar isotope abundances resemble "CHON" particles (Fig. 2b) (Bradley et al., 1996; Keller et al, 1997). Brownlee et al. (1987) proposed a simple but definitive scheme to compare Halley s dust with chondritic IDPs. They compared the ratios of Fe/(Fe+Mg) of measured in Halley s dust with IDPs and meteorites. Fe and Mg were chosen because they are major elements in Halley s dust and IDPs and their submicrometer-scale distribution in grains is strongly influenced by mineral chemistry. Furthermore, the relative abundances of Fe and Mg are unlikely to be strongly influenced by uncertainties in relative ion yields. They found that the distribution of Fe/Fe+Mg) in CP IDPs matches Halley's dust.

Bradley et al., (1992), using a similar scheme, found that the distribution of Mg/Si versus Fe/Si in Halley s dust is found to be similar to that of CP IDPs but dissimilar to other classes of IDPs (and meteorites) (Fig. 4).

A final way to distinguish between some cometary and some asteroidal IDPs is to determine the speeds at which they enter the atmosphere. Most cometary IDPs arrive at 1 AU in orbits with high eccentricities, whereas most asteroidal IDPs arrive in almost circular orbits. Consequently, most cometary IDPs enter the atmosphere at speeds >16 km/sec, whereas most asteroidal IDPs enter at speeds < 14 km/sec (Brownlee et al., 1995). The atmospheric entry speed of IDPs can be estimated by noble gas measurements. While in solar orbit, IDPs accumulate near saturation doses of solar wind He. The temperature at which this He is released from IDPs is an indication of their recent thermal histories, specifically frictional heating during atmospheric entry. Using a stepwise helium release heating method pioneered by A. O. Nier (see Nier and Schlutter, 1990; Nier, 1994), it has been estimated that as many as 15-25% of the IDPs collected in the stratosphere are of cometary origin (Brownlee et al., 1995). Almost all of these "high speed" cometary particles are CP IDPs. Some "low speed" CP IDPs have been identified and assigned as asteroidal particles. Liou and Zook (1996) have suggested these particles may actually be cometary particles which, having been trapped in orbital resonances for $\sim 10^5$ years, approach the Earth with velocities that mimic those of asteroidal grains.

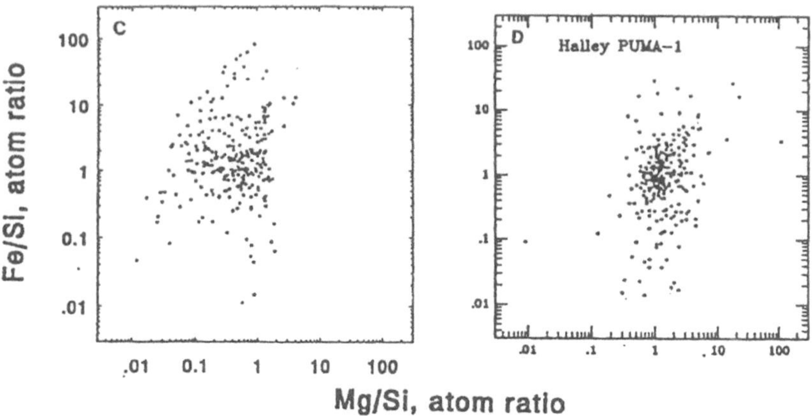


*Figure 4.* Submicrometer-scale distribution of Fe-Mg-Si in a CP IDP (W7027 A11) and comet Halley s dust (data from Bradley et al., 1992).

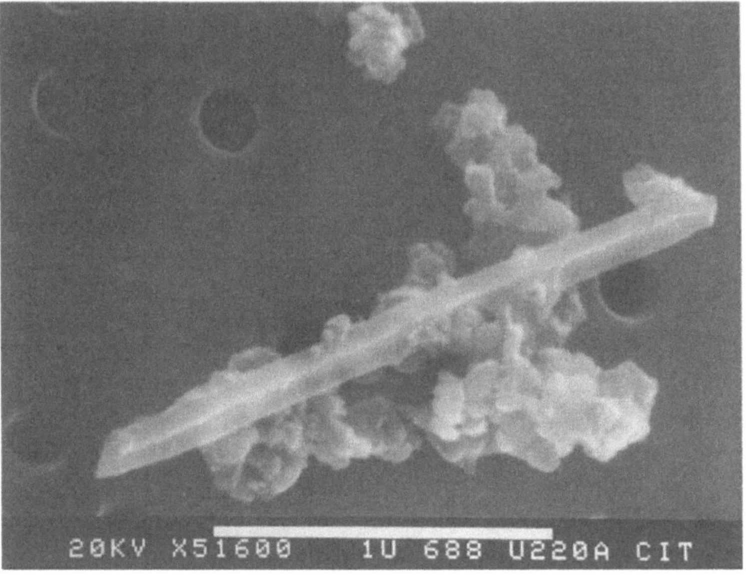


*Figure 5.* SEM image of an enstatite ($MgSiO_3$) "whisker" from CP IDP U220A39. (Adhering material is GEMS-rich chondritic matrix). Transmission electron microscope (TEM) studies of these whiskers indicate that they condensed directly from a (nebular) gas (see Bradley et al., 1983).

The above discussion provides ample evidence that at least some fragile CP IDPs collected in the stratosphere are of cometary origin and, as such, they are presently the only available samples of comets. Comets are unique small bodies formed at the fringes of the solar system. Since they have been isolated at extreme distances from the Sun, they have probably experienced little if any alteration since accretion. In addition to volatiles, the nuclei of comets are likely to contain well-preserved interstellar grains as well as material formed in the solar nebula (Greenberg and Hage, 1990; Mumma et al., 1993). The remainder of this paper focuses specifically on CP IDPs with emphasis on examination of grains that are suspected to have originally formed or been processed in presolar interstellar and circumstellar environments.

## 3. Evidence of pre-solar grain-formation (and destruction) processes in CP IDPs

CP IDPs are more physically, chemically, and isotopically primitive than even the most primitive conventional meteorites. Apart from their extreme porosities and fragile microstructures, most of them contain only

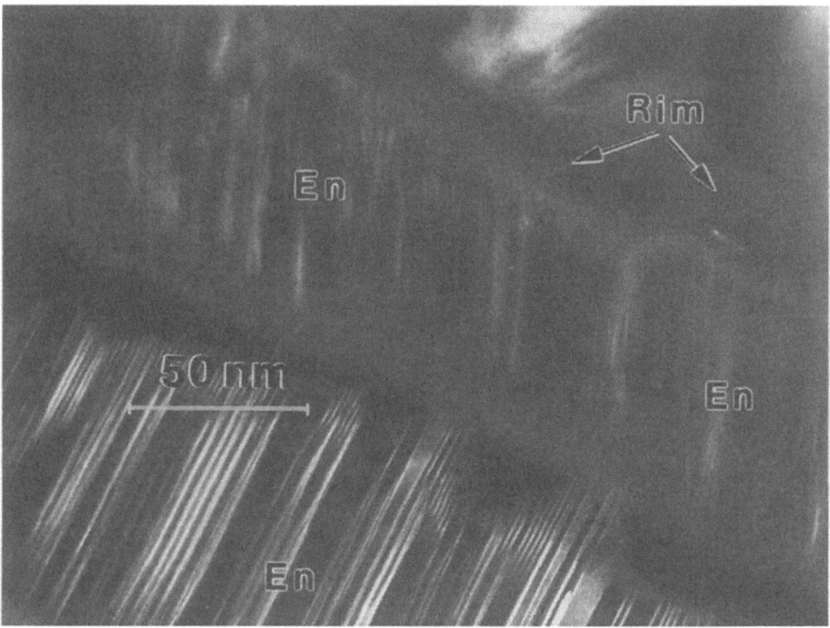


*Figure 6.* Darkfield TEM image of blade-like enstatite "whiskers" in CP IDP U222B42. Some whiskers within CP IDPs have Mg-depleted, glassy rims suggesting that they were pre-accretionally exposed to ionizing radiation.

anhydrous minerals, indicating they have not experienced compaction and aqueous alteration (Bradley and Brownlee, 1986). They contain abundant non-stoichiometric (glass) and disequilibrium mineral assemblages at the nanometer scale (Bradley, 1994). CP IDPs are systematically enriched in volatile elements Na, S, and Zn relative to CI chondritic meteorites (Flynn et al., 1993; Sutton, 1994). (CI chondrites have long been considered as the norm of solar abundances of the elements (Anders and Ebihara, 1982)). CP IDPs also exhibit chemical, structural (crystallographic), and isotopic evidence of primordial grain-forming reactions, much of which has never been observed in meteorites. Since some CP IDPs are almost certainly from comets, these reactions probably reflect presolar interstellar and/or circumstellar processes. Perhaps the most striking feature of all of this evidence is that gas phase reactions are a recurring theme, and gas-to-solid condensation is the fundamental mechanism by which grains are formed throughout space.

## 3.1. CRYSTALLOGRAPHIC EVIDENCE

The earliest SEM and TEM images of CP IDPs revealed a rare but ubiquitous form of the mineral enstatite ($MgSiO_3$) which occurs as highly anisotropic whiskers (Fraundorf, 1981). Figure 5 shows an example of a typical whisker. Enstatite is a common rock-forming mineral but it has never been found with this grossly elongated form either in meteorites or terrestrial rocks. Some of the whiskers contain axial screw dislocations indicating that they grew by spiral growth (Bradley et al., 1993). Whiskers of a variety of minerals have been found in nature. Examples include copper oxide whiskers from hydrothermal deposits (Veblen and Post, 1983), iron oxide (magnetite) from fumeroles (Symonds, 1993), and magnetite in the ALH84001 Martian meteorite (Bradley et al., 1996). In all cases, the whiskers and other coexisting mineral forms condensed directly from vapors and/or supercritical fluids. Whiskers and platelets of more than 30 elements and fifty compounds have been synthesized from vapors in the laboratory (Nadgornyi, 1962).

The significance of enstatite whiskers in CP IDPs is that they provide the first *crystallographic* evidence for the survival of crystals which may have condensed directly from a nebula gas. Some of the condensed enstatite grains have radiation sputtered rims (Fig. 6), indicating that they were exposed to ionizing radiation *prior to* accretion of their cometary parent bodies (Bradley et al., 1997). This observation suggests that they are presolar interstellar or circumstellar silicate grains. However, absolute proof of their presolar origins requires measurement of their isotopic compositions in the ion microprobe. Measurement of a non-solar $^{25}Mg/^{26}Mg$ ratio would unambiguously point towards a presolar origin. The problem with the whiskers is that they are extremely rare, typically on the order of a micrometer in size, and extremely thin with low masses. Recently, the first attempt was made to measure the Mg isotopic composition an unusually large but thin enstatite whisker ($\sim 1 \times 5\,\mu m$ long). As expected, the ion beam rapidly sputtered through the ultrathin ribbon yielding a low secondary ion signal. Nonetheless, the measurement yielded a tantalizing hint of a $^{25}Mg/^{26}Mg$ anomaly (at the $3\sigma$ level).

The potential significance of the enstatite whiskers and other silicates in CP IDPs is very high. Although silicates are among the most abundant grains in interstellar space, (isotopically anomalous) presolar silicates have never been found in meteorites. This is one of the more perplexing questions confronting researchers studying presolar grains in meteoritic materials. Given that the whiskers are vapor phase condensates, and that some of them have irradiation histories that predates comet accretion (Fig. 6), it is likely that they are examples of presolar silicate grains. A major effort is

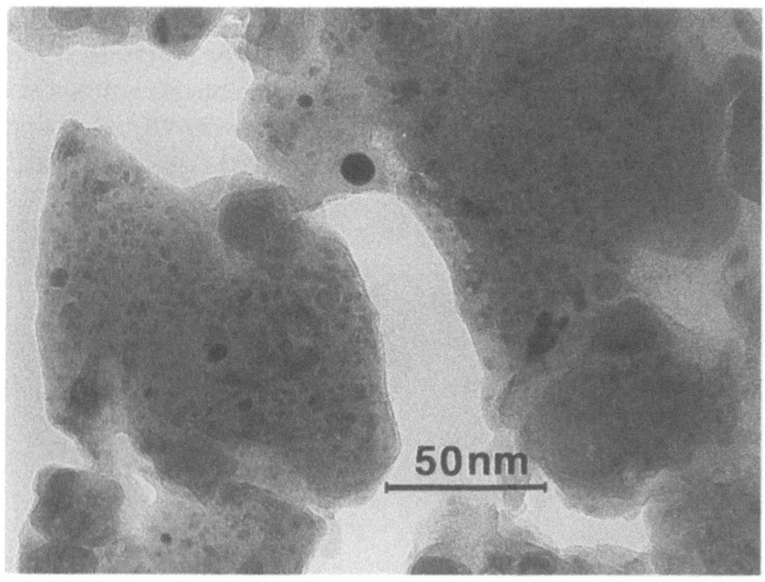


*Figure 7.* Brightfield TEM image of GEMS in CP IDP U222B42.

presently under way to locate enstatites whiskers which are large enough to analyze in the ion microprobe. Measurement of the isotopic compositions of several individual enstatites could determine whether they are solar nebular condensates or presolar grains.

### 3.2. COMPOSITIONAL EVIDENCE

Amorphous carbonaceous material, metal (FeNi), and FeNi carbide crystals are abundant constituents of CP IDPs (Bradley, 1994). The crystal structures of the carbides (in petrographic association with carbonaceous material) have been examined by Bradley et al. (1984b) and Christoffersen and Buseck (1983). A variety of FeNi carbides are found in IDPs. Some of them are almost certainly indigenous while other (high-temperature) carbides probably form as a result of frictional heating during atmospheric entry (Brownlee et al. 1997). Carbides are typically refractory materials but one of the carbides found in IDPs, $\epsilon$-carbide ($[FeNi]_3C$) is a low temperature ($\sim 200\,^{\circ}C$) compound which forms during Fischer-Tropsch type (FTT) catalytic reactions. In the FTT synthesis, catalytically-active metal grains (e.g. FeNi) are exposed to carbon-containing gases (e.g. CO, $CH_4$). Organic and inorganic products are formed and one of the inorganic products ] is elemental carbon which typically precipitates on the surfaces and dissolves into the metal grains and converts them into carbides. The occurrence of $\epsilon$-carbide suggests that at least some of the carbonaceous grains in IDPs may have formed via low temperature catalytic reactions. Hyatsu and Anders (1978) have presented evidence that some of the organic matter in chondritic meteorites (the most chemically similar conventional meteorites to IDPs) were formed via FTT type reactions in the solar nebula.

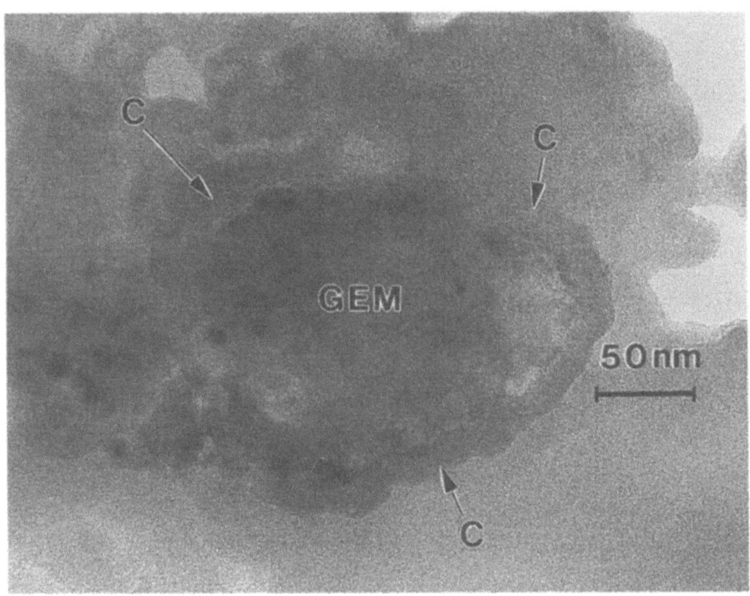


*Figure 8.* Brightfield image of a GEM with a carbonaceous rim (labeled C).

Minor element compositions of olivine and pyroxene grains in CP IDPs are consistent with formation by gas-to-solid condensation. Mg-rich olivine ($Mg_2SiO_4$) and euhedral pyroxene ($MgSiO_3$), i.e. non-whisker forms of enstatite, are common constituents of the fine-grained matrices of CP IDPs. These minerals are also common in meteoritic pyroxenes and olivines, where many of them were likely formed during melting and differentiation on their asteroidal parent body regoliths, but they typically contain $> 1$ wt % Fe and $< 0.5$ wt % Mn. However, in CP IDPs these minerals contain $< 1$ wt % Fe, but Mn (and Cr) can be present in concentrations up to 5 wt %. Klock et al. (1989) describe these unusual minerals as LIME (low iron, manganese enriched) olivines and pyroxenes. They propose that the high Mn contents can best be explained in terms of condensation of the olivine and pyroxene grains from a gas of solar composition.

There is complete solid solution between Mg and Fe in olivines and pyroxenes formed from a molten liquid, but if these cations condense from a nebular gas (of solar composition), Mg partitions strongly into silicates like forsterite and enstatite. Fe condenses as metal but Mn (and Cr), which are not stable as a metals, condense as $Mn_2SiO_4$ (and $Cr_2SiO_4$) in solid solution with forsterite ($Mg_2SiO_4$) and enstatite ($MgSiO_3$). Some enstatite whiskers contain Mn (and Cr) $>$ Fe, i.e. they exhibit morphological evidence (whisker forms), crystallographic evidence (screw dislocations), and trace element evidence [Mn (&Cr) $\gg$ Fe] of formation by vapor phase growth.

### 3.3. ISOTOPIC EVIDENCE

Isotopes are the foundation of the study of presolar grains in meteorites. Most isotope measurements are performed using ion microprobes which are capable of spatial resolution on the order of a micrometer. Measuring the isotopic compositions of individual IDPs in the ion microprobe is more challenging and, since IDPs can be heterogeneous on a scale of nanometers, measuring the isotopic compositions of individual grains *within* IDPs is a formidable challenge. Each nanometer-size grain in an IDP might have an unusual isotopic composition, but the measured signal from many grains could have an average composition not significantly different from the solar value. Thus, existing ion microprobes are only capable of detecting relatively large isotope anomalies in IDPs.

H, C, N, O, Mg, and Si isotope abundances in IDPs have now been measured (McKeegan et al. 1985; Walker, 1994). No large isotopic anomalies outside of terrestrial values have been so far found in C, O, Mg, or Si. However, in addition to the "hint" of a Mg anomaly measured in the enstatite ribbon, the Mg isotopic compositions of 13 IDPs were measured using high precision thermal ionization mass spectrometry (Esat et al. 1979). Evidence of $^{26}Mg$ anomalies in three out of four CP IDPs was observed at the level of three to four parts per thousand, which is close to detection limits. Since the observed effects were small, there has been a tendency to downplay their potential significance (Walker, 1994). However, if real, the origins of these small but measurable anomalies is important. Esat and Taylor (1987) have speculated that the small but measurable Mg anomalies in IDPs could result from intense radiation processing in the interstellar medium.

In contrast to C, O, Mg and Si, major isotopic anomalies in H and N have been measured in multiple chondritic IDPs (Messenger et al., 1996; Staderman et al. 1989; Keller et al., 1997). These particular anomalies bear directly on the question of formation of solids in space in that they may be related to the formation of organic materials formed in ion-molecule reactions in cold interstellar clouds. D/H excesses are observed in meteorites (up to five times the terrestrial value) but excesses as high as fifty times the terrestrial value have been observed in CP IDPs (Messenger et al., 1996).

Furthermore, the D/H in these particles is extremely variable on the scale of $1\mu m$. These D excesses must be the result of chemical mass fractionation, since D is destroyed in stellar nucleosynthesis, and spallation reactions are far too inefficient. The degree of mass fractionation increases with decreasing temperature, as the difference in binding energy between C-H and C-D bonds becomes comparable with the thermal energy. Indeed, extreme D/H fractionation is observed in cold (10 - 100 K) molecular clouds, with some molecules containing 100 - 10000 times the value in H2 gas. At such low temperatures, neutral-neutral reactions are prohibitively slow, but if one of the reactants is ionized there is no activation energy barrier and chemistry proceeds rapidly. This so called ion-molecule chemistry is thought to be the dominant form of chemistry in cold molecular clouds. The D-rich material in meteorites and IDPs likely represents the partial preservation of molecules with higher D excesses formed in presolar molecular clouds. The fact that D excesses in some cluster IDPs reach values observed in interstellar clouds, and the extreme spatial variability in D/H suggests that some IDPs may contain relatively pristine presolar molecular cloud material.

## 4. GEMS

Silicate grains are abundant in circumstellar environments, the interstellar medium, and comets (Tielens, 1990; Mathis, 1993; Hanner, et al., 1994; Dorschner and Henning, 1995; Sandford, 1996). A variety of (isotopically anomalous) interstellar grains have been found in meteorites, but interstellar silicate grains have never been found. If less cosmically abundant grains like silicon carbide and diamond can be found in meteorites, then interstellar silicates should be even easier to find in meteorites and IDPs. This is among the more perplexing questions confronting meteoriticists. The acid dissolution procedures used to concentrate and isolate acid-insoluble presolar grains in meteorites (e.g. graphite, diamond, and silicon carbide) would dissolve silicate grains. However, systematic searches for isotopically anomalous silicates in unprocessed meteorite samples have failed to yield evidence of presolar silicate grains. Alternative explanations are that (1) interstellar silicate grains may have not have survived parent body processing on the parent bodies of meteorites, or (2) their isotopic compositions are simply not significantly different from solar system abundances.

What has been learned about the properties of contemporary interstellar silicate grains from astronomical observations? It is known that interstellar silicates are more glass-like than crystalline, they span a relatively narrow size range (0.1-0.5 $\mu m$ diameter), and their compositions are approximately (within a factor of three) chondritic for most of the major rock-forming elements (Bussoletti and Colangeli, 1990; Mathis, 1993; York, 1994; Martin,

1994). Perhaps their most remarkable property is alignment; while most silicates are dielectrics, interstellar silicate grains align themselves in the galactic magnetic field and cause polarization of starlight. The most likely explanation for grain alignment is that interstellar "amorphous silicates" contain tiny superparamagnetic metal inclusions, a sort of "raisin pudding" of magnetic inclusions in a non-magnetic matrix, as first proposed by Jones and Spitzer (1967). From a materials science viewpoint, the properties of interstellar silicates determined from astronomical measurements appear "exotic" and mutually exclusive of one another.

Assuming CP IDPs are from comets, or "comet-like" outer asteroids (Jewitt and Luu, 1990), then interstellar "amorphous silicate" grains should be an expected component. GEMS (glass with embedded metal and sulfides) are a fundamental component of CP IDPs (Bradley, 1994). They are submicrometer spheroidal grains which contain nanometer-sized FeNi metal and Fe-rich sulfide crystals in silicate glass matrix (Figs. 1b & 7). Their properties appear to have been shaped primarily by exposure to ionizing radiation (Bradley and Ireland, 1996). Since the irradiation occurred *prior to* accretion of the IDPs in which they are encapsulated, the exposure predates formation of the solar system.

Several researchers have commented on the striking similarity between the properties of GEMS and interstellar "amorphous silicates" (Flynn, 1994, Martin, 1995; Goodman and Whittet, 1995; Sandford, 1996). In addition to abundant silicate glass, a prerequisite of interstellar "amorphous silicates", GEMS contain superparamagnetic FeNi inclusions which could explain grain alignment. GEMS have very exotic properties which correspond to those of interstellar silicates. If they are indeed the long sought interstellar silicates then one of the major building blocks of the solar system has been found.

Absolute proof that GEMS are interstellar grains requires measurement of non-solar isotopic compositions. Although GEMS are tiny (submicrometer) objects, the $^{25}Mg/^{26}Mg$ isotopic compositions of four individual GEMS mounted on a gold foil were recently measured in the ion microprobe. The Mg values were not significantly different from solar system abundances (Bradley and Ireland, 1996). This finding does not rule out a presolar interstellar origin for GEMS, since considerable processing of grains occurs in the interstellar medium because grains that were originally isotopically anomalous may become isotopically and chemically homogenized (Nuth, 1988). Recently, a CP IDP with GEMS embedded in carbonaceous material was examined. The carbonaceous material contains significant amounts of nitrogen and ion microprobe measurements revealed non-solar nitrogen abundances (Messenger et al., 1996). Although the isotopic composition of the GEMS in this IDP are unknown, their intimate petrographic association with isotopically anomalous carbonaceous matrix indirectly links them to presolar interstellar environments (Keller et al., 1997). Similarly, GEMS appear to be prominent constituents of deuterium-rich CP IDPs. At least some GEMS exhibit core/mantle structures consistent with the Greenberg model (Fig. 7) (Greenberg, 1984).

## 5. Conclusions

Considerable progress has been made in the study of interplanetary dust particles collected in the stratosphere. Much of the data being collected from IDPs has more relevance to observational astronomy than to mainstream meteoritics. CP IDPs appear more physically, chemically, and isotopically primitive than other known meteoritic materials. There is a very real possibility that some CP IDPs are well-preserved aggregates of circumstellar and/or interstellar materials.

Emerging analytical techniques will continue to play an important role in IDP research. Two particular types of instruments are particularly noteworthy. A new generation "nano-SIMS" ion microprobe presently under development may permit isotopic measurements approaching the nanometer scale. New field emission transmission electron microscopes (FE TEM s) enable compositional mapping of CP IDPs with spatial resolution on the order of five nanometers (50 Å). Given the extreme nanometer-scale heterogeneity within CP IDPs, these instruments will almost certainly play an important role in future IDP research.

Finally, STARDUST is scheduled to return the first actual samples of contemporary interstellar dust and comet grains (from comet Wild 2) in January 2006. These samples will provide groundtruth testing of many of our conclusions and hypotheses concerning IDPs (and meteorites) plus an historic new "motherlode" of particulate meteoritic samples for future generations of meteoriticists and astronomers.

## References

Anders, E. and Ebihara, M. (1982) Solar system abundances of the elements, Geochim. Cosmochim. Acta 46, 2363-2380.

Anders, E. and Zinner, E (1993) Interstellar grains in primitive meteorites: diamond, silicon carbide, and graphite, Meteoritics 28, 490-514.

Arndt, P., Bohsung, J., Maetz, M. and Jessberger, E. K. (1996) The elemental abundances in interplanetary dust particles, Met. Planet. Sci. 31, 817-833.

Binzel, R. P. (1992) The optical spectrum of 5415 Pholus, Icarus 99, 238-240.

Bradley, J. P. (1994a) Nanometer-scale mineralogy and petrography of fine-grained aggregates in anhydrous interplanetary dust particles, Geochim. Cosmochim. Acta 58, 2123-2134.

Bradley, J. P. (1984b) Chemically anomalous, preaccretionally irradiated grains in interplanetary dust from comets, Science 265, 925-929.

Bradley, J. P., Brownlee, D. E. and Snow, T. (1997) GEMS and other pre-accretionally irradiated grains in interplanetary dust particles (IDPs), In: From Stardust to Planetesimals, Eds. Y. J. Pendleton and A. G. G. M. Tielens (Astronomical Society of the Pacific: San Francisco), in press.

Bradley, J. P. and Brownlee, D. E. (1986) Cometary particles: thin sectioning and electron beam analysis, Science 231, 1542-1544.

Bradley, J. P. and Brownlee, D. E. (1992) An interplanetary dust particle linked directly to type CM meteorites and an asteroidal origin, Science 251, 549-552.

Bradley, J. P. and Ireland, T. (1996) The search for interstellar components in interplanetary dust particles. In Physics, Chemistry, and Dynamics of Interplanetary Dust, ASP Conference Series Vol. 104 (Eds. B. A. S. Gustafson and M. S. Hanner), pp. 275-282.

Bradley, J. P., Brownlee, D. E. and Fraundorf, P. (1984a) Discovery of solar flare tracks in interplanetary dust particles, Science 226, 1432-1434.

Bradley, J. P., Brownlee, D. E. and Fraundorf, P. (1984b) Carbon compounds in interplanetary dust: evidence for formation by heterogeneous catalysis, Science 223, 56-58.

Bradley, J. P., Brownlee, D. E. and Veblen, D. R. (1983) Pyroxene whiskers and platelets in interplanetary dust particles, Nature 301, 473-477.

Bradley, J. P., Germani, M. S., and Brownlee, D. E. (1989) Automated thin-film analyses of anhydrous interplanetary dust particles, Earth Planet. Sci. Lett. 93, 1-13.

Bradley, J. P., Harvey, R. P., and McSween, H. Y. Jr. (1996) Magnetite whiskers and platelets in the ALH84001 Martian meteorite: evidence of vapor phase growth, Geochim. Cosmochim. Acta 60, 5149-5155.

Bradley, J. P., Humecki, H. J., and Germani, M. S. (1992) Combined infrared and analytical electron microscope studies of interplanetary dust particles, Ap. J. 394, 643-651.

Bradley, J. P., Keller, L. P., Brownlee, D. E., and Thomas, K. L. (1996) Reflectance spectroscopy of interplanetary dust particles, Met. Planet. Sci. 31, 394-402.

Brownlee, D. E., Joswiak, D. and Bradley, J. P. (1997) Vesicular carbon in strongly heated interplanetary dust particles, Lunar Planet Sci. 28, (in press).

Brownlee, D. E. (1985) Cosmic dust: collection and Research, Ann. Rev. Earth Planet. Sci. 13, 147-173.

Brownlee, D. E. (1994) The origin and role of dust in the early solar system. In Analysis of Interplanetary Dust - (Eds. M. E. Zolensky, T. L. Wilson, F. J. M. Rietmeijer, G. J. Flynn). AIP Conf. Proc. 310, Am. Inst. Phys., pp. 5-8.

Brownlee, D. E., Joswiak, D. J., Schlutter, D. J., Pepin, R. O., Bradley, J. P., and Love, S. G. (1995) Identification of individual cometary IDPs by thermally stepped He release, Lunar Planet. Sci., 26, 193-184.

Brownlee, D. E., Wheelock, M. M, Temple, S, Bradley, J. P. and Kissel, J. (1987) A quantitative comparison of comet Halley and carbonaceous chondrites at the submicrometer level, Lunar Planet. Sci. 25, 185-186.

Bussoletti, E. and Colangeli, L (1990) Cosmic dust: from space to laboratory, Rivista Del Nuovo Cimento 13, 1-70.

Christoffersen, R. and Buseck, P. R. (1983) Epsilon carbide: a low temperature component of interplanetary dust particles, Science 222, 1327-1328.

Dermott, S. F., Jayaraman, S., Xu, Y. L., Gustafson, B. A. S. and Liou, J. C. (1994) A circumstellar ring of asteroidal dust in resonant lock with the Earth, Nature 369, 719-723.

Dorschner, J. and Henning, T. (1995) Dust metamorphosis in the galaxy, Astron. Astrophys. Rev. 6, 271-333.

Esat, T. M., Brownlee, D. E., Papanastassiou, D. A., and Wasserburg, G. J. (1979) Magnesium isotopic compositions of interplanetary dust particles, Science 206, 190-197.

Esat, T. M. and Taylor, S. R. (1987) Mg isotopic compositions of some interplanetary dust particles, Lunar Planet. Sci. 18, 269-270.

Farinella, P., Gonczi, R., and Froeschle, C. (1993) The injection of asteroid fragments

into resonances, Icarus 101, 174-187.
Flynn, G. J. (1990) The near-earth enhancement of asteroidal over cometary dust, Proc. 20th Lunar Plant. Sci. Conf. (LPI, Houston), pp. 363-371.
Flynn, G. J. (1994) The common stuff of stardust, Nature 371, 287-288.
Flynn, G. J., Sutton, S. R., and Bajt,, S. (1993) Trace element content of chondritic cosmic dust: volatile enrichments, thermal alterations, and the possibility of contamination, Lunar Planet Sci. 26, 495-496.
Fomenkova, M. N., Chang, S. and Mukhin, L. M. (1994) Carbonaceous components in the comet Halley dust, Geochim. Cosmochim. Acta 58, 4503-4512.
Fraundorf, P. (1981) Transmission electron microscopy of clinoenstatite "whiskers" in stratosphere-collected interplanetary dust, Lunar Planet. Sci. 12, 292-294.
Gaffey, M. J., Burbine, T. H., and Binzel, R. P. (1993) asteroid spectroscopy: progress and perspectives, Meteoritics 28, 161-187.
Germani, M. S., Bradley, J. P., and Brownlee, D. E. (1990) Automated thin-film analyses of hydrated interplanetary dust particles in the analytical electron microscope, Earth Planet. Sci. Lett. 101, 162-179.
Goodman, A. A. and Whittet, D. C. B. (1995) A point in favor of the superparamagnetic grain hypothesis, Ap. J. 455, L181-L185.
Greenberg, J. M. (1984) Structure and evolution of interstellar grains, Sci. Am. 250, 96-107.
Greenberg, J. M. and Hage, J. I. (1990) From interstellar dust to comets: a unification of observational constraints, Ap. J. 361, 260-274.
Hanner, M. S., Lynch, D. K., and Russel, R. W. (1994) The 8-13 micron spectra of comets and the composition of silicate grains, Ap. J. 425, 274-285.
Hyatsu, R. and Anders, E. (1981) Organic compounds in meteorites and their origins, Top. Curr. Chem. 99, 1-37.
Jessberger, E. K., Christoforidis, A., and Kissel, J. (1988) Aspects of the major element composition of Halley's dust, Nature 332, 691-695.
Jewitt, D. C. and Luu, J. X. (1990) CCD spectra of asteroids II. The Trojans as spectral analogues of cometary nuclei. Astron. J. 100, 933-944.
Jones, R. V. and Spitzer L. (Jr) (1967) Magnetic alignment of interstellar grains, Ap. J. 147, 943-964.
Keller, L. P., S. Messenger, M.M. Miller, K. L. Thomas (1997) Nitrogen speciation in a $^{15}$N-enriched interplanetary dust particle, Lunar Planet. Sci. 28, (in press).
Keller, L. P., Thomas, K. L., and McKay, D. S. (1992) An interplanetary dust particle with links to CI chondrites, Geochim. Cosmochim. Acta 56, 1409-1412.
Kissel, J. and Krueger, F. R. (1987) The organic component of dust from comet Halley as measured by the PUMA mass spectrometer on board Vega 1, Nature 326, 755-760.
Klock, W., Thomas, K. L., McKay, D. S., and Palme, H. (1989) Unusual olivine and pyroxene composition in interplanetary dust and unequilibrated ordinary chondrites, Nature 339, 126-128.
Lawler, M. E., Brownlee, D. E., Temple, S. and Wheelock, M. M. (1989) Iron, magnesium, and silicon in dust from comet Halley, Icarus 80, 225-242.
Liou, J. C., Dermott, S. F. and Xu, Y. L. (1995) The contribution of cometary dust to the zodiacal cloud, Planet. Space Sci. 43, 717-722.
Liou, J. C. and Zook, H. A. (1996) Comets as a source of low eccentricity and low inclination interplanetary dust particles, Icarus 123, 491-502.
Martin, P. G. (1995) On the value of GEMS (glass with embedded metal and sulfides), Ap. J. 445, L63-L66.
Mathis, J. S. (1993) Observations and theories of interstellar dust, Rep. Prog. Phys. 56, 605-652.
McKeegan, K. D., Walker, R. M., and Zinner, E. (1985) Ion microprobe isotopic measurements of individual interplanetary dust particles, Geochim. Cosmochim. Acta 49, 1971-1987.
Messenger, S., Keller, L. P., Thomas, K.L. and Walker, R. M. (1996) Nitrogen petrogra-

phy in two $^{15}$N-rich IDPs, Meteoritics 31, A88.
Mumma, M. J., Stern, S.A., and Weissman, P. R. (1993) Comets and the origin of the solar system: reading the Rosetta stone. In Protostars and Planets III, eds. E. H. Levy et al. (University of Arizona Press), pp. 1175-1252.
Nadgornyi, E. M. (1962) The properties of whiskers, Sov. Phys. Usp. 5, 462-477.
Nier, A. O. (1994) Helium and Neon in interplanetary dust particles. In Analysis of Interplanetary Dust (Eds. M. E. Zolensky, T. L Wilson, F. J. M. Rietmeijer, G. J. Flynn). AIP Conf. Proc. 310, Am. Inst. Phys. pp. 115-126.
Nier, A. O. and Schlutter, D. J. (1990) Helium and neon isotopes in stratospheric particles, Meteoritics 25, 263-267.
Nuth, J. A. (1988) Astrophysical implication of presolar interstellar grains. In Meteorites and the Early Solar System, Eds. J. Kerridge and M. S. Mathews (University of Arizona Press, Tucson) pp. 457-497.
Ott, U (1993) Interstellar grains in meteorites, Nature 364, 25-33.
Rajan, R. S., Brownlee, D. E., Tomandl, D., Hodge, P. W., Farrar, H. Britten, R. A. (1977) Detection of 4He in stratospheric particles gives evidence of extraterrestrial origin, Nature 267, 133-134.
Rietmeijer, F. J. M. (1996) CM-like interplanetary dust particles in lower stratosphere during 1989 October and 1991/July, Met. Planet. Sci. 31, 278-288.
Sandford, S. A. (1987) The collection and analysis of extraterrestrial dust particles, Fund. Cosmic Phys. 12, 1-73.
Sandford, S. A. (1996) The inventory of interstellar materials available for the formation of the solar system, Met. Planet. Sci. 31, 449-476.
Sandford, S. and Walker, R. M. (1985) Laboratory infrared transmission spectra of individual interplanetary dust particles from 2.5 to 25 microns, Ap. J. 291, 838-851.
Schramm, L. S., Brownlee, D. E., and Wheelock, M. M. (1989) Major element compositions of stratospheric micrometeorites, Meteoritics 24, 99-112.
Staderman, F. J., Walker, R. M. and Zinner, E.(1989) Ion microprobe measurements of nitrogen and carbon isotopic variations in individual IDPs, Meteoritics 24, 327.
Sutton, S. R. (1994) Chemical compositions of primitive solar system particles, in Analysis of Interplanetary Dust (Eds. M. E. Zolensky, T. L. Wilson, F. J. M. Rietmeijer, G. J. Flynn). AIP Conf. Proc. 310, Am. Inst. Phys. pp145-157.
Symonds, R. (1993) Scanning electron microscope observations of sublimates from Merapi, Indonesia, Geochem. J. 26, 337-350.
Tielens, A. G. G. M. (1990) Towards a circumstellar silicate mineralogy. In From Miras to PN: Which path for stellar evolution? (Eds. M. O. Mennessier and A. Omont) pp. 186-200. Editions Frontiers, Gif-sur-Yvette, France.
Veblen, D. R. and Post, J. E. (1983) A TEM study of fibroud cuprite (chalcotrichite): microstructures and growth mechanisms, Am. Min. 68, 790-803.
Verniani, F. (1969) Structure and fragmentation of meteoroids, Space Sci. Rev. 10, 230-261.
Walker, R. M. (1994) Isotopic constraints on interstellar material in chondritic IDPs. In Analysis of Interplanetary Dust (Eds. M. E. Zolensky, T. L Wilson, F. J. M. Rietmeijer, G. J. Flynn). AIP Conf. Proc. 310, Am. Inst. Phys. pp. 245-253.
Warren, J. L. and Zolensky, M. E. (1994) Collection and curation of interplanetary dust particles recovered from the stratosphere by NASA. In Analysis of Interplanetary Dust (Eds. M. E. Zolensky, T. L. Wilson, F. J. M. Rietmeijer, G. J. Flynn). AIP Conf. Proc. 310, Am. Inst. Phys. pp. 245-253.
York, D. G. (1994) The elemental composition of interstellar dust, Science 265, 191-192.

# THE DUST AROUND VEGA-LIKE STARS: A BRIDGE FROM YSOS TO COMETS

HAROLD M. BUTNER
*SMTO-Steward Observatory*

HELEN J. WALKER
*CLRC RAL*

AND

DIANE H. WOODEN
*NASA Ames Research Center*

**Abstract.** Vega-like stars are infrared excess stars either nearing or already on the main sequence. The infrared excess is interpreted as evidence for the presence of extended dusty disks surrounding the stars. In a few cases, the disks have been directly imaged. While most of the disks are difficult to observe at optical and near-infrared wavelengths, a few disks are bright enough to study at 10 microns. Mid-infrared spectra from several of these mid-infrared bright Vega-like stars reveals a range of silicate emission features. Such features arise from the thermal emission of silicate grains in the disks. In some cases, the silicate features seen are similar to those seen towards some solar-system comets and interplanetary dust particles.

## 1. Introduction

IRAS observations in 1984 revealed a new class of infrared excess stars, the "Vega-like" stars. The class was discovered through observations of Vega, which was expected to serve as a standard star for IRAS observations out to 100 $\mu$m. Instead, Vega proved to have a significant infrared excess at 25, 60, and 100 $\mu$m over the expected stellar blackbody continuum. In addition to Vega, three other nearby stars were identified in the initial search of the IRAS database: $\beta$ Pic, $\epsilon$ Eri, and $\alpha$ PsA. Backman and Paresce (1993-BP93) provide an excellent overview of the early observations and models of these sources. These main-sequence stars were unusual because they had



*J. M. Greenberg and A. Li (eds.), Formulation and Evolution of Solids in Space,* 505–512.

significant infrared excess emission in the IRAS bands at 25, 60 and 100 $\mu$m, significantly above that expected from the star alone. The lack of visual extinction towards the star combined with infrared colors consistent with thermal emission from dust seemed most consistent with models where a warm (few hundred K) dusty disk surrounded the star.

Visual images (0.89 $\mu$m) of $\beta$ Pic dramatically confirmed the disk hypothesis, revealing a nearly edge-on disk with a radius in excess of 450 AU (BP93). More recently, $\beta$ Pic has been imaged at a variety of wavelengths ranging from the optical (Burrows *et al.* 1996) to 10 $\mu$m (Lagage and Pantin 1994). These observations have been used to model for the grain size and disk density structure (Artymowicz 1996, Greenberg and Li 1996, Li and Greenberg 1998).

Other Vega-like stars have been identified in the IRAS database (BP93). The search is done by looking for IRAS excesses in a large sample of stars (such as all A-stars within 25 parsecs). Most of the stars identified this way have turned out to have very small excesses compared to the 4 prototypes. Infrared excesses, while small, do appear to be common among many nearby stars. A substantial fraction of A, F, G, and K stars in the Gliese catalog shows evidence for a small infrared excess. Most, however, have no 12 $\mu$m excess, forcing one to use either airborne or spaceborne instruments to study their disks at wavelengths shorter than 350 microns. The exact age of many of these systems is uncertain, with some stars possibly as young as 10 million years rather than being on the main sequence (Jura *et al.* 1993). To understand the disk properties of these infrared excess systems, most studies have focused on the "big 3" sources: Vega, $\beta$ Pic, and $\alpha$ PsA. These can all be detected at FIR and millimeter wavelengths, but only $\beta$ Pic can be studied at shorter wavelengths.

The disks around the Vega-like stars are optically thin and thus thought to have very low masses. They appear to be the remnants of the more massive disks which are commonly seen around many young stellar objects. Young stellar disks have masses typically of order 0.02 solar masses, whereas the Vega-like disks have dust amounting to only a few lunar masses. If the bulk of the Vega-like systems are late pre-main-sequence stars or main sequence stars, these systems offer a way to study the details of how disks evolve with time after the surrounding infall envelope has been dispersed by the star.

## 2. $\beta$ Pic-like Systems

One of the great difficulties in studying Vega-like systems is that by and large the disks are not very massive. In addition, even among the prototypes, 3 out of 4 have excess emission only at wavelengths longer than 20

$\mu$m. The lack of 12 $\mu$m emission indicates that the dust is very cold and far away from the star. $\beta$ Pic is the obvious exception where the disk is massive, extends close to the central star, and is oriented edge-on (all factors favorable for study by earth-based observers). It is thus no surprise that searches for other disk systems with ground-based telescopes has proved difficult. For example, numerous optical searches have failed to find counterparts among other suspected Vega-like stars to the disk seen around $\beta$ Pic (BP93). Longer wavelength observations, such as the submillimeter studies of Holland et al (1998), have managed to image several of the prototype disk systems at 850 $\mu$m. However, to study even the 4 prototypes at optical and mid-infrared wavelengths is extremely challenging.

One possibility is to restrict one's observational program to systems that are "$\beta$ Pic-like" , not just "Vega-like". Walker and Wolstencroft (1988-henceforth WW) identified a number of candidates with significant infrared excesses similar to $\beta$ Pic. Most of these are at distances further away than $\beta$ Pic itself, so that they are less well studied than $\beta$ Pic and probably include a mixture of ages. Zuckerman *et al.* (1995) estimate that a number of the objects in the WW list are only a few million years old. The larger distances imply that the disks surrounding these stars are likely to be more massive than $\beta$ Pic. As part of a multi-wavelength program, we have selected a number of the WW-stars for followup studies. (See also Sylvester *et al.* 1996).

## 3. Silicate Emission

Of particular interest is the 8 - 13 $\mu$m region. This wavelength region is accessible from the ground and has a number of spectral features that can be used to study the dust's composition. One of the most important features is a broad diffuse band centered at about 10 $\mu$m. The 10 $\mu$m band commonly is attributed to vibrational fundamentals of the Si-O bond found in silicates. Such 10 $\mu$m features have been seen in either absorption or emission towards a number of astronomical objects, including M-stars, young stars, and comets. When seen in absorption the feature traces the presence of cold dust projected against a bright background. The feature in emission indicates hot optically thin dust. The detailed structure of the silicate band offers clues to the underlying mineralogical composition of the silicate dust. For example, it has been suggested that the presence of a bump at 11.2 $\mu$m might be attributed to crystalline olivine (Hanner *et al.* 1993 and references therein). Other minerals can produce further modifications of the basic silicate shape.

Among astronomical objects, the silicate emission has a shape often referred to as "astronomical silicates" (Li and Greenberg 1997). This is seen towards many dust enshrouded stars. In contrast, observations of some comets such as P/Halley and P/Bradfield 1987 revealed a twin peaked structure (Bregman *et al.* 1987, Hanner *et al.* 1993). The twin peaked feature has also been seen in laboratory spectra of interplanetary dust particles (IDPs - Bradley *et al.* 1992). Such a feature in general is not seen in most interstellar sources (Draine and Lee 1984, Li and Greenberg 1997). In the case of $\beta$ Pic, silicate emission was first detected in its disk by Telesco and Knacke (1991) using spectrophotometry. They compared the feature they found with the Halley spectra and found a plausible match. More recent spectra of $\beta$ Pic were taken by Knacke *et al.* (1993) and compared to a variety of astronomical and cometary spectra. They found that P/Halley spectra and IDP spectra matched $\beta$ Pic's spectra best.

This raises the possibility that we can tie mineral studies of IDPs to the dust found in $\beta$ Pic-like systems if we could obtain high resolution observations of the silicate feature.

## 4. Observations

To study the mid-infrared spectra of the WW-sample, we used the NASA Ames HIFOGS (High-efficiency Faint Object Grating Spectrometer) at the NASA 3 meter IRTF (Infrared Telescope Facility). HIFOGS works in the region of 8 to 13 $\mu$m. See Witteborn *et al.* (1995) for a full description of the instrument. There are 120 detectors, covering the 8 to 13 $\mu$m atmospheric window, with a spectral resolution of 0.05 $\mu$m/pixel. In June 1995, we observed 7 stars brighter than 1 Jy at 12 $\mu$m from the WW list. By restricting ourselves to sources brighter than 1 Jy, we could obtain a reasonable spectrum with good signal-to-noise per pixel. By careful monitoring of known spectral standards and modeling of the atmospheric transmission (i.e. the 10 $\mu$m ozone feature), we obtained overall flux calibration to better than two percent on our brighter sources.

## 5. Discussion

We observed $\alpha$ PsA, $\beta$ Cas, HD 188037, HD 144432, HD 142666, HD 143006 and 51 Oph. HD 188037 has been reported to have an M star companion, which would explain the observed "astronomical" silicate feature we see. Both $\alpha$ PsA and $\beta$ Cas show no evidence of silicate features but there is a hint of the onset of a thermal dust continuum longwards of 12 $\mu$m. HD 143006 has a significant continuum excess in the mid-infrared, but no obvious silicate emission. HD 144432 has a feature reminiscent of that seen towards GW Ori (a binary young stellar object thought to have a disk). HD 142666 has broad silicate features, similar to that Sylvester *et al.* (1996) report for HD 98800. Our data show no examples of PAH-like emission features, though Sylvester *et al.* report PAH features for three other $\beta$ Pic-like sources (HD 169142, HD135344, and HD 141596).

51 Oph is thought to be a Be star with a dust disk. It was reported by Fajardo-Acosta *et al.* (1993) to have silicate emission features similar to $\beta$ Pic and comet P/Halley. They found that the dust around 51 Oph was warmer than $\beta$ Pic and argued that the grains would be larger than the grains around $\beta$ Pic. Sylvester *et al.* (1996) also observed 51 Oph, but make no comment about the emission other than to note it is silicate emission. Our own data has been carefully reduced, and a 550 K blackbody removed from the continuum. The feature appears to have an unusual broad shoulder in the emission beyond 10 $\mu$m. Both P/Halley (Figure 1- bottom) and one of the glass-rich IDPs (U219C11 - discussed by Bradley *et al.* 1992) have spectra very similar to that of 51 Oph.

We applied similar continuum removal to our spectra of HD 142666 and HD 144432 (Figure 1). We find prominent bumps at 11 $\mu$m in both spectra. In contrast, HD 143006 appears featureless and is well represented by a blackbody continuum. All the continuum temperatures that fit the data lie around 400-500 K. If we compare the 3 silicate emission features (51 Oph, HD 142666, HD 144432) with examples of cometary spectra (Hanner *et al.* 1993), we find that the P/Halley and P/Bradfield 1987 spectra are the most similar. A detailed analysis of the spectral features for each star awaits full models, but it would appear that we can use the comets and IDPs as analogs for the average silicate dust found in some of these systems. In young stellar objects (YSOs), the 11.2 $\mu$m crystalline feature is rare, indicating that the dust in the $\beta$ Pic-like systems may have undergone some additional processing. However, recent observations of Herbig Ae/Be stars indicates that their 10 $\mu$m features also may be more complex than traditionally assumed (Wooden-private communication). We thus can link the YSOs to the Vega-like stars through changes in the 10 $\mu$m feature and overall disk properties. Similarly, it appears that we can link the Vega-like dust properties to our own solar system comets (and IDPs).

The recent flurry of observations of Comet Hale-Bopp has revealed an interesting complication. Harker *et al.* (1998) report on HIFOGS observations of comet Hale-Bopp taken over several epochs, spanning the largest range of orbital radii yet observed at 10 $\mu$m. Wooden *et al.* (1998) examine the spectra in detail. They find three critical points: (1) The comet's spectra varies as the comet approaches the sun and then returns to the original spectra as the comet retreats. (2) Olivines dominant the mid-infrared spectra at large distances, but as the comet approaches perihelion, pyroxenes become more prominent. (3) that a simple model of this behavior can be developed that explains the changes as due not to changes in the underlying mineral composition of the dust grains, but due to the dramatic effect of

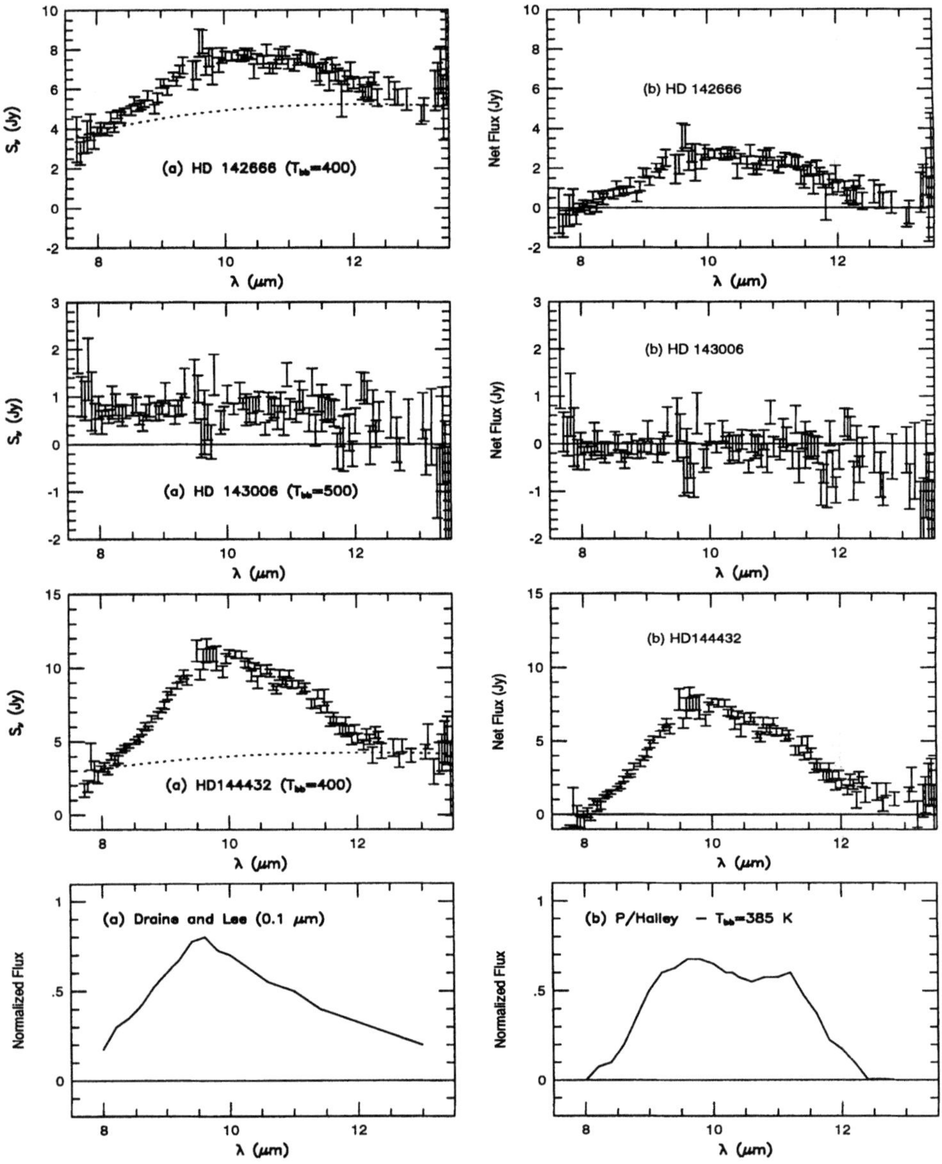


*Figure 1.* Spectra of HD142666, HD143006, and HD144432. (a-left side) For each source, we display the source spectrum (fluxes in Jy) and a blackbody fit superimposed (dashed line). (b-right side) The spectrum of each source consists of the residual flux after subtraction of the displayed blackbody. The blackbodies are 400, 500, and 400 K for HD142666, HD143006, and HD144432 respectively. The bottom two figures are taken from Knacke *et al.* (1993). On the left is the spectra of a Draine and Lee (1984) 0.1 $\mu$m "astronomical silicate". On the right is the spectrum of P/Halley after a 385 K blackbody has been subtracted.

temperature on different grain populations. While Hale-Bopp is the only comet to be so well studied, the role of temperature could be equally important for dust in other comets as well as in $\beta$ Pic-like systems. As our own HIFOGS observations have indicated, the average continuum temperature varies from disk to disk, reflecting the location relative to the star of the material responsible for the mid-infrared emission. We are currently investigating the role such temperature differences might imply for the mid-infrared spectra we observe, and the dust mineral composition we would deduce.

## 6. Conclusions

The results to date suggest that many of the $\beta$ Pic-like stars have dust that has undergone processing similar to that seen in cometary dust. As discussed out by Li and Greenberg (1998) and other authors (see references therein and in BP93), comets explain many of the phenomena identified in $\beta$ Pic. Optical studies indicate the possibility that $\beta$ Pic is undergoing considerable cometary impacts. The comets should be releasing a large amount of dust, that can then be seen in the IR. Since the lifetime of the inner $\beta$ Pic dust disk is quite short, dust replenishment must be on-going. The fact that similar spectra are seen towards a number of "Vega-like" stars suggests that the dust in those systems are undergoing processing similar to $\beta$ Pic's dust.

HMB gratefully wishes to acknowledge the support of the NASA Origins of the Solar System Program (NAGW-4097), the NASA Airborne Astronomy Program (NAG-2-959), the Carnegie Institution of Washington, and the Submillimeter Telescope Observatory (SMTO) for supporting this work. He also wishes to thank the organizers for a travel grant which made his attendance at this NATO ASI possible.

## References

Artymowicz, P. 1996, "Vega-Type Systems", *The Role of Dust in the Formation of Stars*, ed. H. U. Kaufl and R. Siebenmorgen, (Springer: Berlin), p. 137-148.

Backman, D. E., and Paresce, F. 1993 (BP93), "Main Sequence Stars With Circumstellar Solid Material: The Vega Phenomenon", in *Protostars and Planets III*, eds. E. H. Levy, J. I. Lunine, and M. S. Matthews, (Univ. of Arizona: Tucson), p. 1253-1304.

Bradley, J. P., Humecki, H. J., and Germani, M. S. 1992, "Combined Infrared and Analytical Electron Microscope Studies of Interplanetary Dust Particles", *Ap. J.*, **394**, 643-651.

Bregman, J. D., Campins, H., Witteborn, F. C., Wooden, D. H., Rank, D. M., Allamandola, L. J., Cohen, M., and Tielens, A. G. G. M. 1987, "Airborne and Groundbased Spectrophotometry of Comet P/Halley from 5-13 Micrometers", *Astr. Ap.*, **187**, 616-620.

Burrows, C. J., Krist, J. E., Stapelfeldt, K. R., WFPC2 Investigation Team 1995, "HST Observations of the Beta Pictoris Circumstellar Disk", *BAAS*, **187**, 3205.

Draine, B. T., and Lee, H. M. 1984, "Optical Properties of Interstellar Graphite and Silicate Grains", *Ap. J.*, **285**, 89-108.

Fajardo-Acosta, Telesco, C. M., and Knacke, R. F. 1993, "Detection of Silicates in the 51 Ophiuchi System", *Ap. J. (Letters)*, **417**, L33-L36.

Greenberg, J. M., and Li, A. 1996, "Comets as a Source of the Dust in the $\beta$ Pictoris Disk", *The Role of Dust in the Formation of Stars*, ed. H. U. Kaufl and R. Siebenmorgen, (Springer: Berlin), p. 161-174.

Hanner, M. S., Lynch, D. K., and Russell, R. W. 1994, "The 8-13 micron Spectra of Comets and the Composition of Silicate Grains", *Ap. J.*, **425**, 274-285.

Harker, D. E., Wooden, D. H., Woodward, C. E., McMurtry, C. W., Butner, H. M., Goetz, J. A., Pipher, J. L., Forrest, W. J., Koike, C., Rudy, R. J. 1997, "Infrared Imaging and Spectrophotometry of Comet Hale-Bopp", *B. A. A. S.*, **191**, 3402.

Holland, W. S., Greaves, J. S., Zuckerman, B., Webb, R. A., McCarthy, C., Coulson, I. M., Walther, D. M., Dent, W. R. F., Gear, W. K., and Robson, I. 1998, "Submillimetre images of dusty debris around nearby stars", *Nature*, **392**, 788-791.

Knacke, R. F., Fajardo-Acosta, S. B., Telesco, C. M., Hackwell, J. A., Lynch, D. K., and Russell, R. W. 1993, "The Silicates in the Disk of beta Pictoris", *Ap. J.*, **418**, 440-450.

Jura, M., Zuckerman, B., Becklin, E. E., and Smith, R. C. 1993, "Constraints on the Evolution of Remnant Protostellar Dust Debris Around HR 4796", *Ap. J. (Letters)*, 418, L37-L40.

Lagage, P. O. and Pantin, E. 1994, "Dust Depletion in the Inner Disk of beta Pictoris as a Possible Indicator of Planets", *Nature*, **369**, 628-630.

Li, A., and Greenberg, J. M. 1997, "A Unified Model of Interstellar Dust", *Astr. Ap.*, **323**, 566-586.

Li, A., and Greenberg, J. M., 1998, "A Comet Dust Model for the beta Pictoris Disk", *Astr. Ap.*, **331**, 291-313.

Sylvester, R. J., Barlow, M. J., and Skinner, C. J. 1995, "Optical, infrared, and millimetre-wave properties of Vega-like systems", *M.N.R.A.S.*, **279**, 915-939.

Telesco, C. M., and Knacke, R. F. 1991, "Detection of Silicates in the Beta Pictoris Disk", *Ap. J.*, **372**, L29-L31.

Walker, H. J., and Wolstencroft, R. D. 1988 (WW), "Cool Circumstellar Matter Around Nearby Main-Sequence Stars", *Pub. A. S. P.*, **100**, 1509-1521.

Witteborn, F., C., Cohen, M., Bregman, J. D., Heere, K. R., Greene, T. P., and Wooden, D. H. 1995, "HIFOGS, Its Design, Operation, and Calibration" in *ASP Conference Series, vol 73: Airborne Astronomy Symposium on the Galactic Ecosystem: ¿From Gas to Stars to Dust*, ed. M. R. Hass, J. A. Davidson, and E. F. Erickson, (Astronomical Society of the Pacific: San Francisco), p. 573-578.

Wooden, D. H., Harker, D. E., Woodward, C. E., Butner, H. M., Koike, C., Witteborne, and McMurtry, C. M. 1998, "Silicate Mineralogy of the Dust in the Inner Coma of Comet C/1995-O1 (Hale-Bopp) Pre- and Post-Perihelion", *Ap. J.*, in press.

Zuckerman, B., Forveille, T., Kastner, J. H. 1995, "Inhibition of giant-planet formation by rapid gas depletion around young stars", *Nature*, **373**, 494-496.

# ACCRETING GAS AND DUST IN PRE-MAIN SEQUENCE DISK SYSTEMS

M.L. SITKO

*University of Cincinnati*
*Dept. of Physics, Cincinnati, OH 45221-0011, USA*

C.A. GRADY

*Eureka Scientific*
*2452 Delmer St., Suite 100, Oakland, CA 94602, USA*

D.K. LYNCH AND R.W. RUSSELL

*Aerospace Corporation*
*Los Angeles, CA 90009, USA*

AND

M.S. HANNER

*Jet Propulsion Laboratory*
*California Institute of Technology, Pasadena, CA 91109, USA*

**Abstract.** Infrared spectroscopy of pre-main sequence stars with dusty protostellar disks provide information about the evolution of refractory materials in such systems. These systems exhibit varying degrees of strength and structure in the silicate emission band situated at 10 $\mu$m wavelength. Band strength is affected by the mean grain size, while band structure is determined by the chemical composition and degree of crystallinity. In some objects, the silicate band is strong and featureless, similar to that seen in the interstellar medium. In others, the band is often weaker, and exhibits structure consistent with the presence of crystalline olivine. In these latter objects, the band is similar to that of some solar system comets. The strength and structure of the silicate band may be related to the processing history of the system.



*J. M. Greenberg and A. Li (eds.), Formulation and Evolution of Solids in Space,* 513–520.

## 1. Introduction

The study of protostellar disks underwent a major advance in the early 1980's with the discovery, using the IRAS satellite, that a large fraction of all main sequence stars possessed excess infrared emission. By far the most spectacular object in this class of "Vega-type stars" was $\beta$ Pictoris, where coronagraphic imaging showed the presence of a disk-like structure (Smith & Terrile, 1987). That the disk was due to dust was clearly demonstrated with the detection of the silicate emission band at 10 $\mu$m (Knacke *et al.*, 1993). In fact, the emission band is strikingly similar to that of some solar system comets, and confirms the basic assumption that it has dusty material similar to that which went into the formation of planetesimals in our solar system. Later, it was also discovered that this star experienced episodes of infalling gas, presumably the result of infalling solid bodies that were heated by the star - essentially asteroids or comets (Vidal-Madjar *et al.*, 1994).

The Vega-type stars must have evolutionary precursors. Based on their location in the HR diagram, presence of infrared excesses, and other criteria, these are most likely the T Tauri stars (lower mass) and Herbig Ae/Be (HAEBE) stars (higher mass). As for $\beta$ Pic itself, a low-mass HAEBE is indicated. Yet this star is unusual among the Vega-type class in that its disk mass seems to be much higher than any other known. Its low space velocity places it far from a star-forming region in the past few x $10^7$ years. Thus the precursor of $\beta$ Pic should have had a relatively massive disk, and may have been relatively isolated.

## 2. Accreting Material in Disks

A subset of the HAEBEs that are likely evolutionary precursors to stars like $\beta$ Pic has recently been identified. These generally are isolated HAEBEs or older "post-HAEBEs" ("PHAEBEs") with evidence for massive disks of gas and dust seen edge-on. These disks produce large-amplitude flux and polarization variations due to variable obscuration, wavelength-dependent polarization, spectroscopic evidence for infalling gas, and a variety of other phenomena (Grady *et al.*, 1996). These objects are excellent candidates for the study of the protostellar disk phenomenon because they are uncontaminated by extended dust emission, possess infalling gas, and in general they have many the characteristics expected of a pre-main sequence system entering an evolutionary phase where planetesimals are expected to form.

## 3. Dust in HAEBEs and Post-HAEBEs

### 3.1. SILICATE BAND STRENGTH

One of the goals of the study of pre-main sequence systems is to understand the chemical and physical nature of the materials in them at various stages of evolution. In this way, we may be able to chart the overall evolution of planetary systems from molecular cloud to star-plus-planets (if present). It is generally assumed that our own solar system is not unique, and that other planetary systems must undego somewhat similar evolutionary processes as our own did. The similarity in the IR spectra of $\beta$ Pic dust and solar system comet dust supports this assumption.

Dust grains found in interstellar clouds are generally found to possess a well-defined 10 $\mu$m absorption band indicative of the presence of small silicate particles. Such grains, when located near a star, produce an emission band of similar structure. In the "standard" Trapezium-like environments, the observed emission band/continuum ratio is about 3. The feature seen in solar system comets has a smaller ratio, usually less than 2 (Hanner *et al.*, 1994). That observed in the zodiacal light of the solar system has a ratio of about 1.15 (Reach *et al.*, 1996). Such a trend toward a smaller ratio is expected as the spectrum becomes dominated by larger and larger particles. And while the "typical" interstellar band is essentially featureless, that of solar system comets often exhibits structure (when the band is strong enough to measure accurately).

When we actually examine the IR spectra of the accreting edge-on HAEBEs, we see a band/continuum ratio that runs from about 3 to nearly 1 (i.e., no band). The smaller band ratios can be understood as the result of the combined effects of grain agglomeration (removal of small grains onto the surfaces of larger bodies), and radiation pressure. The latter ejects grains smaller than 10 $\mu$m on time scales shorter than a century in these objects (Sitko *et al.*, 1994). The existence of a band at all requires the continual replenishment of small grains.

There is a slight indication that the band/continuum ratio may be related to the age of the systems, but the situation is not clear-cut by any means. Processing of the dust in the disk will be a function of the age, initial environment, efficiency of small-grain removal by radiation pressure, and perhaps a host of other processes.

### 3.2. SILICATE BAND STRUCTURE

If the dust in these objects is truly like that in our solar system, then there is much that can be learned about them from both spectroscopy of solar system comets as well as laboratory analysis of solar system dust. Solar system comets often exhibit an emission feature due to silicates, and the structure of the band suggests the presence of at least some crystalline olivine (Hanner *et al.*, 1994). Primitive interplanetary dust particles of probable cometary origin are found to be aggregates of glassy silicates, often with embedded submicrometer crystals, in a carbonaceous matrix. Combined chemical and spectral analysis of these particles allow the IR spectral features to be correlated with the presence of specific minerological groups, such as olivines and pyroxenes, and a few possess IR spectral features very similar to the emission feature seen in comets (Bradley *et al.*, 1992). In this manner, the spectral information obtained on comets can potentially be modeled using laboratory data on real cometary grains of known composition. We would like to be able to do the same for the dust in the HAEBEs and PHAEBEs.

Up until recently, the band structure that has been seen in both solar system comets and $\beta$ Pic was not readily observed in the HAEBEs and PHAEBEs, even though it would be expected if they are the precursors of $\beta$ Pic objects. However, this situation is changing with the systematic investigation of the isolated edge-on systems. Many (if not most) exhibit a band structure similar to that of $\beta$ Pic and comets, particularly in those objects with a small band/continuum ratio. Under such conditions, the structure is difficult to see at first, due to the strength of the underlying continuum. But this underlying continuum, which can be dominated by the presence of featureless (non-silicate or large silicate) grains, should not *a priori* be expected to be totally identical for any specific solar system comet observed at a single heliocentric distance and all the stellar disk grains distributed over a range of distances in the HAEBEs and PHAEBEs. Specific comparisons of the band structures is more properly made by subtracting the continuum first. When this is done, the silicate bands in these HAEBEs are often very similar in overall structure to that of solar system comets.

In Figure 1 we show the spectra of HD 163296 and Comet Hale-Bopp, obtained with the same instrument, through the same column of terrestrial atmosphere, within a half-hour of one another, and processed the same way, so as to minimize any systematic effects entering into the comparison. These bands are nearly identical, except that the stellar dust has a slightly greater emission near 9.4 microns, a feature probably due to a slightly greater fraction of glassy pyroxenes (Stephens & Russell, 1979) in the stellar grains. Both have similar crystalline olivine features at 11.2 $\mu$m.

In the objects studied, there is a slight indication that, on the average, the more evolved the disk (based on band strength and short-wavelength flux variability), the less "Trapezium-like" and the more "comet-like" the band appears. Figure 2 shows the feature in three objects: RY Tau, which still possesses a thick disk with variable obscuration and a large band/continuum ratio, HD 163296, where the dust emission is weaker, and HD 53187, which

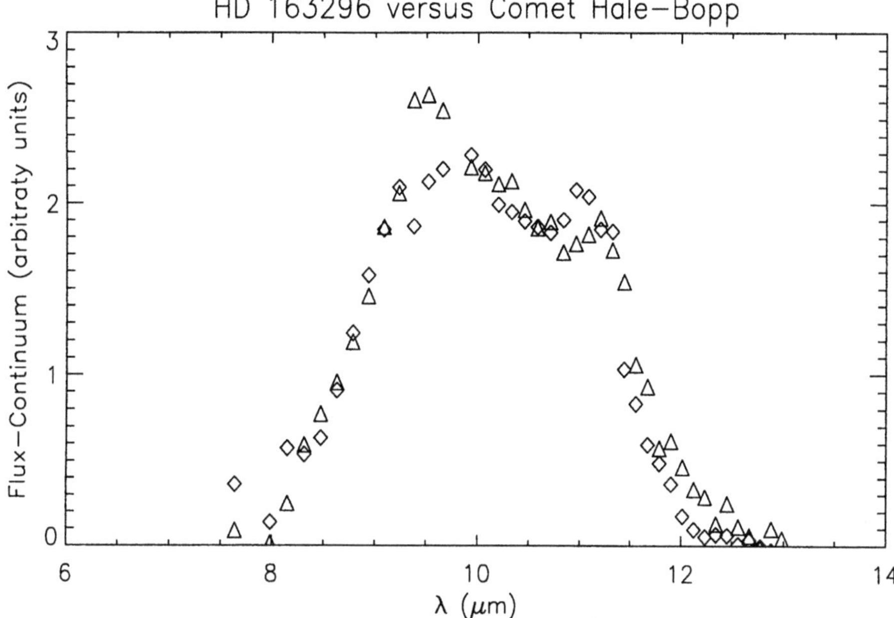


*Figure 1.* The silicate band of HD 163296 (triangles) compared to that of Comet Hale-Bopp (diamonds). The spectra were obtained on 14 October 1996 UT using the Aerospace Corp. BASS spectrometer on the NASA IRTF.

has a very weak band. It would be important to confirm if this is an evolutionary effect, since we do not currently know where or how the material that produces this feature arises. Is it produced in the outflows of evolved stars and "hidden" in the bulk of interstellar grains? Is it partially annealed during disk evolution, or does the removal of coatings allow it to be seen better? Is it partially annealed as dust is carried in toward the star?

In one of these objects, HD 100546, the infalling gas contains low-ionization metals and neutral C, O, and N, is seen at speeds consistent with free-fall from (essentially) infinity to within 0.2 AU of the star, and suggests the presence of stargrazing comets with evaporating silicate and CHON particles (Grady *et al.*, 1997). HD 100546 also possesses IR emission features due to organics as well as at least some Mg-rich crystalline olivine (Waelkens *et al.*, 1996). The infalling gas is also slightly Mg-rich (Grady *et al.*, 1997). Are we seeing the effects of the ablation of the grains? Since many of the primitive glassy IDPs contain crystalline seed cores, perhaps the removal of the outer material allows the presence of such cores to be seen more clearly.

As yet there are many unanswered questions regarding these grains. We

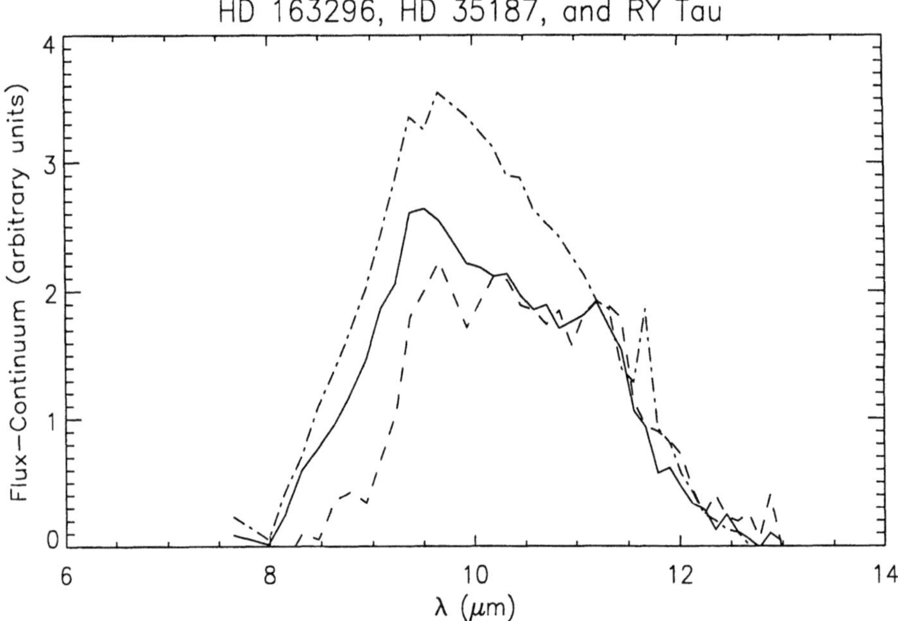


*Figure 2.* The silicate band of HD 163296 (solid line) compared to that of HD 35187 (dashed line) and RY Tau (dot-dashed line). The vertical scaling arbitrary. For the purpose of comparison, the bands are normalized at 11.2 μm.

do not understand their evolution. We do not understand why so many of these objects exhibit spectral evidence for olivines, while the primitive IDPs that have been analyzed in the laboratory so far seem to be dominated by pyroxenes. Further work on both the IDPs and the disk spectra will hopefully result in a (relatively) unified picture of how this material was formed and how it has evolved into its present state.

## References


Bradley, J.P., Humecki, H.J., and Germani, M.S. (1992) Combined Analytical Electron Microscope Studies of Interplanetary Dust Particles, *ApJ*, **394**, pp. 643–651.

Grady, C.A., Pérez, M.R., Talavera, A., Bjorknman, K.S., de Winter, D., Th, P.S., Molster, F., van der Anker, M., Sitko, M.L., Grinin, V.P., Calvet, N., Morrison, N., McCollum, B., and Castelaz, M. (1996) The β Pictoris Phenomenon among Herbig Ae/Be Stars *AstrAp*, **120**, pp. 157–177.

Grady, C.A., Sitko, M.L., Bjorkman, K.S., Pérez, M.R., Lynch, D.K., Russell, R.W., and Hanner, M.S. (1997) The Star-Grazing Extra-Solar Comets in the HO 100546 System *ApJ*, **483**, pp. 449–456.

Hanner, M.S., Lynch, D.K., and Russell, R.W. (1994) The 8-13 Micron Spectra of Comets and the Composition of Silicate Grains *ApJ*, **425**, pp. 274–285.

Knacke, R.F., Fajardo-Acosta, S.B., Telesco, C.M., Hackwell, J.A., Lynch, D.K., and Russell, R.W. (1993) Silicates in the Disk of β Pictoris it ApJ, **418**, pp. 440–450.

Reach, W.T., Abergel, A., Boulanger, F., Dèsert, F.-X., Perault, M., Bernard, J.-P., Blommaert, J., Cesarsky, C., Cesarsky, D., Metcalfe, L., Puget, J.-L., Sibille, F., and Vigroux, L. (1996) Mid-Infrared Spectrum of the Zodiacal Light *A&A*, **315**, pp. L381–L384.

Sitko, M.L., Halbedel, E.M., Lawrence, G., Smith, J.A., and Yanow, K. (1994) Variable Extinction in HD 45677 and the Evolution of Dust Grains in Pre-Main-Sequence Disks *ApJ*, **432**, pp. 753–762.

Smith, B.A., & Terrile, R.J. (1987) A Circumstellar Disk around $\beta$ Pictoris *Science*, **226**, pp. 1421–1424.

Stephens, J.R., & Russell, R.W. (1979) Emission and Extinction of Ground and Vapor-Condensed Silicates from 4 to 14 Microns and the 10 Micron Silicate Feature *ApJ***226**, pp. 780–786.

Vidal-Madjar, A., Lagrange-Henri, A.-M., Feldman, P.D., Beust, H., Lissauer, J.J., Deleuil, M. Ferlet, R., Gry, C., Hobbs, L.M., McGrath, M.A., McPhate, J.B., and Moos, H.W. (1994) HST-GHRS Observations of $\beta$ Pictoris: Additional Evidence for Infalling Comets *A&A*, **290**, pp. 245–258.

Waelkens, C., Waters, L.B.F.M., de Graauw, M.S., Huygen, E., Malfait, K., Plets, H., Vandenbussche, B., Beintema, D.A., Boxhoorn, D.R., Habing, H.J., Heras, A.M., Kester, D.J.M., Lahuis, F., Morris, P.W., Roelfsema, P.R., Salama, A., Siebenmorgen, R., Trams, N.R., van der Bliek, N.R., Valentijn, E.A., and Wesselius, P.R. (1996) SWS Observations of Young Main-Sequence Stars with Dusty Circumstellar Disks *A&A*, **315**, pp L245–L248.

# OBSERVATIONS OF DUST AND MOLECULES IN THE DISKS AND ENVELOPES OF YOUNG STELLAR OBJECTS

MICHIEL R. HOGERHEIJDE AND EWINE F. VAN DISHOECK
*Sterrewacht Leiden*
*P.O. Box 9513, 2300 RA Leiden, The Netherlands*

AND

GEOFFREY A. BLAKE
*Division of Geological and Planetary Sciences, California Institute of Technology, MS 150-21, Pasadena, CA 91125*

**Abstract.** Observations of dust and gas-phase molecules are compared as probes of the embedded phase of low-mass star formation. Results are presented of a detailed study of a sample of low-mass young stellar objects in Taurus with single-dish and interferometer millimeter observations. At least 67% of the objects have compact circumstellar disks, in addition to extended envelopes. $HCO^+$ is found to be an excellent tracer of envelope structure. Interferometer observations of its 1–0 line, and of $^{13}CO$ and $C^{18}O$, reveal the kinematics of the equatorial regions of the envelopes, and the interaction with the bipolar outflows.

## 1. Introduction

Emission from dust has been widely used as a tracer of star formation, as discussed in a recent overview by Chandler & Sargent (1997). An evolutionary classification has been based on the infrared spectral energy distributions of young stellar objects (YSOs), ranging from the deeply embedded class 0 and I phases, via the class II T Tauri phase, to class III pre–main-sequence stars (Lada 1987). Millimeter continuum observations have revealed the presence of circumstellar disks around $\sim 50\%$ of T Tauri stars (Beckwith *et al.* 1990; Osterloh & Beckwith 1995), together with more extended envelopes around many embedded YSOs (Keene & Masson 1990; Terebey, Chandler & André 1993; André & Montmerle 1994).



*J. M. Greenberg and A. Li (eds.), Formulation and Evolution of Solids in Space*, 521–526.

TABLE 1. Source sample

| Source | IRAS PSC | Visible/ Embedded | $L_{bol}$ $L_\odot$ | $M_\star$ [a] $M_\odot$ | $M_{disk}$ $M_\odot$ | $M_{env}$ $M_\odot$ |
|---|---|---|---|---|---|---|
| L1489 IRS | 04016+2610 | Emb | 3.70 | 0.4 | $\leq 0.004$ | 0.016–0.025 |
| T Tau | 04190+1924 | Vis+Emb | 25.50 | 2.7 | 0.023 | 0.029 |
| Haro 6–10 | 04263+2426 | Vis+Emb | 6.98 | 0.9 | 0.010 | $< 0.004$ |
| L1551 IRS 5 | 04287+1801 | Emb | 21.90 | 2.6 | 0.073 | 0.26 |
| L1535 IRS | 04325+2402 | Emb | 0.70 | 0.15 | $\leq 0.005$ | $< 0.010$ |
| TMR 1 | 04361+2547 | Emb | 2.90 | 0.3 | 0.009 | 0.007 |
| TMC 1A | 04365+2535 | Emb | 2.20 | 0.3 | 0.020 | 0.018 |
| L1527 IRS | 04368+2557 | Emb | 1.30 | 0.2 | 0.017 | 0.031 |
| TMC 1 | 04381+2540 | Emb | 0.66 | 0.15 | $\leq 0.005$ | 0.005–0.016 |

[a] Upper limit of stellar mass assuming objects are located on the birthline.

In addition to the spatial distribution of material and the temperature, molecular line observations also probe the density and the velocity field of the gas. This is especially interesting during the embedded phase, when the object continues to accrete matter through infall and at the same time clears its surroundings by the action of a bipolar outflow. The composition of the gas and the dust is affected by these energetic processes, evaporating icy grain mantles and destroying dust particles, thereby influencing the make-up of the primitive solar system.

## 2. Source sample and observations

Our study is based on a sample of nine embedded, low-mass YSOs in the Taurus–Auriga star-forming region ($d$ = 140 pc), chosen by their $HCO^+$ $J$=3–2 emission from the IRAS flux- and colour-limited sample of Tamura *et al.* (1991). Single-dish (sub-)millimeter observations of this sample were obtained in several rotational transitions of $^{12}CO$, $^{13}CO$, $C^{18}O$, $HCO^+$, and $H^{13}CO^+$ using the JCMT[1], the CSO[2], and the IRAM 30m telescope (Hogerheijde *et al.* 1997). Regions of up to $60'' \times 60''$ (10,000 AU) were mapped with beam sizes between $10''$ and $30''$ (1000–4000 AU). The observed lines trace hydrogen number densities of $10^3$–$10^7$ cm$^{-3}$ and kinetic temperatures 10–100 K. Interferometric observations of $HCO^+$, $^{13}CO$, and

[1]The James Clerk Maxwell Telescope is operated by the Joint Astronomy Centre on behalf of the UK Particle Physics and Astronomy Research Council, the Netherlands Organization for Scientific Research and the National Research Council of Canada.

[2]The Caltech Submillimeter Observatory is operated by the California Institute of Technology under funding from the U.S. National Science Foundation (#AST93–13929).

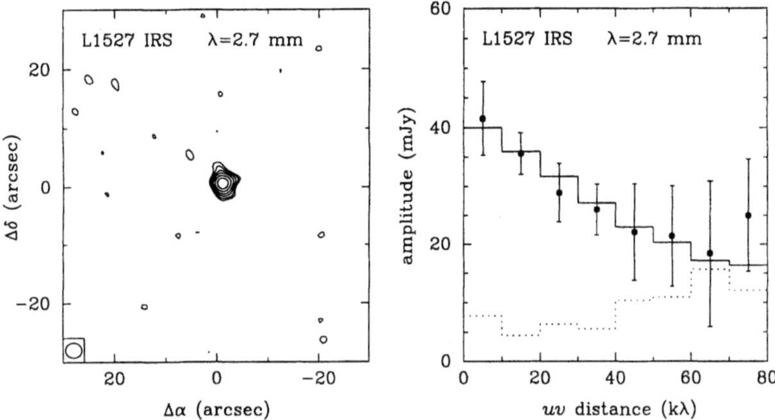


*Figure 1.* Left: cleaned image of 2.7 mm continuum emission of L1527 IRS from OVRO. Contours are drawn at 3 mJy bm$^{-1}$ intervals. Right: 2.7 mm continuum vector-averaged amplitudes of L1527 IRS. The dashed line shows the expectation value in absence of real signal; the solid line is a model fit of a point source plus an extended envelope.

$C^{18}O$ 1–0, as well as $\lambda$=3.4 and 2.7 mm continuum emission were obtained at OVRO[3], with typical synthesized beams of 5″ (700 AU).

## 3. Occurrence of circumstellar disks

Interferometric observations are well suited to investigate the occurrence and growth of disks in the embedded class I phase (cf. Keene & Masson 1990; Terebey *et al.* 1993). Fig. 1 presents the OVRO 2.7 mm continuum measurements of the source L1527 IRS; the left panel shows the cleaned image, the right panel the vector-averaged visibility amplitudes. These amplitudes correspond to the Fourier components of the circular-symmetric part of the sky brightness distribution with respect to the source position, and are especially useful to separate extended and unresolved (i.e., $< 3'' \approx 420$ AU) emission. The flux of the former decreases with increasing baseline separation, while that of the latter stays constant. Both components are clearly present in the measurements of L1527 IRS.

Six of our nine sources show point-source emission, presumably from circumstellar disks, with 2.7 mm fluxes ranging between 11 and 100 mJy. The flux levels toward the three remaining sources were insufficient to constrain the spatial distribution ($< 7$ mJy bm$^{-1}$). The fact that $\geq 67\%$ of embedded YSOs show point-source emission implies that disks are formed very early in te star forming process, $\leq 10^4$ yr after the onset of collapse. No evidence for disk growth during the embedded phase of $\sim 10^5$ yr is found from our data, when comparing the results to the average 2.7 mm

[3]The Owens Valley Millimeter Array is operated by the California Institute of Technology under funding from the U.S. National Science Foundation (#AST93–14079).

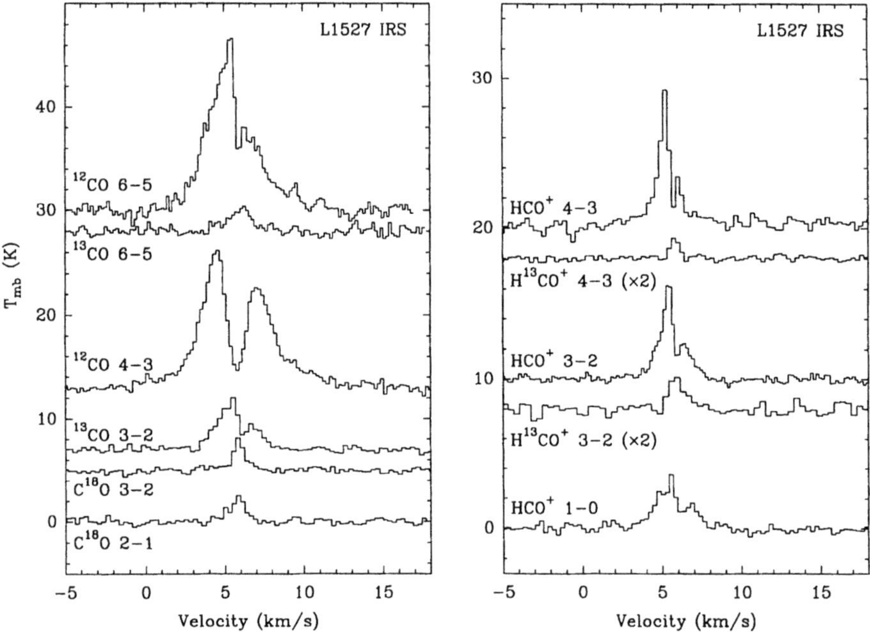


*Figure 2.* Single-dish $^{12}$CO, $^{13}$CO, C$^{18}$O, HCO$^+$ and H$^{13}$CO$^+$ spectra of L1527 IRS.

flux of $\sim$ 24 mJy observed toward more evolved T Tauri stars with ages $\geq 10^6$ yr (Dutrey *et al.* 1996).

The disks are found to contribute 30–75% of the 1.1 mm single-dish flux measured by Moriarty-Schieven *et al.* (1994) in the 19″ JCMT beam. The disk fluxes have been extrapolated using a spectral index of $\alpha = 2.5$, consistent with measurements of L1551 IRS 5, HL Tau and T Tau (Lay *et al.* 1994; Mundy *et al.* 1996; van Langevelde *et al.* 1997). Assuming an average dust temperature of 30 K and optically thin conditions, disk and envelope masses of 0.004–0.073 $M_\odot$ and 0.004–0.26 $M_\odot$, respectively, are inferred. Since the millimeter emission from the disks is likely to be moderately optically thick, it is concluded that the disks and the envelopes are typically equally massive in this phase, with each carrying $\sim$ 10% of the mass of the central star (cf. Table 1).

## 4. Molecular observations of envelopes and outflows

Single-dish molecular line spectra of L1527 IRS are shown in Fig. 2. The line profiles are characterized by a line core of $\Delta V \approx 3$ km s$^{-1}$, line wings originating in the bipolar outflow extending up to 10 km s$^{-1}$ toward the red and the blue, and a $\sim$ 1 km s$^{-1}$ wide self-absorption feature due to cool material in the outer regions of the envelope. In the rarer isotopic species like H$^{13}$CO$^+$ and C$^{18}$O only the line core is sampled. Such spectra

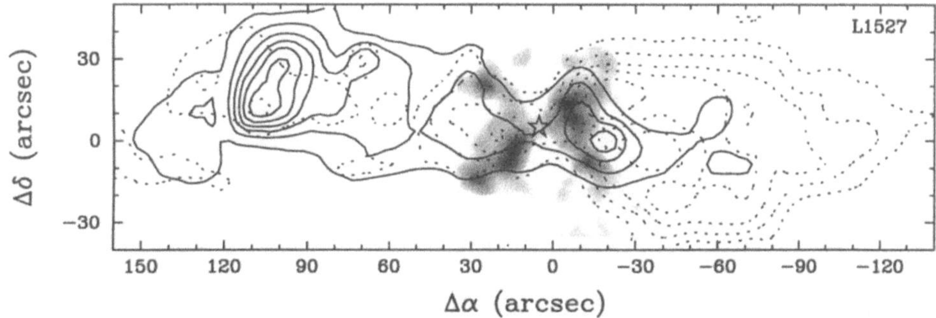


*Figure 3.* Integrated $HCO^+$ 1–0 intensity observed toward L1527 IRS with OVRO (grey scale) superposed on the red (dashed contours) and blue (solid contours) $^{12}CO$ 3–2 outflow lobes (from MacLeod *et al.* 1994).

can be used to test models of protostellar collapse (e.g., Terebey, Shu & Cassen 1984; Galli & Shu 1993; Foster & Chevalier 1993) through a careful treatment of the radiative transfer and the molecular excitation (cf. Zhou *et al.* 1993; Walker, Narayanan & Boss 1994; Choi *et al.* 1995).

Comparison between the millimeter continuum fluxes and the integrated $HCO^+$ and $H^{13}CO^+$ line intensities shows that $HCO^+$ 3–2 and 4–3 are particularly good tracers of the envelopes of our sources. The envelope density distribution is found to follow a radial power law with a slope between 1 and 3, consistent with collapse models.

In the OVRO beam, the $HCO^+$, $^{13}CO$, and $C^{18}O$ emission traces rotation in the equatorial regions of the envelopes, as well as material closely connected with the cavities cleared by the outflow. The best example of this is offered by L1527 IRS (Fig. 3), where the integrated $HCO^+$ emission outlines the red and blue outflow lobes. Whether the outflow actively compresses the material along the cavity walls, or merely provides a low-opacity pathway for heating radiation from the central object to reach into the envelope (cf. Spaans *et al.* 1995) remains subject of further investigation.

## 5. Conclusions

The emission of dust and molecules provides complementary views of the young stellar environment. A detailed study of nine embedded, low-mass YSOs shows that (1) circumstellar disks are formed early in the star formation process and occur around at least two-thirds of all YSOs, (2) $HCO^+$ 3–2 and 4–3 are excellent tracers of the envelopes, and (3) the bipolar outflow has a profound influence in shaping the appearance of the envelopes on 1000 AU scales.

MRH gratefully acknowledges financial support from the *Stimuleringsfonds Internationalisering* of the Netherlands Organization for Scientific Research (NWO) and from the organizers of this School.

## References

1. André, Ph., & Montmerle, T. (1994) From T Tauri stars to protostars: Circumstellar material and young stellar objects in the $\rho$ Ophiuchi cloud. *ApJ*, **420**, pp. 837–862
2. Beckwith, S. V. W., Sargent, A. I., Chini, R. S., & Güsten, R. (1990) A survey for circumstellar disks around young stellar objects. *AJ*, **99**, pp. 924–945
3. Chandler, C. J., Sargent, A. I. (1997) The role of dust in star and planet formation: observations. in *From dust to Planetesimals*, eds. Y. J. Pendleton, A. G. G. M. Tielens, (San Francisco: ASP), in press.
4. Choi, M. H., Evans II, N. J., Gregersen, E. M., & Wang, Y. S. (1995) Modeling line profiles of protostellar collapse in B335. *ApJ*, **448**, pp. 742–747
5. Dutrey, A., Guilloteau, S., Duvert, G., Prato, L., Simon, M., Schuster, K., Menard, F. (1996) Dust and gas distribution around T Tauri stars in Taurus–Auriga. I: Interferometric 2.7 mm continuum and $^{13}$CO $J$=1–0 observations. *A&A*, **309**, pp. 493–504
6. Foster, P. N., & Chevalier, R. A. (1993) Gravitational collapse of an isothermal sphere. *ApJ*, **416**, pp. 303–311
7. Galli, D., & Shu, F. H. (1993) Collapse of magnetized molecular cloud cores. I: Semi-analytical solution. *ApJ*, **417**, pp. 220–242
8. Hogerheijde, M. R., van Dishoeck, E. F., Blake, G. A., & van Langevelde, H. J. (1997) Tracing the envelopes around embedded low-mass young stellar objects with $HCO^+$ and millimeter-continuum observations. *ApJ*, in press.
9. Keene, J., & Masson, C. R. (1990) Detection of a 45 AU radius source around L1551 IRS 5: A possible accretion disk. *ApJ*, **355**, pp. 635–644
10. Lada, C. J. (1987) Star formation: from OB associations to protostars. in *Star Forming Regions*, eds. M. Peimbert, J. Jugaku (Dordrecht: Reidel), pp. 1–17
11. Lay, O. P., Carlstrom, J. E., Hills, R. E., & Phillips, T. G. (1994) Protostellar accretion disks resolved with the JCMT–CSO interferometer. *ApJ*, **434**, pp. L75–L78
12. MacLeod, J., Lorne, A., Harris, A., Tacconi, L., & Schuster, K. (1994) Extended CO 6–5 emission from a nearby outflow source. *JCMT Newsletter*, Sept./Oct. 1994
13. Moriarty-Schieven, G. H., Wannier, P. G., Keene, J., & Tamura, M. (1994) Circumprotostellar environments II. *ApJ*, **436**, pp. 800–806
14. Mundy, L. G., Looney, L. W., Erickson, W., Grossman, A., Welch, W. J., Forster, J. R., Wright, M. C. H., Plambeck, R. L., Lugten, J., & Thornton, D. D. (1996) Imaging the HL Tauri Disk at $\lambda = 2.7$ mm with the BIMA array. *ApJ*, **464**, pp. L169–L173
15. Osterloh, M., & Beckwith, S. V. W. (1995) Millimeter-wave continuum measurements of young stars. *ApJ*, **439**, pp. 288–302
16. Spaans, M., Hogerheijde, M. R., Mundy, L. G., van Dishoeck, E. F. (1995) Photon heating of YSO envelopes as an explanation for CO 6–5. *ApJ*, **455**, pp. L167–L170
17. Tamura, M., Gatley, I., Waller, W., & Werner, M. W. (1991) Two micron morphology of candidate protostars. *ApJ*, **374**, pp. L25–L28
18. Terebey, S., Shu, F. H., & Cassen, P. (1984) The collapse of the cores of slowly rotating isothermal clouds. *ApJ*, **286**, pp. 529–551
19. Terebey, S., Chandler, C. J., & André, Ph. (1993) The contribution of disks and envelopes to the millimeter continuum emission from very young low-mass stars. *ApJ*, **414**, pp. 759–772
20. van Langevelde, H. J., van Dishoeck, E. F., & Blake, G. A. (1997) Dust and CO in the T Tauri circumstellar matter. in *CO: Twenty-five Years of Millimeter-wave Spectroscopy, eds. W. B. Latter et al.*(Kluwer, Dordrecht), p. 469
21. Walker, C. K., Narayanan, G., & Boss, A. P. (1994) Spectroscopic signatures of infall in young protostellar systems. *ApJ*, **431**, pp. 767–782
22. Zhou, S., Evans II, N. J., Kömpe, C., & Walmsley, C. M. (1993) Evidence for protostellar collapse in B335. *ApJ*, **404**, pp. 232–246

# RELEASE MECHANISMS OF VOLATILE MOLECULES IN AMORPHOUS $H_2O$ ICE

SIN-ITI SIRONO AND TETSUO YAMAMOTO
*Department of Earth and Planetary Sciences,*
*Hokkaido University, Sapporo 060, Japan*

**Abstract.** Data of the sublimation experiment of $H_2O$-CO ice are analyzed. The following two possible processes that govern CO release from amorphous $H_2O$ ice are examined separately; one is diffusion of CO molecules in amorphous ice, and the other is CO release associated with a phase transition of amorphous $H_2O$ ice. The peaks in the flux of evolved CO molecules with increasing temperature observed in the experiment are reproduced theoretically. We derive the relation between the temperature at the peak and the activation energy for both processes, and estimate the activation energies of CO diffusion in amorphous ice and of a phase transition of amorphous ice. Brief discussion is given on astrophysical implications.

## 1. Introduction

It is considered that a grain in molecular clouds consists of a silicate core, an organic mantle and an ice mantle (Greenberg et al. 1989). The ice mantle is a mixture of various species such as $H_2O$, CO, and $CO_2$ as clearly revealed by the ISO observations (see Whittet et al. 1996). The amount of the volatile molecules in the ice would depend on the formation conditions of the ice and on the temperature of the grains (Tielens et al. 1991). Laboratory experiments show that volatile molecules are released from the surface of the ice film containing various species with increasing temperature (Kouchi, 1990, Bar-Nun et al. 1985, Sandford and Allamandola 1988). It is remarkable that degassing of volatile molecules do not occur at a constant rate but is observed at several temperatures. We examine two possible release mechanisms of volatile molecules. One is diffusion of volatile molecules through $H_2O$ ice, and the other is squeezing out of volatile molecules associated with phase transitions of $H_2O$ ice.



*J. M. Greenberg and A. Li (eds.), Formulation and Evolution of Solids in Space,* 527–534.

The activation energies in both processes are a crucial physical quantity that determines the release rate of volatile molecules and the temperatures at which remarkable degassing is observed. In astrophysical context the activation energies play a key role in determining the composition of the ice mantle of a grain in molecular clouds, and are also relevant to the thermal evolution of cometary nucleus since the released volatile molecules acts as a cooling agent of the nucleus (Prialnik and Podolak 1995).

The aim of this paper is to analyze the experimental data and to present a method of deriving the activation energy of diffusion of volatile molecules in amorphous $H_2O$ ice and that of a phase transition of amorphous $H_2O$ ice.

## 2. Activation energy of diffusion

Figure 1 shows the result of the experiment (Kouchi 1990), where amorphous $H_2O$ ice containing 20 % of CO is heated from 10 K at a rate of $1\,\mathrm{K\,min^{-1}}$. Note the peaks in the flux of CO molecules released into a vacuum chamber at several temperatures. Different peaks are considered to result from different preservation mechanisms of CO in the ice. Kouchi (1990) suggests that the peak $i$ is due to a phase transition of amorphous $H_2O$ ice to crystalline ice Ic and and the peak $f$ due to a transition from $H_2O_{as}$(I) ice to $H_2O_{as}$(II). Sandford and Allamandola (1988), on the other hand, consider that the peak $i$ results from diffusion of CO in amorphous $H_2O$ ice. The peak $c$ is due to sublimation of pure CO frozen on the $H_2O$ ice. We examine the peaks $i$ and $f$. In this section we determine activation energy of CO diffusion under the assumption that diffusion of CO in amorphous $H_2O$ ice leads to the peaks $i$ and $f$ in the released CO flux. The case of the phase transition will be discussed separately in the next section. Although the discussion is devoted to CO molecules, the same argument can be applied to other molecular species as well.

Let an amorphous $H_2O$ ice film of thickness $l$ containing CO molecules in it be placed at $0 \leq x \leq l$, where $x = 0$ is the surface of a substrate on which the ice is deposited. The number density of CO molecules contained in the ice film at position $x$ at time $t$ is denoted by $n(t, x)$, which is determined by the diffusion equation given by

$$\frac{\partial n}{\partial t} = D(T)\frac{\partial^2 n}{\partial x^2}, \tag{1}$$

where we have ignored the temperature gradient in the film and heat of desorption of CO molecules. $D(T)$ is the diffusion coefficient as a function of temperature $T$, and is written as

$$D(T) = D_0 e^{-E_{\mathrm{df}}/T}, \tag{2}$$

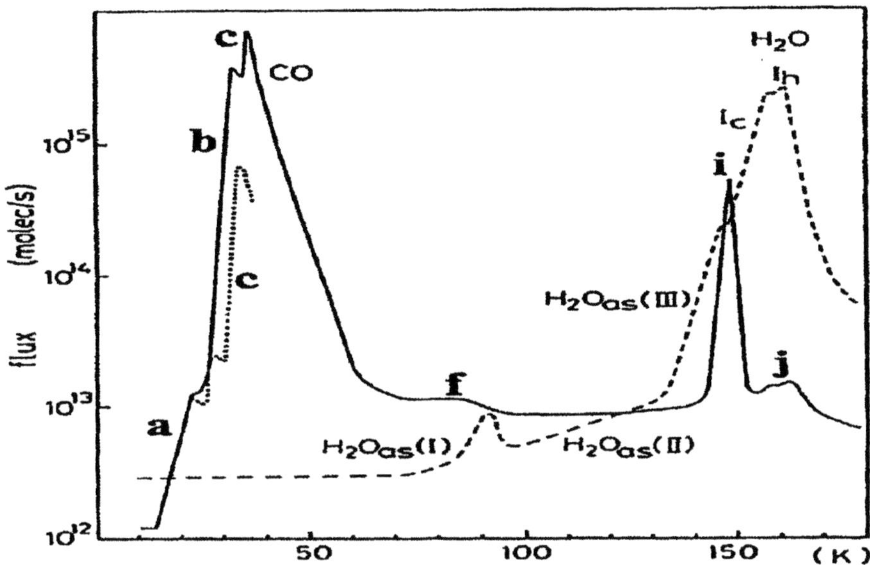


*Figure 1.* Plot of evolved CO from the surface of an $H_2O$ ice film (solid curve). The flux of $H_2O$ molecules is shown by a dashed curve. A dotted curve shows the flux of CO deposited after the formation of the $H_2O$ ice film (from Kouchi, 1990).

where $E_{\rm df}$ is the activation energy of diffusion. The temperature is raised at a constant rate $\alpha$ as in the experimental conditions:

$$T = T_0 + \alpha t. \tag{3}$$

It is assumed that CO molecules are distributed uniformly in the ice film initially, namely $n(0, x) = n_0$ at $t = 0$. The boundary condition at the bottom of the ice film ($x = 0$) is given by

$$\partial n/\partial x|_{x=0} = 0. \tag{4}$$

The boundary condition at the surface ($x = l$) is determined by the balance of the flux of CO molecules diffusing out to the surface of the ice film and that of desorption:

$$J = -D(T) \left.\frac{\partial n}{\partial x}\right|_{x=l} = n(t, l) d\nu e^{-E_{\rm des}/T}, \tag{5}$$

where $E_{\rm des}$ is desorption energy of CO from amorphous $H_2O$ ice, $\nu$ is vibrational frequency of CO perpendicular to the ice surface, and $d$ is diameter of an $H_2O$ molecule. From eqs. (5) and (2), we have

$$-\left.\frac{\partial n}{\partial x}\right|_{x=l} = \frac{n}{l}\frac{l d\nu}{D_0} e^{(E_{\rm df} - E_{\rm des})/T}. \tag{6}$$

In general $E_{df} > E_{des}$ and thus $ld\nu e^{(E_{df}-E_{des})/T}/D_0 \gg 1$ at relevant temperatures, implying that molecules diffusing out to the surface of the ice film desorb quickly compared with the diffusion time scale. Furthermore, the left-hand side $\partial n/\partial x|_{x=l}$ is on the order of $n_0/l$ at relevant time of $t \sim l^2/D$, as is confirmed later (see eq.(11)). Thus $n(t,l)$ is much less than $n_0$ and may be put as

$$n(t,l) = 0. \tag{7}$$

Let us express the time $t$ in terms of $\tau$ defined by

$$\tau = \int_0^t D(\tau')d\tau', \tag{8}$$

where the diffusion constant is a function of time through eq. (2) with eq. (3). The diffusion equation (1) is then expressed by

$$\frac{\partial n}{\partial \tau} = \frac{\partial^2 n}{\partial x^2}. \tag{9}$$

Carrying out Laplace transformation of eq. (9) and integrating its result with the boundary conditions (4) and (7), we have

$$-\frac{d\hat{n}(p,l)}{dx} = \frac{n_0}{\sqrt{p}}\tanh(\sqrt{p}l), \tag{10}$$

where $\hat{n}(p,l) = \int_0^\infty d\tau e^{-p\tau} n(\tau,l)$ is the Laplace transformation of $n(\tau,l)$. The inverse transformation of eq. (10) gives the flux $J$ of CO as

$$J = -D(T)\left.\frac{\partial n}{\partial x}\right|_{x=l} = \frac{D(T)n_0}{l}\sum_{n=0}^{\infty} e^{-(2n+1)^2\pi^2\tau/4l^2}. \tag{11}$$

It is sufficient to retain the first term in the right-hand side of eq. (11) since we are concerned with the time around $\tau/l^2 \sim 1$. The flux $J$ is thus approximated by

$$J = \frac{D_0 n_0}{l}\exp\left(-\frac{E_{df}}{T} - \frac{\pi^2}{4l^2}\int_0^t D(\tau)d\tau\right). \tag{12}$$

It should be pointed out that this expression of $J$ can be applied not only to the laboratory conditions but also to the evolution of cometary or interstellar grains heated up gradually.

Equation (12) indicates that, with increasing temperature, the flux $J$ reaches a peak at some temperature $T_p$. Differentiating the exponent of eq. (12) with respect to $t$, one obtains the relation between the peak temperature $T_p$ and the activation energy $E_{df}$ of diffusion as

$$\frac{E_{df}\alpha}{T_p^2} = \frac{\pi^2 D_0}{4l^2} e^{-E_{df}/T_p} \tag{13}$$

The width $\Delta T$ of the peak is estimated by expanding the exponent of eq. (12) to the second order, and is written as

$$\Delta T = \frac{\sqrt{2} T_p^2}{E_{df}}. \tag{14}$$

¿From eqs. (13) and (14), one may obtain $E_{df}$ and $D_0$ in the diffusion constant given by eq. (2) using $T_p$ and $\Delta T$ observed in the experiment. Actually, large uncertainty is involved in determining the width $\Delta T$ of the peaks from the experimental data. Here we tentatively adopt $D_0 = 330\,\mathrm{cm^2\,s^{-1}}$ (Sandford and Allamandola 1988) and estimate the activation energy of diffusion $E_{df}$ from eq. (13). Figure 2 shows $E_{df}$ as a function of the peak temperature $T_p$ (eq. (13)), where we put $\alpha = 1\,\mathrm{K\,min^{-1}}$ and $l = 0.5\,\mu\mathrm{m}$ (Kouchi, 1990). It can be seen that $E_{df}$ varies almost linearly with $T_p$. For $D_0 = 330\,\mathrm{cm^2\,s^{-1}}$, one obtains $E_{df} = 4700\,\mathrm{K}$ for the peak $i$ ($T_p = 145\,\mathrm{K}$) and 2700 K for the peak $f$ ($T_p = 80\,\mathrm{K}$). The value of $E_{df}$ for the peak $i$ is lower than the value of $E_{df} = 5500\,\mathrm{K}$ obtained by Sandford and Allamandola (1988).

## 3. Phase transition

The coincidence between the peak temperatures of evolved CO and the temperatures at which the phase transitions of amorphous $H_2O$ ice occur (Fig. 1) suggests that release of CO molecules from amorphous ice is somehow associated with phase transitions of the ice at least for the peak $i$. In this section, we shall estimate the activation energy of the phase transition assuming that the CO release is due to the phase transitions.

We assume that the flux of CO molecules released is proportional to the transition rate $\partial\xi/\partial t$ of amorphous $H_2O$ ice, where $\xi$ is the mass fraction of the phase after the transition. The flux $J$ of the evolved CO is then written as

$$J \propto \frac{\partial \xi}{\partial t} = \frac{1-\xi}{A} e^{-E_c/T}, \tag{15}$$

where $E_c$ is the activation energy of the phase transition, and $A$ is time scale of molecular motion such as a period of lattice vibration. Integrating eq. (15), one has $J$ as

$$J = C \exp\left(-\frac{E_c}{T} - \int_0^t e^{-E_c/T} d\tau\right), \tag{16}$$

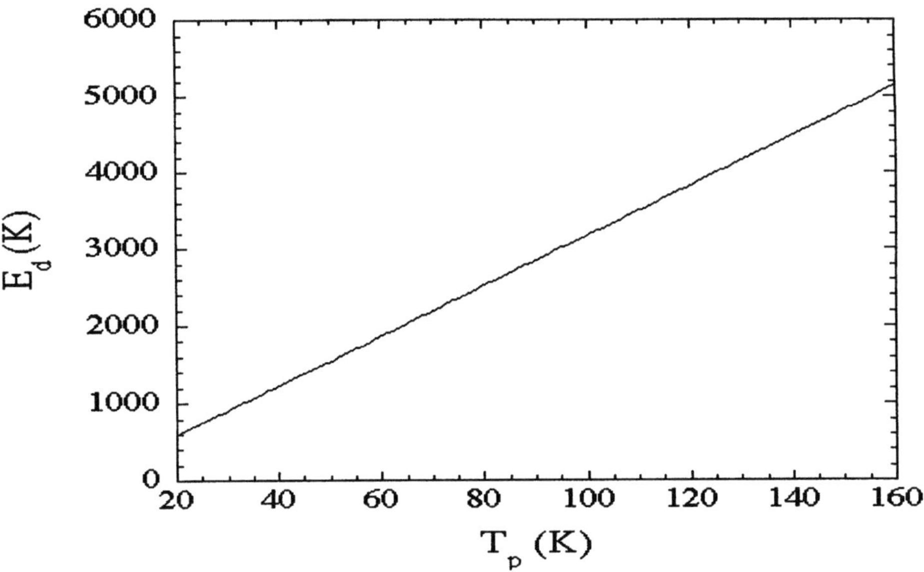


*Figure 2.* The relation between $E_{df}$ and $T_p$ (eq. (13)), where $D_0 = 330\,cm^2\,s^{-1}$ (Sandford and Allamandola 1988), $\alpha = 1\,K\,min^{-1}$, and $l = 0.5\,\mu m$ (Kouchi, 1990).

where $C$ is a constant. Note that this $J$ is of the same form as the $J$ given by eq. (12). Differentiating the exponent of eq. (16) with respect to $t$, the relation between the activation energy $E_c$ and the peak temperature $T_p$ is obtained to be

$$\frac{E_c\alpha}{T_p^2} = \frac{1}{A}e^{-E_c/T_p}. \tag{17}$$

The relation (17) is identical to eq. (13) except that the time scale of diffusion $4l^2/\pi^2 D_0$ is replaced by the time scale $A$. In deriving eq. (17) we have neglected the latent heat that might be deposited at the phase transition. This approximation is valid for the peak $i$, since the time scale of heat conduction $\tau_{cond} = l^2/\chi$ is much shorter than the time scale of heating due to the latent heat deposition $\tau_{heat} = c_p T/L(\partial\xi/\partial t)$ for $T < 190\,K$ in crystallization of amorphous ice, where $\chi$ is thermal diffusivity of amorphous ice, $c_p$ is its specific heat, and $L$ is latent heat of crystallization.

The activation energy $E_c$ for the peak $i$ ($T_p = 145\,K$) is estimated from eq. (17) to be $E_c = 5200\,K$ for $A = 9.54 \times 10^{-14}\,s$ (Schmitt et al. 1989). This is in good agreement with $E_c = 5370\,K$ measured by Schmitt et al. (1989).

## 4. Implications for interstellar ices

From the activation energy $E_{df}$ obtained in §2, we calculate residence time of CO in an amorphous $H_2O$-ice grain mantle with thickness of $0.1\,\mu m$ in

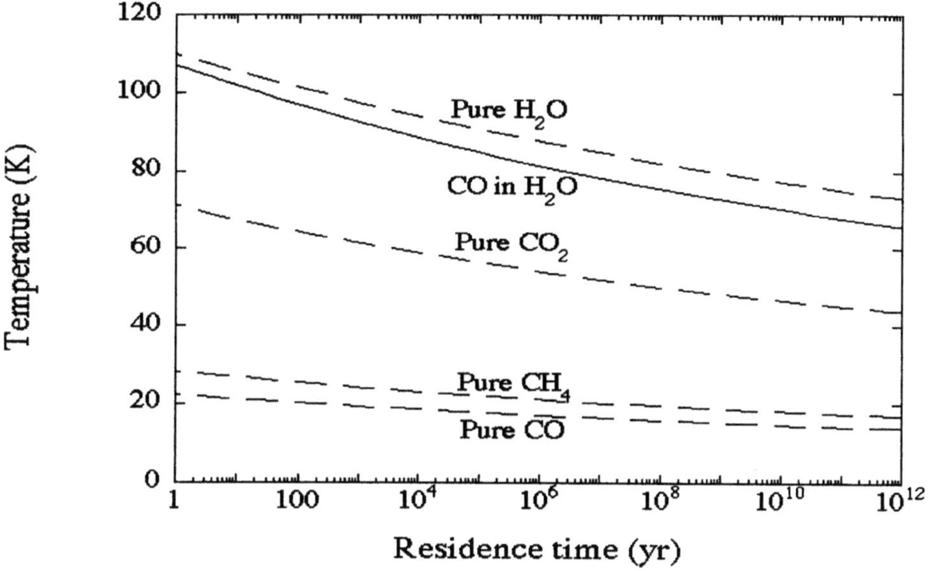


*Figure 3.* Residence time of CO molecules in amorphous $H_2O$ ice of 0.1 $\mu$m-thickness as a function of temperature (solid curve), compared with the residence time for pure CO, $CO_2$, $CH_4$ and $H_2O$ ices (dashed curve). The data of vapor pressure are taken from Yamamoto et al. (1983).

molecular clouds. The result is shown in Fig. 3, in which sublimation time scales of pure CO, $CO_2$, $CH_4$, and $H_2O$ ices are also plotted. It can be seen that the residence time of CO in the amorphous $H_2O$ ice of 80 K is $3.1 \times 10^6$ yr, which is on the same order of magnitude as typical time scale of molecular clouds. Namely an upper limit of the temperature for the presence of CO in polar ice in molecular clouds is 80 K. Furthermore, it should be noted that the residence time of CO in polar ice is longer than the sublimation time scale of pure $CO_2$, which is much less volatile than pure CO. This implies that, if both apolar ice dominated by $CO_2$ and polar ice with 10% of CO are heated gradually, the apolar ice is disappeared first and then the release of CO occurs, and finally $H_2O$ ice is sublimated. The grain temperature may thus be estimated by comparing the observations with Fig. 3. For example, if we observe the grains with polar ice containing CO and without apolar ice dominated by $CO_2$, the grain temperature is constrained between 60 and 80 K.

## 5. Conclusions

The results of the experiment of evolved CO molecules from amorphous $H_2O$ ice are analyzed. It has been shown that the peaks of evolved CO from amorphous $H_2O$ ice are reproduced theoretically by taking account of either diffusion of CO in amorphous $H_2O$ ice or squeezing out of CO associated with phase transitions of amorphous $H_2O$ ice. The microphysics of the release mechanism is still uncertain to determine which is the actual case. We have derived the relation between the activation energy and the temperatures at the peaks for both cases, for which the relations are of the same form. The activation energy of crystallization of amorphous $H_2O$ ice derived from the peak $i$ is in good agreement with the measurement of Schmitt et al. (1989). The residence time scale of CO in amorphous $H_2O$ ice is calculated with the use of the result obtained in the present study. It is shown that the residence time scale of CO in amorphous $H_2O$ ice is longer than that of sublimation of pure $CO_2$. It is suggested that the temperature of interstellar ices can be estimated from comparison of the observational data with the time scale thus calculated.

## Acknowledgment

The authors thank A. Kouchi for enlightening discussion and Mayo Greenberg for inviting them to the NATO ASI on Space Chemistry, where this paper was presented. S.S acknowledges financial support from JSPS. This work was supported in part by Grant-in-Aid for Scientific Research from the Japanese Ministry of Education, Science and Culture.

## References

1. Bar-Nun, A., Herman, G., Laufer, D. and Rappaport, M. L. Trapping and release of gases by water ice and implications for icy bodies. *Icarus*, **63**, 317-332, 1985.
2. Greenberg, J. M., Zhao, N. S. and Hage, J. I. The interstellar dust mode of comet dust constrained by 3.4 $\mu$m and 10 $\mu$m emission. *Adv. Space. Res.* **9**(3), 3-11, 1989.
3. Kouchi, A. Evaporation of $H_2O$-$CO_2$ ice and its astrophysical implications. *J. Cryst. Growth*, **99**, 1220-1226, 1990.
4. Prialnik, D., and Podolak, M. Radioactive heating of porous comet nuclei. *Icarus* **117**, 420-430, 1995.
5. Sandford, S. A. and Allamandola, L. J. The condensation and vaporization behavior of $H_2O$:CO ices and implications for interstellar grains and cometary activity. *Icarus*, **76**, 201-224, 1988.
6. Schmitt, B., Espinasse, S., Grim, R. A. J., Greenberg, J.M. and Klinger, J. Laboratory studies of cometary ice analogues. *ESA SP* 278, pp. 613-619, 1989.
7. Tielens, A. G. G. M., Tokunaga, A. T., Geballe, T. R. and Bass, F. Interstellar solid CO: Polar and nonpolar interstellar ices. *Astroph. J.*, **381**, 181-199, 1991.
8. Whittet, D. C. B., Schutte, W. A., Tielens, A. G. G. M., Boogert, A. C. A., de Graauw, Th., Ehrenfreund, P., Gerakines, P. A., Helmich, F. P., Prusti, T. and van Dishoeck, E. F. An ISO SWS view of interstellar ices–First results. *Astron. Astroph.* **53**, 197-214, 1996.
9. Yamamoto, T., Nakagawa, N. and Fukui, Y. The chemical composition and thermal history of the ice of a cometary nucleus. *Astron. Astroph.*, **122**, 171-176, 1983.

# SCATTERING BY COMPLEX SYSTEMS

## *I: METHODS*

B. Å. S. GUSTAFSON
*Dept. of Astronomy, University of Florida*
*211 SSRB, Gainesville, Fl 32611, USA*

## 1. Introduction

In combination with the long range and complex interaction with matter, the ability to cross empty space makes electromagnetic radiation literally a low cost "space probe". While, much of the message carried by scattered light can be routinely understood, the part that is due to scattering by "complex" structures is hard to decipher. This is to a large extent due to the infinite possible combinations of the relevant particle parameters that include internal structure as well as the commonly considered parameters of refractive index, size, shape, and orientation. The lack of a comprehensive theoretical solution is another reason. This text is in part motivated by the realization that material morphology, on scales comparable to the wavelength of visible light, may bear witness of the conditions during grain growth in a broad variety of cosmic environments.

We first illustrate some fundamental properties of the scattering process by considering scattering by one of the simplest systems - a single atom. We then outline the assumptions made and principles used to derive two commonly used classes of solutions. One of them is exemplified by Mie's [1] classic "exact" solution to the scattering by a homogeneous sphere. The other is the widely used Coupled Dipole Approximation by Purcell and Pennypacker [2]. Finally, we describe the microwave analog method to investigate electromagnetic scattering experimentally. The objective is to convey a sense of the physics of scattering and how it is represented in methods that are commonly used to investigate light scattering. The text is limited to broad principles and there is no claim of completeness.

## 2. Light and the Scattering System

The ability of electromagnetic radiation to propagate through empty space is a consequence of the fact that charges cannot be created and cannot disappear in empty space. While stationary charges and charges moving at a constant rate create a rapidly decaying electric field, use of the Lorentz transformation of General Relativity shows that accel-



*J. M. Greenberg and A. Li (eds.), Formulation and Evolution of Solids in Space,* 535–547.

erating charges create fields at a distant point that are proportional to the angular acceleration as seen from that vantage point. The fields are therefore inversely proportional to the distance which explains the long range of electromagnetic radiation. The intensity which is proportional to the square of the field strength decreases with the inverse square of the distance, as it must to conserve total energy.

When a charge is illuminated, it accelerates in response to the incident field and thus emits light; scattered light. Because the acceleration depends on the bond to nearby charges and because scattering from any nearby charges interfere, a record of the environment is carried across space encoded in the scattered radiation. We start with one of the simplest natural scattering systems; an isolated atom.

## 2.1. SCATTERING BY A SIMPLE SYSTEM; AN ISOLATED ATOM

We get immediately into trouble when we attempt to describe the interaction of electromagnetic radiation and matter using classical electrodynamics. This is because matter consists of atoms and atoms could not exist classically. Realizing that the electron moving through the electric field surrounding a proton is constantly accelerated, we know that it should emit light. In doing so, it loses energy and should eventually fall onto the proton. Quantum mechanics replaces this paradox using concepts such as the famous Heisenberg uncertainty relation between location and momentum of the electron. That discussion is covered in many text books and needs not be repeated in this text. Our starting point is thus that atoms exist and possess given properties as an external field is applied.

Atoms, where the negative and positive charges balance each other on a macroscopic scale, i.e., atoms with overall charge neutrality, still possess an instantaneous microscopic charge imbalance due to an average separation $l$ between charges. This separation averages out over time for single atom molecules unless the electron cloud is distorted by an external field or by interference from nearby charges. An external field can be of any origin including electromagnetic radiation. Although the electron response lags a time varying external field by a finite amount due to inertia and while quantum processes also affect the response, these effects usually only become important at high frequencies. We consider the frequencies of visual light or lower to avoid these complications and so that the field does not vary appreciably across the atom. Experiments suggest that a restoring force acts on the electron cloud and increases linearly with distance from the nucleus under these conditions. This, together with their mass $m$ makes electrons behave like harmonic oscillators with a resonant frequency $\omega_o$, so that charges in an atom can be treated as though they are linked by springs. The equation for the electron's motion may then be written

$$m\left(\frac{d^2 l}{dt^2}+\omega_o^2 l\right)=F \tag{1}$$

where $l$ is the displacement due to the force $F=q_eE$ acting on the electron's charge $q_e$. We use the complex notation $E=E_oe^{i\omega t}$ for the incident electromagnetic wave where $E_o$ is the strength or amplitude of the electric field, $\omega$ the frequency and $t$ the time. Substituting this force, the solution to Eq. 1 is

$$l=\frac{q_eE_o}{m\left(\omega_o^2-\omega^2\right)}e^{i\omega t} \tag{2}$$

We see that the displacement $l$ is proportional to $E_o$ and depends on the charge to mass ratio of the electron, the resonant frequency of the oscillator and the frequency of the incident field. The displacement varies with the same frequency as the incident field.

Note that a force of the same magnitude as that acting on the electron but opposite in direction acts on the nucleus so that the total force on the atom due to the interaction with the electric field averages out at all times. The displacement is inversely proportional to the mass so that the motion of the nucleus is negligible compared to that of the electron cloud.

Because the separation between charges varies with the incident field, the motion of these charges mimics the acceleration of whatever charges are the source of the incident radiation. Therefore their motion is such that they also emit radiation. The intensity of emitted radiation is proportional to the incident intensity. The frequency is the same as the emission from the distant charges. This scattering phenomenon in which the frequency is conserved is called elastic scattering and is the primary subject of this text.

As the charge separation oscillates in response to the incident field, the direction to the charges undergoes angular accelerations that vary depending on the vantage point. It is easy to see that the response to the E-field produces a field of scattered radiation that depends on the projection of the incident E-field. The amplitude is proportional to the incident field and to the sine of the angle between the direction vector and the incident E-field. Equal field strength surfaces are at a distance given by the sine of the angle between the direction vector and $\boldsymbol{l}$ and are thus rotational symmetric about the vector $\boldsymbol{l}$. This is precisely the description of a dipole field. If the charge difference between the nucleus and its set of $n$ electrons is $2nq_e$ and if the charges are separated by the distance $l$, they create a dipole with a dipole moment $p=2nq_el$.

We consider the interaction with the incident magnetic field before we proceed to discuss the scattered radiation by sets of atoms. The changing electric field represented by the vector $\boldsymbol{E}$ cannot exist without an associated magnetic field

$$\boldsymbol{B}=\boldsymbol{k}\times\boldsymbol{E}/c \tag{3}$$

where the Poynting vector $\boldsymbol{k}$ is the unit vector along the direction of propagation. The field $\boldsymbol{B}$ is normal to the electric field and thus normal to the velocity vector $\boldsymbol{v}$ acquired by the charge as it responds to the electric field. The force $\boldsymbol{B}\times\boldsymbol{v}q_e$ acting on the electron is a factor $v/c$ weaker than that due to the E-field and is usually neglected when electrons move at sub-relativistic speeds, as in the atom. We notice that the direction of the electron's induced motion from the E-field is antiparallel to ***E*** and normal to ***k***. As ***E*** reverses direction, ***v*** and ***B*** also reverse. Consequently, the force induced by the magnetic field always acts along the vector ***k*** except at the very moment of motion reversal when the force vanishes. The force acting on the nucleus is weaker because the nucleus' motion in response to the electric field is slower than that of the lower mass electron. Although small, the force acts in the same direction as that on the electron and adds to it.

Weak as the magnetic force may be, it becomes important because the force consistently acts along a given direction. This can be viewed as the physical process whereby momentum is transferred to the atom. One may claim that because scattering changes the momentum of the incident light, no scattering could take place unless the magnetic force transferred momentum to the atom. Interaction with the magnetic field allows the total momentum to be conserved. However, it may be better to view the electric and magnetic interaction as inseparable descriptions of a single physical process. Note that radiation force components in other directions can also result as the scattering body becomes more complex or whenever the oscillator strength is anisotropic, e.g., ref. [3].

We have seen that the scattering of visual light by a single atom creates a dipole scattered field of a frequency dependent magnitude. In addition, a radiation pressure force acts on the scatterer. The scattering of light at a given frequency is thus generally associated with energy transfer to radiation pressure. As a result, the energy at a given frequency is not conserved during the scattering process. Stated differently, all scattering bodies have a finite absorption and distribute the absorbed energy into motion or heat and into radiation at other frequencies. We have also seen that the energy distribution must vary with the frequency of the incident radiation and that this dependency is predictable.

## 2.2. SCATTERING BY A PARTICLE

At the atomic level, all matter consists of positive and negative charges punctuating a volume that is mostly vacuum. As before, we describe each charge as a harmonic oscillator, in a classic representation of matter following Lorentz. Many textbooks describe the electrodynamic properties of bulk material based on interacting and energy dissipating oscillators. This leads to a phenomenological interpretation of bulk matter and of the complex optical constants used to describe bulk matter, e.g., ref. [4], Chapter 9. For our purposes it is sufficient to notice that charges interact in what can be viewed as electromagnetic coupling or alternatively as mutual illumination. Because the speed of light is finite, this coupling or higher order illumination acts as a retardation of the response to the incident field. The scattering by a particle therefore depends, not only on the oscillator strength associated with its individual charges but also on the oscillator strength and configuration of nearby charges. This is one reason for the dependency of scattering on the shape, size, and any internal structure such as cavities or inclusions of other materials. The properties of the scatterer on the scale of the interactions, i.e., on the scale of the wavelength, is thus encoded in the scattered light.

2.2.1. *The refractive Index*

The optical constants that can be obtained through the Lorentz representation using harmonic oscillators have properties that are formally identical to those of quantum mechanical treatments, although the interpretation is not always the same [4, 5, 6]. The main results are that the electrodynamic properties of bulk material at a given frequency can be described using a single complex number. This is either the refractive index $m = n - in'$, or equivalently, the dielectric constant $\varepsilon = \varepsilon' - i\varepsilon''$. (The components for nonmagnetic materials are related through $\varepsilon' = n^2 - n'^2$ and $\varepsilon'' = 2nn'$.) In addition, this treatment leads to definite relations between the frequency dependence of the real and imaginary components known as the Kramers-Kronig relations.

We seek a theoretical justification for representing the optical properties of a material using the refractive index by inquiring as to how the presence of matter might modify Maxwell's equations. We like to know if this change can be uniquely described using a complex number.

The charges associated with individual atoms can be described as charge and current densities that vary over atomic distances, i.e., $10^{-10}$ m, e.g., Scharf, [5]. We note that $10^{-2}$ micron size and larger bodies contain over $10^6$ atoms so that spatial averaging may be justified although temporal averaging is not justified, due to the long time scale of $10^{-13}$ sec associated with thermal motion. Under these conditions, a set of equations that resemble Maxwell's equations can indeed be derived for the fields inside a material accounting for the charge density and the current density, ref. [5], Chapter 4. The charge density depends on the microscopic separation of charges in response to the external field, i.e., on the electric dipole moment per unit volume. A field known as the electric displacement is then established inside the material because the macroscopic electric field is displaced by the polarization as given by simple vector addition. A corresponding "displacement" of the magnetic field due to magnetization of the material (usually referred to as the macroscopic magnetic moment density) is, by way of the Lorentz equation, associated with the macroscopic current density. Because these two new fields appear in the field equations inside the material, the system is no longer closed the way it is when Maxwell's equations are written for empty space. To close the system, we need a description of how the dipole moment and magnetic moment depend on the applied electric and magnetic fields.

If the fields are weak so that the perturbation of the restoring force acting on electrons due to surrounding atoms is negligible and the material is free of permanent electric and magnetic polarization, i.e., not ferroelectric or ferromagnetic, a simple linear relationship is usually assumed. This means that both the polarization and magnetization vanish in the absence of an external field and increase linearly with the applied external field as is observed. These proportionality factors can be written as a complex number. Properly scaled, this number can be interpreted as the refractive index.

We outlined the reasoning behind a description of the optical properties of a material in terms of the effects of the *volume averaged* charges and currents associated with atoms and molecules. This is an approximation to the *discretely* distributed charges and the result can be represented using a complex number. When this number is written as a refractive index, it must have a finite imaginary part corresponding to absorption due to transfer of energy to Brownian motion, radiation pressure, and to other frequencies. This complex number must depend on the frequency and is also expected to change at sufficiently high field strengths, or the representation may break down.

A few comments are in order before we leave this subject. It is well known that interactions between the atoms or molecules in a material strongly affect the polarizability or creation of a dipole moment in response to an external field, and that coupling to thermal lattice vibrations leads to increased absorption. Attempts to derive these interactions using an effective internal field to describe the effect of all the other dipoles on a given dipole, such as the Clausius-Mossotti formula that includes a temperature dependent term, have been only partially successful. However, relations between the change of the real and imaginary part with frequency, the Kramers-Kronig relations, can be derived based on causality, as described in ref. [5], Chapter 4, and are found to be robust. For example, as electrons pick up velocity from the applied field, currents increase and collisions become more frequent. This leads to greater absorption until an equilibrium that limits the current is established. Another observation is that the displaced field can only depend on the applied fields at earlier times so that there is a finite time-lag.

Anisotropic materials that are composed of ordered molecules (crystals) such as Quartz and Graphite have polarizability tensors that depend on the direction of the applied field. Such materials can be described using two or more refractive indexes. We note that the displaced fields in multirefrengent materials are not necessarily parallel to the externally applied fields.

Materials where the molecular bonds are important often have oscillator resonance frequencies in the infrared range of the spectrum that show up as absorption bands and makes this part of the spectrum important for identifying materials. Molecules of dissimilar atoms may possess a permanent finite separation. A good example is the water molecule where both hydrogen atoms tend to reside on the same side of the oxygen atom and the different electron affinities of the atoms create a net dipole moment. The permanent dipoles can be aligned by an external field and also act as harmonic oscillators. Molecules may thus have more than one resonant frequency. At the frequencies of visual light, the induced deformation of the electron cloud dominates, while the realignment of permanent dipoles becomes important at lower frequencies. We thus see how the frequency dependence of scattering can reflect material properties. See ref. [5] for a phenomenological treatment of electrodynamics in simple matter and ref. [6] for a more detailed treatment.

## 3. Methods to Solve the Scattering Problem

In classic electrodynamics, the electric field in any physical circumstance can be represented by the vector sum of the fields from all charges in the universe. Known as the principle of superposition, this makes the evaluation of the scattered field from a particle a straightforward vector addition of the scattered fields from all its dipole charges. However, we must know the field and oscillator strength at each dipole. One family of solutions to the scattering problem attempts to directly evaluate these fields. Another set of solutions use the refractive index to represent the charges in terms of the phase velocity and attenuation of light as it passes through bulk material. The fields inside and outside the particle, including the scattered field, are then related at the surface using a set of boundary conditions. However, the test of validity of any theoretical solution is experimental verification. We introduce the basic ideas behind these groups of theoretical solutions and the microwave analog method. We then compare sample numerical computations with experiment.

### 3.1. DIRECT SOLUTION TO THE SCATTERING BY SETS OF DIPOLES

The representation of a particle using a set of dipoles is conceptionally attractive and can, given a sufficiently large number of dipoles, represent the distribution of charges and currents in discrete volumes near the atom. In addition, the coupled dipole approximation proposed by Purcell and Pennypacker [2] (also known as the discrete dipole approximation) is a highly flexible method. However there is a complication; we must describe the interaction between dipoles. The field incident on a given dipole is the vector addition of the fields scattered by all the other dipoles to the original incident field. Because this applies to all dipoles, the problem is usually solved iteratively, although problems involving sufficiently few dipoles also can be solved through direct matrix inversion. A set of computer codes based on iteration are available from Draine [7].

Numerical tests confirm that the polarizability due to individual charges within sufficiently small volumes can be combined and represented by a single dipole. In practice, each dipole in a numeric solution may therefore represent more than $10^6$ nearby charges. The condition is that the incident field does not change appreciably over the volume. Rayleigh-Gans scattering [8, 9] can be viewed is the simplified approximation where the interaction between dipoles is ignored while the well-known Rayleigh scattering [8, 9] is the small particle limit when the whole particle may be described using a single dipole.

The Rayleigh scattering approximation works well for most materials as long as the particle size across the longest dimension is less than ~0.1 times the wavelength both outside and inside the material. This range can be extended using the Rayleigh-Gans approximation. However, given today's computing power, it is quite practical to use the coupled dipole approximation on these and several times larger systems. As noted above, the size range of interest is of the order of the wavelength. However larger scattering systems are also of interest if they have surface or internal structure on this scale.

### 3.2. BOUNDARY CONDITION METHODS

A widely used family of solutions is based on an elegant method that uses the refractive index, or equivalently, the dielectric constant, as a macroscopic quantity to describe the particle material. This classical method, in use since more than a century, relies on the fact that charges cannot be created or removed neither inside, outside, nor on the boundary of the particle. When solutions are found, they can usually be applied to larger particles than the coupled dipole approximations.

Any generalization of electrostatics to electrodynamics involves moving charges, i.e., a current. In vacuum, this generalization leads directly to Maxwell's equations. They can be derived using the Lorentz force law to deduce the basic electromagnetic field tensor and use of the Lorentz transformation of General Relativity to translate these equations to a moving reference system, e.g., ref. [5], Chapter 2. This family of solutions to the scattering problem uses Maxwell's equations to describe the electromagnetic radiation outside the particle, an analogue set that involves the refractive index to describe the internal field, and a boundary condition to assure that charges are not created or removed at the interface. We refer to these solutions as "boundary condition" solutions. The best known example is Mie's solution to the scattering by an arbitrary single homogeneous sphere [1, 8, 9].

### 3.3. MICROWAVE ANALOG LABORATORY METHOD

Incomplete characterization of the scattering bodies on the scale of the wavelength and of the geometry of the scattering problem hampers most existing light scattering experiments. However, electromagnetic scattering can be investigated with higher accuracy in the microwave range. Further, the fact that dimensions in Maxwell's equations are only encountered as a ratio to the wavelength, show that as long as these equations properly describe the scattering process, it may be scaled to any wavelength. This is known as the principle of electromagnetic similitude. Because the oscillator strength and loss are frequency dependent, an analog material must be used. For example acrylic, with a refractive index near 1.605 -*i*0.003 throughout the microwave range of frequencies, is an analog for silicate materials at visual wavelengths.

The microwave scattering laboratory at the University of Florida [10], is only the latest in a series of facilities, to take advantage of this principle. The first laboratory in the series was built by Prof. J. M. Greenberg in the 1950s. Our laboratory is a modern broad band facility. Because the resonant frequencies of most materials are far removed from microwave frequencies, their refractive index changes only slowly across the microwave range and is usually constant within the w-waveband (75 to 110 GHz) used in Florida. The broad band facility thus allows us to study the dependency on frequency while the refractive index remains constant. In addition, stepping the wavelength is equivalent to studying the scattering from size bins in a size distribution.

## 4. Comparison Between Theory and Experiment

We show examples of angular distributions of the scattered radiation from the forward direction, 0 degrees, to 165 degrees deflection from the direction of the incident radiation. At each angle of scattering the effect of the particle is usually described using a 4 by 4 scattering matrix although all elements are not independent. Equivalently, all observable quantities in an ideally controlled scattering experiment can be described using four intensities and four phase shifts. One intensity and one phase shift each is for the four combinations of polarization with the incident respectively scattered electric vector polarized perpendicular to the plane of scattering (subscript=1) and in the plane of scattering (subscript=2). The first subscript will describe the state of the incident radiation and the second that of the scattered beam. The intensity is normalized to the incident radiation and is given in standard dimensionless units, ref. [8] sect. 2.1, while the phase shift is related to an imaginary light beam scattered at an arbitrary but specified point.

To compare independent solutions to the scattering problem, we attempt to describe the same case in each system. We note that there always remain experimental uncertainties in the refractive index, size, shape, homogeneity and positioning of any experimental target. To eliminate the experimental error associated with orientation, we make the first comparison to the scattering by a sphere.

## 4.1. A SINGLE HOMOGENEOUS SPHERE

The scattering by a homogeneous spherical particle has emerged as a standard test in light scattering since the boundary condition solution, now known as Mie-theory [1, 8, 9], was found about a century ago. In Figures 1a and 1b, we compare the experimentally determined angular distribution of the scattered intensities $I_{1,1}$ and $I_{2,2}$ from a 9.998 mm diameter Bk7 glass sphere at 80 GHz to the Mie-solution. Bk7 glass has a refractive index near 2.5155 -$i$0.0213 in this frequency range. The amplitude and phase of radiation scattered in the angular interval from 5 to 165 degrees in steps of 0.1 degrees and at several frequencies were recorded in an automated procedure taking approximately 8 hours. The process was then repeated with the particle removed. The vector difference between the two records is the scattering by the particle. Because of the great sensitivity to alignment at 0 degrees, a separate run was made for that angle so that the alignment of the receiving antenna with the transmitting antenna could remain stationary over the course of the data logging. No data could be obtained at angles exceeding 165 degrees due to the physical dimensions of the antennae. However, it is possible to add a receiver to the transmitting antenna and thus record scattering at 180 degrees, an option that is not yet implemented. The agreement is very good considering remaining uncertainties in

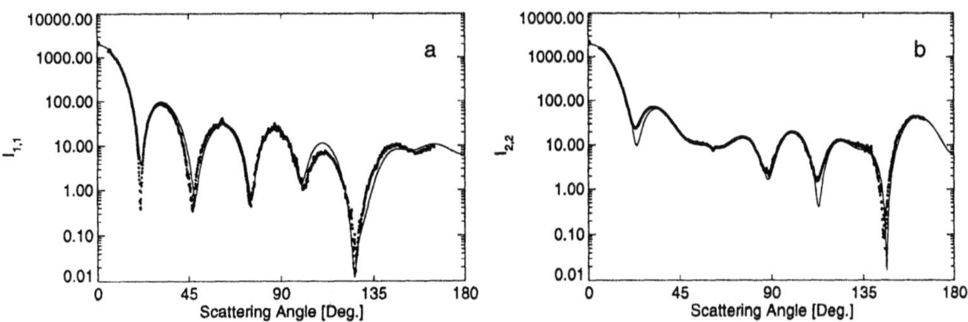


Figure 1a and b. Scattering by a sphere. Intensity of the components polarized perpendicular to the plane of scattering respectively parallel to the plane. Solid line is the Mie solution and dots are for microwave laboratory data.

the refractive index and the geometry of the target. Mie theory predicts no cross polarization components, $I_{1,2}$ or $I_{2,1}$, due to the symmetry of the sphere. The experiment returned cross polarized values below 1 at all scattering angles, and below 0.01 at most angles, in agreement with the theory.

For the polarizations 1,1 respectively 2,2, Figure 1c and 1d, show the phase of the scattered radiation relative the geometric center of the sphere. The agreement is excellent, especially when considering that a 100 micron error in the placement of the sphere corresponds to nearly 20 degree in phase shift at back-scattering angles. To the best of my knowledge, the phase of the scattered radiation has never before been accurately measured except at 0 degrees. This measurement is a new test and confirmation of the correctness of both the experimental data and of Mie-theory. The coupled dipole approximation for the same sphere, using the Discrete Dipole Approximation (DDA) code by Drain [7], is indistinguishable from the Mie solution on the scale of these figures.

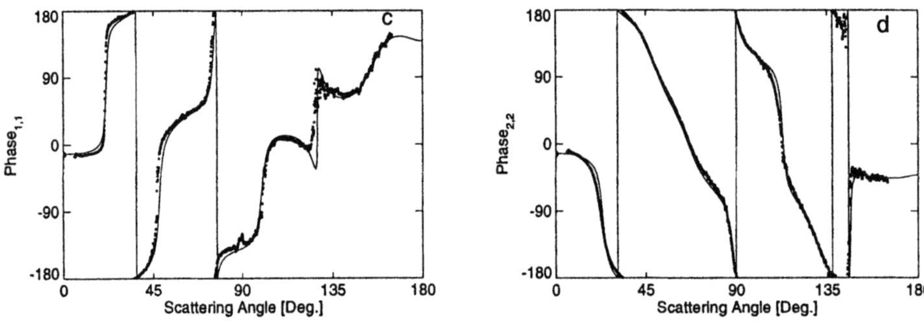


Figure 1c and d. Phase of the scattered components polarized perpendicular to the plane of scattering respectively parallel to the plane. Solid line is the Mie solution and dots are the micr owave data.

## 4.2. CUBE

Boundary condition methods depend on the solution of Maxwell's equations on the surface of the scattering body. When the perpendicular to the surface is undefined and discontinuous as it is along the edges of a cube, a solution cannot be obtained directly. Although it might be possible to circumvent this problem, a literature search found no solution of this type. However, the coupled dipole approximation has no such restrictions and Figures 2a and b show a solution that agrees well with microwave laboratory data.

The 1.27 cm long edges of the acrylic cube of refractive index ~1.605-*i*0.003, are aligned with the ***k***, ***E***, and ***B*** vectors of the incident radiation. The dipoles were place on a similarly oriented cubic lattice so that each edge is represented by a linear array of dipoles. The microwave frequency was 75 GHz.

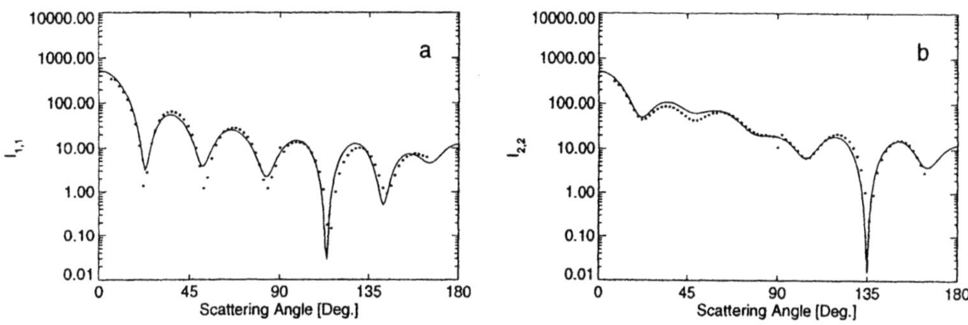


Figure 2a and b. Scattering by a cube. The solid line is for the DDA solution and dots are microwave data.

## 4.3. TWO SPHERES IN CONTACT

Two contacting 9.998 diameter Bk7 glass spheres, similar to the one used in Figure 1, were aligned so that the system's symmetry axis is in the scattering plane and perpendicular to the incident direction. While the agreement between theory and experiment has been good in the previous comparisons, Figures 3a and 3b show significant discrepancy between the experiment and the calculated scattering using the DDA (DDSCAT.5a.8). At the frequency of 75 GHz, each sphere was represented using 65,752 dipoles. The calculations converged to the recommended criterion and all cross sections, also converged after ~300 iterations.

While the DDA works well at smaller sizes, or with lower refractive indexes, this case appears to exemplify when the approximation starts to break down as the coupling becomes sufficiently strong [11]. However, we do not rule out the possibility that a better solution can be found, for example if a better starting approximation is found to the internal field acting on the dipoles, and if care is taken in the placement of dipoles. We note that at sufficiently small sizes the dipole placement becomes particularly important

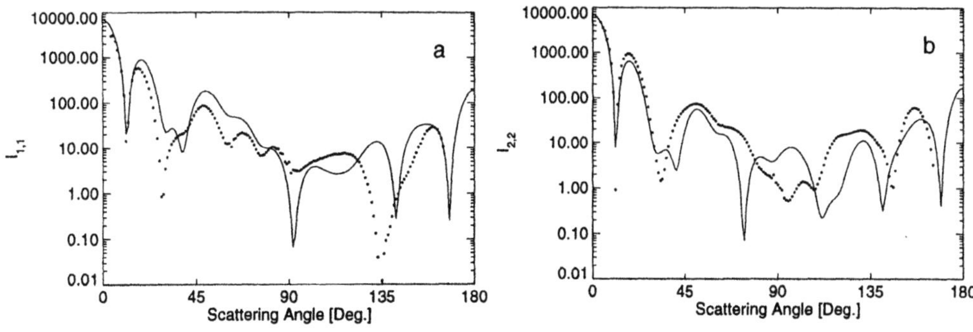


Figure 3a and b. Scattering by two spheres. The solid curve is the DDA solution for the scattering by two spheres in contact. The dots are microwave data.

and the DDA may become inaccurate for that reason. One way to alleviate this problem is to increase the number of dipoles. The translation from refractive index to polarizability is another possible source of error.

Several solutions for the scattering by interacting spheres based on the boundary condition have been proposed. However, the large particle size is insurmountable for most solutions. By comparison, the recently developed solution by Xu [12, 13] solves the system using approximately 1/100 the number of CPU cycles required by the DDA and Figures 4a and 4b show that the solution agrees with the microwave experimental data.

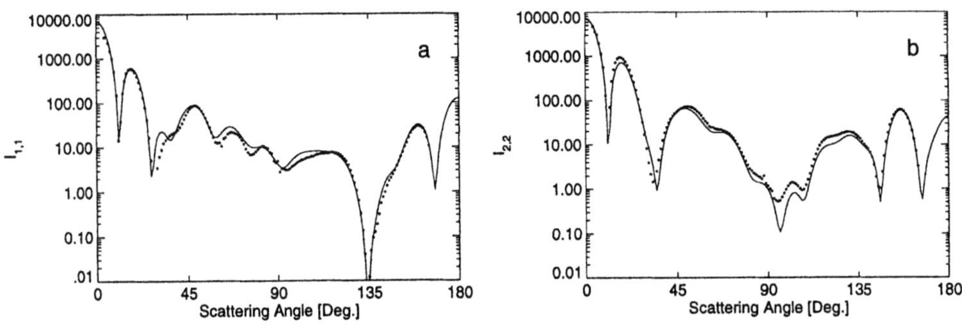


Figure 4a and b. The solid curve is the boundary condition solution to the scattering by two spheres in contact. The dots are the same microwave data as in Fi gures 3a and b.

## 5. Discussion

It is usually stated that the light scattering properties of an arbitrary body depend on the material properties represented by the complex refractive index and the geometry of the scatterer's boundary. The boundary may be defined by the scatterer's size, shape and orientation with respect to the incident and scattered light. For some of the simplest geometries, such as a homogeneous sphere, it is claimed that exact solutions can be obtained and these Mie-solutions are widely used in practice.

The proper statement is that the analytic expressions in the Mie-solution are exact provided that the scattering body is a perfect and homogeneous sphere embedded in an infinite and homogeneous medium and provided that Maxwell's equations apply both inside and outside the particle when the optical properties of the material and the surrounding medium are described by their refractive indices. We note that the medium surrounding the scatterer can be vacuum or any material as long as it is infinite in extent and Maxwell's equations apply.

However, all materials are made of atoms. To be accurate, the description of a material body as a homogeneous medium with a sharp boundary is only an approximation to actual morphologies. Also, the distance to neighboring particles is always finite. Further, the Mie-solution and related solutions for other particle shapes include infinite series that must be truncated in any practical calculation. This is to emphasize that the Mie-solution is an exact formulation to an idealized scattering problem. However, by comparing a sample calculation to experimental data, we see in Figure 1 that the Mie-solution is highly accurate in practice, at least in the microwave range of frequencies. Any deviation is easily explained through uncertainties in the refractive index and imperfections in the geometry of the scattering system. This confirms that the Mie-solution can also be applicable to real conditions.

A boundary condition solution to interacting spheres that is both accurate and fast in comparison to the DDA now exists [12, 13]. The boundary condition solutions are however limited to the finite number of geometries for which specific solutions have been found.

The coupled dipole approximation is also accurate when applied to particles with moderate interaction between dipoles. Its greatest advantage is the versatility. However, as the interactions increase in importance, the approximation appears to break down.

An accurate experimental facility to test theoretical solutions exists at the University of Florida. This facility is able to accurately measure all components of the scattering matrix. Perhaps an equally important role is its ability to fill the gaps between theoretical solutions. This is the subject of the text "Scattering by Complex Systems II: Results from Microwave Measurements" in this volume.

**Acknowledgment:** This work was supported by NASA's Planetary Atmospheres Program through grant NAGW-2482.

## References

1. Mie, G. (1908) *Ann. Physik*, **25**, 377.
2. Purcell, E.M. and Pennypacker, C.R. (1973) Scattering and absorption of light by nonspherical dielectric grains, *Ap. J.* **186**, 705-714.
3. Gustafson, B.Å.S. (1989) Comet ejection and dynamics of nonspherical dust particles and meteoroids, *Ap. J.*, **337**, 945-949.
4. Bohren, C.F. and Huffman, D.R. (1983) *Absorption and Scattering of Light by Small Particles,* John Wiley & Sons, New York.
5. Scharf, G. (1994) *From Electrostatics to Optics A Concise Electrodynamics Course,* Springer-Verlag, Berlin.
6. Hummel, R. E. (1993) *Electric Properties of Materials, second edition,* Springer-Verlag, Berlin.
7. Draine, B.T. (1988) The discrete-dipole approximation and its application to interstellar graphite grains, *Ap. J.*, **333**, 848-872.
8. van de Hulst, H.C. (1957) *Light Scattering by Small Particles,* Wiley, New York.
9. Kerker, M. (1969) Scattering of Light and other electromagnetic radiation, Academic Press, New York.
10. Gustafson, B.Å.S. (1996) Microwave analog to light scattering measurements; A modern implementation of a proven method to achieve precise control, *J. Quant. Spect. Rad. Transf.*, **55:5,** 663-672.
11. Xu, Y.-l. and Gustafson, B.Å.S. (in preparation).
13. Xu, Y.-l. (1995) Electromagnetic scattering by an aggregate of spheres, *Appl. Opt.*, **34,** No. 21, 4573-4588.
13. Xu, Y.-l. (1997) Electromagnetic scattering by an aggregate of spheres: far field, *Appl. Opt.*, **36,** No. 36, 9496-9508.

# SCATTERING BY COMPLEX SYSTEMS

## *II: RESULTS FROM MICROWAVE MEASUREMENTS*

B.Å.S. GUSTAFSON, L. KOLOKOLOVA, J.E. THOMAS-OSIP,
K.W.T. WALDEMARSSON, J. LOESEL, and Y.-l. XU.
*Dept. of Astronomy, University of Florida*
*211 SSRB, Gainesville, Fl 32611, USA*

## 1. Introduction

Our group initiated a systematic exploration of the scattering properties of complex systems using the microwave analogue method [1, 2]. In this article, we select examples that are of interest in astronomy and discuss the angular distribution of the scattering intensity, polarization, and color. This work can be viewed as an extension to the investigation into the scattering by "Bird's-Nest" type particles [3, 4], a model proposed by Greenberg and Gustafson to represent cometary dust.

The scattering and absorption of light by complex systems such as these is a difficult problem to evaluate and no general solution has been found. Solutions that can account for complex structure such as the coupled dipole approximation [5] restrict studies to small size systems. The complex nature of particles is therefore often neglected and the Mie-solution [6, 7] for a homogeneous sphere used in the hope that this will approximate the scattering by non-spheres. Our microwave analog laboratory facility allows us to study the large complex systems discussed in this article. First, we summarize the main features of the microwave laboratory.

## 2. The Microwave Analog Facility at the University of Florida

The microwave analog to light scattering method takes advantage of the principle of electromagnetic similitude to scale the particles to a convenient dimension for manufacturing and handling in the laboratory. This was the basis for a series of highly successful, now defunct, single frequency microwave analog facilities. Prof. J. M. Greenberg built the first unit in the 1950s using the X-band (3.18 cm wavelength) and Prof. R. H. Giese built a Ka-band (8 mm) facility in the 1970s [8]. The University of Florida w-band facility became operational in 1995 and can step through the 75 to 110 GHz waveband, or 2.7 to 4 mm wavelength interval, in up to 512 equally spaced frequencies. A scale factor of 6,700 translates the w-band to the visual range so that a laboratory model in the centimeter range represents a micron size particle scattering



*J. M. Greenberg and A. Li (eds.), Formulation and Evolution of Solids in Space,* 549–563.

optical light. Because electromagnetic properties are necessarily wavelength dependent over such a broad range, an analog target material must be used [1]. We mix a variety of plastics with pigments to represent rocks, ices, and organics.

Figure 1 shows the layout of the Florida facility projected on the scattering plane, i.e., the plane containing the direction (or Poynting) vector **S** of both the incident and the measured scattered radiation. A lens and polarizer equipped antenna illuminates a fixed target with a plane wave that is linearly polarized either perpendicular to the scattering plane or parallel to the plane. The Fresnel lens produces a planar wavefront within a volume 50 to 80 wavelengths to the side centered on the target. An identical receiving antenna assembly can be positioned at any scattering angle, θ, from 0 to 168 degrees counted from the direction of incidence. Like the transmission side, the receiver assem-

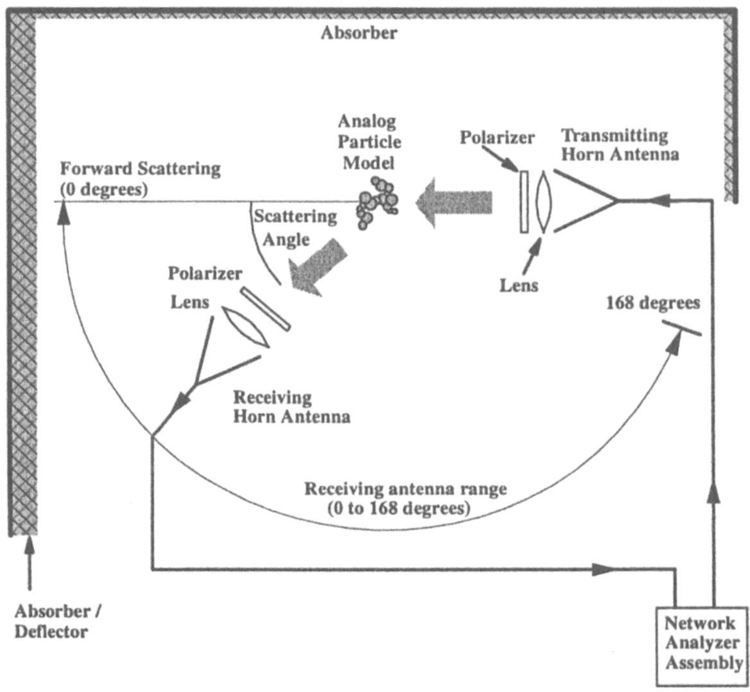


Figure 1. Sketch of the microwave laboratory facility at the University of Florida. The Network Analyzer assembly nominally generates a set of 85 ultra stable wavelengths from 2.7 to 4 millimeters and can generate up to 512 distinct wavelengths. A lens and polarization filter equipped transmitting horn antenna assembly creates a ~30 cm diameter linearly polarized parallel microwave beam illuminating a scale model of a scattering body. Particles scattering visual light are magnified by a factor ~6,000 so that all relevant parameters can be precisely controlled. A mobile receiving antenna assembly identical to the transmitting antenna records the scattered radiation. The plane of polarization of both units can be rotated independently to become parallel or perpendicular to the plane of scattering. Both the intensity loss and phase shift suffered by the scattered wave in comparison to a direct beam is recorded under computer control for sequences of any of the four combinations of polarization and for any scattering angle from 0 to 168 degrees.

bly can be adjusted for either perpendicular or parallel polarization. The scattered far field components are thus measured in four combinations of polarization and with an effectively infinite angular resolution. The counterintuitive property of infinite resolution is illustrated in the figures of ref. [1] (this volume) in which practically every feature of the sharp minima in intensity and in the phase are faithfully reproduced in the laboratory data. There is no apparent degradation of the resolution when compared to theory.

Because we measure the change in both intensity and phase suffered by a wave scattered by our laboratory target in all four polarization combinations, we can characterize the scattering process completely. This is easier to visualize using a geometric description of the polarization ellipse for arbitrarily polarized light than it is using the Stokes parameters directly. We recall that the Stokes parameters [7] are used to describe the intensity and polarization state in nearly all modern works because they lead to a convenient mathematical representation but a geometrical representation in terms of the shape and orientation of the polarization ellipse is equally valid. To describe an arbitrary electromagnetic plane-wave propagating through empty space along the z-axis of a Cartesian coordinate system, it is sufficient to describe the amplitude components of the electric field

$$\begin{aligned} E_x &= A\cos(\tau+\delta_1), \\ E_y &= B\cos(\tau+\delta_2), \\ E_z &= 0. \end{aligned} \tag{1}$$

The phase factors consist of a variable part, $\tau = \omega t - kz$, where $\omega$ is the frequency, $t$ the time, $k=2\pi\omega/c$ the propagation constant or wavenumber, and $c$ the speed of light. The constant parts are the phases $\delta_1$ and $\delta_2$. The magnetic field $\mathbf{H}$ follows from $\mathbf{S} = \mathbf{E} \times \mathbf{H}$. The corresponding Stokes vector is

$$\begin{aligned} I &= A^2 + B^2, \\ Q &= A^2 - B^2, \\ U &= 2AB\cos(\delta_1 - \delta_2), \\ V &= 2AB\sin(\delta_1 - \delta_2). \end{aligned} \tag{2}$$

The squared amplitudes are directly measured intensities and the phases are obtained using a network analyzer that compares the received signal to a stabilized reference signal of known phase. In the laboratory experiments, the signal originates in the network analyzer so that the transmitted signal also can provide the phase reference. We note that there are only three independent parameters describing the wave; the two intensities and the phase difference $\delta_1 - \delta_2$. There exists the relation $I^2=Q^2+U^2+V^2$ for a strictly monochromatic wave so that the Stokes vector also has only three independent parameters. When both the incident and the scattered wave are described by their respective Stokes parameters, the scattering can be described by a linear transformation, mathematically a $4 \times 4$ matrix $\mathbf{F}$;

$$(I, Q, U, V) = \mathbf{F} \cdot (I_0, Q_0, U_0, V_0), \tag{3}$$

where subscripts denote the incident beam parameters. The 16 element **F**-matrix contains only seven independent parameters. Eight parameters are found in the intensity and phase of the mutually perpendicular components measured in the laboratory, or equivalently in the corresponding complex amplitudes. Of these eight, only the phase differences, not the absolute phases, are relevant [7], which reduces the number to seven.

The principle of optical equivalence [7] states that the Stokes parameters contain the complete set of quantities needed to characterize the intensity and state of polarization of a beam of light, in practical analysis. It follows that the transformation matrix **F** between the incident and scattered Stokes vectors fully describe the scattering process to the same level of detail. This principle is based on the fact that optical measurements involve linear transformations only.

In ref. [1] (this volume), we give an example of phase measurements that compare favorably with theory. However, in most practical situations, particles are illuminated by natural light where $Q=U=V=0$ so that there is no definite phase relation between the four polarization components. We see from the form of Eq. 3 that we may concentrate on the upper left quarter of the **F**-matrix as this is sufficient to fully characterize the scattering of natural light. We may rewrite this part of the matrix in terms of quantities measured directly in the laboratory. These are the intensities $I_1$ polarized perpendicular to the scattering plane and $I_2$ polarized along the plane. Also known are the corresponding incident intensities $I_{1,0}$ and $I_{2,0}$. We may then define the transformation matrix

$$\begin{bmatrix} I_2 \\ I_1 \end{bmatrix} = \begin{bmatrix} i_{2,2} & i_{2,1} \\ i_{1,2} & i_{1,1} \end{bmatrix} \times \begin{bmatrix} I_{2,0} \\ I_{1,0} \end{bmatrix}. \tag{4}$$

From these matrix elements, we compute the total scattered intensity in units of the scattering matrix, $1/k^2r^2$, where $r$ is the distance

$$I(\theta) = \frac{i_{1,1} + i_{1,2} + i_{2,1} + i_{2,2}}{2}, \tag{5}$$

and the degree of linear polarization

$$P(\theta) = \frac{(i_{1,1} + i_{1,2}) - (i_{2,1} + i_{2,2})}{i_{1,1} + i_{1,2} + i_{2,1} + i_{2,2}}, \tag{6}$$

both as a function of the scattering angle. All measurements reported in this article were repeated in 84 discrete equally spaced frequencies from 75 to 110 GHz. We integrate the total intensity obtained from Eq. (5) across the wave band which is represented by the sum over intensities normalized to the width of the band, $\Delta\nu$, so that the plotted values

$$\frac{\sum I(\theta)k^{-2}}{\Delta\nu} \tag{7}$$

are proportional to the incident intensity and inversely proportional to the distance squared (in meters). To give a quantitative measure of the color in intensity and in polarization, we split the laboratory spectrum into "blue" and "red" halves (equal frequency intervals). We represent the integral of matrix elements in Eq. (4) illuminated by the solar spectrum across the $[\nu_a, \nu_b]$ interval using;

$$<i>_{[\nu_a,\nu_b]} = \frac{\sum_{freq=\nu_a}^{\nu_b} I_{o,freq} i_{freq} k^{-2}}{\sum_{freq=\nu_a}^{\nu_b} I_{o,freq}} \tag{8}$$

where the indexes indicating the polarization have been dropped and $I_{o,freq}$ is the solar illumination integrated across the small frequency interval corresponding to one frequency step $\nu$-½ to $\nu$+½. This mimics the usual format of reported observations where the nominator is the observed flux across the waveband and the denominator is a normalization to the solar spectrum. We used a simple Planck function to represent the sun. The matrix elements are $<i>_{red}$ where $\nu_a$=1 and $\nu_b$=42 respectively $<i>_{blue}$ for the interval $\nu_a$=43 and $\nu_b$=84. The averages $<I>_{color}$ and $<P>_{color}$ can now be computed by substituting the corresponding averages in Eqs. 5 and 6. In addition, we quantify color through the ratio

$$Color(\theta) = \frac{I_{red} - I_{blue}}{I_{red} + I_{blue}}. \tag{9}$$

To simulate the scattering by a tenuous cloud of randomly oriented particles, all scattering data were averaged over twelve orientations for our largest aggregate and 36 orientations for the others. A more detailed description of the laboratory including the signal processing procedures is given in ref. [2].

## 3. Results

We show results for ensembles of spheres and circular cylinders representing aggregated interstellar grains. These laboratory models are based on the evolutionary sequence that interstellar grains may have undergone as they became incorporated in primitive solar system bodies and eventually were released as cometary and interplanetary dust. This sequence is also the basis for the original "Bird's-Nest" models [3] and is intended to remain consistent with the modeling by Greenberg, in this volume. As the solar system formed, refractory grains were encased in volatiles and trapped inside comets and possibly C- and D- type asteroids. The size and shape of the refractory grains are taken to be those of the classical interstellar grains with dimensions of the order of $10^{-1}$ micron. Because the grains are known from observations of interstellar polarization to be elongated, each grain is either represented by a cylindrical segment or by a set of two to three spheres. The volume ratio of volatiles to refractories as the grains aggregate is given by cosmic abundances, see Greenberg this volume and ref. [9]. No significant compaction or other physical or chemical processing is assumed to have taken place except for the sublimation of volatiles as these comet fragments are exposed to the sun in the inner solar system. These models therefore have several similarities to the original "Bird's-Nest" models [3], including packing factors of the order of 0.1 (or 90% void). The biggest differences are that we have extended the study to larger aggregates and are now able to report color effects.

## 3.1. AGGREGATES

### 3.1.1 *Past results; coherent scattering by small aggregates*

We recall the main results reported on the scattering by "Bird's-Nest" type structures at visual wavelengths. The original paper is an investigation into the scattering by 250 to 500 individual grains assembled into an aggregate based primarily on microwave data obtained in the X-band facility, then at the State University of New York at Albany (SUNYA). A good agreement is found between the measurements and the coherently added scattering by all constituent grains. When the scattering from $N$ identical grains is added coherently, the angular distribution of scattered intensities peaks at $N^2$ times the scattering by the individual grains in the forward direction ($\theta=0$) where the scattering from all grains is in phase. In the backscattering hemisphere where the scattering adds at nearly random phase, the intensity is of the order of $N$ times the scattering by individual grains. The requirement is that the aggregates are at least a few wavelengths across, as was the case. The intensity goes through a series of interference minima in the transition from forward to backscattering. Because coherent addition is found to be a good approximation for all polarization components, their relative intensities are nearly unaffected by the fact that the grains are in an aggregate. The degree of linear polarization (Eq. 6) therefore closely mimics that of individual grains at all scattering angles. That is, when normalized to the scattering by individual grains, the forward scattering peaks at the value $N$ and then drops to near unity. The transition takes places at smaller scattering angles the larger the aggregate is. For 250 to 500 aggregated grains at 10% packing factor, the forward interference pattern extends to 30 to 60 degrees scattering angle. When the degree of linear polarization is similarly normalized to the polarization produced by the constituent particles, the ratio is near unity in all scattering directions. This ratio is unity by definition in the coherent scattering approximation.

Measurements using similar targets and with the same main result were performed in the ka-band facility at the Bochum University [10]. These targets were then coated by a simulated organic refractory mantle and the measurements repeated as successive mantles were grown. A transition from coherent scattering to a mode where a more complex interaction takes place occurs as the mantles grow sufficiently thick. However, the transition occurs for thicker mantles than the values we expect to find based on cosmic abundances.

Loosely packed aggregates of ~500 or fewer interstellar grains of the "Bird's-Nest" type thus have polarization properties similar to those of independently scattering interstellar grains while their intensity distribution is reminiscent of much larger structures. These properties are in rough agreement with the properties observed in the zodiacal light, comet comae and dust tails. The primary difference is the much lower peak polarization than that seen for the model particles. This difference might be overcome with a small amount of scattering from other structures that primarily "contaminates" the low intensities polarized along the scattering plane. This would have negligible effect outside this intensity minimum where the aggregates are good scatterers. The net result could be a reduction of the peak polarization while the overall shape of the polarization curve is preserved. However, the variation with color for these structures is another major subject of which very little was known. Polarization and color along with the intensity distribution will be addressed in the next section.

### 3.1.2. *New results; intensity, polarization, and colors from large aggregates*

Remaining important issues include the observed neutral to red color in the zodiacal light and in comets. If coherent scattering applies, neutral to red color is expected at small scattering angles but not at large scattering angles where it is also observed. The color should be close to that of the constituent grains, which usually is blue, in the back scattering region. The color at small scattering angles should be shifted to the red because the interference peak extends to larger scattering angles at longer wavelengths. Another outstanding question is, at what size aggregates the simple coherent scattering approximation breaks down.

The study of these issues was in part the motivation for the design of the microwave laboratory in Florida as a broad band facility able to simulate colors. The choice of a shorter wavelength was in part motivated by the desire to investigate the scattering by larger structures in comparison to the wavelength than before. In this article we report some of our first findings.

Figure 2a shows an aggregate of 1450±20 polystyrene spheres of 0.63±0.35 mm diameter. This size spheres with refractive index near $m$=1.615-$i$0.03 represent classic size interstellar grains without organic mantles in the visual. The aggregate is therefore most closely related to the ones investigated in the SUNYA laboratory [3]. The primary difference is the larger number of aggregated particles so that the dimensions are of the order of 10 wavelengths across or 5 microns in the visual.

In Figure 2b, the angular distribution of intensity is averaged over 36 orientations and over all 85 frequencies. Two sets of theoretical curves are shown for comparison. The heavy curves are for the coherent scattering approximation (dashed line) and the incoherent scattering (solid line) where any phase relations are neglected. In incoherent scattering, the intensities can be added directly. This is the way the spheres would scatter if dispersed in a tenuous cloud. Phase relations due to differences in the geometric path only are accounted for in our coherent scattering comparison.

Both solutions are comparable except at scattering angles smaller than ~20 degrees where definite phase relations are established. Evidently, the aggregate is sufficiently dispersed so that differences in the geometric path lead to nearly random phase differences at the larger scattering angles. The experimental data indicate that constructive interference is a good approximation only close to the forward direction and incoherent scattering is approached in the back-scattering direction. However, unlike results for the aggregates previously reported, the angular distribution of scattered intensities is considerably brighter at all other scattering angles. We believe that this is due to the detailed structure of the aggregate and is not intrinsic to the larger size aggregate. For practical reasons, the spheres were assembled in groups or clumps where the spheres are in contact with their neighbors on all sides. The assembly tends to behave as an aggregate of these groups rather than an aggregate of individual spheres. This is illustrated using the lighter set of curves where the spheres are grouped in sets of 12 and each group is approximated by an equal volume sphere. This approximation probably exaggerates the coupling between closest neighbors while it neglects coupling to distant spheres. As before, the coherent and incoherent solutions differ only at the smallest scattering angles.

We note than when the particles are cooperating in small groups, they scatter significantly more light per particle than when they are dispersed in a cloud. Extraction of scattering and radiation pressure cross sections showed that the average radiation pressure on the aggregated grains is several times higher than for the same grains dispersed. Most, but not all, of the relevant information is contained in Figure 2b so that this result can be anticipated. However we concentrate on the angular distribution of scattering so that cross sections will not be further discussed in this article.

The degree of linear polarization produced by the constituent particles dispersed so that they do not interact (heavy solid curve in Figure 2c) is close to the well known polarization by isotropic Rayleigh particles. Similar to smaller particles, the polarization is always positive and the maximum near 90 degrees is near unity. There is only a weak asymmetry that shifts the maximum by a small amount to higher scattering angles. The magnitude of this small shift is practically the only color dependency of the polarization across the frequency range and it is too small to be seen in the figure. As the phase shifts due to geometric path differences alone are taken into account, there is no correction for possible optical effective phase velocity differences in the orthogonal polarization directions. The polarization is therefore identical in our coherent and incoherent scattering approximations.

The experimentally determined polarization values are plotted separately for the red and the blue half of the spectral range. While both curves have approximately the same shape as the Rayleigh polarization, the maximum polarization is significantly lower. The decrease in polarization comes primarily from the cross polarization terms $i_{1,2}$ and $i_{2,1}$ in the scattering matrix. By comparison, cross polarization terms vanish in scattering by isotropic particles such as Mie-spheres. The polarization is for the most part slightly more positive in the red than in the blue. Although this color effect is not strong, the red polarization is seen in all experimental data we have obtained for aggregated small particles. Reddening is seen even if the cross polarization terms were ignored. Our interpretation is that coupling between particles in the blue is stronger than in the red, because the particles are larger compared to the wavelength. Stronger interparticle dependence for larger particles is often seen both in the laboratory and in theoretical calculations.

The light dashed and light solid curves corresponding to the coherent and incoherent scattering by the aggregated groups of particles are for the red and blue polarization respectively. They were obtained using the same parameters that generated the light curves in Figure 2b. Because Mie-spheres represent the groups, any cross polarization term vanishes in this approximation as well which helps explain the high polarization. The shift in the maximum polarization with wavelength range is now obvious. We notice that the polarization is red at low scattering angles, but unlike the experimental results, it is blue at high scattering angles. This is due to the well understood shift in the maximum originating in the Mie-calculations.

We have seen that the angular distribution of the intensities can be viewed as coherent scattering by spherical groups of aggregated particles but this approximation does not reproduce the polarization and the color of polarization. However, if the smallest optically important constituents are the groups, these results agree with conclusions drawn in our earlier works. Intensities normalized to the scattering by the constituent particles depend primarily on the geometry of the aggregate while the polarization is close to that from the groups of constituent particles scattering independently. The polarization produced by the irregularly shaped groups is not well represented by spheres. One effect is that the groups produce finite cross polarization. This lowers the maximum polarization.

A sometimes voiced objection to the "Bird's-Nest" model is that if the coherent scattering approximation applies to intensities, the color of the scattered light might be expected to be close to the color of the constituent particles. Because the constituent particles derive from classical interstellar grains, the color should be blue. However, the color of the zodiacal light is neutral or a faint red. The blue color of the constituent particles is given by the heavy solid line in Figure 2d and the color using the coherent scattering approximation with the constituent spheres is the dashed heavy curve. The oscillations are due to a shift in the interference pattern with wavelength and average out if there was a cloud of similar particles. We therefore ignore the small scale features and concentrate on the broader trend with scattering angle. The shift toward the red at small scattering angles is due to the broader forward peak created by coherent scattering in the red. At high scattering angles the shift is toward the blue for this size aggregate. However, the direction and magnitude of the shift depends on the dimensions of the aggregate in a periodic way. We see that the forward scattering always shifts toward red colors while the backscattering can be shifted in either direction. If we were to observe a cloud of aggregates with a broad size distribution, the forward scattering is shifted toward the red while the shifts in the back scattering region average out and we expect to see the same color as that from the constituent particles in this approximation.

Figure 2a. Aggregate of 1450±20 polystyrene ($m$=1.615-$i$0.03) spheres of 0.63±0.35 mm diameter resting on a thin sheet of expanded plastic of low refractive index used to support the delicate assembly during the laboratory experiments. Sets of two to three spheres represent silicate interstellar dust cores in the visual range.

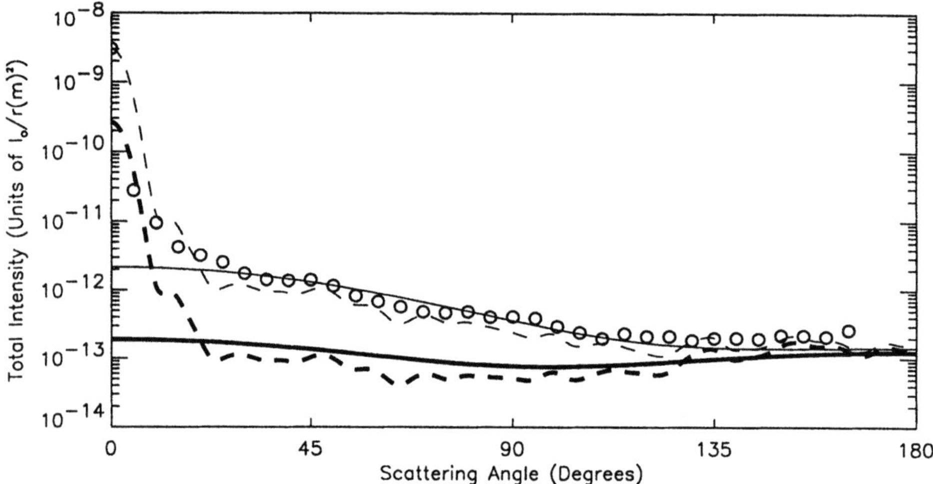


Figure 2b. Angular distribution of scattered intensity averaged over orientations and over frequency. Heavy curves are for the coherent scattering approximation (dashed line) and the incoherent scattering (solid line) where any phase relations are neglected. The lighter set is for the coherent respectively incoherent approximation where spheres are grouped in sets of 12 and each group is approximated by an equal volume sphere.

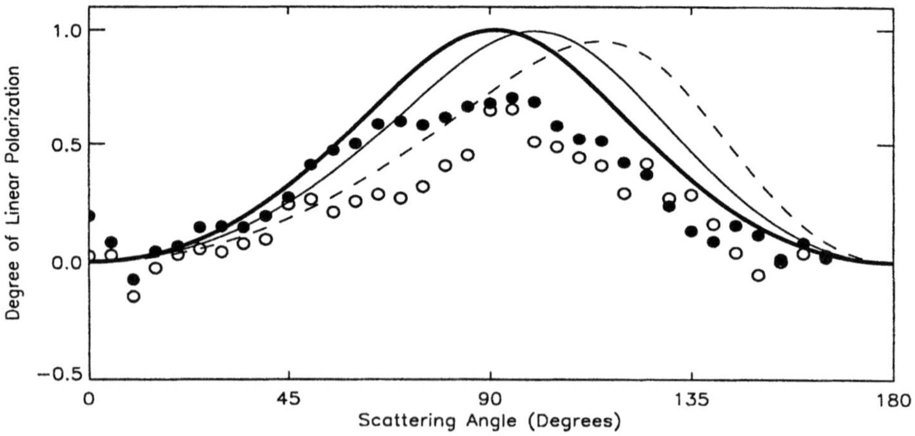


Figure 2c. Angular distribution of the degree of linear polarization. Dots represent experimental results for the red half of the frequency range and circles for the blue half. The heavy solid curve represents both color ranges and both coherent and incoherent scattering by the constituent particles. The light solid curve is for the groups of 12 spheres in the red while the dashed curve is for the blue.

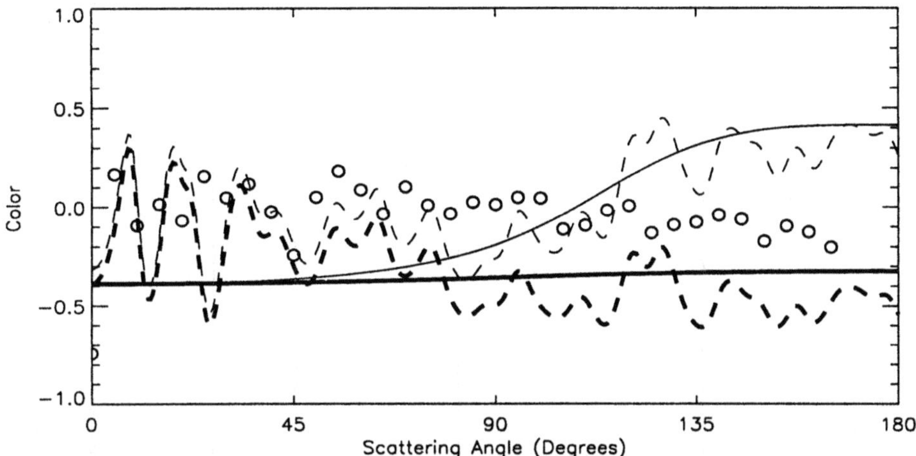


Figure 2d. Color as a function of scattering angle. Experimental data are less blue than the isotropic blue incoherent scattering by constituent spheres (heavy solid line). The solid light line for incoherent scattering by the groups of 12 spheres shows that the reduction in color in the backscattering hemisphere may be due to scattering as groups. The heavy dashed curve is for coherent scattering by the constituent spheres and the light dashed curve is due to the groups of 12 spheres. The reduction in the forward hemisphere may be due to interference due to coherent scattering. Coherent scattering by the groups (light dashed curve) shows reduced color at all angles due to a combination of these e ffects.

However, dots in Figure 2d show that the experimental results are an almost neutral color at all scattering angles. This is much more reminiscent of the color observed in the zodiacal light than the results expected based on the coherent scattering approximation for the constituent particles discussed above. A hint as to the origin of the color is given by the coherent scattering from the groups of particles shown by the light dashed line. We see that the arrangement into groups leads to a redshift at high scattering angles. Together with the shift at small angles from forward coherent scattering, this approaches the color obtained in the experiment. No matter what the explanation for the red shift is, we see that at least this particular "Bird's-Nest" type aggregate shows nearly neutral color and not the blue of its constituent particles.

We note that color can be modified by the color of the bulk material out of which the grains are made. However, we have separated the effects of 1) small constituent spheres that each have blue color due to their size and 2) the effect of aggregation which is to redden the scattering or to reduce the size induced blue color effect. The intrinsic color of the bulk material of actual grains may range from gray (neutral) to brownish yellow or red. We conclude, that based on our interpretation using the coherent scattering by groups of particles to represent the scattering by "Bird's-Nest" type aggregates, the neutral to red color of the zodiacal light might not be in conflict with the "Bird's-Nest" particle model. The angular distribution, magnitude and color of the polarization obtained in the laboratory experiment also are close to the zodiacal values but unlike color, we are unable to argue that these results should hold for a broad range of "Bird's-Nest" type models. To illustrate that the results do hold we show scattering measurements for two other "Bird's-Nest" models.

We used 1.25 mm diameter nylon cylinders to represent the clusters or groups of tightly packed particles and assembled the cylinders into aggregates. Figure 3a shows our "mid size aggregate" consisting of ~1500 cylinders, each ~2 mm in length, making up an approximately 20 wavelengths across assembly (~10 micron in the visual). The packing factor is close to 0.3 in this case. Shown on the same scale in Figure 3b is our 100 wavelengths across "large size aggregate" (~50 micron in the visual) made of a continuous strand of string for practical reasons to reach a lower packing factor using the same amount of material as in the medium aggregate. As for most of our "Bird's-Nest" models, the packing factor is near 0.1. These large aggregates were selected because the zodiacal dust is commonly believed to consist of larger dust grains than the "Bird's-Nests" models that have been discussed previously.

Figure 3c shows the angular distribution of intensity for both aggregates using circles for the mid size aggregate and crosses for the large one. The flat intensity distributions with a forward peak are comparable to that observed in the zodiacal light [11] although contributions from particles scattering along the line of sight makes the deconvolution of observed quantities into dependence on scattering angle uncertain. The laboratory data show a hint of an increase toward backscattering angles where the zodiacal brightness increases to make up the "Gegenschein". Unfortunately, scattering angles above 165 degrees are missing from the laboratory data. The color of the mid size aggregate (Figure 3d) is close to neutral at all scattering angles. However the large aggregate shows blue backscattering approaching that for independent portions of the string. The color for the cylinders or the string alone is near that for the constituent spheres in Figure 2a, i.e., ~ -0.6. Polarization (Figure 3e) is reduced compared to the independent scattering string or cylinders that produce values very close to those for the constituent spheres shown in Fig. 2c. As might be expected, the polarization is closest to the independent values for the large string aggregate in the red. This is because interaction is less important in this low packing factor aggregate and because the string is closer in diameter to a Rayleigh particle in the red. In the blue and for the more tightly packed cylinder aggregate, the polarization is much lower and does no longer resemble Rayleigh polarization. We note that the polarization in the red consistently is more positive than in the blue for both aggregates. By comparison, the observed polarization in the zodiacal light is blue [11] while red polarization is seen in comets [12]. An aggregate made of spheres, but otherwise similar to the mid size aggregate, showed nearly identical scattering in intensity, color, and in polarization.

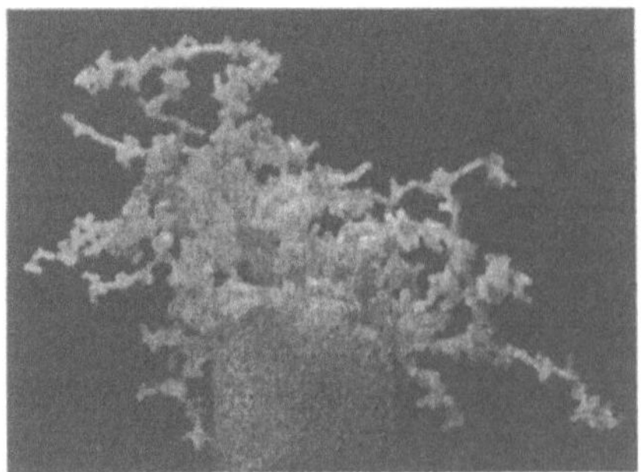

Figure 3a. Mid size aggregate of ~1,500 ~2 mm length cylinders made of 1.25 mm diameter nylon ($m$ ~ 1.74-$i$0.005) string. The model rests on a support of low refractive index expanded Styrofoam. Each cylinder represents the refractory core of an interstellar dust particle.

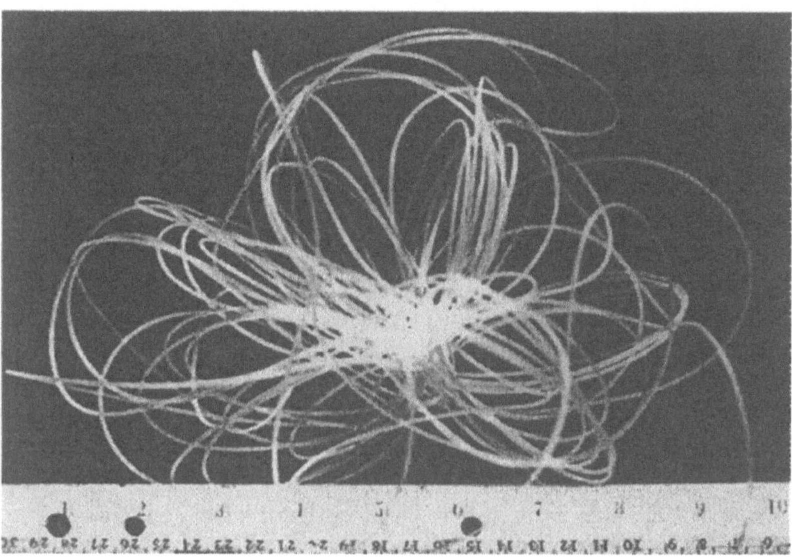

Figure 3b. The large size model representing ~1,500 loosely aggregated interstellar grains is made of identical string to that used to build the mid sized cylinder aggregate shown in Figure 3a. The string was left uncut to achieve a low packing factor. The top scale on the ruler is in inches and the lower scale in centim eters.

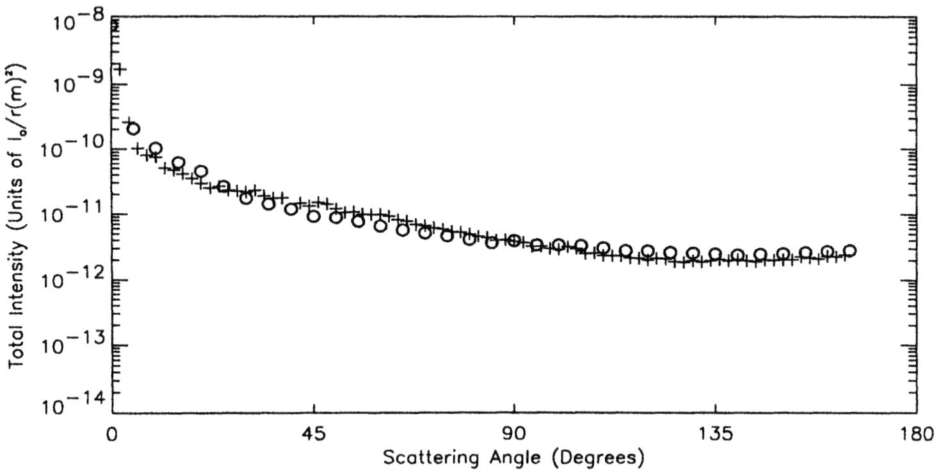


Figure 3c. Angular distribution of scattered intensity averaged over orientations and frequency. Circles are for the mid size cylinder aggregate of Fig. 3a and crosses for the large string aggregate shown in Fig. 3b. The intensities correspond to the scattering by single particles 15 respectively 40 micron across at the unit distance of 1 meter and normalized to the incident intensity. At these sizes the scattering adds incoherently and the sets are therefore nearly equal.

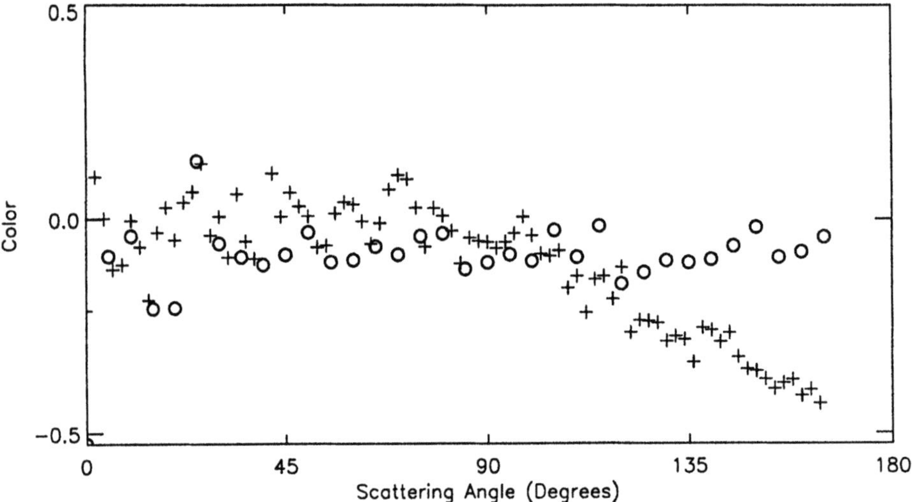


Figure 3d. Angular distribution of the color ratio. The color of the midsize cylinder aggregate (circles) which is more compact is nearly neutral and slightly blue. The large string aggregate (crosses) shows neutral to red color in the forward hemisphere and becomes increasingly blue at higher scattering angles.

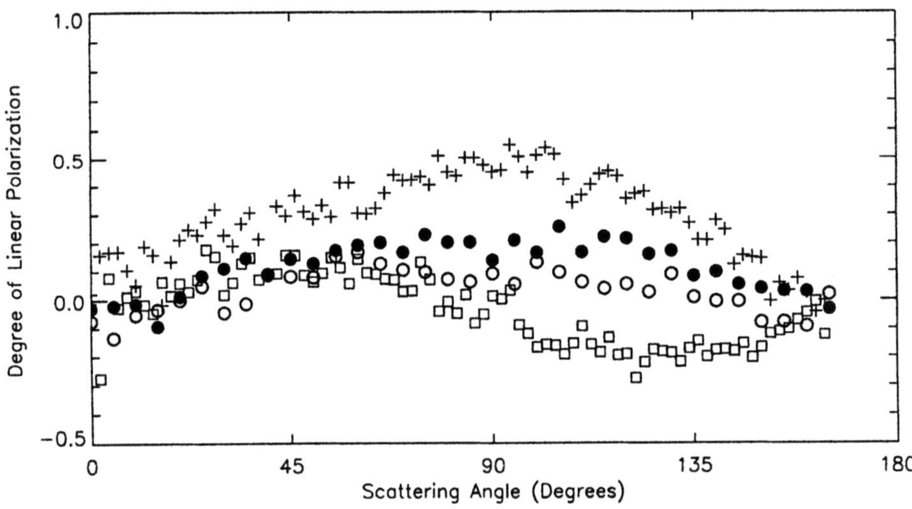


Figure 3e. Polarization plotted versus the scattering angle for the blue and the red halves of the spectrum. Dots are the red for the medium cylinder aggregate and circles the blue. Crosses are for the large string aggregate in the red while squares are for the blue.

We see that while the scattering differs from one model to the other, the general trend is that color and polarization are much lower than they are when produced by independently scattering constituent particles. Both color and polarization is also lower than that due to independent groups of twelve particles that we used to represent clusters of constituent particles. The effect of aggregation is to lower the polarization and a reddening of the forward scattering. Backscattering is also reddened when compared to the color of the constituent particles but the amount varies, probably depending of the importance of coupling to the nearest particles. Unlike the situation for the smaller aggregates investigated previously, polarization is no longer dominated by the constituent particles when the aggregates are ten wavelengths across or larger.

**Acknowledgment**: This work was supported by NASA's Planetary Atmospheres Program through grant NAGW-2482.

## References

1. Gustafson, B.Å.S. (1998) Scattering by Complex Systems I: Methods, this volume.
2. Gustafson, B.Å.S. (1996) Microwave analog to light scattering measurements; A modern implementation of a proven method to achieve precise control, *J. Quant. Spect. Rad. Transf.*, **55:5,** 663-672.
3. Greenberg, J.M. and Gustafson, B.Å.S. (1981) A comet fragment model for Zodiacal light particles, *Astron. Astrophys.* **93,** 35-42.
4. Gustafson, B.Å.S. (1994) Physics of Zodiacal dust, *Ann. Rev. of Earth and Planetary Sciences* **22,** 553-595.
5. Purcell, E.M. and Pennypacker, C.R. (1973) Scattering and absorption of light by nonspherical dielectric grains, *Ap. J.* **186,** 705-714.
6. Mie, G. (1908) Considerations on the optics of turbid media, especially colloidal metal sols, *Ann. Physik.* **25,** pp. 377.
7. van de Hulst, H.C. (1957) *Light Scattering by Small Particles,* Wiley, New York.
8. Zerull, R.H. (1985) Laboratory investigations and optical properties of grains, in R.H. Giese and P. Lamy (eds.), *Properties and Interactions of Interplanetary Dust*, Reidel, Dordrecht, pp. 197-266.
9. Greenberg, J.M., and Hage, J.I. (1990) From interstellar dust to comets: a unification of observational constraints, *Astrophys. J.* **361,** pp. 260-274.
10. Zerull, R.H., Gustafson, B.Å.S., Schulz, K., and Thiele-Corbach, E. (1993) Scattering by aggregates with and without an absorbing mantle: microwave analog experiments, *Applied Optics,* **32:21,** pp. 4088-4100.
11. Lamy, P.L., and Perrin, J.-M. (1986) Volume scattering function and space distribution of the interplanetary dust cloud, *Astron. Astrophys.* **163,** 269-286.
12. Jockers K., Rosenbush, V.K., Bonev, T., and Credner T., *Earth, Moon and Planets*, preprint.

# DYNAMICAL STRUCTURE OF THE ZODIACAL CLOUD

S.F. DERMOTT, K. GROGAN, E. HOLMES AND S. KORTENKAMP*
*University of Florida*
*Department of Astronomy, Gainesville, FL 32611*
**Carnegie Institute of Washington, Washington, D.C. 20015*

**Abstract.** Advances in infrared astronomy and computing power have made detailed study of the structure of the zodiacal cloud possible. An extremely complex cloud has been revealed, quite distinct from the smooth, rotationally symmetric structure assumed prior to the launch of IRAS. The relative contribution of cometary and asteroidal material to the cloud as a whole remains a subject for debate. However, the existence of discrete structures within the cloud, such as the dust bands associated with the Hirayama asteroid families and the Earth's resonant ring, serve as dynamical constraints on the possible sources for the cloud.

## 1. Introduction

The zodiacal cloud, a complex of dust particles that pervades the inner Solar System, is a source of consternation for many astronomers. It dominates the infrared sky and swamps amongst other sources the Cosmic Infrared Background (CIB), the elusive signature of galactic formation in the early Universe. For others, the zodiacal cloud presents a great opportunity for the study of the structure and evolution of our Solar System. Unfortunately, the cloud turns out to be an extremely complicated system which upon inspection raises many dynamical questions. Some of the most important are:

1. What are the sources of the cloud? Both asteroids and comets presumably contribute, but do both contribute significantly or does one dominate over the other? Note that low order secular perturbation theory applies to low eccentricity asteroidal orbits but not to high eccentricity cometary orbits, making asteroidal particle dynamics more tractable.



*J. M. Greenberg and A. Li (eds.), Formulation and Evolution of Solids in Space,* 565–582.

2. If the zodiacal cloud consists of comminution of asteroidal and/or cometary material, what is the size-frequency distribution of the dust? Is there any reason to think that this distribution will vary with distance from the Sun?
3. What is the origin of interplanetary dust particles (IDPs) collected in the atmosphere? Is the source(s) for this material significantly different to that for the zodiacal cloud in general?
4. To what extent has the zodiacal cloud varied over time? How dense was the cloud in the early Solar System, and to what extent could this have affected the climate and the evolution of life on Earth?
5. Can we use the observed structure of our own cloud to detect the presence of unseen planets in exosolar zodiacal clouds?

Historically, the zodiacal cloud has been modeled as a featureless, lenticular structure. Empirical models of this sort have evolved as knowledge of the cloud has increased [1]. However, this approach overlooks the fact that the zodiacal cloud is far from featureless. Sources of structure within the cloud include:

1. The dust bands: comminution products of the Eos, Themis and Koronis asteroid families which appear in the infrared satellite data as 'shoulders' and a 'central cap' superimposed on the broad zodiacal background at low ecliptic latitudes.
2. The symmetry plane of the background cloud.
3. The variation of ecliptic pole brightness with heliocentric distance.
4. The Earth's resonant ring, a structure of asteroidal dust particles trapped in outer mean motion resonances with the planet.

These features represent small, second order effects which serve as dynamical clues and represent our best chance of tackling questions such as those raised above.

## 2. The Dust Bands

The dust bands are a clear indication that the zodiacal cloud is not completely smooth. Discovered by IRAS [2], these features were subsequently attributed to the break up of the major Hirayama asteroid families [3]. It is impossible for a comet to produce a well defined dust band because cometary type orbits have high orbital eccentricities; planetary perturbations produce large variations in these eccentricities and these variations are coupled to those in the inclinations [4]. Therefore even if a group of cometary type orbits initially had identical inclinations, planetary perturbations would disperse those inclinations over a wide range on a timescale of a few precession periods. Figure 1 shows how the dust bands appear in the data.

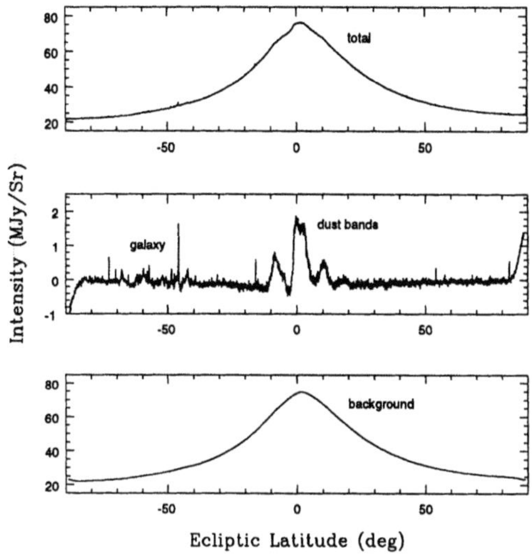


*Figure 1.* A typical IRAS scan (upper panel) is fast-Fourier filtered to produce residuals associated with small scale structure in the cloud (center panel) and the broad zodiacal background (lower panel).

The upper panel shows a typical 'IRAS scan', the variation of zodiacal brightness over the full range of ecliptic latitude (longitude of Earth fixed and solar elongation angle along the line of sight equal to 90°). A fast Fourier filter is applied to this scan, and the resultant residuals and smooth fit are shown in the center panel and lower panel respectively. The residuals show the extent of the small scale structure in the cloud, while the smooth fit is representative of the broad scale zodiacal background.

The dust bands represent the collision products of asteroids, and Jovian perturbations act to precess the longitude of ascending node of the particles such that the nodes are distributed around the ecliptic after approximately $10^6$ years. Since particles in inclined orbits spend a disproportionate amount of time at the extremes of their vertical harmonic oscillations, a set of such orbits with randomly distributed nodes will give rise to two apparent bands of particles symmetrically placed above and below the mean plane of the system. This gives a natural explanation for the 'shoulders' at approximately $\pm 10°$ (the Eos band) and the 'central cap' near the ecliptic plane (a composite Themis and Koronis band).

Figure 2 shows a series of dust profiles viewed in both the leading telescope direction (along the direction of Earth's motion, left panel) and the trailing telescope direction (opposite the direction of Earth's motion, right panel). Each individual profile is produced by coadding a series of raw IRAS scans covering a narrow range of ecliptic longitude to reduce noise. Details to notice in this figure include (1) a sinusoidal variation in the latitudes of peak brightness of the north and south ten degree band pair around the

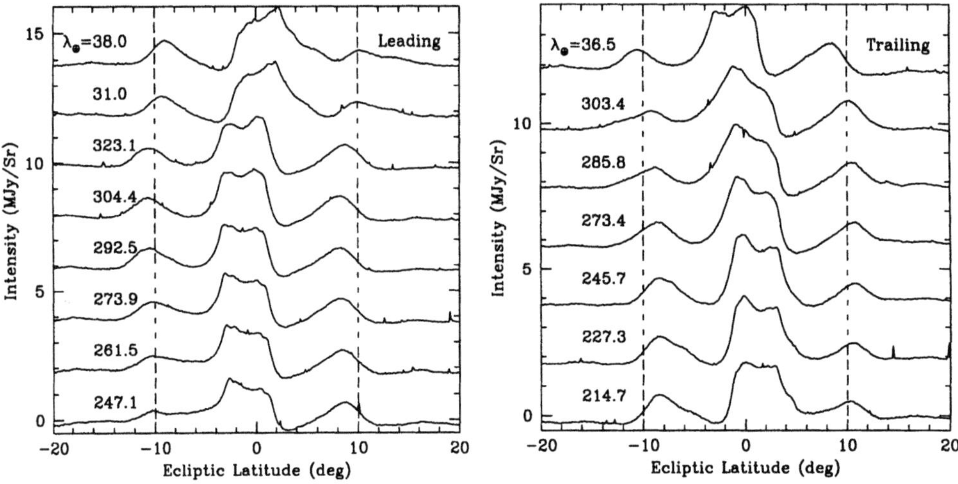


*Figure 2.* Coadded profiles of the dust bands in both the leading direction (along Earth's motion, left panel) and trailing direction (opposite to Earth's motion, right panel). Each profile is taken at a solar elongation angle of 90° and has the associated longitude of Earth labelled.

sky, due to the forced inclinations imposed on the dust particle orbits by planetary gravitational perturbations; (2) the amplitudes of the ten degree band pair vary around the sky, but the north/south components seem to vary opposite in phase to the north/south peaks of the central band. This phase difference is probably due to the small difference between the inclination to the ecliptic of the plane of symmetry of the central dust band and that of the background cloud.

Models of the dust bands produced by the Florida group and constructed from Eos, Themis and Koronis type material have been successful in reproducing these detailed observational features [5]. The approach taken is to produce a fully three-dimensional model of a distribution of dust particles and to perform line of sight integrations through the models to produce brightness profiles which can be directly compared with the observations. The amplitudes of the model profiles are directly proportional to the surface area of material associated with the model, and therefore the amount of material in the dust bands may be directly estimated. Figure 3 shows dust band profiles from two separate models, and the corresponding IRAS observation.

The 2.5° model confines the asteroidal material to the asteroid belt (semi-major axis $>$ 2.0 AU) whereas the 3.5° model allows material to populate the inner Solar System due to Poynting-Robertson drag. The 2.5° and 3.5° refer to the dispersion in proper inclinations of the Eos material in each model required to match the shape of the observed bands. Figure 3 shows that the peak dust band emission from the 2.5° model is approxi-

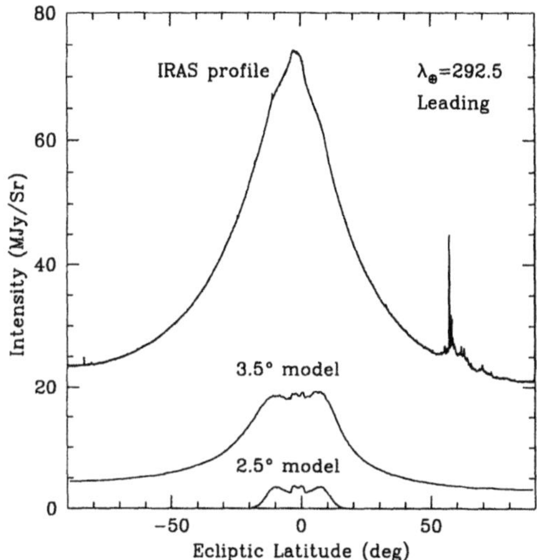


*Figure 3.* Dust band profiles from two separate models, one which confines asteroidal material to the asteroid belt (the 2.5° model) and another which allows material to populate the inner Solar System via Poynting-Robertson drag (the 3.5° model). The corresponding IRAS scan is also shown. Based on these two models, the dust bands contain between 5% and 25% of the total zodiacal cloud, in terms of the material that contributes to the infrared thermal emission in the 25 micron waveband.

mately 5% of the peak zodiacal background emission. However, the emission from the 3.5° model is significantly higher with appreciable emission at the ecliptic poles and a peak intensity approximately 25% that of the zodiacal background. This is because the material populating the inner Solar System is warmer and closer to Earth. It has been calculated that the ratio of dust production by asteroid families to dust production by all asteroids is approximately 1:3 [6]. This would imply that up to 75% of the zodiacal cloud, in terms of material which gives rise to the thermal emission in the 25 micron waveband, may be asteroidal.

The dust band models described above are composed of single sized dust particles, 10 microns in diameter. The dust bands are collision products and will therefore have a size-frequency distribution. The number of particles in such a distribution with a diameter greater than d is given by

$$N(>d) = \frac{1}{3(q-1)} \left( \frac{d_0}{d} \right)^{3(q-1)} \tag{1}$$

where $d_o$ is a constant determined by the total volume of the source material. The associated cumulative area associated with this distribution is given by

$$A = \frac{\pi d_{min}^2}{4(3q-5)} \left( \frac{d_0}{d_{min}} \right)^{3(q-1)} \tag{2}$$

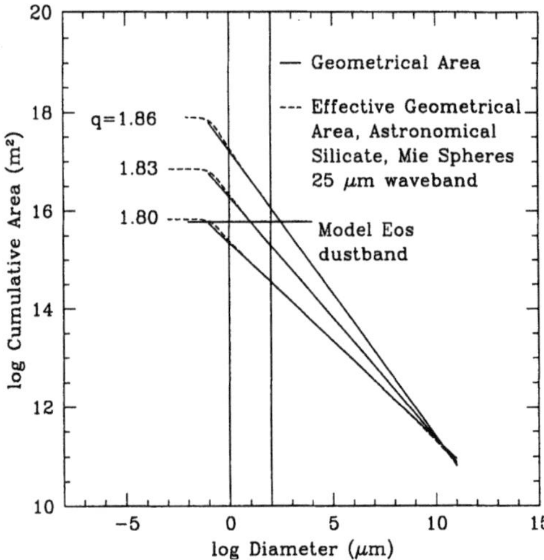


*Figure 4.* The cumulative surface area associated with three different size-frequency distributions centered on the theoretical collisional equilibrium value of q=1.83.

The theoretical solution to a system in collisional equilibrium has a size-frequency index q=1.83 [7]. Figure 4 shows the cumulative surface area of material associated with three different distributions centered on q=1.83. The geometrical area increases to infinity as smaller and smaller particles are included, but the effective surface area of astronomical silicate spheres [8] levels off as the emission efficiency of particles smaller than approximately 0.1 micron approaches zero.

Notice that the surface area given by the single size (10 micron) Eos dust band models is marked on the diagram. This value is remarkably close to that expected for a collisional equilibrium value of q=1.83 and suggests that material in the asteroid belt is close to collisional equilibrium. However, it is expected that the size-frequency distribution will vary with heliocentric distance: Poynting-Robertson drag will act like a sieve, preferentially removing the smaller material. The fact that the model results indicate collisional equilibrium may be indicative of the fact that as material approaches the Earth it appears to move to higher latitudes due to a parallax effect and merges into the background; the filter will therefore be 'biased' towards material in the asteroid belt. Whereas the IRAS observations are concentrated at around 90° solar elongation, the COBE observations have a wide range of solar elongation angle (64° to 124°) and future analysis of the COBE data will provide valuable information on the distribution of material throughout the inner solar system and serve to constrain the dust band models further.

## 3. The Background Cloud

The zodiacal cloud is typically assumed, for the sake of simplicity, to be rotationally symmetric in heliocentric ecliptic longitude. The radial and out-of-ecliptic dependence of the number density distribution of the dust particles are separated and the dust distribution is approximated by an expression of the form

$$n(r,z) \sim r^{-\nu} f(\beta) \tag{3}$$

where z is the height above the ecliptic plane and $\beta = \arcsin(z/r)$ is the heliocentric latitude [1]. The plane of symmetry of the cloud should be dictated by the dominant perturbing forces, the gravitational perturbations produced by the giant planets. Table 1 shows the planes of symmetry with respect to the ecliptic for several features of the Solar System. Figure 5 shows how the plane of symmetry of zodiacal cloud features may be found. The two curves represent the variation of the latitude of peak brightness of the background cloud with ecliptic longitude of Earth, in both the trailing and leading directions. The points at which the latitudes of peak intensity are equal and opposite give the nodes of the cloud; the magnitude of the latitude here gives the inclination of the cloud with respect to the ecliptic. Notice that the result provides evidence for a warp in the disk: the ascending and descending nodes would be exactly 180° apart in the absence of a warp.

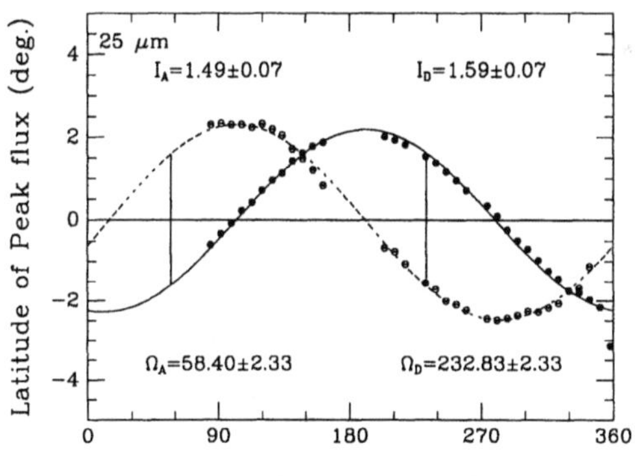


*Figure 5.* The variation of the latitude of peak flux of the zodiacal cloud with ecliptic longitude of Earth in the trailing (solid curve) and leading (dotted curve) directions. The positions at which the trailing and leading latitudes are equal and opposite give the nodal longitudes.

The nodes of the cloud can also be found from the variation of the brightness of the ecliptic poles with ecliptic longitude of Earth, as shown in Figure 6. COBE observations of the north and south ecliptic poles as a function of ecliptic longitude of Earth are shown. When the north and south polar brightnesses are equal the Earth lies in the midplane of the

TABLE 1. Orientation of planes of symmetry with respect to the ecliptic (25 micron waveband)

| Structural Feature | Ascending Node | | Descending Node | |
|---|---|---|---|---|
| | Inclination | Node | Inclination | Node |
| Jupiter | 1.31° | 100.0° | 1.31° | 280.0° |
| Eos family | 1.19° | 97.1° | 1.19° | 277.1° |
| Themis family | 1.22° | 97.8° | 1.22° | 277.8° |
| Ten degree bands | 1.16±0.09° | 100±8° | 1.08±0.06° | 266±8° |
| Zodical Cloud (Ecliptic) | 1.49±0.07° | 58.40±2.33° | 1.59±0.07° | 232.83±2.33° |

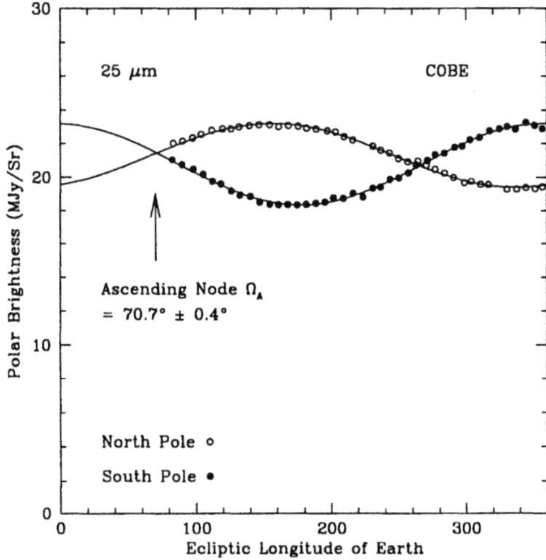


*Figure 6.* The variation of the polar brightness of the zodiacal cloud with ecliptic longitude of Earth. The positions at which the north and south polar brightnesses are equal gives the positions of the nodes.

cloud, the longitude of the ascending/descending node. This method gives an ascending node of 70.7°, and can be thought of as a measurement of the node at 1 AU. This value is significantly different to the values found for the dust bands and the 'in-ecliptic' background, and provides further evidence for a warped zodiacal cloud.

In addition to the conclusions drawn about the structure of the background from these results, the data given in Table 1 provide support for the localized nature of the dust band material separated from the background by the Fourier filter. The observed plane of symmetry of the ten degree band is quite distinct from that of the background cloud, but is closely similar to that of the asteroid families in the outer part of the mainbelt.

## 4. The Polar Flux Gradient

The Earth moves in an eccentric orbit and therefore the heliocentric distance of the Earth varies over the course of a year between aphelion and perihelion. This provides an opportunity to compare the observed variation of the polar flux with heliocentric distance to theoretical values appropriate for asteroidal and cometary distributions, which are markedly different to one another.

We will assume that the polar flux at the heliocentric distance $r$ can be separated into three components in the following manner:

$$B_p(r) = rn(r)f(T) \tag{4}$$

where $n(r)$ is the particle number density in the plane of symmetry of the cloud, $f(T)$ is the variation of the polar flux due to the variation of temperature T with heliocentric distance, and r is a scaling factor (the polar flux will be proportional to the scale height of the cloud). This expression can be rewritten in terms of the polar flux gradient,

$$\frac{r}{B_p(r)}\frac{dB_p(r)}{dr} = 1 + \frac{r}{n(r)}\frac{dn(r)}{dr} + \frac{r}{f(T)}\frac{df(T)}{dr}. \tag{5}$$

If we write the number density as

$$n(r) = \frac{c}{r^\gamma} \tag{6}$$

where c is a constant then the second term on the right hand side of Equation 5 is recognized:

$$\frac{r}{n(r)}\frac{dn(r)}{dr} = -\gamma. \tag{7}$$

However, due to Poynting-Robertson drag,

$$\frac{da}{dt} = -\frac{c}{a}\frac{(2+3e^2)}{(1-e^2)^{3/2}} \tag{8}$$

where c is a constant inversely proportional to the particle radius, and

$$\frac{de}{dt} = -\frac{5c}{2a^2}\frac{e}{(1-e^2)^{1/2}}. \tag{9}$$

Hence,

$$\frac{de}{da} = -\frac{5e}{2a}\frac{(1-e^2)}{(2+3e^2)} \tag{10}$$

$$\frac{a}{\dot{a}}\frac{d\dot{a}}{da} = -\left(1 - \frac{15}{2}e^2\frac{(4+e^2)}{(2+3e^2)^2}\right) \tag{11}$$

and we can show that

$$\gamma = 2 + \frac{a}{\dot{a}}\frac{d\dot{a}}{da}. \tag{12}$$

The final term in Equation 5 represents the variation in polar flux due to temperature alone and can be found from the Planck function,

$$f(T) = B = \frac{2hc}{\lambda^3}\left(e^{T_0/T} - 1\right)^{-1} \tag{13}$$

where

$$T_0 = \frac{hc}{\lambda k}. \tag{14}$$

Consequently,

$$\frac{r}{f(T)}\frac{df(T)}{dr} = \frac{-e^{T_0/T}}{(e^{T_0/T}-1)}\frac{T_0}{2T} = -\alpha. \tag{15}$$

and from Equations 5, 7, 11, 12 and 15, the polar flux gradient becomes

$$\frac{r}{B_p(r)}\frac{dB_p(r)}{dr} = 1 - \gamma - \alpha. \tag{16}$$

While $\alpha$ is a constant in a given waveband, $\gamma$ is strongly dependent on the eccentricity of the dust particles.

Observationally the polar flux gradient can be found from the variation of the mean north/south polar flux with ecliptic longitude of Earth, as shown in Figure 7 in both the 12 and 25 micron wavebands. The ratio of the amplitude to the mean of the sine curve divided by the Earth's eccentricity is a measure of the polar flux gradient. Notice also that the minimum in the polar brightness does not occur at Earth's apocenter, where the local number density of the cloud should be a minimum. This is an indication that the center of the cloud is offset from the Sun.

TABLE 2. The polar flux gradient: observed and theoretical values for single size particles.

| | Predicted | | Observed |
|---|---|---|---|
| | Cometary | Asteroidal | |
| 12 $\mu m$ | -2.56 | -2.05 | -1.98±0.09 |
| 25 $\mu m$ | -1.71 | -1.19 | -1.30±0.07 |

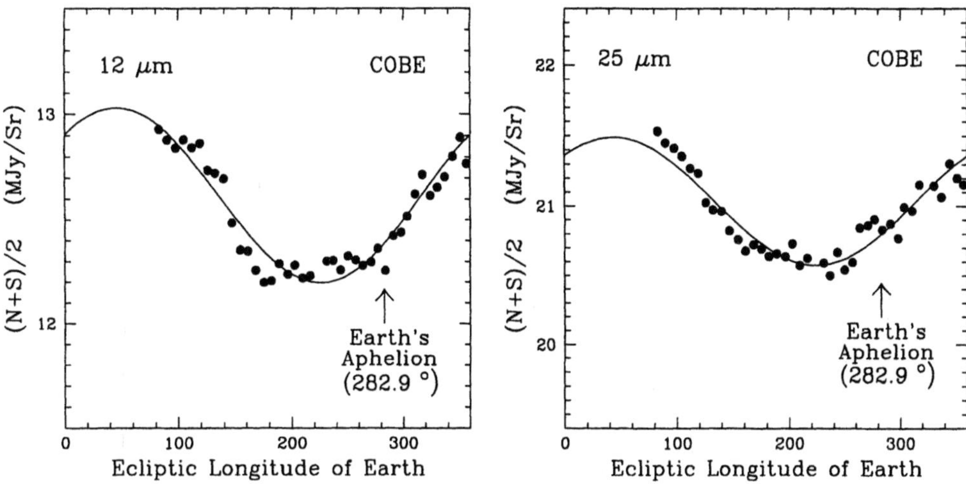


*Figure 7.* The variation of the mean north/south polar brightness as a function of ecliptic longitude of Earth, in both the 12 micron (left panel) and the 25 micron (right panel) wavebands. For each, the ratio of the amplitude to the mean of the sine curve divided by the Earth's eccentricity fis a measure of the polar flux gradient. The minimum in the polar brightness does not occur at Earth's apocenter, where the local number density of the cloud should be a minimum. This is an indication that the center of the cloud is offset from the Sun.

Table 2 compares theoretical values for the flux gradient of cometary type (e $\sim$ 0.3) and asteroidal type (e $\sim$ 0.05) distributions at 1 AU. It is clear that the observed polar flux gradient points to an asteroidal type distribution rather than cometary. However, this conclusion is based on models for asteroidal and cometary distributions of a single size, 10 micron diameter. Figure 8 demonstrates the effect of imposing a size-frequency distribution on the models. The size-frequency index for both model distributions is varied and the polar flux gradient directly obtained by measuring the polar flux as a function of heliocentric distance. These are the curves shown in Figure 8. Also shown are the values for the polar flux gradient in each waveband as obtained from the COBE observations. A high value of q is suggested by this method, at least as high as the value expected for collisional equilibrium. However, without a knowledge of the observed value of q at 1 AU, we are unable to use this method to distinguish between an asteroidal and a cometary source.

Evidence for the actual size-frequency distribution at 1 AU is found from the cratering record on the LDEF satellite [9]. Analysis of micrometeorite craters on the space facing panels of LDEF resulted in an estimate of the total annual mass influx of $3.10^7$ kg yr$^{-1}$. Figure 9 shows this influx in terms of surface area per log mass interval. Rather than the surface area being dominated by the small particle population as expected in collisional equilibrium, the curve is dominated by large particles, peaking at

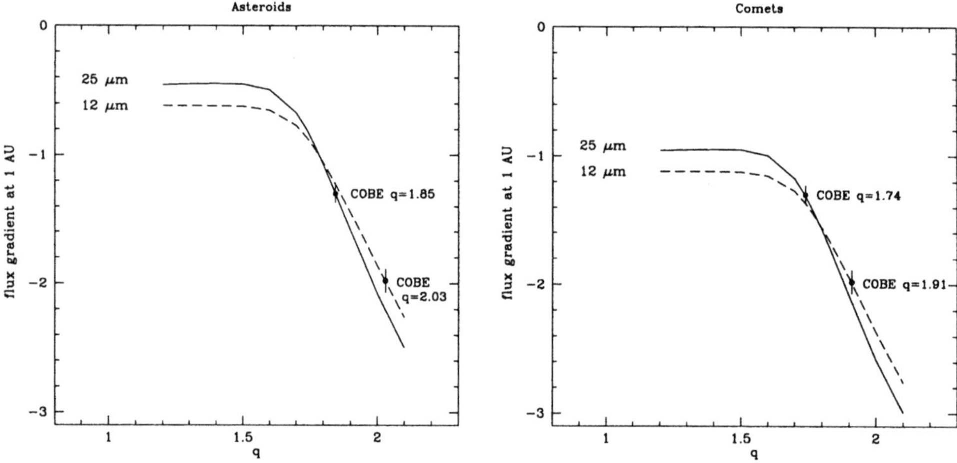


*Figure 8.* The variation of the polar flux gradient with size-frequency index q for both asteroidal (left panel) and cometary (right panel) distributions in the 12 and 25 micron wavebands.

a particle diameter of approximately 100 microns. The linear portion of the curve between 10 and 100 microns indicates a size-frequency index q of approximately 1.1. This result points to the removal of small particles by Poynting-Robertson drag and possibly by collisions as they fragment and are blown out of the Solar System as $\beta$-meteoroids. It follows from this ob-

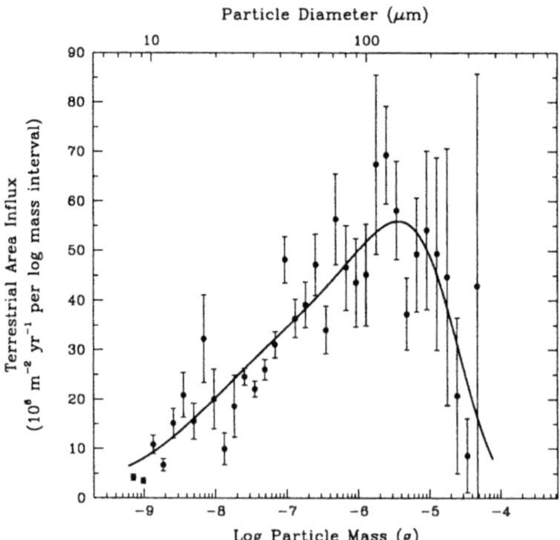


*Figure 9.* The LDEF cratering record. The influx of particles is dominated by the large particles in the distribution rather than the small particles as expected from a distribution in collisional equilibrium. The linear portion of the curve between 10 and 100 microns diameter suggests a size-frequency index q of the zodiacal cloud at 1 AU of approximately 1.1.

servation that while the IRAS dust bands allow us to estimate the input of asteroidal dust to the zodiacal cloud, more information is needed to determine how much of this dust is transported to the Earth. We also note that the LDEF results give information on the size-frequency distribution of the particles accreted by the Earth, but because the Earth preferentially accretes asteroidal particles, and because the Earth's resonant dust ring may have a role in accretion, this distribution may differ from the distribution in the background cloud.

## 5. Earth's Resonant Ring

Numerical integrations have shown that a significant fraction of particles that grind down in the asteroid belt and spiral in towards the Sun via Poynting-Robertson drag are temporarily trapped in corotational resonances external to the orbits of each of the terrestrial planets, particularly Earth [10], [11]. The result is that Earth is embedded in a ring of asteroidal particles (more correctly, low eccentricity particles) that is longitudinally near uniform apart from a cavity that contains the planet. The action of drag introduces a phase lag into the equations of motion of these resonant particles. The physical manifestation of this is a phase difference between the longitude of conjunction of the dust particle and the Earth and the apocenter of the dust particle's orbit, with the result that the position of the Earth within the cabity is markedly asymmetric, with the Earth closer to the edge of the cavity in the trailing orbital direction than in the leading orbital direction.

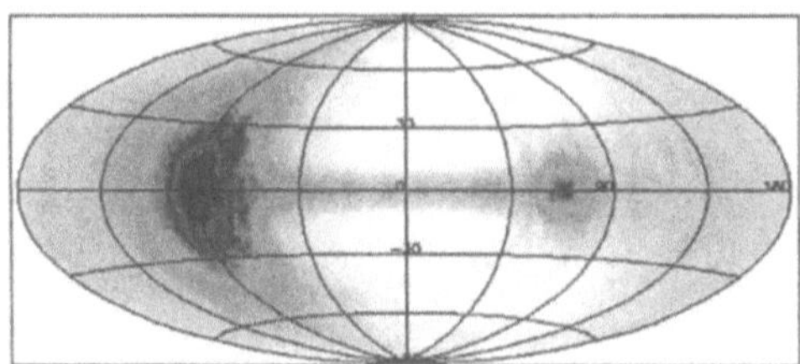

*Figure 10.* Geocentric all-sky view of Earth's resonant ring, trailing direction to the left, leading direction to the right.

This gives a natural explanation for a peculiar asymmetry noted in both the IRAS data and the COBE data, namely that the observations of the peak brightness of the zodiacal cloud in the trailing dirction are consistently greater than those in the trailing direction. Figure 10 shows an all-sky view of a model of the resonant ring [12], the trailing cloud on the left appearing brighter than the leading cloud on the right. A diagram similar to this was shown on the cover of Nature in 1994 [11]. Reach *et al.* 1995 [13] produced a dramatic confirmation of the result, suppressing the background zodiacal cloud by use of an empirical model. The residuals closely matched the structures seen in Figure 10.

## 6. Source of Interplanetary Dust Particles collected at Earth

As the Earth orbits the Sun, it sweeps up particles from the zodiacal cloud. The rate at which particles are captured is dependent on three factors: (1) the average spatial density of particles on Earth-crossing orbits; (2) the geocentric encounter velocity of the particles; (3) the effective gravitational cross-section of the planet. Due to their higher eccentricities, Earth-crossing cometary particles encounter the planet at substantially higher velocities than asteroidal particles and their effective capture cross-section is smaller. Also, the higher eccentricities of the cometary particles result in a lower average spatial density compared with the asteroidal particles. These two factors when combined result in lower capture rates for cometary particles [14], as shown in Figure 11.

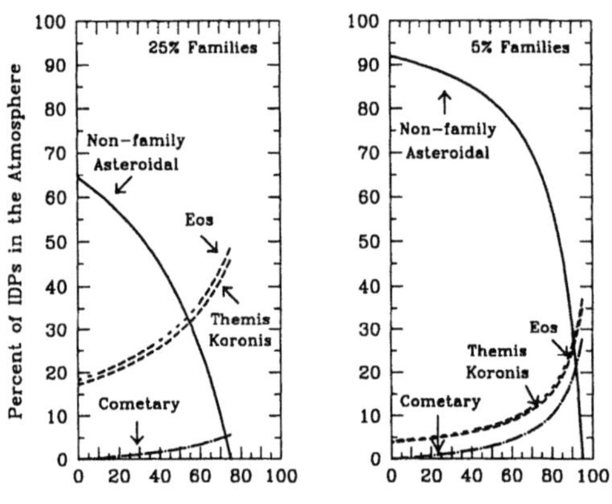


*Figure 11.* Percentages of IDPs in the atmosphere. Using a range of 5% (right panel) to 25% (left panel) for the fraction of asteroid family dust in the background cloud at 1 AU, we combine the relative contributions from each family with their capture rates to determine the percentage of IDPs due to the Eos, Themis and Koronis families. A combination of cometary and non-family asteroidal dust makes up the remainder. In all possible cases, IDPs are predominantly asteroidal.

As discussed earlier, evidence from study of the dust bands suggests that between 5% (right panel) and 25% (left panel) of zodiacal dust originates in the Eos, Themis and Koronis asteroid families. If the ratio of dust production in families to dust production in all asteroids is approximately 1:3 [6] then all asteroids (family and non-family) contribute between 15% and 75%. Comets would then supply the remainder. Although the actual contribution of cometary material to the cloud at 1 AU is unknown, Figure

11 is set up such that any cometary percentage may be chosen, once the estimates for the asteroidal contribtuion taken from the dust band work have been taken into account. Regardless of the cometary contribution to the cloud, certain conclusions are unavoidable from these diagrams. The existence of the dust bands and the evidence that asteroidal dust is transported to the inner Solar System implies that most IDPs are asteroidal. Furthermore, the Themis and Koronis families while contributing between 1% and 4% to the zodiacal cloud may contribute significantly to the atmospheric IDPs due to their low inclinations. Under no circumstances does cometary dust represent more than 28% of IDPs and probably represents less than 10%. Earth predominantly accretes asteroidal dust, and a large, perhaps dominant fraction of this dust comes from the three asteroid families Eos, Themis and Koronis. Samples of these known S-type (Eos, Koronis) and C-type (Themis) asteroids may already exist in our IDP collections.

Muller and MacDonald (1997) [15] suggested that a periodic variation in the accretion rate of IDPs might be driving the Earth's 100,000 year climatic cycle. They proposed that the accretion rate might be linked to the varying inclination of the Earth's orbit with respect to the invariable plane of the Solar System (the plane perpendicular to the total orbital angular momentum vector of the planets). Kortenkamp and Dermott (1998) [16] have reconstructed the orientations of the Earth-crossing portions of the dust bands via direct numerical integrations of individual dust particles back to 1.2 million years ago. At the time when each dust particle became Earth-crossing (approximately 50,000 years after release for a 10 micron diameter particle under the influence of Poynting-Robertson drag), the inclination of the Earth's orbit with respect to the mid-plane of the bands was calculated, and the capture rate of the dust particles calculated (Figure 12).

During periods when the Earth's inclination rises above the mean inclination of the low inclination Themis and Koronis dust bands, Earth is actually outside most of the dust band material for some period of time each year, and the capture rate of the Themis and Koronis material falls. The higher inclination of the Eos dust band ensures that Earth is always within that dust band, and so the capture rate of Eos material is not correlated with the inclination of the Earth. However, the capture rates for all three families are anti-correlated with the Earth's varying orbital eccentricity.

Recent work [18] suggests that asteroids down to a few hundred meters across may be 'rubble piles' - conglomerations of collisional debris held together only by self-gravity. The occasional catastrophic disruption of 10 to 100 km radius rubble pile asteroids would instantaneously liberate a mass of dust several orders of magnitude higher than that associated with the current zodiacal cloud. Within $10^5$ to $10^6$ years Earth would be accreting

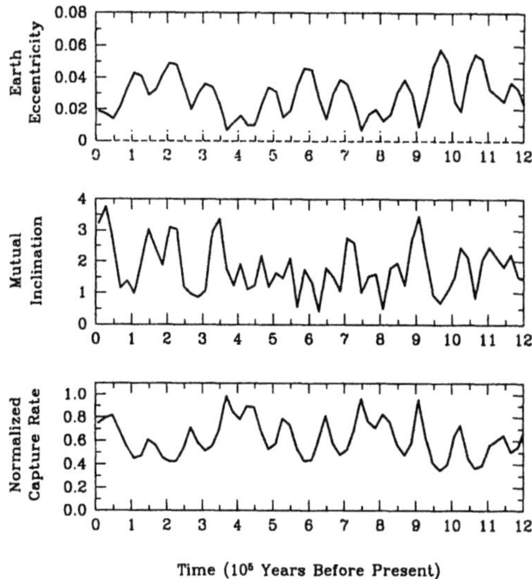


*Figure 12.* Variation in the capture rate of 10 micron diameter Eos particles over the past 1.2 million years. Top panel: Varying eccentricity of the Earth's orbit [17]; Middle panel: Variation in the mutual inclination of the Earth's orbit and the plane of symmetry of the dust particles at 1 AU; Bottom panel: Variation in the capture rate.

dust particles at a rate of approximately $10^{10}$ kg yr$^{-1}$, comparable to the stratospheric loading of dust and aerosols due to explosive volcanic eruptions [19], and may lead to significant changes in Earth's climate.

## 7. Conclusions

Much has been achieved in modeling and interpreting the infrared spacecraft data. Detailed models for the dust bands and the ring have been produced, and asymmetries in the data have been used to interpret the inclination and nodes of the background cloud. However, evidence from both COBE and LDEF suggests that the use of a single-size particle for the components of the zodiacal cloud are oversimplified. In order to unravel such questions as the radial extent of the dust bands and the nature of the dust at 1 AU, a detailed investigation of the size-frequency distribution of the material as a function of heliocentric distance must be made.

Direct evidence for the size-frequency distribution of the zodiacal cloud (albeit at a single heliocentric distance of 1 AU) has been provided by LDEF. However, the precise meaning of the result remains unclear. Earth preferentially captures material with low eccentricites and low inclinations. In addition, most particles trapped in the ring are released from resonance due to close encounters with the Earth, the ring acting as a funnel through which asteroidal particles are deposited into Earth's atmosphere, increasing the efficiency of the transport of carbonaceous material from the asteroid belt to Earth. The extent to which these factors may affect the LDEF results has yet to be determined.

Undertanding the structure of the zodiacal cloud is of the utmost importance, not only in the sense of gaining knowledge of our local environment, but also in that lessons learnt may be directly applied to zodiacal clouds around other stars. Planets not otherwise visible may contribute towards warps in the structure of their own clouds. Dust band and ring structures such as those discussed here may indicate the presence of planets and asteroid belts. Ultimately, the study of these exosolar zodiacal clouds may complement the study of our own in helping us understand the evolution of the Solar System.

## References

1. Giese, R.H., Kneissel, B. and Rittich, U. (1986) Three-dimensional zodiacal dust cloud, a comparative study, *Icarus*, **68**, 395-411.
2. Low, F.J., Beintema, D.A., Gautier, T.N., Gillet, F.C., Beichmann, C.A., Neugebauer, G., Young, E., Aumann, H.H., Boggess, N., Emerson, J.P., Habing, H.J., Hauser, M.G., Houck, J.R., Rowan-Robinson, M., Soifer, B.T., Walker, R.G. and Wesselius, P.R. (1984) Infrared Cirrus: New Components of the Extended infrared Emission, *Astrophys.J*, **278**, L19-L22.
3. Dermott, S.F., Nicholson, P.D., Burns, J.A. and Houck, J.R. (1984) Origin of the Solar System dust bands discovered by IRAS, *Nature*, **312**, 505-509.
4. Liou, J.C., Dermott, S.F. and Xu, Y.L. (1995) The contribution of cometary dust to the zodiacal cloud, *Planet. Space Sci.*, **43**, 717-722.
5. Grogan, K., Dermott, S.F., Jayaraman, S. and Xu, Y.L. (1998) Origin of the Ten Degree Dust Bands, *Planet. Space Sci.*, **45**, 1657-1665.
6. Durda, D.D. and Dermott, S.F. (1997) The Collisional Evolution of the Asteroid Belt and its Contribution to the Zodiacal Cloud, *Icarus*, **130**, 140-164.
7. Dohnanyi, J.S. (1969) Collisional models of asteroids and their debris, *J.Geophys.Res.*, **74**, 2531-2554.
8. Draine, B.T. and Lee, H.M. (1984) Optical properties of interstellar graphite and silicate grains, *Astrophys.J*, **285**, 89-108.
9. Love, S.G. and Brownlee, D.E. (1993) A direct measurement of the terrestrial mass accretion rate of cosmic dust, *Science*, **262**, 550-553.
10. Jackson, A.A and Zook, H.A. (1989) A Solar System dust ring with the Earth as its shepherd, *Nature*, **337**, 629-631.
11. Dermott, S.F., Jayaraman, S., Xu, Y.L., Gustafson, B.Å.S. and Liou, J.C. (1994) A circumsolar ring of asteroidal dust in resonant lock with the Earth, *Nature*, **369**, 719-723.
12. Jayaraman, S. and Dermott, S.F. (1998) Formation, Structure and Observations of the Earth's Resonant Ring, *Icarus*, in press.
13. Reach, W.T., Franz, B.A., Weiland, J.L., Hauser, M.G., Kelsall, T.N., Wright, E.L., Rawley, G., Stemwedel, S.W. and Spiesman, W.J. (1995) Observational Confirmation of a Circumsolar Dust Ring by the COBE satellite, *Nature*, **374**, 521-523.
14. Kortenkamp, S.J. (1996) Ph.D Thesis, University of Florida.
15. Muller, R.A. and MacDonald, G.J. (1997) Glacial Cycles and Astronomical Forcing, *Science*, **277**, 215.
16. Kortenkamp, S.J. and Dermott, S.F. (1998) A 100,000 Year Periodicity in the Accretion Rate of Interplanetary Dust, *Science*, **280**, 874-876.
17. Quinn, T.R., Tremaine, S. and Duncan, M. (1991) A three million year integration

of the Earth's orbit, *Astron. J*, **101**, 2287-2305.
18. Love, S.G. and Aherns, T.J. (1996) Catastrophic Impacts on Gravity Dominated Asteroids, *Icarus*, **124**, 141-155.
19. Rampino, M.R., Self, S. and Stothers, R.B. (1988) *Ann. Rev. Earth Planet. Sci.*, **16**, 73.

# List of Contributors

# List of Participants

**M.V. AKDENIZ**
Middle East Technical University
Dept. of Metallurgical Engineering
ANKARA 06951, Turkey
akdeniz@rorqual.cc.metu.edu.tr

**L. BECKER**
Scripps Institution of Oceanography
Marine Research Department
8602 La Jolla Shares Drive
LA JOLLA, CA 92093-0212, USA
lbecker@ucsd.edu

**J. BLUM**
University of Jena
Astrophysical Institute
Schillergasschen 3
07745 JENA, Germany
blum@astro.uni-jena.de

**A.C.A. BOOGERT**
University of Groningen
Kapteyn Astronomical Institute
P.O. Box 800
9700 AV GRONINGEN
The Netherlands
boog@astro.rug.nl

**J. BOUWMAN**
University of Amsterdam
Astronomical Institute
"Anton Pannekoek"
Kruislaan 403
1098 SJ AMSTERDAM
The Netherlands
jeroenb@astro.uva.nl

**J.P. BRADLEY**
C/O MVA Inc.
5500/200 Oakbrook Parkway
NORCROSS, GA 30093
USA
jbrad@mindspring.com

**J.R. BRUCATO**
Osservatorio Astronomico
di Capodimonte
Laboratorio di Fisica Cosmica
Via Moiariello 16
80131 NAPOLI, Italy
brucato@cosmic.na.astro.it

**H.M. BUTNER**
University of Arizona
SMTO - Steward Observatory
933 N. Cherry Avenue
TUCSON, AZ 85721
USA
hbutner@as.arizona.edu

**J. CAMI**
Space Research Organization
of The Netherlands (SRON)
P.O. Box 800
9700 AV GRONINGEN
The Netherlands
cami@astro.rug.nl

**S. CHARNLEY**
NASA Ames Research Center
Space Science Division, MS 245-3
MOFFETT FIELD, CA 94035-1000
USA
charnley@dusty.arc.nasa.gov

**J.E. CHIAR**
NASA-AMES Research Center
Mail Stop 245-3
MOFFETT FIELD, CA 94035
USA
chiar@misty.arc.nasa.gov

**L. COLANGELI**
Osservatorio Astronomico
di Capodimonte
Via Moiariello 16

80131 NAPOLI, Italy
colangeli@astrna.na.astro.it

**J. CROVISIER**
Observatoire de Paris-Meudon
F-92195 MEUDON
France
crovisie@obspm.fr

**L. D'HENDECOURT**
Institut d'Astrophysique Spatiale
Space Chemistry Department
Campus d'Orsay - Bat 121
ORSAY 91405 Cedex
France
ldh@iaslab.ias.fr

**E. DARTOIS**
Institut d'Astrophysique Spatiale
Space Chemistry Department
Campus d'Orsay, ORSAY 91405
France
dartois@ias.fr

**M.C. DE SANCTIS**
IAS-CNR-Planetologia
Viale dell'Universita' 11
00185 ROMA, Italy
cristina@saturn.ias.fra.cnr.it

**S. DERMOTT**
University of Florida
Astronomy Department
211 Bryant Space Sciences Center
GAINESVILLE, FL 32611
USA
dermott@astro.ufl.edu

**E.F. Van Dishoeck**
Leiden Observatory
P.O. Box 9513, 2300 RA LEIDEN
The Netherlands
ewine@strw.leidenuniv.nl

**C. DOMINIK**
Leiden Observatory
P.O. Box 9513, 2300 RA LEIDEN
The Netherlands
dominik@strw.leidenuniv.nl

**J. DORSCHNER**
University of Jena
Astrophysical Institute
Schillergaesschen 3
D-07745 JENA
Germany
dorsch@astro.uni-jena.de

**H.S. DUZGUN**
Middle East Technical University
Odtu Maden Muh. Bol.
06531 ANKARA, Turkey
duzgun@rorqual.cc.metu.edu.tr

**E. EPIFANI**
Osservatorio Astronomico
di Capodimonte
Laboratorio di Fisica Cosmica
Via Moiariello 16
80131 NAPOLI, Italy
epifani@cosmic.na.astro.it

**F. ESPOSITO**
Osservatorio Astronomico
di Capodimonte
Laboratorio Fisica Cosmica
Via Moiariello 16
80131 NAPOLI, Italy
esposito@cosmic.na.astro.it

**G. FERRINI**
Osservatorio Astronomico
di Capodimonte
Laboratorio di Fisica Cosmica
Via Moiariello 16
80131 NAPOLI, Italy
ferrini@cosmic.na.astro.it

**G. GALAZUTDINOV**
Special Astrophysical Observatory

Coude-Echelle Laboratory
357147 NIZHNIJ ARKHYZ
Russia
gala@sao.ru

**I. GALETICH**
Institute for Low Temperature
Physics and Engineering
Dept. of Molecular Biophysics
47 Lenin Avenue, 310164, Kharzov
Ukraine
galetich@ilt.kharkov.ua

**M. GOLEBIEWSKI**
Torun Center for Astronomy
Gagarina 11, PL-87-100 TORUN
Poland
abyss@astri.uni.torun.pl

**J.M. GREENBERG**
Labratory Astrophysics
University of Leiden
Postbus 9504, 2300 RA Leiden
The Netherlands
greenber@strw.LeidenUniv.nl

**B.A.S. GUSTAFSON**
University of Florida
Astronomy Department
211 SSRB, GAINESVILLE
FL 32611, USA
gustaf@astro.ufl.edu

**H.J. HABING**
Leiden Observatory
P.O. Box 9513, 2300 RA LEIDEN
The Netherlands
habing@strw.leidenuniv.nl

**M.R. HOOGERHEIJDE**
Leiden Observatory
P.O. Box 9513, 2300 RA LEIDEN
The Netherlands
michiel@strw.leidenuniv.nl

**M.A. IATI'**
DIP. Struttura della Materia
e Tecnologie, Fis. Avanzate
S. Agata-Contr. Papardo
Salita Sperone 31
98166 MESSINA, Italy
iati@ortica.unime.it

**A. ILIN**
Pulkovo Observatory
65/1 Pulkovskoe Shosse
196140 ST PETERSBURG
Russia
alexei@ilin.spb.su

**D.A. JOINER**
Rensselaer Polytechnic Institute
Physics & Astronomy Department
15th Street, TROY
NY 12180, USA
joined@rpi.edu

**J. KERRIDGE**
University of California, San Diego
Chemistry Department 0317
LA JOLLA, CA 92093, USA
jkerridg@ucsd.edu

**J. KISSEL**
MPI F. Kernphysik
P/O Box 103980
D-69029 HEIDELBERG
Germany
kissel@kosmo.mpi-hd.mpg.dg

**M. KOLMAN**
Physics Department
Sara Burgerhart straut 25
1055 KV AMSTERDAM
The Netherlands
m.kolman@elsevier.nl

**J. KRELOWSKI**
Torun Center for Astronomy
Gagarina 11, PL-87-100 TORUN

Poland
jacek@astri.uni.torun.pl

**P.O. LAGAGE**
CEA/DSM/DAPNIA/SAP
Centre d'Etude de Saclay
F-91191 GIF-SUR-YVETTE
France
lagage@cea.fr

**F. LAHUIS**
Agencia Espacial Europea
ISO Operations Centre
Apartado de Correos 50727
Villafranca del Castillo
ES-28080 - MADRID Spain
flahuis@iso.vilspa.esa.es
*and* freddy@sron.rug.nl

**A. LI**
Laboratory Astrophysics
University of Leiden
Postbus 9504, 2300 RA LEIDEN
The Netherlands
agli@strw.leidenuniv.nl

**J. LINDHOLM**
Onsala Space Observatory
S-43992 ONSALA
Sweden
jim@oso.chalmers.se

**K. MALFAIT**
University of Leuven
Celestynenlaan 200B
B-3001 HEVERLEE
Belgium
koenm@ster.kuleuven.ac.be

**G. MEEUS**
University of Leuven
Celestynenlaan 200B
B-3001 HEVERLEE
Belgium
gwendolyn@ster.kuleuven.ac.be

**A. MEGIER**
Nicolaus Copernicus University
Centre of Astronomy
Gagarina 11, PL-87-100 TORUN
Poland
anmeg@astri.uni.torun.pl

**F. MOLSTER**
University of Amsterdam
Astronomical Institute
"Anton Pannekoek"
Kruislaan 403
1098 SJ AMSTERDAM
The Netherlands
frankm@astro.uva.nl

**H.J. MUTSCHKE**
University of Jena
Astrophysical Institute
Schillergaesschen 3
D-07745 JENA, Germany
mutschke@astro.uni-jena.de

**A. NILSSON**
Onsala Space Observatory
S-439/92 ONSALA
Sweden
an@oso.chalmers.se

**R. OROSEI**
University of Rome "La Sapienza"
INFOCOM Department
Via Eudossiana 18
00184 ROMA, Italy
orosei@saturn.ias.fra.cnr.it

**E. PALOMBA**
Osservatorio Astronomico
di Capodimonte
Laboratorio di Fisica Cosmica
Via Moiariello 16
80131 NAPOLI, Italy
palomba@astrna.na.astro.it

**M.G. PARISI**
Observatorio Astronomico
Paseo del Bosque S/N
1900 LA PLATA
Argentina
gparisi@fcaglp.fcaglp.unlp.edu.ar

**V. PIRRONELLO**
Istituto di Fisica
Universita di Catania
Catania, Sicily, Italy
*and* Department of Physics
Syracuse University
Syracuse, NY 13244-1130
USA

**T. POPPE**
University of Jena
Schillergasschen 3
07743 JENA, Germany
poppe@astro.uni-jena.de

**S. RAUDELIUNAS**
University of Vilnius
Astronomical Observatory
Ciurlionio 29, 2009 VILNIUS
Lithuania
raudel@astr.vno.osf.lt

**M.G. RAWLINGS**
University of Central Lancashire
Centre for Astrophysics
Corporation Street
PRESTON, LANCS, PR1 2HE
England
m.g.rawlings@uclan.ac.uk

**M.D. REESE**
University of Cincinnati
Physics Department
CINCINNATI, OH 45220
USA
mreese@physunc.phy.uc.edu

**M. ROBINSON**
Osservatorio Astronomico
di Capodimonte
Laboratorio di Fisica Cosmica
Via Moiariello 16
80131 NAPOLI, Italy
robinson@cosmic.na.astro.it

**M.L. SAVAGE**
NASA-AMES Research Center
Space Science Department
MS 245-6
MOFFETT FIELD, CA 94035
USA
savage@ssa1.arc.nasa.gov

**M. SCHNAITER**
University of Jena
Astrophysical Department
Schillergaesschen 3
D-07745 JENA, Germany
schnaiti@astro.uni-jena.de

**W.A. SCHUTTE**
Leiden Observatory
P.O. Box 9513, 2300 RA LEIDEN
The Netherlands
schutte@strwchem.strw.leiden.univ.nl

**O.M. SHALABIEA**
The Ohio-State University
Department of Physics
174 West 18th Avenue
COLUMBUS, OH 43210
USA
osama@mps.ohio-state.edu

**R.Y. SHUPING**
CASA
Uversity of Colorado
Campus Box 389
BOULDER, CO 80303
USA
shuping@casa.colorado.edu

**S. SIRONO**

Hokkaido University
Earth & Planetary Science Dept.
Kita 10, Nishi 8
001 SAPPORO, Japan
sirono@epms.hokudai.ac.jp

**M. SITKO**
University of Cincinnati
Physics Department
CINCINNATI, 45221-0011
USA
sitko@dusty.phy.uc.edu

**T. SMITH**
University of Toledo
Dept. of Physics & Astronomy
Ritter Observatory
Bancroft Street, TOLEDO
OH 43606, USA
tsmith@astro1.panet.utoledo.edu

**T.P. SNOW**
University of Colorado
CAMPUS BOX 389, BOULDER
CO 80309, USA
tsnow@casa.colorado.edu

**P. SONNENTRUCKER**
University of Strasbourg
Solar System Division
Keplerlaan 1
2330, ZH NOORDWIJK
The Netherlands
psonnen@so.estec.esa.nl

**G.S. STRINGFELLOW**
CASA
University of Colorado
Campus Box 389
BOULDER, CO 80309-0389
USA
guy@casa.colorado.edu

**J. THOMAS-OSIP**
University of Florida
Astronomy Department
211 SSRB, GAINESVILLE
FL 32611, USA
jet@astro.ufl.edu

**C. TRUJILLO**
University of Hawaii
Institute for Astronomy
2680 Woodlawn Drive
HONOLULU, HI 96822
USA
chad@ifa.hawaii.edu

**A.G.G.M. TIELENS**
MS 245-3
NASA Ames Research Centre
Moffett Field, CA 94035-1000
USA
tielens@dusty.arc.nasa.gov

**M. Van DEN ANCKER**
University of Amsterdam
Astronomical Institute
Kruislaan 403
1098 SJ, AMSTERDAM
The Netherlands
mario@astro.uva.nl

**V.P. VASILYEV**
Institute of Radio Astronomy
4 Chervonopraporna Str
310002, KHARKIV
Ukraine
rai@ira.kharkov.ua

**K.T. WALDEMARSSON**
University of Florida
Astronomy Department
P.O. Box 112055/211 SSRB
GAINESVILLE, FL 32611
USA
tomas@astro.ufl.edu

**C.M. WRIGHT**
Max-Planck-Institut für

Extraterrestrische Physik
D-85740, GARCHING, Germany
wright@mpe-garching.mpg.de

**G. WURM**
University of Jena
Astrophysical Institute
Schillergbüchen 3
D-07795, JENA, Germany
wurm@astro.uni-jena.de

**T. YAMAMOTO**
Hokkaido University
Earth & Planetary Dept.
N10 W8 060, SAPPORO
Japan
yamamoto@epms.hokudai.ac.jp

**V. ZUBKO**
Nicolas Copernicus University
Center of Astronomy
Gagarina 11, PL-87-100
TORUN, Poland
zubko@astri.uni.torun.pl

# INDEX[1]



[1]restricted to the first page of a paper.

MIX
Papier aus verantwortungsvollen Quellen
Paper from responsible sources
FSC® C105338

If you have any concerns about our products, you can contact us on
**ProductSafety@springernature.com**

In case Publisher is established outside the EU, the EU authorized representative is:
**Springer Nature Customer Service Center GmbH**
**Europaplatz 3, 69115 Heidelberg, Germany**

Printed by Libri Plureos GmbH
in Hamburg, Germany